Danuta Przeworska-Rolewicz
*Institute of Mathematics, Polish Academy of Sciences,
Warsaw, Poland*

Algebraic Analysis

D. Reidel Publishing Company

A MEMBER OF THE KLUWER ACADEMIC PUBLISHERS GROUP

Dordrecht / Boston / Lancaster / Tokyo

PWN-Polish Scientific Publishers

Warszawa

Library of Congress Cataloging-in-Publication Data

CIP

Przeworska-Rolewicz, Danuta.
Algebraic analysis.

(Mathematics and its applications. East European series)
Bibliography: p.
Includes index.
1. Mathematical analysis.   2. Algebra. I. Title.
II. Series: Mathematics and its applications
(D. Reidel Publishing Company).   East European series.
QA300.P99 1987   515   87-4538

ISBN-13: 978-94-010-7139-0   e-ISBN-13: 978-94-009-1427-8
DOI: 10.1007/978-94-009-1427-8

Published by PWN—Polish Scientific Publishers—Warszawa
in co-edition with
D. Reidel Publishing Company, Dordrecht, Holland.

Sold and distributed in the U.S.A. and Canada
by Kluwer Academic Publishers,
101 Philip Drive, Norwell, MA 02061, U.S.A.

Sold and distributed in Albania, Bulgaria, Chinese People's Republic, Cuba, Czecho
slovakia, German Democratic Republic, Hungary, Korean People's Democratic
Republic, Mongolia, Poland, Rumania, the U.S.S.R., Vietnam, and Yugoslavia
by PWN—Polish Scientific Publishers, Warszawa, Poland.

In all remaining countries by
Kluwer Academic Publishers Group,
P.O. Box 322, 3300 AH Dordrecht, Holland.

Contents

Editor's Preface

Approach your problems from the right end and begin with the answers. Then, one day, perhaps you will find the final question.

'The Hermit Clad in Crane Feathers' in R. van Gulik's *The Chinese Maze Murders*.

It isn't that they can't see the solution It is that they can't see the problem.

G. K. Chesterton *The Scandal of Father Brown* 'The point of a Pin'.

Growing specialization and diversification have brought a host of mon o graphs and textbooks on increasingly specialized topics. However, the "tree" of knowledge of mathematics and related fields does not grow only by putting forth new branches. It also happens, quite often in fact, that branches which were thought to be completely disparate are suddenly seen to be related.

Further, the kind and level of sophistication of mathematics applied in various sciences has changed drastically in recent years: measure theory is used (non-trivially) in regional and theoretical economics; algebraic geometry interacts with physics; the Minkowsky lemma, coding theory and the structure of water meet one another in packing and covering theory; quantum fields, crystal defects and mathematical programming profit from homotopy theory; Lie algebras are relevant to filtering; and prediction and electrical engineering can use Stein spaces. And in addition to this there are such new emerging subdisciplines as "experimental mathematics", "CFD", "completely integrable systems", "chaos, synergetics and large-scale order", which are almost impossible to fit into the existing classification schemes. They draw upon widely different sections of mathematics. This programme, Mathematics and Its Applications, is devoted to new emerging (sub)disciplines and to such (new) interrelations as exempla gratia:

— a central concept which plays an important role in several different mathematical and/or scientific specialized areas;

— new applications of the results and ideas from one area of scientific endeavour into another;

— influences which the results, problems and concepts of one field of enquiry have and have had on the development of another.

The Mathematics and Its Applications programme tries to make available a careful selection of books which fit the philosophy outlined above. With such books, which are stimulating rather than definitive, intriguing rather than encyclopaedic, we hope to contribute something towards better communication among the practitioners in diversified fields.

Because of the wealth of scholarly research being undertaken in the Soviet Union, Eastern Europe, and Japan, it was decided to devote special attention to the work emanating from these particular regions. Thus it was decided to start three regional series under the umbrella of the main MIA programme.

There is, and in fact always has been, a substantial algebraic part to analysis. Normally this is not emphasized, but there is definitely much to be said in favour of stressing these aspects and making full use of them.

Though the remark is not immediately germane to this book by itself, it is clear that there is at present a definite algebraic trend in large parts of analysis. Strong algebraic structures carry enormous implications. This is the case in pseudo-differential operator theory for instance. In the case of the so-called completely integrable systems of (partial) differential equations these appear to be so strong that the miracle of solitons is in fact forced by purely formal algebraic properties.

Thus it makes eminent sense to explore to what extent algebraic thinking can help in structuring analysis and giving algebraic analysis its proper place as a foundation stone. In the author's words, algebraic analysis is the theory of right invertible operators in a linear space—think of indefinite integrals. This is an ongoing, vigorous field. And this book constitutes a unique state-of-the-art description of it.

The unreasonable effectiveness of mathematics in science...

> Eugene Wigner

Well, if you know of a better'ole, go to
it.

> Bruce Bairnsfather

What is now proved was once only
imagined.

> William Blake

Bussum, *December* 1986

As long as algebra and geometry proceeded along separate paths, their ad-
vance was slow and their applications
limited.

But when these sciences joined company they drew from each other fresh
vitality and thenceforward marched on
at a rapid pace towards perfection.

> Joseph Louis Lagrange

Michiel Hazewinkel

Introduction

The name "Algebraic Analysis" was used initially by Lagrange: the second revised and enlarged edition of his "Théorie des Fonctions Analytiques" ($M^{me} V^e$ Courcier, Imprimeur–Libraire pour les Mathématiques, Paris, 1813) has a subtitle "Les Principes du Calcul différentiel, dégagés de toute considération d'infiniment petits, d'évanouissans de limit et de fluxions, et réduit à l'analyse algébrique de quantités finies". Note that the publisher of this book was a woman. This title could be explained by the fact that at that time the notions of limit, convergence, and so on, were not precise. However, the same subtitle was used by Cauchy in 1821 in his "Cours d'analyse de l'École Royale Politechnique, 1re partie, Analyse algébrique". In his introduction he wrote: "As to methods, I have sought to make them as rigorous as those of geometry, so as never to have recourse to justifications drawn from the generality of algebra".

This may provoke the idea that the name "algebraic analysis" emphasised that the analyses under consideration were more or less "different" from other concepts of analyses at their time.

Lagrange's book was used as a handbook for several years without regard to Cauchy's ideas. For instance, the young Karl Marx at Bonn University was using this book, as it is known from his letters (cf. H. C. Kennedy, Karl Marx and the foundations of differential calculus, Historia Math. 4 (1977), 303–318).

The term "algebraic analysis" was used in the books: "Istituzioni di Analysi Algebrica" by A. Capelli (Napoli, 1894); "Corso di Analysi Algebrica con Introduzione al Calcolo Infinitesimale" by E. Cesàro (Torino, 1894); "Elementares Lehrbuch der Algebraischen Analysis und der Infinitesimal Rechnung", also by E. Cesàro (Leipzig, 1904); "Course of Algebraic Analysis" (in Russian, Kiev, 1911) and "Treatise on Algebraic Analysis" (in Russian and Ukrainian, Kiev, 1938–1939) by D. O. Grave. The Capelli's book concerns algebraic curves, also the Grave's books are devoted to algebraic problems. The only Cesàro gives an attempt to a common treatment of Algebra, Linear Algebra, Calculus and Differential Equations. Recently, "Foundations of Algebraic Analysis" (Princeton, 1986) by M. Kashiwara, T. Kawai, T. Kimura appeared. This book, however, is concerned with the microlocal analysis.

What led me to "Algebraic Analysis"?

I have written some books devoted to different algebraic methods in analysis (the first one—together with S. Rolewicz, cf. Bibliography, 1968, 1973).

In February of 1972 I found in the journal "Jour de France" the following personal horoscope:

"Après quelques hésitations vous allez découvrir la formule que vous cherchez. Ne craignez pas de l'avance sur l'époque, cette nouveauté plaira".

And, indeed, the same February I found the proper definition of initial operators for right invertible operators and their fundamental properties. Four papers containing these results and their applications to functional-differential equation were submitted for publication that month. This was the first beginning of Calculus in Algebraic Analysis by which I meant the theory of right invertible operators in linear space (without any topology, in general).

The year 1972 was, in fact, good for this theory. A break-through occured when I was asked to read lectures for the first and second year students in the Cybernetic Faculty of the Military Engineering Academy in Warsaw. A new programme of Mathematics based on ideas of Algebraic Analysis was prepared by mathematicians and engineers from this school working in Operations Research Department. Also we prepared new textbooks. This programme was so revolutionary that I never imagined something like this could be realized! However, we did it with extraordinary good results in academic years 1973/74 and 1974/75. I should point out that almost all our students passed all examinations and that marks from physics and engineering subjects were strictly commensurable with these in mathematics.

Words cannot express my gratitude for the support of my work at that time and in the next few years, of the late General Sylwester

Kaliski, professor of Physics, then Commander of the Military Engineering Academy in Warsaw.

A book in Polish entitled "Introduction to Algebraic Analysis and its Applications" was published in 1979.

The next book (in English) "Shifts and Periodicity for Right Invertible Operators" was published in 1980 in London.

The development of this theory still proceeded, but not in such a smooth way, as may follow from the story below. However, always I have found some support. In particular, several mathematicians from different countries have sent me their unpublished results. I am very indebted to all who did so.

The present book is a result of all these efforts. I should point out that a very large part of this book has been written in time of my stay at Monash University in Clayton*, Victoria, Australia, i.e. during the period May 13—July 29, 1984. I would like to express my thanks to the Department of Mathematics at the Monash University for very good conditions of work and to all these members of staff who made several remarks and observations of great value, concerning Algebraic Analysis. This was possible due to a series of lectures I presented on Algebraic Analysis.

This short story of Algebraic Analysis "on lines" shows that this book cannot be treated as a monograph in the field of mathematics more or less already closed. It is rather a guide for some (may be not always classical) routine methods with a principal purpose: to indicate that new possibilities can be found in well-known old results.

To understand this book it is enough to know such notions as convergence, continuity and differentiability (all are included in the new programme of mathematics for high schools in Poland).

The proposed approach to analysis could be also a game for mathematicians lecturing in Calculus and Differential Equations for several years, as it was in the case of myself and my collaborators and, I think, also for students. However, it should be pointed out that although the programme of Mathematics I have written about, and this book are not identical, they have a non-empty intersection.

---

* Clayton is a suburb of Melbourne.

I should also mention that the bibliography which is relatively large, is far from complete with respect to such a large area and the very new subject covered by this book. By consulting the bibliography the reader will find several works related in a sense to Algebraic Analysis which, I hope, may extend the knowledge of the main topic of this book.

Finally, I would like to express my appreciation and gratitude to Professor Roman Ger for his valuable remarks, suggestions and corrections of my manuscript. I wish also to thank to Mrs. Hanna Kalinowska for her help in the preparation of this manuscript.

Warsaw–Clayton, 1984–1985

DANUTA PRZEWORSKA–ROLEWICZ

ACKNOWLEDGMENT

I shall always be indebted to all my physicians and friends who helped me to survive. My gratitude should be extended to the editorial staff of PWN–Polish Scientific Publishers for their unusual aid in the preparation of this book to print.

Warszawa, October 1986   *Danuta Przeworska-Rolewicz*

Chapter

Fundamental properties of linear operators in linear spaces

This chapter contains auxiliary notions and theorems which will be useful in development of Algebraic Analysis.

We assume the reader to be acquainted with notions of sets and mappings and with standard operations on sets. As usually, we denote by $A \cup B$ the union (sum) of the sets A and B, by $A \setminus B$ the difference and by $A \cap B$ the intersection (product, common part) of the sets A and B. The empty set is denoted by $\emptyset$. A set containing one element x only is denoted by $\{x\}$. A set containing a finite numbers of elements $x_1, \ldots, x_n$ is denoted by $\{x_1, \ldots, x_n\}$. We assume that the term "point" has the same meaning as "element" and the terms "mapping" and "transformation" have the same meaning.

Suppose, we are given a family $\mathfrak{A}$ of sets. By $\bigcup_{A \in \mathfrak{A}} A$ and $\bigcap_{A \in \mathfrak{A}} A$ we denote the union and the intersection of all sets belonging to the family $\mathfrak{A}$.

Recall that two sets A and B are of the same cardinality if there exists a one-to-one mapping of A onto B; if that is the case, we write: $\overline{\overline{A}} = \overline{\overline{B}}$. Sets of the same cardinality as the set $\mathbf{N}$ of all positive integers are called countable.

All properties, propositions and theorems in this Chapter are given without proofs. If no reference is indicated then these proofs are assumed to be either trivial or to be found in text-books concerning subjects. The part about algebraic operators and Section 1.5 can be omitted in the first reading.

1.1. GROUPS AND RINGS

A nonempty set X is called a *semigroup* if to every $x \in X$, $y \in X$ there corresponds a unique element $z \in X$ denoted by $z = x \odot y$ satisfying the following conditions:

1

(a) the operation $\odot$ is associative, i.e.

$$x \odot (y \odot z) = (x \odot y) \odot z \quad \text{for all } x, y, z \in X. \qquad (1.1.1)$$

A nonempty X is called a *group* if to every $x \in X$, $y \in X$ there corresponds a unique element $z \in X$, denoted by $z = x \odot y$ satisfying (1.1.1) and the following conditions:

(b) there exists an element $e \in X$ such that

$$e \odot x = x \odot e = x \quad \text{for every } x \in X; \qquad (1.1.2)$$

(c) for every $x \in X$ there exists an element $y \in X$ such that

$$x \odot y = y \odot x = e. \qquad (1.1.3)$$

The operation $\odot$ is called *multiplication*. The element e satisfying Condition (1.1.2) is uniquely determined, and is called either *neutral element of X* or *unit of X*. A semigroup satisfying Condition (1.1.2) is said to be a *semigroup with unit* (or: *neutral element*).

For every $x \in X$ an element $y \in X$ satisfying Condition (1.1.3) is uniquely determined. The unique element $y \in X$ satisfying Condition (1.1.3) is called *inverse of x* and is denoted by x^{-1}.

A nonempty subset Y of a group X is said to be a *subgroup* if Y is a group with respect to the same operation. In other words, $Y \subset X$ is a subgroup of a group X if and only if $e \in Y$, for every $x, y \in Y$ we find $x \odot y \in Y$ and $x^{-1} \in Y$. One can prove that Y is a subgroup X if the condition $x, y \in Y$ implies $x \odot y^{-1} \in Y$. By definition, the sets $\{e\}$ and X itself are subgroups of X.

If X is a group then the cardinality $\overline{\overline{X}}$ of the set X is called the *order* of X. If the order of group X is a positive integer, then X is called a *finite group*.

A group (semigroup) X is said to be *commutative* if

$$x \odot y = y \odot x \quad \text{for all } x, y \in X. \qquad (1.1.4)$$

Traditionally, a commutative group is called an *Abelian group*. The operation $\odot$ in an Abelian group is denoted by $+$ and called *addition*, the inverse element is denoted by $-x$ and the neutral element is denoted by 0 and called the *zero* of the group X. We also write $x + (-y) = x - y$ for all $x, y \in X$ and we call the element $x - y$ the *difference* of x and y.

An Abelian group X (with respect to an operation $+$) is called a *ring* if in X there is defined a second operation $\cdot$ called *multiplication*, such that for every $x, y, z \in X$ we have $x \cdot y \in X$ and

$$x \cdot (y \cdot z) = (x \cdot y) \cdot z, \tag{1.1.5}$$

$$x \cdot (y+z) = x \cdot y + x \cdot z,$$
$$(x+y) \cdot z = x \cdot z + y \cdot z. \tag{1.1.6}$$

It means that the multiplication in a ring is associative and distributive with respect to the addition. The Conditions (1.1.6) are called the *distribution laws*.

In the sequel we shall often write xy instead of $x \cdot y$.

A ring X is said to be *commutative* if

$$xy = yx \quad \text{for all } x, y \in X. \tag{1.1.7}$$

We say that a ring X is a *ring with unit e* if there is an element $e \in X$ such that

$$ex = xe = x \quad \text{for every } x \in X. \tag{1.1.8}$$

If a ring X has a unit then this unit is uniquely determined. It is not difficult to prove that every ring can be extended to a ring with a unit.

A ring $X \neq \{0\}$ *has no divisors of zero* if the condition $xy = 0$ implies either $x = 0$ or $y = 0$.

Suppose that X is a ring with a unit e. An element $x \in X$ is said to be *left (right) invertible* if there exists an element $y \in X$ (resp. $z \in X$) such that

$$yx = e \quad (\text{resp. } xz = e). \tag{1.1.9}$$

The element y (resp. z) is called *left (right) inverse of x*. If for an element $x \in X$ there exists an element $y \in X$ such that

$$xy = yx = e \tag{1.1.10}$$

we say that x is *invertible*. An element $y \in X$ satisfying Equalities (1.1.10) s uniquely determined (provided that exists), is called *inverse of x* and is denoted by x^{-1}.

If an element x belonging to a ring X with a unit e is simultaneously left and right invertible then is invertible.

In other words, if an element x is invertible then a left inverse and a right inverse are both equal to the inverse x^{-1}.

The following characterization of right inverses will be useful in our subsequent considerations:

PROPOSITION 1.1.1 (Arens, 1958). *Suppose that X is a ring with unit e and that $x \in X$ has a right inverse y_0. Then $y \in X$ is a right inverse of x if and only if y is of the form*

$$y = a + y_0(e - xa), \quad \text{where } a \in X \text{ is arbitrary}. \qquad (1.1.11)$$

Observe that we can write Formula (1.1.11) in another form:

$$y = (e - y_0 x)a + y_0, \quad \text{where } a \in X \text{ is arbitrary}. \qquad (1.1.11')$$

A subset Y of a ring X is called a *subring* if Y is a ring with respect to the same operations as X, i.e. if for every $x, y \in Y$ we have $x + y \in Y$, $xy \in Y$, $-x \in Y$, $0 \in Y$. One can prove that $Y \neq \emptyset$ is a subring of a ring X if $x - y \in Y$ and $xy \in Y$ for all $x, y \in Y$.

A nonempty subset Y of a ring X is called a *left* (*right*) *ideal* if for every $y, y_1, y_2 \in Y$, $x \in X$ we have

$$y_1 - y_2 \in Y, \quad xy \in Y \text{ (resp. } yx \in Y). \qquad (1.1.12)$$

An ideal Y in a ring X, which is simultaneously a left and a right ideal, is called a *two-sided ideal* (or simply: *ideal*). This means that $Y \neq \emptyset$ is an ideal in a ring X if for all $y, y_1, y_2 \in Y$ and $x \in X$ we have $y_1 - y_2 \in Y$, $xy \in Y$, $yx \in Y$.

Observe that an ideal $Y \subset X$ is a subring of the ring X. If X is a commutative ring then every left ideal Y is a right ideal. Thus Y is an ideal.

In an arbitrary ring X two sets: $\{0\}$ and X are ideals. These ideals are called *trivial* (or *improper*) *ideals*. All other ideals in X are called *proper* (or *non-trivial*) ideals.

Suppose that X is a ring with a unit e. Then an ideal $Y \subset X$, $Y \neq \{0\}$, is proper if and only if $e \notin Y$.

Suppose that X is an Abelian group and we are given two subsets

$A, B \subset X$. An *algebraic sum* of sets A and B, is a subset of X, denoted by $A+B$, defined as follows:

$$A+B = \{a+b: a \in A, b \in B\}. \tag{1.1.13}$$

If $B = \{b\}$, we write briefly $A+b$ instead of $A+\{b\}$.

Now, suppose that X is a ring and we are given two subsets $A, B \subset X$. We define an algebraic sum $A+B$ by means of the equality (1.1.13). We define also an *algebraic product* of sets A and B, denoted by AB, as follows:

$$AB = \{ab: a \in A, b \in B\}. \tag{1.1.14}$$

If $B = \{b\}$ we write briefly Ab instead of $A\{b\}$ and bA instead of $\{b\}A$. Clearly, $AB \subset X$.

A commutative ring X with a unit e and with the cardinality $\overline{\overline{X}} > 1$ is said to be a *field* if every element $x \in X \setminus \{0\}$ is invertible. Any field is a commutative ring without zero divisors.

The set $X^0 = X \setminus \{0\}$ is a group with respect to the multiplication.

A nonempty subset Y of a field X is called a *subfield* if Y is a field with respect to the same operations, i.e. if for every $x, y \in Y$ we have $x-y \in Y$, $xy \in Y$ and if $y \neq 0$ then $y^{-1} \in Y$.

In the sequel, an arbitrary field of numbers will be called a *field of scalars* and elements of this field will be called either *numbers* or *scalars*. **R** will stand for the field of real numbers.

In a commutative ring X we can define a determinant in the following way. Write

$$\operatorname{sgn} a = \begin{cases} 1, & \text{if } a > 0, \\ 0, & \text{if } a = 0 \quad \text{for } a \in \mathbf{R}, \\ -1, & \text{if } a < 0 \end{cases} \tag{1.1.15}$$

(read: sign of the number a). It is easy to check by induction that for $a_1, \ldots, a_n \in \mathbf{R}$

$$\operatorname{sgn}(a_1 a_2 \ldots a_n) = (\operatorname{sgn} a_1)(\operatorname{sgn} a_2) \ldots (\operatorname{sgn} a_n). \tag{1.1.16}$$

Denote by $\{p_1, \ldots, p_n\}$ an arbitrary permutation of the numbers $1, \ldots, n$. The number

$$\operatorname{sgn}\{p_1, \ldots, p_n\} = \prod_{j,k=1, j \neq k}^{n} \operatorname{sgn}(p_j - p_k) \tag{1.1.17}$$

is called *sign of the permutation* $\{p_1, \ldots, p_n\}$. By definition, sgn $\{p_1, \ldots \ldots, p_n\} \neq 0$. A permutation is said to be *even* if its sign is $+1$. If its sign is -1, a permutation is said to be *odd*.

Suppose, we are given n^2 elements a_{jk} belonging to a commutative ring X. Write

$$
\begin{vmatrix}
a_{11} & a_{12} & \cdots & a_{1n} \\
a_{21} & a_{22} & \cdots & a_{2n} \\
\cdots\cdots\cdots\cdots\cdots \\
a_{n1} & a_{n2} & \cdots & a_{nn}
\end{vmatrix}
$$

$$
= \sum_{\{p_1, \ldots, p_n\}} \operatorname{sgn} \{p_1, \ldots, p_n\} a_{1p_1} \cdots a_{np_n} \qquad (1.1.18)
$$

where the summation is extended on all permutations of numbers $1, \ldots, n$. The element of the ring X determined by means of the equality (1.1.18) is called a *determinant* of the elements a_{jk} $(j, k = 1, \ldots, n)$. Sometimes we shall write briefly:

$$
\det (a_{jk})_{j, k=1, \ldots, n} = \sum_{\{p_1, \ldots, p_n\}} \operatorname{sgn} \{p_1, \ldots, p_n\} a_{1p_1} \cdots a_{np_n}.
$$

$$
(1.1.19)
$$

It is easy to observe that determinants defined by Formula (1.1.18) have the same properties as usual determinants, where a_{jk} are complex or real scalars.

Consider now a system of n equations with n unknowns $x_1, \ldots, x_n$ and with coefficients $a_{jk} \in X$ $(j, k = 1, \ldots, n)$, where X is a commutative ring:

$$
\sum_{j=1}^{n} a_{jk} x_j = y_k \quad (k = 1, \ldots, n) \qquad (1.1.20)
$$

$y_1, \ldots, y_n \in X$ are given. In the same way, as in the scalar case, we obtain Cramer Formulae. Namely, if the element $A = \det (a_{jk})_{j, k=1, \ldots, n}$ is invertible, then the system (1.1.20) has a unique solution, which is of the form

$$
x_j = A^{-1} A_j = A^{-1} \sum_{k=1}^{n} A_{jk} y_k \quad (j = 1, \ldots, n) \qquad (1.1.21)
$$

where the determinant A_j is obtained from the determinant A if we put instead of the j-th column $a_{j1}, \ldots, a_{jn}$ the column $y_1, \ldots, y_n$. A_{jk} are algebraic complements of elements a_{jk}, respectively, i.e. $A_{jk} = (-1)^{j+k}M_{jk}$, where M_{jk} is a minor determinant obtained by cancelling the j-th column and the k-th row in the determinant A.

The Cramer Formulae imply

PROPOSITION 1.1.2. *Suppose that X is a commutative ring, $a_{jk} \in X$ for $j, k = 1, 2, \ldots, n$. If the homogeneous system*

$$\sum_{j=1}^{n} a_{jk}x_j = 0 \quad (k = 1, \ldots, n) \tag{1.1.22}$$

has a non-trivial solution $x_1, \ldots, x_n$, i.e. not all y_j vanish simultaneously, then the element $A = \det(a_{jk})_{j, k=1,\ldots, n}$ is not invertible.

Indeed, suppose that A is invertible. Since $y_1 = \ldots = y_n = 0$, applying Cramer Formulae (1.1.21), we obtain $x_1 = \ldots = x_n = 0$, which contradicts our assumption.

A field X is said to be *algebraically closed* if every polynomial $w(t)$ $= \sum_{k=0}^{n} a_k t^k$ with coefficients $a_0, \ldots, a_n \in X$ has n and only n roots $t_1, \ldots, t_n \in X$, i.e. if

$$w(t) = a \prod_{m=1}^{n} (t - t_m), \quad \text{where } a, t_1, \ldots, t_n \in X. \tag{1.1.23}$$

For instance, the field $\mathbf{R}$ of reals is not algebraically closed, the field $\mathbf{C}$ of complexes is algebraically closed.

A mapping f of a group X onto a group Y is said to be an *isomorphism* if it is one-to-one and preserves the group operations, i.e. if for all x, $y \in X$ we have

$$f(x \odot y) = f(x) \odot f(y), \quad f(x^{-1}) = [f(x)]^{-1} \quad \text{and} \quad f(e_X) = e_Y,$$

where e_X, e_Y are the units in X and Y, respectively. We denote by the same symbol $\odot$ the group operations in both groups X and Y, because it does not lead to any misunderstanding. If $X = Y, f(X) = Y$, then we say that f is an *automorphism*.

In a similar way we define an isomorphism of two rings. For instance, suppose that we have two commutative rings X and Y with the neutral elements 0_X, 0_Y and the units e_X, e_Y, respectively. A mapping f is said to be an *isomorphism of X onto Y* if is one-to-one and preserves the ring operations, i.e. for all $x, y \in X$

$$f(x+y) = f(x)+f(y), \quad f(0_X) = 0_Y, \quad f(-x) = -f(x),$$
$$f(xy) = f(x)f(y), \quad f(e_X) = e_Y \quad \text{and} \quad f(x^{-1}) = [f(x)]^{-1},$$

provided that x is invertible.

If there is an isomorphism of a ring X onto Y we say that X and Y are *isomorphic*.

A field X *has the characteristic zero* if the intersection of all its sub-fields (which is again a subfield of X) is isomorphic with the field $\mathbf{Q}$ of all rational numbers. For instance, $\mathbf{Q}$ itself, the field $\mathbf{R}$ of reals and the field $\mathbf{C}$ of complexes are fields of characteristic zero. All calculations in a field of characteristic zero are going in the same manner as in these fields. In the sequel we shall consider only fields of characteristic zero.

Examples and Exercises

EXAMPLE 1.1.1. A mapping h of a ring X into a ring Y is called a *homomorphism* of X into Y if and only if

$$h(x+y) = h(x)+h(y) \quad \text{and} \quad h(xy) = h(x)h(y)$$
$$\text{for all } x, y \in X \qquad (1.1.24)$$

where on the left-hand side of Equalities (1.1.24) we have the addition and the multiplication in the ring X and on the right-hand side of Equalities (1.1.24) we have the addition and the multiplication in the ring Y. A homomorphism of a ring X into itself is called an *endomorphism* of X. A one-to-one homomorphism of X onto itself is called *automorphism* of X. Suppose, we are given the following mapping of the set $\mathbf{R}$ of all reals into itself:

$$h(a) = -a \quad \text{for all } a \in \mathbf{R}. \qquad (1.1.25)$$

Consider $\mathbf{R}$ as an Abelian group (with respect to the addition of reals). Then the mapping h defined by (1.1.25) is an automorphism of $\mathbf{R}$.

Consider now **R** as a ring (with respect to the addition and the multiplication of reals). Then h is not an endomorphism of **R**. Indeed $h(1 \cdot 1) = h(1) = -1$ and $h(1)h(1) = (-1)(-1) = 1 + h(1 \cdot 1)$.

EXAMPLE 1.1.2. Suppose that $h \neq 0$ is a homomorphism of a ring X into a ring Y. Then the set

$$\ker h = \{x \in X : h(x) = 0\} \tag{1.1.26}$$

(read: *kernel* of h) is an ideal in X. Indeed, suppose that $y, y_1, y_2 \in \ker h$, $x \in X$. Then $h(y) = h(y_1) = h(y_2) = 0$ and

$$h(y_1 - y_2) = h(y_1) - h(y_2) = 0 - 0 = 0, \quad \text{hence } y_1 - y_2 \in \ker h,$$

$$h(xy) = h(x)h(y) = h(x) \cdot 0 = 0, \qquad \text{hence } xy \in \ker h,$$

$$h(yx) = h(y)h(x) = 0 \cdot h(x) = 0, \qquad \text{hence } yx \in \ker h.$$

If the ring X has a unit e then $e \notin \ker h$ and $\ker h \neq X$. Indeed, suppose that $e \in \ker h$. Then $h(e) = 0$. By definition of a homomorphism, for arbitrary $x \in X$ we have $h(x) \neq 0$ and $h(x) = h(ex) = h(e)h(x) = 0 \cdot h(x) = 0$, which contradicts our assumption that $h \neq 0$. Thus $e \notin \ker h$. We therefore conclude that $\ker h \neq X$.

EXERCISE 1.1.1. Suppose that X is a commutative ring with a unit e and that $a \in X$ is arbitrarily fixed. Prove that:

(a) the set aX is an ideal in X (the set aX is defined by Formula (1.1.14));

(b) $aX = X$ if and only if a is invertible;

(c) $aX = -aX$.

EXERCISE 1.1.2. Suppose that X is a commutative ring with a unit e and without zero divisors. Prove that there exists a field Z such that X is a subring of Z.

EXERCISE 1.1.3. Suppose that X is a ring. Prove that:

(a) a left (right) invertible element $x \in X$ cannot belong to a left (right) ideal $Y \subset X$, $Y \neq X$;

(b) an invertible element $x \in X$ cannot belong to a proper ideal $Y \subsetneq X$.

EXERCISE 1.1.4. Suppose that X is a ring with unit e and x has a left inverse y_0. Prove that $y \in X$ is a left inverse of x if and only if $y = a + (e-ax)y_0$ (or: $y = a(e-xy_0)+y_0$), where $a \in X$ is arbitrary (cf. Proposition 1.1.1 and Formula (1.1.11')).

EXERCISE 1.1.5. Suppose that X is a ring with unit e and an element $x \in X$ has two different right (left) inverses. Prove that x is a zero divisor.

EXERCISE 1.1.6. An element a of a ring X with unit is called *idempotent* if $a^2 = a$. Suppose that $x \in X$ has a right inverse y. Prove that $a = e-yx$ is an idempotent. Is a similar statement for left inverses also true?

EXERCISE 1.1.7. An element a of a ring X with unit is called *nilpotent* if there is a positive integer $n > 1$ such that $a^n = 0$, but $a^{n-1} \neq 0$. Prove that the set of all nilpotent elements in a commutative ring X is an ideal in X.

1.2. LINEAR SPACES

An Abelian group X is said to be a *linear space over a field $\mathscr{F}$ of scalars* if in X there is defined an operation of multiplication of elements by scalars satisfying the following conditions:
  for every $x, y \in X$ and $t, s \in \mathscr{F}$ we have $tx \in \mathscr{F}$ and

$$t(x+y) = tx+ty; \quad (t+s)x = tx+sx;$$
$$(ts)x = t(sx); \quad 1 \cdot x = x. \tag{1.2.1}$$

These conditions imply that

$$0 \cdot x = 0 \quad \text{for all } x \in X \tag{1.2.2}$$

(in the equality $0 \cdot x = 0$ on the right-hand side we have the neutral element of the group X. We denote this element by 0, because it does not lead to any misunderstanding).
  Conditions (1.2.1) imply also that

$$\text{if } tx = 0 \quad \text{and } x \neq 0 \quad \text{then } t = 0. \tag{1.2.3}$$

Suppose that X is a linear space over a field $\mathscr{F}$ of scalars. A non-empty subset $Y \subset X$ is called a *linear subspace* (or *linear subset*) of X, if Y is a linear space with respect to the same addition and multiplication by scalars belonging to $\mathscr{F}$. It means that a set $Y \subset X$ is a linear subspace (linear subset) of X if for every $x, y \in Y$ and $t \in \mathscr{F}$ we have $x+y \in Y$ and $tx \in Y$.

Suppose that Y is an arbitrary nonempty subset of a linear space over a field $\mathscr{F}$. The smallest linear subset containing Y is called the *linear span of Y* or the *subspace spanned by the set Y* and is denoted by lin Y.

THEOREM 1.2.1. *If X is a linear space over a field $\mathscr{F}$ and $Y \subset X$ then*

$$\operatorname{lin} Y = \left\{ x \in X \colon x = \sum_{j=1}^{n} t_j x_j, \text{ where } t_j \in \mathscr{F}, x_j \in Y \right.$$
$$\left. (j = 1, ..., n) \right\}. \tag{1.2.4}$$

Elements of the form $\sum_{j=1}^{n} t_j x_j$, where $t_j \in \mathscr{F}$, $x_j \in X$, are called *linear combination of elements $x_1, ..., x_n$*.

We say that an element $x \in X$ is *linearly dependent on a set $Y \subset X$* (or: *on elements of Y*) if $x \in \operatorname{lin} Y$, i.e. if there exist $x_1, ..., x_n \in Y$ and $t_1, ..., t_n \in \mathscr{F}$ such that $x = \sum_{j=1}^{n} t_j x_j$. The elements of a set $Y \subset X$ are said to be *linearly independent* if $x \notin \operatorname{lin}(Y \setminus \{x\})$ for every $x \in Y$.

From the definition of the linear independence follows that elements $x_1, ..., x_n \in X$ are linearly independent if the equality

$$t_1 x_1 + t_2 x_2 + ... + t_n x_n = 0 \quad (t_1, ..., t_n \in \mathscr{F})$$

implies $t_1 = t_2 = ... = t_n = 0$.

In other words, elements $x_1, ..., x_n$ are linearly independent if the only vanishing linear combination is a linear combination with all coefficients equal to zero.

A linear space X over a field $\mathscr{F}$ is said to be *n-dimensional* if n is a minimal number of linearly independent elements $x_1, ..., x_n \in X$ such that $\operatorname{lin}\{x_1, ..., x_n\} = X$. X is said to be *finite dimensional* if it is n-dimensional for some positive integer n.

In this case we write: $\dim X = n$ (resp: $\dim X < +\infty$). The number n is called a *dimension* of the space X. If elements $x_1, \ldots, x_n$ are linearly independent then

$$\dim \text{lin } \{x_1, \ldots, x_n\} = n.$$

If $X = \{0\}$ then, by definition, $\dim X = 0$.

If the space X is not finite dimensional, then we say that X is *infinite dimensional* and we write: $\dim X = +\infty$.

Observe that a finite dimensional linear space X considered over the field $\mathbf{R}$ of reals has a dimension twice greater than $\dim X$ considered over the field $\mathbf{C}$ of complexes. This follows from the fact that the set $\mathbf{C}$ considered as a linear space over $\mathbf{R}$ (i.e., with a usual addition, but with multiplication only by reals) has dimension 2. We therefore assume that if a space X is considered over the field $\mathbf{C}$ then the dimension is to be understood as a dimension of X over this field.

A set (a system) $\mathscr{B}$ of elements of a linear space X over a field $\mathscr{F}$ is called a *basis* if every element $x \in X$ can be written in a unique way as a linear combination of elements belonging to $\mathscr{B}$. The uniqueness of this representation implies that $\mathscr{B}$ consists of linearly independent elements.

PROPOSITION 1.2.1. *If X is a linear space over the field $\mathscr{F}$ and $Y \subset X$ is a set of linearly independent elements, then Y is a basis in the linear span of Y.*

Suppose, we are given linear spaces $X_1, \ldots, X_n$ all over the same field $\mathscr{F}$ of scalars. A set X which consists of all ordered n-tuples $x = (x_1, \ldots, x_n)$ such that $x_j \in X_j$ $(j = 1, \ldots, n)$ is called a *Cartesian product* of spaces $X_1, X_2, \ldots, X_n$ and is denoted by $X_1 \times X_2 \times \ldots \times X_n$. Define the addition and the multiplication by scalars as follows:

$$(x_1, \ldots, x_n) + (y_1, \ldots, y_n) = (x_1 + y_1, \ldots, x_n + y_n), \qquad (1.2.5)$$

$$t(x_1, \ldots, x_n) = (tx_1, \ldots, tx_n), \qquad (1.2.6)$$

for $t \in \mathscr{F}$, $x_j, y_j \in X_j$ $(j = 1, \ldots, n)$.

It is easy to verify that the set $X_1 \times X_2 \times \ldots \times X_n$ with the addition and the multiplication by scalars defined by Formulae (1.2.5), (1.2.6) is a linear space over the field $\mathscr{F}$. Thus the Cartesian product of linear spaces is again a linear space (over the same field of scalars).

Suppose now, that $X_1, ..., X_n$ are subspaces of a linear space X over a field $\mathscr{F}$. The algebraic sum of $X_1, ..., X_n$, i.e. the set $X_1 + X_2 + + ... + X_n = \{x_1 + x_2 + ... + x_n: x_j \in X_j \ (j = 1, 2, ..., n)\}$ is called a *direct sum* of $X_1, ..., X_n$ if $X_j \cap \sum_{\substack{k \neq j}}^{n} X_k = \{0\}$ for $j = 1, 2, ..., n$. The direct sum of subspaces $X_1, ..., X_n$ is denoted by $X_1 \oplus ... \oplus X_n$. Observe that the condition $X_j \cap \sum_{\substack{k \neq i}}^{n} X_k = \{0\}$ implies that every element $x \in X_1 \oplus \oplus ... \oplus X_n$ can be written in a unique way in the form: $x = x_1 + ... + x_n$, where $x_j \in X_j \ (j = 1, ..., n)$. If $X = X_1 \oplus ... \oplus X_n$, we say that X is *decomposed onto the direct sum of spaces* $X_1, ..., X_n$.

Suppose that X is a linear space over the field $\mathbf{R}$ of reals. This space can be embedded in a natural way in a linear space over the field $\mathbf{C}$ of complexes. Indeed, consider the set of all ordered pairs (x, y), where $x, y \in X$, with the addition and the multiplication by scalars defined as follows:

$$(x_1, y_1) + (x_2, y_2) = (x_1 + x_2, y_1 + y_2) \quad \text{for } x_1, x_2, y_1, y_2 \in X, \tag{1.2.7}$$

$$(a + ib)(x, y) = (ax - by, ay + bx) \quad \text{for } x, y \in X, a + ib \in \mathbf{C}. \tag{1.2.8}$$

It is easy to check that these operations satisfy the distribution laws given in the definition of a linear space and that the multiplication by scalars is associative.

A linear space over the field $\mathbf{C}$ introduced in this way will be denoted by $X + iX$. The embedding $x \rightarrow (x, 0)$ permits us to consider X as a subspace of $X + iX$.

Suppose that X is a linear space over a field $\mathscr{F}$ and that Y is a subspace of X. To every element $x \in X$ there correspond the set

$$[x] = \{x + y: y \in Y\} = x + Y. \tag{1.2.9}$$

The set $[x]$ is called a *coset determined by the element* $x \in X$.

The coset $[x]$ can be also written in the following, equivalent form:

$$[x] = \{z \in X: x - z \in Y\}. \tag{1.2.10}$$

Two cosets are either disjoint or equal, i.e. if $x_1, x_2 \in X$ then either $[x_1] = [x_2]$ or $[x_1] \cap [x_2] = \emptyset$.

Consider the set of all cosets determined by elements $x \in X$:

$$X/Y = \{[x]: x \in X\}, \quad \text{where } [x] = x + Y. \tag{1.2.11}$$

Define in the set X/Y the addition and the multiplication by scalars as follows:

$$[x] + [y] = [x + y], \quad t[x] = [tx] \quad \text{for } x, y \in X, t \in \mathscr{F}. \tag{1.2.12}$$

The set X/Y with the addition and the multiplication by scalars defined by Formulae (1.2.12) is a linear space over the field $\mathscr{F}$. By definition, $[0] = Y$. The space X/Y is called a *quotient space*.

Consider a subspace Y of a linear space X over a field $\mathscr{F}$ of scalars. The *codimension* (or a *defect*) of subspace Y is the dimension of the quotient space X/Y:

$$\operatorname{codim} Y = \dim X/Y. \tag{1.2.13}$$

Suppose now, that X is a linear space over the field $\mathbf{R}$ of reals. A *convex combination* of elements $x_1, \ldots, x_n \in X$ is an element $x \in X$ of the form

$$x = \sum_{j=1}^{n} a_j x_j, \quad \text{where } a_j \geqslant 0 \text{ and } \sum_{j=1}^{n} a_j = 1. \tag{1.2.14}$$

A *closed interval* $[x, y]$, where $x, y \in X$, is the set of all convex combinations of x and y, i.e. the set

$$[x, y] = \{tx + (1-t)y: 0 \leqslant t \leqslant 1\}. \tag{1.2.15}$$

A subset Y of a linear space X over $\mathbf{R}$ is said to be a *convex set* if $x, y \in Y$ implies $[x, y] \subset Y$. In other words, a set Y is *convex* if Y contains all convex combinations of every pair of elements $x, y \in Y$.

The *convex hull* of a subset Y in a linear space X over $\mathbf{R}$ is the smallest convex set containing the set Y. The convex hull of a set Y is denoted by $\operatorname{conv} Y$. We can prove that

$$\operatorname{conv} Y = \left\{ x \in X: x = \sum_{j=1}^{n} a_j x_j, \ x_j \in Y, \ a_j \geqslant 0, \ \sum_{j=1}^{n} a_j = 1 \right\}. \tag{1.2.16}$$

If a linear space X over a field $\mathscr{F}$ is a ring (with respect to the same addition) and $a(xy) = (ax)y = x(ay)$ for all $a \in \mathscr{F}$ and $x, y \in X$ then we call X a *linear ring* (or *algebra*). A nonempty subset Y of a linear ring X is called a *linear subring* (*subalgebra*) if Y is a linear ring with respect to the same addition, multiplication of elements and multiplication by scalars. A nonempty subset Y of a linear ring over a field $\mathscr{F}$ is a linear subring if for every $x, y \in Y$, $t \in \mathscr{F}$ we have $x - y \in Y$, $xy \in Y$, $tx \in Y$.

Suppose that the subring Y of a linear ring X over a field $\mathscr{F}$ is a proper ideal in X. Consider the set $X/Y = \{[x] : x \subset X\}$, where, as before, $[x] = x + Y$, with the addition and the multiplication by scalars defined by Formulae (1.2.12). Define the multiplication of cosets as follows:

$$[x][y] = [xy] \quad \text{for } x, y \in Y. \tag{1.2.17}$$

It is easy to verify that the set X/Y with the operations on cosets defined by Formulae (1.2.12) and (1.2.17) is a linear ring over the field $\mathscr{F}$. This ring is called a *quotient linear ring* (*quotient algebra*).

One can prove that every linear ring can be extended to a linear ring with unit.

A subset Y of a linear space X over the field $\mathscr{F}$ is said to be a *linear manifold* if $x, y \in Y$, $a, b \in \mathscr{F}$ and $a + b = 1$ imply $ax + by \in Y$.

Examples and Exercises

EXAMPLE 1.2.1. Define $\mathbf{R}^1 = \mathbf{R}$ and $\mathbf{R}^{k+1} = \mathbf{R} \times \mathbf{R}^k$ for $k = 1, 2, ..., \mathbf{R}^n$ is a linear space over $\mathbf{R}$ with respect to the usual addition of vectors and multiplication of a vector by a scalar: if $x = (x_1, ..., x_n), y = (y_1, ..., y_n) \in \mathbf{R}^n$, $t \in \mathbf{R}$, then $(x_1, ..., x_n) + (y_1, ..., y_n) = (x_1 + y_1, ..., x_n + +y_n)$, $t(x_1, ..., x_n) = (tx_1, ..., tx_n)$.

The set of all unit vectors, i.e. vectors of the form

$$(1, 0, ..., 0), (0, 1, 0, ..., 0), ..., (0, 0, ..., 0, 1),$$

is a basis in $\mathbf{R}^n$.

Of course, $\dim \mathbf{R}^n = n$. Observe that we can write also:

$$\mathbf{R}^n = \underbrace{\mathbf{R} \oplus ... \oplus \mathbf{R}}_{n\text{-fold}}.$$

EXAMPLE 1.2.2. The space $\mathbf{R}^2$ over $\mathbf{C}$ with the usual addition and multiplication by complex scalars, i.e. simply the plane $\mathbf{C}$ of complex numbers, is a linear space over $\mathbf{C}$ with a basis $\{1\}$. Of course dim $\mathbf{R}^2 = 1$. The same space $\mathbf{R}^2$, if we write $\mathbf{R}^2 = \mathbf{C} = \mathbf{R} + i\mathbf{R}$, has the basis $\{1, i\} = \{(1, 0), (0, 1)\}$ and dim $(\mathbf{R} + i\mathbf{R}) = 2$.

EXAMPLE 1.2.3. The set $\mathscr{F}_n[t]$ ($t \in \mathbf{R}$ or $t \in \mathbf{C}$) of all polynomials of degree n with coefficients from a field $\mathscr{F}$: $w_n(t) = \sum_{k=0}^{n} a_k t^k$, where $a_0, \ldots$
$\ldots, a_n \in \mathscr{F}$ is a linear space over $\mathscr{F}$ if we define the addition and the multiplication by a scalar as follows:

$$w_n(t) + v_n(t) = \sum_{k=0}^{n} (a_k + b_k) t^k, \quad \text{where } v_n(t) = \sum_{k=0}^{n} b_k t^k,$$

(1.2.18)

$$a w_n(t) = \sum_{k=0}^{n} (a a_k) t^k, \quad \text{where } a \in \mathscr{F}.$$

A basis for the space $\mathscr{F}_n[t]$ is the set $\{1, t, \ldots, t^n\}$ and dim $\mathscr{F}_n[t] = n+1$.

EXAMPLE 1.2.4. The set $\mathscr{F}[t]$ of all polynomials with coefficients in $\mathscr{F}$ is a linear space over $\mathscr{F}$, if the multiplication by a scalar is defined as in Example 1.2.3 and the addition as follows: let $w_n(t) = \sum_{k=0}^{n} a_k t^k$ and
$v_m(t) = \sum_{k=0}^{m} b_k t^k$, where $a_k, b_k \in \mathscr{F}$, $n \leqslant m$. Then

$$w_n + v_m = u_N, \quad \text{where } N = \max(n, m) = m$$

and

$$u_N(t) = \sum_{k=0}^{N} c_k t^k,$$

$$\text{where } c_k = \begin{cases} a_k + b_k & \text{for } k = 0, 1, \ldots, \min(n, m) = n, \\ b_k & \text{for } k = n+1, \ldots, m. \end{cases}$$

The basis in the space $\mathscr{F}[t]$ is the set $\{1, t, t^2, \ldots\}$ and dim $\mathscr{F}[t] = +\infty$.

EXAMPLE 1.2.5. The set X of all functions defined on a set Ω and with values in a field $\mathscr{F}$ of scalars is a linear space over $\mathscr{F}$ if the addition and the multiplication by a scalar are defined as follows:

$$(x+y)(t) = x(t)+y(t); \quad (ax)(t) = ax(t)$$

$$\text{for } x, y \in X, t \in \Omega, a \in \mathscr{F}. \quad (1.2.19)$$

EXAMPLE 1.2.6. The set X of all real-valued functions defined and bounded on a set Ω with the addition and the multiplication by scalars defined by Formulae (1.2.19) is a linear space over $\mathbf{R}$. Indeed, a sum of two bounded functions and a product of a bounded function by a scalar are again bounded functions. Of course, dim $X = +\infty$.

EXAMPLE 1.2.7. The set $C[a, b]$ of all real-valued functions defined and continuous on a closed interval $[a, b]$ is a linear space over $\mathbf{R}$ if the addition and the multiplication by a scalar are defined by Formulae (1.2.19). Indeed, a sum of two continuous functions and a product of a continuous function by a scalar are again continuous functions. Of course, dim $C[a, b] = +\infty$.

EXAMPLE 1.2.8. The set X of all real-valued functions defined and continuous on a closed interval $[a, b]$ and differentiable at each point $t \in [a, b]$ is a linear space over $\mathbf{R}$ if the addition and the multiplication are defined by Formulae (1.2.19). This is a simple consequence of the following fact: if $x, y \in X$, then there exist the derivatives x' and y' for every $t \in [a, b]$ and

$$(x+y)' = x'+y', \, . \, (\lambda x)' = \lambda x' \quad \text{for } \lambda \in \mathbf{R}. \quad (1.2.20)$$

EXAMPLE 1.2.9. The set $C^1[a, b]$ of all real-valued functions defined on a closed interval $[a, b]$ and having a continuous derivative in (a, b) is a linear space over $\mathbf{R}$ with the addition and the multiplication by a scalar defined by Formulae (1.2.19), if we assume that there exist limits: $x'(a) = \lim_{t \to a+0} x'(t)$ and $x'(b) = \lim_{t \to b-0} x'(t)$. Indeed, if $x, y \in C^1[a, b]$, $\lambda \in \mathbf{R}$, then Formulae (1.2.20) imply that $x'+y' = (x+y)'$ and $(\lambda x)' = \lambda x'$ are continuous functions. Of course, dim $C^1[a, b] = +\infty$.

EXERCISE 1.2.1. Prove that:

(a) the set $C^n[a, b]$ of all real-valued functions defined on a closed interval $[a, b]$ and having a continuous derivative of the order n in (a, b) is a linear space over $\mathbf{R}$ (we admit here a similar convention as in Example 1.2.9 concerning $x^{(k)}(a)$ and $x^{(k)}(b)$ $(k = 0, 1, ..., n))$;

(b) the space $C^{n+1}[a, b]$ is a linear subspace of the space $C^n[a, b]$ $(n = 0, 1, 2, ...)$, where we write $C^0[a, b] = C[a, b]$;

(c) the set $C^\infty[a, b]$ of all real-valued functions defined on a closed interval $[a, b]$ and having continuous derivatives of an arbitrary order in (a, b) is a linear space over $\mathbf{R}$;

(d) $C^\infty[a, b]$ is a linear subspace of each of spaces $C^n[a, b]$ $(n = 0, 1, 2, ...)$.

EXERCISE 1.2.2. Suppose, we are given two disjoint sets E_1 and E_2. If E is an arbitrary set then we denote by X_E a linear space of all functions defined on the set E and with values in a field $\mathscr{F}$ of scalars. Prove that $X_{E_1 \cup E_2} = X_{E_1} \oplus X_{E_2}$.

EXAMPLE 1.2.10. The linear space $C[a, b]$ determined in Example 1.2.7 is a linear ring over $\mathbf{R}$ if we define the multiplication of functions in the usual way:

$$(xy)(t) = x(t)y(t) \quad \text{for } x, y \in C[a, b], t \in [a, b]. \qquad (1.2.21)$$

For an arbitrary fixed $t_0 \in [a, b]$ the set

$$X_0 = \{x \in C[a, b]: x(t_0) = 0\}$$

is a proper ideal in the ring $C[a, b]$. Indeed, if $x, y \in X_0$ then $x(t_0) - -y(t_0) = 0$, hence $x - y \in X_0$. If $x \in X$, $y \in X_0$, then $x(t_0)y(t_0) = 0$, $y(t_0)x(t_0) = 0$, hence $xy, yx \in X_0$. Moreover, observe that the function $e(t) \equiv 1$, which is the unit of the ring under consideration, does not belong to X_0. Indeed, $e(t_0) = 1 \neq 0$. The quotient linear ring $C[a, b]/X_0$ can be identified with the set of all functions constant on the interval $[a, b]$. This follows from the fact that $y \in [x]$ for an $x \in C[a, b]$ if and only if $x - y \in X_0$, i.e. $x(t_0) = y(t_0)$. We therefore conclude that codim $X_0 = \dim C[a, b]/X_0 = 1$.

EXERCISE 1.2.3. Prove that:
(a) the set
$$X_1 = \{x \in C[a, b]: x(t_0) = 0, x(t_1) = 0 \text{ for } a < t_0 < t_1 < b\}$$
is a proper ideal in the linear ring $C[a, b]$ (cf. Example 1.2.10) where t_0, t_1 are arbitrarily fixed in (a, b);
(b) codim $X_1 = \dim C[a, b]/X_1 = 2$. The quotient ring $C[a, b]/X_1$ can be identified with the set $\{x \in C[a, b]: x(t) = \alpha t + \beta \text{ for } \alpha, \beta \in \mathbf{R}\}$.

EXERCISE 1.2.4. Prove that:
(a) the set
$$Y = \{x \in C[a, b]: x(t) = 0 \text{ for } a \leqslant t_0 \leqslant t \leqslant t_1 \leqslant b\}$$
is a proper ideal in the linear ring $C[a, b]$ if $t_1 \neq t_0$ and either $t_0 \neq a$ or $t_1 \neq b$;
(b) if the conditions of point (a) are satisfied then the quotient ring $C[a, b]/Y$ can be identified with the space $C[t_0, t_1]$; thus codim $Y = \dim C[t_0, t_1]/Y = +\infty$.

EXAMPLE 1.2.11. A convex hull of two points $a, b \in \mathbf{R}$ is the closed interval $[a, b]$. A convex hull of 3 linearly independent points in $\mathbf{R}^2$ is a triangle with vertices in these points. A convex hull of $n+1$ linearly independent points in $\mathbf{R}^n$ is an n-dimensional simplex with vertices at these points ($n = 2, 3, ...$).

EXERCISE 1.2.5. Determine a convex hull of n points in $\mathbf{R}^2$ and $\mathbf{R}^3$ ($n = 2, 3, ...$).

EXERCISE 1.2.6. Are the space X defined in Example 1.2.8, the spaces $C^n[a, b]$ ($n = 0, 1, 2, ...$) and the space $C^\infty[a, b]$ linear rings with respect to the multiplication defined by Formula (1.2.21)?

EXERCISE 1.2.7. A multiplication in a ring X is called *trivial* if $xy = 0$ for all $x, y \in X$. Prove that every linear space is a linear ring with a trivial multiplication.

EXERCISE 1.2.8. Suppose that X is the set of all sequences $a = \{a_n\}$, where $a_n \in \mathbf{R}$ $(n = 0, 1, 2, ...)$. If $a = \{a_n\}$, $b = \{b_n\} \in X$, then the addition and multiplication by a scalar $\lambda \in \mathbf{R}$ is defined as the addition and multiplication by λ of coordinates, i.e.

$$a+b = \{a_n+b_n\}, \quad \lambda a = \{\lambda a_n\}. \tag{1.2.22}$$

A *convolution* $a*b$ is, by definition, a sequence

$$a*b = \left\{\sum_{k=0}^{n} a_k b_{n-k}\right\}. \tag{1.2.23}$$

Prove that X is a commutative linear ring over $\mathbf{R}$ with a unit if the addition and multiplication by a scalar are defined by Formulae (1.2.22) and the multiplication of two elements $a, b \in X$ by its convolution, i.e. by Formula (1.2.23). Does this ring have zero divisors?

EXERCISE 1.2.9. Determine linear manifolds in $\mathbf{R}$, $\mathbf{R}^2$, $\mathbf{R}^3$.

EXERCISE 1.2.10. Suppose that X is a linear space over a field $\mathscr{F}$, Y is a linear manifold in X and $x_0 \in Y$ is arbitrarily fixed.

(a) Prove that the set $Y_0 = Y - x_0 = \{y - x_0 : y \in Y\}$ is a linear subspace of X which is independent of the choice of $x_0 \in Y$.

(b) We say that a linear manifold Y *has the codimension* 1 if the quotient space X/Y, where Y is defined in Point (a) has the dimension 1. A linear manifold of codimension 1 is called a *hyperplane*. Determine hyperplanes in $\mathbf{R}$, $\mathbf{R}^2$, $\mathbf{R}^3$.

EXERCISE 1.2.11. Prove that the set $\mathscr{F}[t]$ determined in Example 1.2.4 is a linear ring if we define multiplication of two polynomials as follows:

$$\left(\sum_{k=0}^{n} a_k t^k\right)\left(\sum_{j=0}^{m} b_j t^j\right) = \sum_{k=0}^{n} \sum_{j=0}^{m} a_k b_j t^{j+k}.$$

1.3. LINEAR OPERATORS AND LINEAR FUNCTIONALS

Suppose, we are given two linear spaces X and Y, both over the same field $\mathscr{F}$ of scalars. A mapping A of a linear subset dom A of X into Y is said to be a *linear operator* if

$$A(x+y) = Ax + Ay \quad \text{for all } x, y \in \text{dom } A, \tag{1.3.1}$$

$$A(tx) = tAx \quad \text{for all } x \in \text{dom } A, t \in \mathscr{F}. \tag{1.3.2}$$

The set dom A is called the *domain* of the operator A. Precisely, a linear operator is a pair $(A, \text{dom } A)$, because it is determined by its domain and by the form of the mapping A. However, in our subsequent considerations we shall write traditionally A instead of $(A, \text{dom } A)$, since this notation does not lead to any misunderstanding.

Suppose that $G \subset \text{dom } A$. Write

$$AG = \{Ax : x \in G\}. \tag{1.3.3}$$

By definition, $AG \subset Y$. The set AG is called the *image* of the set G. The set $A \text{ dom } A$ is called the *range* of operator A (the *set of values of A*).

A *graph* of a linear operator A is a subset of the Cartesian product $X \times Y$ defined as follows:

$$\text{graph } A = \{(x, y) : x \in \text{dom } A, y = Ax\}. \tag{1.3.4}$$

A set of all linear operators with the domain contained in the space X and the range contained in the space Y will be denoted by $L(X \to Y)$.

An *identity operator* in the space X is an operator I_X defined by means of the equality:

$$I_X x = x \quad \text{for all } x \in X. \tag{1.3.5}$$

Sometimes, if it does not lead to any misunderstanding, we shall write I instead of I_X.

If an operator $A \in L(X \to Y)$ is one-to-one we can define an *inverse operator* A^{-1} in the following way: for all $y \in A \text{ dom } A$

$$A^{-1}y = x, \quad \text{where } x \in \text{dom } A \text{ and } y = Ax. \tag{1.3.6}$$

Observe that, by our assumption, to every y there corresponds a unique $x \in \text{dom } A$ and, by definition,

$$\text{dom } A^{-1} = A \text{ dom } A \subset Y, \quad A^{-1} \text{dom } A^{-1} = \text{dom } A \subset X. \tag{1.3.7}$$

For every $x \in \text{dom } A$, if $y = Ax$ then

$$(A^{-1}A)x = A^{-1}(Ax) = A^{-1}y = x,$$
$$(AA^{-1})y = A(A^{-1}y) = Ax = y.$$

Thus

$$A^{-1}A = I_{\text{dom }A}, \quad AA^{-1} = I_{A\text{dom }A}. \tag{1.3.8}$$

Hence A^{-1} is a uniquely determined inverse of A. It is easy to verify that A^{-1} is also a linear operator.

If an operator $A \in L(X \to Y)$ has an inverse operator, we say that A is *invertible*.

An operator $A \in L(X \to Y)$ is called an *isomorphism* if dom $A = X$, A dom $A = Y$ and if A is one-to-one. By definition, if A is an isomorphism then it is invertible, the inverse operator A^{-1} is one-to-one and dom $A^{-1} = Y$, A^{-1}dom $A^{-1} = X$. Hence A^{-1} is also an isomorphism.

Two spaces X and Y are *isomorphic* if there is an isomorphism A mapping X onto Y.

We define the *sum of two operators* $A, B \in L(X \to Y)$ and the *product of an operator by a scalar* as follows: $\text{dom}(A+B) = \text{dom } A \cap \text{dom } B$ and

$$\begin{aligned}
(A+B)x &= Ax+Bx \quad \text{for } x \in \text{dom } A \cap \text{dom } B, \\
(tA)x &= t(Ax) \quad \text{for } x \in \text{dom } A, t \in \mathscr{F}.
\end{aligned} \tag{1.3.9}$$

If dom A = dom B = dom C then $(A+B)+C = A+(B+C)$ and $A+B = B+A$.

Observe, that an operator C such that $A+C = B$ for $A, B \in L(X \to Y)$ does not necessarily exist. For instance, it is so in the case when dom A $\cap$dom $B = \{0\}$. If such operator C exists, we write $C = B-A$ and we call C the *difference* of the operators B and A. The operation $-$ is called *subtraction* of operators. By definition, if $B-A$ is well-determined then

$$B-A = B+(-A) \quad \text{on dom } A \cap \text{dom } B.$$

Write

$$L_0(X \to Y) = \{A \in L(X \to Y): \text{dom } A = X\}.$$

Since the addition of two arbitrary operators belonging for $L_0(X \to Y)$ is well-determined, associative, commutative and to every operators $A, B \in L_0(X \to Y)$ there exists an operator $C = B-A$, we conclude that $L_0(X \to Y)$ is an Abelian group. A neutral element of this group is an operator Θ such that $\Theta x = 0$ for every $x \in X$. In the sequel we shall denote this "zero operator" Θ by 0, since it does not lead to any misunderstanding. Formulae (1.3.9) imply that the Abelian group $L_0(X \to Y)$ is a linear space over the field $\mathscr{F}$.

Suppose now that X, Y, Z are linear spaces over a field of scalars, $A \in L(Y \to Z)$, $B \in L(X \to Y)$ and $B \operatorname{dom} B \subset \operatorname{dom} A \subset Y$. A *superposition* (*product*) AB of operators A and B is defined as follows:

$$(AB)x = A(Bx) \quad \text{for all } x \in \operatorname{dom} B. \tag{1.3.10}$$

By definition, $AB \in L(X \to Z)$, $\operatorname{dom} AB = \operatorname{dom} B$, $AB \operatorname{dom} AB = A \operatorname{dom} B$. The superposition (if is well-determined) is distributive with respect to the addition of operators. The superposition (if is well-determined) is also associative.

Two operators A and B are said to be *commutative* if both superpositions AB, BA exist and

$$AB = BA \quad \text{on } \operatorname{dom} A \cap \operatorname{dom} B. \tag{1.3.11}$$

Write

$$L(X) = L(X \to X)$$

and

$$L_0(X) = L_0(X \to X) = \{A \in L(X): \operatorname{dom} A = X\}.$$

Formulae (1.3.10) imply that $L_0(X)$ is not only a linear space but also a linear ring with respect to the multiplication of two operators $A, B \in L_0(X)$ defined as their superposition AB. Indeed, if $A, B \in L_0(X)$ then $B \operatorname{dom} B \subset \operatorname{dom} A = X$. Hence AB is well-determined for all $A, B \in L_0(X)$. The linear ring $L_0(X)$ has a unit, namely the identity operator $I_X = I$. However, $L_0(X)$ is a non-commutative ring with zero divisors.

An operator $P \in L_0(X)$ is called a *projection* (*projection operator, projector*) if $P^2 = P$, where we write $P^2 = P \cdot P$. If $P \in L_0(X)$ is a projection then also $I - P$ is a projection.

THEOREM 1.3.1. *A projection $P \in L_0(X)$ determines a decomposition of the space X onto a direct sum $X = Y \oplus Z$, where*

$$\begin{aligned} Y &= \{x \in X: Px = x\} = PX, \\ Z &= \{x \in X: Px = 0\} = (I - P)X. \end{aligned} \tag{1.3.12}$$

Conversely, if $X = Y \oplus Z$, then there exists a projection $P \in L_0(X)$ such that $Y = PX$, $Z = (I-P)X$.

Having already proved this one-to-one correspondence between projections and decompositions of a space X onto direct sum, we can say that "Y is a projection of X in the direction Z" or that the operator P projects X onto Y in "the direction Z". Observe that, if P projects X onto Y in the direction Z then $I-P$ projects X onto Z in the direction Y.

Suppose that X_0 is a linear subspace of a linear space X. Then every operator $A \in L_0(X \rightarrow Y)$ such that $AX_0 \subset X_0$, induces a linear operator $[A] \in L_0(X/X_0 \rightarrow Y/AX_0)$ by means of the equality:

$$[A][x] = [Ax] \quad \text{for } x \in X, \tag{1.3.13}$$

where $[x]$ is a coset in the quotient space X/X_0 determined by an element $x \in X$, i.e. $[x] = x+X_0$, $[Ax]$ is a coset in the quotient space Y/AX_0, i.e. $[Ax] = Ax+AX_0$.

If X_0 is a subspace of a linear space X and $A \in L(X \rightarrow Y)$ then a *restriction* $A|_{X_0}$ of A to the subspace X_0 is defined as follows:

$$A|_{X_0} x = Ax \quad \text{for } x \in X_0 \cap \text{dom } A. \tag{1.3.14}$$

An operator $A_1 \in L_0(X \rightarrow Y)$ is called an *extension* of an operator $A \in L_0(X_0 \rightarrow Y)$ to the space X, where X_0 is a linear subspace of X, if

$$A_1 x = Ax \quad \text{for } x \in X_0,$$

i.e. if A is a restriction of the operator A_1 to the subspace X_0.

Suppose that $A \in L(X \rightarrow Y)$. Write

$$\ker A = \{x \in \text{dom } A : Ax = 0\}. \tag{1.3.15}$$

The set $\ker A$ is called the *kernel* of the operator A and is a linear subspace of X.

The dimension of the kernel of an operator $A \in L(X \rightarrow Y)$ is called the *nullity* of A and is denoted by α_A, i.e.

$$\alpha_A = \dim \ker A. \tag{1.3.16}$$

The *cokernel* (or a *defect space*) of an operator $A \in L(X \rightarrow Y)$ is the

quotient space Y/A dom A. The *deficiency* β_A of an operator $A \in L(X \to Y)$ is defined by the equality:

$$\beta_A = \dim Y/A \text{ dom } A = \text{codim } A \text{ dom } A. \tag{1.3.17}$$

By definition the deficiency of an operator A is equal to the defect of its range.

If at least one of the numbers α_A, β_A is finite we define the *index* $\varkappa_A$ of an operator $A \in L(X \to Y)$ in the following way:

$$\varkappa_A = \begin{cases} \beta_A - \alpha_A & \text{if } \alpha_A < +\infty, \beta_A < +\infty, \\ +\infty & \text{if } \alpha_A < +\infty, \beta_A = +\infty, \\ -\infty & \text{if } \alpha_A = +\infty, \beta_A < +\infty. \end{cases} \tag{1.3.18}$$

THEOREM 1.3.2. *If in a linear space X the maximal number of linearly independent elements is n then* $\dim X = n$.

An immediate consequence of Theorems 1.2.1 and 1.3.2 is

COROLLARY 1.3.1. *If X is an n-dimensional linear space then every set of n linearly independent elements $\{x_1, \ldots, x_n\} \subset X$, is a basis in X.*

THEOREM 1.3.3. *If Y is a subspace of a linear space X and the codimension of Y is finite then there exists a subspace $Z \subset X$ such that $X = Y \oplus Z$ and* $\dim Z = \text{codim } Y$.

The construction of the subspace Z is the following. By assumptions there exists n linearly independent cosets $[x_1], \ldots, [x_n] \in X/Y$ and the coset $[x] \in X/Y$ can be written in a unique way in the form:

$$[x] = \sum_{j=1}^{n} t_j [x_j], \quad \text{where } t_1, \ldots, t_n \text{ are scalars.}$$

Let $y_1, \ldots, y_n \in X$ be arbitrary elements such that $y_j \in [x_j]$ $(j = 1, 2, \ldots, n)$. The elements $y_1, \ldots, y_n$ are linearly independent. Thus every element $x \in Y$ can be written uniquely in the form

$$x = y + \sum_{j=1}^{n} t_j y_j, \quad \text{where } y \in Y.$$

Writing $Z = \lin\{y_1, \dots, y_n\}$ we conclude that $X = Y \oplus Z$ and $\dim Z = n$.

If we assume that the *axiom of choice* is satisfied, Theorem 1.3.3 is true without the assumption that $\operatorname{codim} Y < +\infty$. Namely, we have

THEOREM 1.3.4. *If Y is an arbitrary subspace of a linear space X then there exists a linear subspace $Z \subset X$ such that $X = Y \oplus Z$.*

A consequence of Theorem 1.3.4 (or Theorem 1.3.3 if $\operatorname{codim} Y < +\infty$) is

COROLLARY 1.3.2. *Let X_0 be a subspace of a linear space X. Then every linear operator $A \in L(X \to Y)$, such that $\operatorname{dom} A = X_0$ can be extended to an operator A_1 such that $\operatorname{dom} A_1 = X$ and $A_1 \operatorname{dom} A_1 = A_1 X = A \operatorname{dom} A$, i.e. the range is preserved.*

Suppose now that X is a linear space over a field $\mathscr{F}$ of scalars and that either $\mathscr{F} = \mathbf{R}$ or $\mathscr{F} = \mathbf{C}$.

An operator $f \in L_0(X \to \mathscr{F})$ is called a *linear functional*.

Write $X' = L_0(X \to \mathscr{F})$. By definition, X' is a linear space over $\mathscr{F}$ which consists of all linear functionals determined on the space X. We say that X' is a *space conjugate* to X.

If X is an n-dimensional space over $\mathscr{F}$ (where either $\mathscr{F} = \mathbf{R}$ or $\mathscr{F} = \mathbf{C}$) with a basis $\{x_1, \dots, x_n\}$ then every linear functional f on the space X is of the form

$$f(x) = \sum_{j=1}^{n} t_j a_j, \quad \text{where } x = \sum_{j=1}^{n} t_j x_j \in X, \quad t_1, \dots, t_n \in \mathscr{F}$$

$$(1.3.19)$$

and $a_j = f(x_j)$ $(j = 1, \dots, n)$, i.e. f is uniquely determined by its values on the elements of the basis in X.

THEOREM 1.3.5. *Suppose that X is an n-dimensional linear space over $\mathscr{F}$ (where either $\mathscr{F} = \mathbf{R}$ or $\mathscr{F} = \mathbf{C}$) with a basis $\{x_1, \dots, x_n\}$. Then there exist functionals $f_1, \dots, f_n$ which are linearly independent and such that*

$$f_j(x_m) = \delta_{jm} \quad (j, m = 1, 2, \dots, n),$$

$$(1.3.20)$$

where δ_{jm} is the so-called Kronecker symbol:

$$\delta_{jm} = \begin{cases} 1 & \text{if } j = m, \\ 0 & \text{if } j \neq m. \end{cases} \tag{1.3.21}$$

Namely, $f_j(x) = t_j$ $(j = 1, ..., n)$ for every $x = \sum_{j=1}^{n} t_j x_j \in X$.

COROLLARY 1.3.3. *If X is an n-dimensional space then* $\dim X' = \dim X$ $= n$ *and the set* $\{f_1, ..., f_n\}$ *(where the functionals $f_1, ..., f_n \in X'$ are determined in Theorem 1.3.5) is a basis in X'.*

THEOREM 1.3.6. *Suppose that X is a linear space (over $\mathbf{R}$ or $\mathbf{C}$) and that $g, f_1, ..., f_n \in X'$. If*

$$f_j(x) = 0 \text{ for } j = 1, 2, ..., n \quad \text{implies} \quad g(x) = 0 \tag{1.3.22}$$

then the functional g is linearly dependent on the functionals $f_1, ..., f_n$.

Suppose that X is an n-dimensional linear space with a basis $\{x_1, ..., x_n\}$ and Y is an m-dimensional linear space with a basis $\{y_1, ..., y_m\}$, both over the same field $\mathscr{F}$ of scalars. Let $A \in L_0(X \to Y)$. Let $x = \sum_{j=1}^{n} t_j x_j \in X$, where $t_1, ..., t_n \in \mathscr{F}$, be arbitrary. Then

$$Ax = A \sum_{j=1}^{n} t_j x_j = \sum_{j=1}^{n} t_j A x_j.$$

On the other hand, since $Ax \in Y$, we can find $c_1, ..., c_m$ such that $Ax = \sum_{k=1}^{n} c_k y_k$. Indeed, since $Ax_j \in Y$, we have $Ax_j = \sum_{k=1}^{m} a_{jk} y_k$, where $a_{jk} \in \mathscr{F}$ $(j = 1, 2, ..., n; \ k = 1, 2, ..., m)$. Thus

$$Ax = \sum_{j=1}^{n} t_j A x_j = \sum_{j=1}^{n} t_j \sum_{k=1}^{m} a_{jk} y_k = \sum_{k=1}^{m} \left(\sum_{j=1}^{n} t_j a_{jk} \right) y_k.$$

We therefore conclude that

$$c_k = \sum_{j=1}^{n} t_j a_{jk} \quad (k = 1, 2, ..., m).$$

The coefficients a_{jk} determine the transformation of the basis $\{x_1, \ldots, x_n\}$ into the basis $\{y_1, \ldots, y_n\}$ by the operator A. Thus there is a *one-to-one correspondence* between the operators $A \in L_0(X \to Y)$ and the matrices

$$\begin{bmatrix} a_{11} & a_{21} & \ldots & a_{n1} \\ a_{12} & a_{22} & \ldots & a_{n2} \\ \cdots\cdots\cdots\cdots\cdots \\ a_{1m} & a_{2m} & \ldots & a_{nm} \end{bmatrix} = (a_{jk})_{\substack{j=1,\ldots,n \\ k=1,\ldots,m}}$$

of dimension $m \times n$. We shall denote the operator A and its matrix by the same letter A. Sometimes we shall write briefly

$$A = (a_{jk})_{\substack{j=1,\ldots,n \\ k=1,\ldots,m}}.$$

THEOREM 1.3.7. *Suppose that X, Y, Z are linear spaces over a field $\mathscr{F}$, $\dim X = n$, $\dim Y = m$ and $\dim Z = p$. If*

$$(a_{jk})_{\substack{j=1,\ldots,m \\ k=1,\ldots,p}} = A \in L_0(Y \to Z),$$

$$(b_{lj})_{\substack{l=1,\ldots,n \\ j=1,\ldots,m}} = B \in L_0(X \to Y)$$

then AB exists and

$$AB = (c_{lk})_{\substack{l=1,\ldots,n \\ k=1,\ldots,p}}, \quad \text{where } c_{lk} = \sum_{j=1}^{m} a_{jk} b_{lj}. \tag{1.3.23}$$

An operator $A \in L_0(X \to Y)$ is said to be *finite dimensional* if its range is finite dimensional. If $\dim A \operatorname{dom} A = n$, we say that A is an *n-dimensional operator*. Clearly, if $A \in L_0(X \to Y)$ and $\dim Y < +\infty$ then A is a finite dimensional operator.

PROPOSITION 1.3.1. *An operator $K \in L_0(X \to Y)$ is n-dimensional if and only if there exist $f_1, \ldots, f_n \in X'$ and linearly independent $y_1, \ldots, y_n \in Y$ such that*

$$Kx = \sum_{j=1}^{n} f_j(x) y_j \quad \text{for all } x \in X. \tag{1.3.24}$$

Suppose that we are given a class $\mathscr{E}$ (non-necessarily linear) of linear operators. An operator B is said to be an $\mathscr{E}$-*perturbation of an operator* $A \in \mathscr{E}$ if $A + B \in \mathscr{E}$. An operator B is said to be a *perturbation of the class* $\mathscr{E}$ if $A + B \in \mathscr{E}$ for every $A \in \mathscr{E}$. By $\Pi_{\mathscr{E}}$ we shall denote the set of all perturbations of the class $\mathscr{E}$. This set is additive, i.e. if A_1, A_2 are perturbations of the class $\mathscr{E}$ then $A_1 + A_2$ is again a perturbation of $\mathscr{E}$. If $\mathscr{E}$ is homogeneous, i.e. $\alpha A \in \mathscr{E}$ for every scalar α and every $A \in \mathscr{E}$, then the set $\Pi_{\mathscr{E}}$ is linear (cf. the author and S. Rolewicz, 1968).

THEOREM 1.3.8. *Every finite dimensional operator* $K \in L(X \to Y)$ *is an index preserving perturbation of the class of all operators* $A \in L(X \to Y)$ *with a finite nullity (deficiency), i.e.* $\varkappa_{A+K} = \varkappa_A$.

THEOREM 1.3.9. *Suppose that the class of all operators belonging to* $L_0(X \to Y)$ *with a finite nullity (deficiency) is non-empty. Then an operator* $K \in L_0(X \to Y)$ *is a perturbation of this class if and only if* K *is finite dimensional.*

An operator $A \in L_0(X \to Y)$ is said to be *right (left) invertible* if there is an operator $B \in L_0(Y \to X)$ such that $AB = I_Y$ (respectively: $BA = I_X$). One can prove that

(a) A is right invertible if and only if is a mapping onto, i.e. its deficiency $\beta_A = 0$,

(b) A is left invertible if $\ker A = \{0\}$, i.e. its nullity $\alpha_A = 0$,

(c) If A is simultaneously left and right invertible then A is invertible.

THEOREM 1.3.10. *An operator* $A \in L_0(X \to Y)$ *has a finite nullity (deficiency) if and only if* $A = S + K$, *where* $S \in L_0(X \to Y)$ *is left (right) invertible and* K *is finite dimensional.*

COROLLARY 1.3.4. *An operator* $A \in L_0(X \to Y)$ *with finite nullity and deficiency has the index zero:* $\varkappa_A = 0$ *if and only if* $A = S + K$, *where* S *is invertible and* K *is finite dimensional.*

Examples and Exercises

EXAMPLE 1.3.1. Consider the space **C** of all complex numbers with usual addition and multiplication by reals. Define an operator A by

$$Az = az + b\bar{z}, \quad \text{where } a, b \in \mathbf{C} \ (z \in \mathbf{C}).$$

It is easy to check that A is a linear operator acting on **C**. Write: $f(z) = \text{Re } z$, $g(z) = \text{Im } z$, $h(z) = |z|$, where $z \in \mathbf{C}$. It is easy to check that f and g are linear functionals (because they map linearly **C** into **R**). Observe that h is not a linear functional. Indeed,

$$h(1+i) = |1+i| = \sqrt{2}\,; \quad h(1) + h(i) = |1| + |i| = 2 \neq h(1+i).$$

EXAMPLE 1.3.2. Consider the space $X = C[a, b]$. Suppose, we are given a function $g \in C[a, b]$ such that $g(t) \neq 0$ for $a \leqslant t \leqslant b$. Write: $(Ax)(t) = g(t)x(t)$ for $x \in C[a, b]$. It is easy to check that $A \in L_0(X)$. Moreover, A is invertible and $(A^{-1}x)(t) = x(t)/g(t)$.

Suppose now that $h \in C^1[a, b]$, $a \leqslant h(t) \leqslant b$, $h(a) = a, h(b) = b$ and $h'(t) > 0$ for $a \leqslant t \leqslant b$. Write

$$(Bx)(t) = x\big(h(t)\big) \quad \text{for } x \in C[a, b], t \in [a, b].$$

It is easy to check that $B \in L_0(X)$ and that B is an invertible operator, namely

$$(B^{-1}x)(t) = x\big(h^{-1}(t)\big) \quad \text{for } x \in C[a, b], t \in [a, b],$$

where h^{-1} denotes the inverse function for h, which exists by our assumption. Write: $f(x) = \sum_{j=1}^{n} a_j x(t_j)$ for $x \in C[a, b]$, where $a_1, \ldots, a_n \in \mathbf{R}$; $t_1, \ldots, t_n \in [a, b]$ are arbitrarily fixed. It is easy to verify that f is a linear functional over $C[a, b]$.

EXAMPLE 1.3.3. If linear spaces $X_1, \ldots, X_n$ are finite dimensional then we have obvious equalities:

$$\dim X_1 \times X_2 \times \ldots \times X_n = \dim X_1 + \dim X_2 + \ldots + \dim X_n,$$

$$(1.3.25)$$

$$\dim X_1 \oplus X_2 \oplus \ldots \oplus X_n = \dim X_1 + \dim X_2 + \ldots + \dim X_n.$$

If at least one space X_j is infinite dimensional then both, the Cartesian product $X_1 \times \ldots \times X_n$ and the direct sum $X_1 \oplus \ldots \oplus X_n$ are infinite dimensional. We therefore conclude that Formulae (1.3.25) hold also in the case of infinite dimensional spaces.

EXERCISE 1.3.1. Prove that a shift of points in $\mathbf{R}^n$ ($n = 1, 2, \ldots$) is not a linear operator.

EXERCISE 1.3.2. Give examples of projections in $\mathbf{R}^n$, $C[a, b]$ and $\mathscr{F}_n[t]$ (cf. Exercises 1.2.3, 1.2.5).

EXERCISE 1.3.3. Prove that a rotation through an angle φ in $\mathbf{R}^2$ is a linear operator. Determine the matrix of this operator.

EXERCISE 1.3.4. Suppose that X is a linear space over $\mathscr{F}$ (where either $\mathscr{F} = \mathbf{R}$ or $\mathscr{F} = \mathbf{C}$) and f is a linear functional on X. Prove that: (a) the set $H_f = \{x \in X : f(x) = 1\}$ is a hyperplane which does not contain the point zero; (b) if H is a hyperplane in X not containing zero then there exists a linear functional f such that $H = \{x \in X : f(x) = 1\}$ (cf. Exercise 1.2.10).

EXERCISE 1.3.5. Give examples of linear functionals in $\mathbf{R}^n$ and determine corresponding hyperplanes (cf. Exercise 1.3.4). *Example*: if $f(x) = x_1 + \ldots + x_n$ then $H_f = \{x \in \mathbf{R}^n : x_1 + \ldots + x_n = 1\}$, where $x = (x_1, \ldots \ldots, x_n)$.

EXERCISE 1.3.6. Give examples of linear functionals in $\mathscr{F}_n[t]$, $\mathscr{F}[t]$ and $C[a, b]$ and determine corresponding hyperplanes (cf. Exercises 1.2.3, 1.2.4, 1.2.5).

EXERCISE 1.3.7. Mappings preserving all linear manifolds are called *affinities* (or *affine transformations*). Describe affinities in $\mathbf{R}^2$ and $\mathbf{R}^3$.

EXERCISE 1.3.8. Prove that the set of all square matrices of dimension n is a linear ring with unit. Give examples of divisors of zero in the ring.

EXERCISE 1.3.9. Prove that the set of all finite dimensional operators acting in a linear space X is an ideal in the linear ring $L_0(X)$ and that this ideal is proper if X is of infinite dimension.

EXERCISE 1.3.10. Suppose that X is an n-dimensional linear space over $\mathscr{F}$ (where either $\mathscr{F} = \mathbf{R}$ or $\mathscr{F} = \mathbf{C}$) and $A \in L_0(X)$. Every one-to-one mapping B of a basis in X into another basis in X is called a *change of the coordinate system*. Any such B has an extension on the whole space X which is invertible and $B, B^{-1} \in L_0(X)$. Prove that a change of the coordinate system B maps every operator A into the operator $B^{-1}AB$. Give examples in $\mathbf{R}^n$.

EXERCISE 1.3.11. A square matrix $A = (a_{jk})_{j,k=1,\ldots,n}$ is called *diagonal* if $a_{jk} = 0$ for $k \neq j$ $(j, k = 1, \ldots, n)$. Prove that:
(a) If A is diagonal then A^m is diagonal $(m = 1, 2, \ldots)$.
(b) The set of all diagonal matrices of dimension n is a subring of the linear ring described in Exercise 1.3.8. Does this subring have zero divisors?

EXERCISE 1.3.12. An operator $A \in L_0(X)$ is said to be *nilpotent of order* k if $A^k = 0$ but $A^{k-1} \neq 0$. Suppose that $A \in L_0(X)$ is nilpotent of order k. Prove that there exists an $x_0 \in X$ such that $x_0 \neq 0$ and the elements $x_0, Ax_0, \ldots, A^{k-1}x_0$ are linearly independent.

EXERCISE 1.3.13. Suppose that $K \in L_0(X)$ is a finite dimensional operator. Prove that $\alpha_{I+K} = \beta_{I+K} = n-k$, where $n = \dim KX$, k is the rank of matrix $I+K = (f_{ji}(y_i))_{i,j=1,2,\ldots,n}$, $Kx = \sum_{i=1}^{n} f_i(x)y_i$ for $x \in X$.

1.4. EIGENSPACES AND PRINCIPAL SPACES. ALGEBRAIC OPERATORS. VOLTERRA OPERATORS

Suppose that X is a linear space over an algebraically closed field $\mathscr{F}$ and $A \in L_0(X)$. A scalar $\lambda \in \mathscr{F}$ is said to be a *regular value* of A if the operator $A - \lambda I$ is invertible. The set of all scalars λ which are not regular values of A is called *spectrum* of A and is denoted by spectr A. Evidently, spectr $A \subset \mathscr{F}$. If $\lambda \in$ spectr A and there exists an $x \in X$ such that $x \neq 0$ and $(A - \lambda I)x = 0$, i.e. if there exists a non-trivial solution of the equation $Ax = \lambda x$ then λ is called an *eigenvalue* of A and x is called an *eigenvector* corresponding to the eigenvalue λ. The linear span of all eigenvectors corresponding to an eigenvalue λ is called an *eigenspace* of the operator A corresponding to the eigenvalue λ. By definition, every eigenspace is of the form

$$\{x \in X : Ax = \lambda x\} = \ker(A - \lambda I) \tag{1.4.1}$$

hence is a linear subspace of X.

A *principal vector* corresponding to a $\lambda \in \mathscr{F}$ is an element $x \in X \setminus \{0\}$ such that there exists a positive integer n for which $(A - \lambda I)^n x = 0$, i.e. $x \subset \ker(A - \lambda I)^n$. A linear span of all principal vectors corresponding to the value λ is called a *principal space* of the operator A corresponding to the value λ. By definition, a principal space corresponding to a value λ is of the form

$$\bigcup_{n=1}^{\infty} \{x \in X : (A - \lambda I)^n x = 0\} = \bigcup_{n=1}^{\infty} \ker(A - \lambda I)^n. \tag{1.4.2}$$

The dimension of a principal space corresponding to an eigenvalue λ_0 of an operator A is called the *multiplicity* of the eigenvalue λ_0.

If there exist principal vectors corresponding to a scalar λ_0 then there exist eigenvectors corresponding to this value, namely, $x_0 = (A - \lambda I)^{n-1} x$, where n is the smallest positive integer such that $0 \ne x \in \ker(A - \lambda I)^n$. On the other hand, every eigenvector x corresponding to an eigenvalue λ_0 is a principal vector. We therefore conclude that the dimension of a principal space is not smaller than the dimension of the eigenspace corresponding to the eigenvalue.

PROPOSITION 1.4.1. *All eigenvalues of a square matrix $A = (a_{jk})_{j, k=1, \ldots, N}$ are roots of the equation*

$$\det(A - \lambda I) = 0. \tag{1.4.3}$$

The determinant $\det(A - \lambda I)$ is called the *characteristic determinant* of the matrix A. (We admit here that a unit matrix E corresponding to the identity operator I is denoted also by I). Observe that the determinant $\det(A - \lambda I)$ is a polynomial in λ.

We shall consider now a class of operators for which it will be easy to determine principal spaces and eigenspaces. In particular, we shall see that to this class belong all operators transforming finite dimensional spaces into themself.

Suppose that $\mathscr{F} = \mathbf{C}$. We say that an operator $A \in L_0(X)$ is *algebraic* if there exists a polynomial $P(t) = p_0 + p_1 t + \ldots + p_N t^N$ with $p_0, \ldots$ $\ldots, p_N \in \mathbf{C}$ such that $P(A) = 0$ on X. Without loss of generality we can assume that the polynomial $P(t)$ is *normed*, i.e. $p_N = 1$. We say that an algebraic operator $A \in L_0(X)$ *is of order* N if there does not exist a normed polynomial $Q(t)$ of degree $m < N$ such that $Q(A) = 0$ on X, i.e. if N is a minimal degree of a polynomial identity $P(A) = 0$ satisfied by the operator A. Such a minimal polynomial $P(t)$ is called the *characteristic polynomial* of A and its roots are called the *characteristic roots* of A. We shall see that having already known the characteristic polynomial and characteristic roots of an algebraic operator A we will be able to determine principal spaces and eigenspaces of A.

LEMMA 1.4.1 (Hermite interpolation formula with multiple knots). *There exists a unique polynomial $W(t)$ of degree $N-1$ which together with its derivatives admits the given values y_{ki} in the given different points $t_1, \ldots, t_n$. More precisely,*

$$W^{(k)}(t_i) = y_{ki}$$
$$(k = 0, 1, \ldots, r_i - 1; i = 1, 2, \ldots, n; r_1 + \ldots + r_n = N),$$

where we write

$$W^{(0)} = W, \quad W^{(k)} = \frac{\mathrm{d}^k W}{\mathrm{d} t^k} \quad (k = 1, 2, \ldots).$$

The polynomial $W(t)$, we are looking for, is of the form

$$W(t) = \sum_{i=1}^{n} \frac{P(t)}{(t-t_i)^{r_i}} \sum_{k=0}^{r_i-1} y_{ki} \left\{ \frac{(t-t_i)^{r_i}}{P(t)} \right\}_{(r_i-1-k;\, t_i)} \frac{(t-t_i)^k}{k!},$$

$$(1.4.4)$$

where we write

$$P(t) = \prod_{m=1}^{n} (t-t_m)^{r_m}; \quad \{f(t)\}_{(k,\, s)} = \sum_{m=0}^{k} \frac{\mathrm{d}^m f(t)}{\mathrm{d} t^m} \bigg|_{t=s} \frac{(t-s)^m}{m!}$$

$$(1.4.5)$$

for any function f k-times differentiable in a neighbourhood of the point s.

If $t_1, \ldots, t_n$ are single knots, then Hermite interpolation formula yields the *Lagrange interpolation formula*:

$$W(t) = \sum_{i=1}^{n} y_i \prod_{m=1, m\neq i}^{n} \frac{t-t_m}{t_i-t_m}.$$ (1.4.6)

LEMMA 1.4.2 (Partition of unit). *If*

$$p_i(t) = q_i(t) \prod_{m=1, m\neq i}^{n} (t-t_m)^{r_m};$$

$$q_i(t) = \left\{ \frac{(t-t_i)^{r_i}}{P(t)} \right\}_{(r_i-1; t_i)} \quad (i = 1, 2, \ldots, n)$$ (1.4.7)

then

$$\sum_{i=1}^{n} p_i(t) \equiv 1$$ (1.4.8)

and this representation holds for arbitrary fixed points $t_1, \ldots, t_n$ and positive integers $r_1, \ldots, r_n$.

In the case of single knots Formula (1.4.8) is the following:

$$\sum_{i=1}^{n} \prod_{m=1, m\neq i}^{n} \frac{t-t_m}{t_i-t_m} \equiv 1.$$ (1.4.9)

THEOREM 1.4.1. *If $A \in L_0(X)$ then the following conditions are equivalent:*

(a) *A is an algebraic operator with a characteristic polynomial*

$$P(t) = \prod_{m=1}^{n} (t-t_m)^{r_m} \quad \text{of order } N = r_1 + \ldots + r_n;$$

(b) *There exist n operators $P_1, \ldots, P_n \in L_0(X)$ such that*

$$P_j P_k = \begin{cases} P_k & \text{for } j = k, \\ 0 & \text{for } j \neq k, \end{cases}$$

$$\sum_{j=1}^{n} P_j = I \quad \text{and} \quad (A - t_j I)^{r_j} P_j = 0 \quad (j, k = 1, 2, \ldots, n)$$

namely, $P_j = p_j(A)$, where the polynomials p_j are determined by Formula (1.4.7) (i.e. P_j are disjoint projectors giving a partition of unit);

(c) *The space X is a direct sum of n principal spaces of the operator A corresponding to the eigenvalues $t_1, ..., t_n$ with multiplicities $r_1, ..., r_n$, respectively, i.e.*

$$X = X_1 \oplus X_2 \oplus ... \oplus X_n,$$

where $X_j = \ker (A - t_j I)^{r_j}, j = 1, ..., n$.

Putting $r_j = 1$ for $j = 1, ..., n$ we have $n = N$ and we obtain immediately

COROLLARY 1.4.1. *If $A \in L_0(X)$ then the following conditions are equivalent:*

(a′) *A is an algebraic operator of order N and the characteristic polynomial $P(t)$ of A has single roots only, i.e.*

$$P(t) = \prod_{j=1}^{N} (t - t_j), \quad \text{where } t_i \neq t_j \text{ for } i \neq j;$$

(b′) *There exist N operators $P_j \in L_0(X)$ such that $P_j P_k = 0$ for $j \neq k$,*
$P_j^2 = P_j, \sum_{j=1}^{N} P_j = I$ *and $AP_j = t_j P_j$, namely*

$$P_j = \prod_{m=1, m \neq i}^{N} (t_j - t_m)^{-1}(A - t_m I) \quad (j, k = 1, 2, ..., N);$$

(c′) *The space X is a direct sum of N eigenspaces X_j of the operator A corresponding to the eigenvalues $t_1, ..., t_n$ respectively, i.e.*

$$X = X_1 \oplus ... \oplus X_N, \quad \text{where } X_j = \ker(A - t_j I), \quad j = 1, ..., n.$$

In other words, $x \in X_j$ if and only if $Ax = t_j x$ $(j = 1, 2, ..., N)$.

COROLLARY 1.4.2 (Cayley–Hamilton Theorem). *If $\dim X < +\infty$ then every operator $A \in L_0(X)$ is an algebraic operator and its characteristic polynomial is a divisor of the polynomial $Q(\lambda) = \det(A - \lambda I)$.*

It is an immediate consequence of the fact that to every operator $A \in L_0(X)$ there corresponds a square matrix $A = (a_{jk})_{j,k=1,...,\dim X}$, of Proposition 1.4.1 and of (c) of Theorem 1.4.1.

Suppose that $\dim X < +\infty$, $A \in L_0(X)$ and the characteristic polynomial of A is of the form $P(t) = \prod\limits_{m=1}^{n} (t-t_m)^{r_m}$.

A *Jordan matrix* corresponding to the characteristic root t_m is defined as a square matrix of dimension $k \leqslant r_m$ which is of the form

$$J_{m,k} = \begin{cases} (t_m) & \text{if } r_m = 1, \\ \begin{bmatrix} t_m & 1 & & 0 \\ & \ddots & \ddots & \\ & & t_m & 1 \\ & & & \ddots & \ddots \\ 0 & & & & t_m \end{bmatrix} & \begin{array}{l} (k = 1, 2, \ldots, r_m, \\ m = 1, 2, \ldots, n) \\ \text{if } r_m \geqslant 2. \end{array} \end{cases} \qquad (1.4.10)$$

THEOREM 1.4.2 (Jordan Theorem). *Suppose that* $\dim X < +\infty$ *and a square matrix* $A = (a_{jk})_{j,k=1,2,\ldots,\dim X}$ *has the characteristic polynomial* $P(t) = \prod\limits_{m=1}^{n} (t-t_m)^{r_m}$, $N = r_1 + \ldots + r_n$. *Then by a suitable change of basis in* X *the matrix* A *can be represented in the form:*

$$A = [\![J_{m,k}]\!]_{\substack{m=1,2,\ldots,n \\ k=0,1,\ldots,r_m}},$$

where we admit the following convention: If $A_1, \ldots, A_M$ *are square matrices of dimensions* $n_1, \ldots, n_M$, *respectively, then we denote by* $[A_i]_{i=1,\ldots,M}$ *the square matrix*

$$\begin{bmatrix} A_1 & & 0 \\ & \ddots & \\ 0 & & A_M \end{bmatrix}$$

of dimension $n_1, \ldots, n_M$; $J_{m,k}$ *are Jordan matrices corresponding to the characteristic root* t_m, *i.e. for* $k \leqslant r_m$ *we have* $J_{m,k} = (t_m)$ *if* $r_m = 1$ *and*

$$J_{m,k} = \begin{bmatrix} t_m & 1 & & 0 \\ & \ddots & \ddots & \\ & & t_m & 1 \\ & & & \ddots & \ddots \\ 0 & & & & t_m \end{bmatrix} \quad \text{if } r_m \geqslant 2$$

and some of $J_{m,k}$ *appear in the matrix* A *several times, some—zero times.*

The simple proof based on properties of algebraic operators has been given by Przeworska-Rolewicz, 1977.

PROPOSITION 1.4.2. *If $A \in L_0(X)$ is an algebraic operator with the characteristic polynomial $P(t) = \prod\limits_{m=1}^{n} (t-t_m)^{r_m}$ and*

$$Q_k = (A - t_k I) P_k \quad (k = 1, 2, ..., n), \tag{1.4.11}$$

where $P_1, ..., P_n$ are defined in Theorem 1.4.1, then

$$Q_j Q_k = 0 \quad for \ j \neq k, \tag{1.4.12}$$

$$P_k Q_j = Q_j P_k = \begin{cases} 0 & for \ j \neq k \\ Q_j & for \ j = k \end{cases} \quad (j, k = 1, 2, ..., n), \tag{1.4.13}$$

$$Q_k^m = (A - t_k I)^m P_k \quad for \ k = 1, 2, ..., n; m = 1, 2, ... \tag{1.4.14}$$

In particular

$$Q_k^{r_k} = 0 \quad (k = 1, 2, ..., n). \tag{1.4.15}$$

THEOREM 1.4.3. *Suppose that $A \in L_0(X)$ is an algebraic operator with the characteristic polynomial $P(t) = \prod\limits_{m=1}^{n} (t-t_m)^{r_m}$. Then*

$$A^m = \sum_{j=1}^{n} \left[t_j^m I + \sum_{k=1}^{m} \binom{m}{k} t_j^{m-k} (A - t_j I)^k \right] P_j \quad for \ m = 1, 2, ... \tag{1.4.16}$$

COROLLARY 1.4.3. *If A satisfies the assumption of Theorem 1.4.3, then*

$$A^m P_i = t_i^m P_i + \sum_{k=1}^{m} \binom{m}{k} t_i^{m-k} (A - t_i I)^k P_i \tag{1.4.17}$$

for $i = 1, 2, ..., n; m = 1, 2, ...$

COROLLARY 1.4.4. *If $A \in L_0(X)$ is an algebraic operator with single roots only, i.e. with the characteristic polynomial $P(t) = \prod\limits_{i=1}^{n} (t-t_i)$, then*

$$A^m = \sum_{j=1}^{n} t_j^m P_j, \qquad (1.4.18)$$

$$A^m P_i = t_i^m P_i \quad for \ i = 1, 2, ..., n \qquad (1.4.19)$$

$(m = 1, 2, ...)$.

For an arbitrary polynomial $Q(t)$ with complex coefficients

$$Q(A) = \sum_{j=1}^{n} Q(t_j) P_j. \qquad (1.4.20)$$

THEOREM 1.4.4. If $A \in L_0(X)$ is an algebraic operator with the characteristic polynomial $P(t) = \prod_{j=1}^{n} (t-t_j)^{r_j}$ then for every $\lambda \neq t_1, ..., t_n$ the operator $A - \lambda I$ is invertible and

$(A - \lambda I)^{-1}$

$$= \sum_{j=1}^{n} \left[\frac{1}{t_j - \lambda} I + \sum_{m=1}^{r_j-1} \frac{(-1)^{m+1}}{(t_j - \lambda)^{m+1}} (A - t_j I)^m \right] P_j. \qquad (1.4.21)$$

Theorems 1.4.3 and 1.4.4 in the case of single roots (and $\lambda = 0$) were proved by the author, 1962, also 1973. These theorems and Proposition 1.4.2 in the case of multiple roots were proved by Tasche, 1975 using analytic properties of algebraic operators in Banach spaces.

The reader can find another properties of algebraic operators either in the monography of the author and S. Rolewicz, 1968, or in the monography of the author, 1973, devoted to various applications of algebraic operators. For some other applications of algebraic operators see: Fortuna, 1974, Meyer, 1975, Włodarska-Dymitruk, 1975, Voiculescu, 1974, Meyer and Silbermann, 1977, Reynerts, 1979, Yaohua, 1981, Nguyen Van Mau, 1983, 1983a, 1984, Roach, 1983.

Tasche, 1975, has given another form of projectors P_k appearing in Theorem 1.4.1 by application of analytic properties of algebraic operators in Banach spaces. We will show this form using the following lemma about decomposition of rational function onto vulgar fractions:

LEMMA 1.4.3. Suppose that

$$P(t) = \prod_{m=1}^{n} (t-t_m)^{r_m}. \qquad (1.4.22)$$

Then

$$\frac{1}{P(t)} = \sum_{j=1}^{n} \sum_{k=0}^{r_j-1} \frac{d_{jk}}{k!} \frac{1}{(t-t_j)^{r_j-k}},$$ (1.4.23)

where

$$d_{jk} = \left\{ \frac{\mathrm{d}^k}{\mathrm{d}t^k} [p_j(t)]^{-1} \right\}_{t=t_j}, \quad p_j(t) = (t-t_j)^- \, P(t)$$ (1.4.24)

$(j = 1, 2, ..., n; k = 0, 1, ..., r_j-1).$

COROLLARY 1.4.5. *Suppose that* $A \in L_0(X)$ *is an algebraic operator with the characteristic polynomial* $P(t) = \prod_{m=1}^{n} (t - t_m)^{r_m}$. *Then*

$$P_J = \hat{p_j}(A), \quad where \; \hat{p_j}(t) = p_j(t) \sum_{k=0}^{r_j-1} \frac{d_{jk}}{k!} (t-t_j)^k,$$ (1.4.25)

where polynomials $p_j(t)$ *and coefficients* d_{jk} *are defined by Formulae* (1.4.24) $(j = 1, 2, ..., n)$.

An operator $A \in L_0(X)$ is said to be a *Volterra operator* if the operator $I - \lambda A$ is invertible for every scalar λ. The set of all Volterra operators belonging to $L_0(X)$ will be denoted by $V(X)$. If $A \in V(X)$ then the homogeneous equation $(I - \lambda A)x = 0$ has only zero as a solution for every scalar λ.

THEOREM 1.4.5 (cf. Przeworska-Rolewicz, 1980). *Let* X *be a linear space over an algebraically closed field* $\mathscr{F}$ *and let* $A \in V(X)$. *Consider the operators* $A_\lambda = (I - \lambda A)^{-1}$ *for all* $\lambda \in \mathscr{F}$. *Let* $\lambda \neq 0$ *be arbitrarily fixed. Then*
  (i) *if* dim ker $A = 0$ *then* A_λ *has no eigenvalues;*
  (ii) *if* dim ker $A > 0$ *then* A_λ *has at most one eigenvalue* $\mu = 1$, *i.e.* A_λ *are not Volterra operators.*

Proof. Suppose that $\lambda \neq 0$ is arbitrarily fixed and there exists a $u \neq 0$ such that for a $\mu \neq 0$ we have $A_\lambda u = \mu u$. (The case $\mu = 0$ is a trivial one). This implies $(I- \lambda A)^{-1}u = A_\lambda u = \mu u$ and $u = \mu(I- \lambda A)u$. Thus we have the equality

$$(\mu - 1)u = \mu \lambda Au. \tag{1.4.26}$$

If $\mu \neq 1$ this equality can be written as follows:

$$\left(I- \frac{\mu \lambda}{\mu - 1} A\right)u = 0.$$

This, and our assumption that $A \in V(X)$ together imply that $u = 0$ (because the operator $I- [\mu \lambda/(\mu - 1)]A$ is invertible), which contradicts our assumption that $u \neq 0$. We therefore conclude that a number $\mu \neq 1$ cannot be an eigenvalue of each operator A_λ ($\lambda \neq 0$).

Consider now the case $\mu = 1$. From Equation (1.4.26) we have $- \lambda Au = 0$. Since $\lambda \neq 0$ we find $Au = 0$. If dim ker $A = 0$ we have $u = 0$ which also contradicts our assumptions. Then $\mu = 1$ is not an eigenvalue of A_λ. If dim ker $A > 0$ then there is a $u \neq 0$ such that $Au = 0$. This implies that u is an eigenvector of A_λ corresponding to the eigenvalue $\mu = 1$ and $A_\lambda \notin V(X)$. ∎

Examples and Exercises

EXAMPLE 1.4.1. Every finite dimensional operator $A \in L_0(X)$, where X is an infinite dimensional space, is algebraic. Indeed, by definition, the space AX is finite dimensional. Denote by A_0 the matrix corresponding to the operator A restricted to the space $AX \subset X$. Then, by the Cayley–Hamilton Theorem (Corollary 1.4.2) the operator A_0 is algebraic and its characteristic polynomial $P(\lambda)$ is a divisor of the polynomial $\det(A_0 - \lambda I)$. Hence the characteristic polynomial of the operator A on the space X is $\lambda P(\lambda)$. Indeed, $AP(A)X = P(A)AX = P(A_0)AX = 0$.

Observe that 0 is a characteristic root of the operator A. The operator A is not invertible, because the equation $Ax = 0$ has non-trivial solutions of the form $x = P(A)y$, where $y \in X\setminus AX \neq \{0\}$.

EXAMPLE 1.4.2. An operator $A \in L_0(X)$ is said to be an *involution of order n* if $A^n = I$ and $A^k \neq I$ for $k = 1, 2, ..., n-1$. If $n = 2$ we say briefly that A is an *involution*. By definition, every involution of order n

is an algebraic operator with the characteristic polynomial $t^n - 1$ and with characteristic roots $1, \varepsilon, \ldots, \varepsilon^{n-1}$, where $\varepsilon = \exp 2\pi i/n$.

EXERCISE 1.4.1. Suppose that $A \in L_0(X)$ is an involution of order n. Prove that the projectors P_k determined in Theorem 1.4.1 are of the form

$$P_k = \frac{1}{n} \sum_{j=0}^{n-1} \varepsilon^{-jk} A^j, \qquad (1.4.27)$$

where $\varepsilon = \exp 2\pi i/n$ $(k = 1, 2, \ldots, n)$.

In particular, if A is an involution, then

$$P_1 = \frac{1}{2}(I - A), \quad P_2 = \frac{1}{2}(I + A) \qquad (1.4.28)$$

(cf. Przeworska-Rolewicz, 1961, p. 61–62, also 1973).

EXAMPLE 1.4.3. Let X be an arbitrary linear space of functions determined on the real line. Define a *reflection* of functions: $(Sx)(t) = x(-t)$ for $x \in X$. Then $S^2 = I$. We find $X = X_1 \oplus X_2$, where X_1 is the space of all odd functions, i.e. functions $x(t)$ satisfying the condition $x(-t) = -x(t)$, and X_2 is the space of all even functions, i.e. functions $x(t)$ such that $x(-t) = x(t)$. Hence it follows that every function can be written uniquely as the sum of an even and an odd function (Fig. 1.1), namely, as $x = x^+ + x^-$, where $x^+(t) = \frac{1}{2}[x(t) + x(-t)]$, $x^-(t) = \frac{1}{2}[x(t) - x(-t)]$.

EXERCISE 1.4.2. Let X be the space of all square matrices of dimension with real elements. Suppose we are given a matrix $A = (a_{jk})_{\substack{j=1,\ldots,n \\ k=1,\ldots,m}}$. Recall that the matrix

$$A^T = (a_{kj})_{\substack{k=1,\ldots,m \\ j=1,\ldots,n}}$$

is called the *transpose* of A. This means that we obtain a transposed matrix by changing of roles of columns and rows in the matrix A. Prove that:

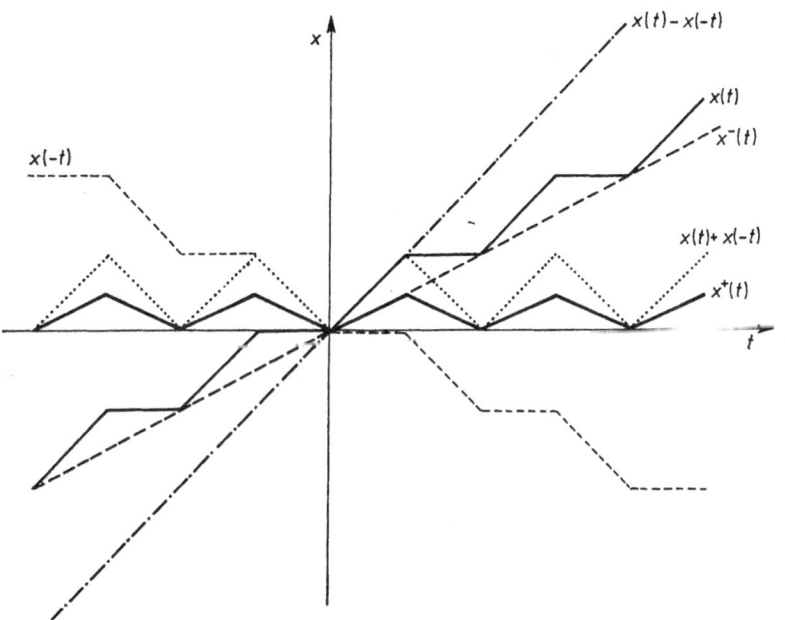

Fig. 1.1. Decomposition of a function into even and odd parts

(a) the operation of transposition of a matrix is an involution,

(b) every matrix can be written uniquely as a sum of a symmetric matrix and an antisymmetric matrix. Recall that a matrix $(a_{jk})_{j,\,k=1,\ldots,n}$ is *symmetric* if $a_{kj} = a_{jk}$ and is *antisymmetric* if $a_{kj} = -a_{jk}$.

EXERCISE 1.4.3. Let X be an arbitrary linear space of functions determined on the complex plane. Let $S \in L_0(X)$ be an operator of rotation through the angle $\alpha = 2\pi/N$, where N is an integer greater than 1, i.e. $(Sx)(t) = x(\exp(2\pi i/N)t)$ for $t \in \mathbf{C}$, $x \in X$. Prove that S is an involution of order N and determine components given by the partition of unit.

EXERCISE 1.4.4. Let X be the linear space of all functions defined for all $t \in \mathbf{R}$ and periodic with a period ω. Prove that the shift operator defined as follows: $(Sx)(t) = x(t - \omega/N)$, where N is an arbitrary integer greater than 1, is an involution of order N and determine components given by the partition of unit.

EXERCISE 1.4.5. Suppose that X is an arbitrary linear space, $A, B \in L_0(X)$ and A is algebraic. Prove that a complex number $\lambda \neq 0$ such that the *commutator* $AB - BA = \lambda I$ does not exists for every $B \in L_0(X)$ (cf. the author, 1973).

EXERCISE 1.4.6. Consider the linear space $\mathcal{F}_n[t]$ of all polynomials of degree n having coefficients belonging to a field $\mathcal{F}$ (cf. Exercise 1.2.3). Define for polynomials $w(t) = \sum\limits_{k=1}^{n} w_k t^k$, where $w_k \in \mathcal{F}$ the operator D by means of the equality

$$(Dw)(t) = \sum_{k=1}^{n} k w_k t^{k-1}. \tag{1.4.29}$$

Prove that D is nilpotent of degree $n+1$ on the space $\mathcal{F}_n[t]$, i.e. $D^{n+1} = 0$. It means that D is an algebraic operator having only a root 0 with multiplicity $n+1$.

EXERCISE 1.4.7. Let S be an algebraic operator with simple roots $t_1, \ldots$ $\ldots, t_n$ only. Let A, B be polynomials with scalar coefficients. The function Φ_A is said to be the *symbol* of the operator $A(S)$ if $\Phi_A(t) = A(t)$, where $t \in \Omega_A = \{t_1, \ldots, t_n\}$. Prove that $\Phi_{A+B} = \Phi_A + \Phi_B$ and $\Phi_{AB} = \Phi_A \Phi_B$.

1.5. CONJUGATE SPACES AND CONJUGATE OPERATORS

In the present section we shall consider conjugate spaces and conjugate operators in the same way as these used by Przeworska-Rolewicz and S. Rolewicz, 1968.

In Section 1.3 we have denoted by X' the space conjugate to the space X, i.e. the space of all linear functionals defined on X. A subspace $\Xi \subset X'$ is said to be *total* if $\xi(x) = 0$ for all $\xi \in \Xi$ implies $x = 0$, where $x \in X$. It is easy to verify that X' itself is total.

In the sequel any total subspace of X' will be called also a *conjugate space* to X.

Observe that elements $x \in X$ can be treated as functionals defined on a conjugate space $\varXi \subset X'$ for the following relation:

$$F_x(\xi) = \xi(x) \quad \text{for } x \in X \text{ and } \xi \in \varXi.$$

Thus, if we denote by $\varXi'$ the space of all linear functionals defined on $\varXi$ then the space X is transformed into $\varXi'$ in a uniquely determined way. This mapping is called the *canonical embedding* and denoted by $\varkappa$. The image $\varkappa X$ of the space X is a conjugate space for $\varXi$ since the condition $\xi(x) = 0$ for all $x \in X$ implies $\xi = 0$.

Suppose that we are given two linear spaces X and Y, both over the same field of scalars. Suppose that $H \subset Y'$ is a conjugate space.* Then to every operator $A \in L(X \to Y)$ there corresponds an operator ηA such that $\mathrm{dom}\, \eta A = H$, the range of ηA is X', defined by means of the equality:

$$(\eta A)x = \eta(Ax) \quad \text{for all } x \in \mathrm{dom}\, A, \, \eta \in H. \tag{1.5.1}$$

The operator ηA is said to be a *conjugate operator* for A and will be denoted by A' (or by A^*). Therefore, by definition, $A'\eta = \eta A$ for all $\eta \in H$. It is easy to verify that $I' = I$ and that $(A+B)' = A'+B'$ if the sum $A+B$ is well determined.

Let $\varXi \subset X'$ be an arbitrary conjugate space. Consider an operator A' as defined for all these $\eta \in H$ for which $A'\eta = \eta A \in \varXi$. In that way to every operator $A \in L(X \to Y)$ there corresponds an operator A' $\in L(H \to \varXi)$.

This general definition admits also conjugate operators defined on the set $\{0\}$ only. We therefore shall consider in the sequel only such operators $A \in L_0(X \to Y)$ for which $A' \in L_0(H \to \varXi)$, i.e. operators $A \in L_0(X \to Y)$ such that $A'\eta \in \varXi$ for all $\eta \in H$. The set of all these operators will be denoted by $L_0(X \to Y, H \to \varXi)$. By definition

$$L_0(X \to Y, H \to \varXi)$$
$$= \{A \in L_0(X \to Y): A'\eta = \eta A \in \varXi \text{ for all } \eta \in H\}, \tag{1.5.2}$$

$$L_0(X, \varXi) = L_0(X \to X, \varXi \to \varXi). \tag{1.5.3}$$

For operators belonging to $L(X)$ we admit the following convention: we shall consider only such conjugate spaces $\varXi \subset X'$ that $\eta(Ax)$

---

* Here and in the sequel H is the Greek capital "eta".

$= (A'\eta)(x)$ for $x \in \mathrm{dom}\, A$, $\eta \in \varXi$ and η has a unique extension to a $\xi \in \varXi$. It means that every functional $\xi \in \varXi$ is uniquely determined by its restriction to $\mathrm{dom}\, A \subset X$, i.e. $A'\eta = \xi|_{\mathrm{dom}A}$ and we can identify ξ and $\xi|_{\mathrm{dom}A}$. The set of all operators satisfying these conditions will be denoted by $L(X, \varXi)$.

Evidently, $L_0(X \to Y, H \to \varXi)$ is a linear space. The space $L_0(X, \varXi)$ is an algebra (a linear ring) and $(AB)' = B'A'$ for all $A, B \in L_0(X, \varXi)$. If $A \in L_0(X, \varXi)$ then we say that A *preserves the conjugate space*.

THEOREM 1.5.1. *A finite dimensional operator K defined by means of the equality*

$$Kx = \sum_{j=1}^{n} f_j(x) y_j \quad \text{for } x \in X$$

(where $f_1, \dots, f_n \in X'$ and $y_1, \dots, y_n \in Y$ are linearly independent), belongs to $L_0(X \to Y, H \to \varXi)$ if and only if $f_1, \dots, f_n \in \varXi$.

THEOREM 1.5.2. *If $A \in L_0(X \to Y, H \to \varXi)$ is right invertible (left invertible, invertible) then the conjugate operator A' is left invertible (right invertible, invertible, respectively).*

THEOREM 1.5.3. *If $A \in L_0(X \to Y, Y' \to X')$ then the nullity of A' is equal to the deficiency of A and the nullity of A is equal to the deficiency of A':*

$$\alpha_{A'} = \beta_A, \quad \alpha_A = \beta_{A'}. \tag{1.5.4}$$

In general, we have

THEOREM 1.5.4. *If $A \in L_0(X \to Y, H \to \varXi)$ then $\alpha_A \leqslant \beta_{A'}$.*

Operators belonging to $L_0(X \to Y, H \to \varXi)$ with finite nullity and deficiency and satisfying Conditions (1.5.4) are called $\varPhi_H$-*operators*.

THEOREM 1.5.5. *If $K \in L_0(X, \varXi)$ is finite dimensional then $I + K$ is a $\varPhi_{\varXi}$-operator and $\varkappa_{I+K} = 0$.*

An operator $T \in L_0(X \to Y, H \to \varXi)$ *satisfies the Fredholm alternative* if the following three conditions are satisfied:

(i) $\alpha_{I+T} = \dim \ker (I+T) < +\infty$, i.e. the homogeneous equation $(I+T)x = 0$ has a finite number k of linearly independent solutions;

(ii) $\alpha_{I'+T'} = \alpha_{I+T}$, i.e. the conjugate homogeneous equation has a finite number k' of linearly independent solutions and $k' = k$;

(iii) the equation $(I+T)x = y$, $y \in Y$ has a solution if and only if $\xi(y) = 0$ for every $\xi \in \ker(I'+T')$.

This alternative, i.e. the conditions (i), (ii), (iii), have been proved firstly for a class of integral equations by Swedish mathematician Ivar Fredholm in the years 1900–1903.

THEOREM 1.5.6. *An operator $T \in L_0(X \to Y, H \to \Xi)$ satisfies the Fredholm Alternative if and only if $I+T$ is a Φ_H-operator with the index zero:* $\varkappa_{I+T} = 0$.

An operator $T \in L_0(X \to Y, H \to \Xi)$ is said to be *of the Noether type (Noetherian)* if $I+T$ is a Φ_H-operator. In this case we do not assume that $\varkappa_{I+T} = 0$.

All finite dimensional operators satisfy the Fredholm Alternative (cf. Theorem 1.5.5).

If an operator $A \in L_0(X \to Y, Y' \to X')$ has finite nullity and deficiency then the operator $T = A - I$ is of the Noether type.

THEOREM 1.5.7. *If $A \in L_0(X \to Y)$ has finite nullity and deficiency then for every conjugate space $\Xi \subset X'$ there exists a Φ_Ξ-operator $B \in L_0(Y \to X)$ such that the operators $AB - I$ and $BA - I$ are finite dimensional in the spaces Y and X, respectively.*

THEOREM 1.5.8. *If $B \in L_0(X \to Y)$ is a Φ_H-operator, $A \in L_0(Y \to Z)$ is a Φ_Σ-operator and $HA \subset \Sigma$ then the superposition AB is a Φ_Σ-operator and $\varkappa_{AB} = \varkappa_A + \varkappa_B$.*

THEOREM 1.5.9. *If an algebra $\mathfrak{X}(X)$ of linear operators contains all finite dimensional operators $K \in L_0(X, \Xi)$ such that $I+K$ are Φ_Ξ-operators then $\mathfrak{X}(X) \subset L_0(X, \Xi)$.*

THEOREM 1.5.10. *If $A \in L_0(X \to Y)$ is a Φ_H-operator then every finite dimensional operator $K \in L_0(X \to Y, H \to \Xi)$ is a Φ_H-perturbation of A.*

THEOREM 1.5.11 (First theorem on reduction). *Let* $\Xi \subset X'$ *be a conjugate space. Suppose that* $T \in L_0(X, \Xi)$ *and* $\alpha_{I+T} < +\infty$, $\alpha_{I^{\bullet}+T'} < +\infty$. *Let* X_0 *be a subspace of* X *containing* TX *and let* Ξ_0 *be a subspace of* Ξ *containing* $\Xi T = T'\Xi$:

$$TX \subset X_0 \subset X; \quad \Xi T \subset \Xi_0 \subset \Xi.$$

Let A *be the restriction of* $I+T$ *to the space* X_0: $A = I+T|_{X_0}$. *Then* $A \in L_0(X_0, \Xi_0)$ *and* $\alpha_A = \alpha_{I+T}$, $\alpha_{A'} = \alpha_{I+T'}$. *In other words:*

$$\ker(I+T) \subset X_0, \quad \ker(I'+T') \subset \Xi_0.$$

THEOREM 1.5.12 (Second theorem on reduction). *Let* X_0 *be a subspace of linear space* X *and let* Ξ_0 *be a subspace of a conjugate space* $\Xi \subset X'$. *Suppose that for an operator* $A \in L_0(X_0, \Xi_0)$ *there exists an operator* B *such that*

$$AB = I+T_1, \quad BA = I+T_2,$$

where T_1, T_2 *have extensions* $\tilde{T}_1, \tilde{T}_2 \in L_0(X, \Xi)$ *such that* $I+\tilde{T}_1, I+\tilde{T}_2$ *are* Φ_Ξ-*operators. Then* A *is a* Φ_Ξ-*operator.*

Exercises

EXERCISE 1.5.1. Let X, Y be linear spaces (over $\mathbf{C}$ or $\mathbf{R}$) and let dim $X = n$, dim $Y = m$. If $A \in L_0(X \to Y)$ then $A' = A^T$, where A^T denotes the transpose of the matrix $A = (a_{jk})_{\substack{j=1,\ldots,n \\ k=1,\ldots,m}}$, i.e. $A^T = (a_{kj})_{\substack{k=1,\ldots,m \\ j=1,\ldots,n}}$.

It means that an operator conjugate to A is determined by the transpose of the matrix A. Prove that:

1° If A is diagonal then $A^T = A$;

2° $(A^T)^T = A$, $(\lambda A)^T = \lambda A^T$ for all scalars λ, $(A+B)^T = A^T+B^T$, $(AB)^T = B^T A^T$.

EXERCISE 1.5.2. Suppose that $A \in L_0(X, \Xi)$ is an algebraic operator with the characteristic polynomial $P(t)$. Prove that the conjugate operator A' is again algebraic with the same characteristic polynomial.

EXERCISE 1.5.3. Suppose that either $\mathscr{F} = \mathbf{C}$ or $\mathscr{F} = \mathbf{R}$ and $A = (a_{jk})_{j,k=1,\ldots,n}$, where $a_{n-k+1,k} = \lambda \in \mathscr{F}$ and $a_{jk} = 0$ for $j \neq n-k+1$

$(j, k = 1, ..., n)$. Prove that $A^T = A$. Does $A^T = A$ if $a_{n-k+1, k} = \lambda_k \in \mathscr{F}$ and $\lambda_k \neq \lambda_j$ for $k \neq j$?

EXERCISE 1.5.4. Suppose that A is a non-singular square matrix, i.e. $\det A \neq 0$. Prove that

$$(A^T)^{-1} = (A^{-1})^T, \quad \det A^{-1} = (\det A)^{-1}.$$

EXERCISE 1.5.5. Suppose that $\dim X = n$, $\dim Y = m \neq n$ and that $A \in L_0(X \to Y)$. Then $A^T \in L_0(Y \to X)$ and $A^T A \in L_0(X)$ is a square matrix of dimension n. Prove that:

(a) If $\det A^T A \neq 0$ then the equation $Ax = y$, $y \in Y$, has a unique solution $x = (A^T A)^{-1} A^T y$, i.e. the homogeneous equation $Ax = 0$ has only zero as a solution;

(b) If there is $x \neq 0$ such that $Ax = 0$ then $\det A^T A = 0$;

(c) If $y \neq 0$, $A^T y = 0$ and $\det A^T A \neq 0$ then the equation $Ax = y$ has no solutions;

(For the proofs act on both sides of the equation $Ax = y$ by A^T.)

(d) Give conditions of solvability of the equation $Ax = y$ in the case $\det AA^T \neq 0$ (by substitution: $x = A^T u$);

(e) Explain why the results (a), (b), (c) are more useful when $n > m$ and results of the point (d) are more useful when $m > n$.

EXERCISE 1.5.6. An operator B is said to be a *generalized inverse* of A if

$$ABA = A, \quad BAB = B. \tag{1.5.5}$$

Prove that:

$1°$ If A has a generalized inverse B then B has a generalized inverse A;

$2°$ The operators A and B determined in Theorem 1.5.7 are generalized inverses, each to another;

$3°$ If $A \in L_0(X \to Y)$ has finite nullity and deficiency then for every conjugate space $\Xi \subset X'$ there exists a Φ_Ξ-operator $B \in L_0(Y \to X)$ which is a generalized inverse of A.

EXERCISE 1.5.7. Suppose that $T \in L_0(X)$ and that there is a positive integer m such that $I - T^m$ is a Φ_Ξ-operator, where $\Xi \subset X'$ is a conjugate space. Prove that $I - T$ is a Φ_Ξ-operator.

Chapter 2

Calculus of right invertible operators

2.1. PROPERTIES OF RIGHT INVERSES. INDEFINITE INTEGRALS

Suppose that X is a linear space over a field $\mathscr{F}$ of scalars. An operator $D \in L(X)$ is said to be *right invertible* if there is an operator $R \in L_0(X)$ such that $RX \subset \text{dom } D$ and

$$DR = I \qquad (2.1.1)$$

(cf. Exercise 1.3.13).

The operator R is called a *right inverse* of D. The set of all right invertible operators will be denoted by $R(X)$. The set of all right inverses of an operator $D \in R(X)$ will be denoted by $\mathscr{R}_D$. Sometimes we shall write also $\mathscr{R}_D = \{R_\gamma\}_{\gamma \in \Gamma}$.

As a matter of fact, the assumption dom $R = X$ is not essential but simplifies our subsequent considerations. Moreover, this assumption is satisfied in applications.

Suppose that x is an arbitrarily fixed element of the space X. Let $D \in R(X)$. The set

$$\mathscr{R}_D x = \{R_\gamma x\}_{\gamma \in \Gamma}$$

is called the *indefinite integral* of x. Each of elements $R_\gamma x$, where $\gamma \in \Gamma$, is called a *primitive* for x. By definition, if y is a primitive for x then $Dy = x$. Indeed, there is an index $\gamma \in \Gamma$ such that $y = R_\gamma x$. This implies $Dy = DR_\gamma x = x$, because $DR_\gamma = I$.

Suppose that $D \in R(X)$. The kernel of the operator D is called the *space of constants* for D. Each element $z \in \ker D$ will be called a *constant*. Observe that, by definition, an element $z \in X$ is a constant for D if and only if $Dz = 0$.

PROPOSITION 2.1.1. *If $D \in R(X)$, $R \in \mathscr{R}_D$ then $D^k R^k = I$ for $k = 1, 2, \ldots$*

Proof (by induction). For $k = 1$ we have $DR = I$ by the definition of R. Suppose now that $D^k R^k = I$ for an arbitrary positive integer k. Then $D^{k+1} R^{k+1} = D(D^k R^k) R = DR = I$, which completes the proof. ∎

PROPOSITION 2.1.2. *Suppose that* $D \in R(X)$, $R_1, R_2 \in \mathscr{R}_D$ *and* $y_1 = R_1 x$, $y_2 = R_2 x$, *where* $x \in X$ *is arbitrary. Then* $y_1 - y_2 \in \ker D$.

Indeed, $D(y_1 - y_2) = Dy_1 - Dy_2 = DR_1 x - DR_2 x = x - x = 0$, hence $y_1 - y_2 \in \ker D$. ∎

In other words: *A difference of two primitives for a given element* $x \in X$ *is a constant.*

This implies that an indefinite integral is well-determined if we know at least one right inverse. Indeed, we know in this case at least one primitive and all other primitives are obtained by addition of a constant. We therefore obtain

PROPOSITION 2.1.3. *If* $D \in R(X)$ *and* $R \in \mathscr{R}_D$ *then the indefinite integral of an element* $x \in X$ *is of the form*

$$\mathscr{R}_D x = \{Rx + z : z \in \ker D\} = Rx + \ker D. \tag{2.1.2}$$

This property can be also formulated in the following way: *The indefinite integral of an element* $x \in X$ *is a sum of a primitive element and an arbitrary constant.*

PROPOSITION 2.1.4. *If* $D \in R(X)$ *then for every* $R \in \mathscr{R}_D$ *we have*

$$\operatorname{dom} D = RX \oplus \ker D. \tag{2.1.3}$$

Proof. If $x \in \operatorname{dom} D$ then we can write: $x = u + w$, where $u = RDx$, $w = (I - RD)x$. By definition, $u \in RX$ and $Dw = D(I - RD)x = Dx - (DR)Dx = Dx - Dx = 0$, hence $w \in \ker D$ and $x = u + w \in RX + \ker D$. Suppose now that $x \in X$, $z \in \ker D$ and $R \in \mathscr{R}_D$ are arbitrarily fixed. Then $y = Rx + z \in \operatorname{dom} D$ for $Dy = DRx + Dz = x$. Suppose that $u \in RX \cap \ker D$. Then there is a $w \in X$ such that $u = Rw$. On the other hand, $u \in \ker D$, hence $Du = 0$. But $Du = DRw = w$, which implies $w = 0$ and $u = Rw = 0$. This means that $\operatorname{dom} D$ is a direct sum of RX

and ker D. By Proposition 2.1.2 the existence of the decomposition (2.1.3) is independent of the choice of the right inverse R.   ∎

THEOREM 2.1.1. *Suppose that $D \in R(X)$ and $R_1 \in \mathscr{R}_D$. Then every right inverse of D is of the form*

$$R = A + R_1(I - DA) = R_1 + (I - R_1 D)A, \qquad (2.1.4)$$

where $A \in L_0(X)$, $AX \subset \mathrm{dom}\, D$, i.e.

$$\mathscr{R}_D = \{R_1 + (I - R_1 D)A : A \in L_0(X), AX \subset \mathrm{dom}\, D\} \qquad (2.1.5)$$

(cf. Proposition 1.1.1 for rings).

Proof. Suppose that $R_1 \in \mathscr{R}_D$ and R is of the form (2.1.4). Since $DR = I$, we have

$$DR = D[A + R_1(I - DA)] = DA + DR_1(I - DA)$$
$$= DA + I - DA = I.$$

Thus $R \in \mathscr{R}_D$. Observe that

$$R = A + R_1(I - DA) = A + R_1 - R_1 DA = R_1 + (I - R_1 D)A.$$

Conversely, suppose that $R_1 \in \mathscr{R}_D$ is given and $R \in \mathscr{R}_D$ is arbitrarily chosen. Write: $A = R - R_1$. By definition, $A \in L_0(X)$ and $AX \subset \mathrm{dom}\, D$. Since $DR_1 = I$, $DR = I$, we find

$$R_1 + (I - R_1 D)A = R_1 + (I - R_1 D)(R - R_1)$$
$$= R_1 + R - R_1(DR) - R_1 + R_1(DR_1)$$
$$= R_1 + R - R_1 - R_1 + R_1 = R$$

which was to be proved.   ∎

Observe that if $D \in R(X)$, $R \in \mathscr{R}_D$ and $x \in X$, then $Rx = 0$ implies $x = 0$. Indeed, $x = DRx = 0$.

Examples and Exercises

Before we pass to examples we will give lemmas which will be applied in these examples.

Recall that for functions belonging to $C[a, b]$ we have the following statements (cf. for instance Kuratowski, 1969):

1° Every continuous function on $[a, b]$ is a limit of a uniformly convergent sequence of piecewise linear functions.

2° Suppose we are given a sequence $\{x_n\}$ of functions defined for $t \in [a, b]$, having continuous derivatives on $[a, b]$ and such that the sequence $\{x_n'\}$ is uniformly convergent to a function $y \in C[a, b]$, where we write $x'(t) = \dfrac{d}{dt} x(t)$ for an arbitrary differentiable function, when $a < t < b$ and $x'(a)$ for the right-hand derivative, $x'(b)$ for the left-hand derivative. Suppose moreover, that there is a $c \in [a, b]$ such that the sequence $\{x_n(c)\}$ is convergent. Then the sequence $\{x_n(t)\}$ is convergent for every $t \in [a, b]$. Moreover, if $x(t) = \lim_{n \to \infty} x_n(t)$, then $x'(t) = y(t)$ $= \lim_{n \to \infty} x_n'(t)$. We say that a function ξ defined in $[a, b]$ is *primitive* for a function $x \in C[a, b]$ if $\xi'(t) = x(t)$ for $a \leqslant t \leqslant b$.

LEMMA 2.1.1. *Suppose we are given a point $t_0 \in [a, b]$ and an arbitrary real number c. If a function $x(t)$ defined on the interval $[a, b]$ has a primitive function $\xi(t)$ then there exists a primitive function $\eta(t)$ for x such that $\eta(t_0) = c$.*

Proof. Write $\eta(t) = \xi(t) - \xi(t_0) + c$. By our assumption the derivative η' exists and $\eta'(t) = \xi'(t) = x(t)$ for $a \leqslant t \leqslant b$. Thus η is a primitive function for x. Moreover, $\eta(t_0) = \xi(t_0) - \xi(t_0) + c = c$. Hence η satisfies the required conditions. ∎

LEMMA 2.1.2. *If a sequence $\{x_n\} \subset C[a, b]$ is uniformly convergent to a function x and all functions $x_n(t)$ have primitive functions $\xi_n(t)$ then the function x has a primitive function.*

Proof. Lemma 2.1.1 implies that the functions $\xi_n(t)$ can be chosen in such a way that $\xi_n(t_0) = 0$ for a $t_0 \in [a, b]$ $(n = 1, 2, \ldots)$. Having already chosen functions ξ_n, we conclude that the sequence $\{\xi_n(t_0)\}$, as a sequence of zeros, is convergent to zero. Thus the second of statements cited above implies that the sequence $\{\xi_n(t)\}$ is convergent for every $t \in [a, b]$ to a function $\xi(t)$ such that $\xi'(t) = x(t)$. This proves that x has a primitive function. ∎

LEMMA 2.1.3. *Every function continuous on a closed interval has a primitive function on this interval.*

Proof. Since every function $x \in C[a, b]$ is a limit of uniformly convergent sequence of piecewise linear functions, Lemma 2.1.2 implies that it is enough to prove that every function piecewise linear has a primitive function. Suppose now that y is a piecewise linear function on $[a, b]$. This means that there exists a finite system of points $a_0, ..., a_n$ such that $a = a_0 < a_1 < ... < a_n = b$ and real numbers $c_1, ..., c_n, d_1, ..., d_n$ such that

$$x(t) = c_k t + d_k \quad \text{for } a_{k-1} \leqslant t \leqslant a_k \ (k = 1, 2, ..., n).$$

Write

$$\xi_k(t) = \frac{1}{2} c_k t^2 + d_k t + e_k \quad \text{for } a_{k-1} \leqslant t \leqslant a_k \ (k = 1, 2, ..., n),$$

where

$$e_1 = 0, \quad e_{k+1} = \frac{1}{2} c_k a_k^2 + d_k a_k + e_k - \left(\frac{1}{2} c_{k+1} a_k^2 + d_{k+1} a_k \right)$$

$$(k = 1, 2, ..., n-1).$$

This definition implies that we have

$$\xi_k(a_k) = \xi_{k+1}(a_k) \quad (k = 1, 2, ..., n).$$

Thus the function ξ determined by

$$\xi(t) = \xi_k(t) \quad \text{for } a_{k-1} \leqslant t \leqslant a_k \ (k = 1, 2, ..., n)$$

is well-determined in $[a, b]$. Moreover, this function is differentiable in $[a, b]$, because we find

$$\xi'(t) = \xi'_k(t) = c_k t + d_k = x(t) \quad \text{for } a_{k-1} \leqslant t \leqslant a_k$$

$$(k = 1, 2, ..., n).$$

Thus ξ is a primitive function for x.   ∎

EXAMPLE 2.1.1. Let $X = C[a, b]$. Write $D = \mathrm{d}/\mathrm{d}t$. Lemmas 2.1.1 and 2.1.3 together imply that for every function $x \in C[a, b]$ there exists a unique primitive function ξ such that $\xi(t_0) = 0$, where t_0 is arbitrarily fixed in $[a, b]$. Observe that ξ, as a function having a continuous derivative

$x \in C[a, b]$, belongs to $C[a, b]$ itself. We shall write this unique primitive function as follows:

$$\xi(t) = \int_{t_0}^{t} x(s) \, ds$$

(read: $\xi(t)$ is equal to the *integral* from t_0 to t of $x(s)$ with respect to s). The numbers t and t_0 are called *upper* and *lower limits of the integral*. Observe that the variable s under the sign of the integral is an *apparent variable*. Indeed, if we put instead of s an arbitrary letter we obtain the same result:

$$\int_{t_0}^{t} x(u) \, du = \xi(t) = \int_{t_0}^{t} x(s) \, ds.$$

Write

$$(Rx)(t) = \int_{t_0}^{t} x(s) \, ds \quad \text{for } x \in C[a, b]. \tag{2.1.6}$$

From the definition it follows that R is well-defined for all continuous functions. Moreover, R is a linear operator. Indeed, let $x, y \in C[a, b]$ and $\lambda, \mu \in \mathbf{R}$ be arbitrary. Then $\lambda x + \mu y \in C[a, b]$. Write

$$\xi(t) = \int_{t_0}^{t} x(s) \, ds, \quad \eta(t) = \int_{t_0}^{t} y(s) \, ds,$$

$$\zeta(t) = \int_{t_0}^{t} [\lambda x(s) + \mu y(s)] \, ds.$$

Since $\zeta' = \lambda x + \mu y = \lambda \xi' + \mu \eta'$ and $\zeta(t_0) = 0 = \lambda \xi(t_0) + \mu \eta(t_0)$, we conclude that $\zeta = \lambda \xi + \mu \eta$. Thus $R(\lambda x + \mu y) = \zeta = \lambda \xi + \mu \eta = \lambda Rx + \mu Ry$. Hence $R \in L_0(X)$.

The definition of R implies that

$$(DRx)(t) = \frac{d}{dt} \xi(t) = x(t) \quad \text{for all } x \in C[a, b].$$

Thus $RX \subset \operatorname{dom} D = C^1[a, b]$ and $DR = I$. This last fact can be also formulated as follows: *A derivative of an integral with respect to its*

*upper limit exists and is equal to the value of the integrated function
at this upper limit, i.e.*

$$\frac{\mathrm{d}}{\mathrm{d}t} \int_{t_0}^{t} x(s)\,\mathrm{d}s = x(t) \quad \text{for all } x \in C[a, b], a \leqslant t \leqslant b. \qquad (2.1.7)$$

A function $x(t)$ is said to be *constant on* $[a, b]$ if there is a scalar c
such that $x(t) = c$ for all $t \in [a, b]$. Since there is a one-to-one correspondence between all constant functions belonging to $C[a, b]$ and scalars
$c \in \mathbf{R}$ we can identify in the example constants with scalars if it does not
lead to a misunderstanding. A correspondence of this type does not
hold in some cases which will be considered.

Since in our case ker $D = \{x \in C^1[a, b]\colon x'(t) = 0 \text{ for } a \leqslant t \leqslant b\}$
is the space of all functions constant in $[a, b]$, we have dim ker $D = 1$.

The operator $D = \mathrm{d}/\mathrm{d}t$ is right invertible but not invertible. Indeed,
by definition

$$(RDx)(t) = \int_{t_0}^{t} x'(s)\,\mathrm{d}s = x(t) - x(t_0) \quad \text{for } x \in \text{dom } D$$

(because $x(t) - x(t_0)$ is the unique primitive function for $x'(t)$, which
has the value 0 at the point t_0). Thus, if $x(t_0) \neq 0$ we find $RDx \neq x$.
Hence $RD \neq I$.

An indefinite integral of a function $x \in C[a, b]$ is denoted by

$$\int x(t)\,\mathrm{d}t$$

(read: integral of $x(t)$ with respect to t). Thus, by definition,

$$\int x(t)\,\mathrm{d}t = \left\{ \int_{t_0}^{t} x(s)\,\mathrm{d}s + c\colon c \in \mathbf{R} \right\} \qquad (2.1.8)$$

for an arbitrarily fixed $t_0 \in [a, b]$.

EXAMPLE 2.1.2. Let

$$\Omega = \{(t, s) \in \mathbf{R}^2\colon a \leqslant t, s \leqslant b\}.$$

Denote by $C(\Omega)$ the space of all real-valued functions $x(t, s)$ defined
and continuous for $(t, s) \in \Omega$. If for an $x \in C(\Omega)$ the limit

$$\lim_{h \to 0} \frac{x(t_0+h, s)-x(t_0, s)}{h}$$

exists, we call this limit a *partial derivative of x with respect to the variable* t at t_0 and we write

$$x'_t(t_0) = \lim_{h \to 0} \frac{x(t_0+h, s)-x(t_0, s)}{h}.$$

We write also $\partial x/\partial t$ instead of x'_t. Write

$$D = \partial/\partial t,$$

$$\text{dom } D = \{x \in C(\Omega): x(t, s_0) \in C^1[a, b] \text{ for every fixed } s_0 \in [a, b]\},$$

$$(Rx)(t, s) = \int_a^t x(u, s)du \quad \text{for } (t, s) \in \Omega.$$

In a similar way as in Example 2.1.1 we conclude that $D \in R(X)$, where $X = C(\Omega)$, and that R is a right inverse of D. Moreover, in our case

$$\text{ker } D = \{x = x(t, s): x(t, s) = \varphi(s), \varphi \in C[a, b]\}. \tag{2.1.9}$$

Indeed, $\partial \varphi(s)/\partial t = 0$ for all $s \in [a, b]$. On the other hand, if $\partial x(t, s)/\partial t = 0$ for an $x \in C(\Omega)$ then $x(t, s) = \varphi(s)$ for all $(t, s) \in \Omega$.

The form of the kernel of D implies that dim ker $D = +\infty$. The operator $D = \partial/\partial t$ is not invertible in $C(\Omega)$, because, in general,

$$(RDx)(t) = \int_a^t x'_t(u, s)ds = x(t, s)-x(a, s) \neq x(t, s)$$

$$\text{for } x \in \text{dom } D.$$

An indefinite integral of a function $x \in C(\Omega)$ is then of the form:

$$\mathscr{R}_D x = \left\{ \int_a^t x(u, s)du + \varphi(s): \varphi \in C[a, b] \right\}.$$

In a similar way we can consider a partial derivative with respect to s (denoted by either $\partial x/\partial s$ or x'_s) as a right invertible operator.

EXAMPLE 2.1.3. Suppose that $a \in [0, 1]$ is arbitrarily fixed. Consider the space

$$X = \{x \in C^\infty[0, 1]: x^{(k)}(a) = 0 \text{ for } k = 0, 1, 2, \ldots\}$$

(cf. Exercise 1.2.1). Write $D = d/dt$ and $(Rx)(t) = \int_a^t x(s)ds$. It is easy to see that both operators D and R are defined on the whole space X. Let $x \in X$ be arbitrary and write $y = Rx$. Then $y(a) = (Rx)(a) = \int_a^a x(s)ds = 0$. Since $y'(t) = x(t)$, we have $y'(a) \doteq x(a) = 0$. By an easy induction we conclude that $y^{(k+1)} = x^{(k)}$ ($k = 0, 1, 2, \ldots$). Thus $y^{(k+1)}(a) = x^{(k)}(a) = 0$ for $k = 0, 1, 2, \ldots$ This implies that $RX \subset X = \text{dom } D$. Moreover, $DRx = Dy = y' = x$. The arbitrariness of x implies that $DR = I$.

On the other hand, we have

$$(RDx)(t) = \int_a^t x'(s)ds = x(t) - x(a) = x(t) \quad \text{for } x \in X.$$

Thus $RD = I$ and the operator D is invertible in the space X. This implies that $\ker D = \{0\}$ and $\dim \ker D = 0$. Indeed, if $Dx = 0$ then $x = RDx = 0$. We therefore conclude that the indefinite integral of $x \in X$ is a one-element set $\left\{\int_a^t x(s)ds\right\}$.

EXAMPLE 2.1.4. Suppose that X is the space (s) of all sequences $\{x_n\}$, $x_n \in \mathbf{R}$, $n \in \mathbf{N}$, with addition and multiplication by scalars defined as follows: if $x = \{x_n\}, y = \{y_n\}, \lambda \in \mathbf{R}$ then $x + y = \{x_n + y_n\}, \lambda x = \{\lambda x_n\}$ (cf. Exercise 1.2.8). Define the operator D by the equality $Dx = \{x_{n+1} - x_n\}$. We have: $\text{dom } D = X$. Write now $Rx = y$ for $x \in X$ where $y = \{y_n\}, y_1 = 0, y_{n+1} = \sum_{k=1}^n x_k$ for $n \geqslant 1$. Thus $DRx = Dy = \{y_{n+1} - y_n\} = \{x_n\} = x$ for $x \in X$ and $DR = I$. It is easy to verify that the operator D is not invertible. The space of constants has the form

$$\ker D = \{z = \{z_n\}: z_{n+1} - z_n = 0 \text{ for } n \in \mathbf{N}\}$$
$$= \{z = \{z_n\}: z_n = c, n \in \mathbf{N}, c \in \mathbf{R}\}.$$

We therefore conclude that the indefinite integral of an element $x \in X$ has the form

$$\mathscr{R}_D x = \{Rx + z : z \in \ker D\}$$

$$= \left\{ y = \{y_n\} : y_1 = c, y_{n+1} = \sum_{k=1}^{n} x_k + c \text{ for } n \in \mathbf{N}, c \in \mathbf{R} \right\}.$$

We shall prove now that R is a Volterra operator. Indeed, consider the equation $(I - \lambda R)x = y$, where $y \in X$ and $\lambda \in \mathbf{R}$ are arbitrarily fixed. From the definition of the operator R we obtain a sequence of equalities

$$x_1 = y_1, \quad x_{n+1} - \lambda \sum_{k=1}^{n} x_k = y_{n+1} \quad (n = 1, 2, ...).$$

A solution of our equation is then of the form:

$$x_1 = y_1, \quad x_{n+1} = y_{n+1} + \lambda \sum_{k=1}^{n} (\lambda+1)^{k-1} y_{n+1-k}$$

$$(n = 1, 2, ...). \quad (2.1.10)$$

Proof (by induction). For $n = 1$ we have $x_2 = y_2$. Suppose that Formulae (2.1.10) are true for an arbitrary positive integer n. Then, by this assumption, and from the form of our equation we have

$$x_{n+2} = y_{n+2} + \lambda \sum_{k=1}^{n+1} x_k = y_{n+2} + \lambda \sum_{k=1}^{n} x_k + \lambda x_{n+1}$$

$$= y_{n+2} + x_{n+1} - y_{n+1} + \lambda x_{n+1} = y_{n+2} + (\lambda+1)x_{n+1} - y_{n+1}$$

$$= y_{n+2} + (\lambda+1) \left[y_{n+1} + \lambda \sum_{k=1}^{n} (\lambda+1)^{k-1} y_{n+1-k} \right] - y_{n+1}$$

$$= y_{n+2} + (\lambda+1)y_{n+1} - y_{n+1} + \lambda \sum_{k=1}^{n} (\lambda+1)^k y_{n+1-k}$$

$$= y_{n+2} + \lambda y_{n+1} + \lambda \sum_{m=2}^{n+1} (\lambda+1)^{m-1} y_{n+2-m}$$

$$= y_{n+2} + \lambda \sum_{m=1}^{n+1} (\lambda+1)^{m-1} y_{n+2-m}$$

which was to be proved.

Thus $x = \{x_n\} = 0$ if and only if $y = \{y_n\} = 0$ and the operator $I - \lambda R$ is invertible for every $\lambda \in \mathbf{R}$. Hence R is a Volterra operator. Moreover, if $(I - \lambda R)^{-1} y = u$, where $u = \{u_n\}$, then $u_1 = y_1$ and

$$u_{n+1} = y_{n+1} + \lambda \sum_{k=1}^{n} (\lambda + 1)^{k-1} y_{n+1-k} \quad (n = 1, 2, \ldots).$$

EXAMPLE 2.1.5. Suppose that X is a linear space, $D \in R(X)$, $R \in \mathcal{R}_D$, $A \in L(X)$, $B \in L(X)$ and the following equality holds:

$$DA = BAD \quad \text{on dom } D.$$

An immediate consequence of this equality is that

$$RBADx = RDAx \quad \text{for } x \in \text{dom } D. \tag{2.1.11}$$

EXAMPLE 2.1.6. Suppose that X, D, R are defined as in Example 2.1.1. Suppose we are given a function $g \in C^1[a, b]$. Define $A, B \in L(X)$ by

$$(Ax)(t) = x(g(t)) \quad \text{and} \quad (Bx)(t) = g'(t)x(t) \quad \text{for } x \in C[a, b]$$

(cf. Example 1.3.2). Observe that dom $D = C^1[a, b]$ and that for all $x \in C^1[a, b]$ we have

$$(DAx)(t) = \frac{d}{dt} x(g(t)) = g'(t)x'(g(t)) = (BADx)(t).$$

Thus all conditions of Example 2.1.5 are satisfied. Formula (2.1.11) implies that if ξ is a primitive function for $x \in C[a, b]$, i.e. $D\xi = \xi' = x$, we obtain

$$\int_{t_0}^{t} g'(s)x(g(s))ds = (RBAx)(t) = (RBAD\xi)(t)$$

$$= (RDA\xi)(t) = \int_{t_0}^{t} [\xi(g(s))]'ds = \xi(g(t)) - \xi(g(t_0)).$$

Finally we have

$$\int_{t_0}^{t} g'(s)x(g(s))ds = \xi(g(t)) - \xi(g(t_0)), \quad \text{where } \xi' = x. \tag{2.1.12}$$

The last formula is called the *formula for integration by substitution*. Put x instead of ξ in Formula (2.1.12). Since $x\big(g(t_0)\big)$ is constant, we obtain immediately the *formula of integration by substitution for indefinite integrals*:

$$\int g'(t)x'\big(g(t)\big)\mathrm{d}t = x\big(g(t)\big)+c \quad \text{for } x \in C^1[a, b]. \qquad (2.1.13)$$

From Formula (2.1.12) we can derive another form for the formula of integration by substitution for indefinite integrals, namely

$$\int g'(t)x\big(g(t)\big)\mathrm{d}t = \int x(u)\mathrm{d}u, \quad \text{where } u = g(t),\, x \in C[a, b]. \qquad (2.1.14)$$

For instance, if $x(t) = \sin t$ then

$$\int g'(t)\sin g(t)\,\mathrm{d}t = \int \sin u\,\mathrm{d}u|_{u=g(t)} = -\cos u|_{u=g(t)}+c$$
$$= -\cos g(t)+c, \quad \text{where } c \in \mathbf{R} \text{ is arbitrary.}$$

EXAMPLE 2.1.7. A commutative linear ring X is called a *Leibniz algebra* if there is an operator $D \in R(X)$ which satisfies the *Leibniz condition*

$$D(xy) = xDy+yDx \quad \text{for } x, y \in \mathrm{dom}\, D. \qquad (2.1.15)$$

Then the indefinite integral satisfies the following equality

$$\mathscr{R}_D(xDy) = xy-\mathscr{R}_D(yDx) \quad \text{for } x, y \in \mathrm{dom}\, D. \qquad (2.1.16)$$

This formula is called the *formula of integration by parts* for indefinite integrals.

Formula (2.1.3) implies that Formula (2.1.16) can be rewritten as follows:

$$\{R(xDy)+z\colon z \in \ker D\} = \{xy-R(yDx)+z_1\colon z_1 \in \ker D\}, \qquad (2.1.16')$$

for $x, y \in \mathrm{dom}\, D$. Indeed, for primitive elements and $z, z_1 \in \ker D$ we have

$$D[R(xDy)+z] = (DR)(xDy)+Dz = xDy+yDx-yDx$$
$$= D(xy)-yDx = D(xy)-(DR)(yDx)+Dz_1$$
$$= D[xy-R(yDx)+z_1],$$

which proves Equalities (2.1.16') and (2.1.16).

EXAMPLE 2.1.8. Suppose that X, D, R are defined as in Example 2.1.1. We have shown in Example 1.2.10 that $X = C[a, b]$ is a commutative linear ring with the usual multiplication of functions. Moreover, the operator $D = \mathrm{d}/\mathrm{d}t$ satisfies the Leibniz condition (2.1.15), because

$$\frac{\mathrm{d}}{\mathrm{d}t}(xy) = x\frac{\mathrm{d}}{\mathrm{d}t}y + y\frac{\mathrm{d}}{\mathrm{d}t}x \quad \text{for } x, y \in C^1[a, b].$$

Thus the formula of integration by parts for indefinite integrals is of the form

$$\int x(t)y'(t)\,\mathrm{d}t = x(t)y(t) - \int x'(t)y(t)\,\mathrm{d}t \quad \text{for } x, y \in C[a, b].$$

EXAMPLE 2.1.9. Suppose that $X = (s)$ consists of all real sequences $\{x_n\}$, where $n \in \mathbf{N}$ (cf. Example 2.1.4). Write for $x = \{x_n\} \in (s)$

$$S_l\{x_n\} = \{x_{n+1}\}, \quad S_r\{x_n\} = \{x_{n-1}\}, \tag{2.1.17}$$

where we assume $x_0 = 0$ for all $x \in (s)$. It is easy to check that for every $\{x_n\} \in (s)$ the shift operators S_l and S_r satisfy the equalities

$$S_l S_r = I, \quad S_r S_l\{x_n\} = \{0, x_2, x_3, \ldots\} \neq \{x_n\},$$

which means that the operator S_l is right invertible, but not invertible. We have ker $S_l = \{\{x_n\}: x_n = 0 \text{ for } n = 2, 3, \ldots\}$.

REMARK. If we consider two-sided sequences, i.e. $n = \pm 1, \pm 2, \ldots$ then the operator S_l is invertible and $S_l^{-1} = S_r$.

EXAMPLE 2.1.10. Suppose that $\Omega \subset \mathbf{R}^n$ is an arbitrary domain and let $X = C(\Omega)$. Consider the operator

$$D = \sum_{j=1}^{n} a_j \frac{\partial}{\partial t_j}, \quad \text{where } t = (t_1, \ldots, t_n) \in \Omega, \tag{2.1.18}$$

$a_1, \ldots, a_n \in \mathbf{R}$ do not vanish simultaneously (cf. Example 2.1.2). We shall prove that the operator D is right invertible. Indeed, write $u_{1j} = a_j$ $(j = 1, \ldots, n)$ and $u_1 = (u_{11}, \ldots, u_{1n}) \in \mathbf{R}^n$. Denote by $u_j = (u_{j1}, \ldots, u_{jn}) \in \mathbf{R}^n$ $(j = 2, \ldots, n)$ vectors which are orthogonal to u_1 and such that the set $\{u_1, \ldots, u_n\}$ is linearly independent. This implies that $U = \det(u_{jk})_{j,k=1,\ldots,n} \neq 0$.

If we write

$$t_j = \sum_{k=1}^{n} u_{kj} v_k \quad (j = 1, ..., n), v = (v_1, ..., v_n) \in \mathbf{R}^n \quad (2.1.19)$$

we have the following substitution for an arbitrary function $x \in C^1(\Omega)$:

$$y(v) = y(v_1, ..., v_n) = x\left(\sum_{k=1}^{n} u_{k1} v_k, ..., \sum_{k=1}^{n} u_{kn} v_k\right)$$

$$= x(t_1, ..., t_n) = x(t)$$

and

$$\frac{\partial y(v)}{\partial v_1} = \sum_{j=1}^{n} \frac{\partial x(t)}{\partial t_j} \frac{\partial t_j}{\partial v_1} = \sum_{j=1}^{n} u_{1j} \frac{\partial x(t)}{\partial t}$$

$$= \sum_{j=1}^{n} a_j \frac{\partial x(t)}{\partial t} = (Dx)(t).$$

The operator $\partial/\partial v_1$ has a right inverse R, namely

$$(Rw)(v) = \int_{v_1^0}^{v_1} w(\eta, v_2, ..., v_n) d\eta.$$

Equalities (2.1.19) and the Cramer Formulae together imply that

$$v_k = U^{-1} U_k(t) \quad (k = 1, ..., n),$$

where we obtain the determinant U_k if we put instead of the k-th column $(t_1, ..., t_n)$. Thus if $\xi(t) = w(v)$, we find

$$(R_0 \xi)(t) = \int_{U^{-1}U_1(t^0)}^{U^{-1}U_1(t)} \xi[U^{-1} U_1(t'), ..., U^{-1} U_n(t')] d\tau,$$

where $t' = (\tau, t_2, ..., t_n) \in \Omega$, $v_1^0 = U^{-1} U_1(t^0)$, and $t^0 \in \Omega$ is arbitrarily fixed. By construction, the operator R_0 is a right inverse of D in the space $C(\Omega)$.

EXERCISE 2.1.1. A real-valued function of a real variable is said to be *elementary* if is obtained by the following operations: addition, subtraction, multiplication, division, superposition and inversion of polynomials, trigonometric functions, exponential functions, whenever these operations are well-determined. Suppose that X, D, R are defined, as in Example 2.1.1. Give methods for calculating indefinite integrals of elementary functions.

EXERCISE 2.1.2. Suppose that X is a commutative algebra (linear ring) with a non-trivial multiplication, i.e. there exist $x, y \in X$ such that $xy \neq 0$. Suppose also that $D \in R(X)$ satisfies the Leibniz condition (2.1.15), i.e. X is a Leibniz algebra according with the definition in Example 2.1.7. Prove that the operator D^n does not satisfy the Leibniz condition for $n = 2, 3, \ldots$ (cf. Mażbic–Kulma, 1971).

EXERCISE 2.1.3. Suppose that X is a Leibniz algebra with unit e, i.e. there is an operator $D \in R(X)$ which satisfies the Leibniz condition (2.1.15). Prove that:

(a) $D(\lambda e) = 0$ for every scalar λ,

(b) $Dx^n = nx^{n-1}Dx$ for $x \in$ dom D $(n = 1, 2, \ldots)$,

(c) there is $g \in X$ such that $Dg = e$, namely $g = Re$, where $R \in \mathcal{R}_D$ is arbitrary,

(d) if $g = Re$ for an $R \in \mathcal{R}_D$ then $Dg^n = ng^{n-1}$ for $n = 1, 2, \ldots$

EXERCISE 2.1.4. Suppose that X is a linear space over a field $\mathscr{F}$ and $D_1, \ldots, D_m \in R(X)$, $R_k \in \mathcal{R}_{D^k}$ $(k = 1, 2, \ldots, m)$. Prove that:

(a) for arbitrary positive integers $k_1, \ldots, k_m$ the operator $D = D_1^{k_1} \ldots D_m^{k_m}$ is right invertible and has a right inverse $R = R_m^{k_m} \ldots R_1^{k_1}$,

(b) if, in particular, $D_1 = D_2 = \ldots = D_m = D$, $k_1 = k_2 = \ldots = k_m = 1$ then $D^m \in R(X)$ and $R = R_1 \ldots R_m \in \mathcal{R}_{D^m}$.

EXERCISE 2.1.5. Suppose that a linear space X is defined as in Example 2.1.2, $D_1 = \partial/\partial t$, $D_2 = \partial/\partial s$. Example 2.1.2 and Exercise 2.1.4 together imply that the operator $D = D_1 D_2$ is right invertible. Determine a right inverse of D.

EXERCISE 2.1.6. Suppose that X is a Leibniz algebra, i.e. there exists an operator $D \in R(X)$ which satisfies Condition (2.1.15). Prove that

$$D^n(xy) = \sum_{k=0}^{n} \binom{n}{k}(D^k x)(D^{n-k}y)$$

for every $x, y \in$ dom D^n $(n = 1, 2, ...)$,   (2.1.20)

where by $\binom{n}{k}$ we denote the *Newton coefficients* of n-th power of a binomial, i.e.

$$\binom{n}{k} = \frac{n!}{k!(n-k)!} \quad (k = 0, 1, ..., n; n = 1, 2, ...).$$

Formula (2.1.20) is called *Leibniz formula*. We point out that the proof of Leibniz formula is by induction, if we use the following property of Newton coefficients: $\binom{n+1}{k} = \binom{n}{k} + \binom{n}{k-1}$ for $k = 1, 2, ..., n$; $n = 1, 2, ...$

EXERCISE 2.1.7. Suppose that $X = C^\infty[0, 1]$ and $(Dx)(t) = tx(t) - dx(t)/dt$ for $x \in X$. Prove that the operator D is right invertible in X and a right inverse of D is $R = -R_0(I - tR_0)^{-1}$, where $(R_0 x)(t) = \int_0^t x(s)\,ds$.

The operator D, called *Hermite derivative*, was considered by Antosik and Mikusiński, 1968.

2.2. INITIAL OPERATORS. TAYLOR–GONTCHAROV FORMULA. DEFINITE INTEGRALS

The following definition plays a fundamental role in our subsequent considerations:

DEFINITION 2.2.1. An operator $F \in L(X)$ is said to be an *initial operator for an operator* $D \in R(X)$ *corresponding to a right inverse* R *of* D if
   (i) F is a projection onto the space of constants, i.e.

$$F^2 = F, \quad FX = \ker D,$$

   (ii)

$$FR = 0.$$

As a matter of fact, the assumption that dom $F = X$ is not essential. It is enough to assume for some considerations that dom $D \subset$ dom $F \subset X$. However, for simplicity only we assume here and in the sequel that dom $F = X$.

Definition 2.2.1 immediately implies that

$$Fz = z \quad \text{for every } z \in \ker D. \tag{2.2.1}$$

Moreover, we have

$$DF = 0 \quad \text{on } X, \tag{2.2.2}$$

$$\ker F = RX \quad \text{and} \quad \ker D \cap \ker F = \{0\}. \tag{2.2.3}$$

Indeed, by definition $Fx \in \ker D$ for every $x \in X$, hence $DFx = 0$. The arbitrariness of x implies $DF = 0$. The property $FR = 0$ implies that $\ker F = RX$. Suppose now that $z \in \ker D$ and $Fz = 0$. Formula (2.2.1) implies that $z = Fz = 0$ which proves that $\ker D \cap \ker F = \{0\}$.

THEOREM 2.2.1. *Suppose that $D \in R(X)$. A necessary and sufficient condition for an operator $F \in L(X)$ to be an initial operator for D corresponding to an $R \in \mathcal{R}_D$ is that*

$$F = I - RD \quad \text{on dom } D. \tag{2.2.4}$$

Proof. Necessity. Suppose that F is an initial operator for D corresponding to an $R \in \mathcal{R}_D$. Let $x \in$ dom D be arbitrarily fixed. Write $u = RDx$. By definition, $Du = D(RDx) = DR(Dx) = Dx$. Thus $D(u-x) = Du - Dx = Dx - Dx = 0$, which implies that $z = u - x \in \ker D$. Thus, by Formula (2.2.1), $F(u-x) = Fz = z = u - z$. On the other hand, since $FR = 0$, we have $Fu = FRDx = 0$. We therefore conclude that

$$(I - RD)x = x - RDx = x - u = -z = -Fz = -F(u-x)$$
$$= Fu + Fx = Fx.$$

The arbitrariness of $x \in$ dom D implies Formula (2.2.4).

Sufficiency. Suppose that $F = I - RD$ on dom D. Thus we have $F^2 = (I - RD)^2 = I - RD - RD + R(DR)D = I - RD + RD = I - RD = F$ and F is a projection operator. Since $DF = D(I - RD) = D - (DR)D = D - D = 0$, we have F dom $D \subset \ker D$. Moreover, if $z \in \ker D$ then $Dz = 0$. Thus $Fz - (I - RD)z = z - RDz = z$. This implies that F is a projection onto ker D. Since $FR = (I - RD)R = R - R = 0$, we conclude that F is an initial operator for D corresponding to R. ∎

Observe that Theorem 2.2.1 characterizes uniquely the operator F on the domain of the operator D. What is outside of the domain of D does not play any role.

PROPOSITION 2.2.1. *If an operator* $A \in L(X)$ *is invertible then initial operators different than zero do not exist.*

Proof. Indeed, let $B \in L(X)$ be an inverse of A, i.e. $BA = I$, $AB = I$. If we write, according with Formula (2.2.4), $F = I - BA$, we conclude that $F = I - BA = I - I = 0$. ∎

This shows that *non-trivial initial operators do exist only for operators which are right invertible but not invertible.*

Theorem 2.2.1 immediately implies

THEOREM 2.2.2. *The family* $\mathscr{R}_D = \{R_\gamma\}_{\gamma \in \Gamma}$ *of all right inverses of an operator* $D \in R(X)$ *induces in a unique way the family* $\mathscr{F}_D = \{F_\gamma\}_{\gamma \in \Gamma}$ *of initial operators for* D *defined by means of the equality*

$$F_\gamma = I - R_\gamma D \quad \text{on dom } D \text{ for every } \gamma \in \Gamma. \tag{2.2.5}$$

THEOREM 2.2.3 (Taylor–Gontcharov Formula). *Suppose that* $D \in R(X)$ *and* $\mathscr{F}_D = \{F_\gamma\}_{\gamma \in \Gamma}$ *denotes the family of initial operators induced by* $\mathscr{R}_D = \{R_\gamma\}_{\gamma \in \Gamma}$. *Let* $\{\gamma_n\} \subset \Gamma$ *be an arbitrary sequence of indices. Then for every positive integer* N *the following identity holds*:

$$I = F_{\gamma_0} + \sum_{k=1}^{N-1} R_{\gamma_0} \dots R_{\gamma_{k-1}} F_{\gamma_k} D^k + R_{\gamma_0} \dots R_{\gamma_{N-1}} D^N$$

$$\text{on dom } D^N. \tag{2.2.6}$$

Proof (by induction). For $N = 1$ we have from Formula (2.2.5) $I = F_{\gamma_0} + R_{\gamma_0} D$ on the domain of the operator D. Suppose that the identity (2.2.6) holds for every fixed $N \geqslant 1$. Then, by the induction assumption, we have on the domain of the operator D^{N+1},

$$R_{\gamma_0} \ldots R_{\gamma_N} D^{N+1} = R_{\gamma_0} \ldots R_{\gamma_{N-1}} (R_{\gamma_N} D) D^N$$

$$= R_{\gamma_0} \ldots R_{\gamma_{N-1}} (I - F_{\gamma_N}) D^N$$

$$= R_{\gamma_0} \ldots R_{\gamma_{N-1}} D^N - R_{\gamma_0} \ldots R_{\gamma_{N-1}} F_{\gamma_N} D^N$$

$$= I - F_{\gamma_0} - \sum_{k=1}^{N-1} R_{\gamma_0} \ldots R_{\gamma_{k-1}} F_{\gamma_k} D^k - R_{\gamma_0} \ldots R_{\gamma_{N-1}} F_{\gamma_N} D^N$$

$$= I - F_{\gamma_0} - \sum_{k=1}^{N} R_{\gamma_0} \ldots R_{\gamma_{k-1}} F_{\gamma_k} D^k,$$

which was to be proved.   ∎

Putting $R_{\gamma_n} = R$ and $F_{\gamma_n} = F$ for $n = 0, 1, 2, \ldots$ we obtain immediately

COROLLARY 2.2.1 (Taylor Formula). *If $D \in R(X)$ and F is an initial operator for D corresponding to an $R \in \mathcal{R}_D$ then*

$$I = \sum_{k=0}^{N-1} R^k F D^k + R^N D^N \quad on \; \text{dom} \, D^N \; (N = 1, 2, \ldots). \quad (2.2.7)$$

COROLLARY 2.2.2. *Suppose that all assumptions of Theorem 2.2.3 are satisfied. Then for every positive integer N*

$$\ker D^N = \left\{ z = z_0 + \sum_{k=1}^{N-1} R_{\gamma_0} \ldots R_{\gamma_{k-1}} z_k : z_0, \ldots, z_{N-1} \in \ker D \right\}.$$

$$(2.2.8)$$

Proof. Suppose that $z = z_0 + \sum_{k=1}^{N-1} R_{\gamma_0} \ldots R_{\gamma_{k-1}} z_k$, where $z_0, \ldots, z_{N-1}$ $\in \ker D$. Then

$$D^N z = D^N z_0 + \sum_{k=1}^{N-1} D^N R_{\gamma_0} \ldots R_{\gamma_{k-1}} z_k = \sum_{k=1}^{N-1} D^{N-k} z_k = 0$$

which implies $z \in \ker D^N$. Conversely, suppose that $z \in \ker D^N$. Since $D^N z = 0$ then the Taylor–Gontcharov Formula implies that

$$z = F_{\gamma_0} z + \sum_{k=1}^{N-1} R_{\gamma_0} \ldots R_{\gamma_{k-1}} F_{\gamma_k} D^k z.$$

Write $z_k = F_{\gamma_k} D^k z$ for $k = 0, 1, ..., N-1$. By definition, $z_0, ..., z_{N-1}$ $\in \ker D$. Thus z is of the required form.   ■

Putting in Corollary 2.2.2 $R_{\gamma_k} = R$, $F_{\gamma_k} = F$ for $k = 0, 1, 2, ...$ we obtain immediately

COROLLARY 2.2.3. *If $D \in R(X)$ and F is an initial operator for D corresponding to an $R \in \mathcal{R}_D$ then*

$$\ker D^N = \left\{ z = \sum_{k=0}^{N-1} R^k z_k : z_0, ..., z_{N-1} \in \ker D \right\}$$

$$(N = 1, 2, ...). \quad (2.2.9)$$

We still assume that $\mathcal{R}_D = \{R_\gamma\}_{\gamma \in \Gamma}$ is the family of all right inverses for an operator $D \in R(X)$ and that $\mathscr{F}_D = \{F_\gamma\}_{\gamma \in \Gamma}$ is induced family of initial operators for D. Then we have the following properties:

$$F_\alpha F_\beta = F_\beta \quad \text{for all } \alpha, \beta \in \Gamma, \quad (2.2.10)$$

$$F_\beta R_\alpha = R_\alpha - R_\beta \quad \text{for all } \alpha, \beta \in \Gamma. \quad (2.2.11)$$

Indeed, since $DF_\beta = 0$ for every $\beta \in \Gamma$, we find $F_\alpha F_\beta = (I - R_\alpha D) F_\beta$ $= F_\beta - R_\alpha (DF_\beta) = F_\beta$. Moreover, $F_\beta R_\alpha = (I - R_\beta D) R_\alpha = R_\alpha - R_\beta (DR_\alpha)$ $= R_\alpha - R_\beta$.   ■

PROPERTY 2.2.1. *For all $\alpha, \beta, \gamma \in \Gamma$ the operator $F_\beta R_\gamma - F_\alpha R_\gamma$ does not depend on the choice of an operator $R_\gamma \in \mathcal{R}_D$.*

Proof. Indeed, Formula (2.2.11) implies that $F_\beta R_\gamma - F_\alpha R_\gamma = R_\gamma - R_\beta -$ $- (R_\gamma - R_\alpha) = R_\alpha - R_\beta + R_\gamma - R_\gamma = R_\alpha - R_\beta = F_\beta R_\alpha$.   ■

Property 2.2.1 shows us that in fact the operator $F_\beta R_\gamma - F_\alpha R_\gamma$, depends only on indices α, β. This permit us to write

$$I_\alpha^\beta = F_\beta R_\gamma - F_\alpha R_\gamma \quad \text{for } \alpha, \beta, \gamma \in \Gamma. \quad (2.2.12)$$

We say that I_α^β is an *operator of definite integration*. For every $x \in X$ the element $I_\alpha^\beta x$ is called *definite integral of x*. The indices α and β are called *lower* and *upper limits of integration*.

The proof of Property 2.2.1 immediately implies that

$$I_\alpha^\beta = F_\beta R_\alpha \quad \text{for } \alpha, \beta \in \Gamma. \quad (2.2.13)$$

70 2. Calculus of right invertible operators

PROPERTY 2.2.2. *For arbitrary $x \in X$, $\alpha, \beta \in \Gamma$ we have*

$$I_\alpha^\beta x = z, \quad where \ z \in \ker D.$$

In other words: *a definite integral of an arbitrary element is a constant.*

Proof. Indeed, Formula (2.2.13) implies that $DI_\alpha^\beta x = DF_\beta R_\alpha x = 0$, because $DF_\beta = 0$. Thus $z = I_\alpha^\beta x \in \ker D$. ∎

PROPERTY 2.2.3. *For arbitrary $\alpha, \beta \in \Gamma$*

$$I_\beta^\alpha = -I_\alpha^\beta. \tag{2.2.14}$$

In other words: *a change of the role of lower and upper limits of integration changes the sign of the operator of definite integration, and consequently, changes also the sign of the definite integral of an arbitrary element.*

Proof. Indeed, $I_\alpha^\beta + I_\beta^\alpha = F_\beta R_\gamma - F_\alpha R_\gamma + F_\alpha R_\gamma - F_\beta R_\gamma = 0$. ∎

PROPERTY 2.2.4. *For arbitrary $\alpha, \beta, \delta \in \Gamma$*

$$I_\alpha^\delta + I_\delta^\beta = I_\alpha^\beta. \tag{2.2.15}$$

Proof. Indeed, $I_\alpha^\delta + I_\delta^\beta = F_\delta R_\gamma - F_\alpha R_\gamma + F_\beta R_\gamma - F_\delta R_\gamma = F_\beta R_\gamma - F_\alpha R_\gamma = I_\alpha^\beta$. ∎

PROPERTY 2.2.5. *For arbitrary $\alpha, \beta \in \Gamma$*

$$I_\alpha^\beta D = F_\beta - F_\alpha, \tag{2.2.16}$$

i.e.

$$I_\alpha^\beta Dx = F_\beta x - F_\alpha x \quad for \ x \in dom \ D. \tag{2.2.17}$$

Proof. Indeed, Formulae (2.2.10) and (2.2.13) together imply that

$$I_\alpha^\beta D = F_\beta R_\alpha D = F_\beta(I - F_\alpha) = F_\beta - F_\beta F_\alpha = F_\beta - F_\alpha. \quad ∎$$

Any element Fx, where $x \in X$, and F is an initial operator, is said to be an *initial value of the element x.*

Since the element $x \in dom \ D$ is primitive for $y = Dx$, we can reformulate Property 2.2.5 in the following way:

PROPERTY 2.2.6. *If $x \in X$, α, $\beta \in \Gamma$ are arbitrary and $y \in X$ is an arbitrary primitive element for x then*

$$I_\alpha^\beta x = F_\beta y - F_\alpha y. \tag{2.2.18}$$

In other words: *a definite integral is equal to the difference of initial values of an arbitrary primitive element corresponding to lower and upper limits of integration.*

PROPOSITION 2.2.2. *Suppose that $D \in R(X)$, dim ker $D \neq 0$, F and $F_1 \neq F$ are initial operators for D and F corresponds to a right inverse $R \in \mathscr{R}_D$. Then for every $z \in \ker D$ there exists an $x \in X$ such that $F_1 Rx = z$.*

In other words: *for every constant there exists an element such that the definite integral of this element is equal to the given constant.*

Proof. Indeed, since F_1 is an initial operator for D, F_1 is a mapping onto ker D. Hence $F_1 RX = \ker D$, because dom $R = X$. Let $z \in \ker D$ be arbitrarily fixed. Then there exists an $x \in X$ such that $F_1 Rx = z$. ∎

Theorems 2.2.1 and 2.2.2 characterize initial operators by means of right inverses. We shall show that one can also characterize right inverses by means of initial operators. This characterization is of great importance. Indeed, in the sequel we shall see that in many problems initial operators are given, and we have to look for corresponding right inverses.

THEOREM 2.2.4. *Suppose that $D \in R(X)$ and $F \in L_0(X)$ is a projection onto the space of constants. Then F is an initial operator for D corresponding to the right inverse $R = R_1 - FR_1$ for every $R_1 \in \mathscr{R}_D$ and R is uniquely determined, independently of the choice of $R_1 \in \mathscr{R}_D$.*

Proof. Since, by assumptions, we have $DF = 0$ and $DR_1 = I$, we conclude that $DR = D(R_1 - FR_1) = DR_1 - DFR_1 = I$. Thus R is a right inverse for D. Since $F^2 = F$, we have $FR = F(R_1 - FR_1) = FR_1 - F^2 R_1 = FR_1 - FR_1 = 0$. Hence F is an initial operator for D corresponding to R. We still have to prove that the operator R is uniquely determined independently of the choice of $R_1 \in \mathscr{R}_D$. Suppose that $R_2 \in \mathscr{R}_D$ and $R_2 \neq R_1$. Write: $R_3 = R_2 - FR_2$. As before, we prove that R_3 is a right inverse for D and that F is an initial operator for D corresponding to R_2, i.e. $F = I - R_2 D$ on dom D. Then $R_3 - R = R_2 - - FR_2 - (R_1 - FR_1) = (I - F)(R_2 - R_1) = R_2 D(R_2 - R_1) = R_2(DR_2 - - DR_1) = R_2 - R_2 = 0$. Hence $R_3 = R$, which was to be proved. ∎

PROPERTY 2.2.7. *If* $D \in R(X)$ *and* $R, R_1 \in \mathcal{R}_D$ *commute, then* $R_1 = R$.

Proof. Indeed, by our assumption, $R = (DR_1)R = D(R_1 R) = D(RR_1) = (DR)R_1 = R_1$. ∎

PROPERTY 2.2.8. *If* $D \in R(X)$ *and* F, F_1 *are initial operators for* D *which commute, then* $F_1 = F$.

Proof. Indeed, by our assumption and Formula (2.2.10), we have $F_1 = FF_1 = F_1 F = F$.

PROPERTY 2.2.9. *Suppose that* $D \in R(X)$ *and* F_1, F_2 *are initial operators for* D *corresponding to right inverses* R_1, R_2, *respectively. If* $R_1 = R_2$ *then* $F_1 = F_2$. *Conversely, if* $F_1 = F_2$, *then* $R_1 = R_2$.

Proof. Indeed, if $R_1 = R_2$, then $F_1 = I - R_1 D = I - R_2 D = F_2$ on dom D. Conversely, suppose that $F_1 = F_2$. Since $F_1 R_1 = 0$, Theorem 2.2.4 implies that $R_2 = R_1 - F_2 R_1 = R_1 - F_1 R_1 = R_1$. ∎

Properties 2.2.7, 2.2.8 and 2.2.9 show that there is no sense in considering commutative right inverses and initial operators. The case of noncommutative right inverses and initial operators is, indeed, essential and interesting.

THEOREM 2.2.5. *If* $D \in R(X)$ *and* F *is an initial operator for* D *corresponding to a right inverse* R *of* D *then the set* $\mathcal{R}_D$ *of all right inverses of* D *is of the form*

$$\mathcal{R}_D = \{R + FA: A \in L_0(X)\} \tag{2.2.19}$$

and the set $\mathcal{F}_D$ *of all initial operators for* D *is of the form*

$$\mathcal{F}_D = \{F(I - AD): A \in L_0(X)\}. \tag{2.2.20}$$

Proof. If $R_1 = R + FA$ for an $A \in L_0(X)$ then $DR_1 = D(R + FA) = DR +$

$+DFA = I$. Hence $R_1 \in \mathscr{R}_D$. On the other hand, if $R_1 \in \mathscr{R}_D$ then $DR_1 = I$ and $FR_1 = (I-RD)R_1 = R_1-R(DR_1) = R_1-R$. Hence $R_1 = R+FR_1 = R+F(R_1-R)$, since $FR = 0$. Writing $A = R_1-R$ we conclude that $R_1 = R+FA$, where $A \in L_0(X)$. We therefore have proved that the set $\mathscr{R}_D$ is of the form (2.2.19). Suppose now that $R_1 \in \mathscr{R}_D$ is arbitrarily fixed. If F_1 is an initial operator corresponding to R_1 then on dom D we find: $F_1 = (I-R_1 D) = I-(R+FA)D = I-RD-FAD = F-FAD = F(I-AD)$. The arbitrariness of F_1 implies Formula (2.2.20). ∎

THEOREM 2.2.6. *Suppose that $F_0, \ldots, F_m$ are initial operators for $D \in R(X)$ and that $P_1, \ldots, P_m \in L(X)$ preserve the space of constants, i.e. $P_k \ker D \subset \ker D$ $(k = 1, \ldots, m)$. Let*

$$F = F_0 + \sum_{k=1}^{m} P_k F_k D^k. \tag{2.2.21}$$

Then F is an initial operator for D corresponding to a right inverse

$$R = R_0 - \sum_{k=1}^{m} P_k F_k D^{k-1}. \tag{2.2.22}$$

Proof. By our assumptions we have $DP_k = 0$ for $k = 1, \ldots, m$. Thus $DF = DF_0 + D\sum_{k=1}^{m} P_k F_k D^k = DF_0 + \sum_{k=1}^{m} DP_k F_k D^k = 0$. Then F maps X into $\ker D$. Moreover $F^2 = \left(F_0 + \sum_{k=1}^{m} P_k F_k D^k\right)F = F_0 F + \sum_{k=1}^{m} P_k F_k D^k F = F_0 F = (I-R_0 D)F = F-R_0 DF = F$.

Since for an arbitrary $z \in \ker D$ we have $Fz = F_0 z + \sum_{k=1}^{m} P_k F_k D^k z = F_0 z = z$, we conclude that F is a projection onto the space of constants. Theorem 2.2.4 implies that F is an initial operator for D corresponding to a right inverse defined as follows: $R = R_0 - FR_0$. But $F_0 R_0 = 0$. Thus $R = R_0 - FR_0 = \left(F_0 + \sum_{k=1}^{m} P_k F_k D^k\right)R_0 = R_0 - F_0 R_0 -$

$- \sum_{k=1}^{m} P_k F_k D^k R_0 = R_0 - \sum_{k=1}^{m} P_k F_k D^{k-1}$ which was to be proved. ∎

THEOREM 2.2.7. *Suppose that $F_0, \ldots, F_m$ are initial operators for an operator $D \in R(X)$ corresponding to right inverses $R_0, \ldots, R_m$, respectively. Write*

$$F = \sum_{k=0}^{m} a_k F_k,$$

where $a_0, \ldots, a_m$ are scalars not all zero. Then F is an initial operator for D if and only if

$$\sum_{k=0}^{m} a_k = 1. \qquad (2.2.23)$$

If this condition is satisfied then the initial operator F corresponds to the right inverse

$$R = \sum_{k=0}^{m} a_k R^k. \qquad (2.2.24)$$

Proof. Necessity. Suppose that F is an initial operator for D. Then for an arbitrary $z \in \ker D$, $z \neq 0$, we have

$$z = Fz = \sum_{k=0}^{m} a_k F_k z = \sum_{k=0}^{m} a_k z = \left(\sum_{k=0}^{m} a_k\right) z,$$

which implies the condition (2.2.23).

Sufficiency. Suppose that the condition (2.2.23) is satisfied. Then, by our assumption, $FX \subset \ker D$. Let $z \in \ker D$ be arbitrary. Then Fz

$$= \sum_{k=0}^{m} a_k F_k z = \sum_{k=0}^{m} a_k z = \left(\sum_{k=0}^{m} a_k\right) z = z. \text{ Moreover, } F^2 z = F(Fz) = Fz$$

$= z$. Hence F is a projection onto $\ker D$.

Theorem 2.2.4 implies that F is an initial operator for D corresponding to a uniquely determined right inverse

$$R = R_0 - FR_0 = R_0 - \sum_{k=0}^{m} a_k F_k R_0 = R_0 - \sum_{k=0}^{m} a_k (I - R_k D) R_0$$

$$= R_0 - \sum_{k=0}^{m} a_k R_0 + \sum_{k=0}^{m} a_k R_k DR_0 = \left(1 - \sum_{k=0}^{m} a_k\right) R_0 + \sum_{k=0}^{m} a_k R_k$$

$$= \sum_{k=0}^{m} a_k R_k,$$

which was to be proved. ∎

In particular, Theorem 2.2.7 implies that a *convex combination of initial operators is again an initial operator*.

THEOREM 2.2.8. *Suppose that* $D_1, \ldots, D_m \in R(X)$ *and that* F_j *is an initial operator for* D_j *corresponding to* $R_j \in \mathcal{R}_{D_j}$ $(j = 1, 2, \ldots, m)$. *Then the operator* $F = F_m + R_m F_{m-1} D_m + \ldots + R_m \ldots R_2 F_1 D_1 \ldots D_m$ *is an initial operator for the operator* $D = D_1 \ldots D_m$ *(if this superposition exists) corresponding to the right inverse* $R = R_m \ldots R_1$ *of* D.

Proof. It is easy to verify that $DR = D_1 \ldots D_m R_m \ldots R_1 = I$ (cf. Exercise 2.1.4). Thus $D \in R(X)$ and $R \in \mathcal{R}_D$. Formula (2.2.4) implies that on the domain of D we have

$$
\begin{aligned}
F &= F_m + R_m F_{m-1} D_m + \ldots + R_m \ldots R_2 F_1 D_2 \ldots D_m \\
&= (I - R_m D_m) + R_m (I - R_{m-1} D_{m-1}) D_m + \ldots + \\
&\quad + R_m \ldots R_2 (I - R_1 D_1) D_2 \ldots D_m \\
&= I - R_m D_m + R_m D_m - R_m R_{m-1} D_{m-1} D_m + \ldots + \\
&\quad + R_m \ldots R_2 D_2 \ldots D_m - R_m \ldots R_2 R_1 D_1 D_2 \ldots D_m \\
&= I - R_m \ldots R_1 D_1 \ldots D_m = I - RD.
\end{aligned}
$$

Hence F is an initial operator for D corresponding to R. ∎

Theorems 2.2.5 and 2.2.8 immediately imply

COROLLARY 2.2.4. *Suppose that all assumptions of Theorem 2.2.8 are satisfied and* F *is an initial operator for* $D = D_1 \ldots D_m$ *corresponding to an* $R \in \mathcal{R}_D$. *Then*

$$\mathcal{R}_D = \{(R_m + F_m A_m) \ldots (R_1 + F_1 A_1): A_1, \ldots, A_m \in L_0(X)\},$$

$$\mathcal{F}_D = \{F_m (I - A_m D_m) + \ldots + (R_m + F_m A_m) \ldots (R_2 + F_2 A_2) F_1 (I - A_1 D_1) D_2 \ldots D_m: A_1, \ldots, A_m \in L_0(X)\}.$$

We have the following characterization of right invertible operators:

THEOREM 2.2.9. *Suppose that* $A \in L(X)$. *If there exists an operator* $B \in L_0(X)$ *such that* $BX \subset \mathrm{dom}\, A$ *and*:
  (i) $\ker B = \{0\}$,
  (ii) *the operator* $P = I - BA$ *defined on* $\mathrm{dom}\, A$ *is a projection into* $\ker A$,
  (iii) $PB = 0$,
then the operator A *is right invertible and* P *is an initial operator for* A *corresponding to the right inverse* B.

Proof. By definition, $P^2 = P$ and $P\,\mathrm{dom}\, A \subset \ker A$. If $x \in \ker A$ then $Ax = 0$ and $Px = x - BAx = x$. Hence P is a projection onto $\ker A$.

Suppose that $x \in \mathrm{dom}\, A$ is arbitrarily fixed. Then $Px = x - BAx \in \ker A$. Thus $A(Px) = 0$ and $Ax - ABAx = A(I - BA)x = APx = 0$. The arbitrariness of x implies that $A = ABA$ on $\mathrm{dom}\, A$. Hence $AB = ABAB = (AB)^2$ which implies that the operator $U = AB$ is a projection and $\mathrm{dom}\, U = \mathrm{dom}\, AB = \mathrm{dom}\, B = X$. Suppose that $U \neq I$ on X. Then there exists a $y \in X$ such that $y \neq 0$ and $v = Uy - y = 0$. Hence $ABv = Uv = U(Uy - y) = U^2y - Uy = Uy - Uy = 0$ and $BA(Bv) = B(ABv) = 0$, which implies $PBv = Bv - BA(Bv) = Bv$. On the other hand, the equality $ABv = 0$ implies that $Bv \in \ker A$. Thus, Condition (iii) implies that $Bv = P(Bv) = PBv = 0$. Since $Bv = 0$, Condition (i) implies that $v = 0$, which contradicts to our assumption. Thus $ABy = y$ for all $y \in X$, i.e. $AB = I$ on X. We therefore conclude that B is a right inverse for A and that P is an initial operator for A corresponding to B. ∎

Examples and Exercises

EXAMPLE 2.2.1. Suppose, as in Example 2.1.1, that $X = C[a, b]$, $D = \dfrac{d}{dt}$ and $(Rx)(t) = \int_{t_0}^{t} x(s)\,ds$, where $a \leqslant t_0 \leqslant b$ is arbitrarily fixed. By Theorem 2.2.1, if $x \in \mathrm{dom}\, D = C^1[a, b]$ then $(Fx)(t) = (I - RD)x(t) = x(t) - (RDx)(t) = x(t) - \int_{t_0}^{t} x'(s)\,ds = x(t) - x(t) + x(t_0) = x(t_0)$.

Observe that the operator F defined by means of the equality

$$(Fx)(t) = x(t_0) \quad \text{for } a \leqslant t \leqslant b \tag{2.2.25}$$

is defined not only for $x \in C^1[a, b]$, but on the whole space $C[a, b]$. This operator acts in this way that to every continuous function $x(t)$ there corresponds a constant function $z(t)$ which has the value equal to the value of x at the point t_0, i.e. $z(t) = x(t_0)$ for all $t \in [a, b]$.
Consider the set

$$\{R_c\}_{c \in [a, b]}, \quad \text{where } (R_c x)(t) = \int_c^t x(s) \, ds \text{ for } x \in C[a, b].$$

Theorem 2.2.2 and Formula (2.2.25) together imply that the induced family of initial operators has the form

$$\{F_c\}_{c \in [a, b]}, \quad \text{where } (F_c x)(t) = x(c).$$

If y is an arbitrary primitive function for $x \in C[a, b]$, and c_1, c_2 are arbitrarily fixed in $[a, b]$ then, by Property 2.2.6, we find

$$\int_{c_1}^{c_2} x(s) \, ds = y(c_1) - y(c_2), \quad \text{where } y' = x. \tag{2.2.26}$$

Thus the formula of integration by parts has the following form:

$$\int_{c_1}^{c_2} x(s) y'(s) \, ds = [x(s) y(s)]_{c_1}^{c_2} - \int_{c_1}^{c_2} x'(s) y(s) \, ds, \tag{2.2.27}$$

where $x, y \in C^1[a, b]$ and we write

$$[u(s)]_{c_1}^{c_2} = u(c_2) - u(c_1) \quad \text{for } u \in C[a, b], a \leqslant c_1, c_2 \leqslant b.$$

To apply Taylor Formula for the operator $D = d/dt$, we shall prove by induction that

$$(R^k x)(t) = \int_{t_0}^t \frac{(t-s)^{k-1}}{(k-1)!} x(s) \, ds \quad \text{for } x \in C[a, b],$$

$$a \leqslant t_0 \leqslant b \quad (k = 1, 2, \ldots). \tag{2.2.28}$$

For $k = 1$ this formula follows from the definition of the operator R.

Suppose this formula to be true for an arbitrarily fixed $k \geqslant 1$. By the induction assumption and the integration by parts we obtain

$$\int_{t_0}^{t} \frac{(t-s)^k}{k!} x(s) \, ds$$

$$= \left[\frac{(t-s)^k}{k!} \int_{t_0}^{s} x(u) \, du \right]_{t_0}^{t} - \int_{t_0}^{t} -k \frac{(t-s)^{k-1}}{k!} \left[\int_{t_0}^{s} x(u) \, du \, ds \right]$$

$$= \int_{t_0}^{t} \frac{(t-s)^{k-1}}{(k-1)!} \left[\int_{t_0}^{s} x(u) \, du \right] ds = [R^k(Rx)](t) = (R^{k+1}x)(t),$$

which was to be proved.

In particular, if we put $x(t) = c$ for $a \leqslant t \leqslant b$, where $c \in \mathbf{R}$, we obtain

$$(R^k c)(t) = c \frac{(t-t_0)^k}{k!} \quad \text{for } c \in \mathbf{R} \quad (k = 1, 2, ...). \quad (2.2.29)$$

Indeed,

$$(R^k c)(t) = \int_{t_0}^{t} \frac{(t-s)^{k-1}}{(k-1)!} c \, ds = c \int_{t_0}^{t} \frac{(t-s)^{k-1}}{(k-1)!} \, ds$$

$$= c \left[\frac{(t-s)^k}{k!} \right]_{t_0}^{t} = c \frac{(t-t_0)^k}{k!}.$$

This, and Taylor Formula (2.2.7) together, imply that every function $x \in C^N[a, b]$ ($N = 1, 2, ...$) can be presented in the form

$$x(t) = \sum_{k=0}^{N-1} (R^k F D^k x)(t) + (R^N D^N x)(t) = (Fx)(t) +$$

$$+ \sum_{k=1}^{N-1} \int_{t_0}^{t} \frac{(t-s)^{k-1}}{(k-1)!} (Fx^{(k)})(s) \, ds + \int_{t_0}^{t} \frac{(t-s)^{N-1}}{(N-1)!} x^{(N)}(s) \, ds$$

$$= x(t_0) + \sum_{k=1}^{N-1} \int_{t_0}^{t} \frac{(t-s)^{k-1}}{(k-1)!} x^{(k)}(t_0) \, ds +$$

$$+ \int_{t_0}^{t} \frac{(t-s)^{N-1}}{(N-1)!} x^{(N)}(s)\,ds$$

$$= x(t_0) + \sum_{k=1}^{N-1} \frac{(t-t_0)^k}{k!} x^{(k)}(t_0) + \int_{t_0}^{t} \frac{(t-s)^{N-1}}{(N-1)!} x^{(N)}(s)\,ds$$

$$= \sum_{k=0}^{N-1} \frac{(t-t_0)^k}{k!} x^{(k)}(t_0) + \int_{t_0}^{t} \frac{(t-s)^{N-1}}{(N-1)!} x^{(N)}(s)\,ds.$$

Finally we obtain a Taylor Formula for functions $x \in C^N [a, b]$ in the form

$$x(t) = \sum_{k=0}^{N-1} \frac{(t-t_0)^k}{k!} x^{(k)}(t_0) + R_N(t), \qquad (2.2.30)$$

where

$$R_N(t) = \int_{t_0}^{t} \frac{(t-s)^{N-1}}{(N-1)!} x^{(N)}(s)\,ds \quad (N = 1, 2, ...). \qquad (2.2.31)$$

The function $R_N(t)$ is said to be N-th *integral remainder* in the Taylor Formula (2.2.30). To obtain this remainder in another form, we can apply the following classical properties of continuous functions: every function $x \in C[a, b]$ reaches the greatest lower bound $m = \inf\limits_{a \leqslant t \leqslant b} x(t)$ and the least upper bound $M = \sup\limits_{a \leqslant t \leqslant b} x(t)$ in the interval (a, b). Moreover, for any $c \in [m, M]$ there is a $t \in [a, b]$ such that $x(t) = c$, i.e. the *function $x(t)$ admits all intermediate values.*

Suppose we are given a function $x \in C^N[a, b]$ $(N = 1, 2, ...)$. Write: $m = \inf\limits_{a \leqslant s \leqslant b} x^{(N)}(s)$, $M = \sup\limits_{a \leqslant s \leqslant b} x^{(N)}(s)$. Observe that the function $(t-s)^{N-1}/(N-1)!$ is non-negative for $t_0 \leqslant s \leqslant t$. Thus we have the following estimation:

$$m \frac{(t-t_0)^N}{N!} = m \int_{t_0}^{t} \frac{(t-s)^{N-1}}{(N-1)!}\,ds \leqslant R_N(t) \leqslant M \int_{t_0}^{t} \frac{(t-s)^{N-1}}{(N-1)!}\,ds$$

$$= M \frac{(t-t_0)^N}{N!}.$$

By our assumptions we conclude that there exists a number Θ $\in (0, 1)$ such that $s = t_0 + \Theta(t-t_0)$ and

$$R_N(t) = x^{(N)}(t_0 + \Theta(t-t_0)) \frac{(t-t_0)^N}{N!}, \quad 0 < \Theta < 1. \quad (2.2.32)$$

This is the so-called *Lagrange remainder*.

If we write now

$$m(t) = \inf_{a \leqslant s \leqslant b} \frac{(t-s)^{N-1}}{(N-1)!} x^{(N)}(s),$$

$$M(t) = \sup_{a \leqslant s \leqslant b} \frac{(t-s)^{N-1}}{(N-1)!} x^{(N)}(s),$$

we obtain the following estimation:

$$m(t)(t-t_0) \leqslant R_N(t) \leqslant M(t)(t-t_0) \quad \text{for all } t \in [a, b].$$

Then there exists a number $\Theta' \in (0, 1)$ such that

$$R_N(t) = x^{(N)}(t_0 + \Theta'(t-t_0)) \frac{\{t - [t_0 + \Theta'(t-t_0)]\}^{N-1}}{(N-1)!}(t-t_0)$$

$$= x^{(N)}(t_0 + \Theta'(t-t_0)) \frac{(1-\Theta')^{N-1}(t-t_0)^N}{(N-1)!}.$$

Thus

$$R_N(t) = x^{(N)}(t_0 + \Theta'(t-t_0)) \frac{(1-\Theta')^{N-1}(t-t_0)^N}{(N-1)!},$$

$$0 < \Theta' < 1. \quad (2.2.32')$$

This is the so-called *Cauchy remainder*.

If the point $t_0 = 0$ belongs to the interval $[a, b]$ we put in Formula (2.2.30) $t_0 = 0$ and we obtain the so-called *Maclaurin Formula*:

$$x(t) = \sum_{k=0}^{N-1} \frac{t^k}{k!} x^{(k)}(0) + R_N(t), \quad (2.2.33)$$

where

$$R_N(t) = \int_0^t \frac{(t-s)^{N-1}}{(N-1)!} x^{(N)}(s) \, ds \quad \text{or} \quad R_N(t) = x^{(N)}(\Theta t) \frac{t^N}{N!},$$

$$0 < \Theta < 1$$

or

$$R_N(t) = x^{(N)}(\Theta' t) \frac{(1-\Theta')^{N-1} t^N}{(N-1)!}, \quad 0 < \Theta' < 1.$$

Suppose now that

$$x \in C^\infty[a, b] \quad \text{and} \quad \lim_{N \to \infty} R_N(t) = 0 \quad \text{for } t \in [a, b]. \quad (2.2.34)$$

Then, from Formula (2.2.30), it follows that

$$x(t) = \sum_{k=0}^{\infty} x^{(k)}(t_0) \frac{(t-t_0)^k}{k!}. \quad (2.2.35)$$

The convergent series (2.2.35) is called the *Taylor series*.

If Condition (2.2.34) is satisfied, we say that the function $x(t)$ *has an expansion in the Taylor series in the interval* $[a, b]$. In particular, if $t_0 = 0$ and Condition (2.2.34) is satisfied, we say that the function $x(t)$ *has an expansion in the Maclaurin series* of the form

$$x(t) = \sum_{k=0}^{\infty} \frac{t^k}{k!} x^{(k)}(0). \quad (2.2.36)$$

Observe that Taylor and Maclaurin series are *power series*, i.e. are of the form $\sum_{k=0}^{\infty} a_k(t-t_0)^k$, where $a_k \in \mathbf{R}$. Indeed, we can put $a_k = x^{(k)}(t_0)/k!$ for $k = 0, 1, 2, \ldots$

EXAMPLE 2.2.2. Suppose, as in Example 2.2.1, that $X = C[a, b]$, $D = \dfrac{d}{dt}$ and $(Rx)(t) = \int_a^t x(s) ds$. We shall prove that R is a Volterra operator, i.e. the operator $I - \lambda R$ is invertible for every scalar λ. Write

$$S_n(t) = \sum_{k=0}^{n} \frac{t^k}{k!} \quad (n = 0, 1, 2, \ldots).$$

The sequence $\{S_n\}$ is uniformly and absolutely convergent in each closed interval $[a, b]$. We admit the following denotation:

$$e^t = \lim_{n\to\infty} S_n(t) = \sum_{n=0}^{\infty} \frac{t^n}{n!}, \quad t \in \mathbf{R}. \qquad (2.2.37)$$

The function e^t is called the *exponential function*. Observe that, by definition, we have

$$e^0 = 1, \quad e^{t+s} = e^t e^s \quad \text{and} \quad \frac{d}{dt} e^t = e^t \quad \text{for all } t, s \in \mathbf{R}.$$
$$(2.2.38)$$

Indeed,

$$e^{t+s} = \sum_{n=0}^{\infty} \frac{(t+s)^n}{n!} = \sum_{n=0}^{\infty} \frac{1}{n!} \sum_{k=0}^{n} \binom{n}{k} t^k s^{n-k}$$

$$= \sum_{n=0}^{\infty} \sum_{k=0}^{n} \frac{1}{n!} \frac{n!}{(n-k)!k!} t^k s^{n-k} = \sum_{n=0}^{\infty} \sum_{k=0}^{n} \frac{t^k}{k!} \frac{s^{n-k}}{(n-k)!}$$

$$= \left(\sum_{n=0}^{\infty} \frac{t^n}{n!}\right)\left(\sum_{n=0}^{\infty} \frac{s^n}{n!}\right) = e^t e^s.$$

Moreover, since

$$\frac{d}{dt} S_n(t) = \sum_{k=1}^{n} k \frac{t^{k-1}}{k!} = \sum_{k=1}^{n} \frac{t^{k-1}}{(k-1)!} = \sum_{m=0}^{n-1} \frac{t^m}{m!} = S_{n-1}(t),$$

we conclude that

$$\lim_{n\to\infty} \frac{d}{dt} S_n(t) = \lim_{n\to\infty} S_{n-1}(t) = e^t.$$

Since the series of derivatives is also uniformly and absolutely convergent in each interval $[a, b]$, we find $\dfrac{d}{dt} e^t = \dfrac{d}{dt} \lim_{n\to\infty} S_n(t) = \lim_{n\to\infty} \dfrac{d}{dt} S_n(t)$ $= \lim_{n\to\infty} S_{n-1}(t) = e^t.$

If we put $t = 1$, we obtain a number $e = \displaystyle\sum_{n=0}^{\infty} \frac{1}{n!}$.

One can prove that this number is *transcendental* (i.e. a polynomial $w(t)$ with integer coefficients such that $w(e) = 0$ does not exist). An approximative value of e is: $e \approx 2.7182818285 \dots$

Define now an operator B by means of the exponential function:

$$(Bx)(t) = \int_{t_0}^{t} e^{\lambda(t-s)} x(s) \, ds \quad \text{for } x \in C[a, b], \tag{2.2.39}$$

where $t_0 \in [a, b]$ is arbitrarily fixed. We shall prove that

$$(I + \lambda B)(I - \lambda R) = (I - \lambda R)(I + \lambda B) = I \quad \text{for all } \lambda \in \mathbf{R}. \tag{2.2.40}$$

Indeed, without loss of generality we can assume that $\lambda \neq 0$. Then, by integration by parts, we obtain for every $x \in C[a, b]$:

$$[(I + \lambda B)(I - \lambda R)x](t) = [(I + \lambda B - \lambda R - \lambda^2 BR)x](t)$$

$$= [x + \lambda(B - R)x - \lambda^2 BR x](t)$$

$$= x(t) + \lambda \left[\int_{t_0}^{t} e^{\lambda(t-s)} x(s) \, ds - \int_{t_0}^{t} x(s) \, ds \right] -$$

$$- \lambda^2 \int_{t_0}^{t} e^{\lambda(t-s)} \left[\int_{t_0}^{s} x(u) \, du \right] ds$$

$$= x(t) + \lambda \int_{t_0}^{t} [e^{\lambda(t-s)} - 1] x(s) \, ds - \lambda^2 e^{\lambda t} \int_{t_0}^{t} e^{-\lambda s} \left[\int_{t_0}^{s} x(u) \, du \right] ds$$

$$= x(t) + \lambda \int_{t_0}^{t} [e^{\lambda(t-s)} - 1] x(s) \, ds -$$

$$- \lambda^2 e^{\lambda t} \left\{ \left[-\frac{1}{\lambda} e^{-\lambda s} \int_{t_0}^{s} x(u) \, du \right]_{t_0}^{t} - \int_{t_0}^{t} -\frac{1}{\lambda} e^{-\lambda s} x(s) \, ds \right\}$$

$$= x(t) + \lambda \int_{t_0}^{t} [e^{\lambda(t-s)} - 1] x(s) \, ds +$$

$$+ \lambda \int_{t_0}^{t} x(u)\,du - \lambda \int_{t_0}^{t} e^{\lambda(t-s)} x(s)\,ds$$

$$= x(t) + \lambda \int_{t_0}^{t} [e^{\lambda(t-s)} - 1 + 1 - e^{\lambda(t-s)}] x(s)\,ds = x(t).$$

Hence, $(I + \lambda B)(I - \lambda R) = I$. A similar proof for the second equality $(I - \lambda R)(I + \lambda B) = I$. Thus Equalities (2.2.40) imply that the operator R is invertible for every scalar λ and $(I - \lambda R)^{-1} = I + \lambda B$, i.e.

$$[(I - \lambda R)^{-1} x](t) = x(t) + \lambda \int_{t_0}^{t} e^{\lambda(t-s)} x(s)\,ds \quad \text{for } x \in C[a, b].$$

$$(2.2.41)$$

EXAMPLE 2.2.3. Suppose that $X = C[a, b]$ and that $q \in C[a, b]$. Let $q_0 = \int_a^b q(s)\,ds \neq 0$. Define an operator F by means of the equality

$$(Fx)(t) = \frac{1}{q_0} \int_a^b q(s) x(s)\,ds \quad \text{for } x \in C[a, b]. \qquad (2.2.42)$$

We shall prove that F is an initial operator for $D = d/dt$ corresponding to a right inverse defined as follows

$$(Rx)(t) = \int_a^t x(s)\,ds - \frac{1}{q_0} \int_a^b \left[q(s) \int_a^s x(u)\,du \right] ds. \qquad (2.2.43)$$

Indeed, since values of Fx are constants, we have $FX \subset \ker D$. Suppose that $c \in \mathbf{R}$. Then the function $z(t) \equiv c$ is a constant, hence $z \in \ker D$ and

$$(Fz)(t) = \frac{1}{q_0} \int_a^b q(s) z(s)\,ds = \frac{1}{q_0} \int_a^b c q(s)\,ds$$

$$= \frac{c}{q_0} \int_a^b q(s)\,ds = \frac{c}{q_0} q_0 = z(t).$$

Thus $Fz = z$ for $z \in \ker D$, i.e. F is a mapping onto $\ker D$. Let $x \in X$ be arbitrarily fixed. Write $z = Fx$. Then $z \in \ker D$ and $F^2 x = F(Fx) = Fz = z = Fx$. The arbitrariness of $x \in X$ implies that F is a projection onto the space of constants $\ker D$. All assumptions of Theorem 2.2.4 are satisfied. We therefore conclude that F is an initial operator for $D = d/dt$ corresponding to a right inverse $R = R_0 - FR_0$, where we can admit $(R_0 x)(t) = \int_a^t x(s) \, ds$. This implies that R is of the form (2.2.43).

This example in the case $q(t) \equiv 1$ has been given by Tasche, 1974. The operator F has then the form

$$(Fx)(t) = \frac{1}{b-a} \int_a^b x(s) \, ds \quad \text{for } x \in X, \qquad (2.2.42')$$

and the operator R is of the form

$$(Rx)(t) = \int_a^t x(s) \, ds - \int_a^b \frac{b-s}{b-a} x(s) \, ds \quad \text{for } x \in X. \qquad (2.2.43')$$

Indeed, integrating by parts we find

$$(Rx)(t) = [(R_0 - FR_0)x](t) = \int_a^t x(s) \, ds - \frac{1}{b-a} \int_a^b \left[\int_a^s x(u) \, du \right] ds$$

$$= \int_a^t x(s) \, ds - \frac{1}{b-a} \left\{ \left[s \int_a^s x(u) \, du \right]_a^b - \int_a^b sx(s) \, ds \right\}$$

$$= \int_a^t x(s) \, ds - \frac{1}{b-a} \left[b \int_a^b x(s) \, ds - \int_a^b sx(s) \, ds \right]$$

$$= \int_a^t x(s) \, ds - \int_a^b \frac{b-s}{b-a} x(s) \, ds.$$

EXAMPLE 2.2.4. Let $X = C[a, b]$ and let $d \in \mathbf{R}$ be arbitrarily fixed. Define an operator F as follows:

$$(Fx)(t) = dx(a) + (1-d)x(b) \quad \text{for } x \in X. \qquad (2.2.44)$$

Example 2.2.1 shows that the operators $(F_a x)(t) = x(a)$ and $(F_b x)(t) = x(b)$ are initial operators for the operator $D = d/dt$. Thus Theorem 2.2.7 implies that F is again an initial operator for $D = d/dt$ for $F = dF_a + (1-d)F_b$ and the sum of coefficients $d, 1-d$ is equal to 1. The initial operators F_a and F_b correspond to the right inverses R_a and R_b, respectively, defined as follows:

$$(R_a x)(t) = \int_a^t x(s)\,ds, \quad (R_b x)(t) = \int_b^t x(s)\,ds \quad \text{for } x \in X.$$

Then, by Theorem 2.2.7, the initial operator F corresponds to a right inverse $R = dR_a + (1-d)R_b$, i.e.

$$(Rx)(t) = d\int_a^t x(s)\,ds + (1-d)\int_b^t x(s)\,ds \quad \text{for } x \in X.$$

EXAMPLE 2.2.5. Let $X = C[0, 1]$ and $d \in \mathbf{R}$ be arbitrarily fixed. Define an operator F as follows:

$$(Fx)(t) = dx(0) + (1-d)\int_0^1 x(s)\,ds. \tag{2.2.45}$$

Theorem 2.2.7 and Example 2.2.5 (for $a = 0$, $b = 1$) together imply that F is an initial operator for the operator $D = d/dt$ corresponding to a right inverse R defined by means of the equality

$$(Rx)(t) = d\int_0^t x(s)\,ds + (1-d)\left[\int_0^t x(s)\,ds - \int_0^1 (1-s)x(s)\,ds\right]$$

$$= \int_0^t x(s)\,ds + (d-1)\int_0^1 (1-s)x(s)\,ds.$$

EXAMPLE 2.2.6. Let $X = C^1[a, b]$ and let $d \in \mathbf{R}$ be arbitrary. Define an operator F as follows:

$$(Fx)(t) = x(a) + dx'(b) \quad \text{for } x \in X. \tag{2.2.46}$$

Thus $F = F_a + dF_b D$, where $D = d/dt$ and F_a, F_b are defined as in Example 2.2.4. By Theorem 2.2.6, F is an initial operator for $D = d/dt$

corresponding to a right inverse $R = R_a - dF_b$, i.e.

$$(Rx)(t) = \int_a^t x(s)\,ds - dx(b).$$

EXAMPLE 2.2.7. Suppose that X is a linear space, $D \in R(X)$, $\mathscr{R}_D = \{R_\gamma\}_{\gamma \in \Gamma}$, $\mathscr{F}_D = \{F_\gamma\}_{\gamma \in \Gamma}$ (where F_γ is an initial operator for D corresponding to the right inverse R_γ), $B \in L(X)$, B is invertible and the following equality holds: $DA = BAD$ on dom D. Formula (2.1.11) implies that $F_\beta R_\alpha BADx = (F_\beta - F_\alpha)Ax$ for $x \in $ dom D, $\alpha, \beta \in \Gamma$.

Indeed, if $x \subset $ dom D then

$$\begin{aligned} F_\beta R_\alpha BADx &= F_\beta R_\alpha DAx = F_\beta(I - F_\alpha)Ax \\ &= (F_\beta - F_\beta F_\alpha)Ax = (F_\beta - F_\alpha)Ax. \end{aligned}$$

EXAMPLE 2.2.8. Suppose that $X = C[a, b]$ and $D = d/dt$. Write: $(R_c x)(t)$ $= \int_c^t x(s)\,ds$ for $a \leqslant c \leqslant b$. Then an initial operator F_c corresponding to R_c is $(F_c x)(t) = x(c)$ (cf. Example 2.2.1). Put $(Ax)(t) = x(g(t))$, where $g \in C^1[a, b]$. Then $(DAx)(t) = x'(q(t))q'(t) = (BADx)(t)$, where $(Bx)(t) = q'(t)x(t)$. In a similar way, as in Example 2.1.6, we obtain

$$\int_{c_1}^{c_2} g'(s)x(g(s))\,ds = \int_{g(c_1)}^{g(c_2)} x(u)\,du, \tag{2.2.47}$$

for $a \leqslant c_1 \leqslant c_2 \leqslant b$. This is a *formula of integration by substitution for definite integrals*. Indeed, we can write

$$\int_{g(c_1)}^{g(c_2)} x(s)\,ds = \int_c^{g(c_2)} x(s)\,ds - \int_c^{g(c_1)} x(s)\,ds = \xi(g(c_2)) - \xi(g(c_1)),$$

where $\xi(t) = \int_c^t x(s)\,ds$ is a primitive function for x for an arbitrarily fixed $c \in [a, b]$. Formula (2.2.47) is also called a *formula for change of limits of integration*.

EXERCISE 2.2.1. Suppose that $X = C[a, b]$ and $D = d/dt$. Suppose we are given N points $t_0, ..., t_{N-1} \in [a, b]$. Write Taylor–Gontcharov

Formula for initial operators F_k defined as follows: $(F_k x)(t) = x(t_k)$
for $k = 0, 1, ..., N-1$, $x \in C[a, b]$ (cf. Example 2.2.1).

EXERCISE 2.2.2. Suppose that X is a linear space, $D \in R(X)$, $R \in \mathcal{R}_D$,
$A \in L_0(X)$, $AX \subset \operatorname{dom} D$ and, moreover, A is invertible. Prove that:
    (i) the operator $D_1 = DA$ is right invertible and the operator F_1
$= A^{-1}FA$ is an initial operator for D_1 corresponding to $R_1 = A^{-1}R$
$\in \mathcal{R}_{D_1}$, where F is an initial operator for D corresponding to an $R \in \mathcal{R}_D$;
    (ii) the operator $D_2 = A^{-1}D$ is right invertible, the operator $F_2 = F$
is an initial operator for D_2 corresponding to $R_2 = RA \in \mathcal{R}_{D_2}$, where F
and R are as in point (i).

EXERCISE 2.2.3. Suppose that $D \in R(X)$ and $F_0, ..., F_{N-1}$ are initial
operators for D corresponding to right inverses $R_0, ..., R_{N-1}$
$\in \mathcal{R}_D$. Using Taylor–Gontcharov Formula show that the operator
$$F = F_0 + \sum_{k=1}^{N-1} R_0 ... R_{k-1} F_k D^k \text{ is an initial operator for } D^N \in R(X)$$
corresponding to a right inverse $R = R_0 ... R_{N-1} \in \mathcal{R}_{D^N}$.

EXERCISE 2.2.4. Suppose that X, D, R are defined as in Example 2.2.1.
The proof of Lemma 2.1.3 implies that the area P of the domain bounded
by the graph of the curve $x = x(t)$, the coordinate axis $y = 0$ and the
lines $t = a$ and $t = b$ is equal to the definite integral $\int_a^b x(s)ds$ (see
Fig. 2.1).

EXERCISE 2.2.5.   The   following   theorem   is   well-known   as   the

ROLLE THEOREM. If $x \in C^1[a, b]$ and $x(a) = x(b)$ then there exists
a point c, $a < c < b$, such that $x'(c) = 0$.

    Indeed, define X, D, F as in Example 2.2.1. Put $t_0 = a$, $t = b$, $N = 1$
and write the Taylor Formula (2.2.30) with the Lagrange remainder
(2.2.32). Then we obtain $x(b) = x(a)+x'(a+\Theta(b-a))(b-a)$. This implies $x'(a+\Theta(b-a)) = 0$ for a Θ, $0 < \Theta < 1$. Writing $c = a+\Theta(b-a)$
we conclude that $x'(c) = 0$ for a $c \in (a, b)$. Let $x \in C^1[a, b]$. Denote

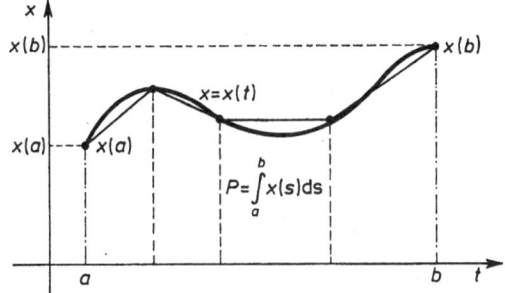

Fig. 2.1. Geometric interpretation of a definite integral

by α the angle of a tangent at a point $c \in (a, b)$ of the curve $x = x(t)$ with the coordinate axis $y = 0$. Then $\tan \alpha = x'(c)$. Thus the Rolle Theorem has the following geometric interpretation: if $x(a) = x(b)$ then there exists an internal point c of the interval $[a, b]$ such that the tangent at point c is parallel to the coordinate axis $y = 0$ (see Fig. 2.2). The Rolle Theorem is a theorem about an intermediate value of a function.

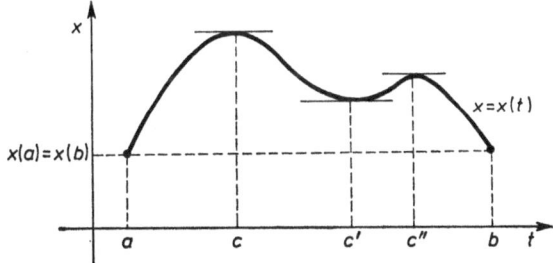

Fig. 2.2. Geometric interpretation of the Rolle Theorem

EXERCISE 2.2.6. The following theorem is well-known as the

LAGRANGE THEOREM. If $x \in C^1[a, b]$ then there exists a number Θ, $0 < \Theta < 1$, such that

$$\frac{x(b) - x(a)}{b - a} = x'(a + \Theta(b - a)).$$ (2.2.48)

Indeed, in the same way, as in the preceding exercise, we obtain from Formulae (2.2.30) and (2.2.32)

$$x(b) = x(a) + x'(a + \Theta(b-a))(b-a) \quad \text{for a } \Theta \in (0, 1).$$

$$(2.2.49)$$

Formula (2.2.48) has the following geometric interpretation: there exists a point $c \in (a, b)$, namely $c = a + \Theta(b-a)$, $0 < \Theta < 1$, such that the tangent at the point c is parallel to the secant passing through points $(a, x(a))$ and $(b, x(b))$ (see Fig. 2.3). This is again a theorem about an intermediate value of a function.

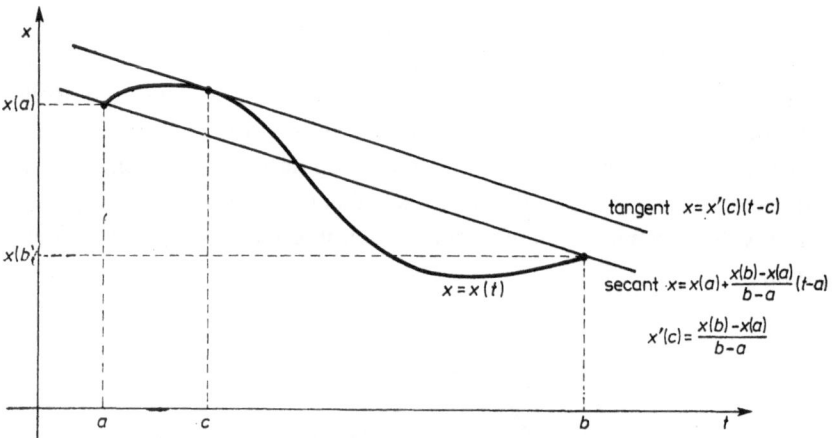

Fig. 2.3. Geometric interpretation of the Lagrange Theorem

EXERCISE 2.2.7. Prove the so-called

CAUCHY THEOREM. *If $x, y \in C^1[a, b]$ and $y'(t) \neq 0$ for $t \in [a, b]$ then there exists a positive number $\Theta < 1$ such that*

$$\frac{x(b) - x(a)}{y(b) - y(a)} = \frac{x'(a + \Theta h)}{y'(a + \Theta h)}, \quad \text{where } h = b - a. \qquad (2.2.50)$$

Indeed, since $y'(t) \neq 0$ in $[a, b]$ then by Rolle Theorem (Exercise 2.2.5) we have $y(b) \neq y(a)$. Write

$$u(t) = x(a) - x(t) + [y(t) - y(a)] \frac{x(b) - x(a)}{y(b) - y(a)}, \quad t \in [a, b].$$

We can prove that the function $u(t)$ satisfies all conditions of the Rolle Theorem and we conclude that there is a number $0 < \Theta < 1$ such that $u'(a+\Theta h) = 0$, which implies the required formula (2.2.50). This is again a theorem concerning an intermediate value of a function.

EXERCISE 2.2.8. Write expansions in the Maclaurin series for the following functions: $\sin x$, $\cos x$, $1/(1+x)$, $1/(1-x)$, $1/(1-x^2)$, $\ln(1+x)$.

EXERCISE 2.2.9. Suppose that X, D, R are defined as in Example 2.1.4. Prove that:

(i) an initial operator F for D corresponding to R is of the form $Fx = \{z_n\}$, where $x = \{x_n\}$, $z_n = x_1$ $(n = 1, 2, ...)$;

(ii) $D^k x = \left\{ \sum\limits_{m=0}^{k} \binom{k}{m} (-1)^m x_{n+k-m} \right\}$, where $x = \{x_n\}$ $(k = 1, 2, ...)$;

(iii) $FD^k x = \{z_{n,k}\}$, where $z_{n,k} = c_k = \sum\limits_{m=0}^{k} \binom{k}{m}(-1)^m x_{k+1-m}$ for k, n
$= 1, 2, ...$;

(iv) if $u = R^{j+1}x$, where $u = \{u_n\}$, $x = \{x_n\}$, $j = 0, 1, 2, ...$ then

$$u_n = \begin{cases} 0 & \text{for } n = 1, 2, ..., j+1, \\ \sum\limits_{m=j}^{n-2} \binom{m}{j} x_{n-1-m} & \text{for } n = j+2, j+3, ... ; \end{cases}$$

(v) if $u = R^{j+1}z$, where $j = 0, 1, 2, ...$, $z = \{z_n\} \in \ker D$, since $z_n = c \in \mathbf{R}$ $(n = 1, 2, ...)$ we have

$$u_n = \begin{cases} 0 & \text{for } n = 1, 2, ..., j+1, \\ c \sum\limits_{m=1}^{n-2} \binom{m}{j} & \text{for } n = j+2, j+3, ... ; \end{cases}$$

(vi) the Taylor Formula is of the form $(N = 1, 2, ...)$

$$x = \left\{ \sum\limits_{k=0}^{N-1} a_{n,k} \sum\limits_{m=0}^{n} \binom{k}{m}(-1)^m x_{k+1-m} \right\} + \{R_{N,n}\},$$

where

$$a_{n,k} = \begin{cases} 0 & \text{for } n = 1, 2, \ldots, k+1, \\ \displaystyle\sum_{j=1}^{n-2} \binom{k}{j} & \text{for } n = k+2, k+3, \ldots, \end{cases}$$

$$R_{N,n} = R^N D^N x$$

$$= \begin{cases} 0 & \text{for } n = 1, 2, \ldots, N, \\ \displaystyle\sum_{m=N-1}^{n-2} \binom{m}{n-1} \sum_{j=1}^{n-1-m} \binom{n-1-m}{j}(-1)^j x_{n-m-j} \\ & \text{for } n = N+1, N+2, \ldots \end{cases}$$

EXERCISE 2.2.10. Suppose that X, D, R are defined as in Example 2.1.4. Let $m > 1$ be an arbitrarily fixed positive integer. Prove (applying Theorem 2.2.4) that the operator F_m defined by means of the equality

$$F_m x = \{z_n\}, \quad z_n = \frac{1}{m}\sum_{k=1}^{m} x_k \quad \text{for } n = 1, 2, \ldots, x = \{x_n\} \in X$$

is an initial operator for the operator D corresponding to the right inverse R_m defined as follows:

$$R_m x = y = \{y_n\}, \quad \text{where } x = \{x_n\} \in X,$$

$$y_1 = -\frac{1}{m}\sum_{j=1}^{m-1}(m-j)x_j,$$

$$y_n = \sum_{j=1}^{n-1} x_j - \frac{1}{m}\sum_{j=1}^{m-1}(m-j)x_j \quad \text{for } n \geqslant 2.$$

EXERCISE 2.2.11. Suppose that X, D, R are defined as in Example 2.1.4. Let $m > 1$ be an arbitrarily fixed positive integer and $\alpha \neq 0$ be an arbitrarily fixed real number. Prove that the operator F defined as follows:

$$Fx = \{z_n\}, \quad z_n = \alpha x_1 + \frac{1-\alpha}{m}\sum_{k=1}^{m} x_k \quad (n = 1, 2, \ldots)$$

is an initial operator for the operator D and determines the corresponding right inverse. (Apply Theorem 2.2.7, Point (i) of Exercise 2.2.9 and Exercise 2.2.10.)

EXERCISE 2.2.12. Suppose that $D_1, D_2 \in R(X)$, $R_i \in \mathcal{R}_{D_i}$ $(i = 1, 2)$ and $D_1 D_2 = D_2 D_1$ on dom $D_1 \cap$ dom D_2. Prove that a necessary and sufficient condition for the operators R_1 and R_2 to be commutative is that there exists an operator $A \in L_0(X)$ such that $F_1 A = 0$, $F_2 D_1 A = 0$, where F_i is an initial operator for D_i corresponding to the right inverse R_i $(i = 1, 2)$. Show that one can admit $A = R_1 R_2 - R_2 R_1$ (cf. Przeworska-Rolewicz, 1981 (1979)).

EXERCISE 2.2.13. Are conditions (i), (ii), (iii) be all essential for the proof of Theorem 2.2.9?

2.3. EXPONENTIALS, SINE AND COSINE OPERATORS AND VOLTERRA RIGHT INVERSES

Suppose that X is a linear space over an algebraically closed field $\mathcal{F}$. For instance, $\mathcal{F} = \mathbf{C}$. If X is a linear space over the reals we can consider instead of X its natural extension to a linear space over the complexes (cf. Section 1.1.2).

DEFINITION 2.3.1. If a number $\lambda \in \mathbf{C}$ is an eigenvalue of the operator D then every eigenvector x_λ corresponding to the eigenvalue λ is said to be an *exponential element* (shortly: an *exponential*). By definition, x_λ is an exponential if and only if $x_\lambda \neq 0$ and $x_\lambda \in \ker(D - \lambda I)$.

PROPOSITION 2.3.1. *Suppose that $D \in R(X)$ and F is an initial operator for D corresponding to an $R \in \mathcal{R}_D$. If $0 \neq x_\lambda \in \ker(I - \lambda R)$ for a $\lambda \in \mathcal{F}$ then $x_\lambda \in \ker(D - \lambda I)$, i.e. is an exponential.*

Proof. Indeed, since $x_\lambda \in \ker(I - \lambda R)$ we have $(I - \lambda R)x_\lambda = 0$ and

$$0 = D(I - \lambda R)x_\lambda = (D - \lambda DR)x_\lambda = (D - \lambda I)x_\lambda$$

which implies $x_\lambda \in \ker(D - \lambda I)$. ∎

PROPOSITION 2.3.2. *Suppose that all assumptions of Proposition* 2.3.1 *are satisfied. If* x_λ *is an exponential and* $Fx_\lambda = 0$ *then* $x_\lambda \in \ker(I - \lambda R)$, *i.e.* R *is not a Volterra operator.*

Proof. Indeed, by our assumptions, we have $0 \neq x_\lambda \in \ker(D - \lambda I)$, i.e. $Dx_\lambda = \lambda x_\lambda$, and $Fx_\lambda = 0$ which implies $x_\lambda = x_\lambda - Fx_\lambda = (I - F)x_\lambda = RDx_\lambda = \lambda Rx_\lambda$. Thus $x_\lambda \in \ker(I - \lambda R)$. Since $x_\lambda \neq 0$ we conclude that R is not a Volterra operator.   ∎

THEOREM 2.3.1. *Suppose that* $D \in R(X)$ *and* $\{\lambda_n\} \subset \mathscr{F}$ *is a sequence of eigenvalues for* D *such that* $\lambda_i \neq \lambda_j$ *for* $i \neq j$. *Then for an arbitrary positive integer* n *the exponentials* $x_{\lambda_1}, \ldots, x_{\lambda_n}$ *are linearly independent.*

Proof (by induction). For $n = 1$ we have from the definition $x_{\lambda_1} \neq 0$. Suppose that for an arbitrarily fixed $n \geq 1$ the exponentials $x_{\lambda_1}, \ldots, x_{\lambda_n}$ are linearly independent, i.e.

$$\sum_{j=1}^{n} a_j x_{\lambda_j} = 0 \quad \text{for } a_1, \ldots, a_n \in \mathscr{F}$$

implies

$$a_1 = \ldots = a_n = 0.$$

Suppose that the exponential $x_{\lambda_{n+1}}$ is linearly dependent on the set $\{x_{\lambda_1}, \ldots, x_{\lambda_n}\}$. This means that there exists a scalar $a_{n+1} \neq 0$ such that

$$\sum_{j=1}^{n+1} a_j x_{\lambda_j} = 0. \tag{2.3.1}$$

Acting on both sides of Equality (2.3.1) by operator D we get

$$0 = D\sum_{j=1}^{n+1} a_j x_{\lambda_j} = \sum_{j=1}^{n+1} a_j Dx_{\lambda_j} = \sum_{j=1}^{n+1} a_j \lambda_j x_{\lambda_j}. \tag{2.3.1'}$$

Multiplying (2.3.1) by λ_{n+1} and subtracting (2.3.1') we find

$$0 = \sum_{j=1}^{n+1} a_j(\lambda_{n+1} - \lambda_j)x_{\lambda_j} = \sum_{j=1}^{n} a_j(\lambda_{n+1} - \lambda_j)x_{\lambda_j}.$$

Since $\lambda_{n+1} \neq \lambda_j$ for $j = 1, \ldots, n$, our induction assumption implies that $a_1 = \ldots = a_n = 0$. This, and Equality (2.3.1) together, imply that $a_{n+1} = 0$ which contradicts our assumption. We therefore conclude that $x_{\lambda_{n+1}}$ is linearly independent of the set $\{x_{\lambda_1}, \ldots, x_{\lambda_n}\}$, what was to be proved. ∎

THEOREM 2.3.2. *Suppose that* $D \in R(X)$ *and* F *is an initial operator for* D *corresponding to an* $R \in \mathcal{R}_D$. *Suppose that* x_λ *is an exponential. Then* x_λ *is an eigenvector for* R *corresponding to the eigenvalue* $1/\lambda$ *if and only if* $Fx_\lambda = 0$.

Proof. Sufficiency. It follows immediately from Proposition 2.3.2.

Necessity. Suppose that $\mu = 1/\lambda$ is an eigenvalue of R and the corresponding eigenvector x_λ is an exponential (corresponding to the eigenvalue λ). Since $Dx_\lambda = \lambda x_\lambda$ and $Rx_\lambda = \mu x_\lambda$ we find $0 = \mu x_\lambda - Rx_\lambda = (1/\lambda)x_\lambda - Rx_\lambda = (1/\lambda)(I - \lambda R)x_\lambda = (1/\lambda)(I - RD)x_\lambda = Fx_\lambda$, which was to be proved. ∎

A consequence of Theorem 1.4.5 is

PROPOSITION 2.3.3. *If* R *is a Volterra right inverse, i.e.* $R \in \mathcal{R}_D \cap V(X)$ *for a* $D \in R(X)$, *then the operators* $e_\lambda = (I - \lambda R)^{-1}$ *(which exist by our assumption) have no eigenvalues for* $\lambda \neq 0$.

Proof. Indeed, suppose that $u \in \ker R$. Then $Ru = 0$ and $u = DRu = 0$. Consequently, $\ker R = \{0\}$. This and Point (i) of Theorem 1.4.5 together imply that the operators e_λ have no eigenvalues. ∎

THEOREM 2.3.3. *Suppose that* $D \in R(X)$, $\dim \ker D \neq 0$, $R \in \mathcal{R}_D$ *and that the operator* $I - \lambda R$ *is invertible for a* $\lambda \in \mathcal{F}$. *Then*:

(i) λ *is an eigenvalue of the operator* D *and the corresponding exponentials are of the form*

$$e_\lambda(z) = (I - \lambda R)^{-1}z, \quad where \ z \in \ker D. \tag{2.3.2}$$

Moreover, the dimension of the eigenspace corresponding to the eigenvalue λ *is equal to the dimension of the space of constants* $\ker D$.

(ii) *If* $\lambda \neq 0$ *then there exist non-trivial exponential elements*: $e_\lambda(z) \neq 0$.

(iii) *If* F *is an initial operator for* D *corresponding to the operator* R *then exponentials* $e_\lambda(z)$ *are uniquely determined by their initial values*:

$$e_\lambda(z) = (I - \lambda R)^{-1}F[e_\lambda(z)], \quad i.e. \quad F[e_\lambda(z)] = z. \tag{2.3.3}$$

Proof. (i) By definition, $(I - \lambda R)e_\lambda(z) = (I - \lambda R)(I - \lambda R)^{-1}z = z$, where $z \in \ker D$. Thus $e_\lambda(z) = z + \lambda Re_\lambda(z)$ and $De_\lambda(z) = Dz + \lambda DRe_\lambda(z) = \lambda e_\lambda(z)$. This proves that the element $e_\lambda(z)$ is an eigenvector of the operator D corresponding to the eigenvalue λ. Since the operator $I - \lambda R$ is invertible, we conclude that $\dim\{e_\lambda(z): z \in \ker D\} = \dim(I - \lambda R)^{-1} \ker D = \dim \ker D \neq 0$.

(ii) Suppose that $\lambda \neq 0$ and $e_\lambda(z) = (I - \lambda R)^{-1}z = 0$. This implies that $z = (I - \lambda R)e_\lambda(z) = 0$ for an arbitrary $z \in \ker D$, which contradicts to our assumption that $\dim \ker D \neq 0$.

(iii) The definition and (i) together imply that $De_\lambda(z) = \lambda e_\lambda(z)$. Thus $Fe_\lambda(z) = (I - RD)e_\lambda(z) = e_\lambda(z) - \lambda Re_\lambda(z) = (I - \lambda R)e_\lambda(z)$, which implies $e_\lambda(z) = (I - \lambda R)^{-1}F[e_\lambda(z)]$, i.e. $F[e_\lambda(z)] = z$. ∎

COROLLARY 2.3.1. *Suppose that $D \in R(X)$, $\dim \ker D \neq 0$, and that there is a right inverse R of D which is a Volterra operator, i.e. $R \in \mathscr{R}_D \cap \cap V(X)$. Then:*

(i) *Every scalar λ is an eigenvalue of the operator D, i.e. for every $\lambda \in \mathscr{F}$ there exist non-trivial exponentials.*

(ii) *There exist $x \in X$ such that $x \neq 0$ and $Dx = x$, namely, $x = e_1(z)$, where $z \in \ker D$.*

Proof. By our assumption the operator $I - \lambda R$ is invertible for every $\lambda \in \mathscr{F}$. Thus Point (i) of Theorem 2.3.3 implies that all numbers $\lambda \in \mathscr{F}$ are eigenvalues of the operator D and the corresponding eigenvectors are $e_\lambda(z) = (I - \lambda R)^{-1}z$, where $z \in \ker D$. This and Point (ii) of Theorem 2.3.3 together imply the existence of non-trivial exponential elements for every $\lambda \in \mathscr{F}$. Putting $\lambda = 1$ we obtain Point (ii) of our Corollary. ∎

COROLLARY 2.3.2. *Suppose that $D \in R(X)$, $\{R_\gamma\}_{\gamma \in \Gamma_0} \subset \mathscr{R}_D$. Let $\mathscr{F}_D^0 = \{F_\gamma\}_{\gamma \in \Gamma_0}$ denotes the induced family of initial operators. If $R_\gamma \in V(X)$ for every $\gamma \in \Gamma_0$ (i.e. R_γ is a Volterra operator for every $\gamma \in \Gamma_0$) then $(F_\beta - F_\alpha)e_\lambda(z) = \lambda F_\beta R_\alpha e_\lambda(z)$, where $z \in \ker D$, $\alpha, \beta \in \Gamma_0$.*

Proof. Indeed, by definition and Corollary 2.3.1 we have

$$\lambda F_\beta R_\alpha e_\lambda(z) = F_\beta R_\alpha[\lambda e_\lambda(z)] = F_\beta R_\alpha De_\lambda(z)$$
$$= F_\beta(I - F_\alpha)e_\lambda(z) = (F_\beta - F_\beta F_\alpha)e_\lambda(z) = (F_\beta - F_\alpha)e_\lambda(z). ∎$$

Corollary 2.3.1 permits us to introduce

DEFINITION 2.3.2. If $D \in R(X)$ and there is an $R \in \mathcal{R}_D \cap V(X)$, then every operator

$$e_\lambda = (I - \lambda R)^{-1}, \quad \lambda \in \mathcal{F} \tag{2.3.4}$$

is called an *exponential operator*.

THEOREM 2.3.4. *Suppose that* $D \in R(X)$, *dim* $\ker D \neq 0$ *and* F *is an initial operator corresponding to an* $R \in \mathcal{R}_D \cap V(X)$. *Suppose that* $F_1 \in L_0(X)$ *is an arbitrary projection onto* $\ker D$, *i.e. an initial operator for* D *corresponding to the right inverse* $R_1 = R - F_1 R$ *(cf. Theorem 2.2.4). Let* $0 \neq \lambda \in \mathcal{F}$. *Then the operator* R_1 *has an eigenvector* $e_\lambda(z)$ *corresponding to the eigenvalue* $1/\lambda$ *if and only if there exists* $0 \neq z \in \ker D$ *such that*

$$F_1 e_\lambda(z) = 0, \quad \text{where } e_\lambda(z) = (I - \lambda R)^{-1} z. \tag{2.3.5}$$

Proof. Exponentials $e_\lambda(z) = (I - \lambda R)^{-1} z$ are well-defined for all $\lambda \in \mathcal{F}$, $z \in \ker D$ by our assumption that $R \in V(X)$. By definition, $D e_\lambda(z) = \lambda e_\lambda(z)$ for all $\lambda \in \mathcal{F}$, $z \in \ker D$. This, and Theorem 2.3.2, together imply the conclusion of our Theorem. ∎

COROLLARY 2.3.3. *Suppose that all assumptions of Theorem* 2.3.4 *are satisfied and Condition* (2.3.5) *holds. Then* R_1 *is not a Volterra operator.*

COROLLARY 2.3.4. *Suppose that all assumptions of Theorem* 2.3.4 *are satisfied. Then* $R_1 \in V(X_0)$, *where* $X_0 = \{x \in X: F_1 R x = 0\}$.

Proof. Indeed, if $x \in X_0$ then $R_1 x = (R - F_1 R)x = Rx - F_1 Rx = Rx$. Since $R \in V(X)$, the arbitrariness of $x \in X_0$ implies $R_1 \in V(X_0)$. ∎

COROLLARY 2.3.5. *Suppose that all assumptions of Theorem* 2.3.4 *are satisfied. Suppose that* F_1 *is a projection onto* $\ker D$. *Then the operator* $F_2 = dFD + F_1$, *where* $0 \neq d \in \mathcal{F}$, *is an initial operator for* D. *Moreover, if there exist* $0 \neq \lambda \in \mathcal{F}$ *and* $0 \neq z \in \ker D$ *such that*

$$F_1 e_\lambda(z) = -\lambda dz, \tag{2.3.6}$$

then the corresponding right inverse $R_2 = R - dF - F_1 R$ *is not a Volterra operator.*

Proof. It is easy to verify that F_2 is a projection onto ker D. Then Theorem 2.2.4 implies that F_2 is an initial operator for D corresponding to the right inverse $R_2 = R - F_2 R = R - dFDR - F_1 R = R - dF - F_1 R$. Suppose now that Condition (2.3.6) holds. Write $u = e_\lambda(z)$ and $\mu = 1/\lambda$. Since $Du = \lambda u$, $Fu = Fe_\lambda(z) = z$, we get

$$\mu u - R_2 u = \frac{1}{\lambda}(I - \lambda R_2)u = \frac{1}{\lambda}(I - R_2 D)u = \frac{1}{\lambda}F_2 u$$

$$= \frac{1}{\lambda}(dFD + F_1)u = \frac{1}{\lambda}(dFDu + F_1 u)$$

$$= \frac{1}{\lambda}(\lambda dFu - \lambda dz) = \frac{1}{\lambda}(\lambda dz - \lambda dz) = 0$$

which proves that u is an eigenvector for R corresponding to the eigenvalue $\mu = 1/\lambda$. Thus $R_2 \notin V(X)$.   ∎

THEOREM 2.3.5. *Suppose that $D \in R(X)$, dim ker $D \neq 0$ and F is an initial operator for D corresponding to an $R \in \mathscr{R}_D$. If there exists $A \in L_0(X)$ such that $F(I - \lambda A)x_\lambda = 0$, where x_λ is an exponential, then the right inverse $R_1 = R + FA$ is not a Volterra operator.*

Proof. Observe that, by definition, $R_1 = R + FA$ is a right inverse for D. By our assumptions $Dx_\lambda = \lambda x_\lambda$ and $Fx_\lambda = \lambda FAx_\lambda$. This implies that

$$(I - \lambda R_1)x_\lambda = x_\lambda - \lambda Rx_\lambda - \lambda FAx_\lambda = x_\lambda - RDx_\lambda - \lambda FAx_\lambda$$
$$= x_\lambda - (I - F)x_\lambda - \lambda FAx_\lambda = x_\lambda - x_\lambda + Fx_\lambda - \lambda FAx_\lambda = 0$$

and $0 \neq x_\lambda \in \ker(I - \lambda R_1)$. We therefore conclude that $R_1 \notin V(X)$.   ∎

Write for a $D \in R(X)$ an $R \in \mathscr{R}_D \cap V(X)$:

$$E(R) = \lim\{e_\lambda(z) = (I - \lambda R)^{-1}z : z \in \ker D, \lambda \in \mathscr{F}\}. \qquad (2.3.7)$$

It is easy to verify that

$$E(R) = \bigcup_{\lambda \in \mathscr{F}} \ker(D - \lambda I). \qquad (2.3.8)$$

PROPOSITION 2.3.4. *Suppose that $R \in \mathscr{R}_D \cap V(X)$ for a $D \in R(X)$ and dim ker $D \neq 0$. Then the set $E(R)$, defined by Formula (2.3.7), of all exponentials is independent of the choice of a right inverse R. Moreover,* *if* $\lambda_1, \ldots, \lambda_n$ *are different each of another then the set* $\{e_{\lambda_j}(z_j)\}_{j=1,\ldots,n}$ *consists of linearly independent elements.*

Proof. Indeed, if $R_1, R_2 \in \mathscr{R}_D \cap V(X)$, $R_1 \neq R_2$ then exponentials $(I - \lambda R_1)^{-1} z$, $(I - \lambda R_2)^{-1} z \in \ker(D - \lambda I)$ for arbitrarily fixed $\lambda \in \mathscr{F}$, $z \in \ker D$. This, and Formula (2.3.8), together imply that $E(R_1) = E(R_2)$. Theorem 2.3.1 implies the linear independence of exponentials. ∎

DEFINITION 2.3.3. If $D \in R(X)$, $\mathscr{F} = \mathbf{R}$, and there is an $R \in \mathscr{R}_D \cap V(X)$ then operators

$$c_\lambda = \frac{1}{2}(e_{\lambda i} + e_{-\lambda i}), \quad s_\lambda = \frac{1}{2i}(e_{\lambda i} - e_{-\lambda i}) \qquad (2.3.9)$$

are called *cosine* and *sine operators*, respectively. The elements $c_\lambda(z)$ and $s_\lambda(z)$, where $z \in \ker D$, are called *cosine* and *sine elements*, respectively.

THEOREM 2.3.6. *If* $D \in R(X)$, $\mathscr{F} = \mathbf{R}$ *and there is an* $R \in \mathscr{R}_D \cap V(X)$ *then the cosine and sine operators have the following properties for* $\lambda, \mu \in \mathbf{R}$:

$$c_\lambda = (I + \lambda^2 R^2)^{-1}, \quad s_\lambda = \lambda R(I + \lambda^2 R^2)^{-1}, \qquad (2.3.10)$$

$$c_\lambda^2 + s_\lambda^2 = e_{\lambda i} e_{-\lambda i}, \qquad (2.3.11)$$

$$c_\lambda s_\mu + c_\mu s_\lambda = \frac{1}{2i}(e_{\lambda i} e_{\mu i} - e_{-\lambda i} e_{-\mu i}), \qquad (2.3.12)$$

$$c_\lambda c_\mu - s_\lambda s_\mu = \frac{1}{2}(e_{\lambda i} e_{\mu i} + e_{-\lambda i} e_{-\mu i}). \qquad (2.3.13)$$

Proof. The definitions (2.3.9) imply that

$$c_\lambda = \frac{1}{2}(e_{\lambda i} + e_{-\lambda i}) = \frac{1}{2}[(I - \lambda i R)^{-1} + (I + \lambda i R)^{-1}]$$

$$= \frac{1}{2}(I - \lambda i R)^{-1}(I + \lambda i R)^{-1}(I + \lambda i R + I - \lambda i R)$$

$$= \frac{1}{2}(I + \lambda^2 R^2)^{-1} 2I = (I + \lambda^2 R^2)^{-1},$$

$$s_\lambda = \frac{1}{2i}(e_{\lambda i} - e_{-\lambda i}) = \frac{1}{2i}[(I - \lambda i R)^{-1} - (I + \lambda i R)^{-1}]$$

$$= \frac{1}{2i}(I - \lambda i R)^{-1}(I + \lambda i R)^{-1}[I + \lambda i R - (I - \lambda i R)]$$

$$= \frac{1}{2i}(I + \lambda^2 R^2)^{-1}(I + \lambda i R - I + \lambda i R)$$

$$= \frac{1}{2i}(I + \lambda^2 R^2)^{-1} 2\lambda i R = \lambda R (I + \lambda^2 R^2)^{-1}.$$

Furthermore we have

$$c_\lambda^2 + s_\lambda^2 = \frac{1}{4}(e_{\lambda i} + e_{-\lambda i})^2 + \frac{1}{4i^2}(e_{\lambda i} - e_{-\lambda i})^2$$

$$= \frac{1}{4}(e_{\lambda i}^2 + 2e_{\lambda i}e_{-\lambda i} + e_{-\lambda i}^2) - \frac{1}{4}(e_{\lambda i}^2 - 2e_{\lambda i}e_{-\lambda i} + e_{-\lambda i}^2)$$

$$= \frac{1}{4} 4e_{\lambda i}e_{-\lambda i} = e_{\lambda i}e_{-\lambda i},$$

$$c_\lambda s_\mu + c_\mu s_\lambda = \frac{1}{2}(e_{\lambda i} + e_{-\lambda i}) \frac{1}{2i}(e_{\mu i} - e_{-\mu i}) +$$

$$+ \frac{1}{2i}(e_{\lambda i} - e_{-\lambda i}) \frac{1}{2}(e_{\mu i} + e_{-\mu i})$$

$$= \frac{1}{4i}(e_{\lambda i}e_{\mu i} - e_{-\mu i}e_{\lambda i} + e_{\mu i}e_{-\lambda i} - e_{-\lambda i}e_{-\mu i} +$$

$$+ e_{\lambda i}e_{\mu i} - e_{-\lambda i}e_{\mu i} + e_{\lambda i}e_{-\mu i} - e_{-\lambda i}e_{-\mu i})$$

$$= \frac{1}{4i}(2e_{\lambda i}e_{\mu i} - 2e_{-\lambda i}e_{-\mu i}) = \frac{1}{2i}(e_{\lambda i}e_{\mu i} - e_{-\lambda i}e_{-\mu i}).$$

The proof of Equality (2.3.13) is similar. ∎

THEOREM 2.3.7. *If* $D \in R(X)$, dim ker $D \neq 0$ *and there is an* $R \in \mathcal{R}_D \cap V(X)$ *then*

$$Dc_\lambda = -\lambda s_\lambda, \quad Ds_\lambda = \lambda c_\lambda \quad for \ \lambda \in \mathbf{R}. \tag{2.3.14}$$

Proof. Indeed, from the Definitions (2.3.9) for every $\lambda \in \mathbf{R}$, $z \in$ ker D we have

$$Dc_\lambda = \frac{1}{2} D(e_{\lambda i} + e_{-\lambda i}) = \frac{1}{2} (\lambda i e_{\lambda i} - \lambda i e_{-\lambda i})$$

$$= \frac{1}{2} \lambda i (e_{\lambda i} - e_{-\lambda i}) = -\frac{\lambda}{2i} (e_{\lambda i} - e_{-\lambda i}) = -\lambda s_\lambda.$$

Since $DR = I$, we have for every $\lambda \in \mathbf{R}$

$$Ds_\lambda = \lambda DR(I + \lambda^2 R^2)^{-1} = \lambda(I + \lambda^2 R^2)^{-1} = \lambda c_\lambda. \qquad \blacksquare$$

THEOREM 2.3.8. *Suppose that all assumptions of Theorem 2.3.7 are satisfied. Let F be an initial operator corresponding to R. Then*:

(i) *For arbitrurily fixed $\lambda \in \mathbf{R}$ and $z \in \ker D$ the element $c_\lambda(z)$ does not vanish. Moreover,*

$$c_0(z) = z, \quad s_0(z) = 0, \quad Fs_\lambda(z) = 0. \qquad (2.3.15)$$

(ii) *For arbitrarily fixed $\lambda \in \mathbf{R}$ and $z \in \ker D$ the element $c_\lambda(z)$ is even with respect to λ and the element $s_\lambda(z)$ is odd.*

(iii)

$$c_\lambda(z) \in E(R), \quad s_\lambda(z) \in E(R) \quad \text{for all } \lambda \in \mathbf{R}, z \in \ker D, \qquad (2.3.16)$$

where the set $E(R)$ of exponentials is defined by Formula (2.3.7).

Proof. (i) Suppose that there exist $\lambda \in \mathbf{R}$ and $z \in \ker D$ such that $z \neq 0$ and $c_\lambda(z) = 0$. Formulae (2.3.10) imply that

$$Fs_\lambda(z) = \lambda FR(I + \lambda^2 R^2)^{-1}z = 0 \quad \text{for all } z \in \ker D,$$
$$z = (I + \lambda^2 R^2)(I + \lambda^2 R^2)^{-1}z = (I + \lambda^2 R^2)c_\lambda(z) = 0 \qquad (2.3.17)$$

which contradicts our assumptions.

If $\lambda = 0$ then Formulae (2.3.10) imply also that $c_0(z) = z$, $s_0(z) = 0$.

(1) Let $c_\lambda(z) = 0$. Suppose that $\lambda = 0$. Then the first of Formulae (2.3.10) implies that $z = c_0(z) = 0$, which contradicts our assumption that $z \neq 0$. Now, suppose that $\lambda \neq 0$. Then the first of Equalities (2.3.17) implies that $z = 0$, for the operator $(I + \lambda^2 R^2)^{-1}$ is invertible. This also contradicts our assumption that $z \neq 0$.

(2) Let $s_\lambda(z) = 0$. If $\lambda = 0$ then the first of Formulae (2.3.10) implies that $c_0(z) = z \neq 0$. Suppose that $\lambda \neq 0$. Then the second of Formulae (2.3.10) implies that $\lambda Rz = 0$. Since $\lambda \neq 0$ we find $Rz = 0$ and $z = DRz = 0$, which contradicts our assumption that $z \neq 0$.

(ii) Let $\lambda \in \mathbf{R}$ and $z \in \ker D$ be arbitrary. Formulae (2.3.10) imply that

$$c_{-\lambda}(z) = [I + (-\lambda)^2 R^2]^{-1} z = (I + \lambda^2 R^2)^{-1} z = c_\lambda(z),$$
$$s_{-\lambda}(z) = -\lambda R [I + (-\lambda)^2 R^2]^{-1} z = -\lambda R (I + \lambda^2 R^2)^{-1} z = -s_\lambda(z).$$

This means that cosine elements are even and sine elements are odd with respect to λ.

(iii) The elements $c_\lambda(z) = \dfrac{1}{2}[e_{\lambda i}(z) + e_{-\lambda i}(z)]$, $s_\lambda(z) = \dfrac{1}{2i}[e_{\lambda i}(z) - e_{-\lambda i}(z)]$, $\lambda \in \mathbf{R}$, $z \in \ker D$, as linear combinations of exponentials belong to the set $E(R)$. ∎

REMARK 2.3.1. Theorems 2.3.3, 2.3.4 and Corollaries 2.3.1, 2.3.2 hold even if R is not a Volterra operator but satisfies a weaker assumption. Namely, denote by $V_L(X)$ the set of operators $A \in L_0(X)$ such that for every scalar λ there exists an operator A_λ satisfying the following conditions:

$$A_\lambda(I - \lambda A) = I \quad \text{but} \quad (I - \lambda A) A_\lambda \neq I \tag{2.3.18}$$

(i.e. the operators $I - \lambda A$ are left invertible, but not invertible).

Suppose that for a given $D \in R(X)$, dim $\ker D \neq 0$, there exists an $R \in \mathcal{R}_D \cap V_L(X)$, i.e. there exists a right inverse R and operators R_λ $(\lambda \in \mathcal{F})$ such that

$$R_\lambda(I - \lambda R) = I \quad \text{and} \quad (I - \lambda R) R_\lambda \neq I \quad \text{for every } \lambda \in \mathcal{F}.$$

Then elements $x_\lambda = R_\lambda z$, where $\lambda \in \mathcal{F}$, $z \in \ker D$ are arbitrary, are exponentials for D. Indeed, if $Dx_\lambda = \lambda x_\lambda$ then by our assumption $0 = (D - \lambda I) x_\lambda = D(I - \lambda R) x_\lambda$, which implies $(I - \lambda R) x_\lambda = z$, where $z \in \ker D$ and $x_\lambda = R_\lambda(I - \lambda R) x_\lambda = R_\lambda z$. Thus $R_\lambda z$ are exponentials. Further considerations are going on the same lines. Similarly, we can define sine and cosine operators by means of operators R_λ (cf. also Exercise 2.3.14).

Examples and Exercises

EXAMPLE 2.3.1. Suppose that $X = C[a, b]$, $D = d/dt$, and $(Rx)(t) = \int_{t_0}^{t} x(s) \, ds$ for $x \in C[a, b]$, where $t_0 \in [a, b]$ is arbitrarily fixed. Formula

(2.2.41) shows that R is a Volterra operator. Moreover, if $z(t) \equiv c \in \mathbf{R}$, i.e. z is a constant, then from the same Formula (2.2.41) we find

$$[e_\lambda(z)](t) = [(I-\lambda R)^{-1}z](t)$$

$$= z(t) + \lambda \int_{t_0}^{t} e^{\lambda(t-s)}z(s)\,ds$$

$$= c + \lambda c \int_{t_0}^{t} e^{\lambda(t-s)}ds = c + c[e^{\lambda(t-s)}]_{t_0}^{t}$$

$$- c - c^0 + ce^{\lambda(t-t_0)} = c - c + ce^{\lambda(t-t_0)}$$

$$= ce^{\lambda(t-t_0)}.$$

Therefore every exponential in the space $C[a, b]$ is an exponential function. In particular, if we put $c = e^{\lambda t_0}$ then we obtain exponential functions of the form $ce^{\lambda t}$.

We know (cf. Example 2.2.1) that an initial operator F corresponding to R is of the form: $(Fx)(t) = x(t_0)$ for $x \in C[a, b]$. Therefore, a unique exponential satisfying the condition $F[e_\lambda(z)] = 1$ is the function $e^{\lambda(t-t_0)}$ $= c_0 e^{\lambda t}$, where $c_0 = e^{-\lambda t_0}$. Indeed, if $1 = [ce^{\lambda(t-t_0)}]_{t=t_0}$ then $1 = ce^0$ $= c$. Thus exponential functions are uniquely determined by their initial values, as the general case.

Having already defined exponentials we find for arbitrary $\lambda \in \mathbf{R}$, $z(t) \equiv c$, $c \in \mathbf{R}$, cosine and sine elements. Namely, we have (cf. Formula (2.3.9) and Definition 2.3.3):

$$[c_\lambda(z)](t) = \frac{1}{2}[e_{\lambda i}(z) + e_{-\lambda i}(z)] = \frac{1}{2}[ce^{\lambda it} + ce^{-\lambda it}]$$

$$= c\,\frac{e^{\lambda it} + e^{-\lambda it}}{2} = \frac{c}{2}\left[\sum_{n=0}^{\infty} \frac{(\lambda it)^n}{n!} + \sum_{n=0}^{\infty} \frac{(-\lambda it)^n}{n!}\right]$$

$$= \frac{c}{2}\sum_{n=0}^{\infty} \frac{\lambda^n t^n i^n[1+(-1)^n]}{n!} = \frac{c}{2}\sum_{n=0}^{\infty} \frac{2\lambda^{2n}t^{2n}i^{2n}}{n!}$$

$$= c\sum_{n=0}^{\infty} \frac{(-1)^n \lambda^{2n}t^{2n}}{(2n)!}.$$

In a similar way we find

$$[s_\lambda(z)](t) = c \sum_{n=0}^{\infty} \frac{(-1)^n \lambda^{2n+1} t^{2n+1}}{(2n+1)!}.$$

On the other hand, if we expand functions $\cos \lambda t$ and $\sin \lambda t$, in Maclaurin series (Exercise 2.2.8), we obtain the same series convergent uniformly in every interval $[a, b]$. Thus

$$[c_\lambda(z)](t) = c \cos \lambda t, \quad [s_\lambda(z)](t) = c \sin \lambda t,$$

and cosine and sine elements are cosine and sine functions, respectively.

EXAMPLE 2.3.2. Suppose that $X = C(\Omega)$, where $\Omega = \{(t, s) \in \mathbf{R}^2 :$ $a \leqslant t, s \leqslant b\}$ and $D = \dfrac{\partial}{\partial t}$, $(Rx)(t, s) = \int_{t_0}^{t} x(u, s)du$ for $x \in C(\Omega)$, $(t, s) \in \Omega$. We have proved in Example 2.1.2 that R is a right inverse of D, ker D is defined by Formula (2.1.9) and dim ker $D = +\infty$.

Arguing, as in Example 2.2.1, we prove that an initial operator F corresponding to R is of the form: $(Fx)(t, s) = x(t_0, s)$. Since constants for the operator $D = \partial/\partial t$ are all functions of one variable s belonging to $C[a, b]$, we conclude that all exponentials are of the form: $[e_\lambda(z)](t, s) = z(s)e^{\lambda(t-t_0)}$, where $z \in C[a, b]$, $\lambda \in \mathbf{C}$ are arbitrary. Similarly, cosine and sine elements are of the form: $z_1(s) \cos \lambda t$, $z_2(s) \sin \lambda t$, where $z_1, z_2 \in C[a, b]$, $\lambda \in \mathbf{R}$ are arbitrary.

EXAMPLE 2.3.3. Suppose that $X = (s)$, $Dx = \{x_{n+1}-x_n\}$ for $x = \{x_n\}$ $\in (s)$ and $Rx = y = \{y_n\}$, where $y_1 = 0$, $y_n = \sum_{k=1}^{n-1} x_k$ for $k = 2, 3, \ldots$ We have proved in Example 2.1.4 that R is a right inverse of D which is a Volterra operator. Moreover,

$$(I-\lambda R)^{-1}y = u, \quad \text{where } y = \{y_n\}, u = \{u_n\} \in (s), \lambda \in \mathbf{C},$$
$$\tag{2.3.19}$$

$$u_1 = y_1, \quad u_n = y_n + \lambda \sum_{k=1}^{n-1} (\lambda+1)^{k-1} y_{n-k} \quad (n = 2, 3, \ldots).$$

We have also ker $D = \{z = \{z_n\} \in (s) : z_n = c, c \in \mathbf{R} \ (n = 1, 2, \ldots)\}$. An initial operator F corresponding to R is then of the form $Fx = z$

(ii) $D\mathrm{ch}_\lambda = \lambda \mathrm{sh}_\lambda$, $\quad D\mathrm{sh}_\lambda = \lambda \mathrm{ch}_\lambda$ $\quad$ for $\lambda \in \mathbf{R}$; $\hfill$ (2.3.25)

(iii) Elements $\mathrm{ch}_\lambda(z)$, $\mathrm{sh}_\lambda(z)$, where $z \in \ker D$, $\lambda \in \mathbf{R}$ are linearly independent;

(iv) If $z_0, z_1 \in \ker D$, $\lambda \in \mathbf{R}$ are arbitrary then

$$u = \mathrm{ch}_\lambda(z_0) + \mathrm{sh}_\lambda(z_1) \in \ker (D^2 - \lambda^2 I). \qquad (2.3.26)$$

EXERCISE 2.3.3 (cf. Tasche, 1976). Let X be the set of all complex-valued functions determined and continuous on the half-axis $[1, +\infty)$ with the usual addition and multiplication by scalars. Prove that:

(i) X is a linear space over $\mathbf{C}$;

(ii) The operator $D = t\dfrac{\mathrm{d}}{\mathrm{d}t}$ is right invertible and has a right inverse

R defined by means of the equality $(Rx)(t) = \int\limits_1^t s^{-1} x(s)\,\mathrm{d}s$ for $x \in X$;

(iii) $\dim \ker D = 1$;

(iv) Every $\lambda \in \mathbf{C}$ is an eigenvalue of D and exponentials are of the form Ct^λ, where $C, \lambda \in \mathbf{C}$ are arbitrary.

EXERCISE 2.3.4 (cf. Dimovski, 1973; Tasche, 1976). Suppose we are given real functions $p(t) > 0$, $q(t)$ continuous on the half-axis $[0, +\infty)$. Write:

$$r(t) = \exp \int\limits_0^t \frac{q(s)}{p(s)}\,\mathrm{d}s, \quad u(t) = \int\limits_0^t \frac{\mathrm{d}s}{p(s)} \qquad (2.3.27)$$

(where we admit: $\exp v = e^v$). By our assumptions there exist $\mathrm{d}u/\mathrm{d}t$ and the inverse function u^{-1}. Denote by X the set of all complex-valued functions defined and continuous on $[0, +\infty)$ with the usual addition and multiplication by scalars. Prove that:

(i) X is a linear space over $\mathbf{C}$;

(ii) The operator $D = \dfrac{\mathrm{d}p}{\mathrm{d}t} + q$ is right invertible and has a right inverse defined as follows:

$$(Rx)(t) = \frac{1}{r(t)} \int\limits_0^t \frac{r(s)}{p(s)}\,x(s)\,\mathrm{d}s \quad \text{for } x \in X;$$

(iii) dim ker $D = 1$;

(iv) Every $\lambda \in \mathbf{C}$ is an eigenvalue of D and exponentials are of the form $\dfrac{C}{r(t)} \exp[\lambda u(t)]$, where $C, \lambda \in \mathbf{C}$ are arbitrary.

EXERCISE 2.3.5 (cf. Tasche, 1976). Let X be the set of all functions $x(t) = t^p \xi(t)$, where ξ are continuous complex-valued functions defined on the half-axis $[0, +\infty)$ and $p > -1$ is arbitrary. Prove that:

(i) X with the usual addition and multiplication by scalars is a linear space over $\mathbf{C}$;

(ii) The operator $D = t^{-1} \dfrac{d}{dt} t \dfrac{d}{dt}$ is right invertible (cf. Ditkin and Prudnikov, 1962) and dim ker $D = 2$;

(iii) The operator F defined by means of the equality $(Fx)(t) = \xi_0 + \xi_1 \ln t$, where

$$\xi_1 = \lim_{t \to +0} tx'(t), \quad \xi_0 = \lim_{t \to +0} [x(t) - \xi_1 \ln t]$$

is an initial operator for D corresponding to the right inverse

$$(Rx)(t) = \frac{1}{4} t^2 \int_0^1 \int_0^1 x\left(t\sqrt{uv}\right) du\, dv \quad \text{for } x \in X;$$

(iv) Every $\lambda \in \mathbf{C}$ is an eigenvalue of D and exponentials are of the form:

$$[e_\lambda(z)](t) = CJ_0(\lambda^{1/2}t), \tag{2.3.28}$$

where $C \in \mathbf{C}$ is arbitrary and J_n denotes the Bessel function of order n, i.e.

$$J_n(t) = \sum_{k=0}^{\infty} \frac{(-1)^k}{k!(n+k)!} \left(\frac{t}{2}\right)^{n+k} \quad (n = 0, 1, 2, \ldots). \tag{2.3.29}$$

EXERCISE 2.3.6. Write $\Omega = [0, +\infty) \times [0, +\infty)$. Let X be the set of all continuous complex-valued functions defined on Ω. Prove that:

(i) X with the usual addition and multiplication by scalars is a linear space over $\mathbf{C}$;

for $x = \{x_n\} \in (s)$, where $z = \{z_n\}$, $z_n = x_1$ for $n = 1, 2, \ldots$ (cf. Exercise 2.2.9).

Formula (2.3.19) implies that exponentials are of the form:

$$e_\lambda(z) = c\{(\lambda+1)^{n-1}\}, \quad \text{where } z = \{z_n\},$$

$$z_n = c \in \mathbf{R} \quad (n = 1, 2, \ldots), \lambda \in \mathbf{C}. \tag{2.3.20}$$

Indeed, by definition $e_\lambda(z) = (I - \lambda R)^{-1}z = u$, where $u = \{u_n\}$, $u_1 = z_1 = c$ and for $n = 2, 3, \ldots$

$$u_n = z_n + \lambda \sum_{k=1}^{n-1} (\lambda+1)^{k-1} z_{n-k} = c + \lambda \sum_{k=1}^{n-1} (\lambda+1)^{k-1} c$$

$$= c\left[1 + \lambda \sum_{k=1}^{n-1} (\lambda+1)^{k-1}\right] = c\left[1 + \lambda \frac{(\lambda+1)^{n-1}-1}{\lambda+1-1}\right]$$

$$= c[1 + (\lambda+1)^{n-1} - 1] = c(\lambda+1)^{n-1}.$$

Formula (2.3.20) implies that cosine and sine elements are of the form:

$$c_\lambda(z) = c\left\{ \sum_{k=0}^{[(n-1)/2]} (-1)^k \binom{n-1}{2k} \lambda^{2k}\right\}, \tag{2.3.21}$$

$$s_\lambda(z) = c\left\{ \sum_{k=0}^{[(n-1)/2]+1} (-1)^k \binom{n-1}{2k+1} \lambda^{2k+1}\right\}, \tag{2.3.22}$$

$$z = \{z_n\}, \quad z_n = c \in \mathbf{R}, \quad \lambda \in \mathbf{C}.$$

Indeed, denote by $[p]$ the integer part of a real p, i.e. the greatest positive integer $n \leqslant p$. Then

$$c_\lambda(z) = \frac{1}{2}[e_{\lambda i}(z) + e_{-\lambda i}(z)] = \frac{c}{2}\{(\lambda i+1)^{n-1}\} +$$

$$+ \frac{c}{2}\{(-\lambda i+1)^{n-1}\} = \frac{c}{2}\{(\lambda i+1)^{n-1} + (-\lambda i+1)^{n-1}\}$$

$$= \frac{c}{2}\left\{\sum_{k=0}^{n-1} \binom{n-1}{k}(\lambda i)^k + \sum_{k=0}^{n-1} \binom{n-1}{k}(-\lambda i)^k\right\}$$

$$= \frac{c}{2} \left\{ \sum_{k=0}^{n-1} \binom{n-1}{k} \lambda^k i^k [1 + (-1)^k] \right\}$$

$$= \frac{c}{2} \left\{ \sum_{k=0}^{[(n-1)/2]} 2 \binom{n-1}{2k} \lambda^{2k} i^{2k} \right\}$$

$$= c \left\{ \sum_{k=0}^{[(n-1)/2]} (-1)^k \binom{n-1}{2k} \lambda^{2k} \right\}.$$

Similarly,

$$s_\lambda(z) = \frac{1}{2i} [e_{\lambda i}(z) - e_{-\lambda i}(z)]$$

$$= \frac{c}{2i} \{ (\lambda i + 1)^{n-1} - (-\lambda i + 1)^{n-1} \}$$

$$= \frac{c}{2i} \left\{ \sum_{k=0}^{n-1} \binom{n-1}{k} \lambda^k i^k [1 - (-1)^k] \right\}$$

$$= \frac{c}{2i} \left\{ \sum_{k=0}^{[(n-1)/2]+1} 2 \binom{n-1}{2k+1} \lambda^{2k+1} i^{2k+1} \right\}$$

$$= c \left\{ \sum_{k=0}^{[(n-1)/2]+1} (-1)^k \binom{n-1}{2k+1} \lambda^{2k+1} \right\}.$$

EXERCISE 2.3.1. Suppose that X is a linear space over $\mathbf{C}$, $D \in R(X)$, dim ker $D \neq 0$ and $R \in \mathscr{R}_D \cap V(X)$. Prove that $u = c_\lambda(z_0) + s_\lambda(z_1)$ $\in$ ker $(D^2 + \lambda^2 I)$ for all $z_0, z_1 \in$ ker D, $\lambda \in \mathbf{R}$.

EXERCISE 2.3.2. Suppose that all assumptions of Exercise 2.3.1 are satisfied. Write

$$\mathrm{ch}_\lambda = (I - \lambda^2 R^2)^{-1}, \quad \mathrm{sh}_\lambda = \lambda R(I - \lambda^2 R^2)^{-1} \quad \text{for } \lambda \in \mathbf{R}.$$
$$(2.3.23)$$

Operators ch_λ, sh_λ are called *cosine hyperbolic* and *sine hyperbolic* operators, respectively. Prove that

(i) $\mathrm{ch}_\lambda = \frac{1}{2} (e_\lambda + e_{-\lambda})$, $\mathrm{sh}_\lambda = \frac{1}{2} (e_\lambda - e_{-\lambda})$ for $\lambda \in \mathbf{R}$;   (2.3.24)

(ii) The operator $D = \dfrac{\partial^2}{\partial s \partial t}$ is right invertible and dim ker D $= +\infty$;

(iii) The operator F defined as $(Fx)(t, s) = x(0, s) + x(t, 0) - x(0, 0)$ for $x \in X$ is an initial operator for D corresponding to the right inverse

$$(Rx)(t, s) = \int_0^t \int_0^s x(u, v) du\, dv \quad \text{for } x \in X;$$

(iv) Every $\lambda \in \mathbf{C}$ is an eigenvalue of D and exponentials are of the form

$$[e_\lambda(z)](t, s) = CJ_0\big(2\sqrt{\lambda ts}\big),$$

where $C \in \mathbf{C}$ is arbitrary and J_0 is the Bessel function of order zero defined by Formula (2.3.29) (cf. Tasche, 1976).

EXERCISE 2.3.7. Suppose that $D \in R(X)$, dim ker $D \neq 0$ and $R \in \mathcal{R}_D \cap \cap V_L(X)$, where $V_L(X)$ is defined in Remark 2.3.1 (cf. Formula (2.3.18)). Let $c_\lambda(z)$ and $s_\lambda(z)$ be cosine and sine elements induced by exponentials $x_\lambda = R_\lambda z$, $z \in \ker D$, $\lambda \in \mathbf{C}$ defined in Remark 2.3.1. Prove that:

(i) $u = c_\lambda(z_0) + s_\lambda(z_1) \in \ker (D^2 + \lambda^2 I)$ for all $z_0, z_1 \in \ker D$, $\lambda \in \mathbf{R}$;

(ii) The operators ch_λ and sh_λ defined in Exercise 2.3.2 can be also defined in a similar way in our case and have properties (ii)–(iv) in that Exercise.

EXERCISE 2.3.8. Suppose that $X = C[0, T]$ and $D = \mathrm{d}/\mathrm{d}t$. Prove that

(i) The operator F defined as

$$(Fx)(t) = x(0) - x(T) + \frac{1}{T}\int_0^T x(s)\mathrm{d}s \quad \text{for } x \in X$$

is an initial operator for D;

(ii) Find a formula for the corresponding right inverse R and prove that R is not a Volterra operator (apply Theorem 2.3.4).

EXERCISE 2.3.9. Let $X = C[a, b]$ and let

$$(Ax)(t) = \int_{t_0}^t \Big[\int_{t_1}^t x(u) \mathrm{d}u\Big] \mathrm{d}s \quad \text{for } x \in X, \text{ where } a \leqslant t_0, t_1 \leqslant b.$$

Observe that A can be written in the form

$$(Ax)(t) = \int_{t_0}^{t_1} (t_0 - s)x(s)\,ds + \int_{t_1}^{t} (t-s)x(s)\,ds.$$

Prove that:

(i) $A \in L_0(X)$ and is a right inverse for the operator d^2/dt^2;

(ii) If $t_1 = t_0$ then $A \in V(X)$;

(iii) If $t_1 \neq t_0$ and $\mu \in \mathbf{R}$ then ker $(I - \mu^2 A) = \{0\}$;

(iv) If $t_1 \neq t_0$ then for an arbitrary integer k all numbers λ_k $= -\mu_k^{-2} = -[\pi/(t_1 - t_0)(2k+1)]^{-2}$ are eigenvalues of A and the corresponding eigenfunctions are $C_1 \cos \mu_k t + C_2 \sin \mu_k t$, where $C_1, C_2 \in \mathbf{R}$. Consequently, $A \notin V(X)$.

EXERCISE 2.3.10. Let $X = C[a, b]$ and let

$$(Ax)(t) = \int_0^t \left\{ \int_{t_1}^s \left[\int_0^u x(v)\,dv \right] du \right\} ds \quad \text{for } x \in X,$$

where $a \leqslant 0$, $t_1 \leqslant b$, $t_1 \neq 0$. Prove that:

(i) $A \in L_0(X)$ and is a right inverse of the operator d^3/dt^3;

(ii) The functions

$$C_1 e^{\mu_k t} + e^{-\mu_k t}\left(C_2 \cos \frac{\sqrt{3}}{2} \mu_k t + C_3 \sin \frac{\sqrt{3}}{2} \mu_k t \right),$$

where $C_1, C_2, C_3 \in \mathbf{R}$, are eigenvectors of the operator A corresponding to the eigenvalues

$$\lambda_k = \left[\frac{2}{\sqrt{3}} t_1 \left(\operatorname{arc\,cot} \frac{2}{3} + k\pi \right) \right]^{-2} = \mu_k^{-2},$$

where k is an arbitrary integer. Consequently $A \notin V(X)$.

EXERCISE 2.3.11. Let $X = C[a, b]$ and let

$$(Ax)(t) = \int_{t_0}^{t} \left\{ \int_{t_1}^{s_1} \left[\int_{t_2}^{s_2} \left(\int_{t_3}^{s_3} x(u)\,du \right) ds_3 \right] ds_2 \right\} ds_1$$

$$\text{for } x \in X, a \leqslant t_0, t_1, t_2, t_3, t_4 \leqslant b.$$

Prove that:
  (i) $A \in L_0(X)$ and is a right inverse of the operators d^4/dt^4;
  (ii) If points t_0, t_1, t_2, t_3 are chosen in the following way: $t_1 \neq t_0$, $t_3 \neq t_2$ and there exist integers j, k such that $\dfrac{t_3 - t_2}{t_1 - t_0} = \dfrac{2j+1}{2k+1}$, then the operator A has eigenfunctions $\sin \mu_k t$ corresponding to the eigenvalues

$$\mu_k = \frac{\pi}{(t_1 - t_0)} (2k+1) = \frac{\pi}{(t_3 - t_2)} (2j+1),$$

where j, k are integers. Consequently, $A \notin V(X)$.

OPEN QUESTION 2.3.1. Are all Volterra right inverses for a $D \in R(X)$ linearly independent?

EXERCISE 2.3.12. Suppose that $D \in R(X)$, F is an initial operator to an $R \in \mathscr{R}_D \cap V(X)$, $e_\lambda = (I - \lambda R)^{-1}$ for all $\lambda \in \mathscr{F}$ and $F_1 \neq F$ is a projection onto ker D, hence F_1 corresponds to a right inverse $R_1 = R - F_1 R$. Write $R^0 = I - \lambda R_1 = I - \lambda R + \lambda F_1 R$. Prove that:
  (i) $DR^0 = D(I - \lambda R)$, $R^0 e_\lambda(\ker D) \subset \ker D$ for all λ;
  (ii) If $F_1 e_\lambda(z) \neq 0$ for all $\lambda \in \mathscr{F}$, $z \in \ker D$ then ker $R^0 = \{0\}$ and R^0 is a mapping onto, hence R_0 is invertible, which implies $R_1 \in V(X)$.

When $X = C[0,1]$, $D = \dfrac{d}{dt}$, $R = \int_0^t$, $(Fx)(t) = x(0)$, $(F_1 x)(t) = x(1)$ for $x \in X$ we have $(R^0 x) A) = x(t) - \lambda \int_0^t x(s)ds + \lambda \int_0^1 x(s)ds = x(t) - \lambda \int_1^t x(s)ds$, $R_0 e^{\lambda t} = e^\lambda \neq 0$ and $F_1 e^{\lambda t} = e^\lambda \neq 0$ for all $\lambda \in \mathbb{C}$.

EXERCISE 2.3.13. Prove that the operator T defined as

$$(Tx)(t) = \begin{cases} \dfrac{1}{t} \displaystyle\int_t^{2t} x(s)ds & \text{for } t \neq 0, \\ x(0) & \text{for } t = 0, \end{cases} \quad x \in C(\mathbb{R}),$$

is a Volterra operator.

EXERCISE 2.3.14. Suppose that $D \in R(X)$, dim ker $D \neq 0$, F is an initial operator for D corresponding to an $R \in \mathscr{R}_D$.

(i) Denote by $\mathscr{F}_r$ the set $\{\lambda \in \mathscr{F}: I - \lambda R \in R(X)\}$. Then for all $\lambda \in \mathscr{F}_r$ there exists an R_λ such that $(I - \lambda R)R_\lambda = I$. Prove that:

(a) $\ker(D - \lambda I) = \ker DR_\lambda = R_\lambda \ker D \oplus \ker(I - \lambda R)$ for all $\lambda \in \mathscr{F}_r$;

(b) Exponentials exist for all $\lambda \in \mathscr{F}_r$ and are of the form: $x_\lambda = R_\lambda z + r_\lambda$, where $r_\lambda \in \ker(I - \lambda R)$ and $z \in \ker D$ are arbitrary;

(c) $Fx_\lambda = z$.

(ii) Denote by $\Lambda(X)$ the set of all left invertible operators belonging to $L(X)$ and by $\mathscr{F}_l$ the set $\{\lambda \in \mathscr{F}: I - \lambda R \in \Lambda(X)\}$. Then for all $\lambda \in \mathscr{F}_l$ there exists an L_λ such that $L_\lambda(I - \lambda R) = I$. Prove that:

(a) $\ker(D - \lambda I) = L_\lambda \ker D$ for all $\lambda \in \mathscr{F}_l$;

(b) Exponentials exist for all $\lambda \in \mathscr{F}_l$ and are of the form: $x_\lambda = L_\lambda z$, where $z \in \ker D$ is arbitrary and $Fx_\lambda = z$.

(iii) Write $\mathscr{F}_0 = \mathscr{F}_r \cap \mathscr{F}_l$ and $\mathscr{F}_D = \mathscr{F}_l \cup \mathscr{F}_r$. Prove that:

(a) If $\lambda \in \mathscr{F}_0$ then $I - \lambda R$ is invertible and exponentials are of the form $e_\lambda(z) = (I - \lambda R)^{-1}$, where $z \in \ker D$ is arbitrary, with the property $Fe_\lambda(z) = z$;

(b) For all $\lambda \in \mathscr{F}_0$ there exist exponentials.

2.4. D-POLYNOMIALS

To begin with we prove the following:

PROPOSITION 2.4.1. *Suppose that $D \in R(X)$ and $R \in \mathscr{R}_D$. Then R is not a nilpotent operator (cf. Exercise 1.3.12).*

Proof. Suppose that R is a nilpotent operator of the order n, i.e. there exists a positive integer $n \geq 2$ such that $R^n = 0$ and $R^{k-1} \neq 0$ for $k = 0, 1, ..., n-1$. Since $DR = I$ we find that $R^{n-1} = DR^n = 0$ which contradicts our assumption. ∎

DEFINITION 2.4.1. Let $D \in R(X)$ and dim ker $D > 0$. A *D-monomial of degree n* is an element of the form: $R^n z$, where $R \in \mathscr{R}_D$, $z \in \ker D$ $(n = 0, 1, 2, ...)$. A *D-polynomial of degree n* is an element of the form

$$\sum_{k=0}^{n} R^k z_k, \quad \text{where } z_0, ..., z_n \in \ker D, R \in \mathscr{R}_D.$$

is a Volterra operator. Conversely, if R^0 is a Volterra operator then R is a Volterra operator*.

Proof. Suppose that $R \in \mathscr{R}_D \cap V(X)$. Observe that all polynomials in D (and in R, also) with scalar coefficients are commutative. Since $R \in V(X)$ the operators $I - \lambda R$ are invertible for all $\lambda \in \mathscr{F}$ which implies that the operator $Q(I, R) = \prod\limits_{m=0}^{n} (I - t_m R)^{r_m}$ is invertible as a superposition of invertible operators. This implies that the operator $Q(D)$ is right invertible and has a right inverse $R_1 = R^N [Q(I, R)]^{-1} - [Q(I, R)]^{-1} R^N$. Indeed,

$$Q(D) R_1 = Q(D) R^N [Q(I, R)]^{-1} = \left(\sum_{k=0}^{N} q_k D^k R^N \right) [Q(I, R)]^{-1}$$

$$= \left[\sum_{k=0}^{N} q_k (D^k R^k) R^{N-k} \right] [Q(I, R)]^{-1}$$

$$= \left(\sum_{k=0}^{N} q_k R^{N-k} \right) [Q(I, R)]^{-1} = Q(I, R) [Q(I, R)]^{-1} = I.$$

Consequently, the operator $P(D) = Q(D) D^M = D^M Q(D)$ is also right invertible and has a right inverse of the form (2.4.5). Indeed, since $Q(D) R_1 = I$ we have $P(D) R^0 = D^M Q(D) [Q(I, R)]^{-1} R^{N+M} = D^M Q(D) R_1 R^M = D^M R^M = I.$

We shall prove that $R^0 \in V(X)$. For an arbitrary $\lambda \in \mathscr{F}$ we have $I - \lambda R^0 = I - \lambda [Q(I, R)]^{-1} R^{N+M} = [Q(I, R)]^{-1} [Q(I, R) - \lambda R^{N+M}]$

$= [Q(I, R)]^{-1} Q_1(R)$. The operator $Q(I, R) - \lambda R^{N+M} = \sum\limits_{k=0}^{n} q_k R^{N-k} - \lambda R^{N+M}$ as a polynomial in R with scalar coefficients of degree $N+M$ can be written in the form $Q_1(R) = \prod\limits_{m=1}^{N+M} (I - a_m R)$, where $a_m \in \mathscr{F}$. Observe that $Q_1(0) = q_N = 1$. We conclude that the operator $I - \lambda R^0$ as a superposition of invertible operators is again invertible. The arbitrariness of $\lambda \in \mathscr{F}$ implies that $R^0 \in V(X)$.

Conversely, suppose that the operator R^0 defined by Formula (2.4.5)

---

* The implication $R^0 \in V(X) \rightarrow R \in V(X)$ has been proved by H. von Trotha in 1978.

is a Volterra operator. Let $\lambda \in \mathscr{F}$ be arbitrarily fixed. Write $\mu = Q(\lambda)$. By our assumption the operator $I - \mu R^0 = I - \mu [Q(I, R)]^{-1} R^{N+M}$ $= [Q(I, R)]^{-1} [Q(I, R) - \mu R^{N+M}]$. Since the operators $[Q(I, R)]^{-1}$ and $Q(I, R) - \mu R^{N+M}$ are commutative, we conclude that the operator $P_\mu(I, R) = Q(I, R) - \mu R^{N+M}$ is invertible, where we write: $P_\mu(t, s)$ $= Q(t, s) - \mu s^{N+M}$, $P_\mu(t) = P_\mu(t, 1)$. By our assumption $P_\mu(\lambda) = P_\mu(\lambda, 1)$ $= Q(\lambda, 1) - \mu = Q(\lambda) - \mu = 0$. We therefore conclude that λ is a root of the polynomial $P_\mu(t)$ and that $P_\mu(t) = (t - \lambda) Q_\mu(t)$, where $Q_\mu(t)$ is a well-defined polynomial. Denote by $Q_\mu(t, s)$ the corresponding polynomial of two variables, i.e. a polynomial $Q_\mu(t, s)$ such that $P_\mu(t, s)$ $= (t - \lambda s) Q_\mu(t, s)$. Since the polynomial $Q(I, R) - \mu R^{N+M} = P_\mu(I, R)$ is invertible, we conclude that the operator $I - \lambda R$ is invertible. The arbitrariness of $\lambda \in \mathscr{F}$ implies that $R \in V(X)$.　∎

THEOREM 2.4.2. *Suppose that* $D \in R(X)$, $\dim \ker D \neq 0$ *and* $R \in \mathscr{R}_D$. *Let* $q(D) = \sum\limits_{j=0}^{N} q_j D^j$ *be an arbitrary polynomial with scalar coefficients. Then* $q(D) = 0$ *if and only if*

$$q(D) R^k z = 0 \quad \text{for all } z \in \ker D \quad (k = 0, 1, 2, \ldots). \qquad (2.4.6)$$

Proof. Necessity is obvious. In order to prove that Condition (2.4.6) implies $q(D) = 0$ observe that $R^n z \in \ker D^{n+1}$ $(n = 0, 1, 2, \ldots)$ for all $z \in \ker D$ (cf. Corollary 2.4.1). Then for all $z \in \ker D$ $(n = 0, 1, 2, \ldots)$ we find

$$0 = q(D) R^n z = \sum_{k=0}^{N} q_k D^k R^n z = \sum_{k=0}^{n} q_k R^{n-k} z + \sum_{k=n+1}^{N} q_k D^{k-n} z$$

since, by definition, $DR = I$ and $Dz = 0$. This, and the linear independence of D-monomials $z, Rz, R^2 z, \ldots$ (Proposition 2.4.2), together imply that $q_0 = \ldots = q_n = 0$ for $n = 0, 1, \ldots, N$, i.e. $q(D) = 0$.　∎

COROLLARY 2.4.2. *Suppose that all assumptions of Theorem 2.4.2 are satisfied. Then* $q(D) = 0$ *on* X *if and only if* $q(D) P(R) = \{0\}$, *where the set* $P(R)$ *is defined by Formula* (2.4.3).

COROLLARY 2.4.3. *Suppose that all assumptions of Theorem 2.4.2 are satisfied. Then* $q(D) = 0$ *if and only if* $q(D) R^k z = 0$ *for all* $z \in \ker D$

and $k = 0, 1, ..., m$, *where* $m = N+1$, *i.e. it is sufficient to admit a finite number of Conditions* (2.4.6).

THEOREM 2.4.3. *Suppose that* X *is a commutative algebra* (*linear ring*) *satisfying the condition*:

if $x, y \in \text{dom } D$ *then* $xy \in \text{dom } D$.

Suppose, moreover, that $\ker D$ *is not an annihilator in* X, *i.e. if* $x \ker D$ $= \{0\}$ *for an* $x \in X$ *then* $x = 0$. *Let* $q(D) = \sum_{j=0}^{n} q_j D^j$ *where* $q_0, ..., q_n \in X$.

Then $q(D) = 0$ *if and only if* $q(D) R^k z = 0$ *for all* $z \in \ker D$ $(k = 0, 1, 2, ...)$, *where* $R \in \mathscr{R}_D$ *is arbitrarily fixed*.

Proof. Necessity is obvious.

Sufficiency will be proved by induction. Assume that $q(D) R^k z = 0$ for all $z \in \ker D$ and $k = 0, 1, 2, ...$ In a similar way, as in the proof of Theorem 2.4.2 we shall obtain $\sum_{j=0}^{k} q_j R^{k-j} z = 0$ for all $z \in \ker D$ $(k = 0, 1, 2)$. Let $k = 0$. Then we have $q_0 R^k z = 0$ and $R^k z \neq 0$ for $z \neq 0$. Indeed, if $Ru = 0$ then $u = DRu = 0$. The arbitrariness of $z \in \ker D$ and our assumptions together imply that $q_0 = 0$. Suppose that $q_0 = ... = q_m = 0$ for an arbitrarily fixed $m \geqslant 0$. Then $0 = \sum_{j=1}^{m+1} q_j R^{m+1-j} z = q_{m+1} z$. The arbitrariness of $z \in \ker D$ implies $q_{m+1} = 0$ which finishes the proof. ∎

Examples and Exercises

EXAMPLE 2.4.1 (cf. Carlitz, 1976; Ince, 1927). Let $P_1(D)$, $P_2(D)$ be *polynomial differential operators*, i.e. finite sums of monomials of the form $A_0(t) DA_1(t) ... A_{n-1}(t) DA_n(t)$, where $D = \mathrm{d}/\mathrm{d}t$, $A_0, ..., A_n$ are given functions, differentiable on $[0, T]$. Using the identity

$$DA_j = A_j D + A_j', \quad \text{where } A_j' = \frac{\mathrm{d}}{\mathrm{d}t} A,$$

we can write these polynomials in the form

$$P_i(D) = \sum_{j=1}^{m} P_j^{(i)}(x) D^i \quad (i = 1, 2).$$

Theorem 2.4.3 implies then that $P_1(D) = P_2(D)$ if and only if $P_1(D)t^k = P_2(D)t^k$ for $k = 0, 1, 2, ...$

Indeed, if we assume that $(Rx)(t) = \int_0^t x(s)ds$ then we conclude that $P(R) = \text{lin}\{t^n: n = 0, 1, 2, ...\}$. In particular, if $P(D)$ is a polynomial differential operator then $P(D) = 0$ if and only if $P(D)t^k = 0$ for $k = 0, 1, 2, ...$

EXAMPLE 2.4.2. For every D-monomial of the order $n \geqslant 1$ there exists an initial operator such that the corresponding initial value is equal zero. Indeed, let $D \in R(X)$, $\dim \ker D \neq 0$ and let F be an initial operator for D corresponding to an $R \in \mathcal{R}_D$. Then

$$FR^k z = 0 \quad \text{for all } z \in \ker D, \, k = 1, 2, ... \qquad (2.4.7)$$

for $FR = 0$.

EXERCISE 2.4.1. Suppose that $X = C[a, b]$, $D = d/dt$ and $d \in \mathbf{R}$ is arbitrarily fixed. Define the operators F_1, F_2, F_3, F_4 as

$$(F_1 x)(t) = x(a),$$

$$(F_2 x)(t) = \frac{1}{b-a}\int_a^b x(s)ds,$$

$$(F_3 x)(t) = dx(a) + (1-d)x(b),$$

$$(F_4 x)(t) = x(a) - x(b) + \frac{1}{b-a}\int_a^b x(s)ds,$$

for $x \in X$.

(i) Prove that F_1, F_2, F_3, F_4 are initial operators for D corresponding to the right inverses

$$(R_1 x)(t) = \int_a^t x(s)ds,$$

$$(R_2 x)(t) = \int_a^t x(s)ds - \frac{1}{b-a}\int_a^b (b-s)x(s)ds,$$

PROPOSITION 2.4.2. *Suppose that* $D \in R(X)$, dim ker $D \neq 0$, $R_1, \ldots, R_n$ $\in \mathcal{R}_D$, $z_1, \ldots, z_n \in$ ker D *and* $z_k \neq 0$ *for* $k = 1, \ldots, n$, *where* n *is an arbitrary fixed positive integer. Then D-monomials* $z_0, R_1 z_1, \ldots, R_n^n z_n$ *are linearly independent.*

Proof (by induction). For $n = 0$ our proposition holds since by our assumption $z_0 \neq 0$. Let $n \geq 1$ be arbitrarily fixed and let *D*-monomials $z_0, R_1 z_1, \ldots, R_n^n z_n$ be linearly independent. Suppose that $0 \neq z_{n+1} \in$ ker D and that the *D*-monomial $R_{n+1}^{n+1} z_{n+1}$ is linearly dependent on the *D*-monomials $z_0, \ldots, R_n^n z_n$. It means that there exist scalars $a_0, \ldots, a_{n+1}$ not vanishing simultaneously and such that

$$\sum_{k=0}^{n+1} a_k R_k^k z_k = 0. \tag{2.4.1}$$

Since $z_0, \ldots, z_{n+1} \in$ ker D, acting on both sides of this equality by the operator D^{n+1} we find

$$0 = D^{n+1} \sum_{k=0}^{n+1} a_k R_k^k z_k = \sum_{k=0}^{n+1} a_k D^{n+1} R_k^k z_k$$

$$= \sum_{k=0}^{n+1} a_k D^{n+1-k} z_k = a_{n+1} z_{n+1}. \tag{2.4.2}$$

By our assumption $z_{n+1} \neq 0$. This implies that $a_{n+1} = 0$. Equality (2.4.2) is then of the form: $\sum_{k=0}^{n} a_k R_k^k z_k = 0$ which contradicts our inductive assumption that the *D*-monomials $z_0, R_1 z_1, \ldots, R_n^n z_n$ are linearly independent. We therefore conclude that the *D*-monomials $z_0, R_1 z_1, \ldots$ $\ldots, R_{n+1}^{n+1} z_{n+1}$ are linearly independent which was to be proved. ∎

In particular, if $R_1 = \ldots = R_n = R \in \mathcal{R}_D$ then we conclude that the *D*-monomials $z_0, R z_1, \ldots, R^n z_n$ are linearly independent for arbitrary $z_0, \ldots, z_n \in$ ker D $(n = 0, 1, 2, \ldots)$.

Write for $D \in R(X)$ and $R \in \mathcal{R}_D$

$$P(R) = \text{lin}\{R^k z : z \in \text{ker } D, k = 0, 1, 2, \ldots\}. \tag{2.4.3}$$

By definition and Proposition 2.4.2 the set $P(R)$ consists of linearly independent elements (provided that ker $D \neq \{0\}$).

PROPOSITION 2.4.3. *Suppose that $D \in R(X)$ and dim ker $D \neq 0$. Then the set $P(R)$ of all D-polynomials defined by Formula (2.4.3) is independent of the choice of a right inverse, i.e. if $R_1, R_2 \in \mathscr{R}_D, R_1 \neq R_2$ then $P(R_1) = P(R_2)$.*

Proof. Indeed, $R_2 - R_1 = F_1 R_2$, where F_1 is an initial operator for D corresponding to R_1, i.e. $R_2 z = R_1 z + z_0$ for all $z \in$ ker D, where $z_0 \in$ ker D. By an easy induction we conclude that

$$R_2^k z = R_1^k z + \sum_{j=0}^{k-1} R_1^j z_j \quad \text{for all } z \in \text{ker } D, k \in \mathbf{N},$$

where $z_0, \ldots, z_{k-1} \in$ ker D. ∎

PROPOSITION 2.4.4. *Suppose that $D \in R(X)$, dim ker $D \neq 0$ and $R \in \mathscr{R}_D$. Then*

$$P(R) = \bigcup_{n \in \mathbf{N}} \text{ker } D^n. \tag{2.4.3'}$$

Proof. Indeed, let $n \in \mathbf{N}$ be arbitrarily fixed. Consider a D-monomial $R^{n-1}z$, where $z \in$ ker D is arbitrarily fixed. Then $D^n(R^{n-1}z) = D(D^{n-1}R^{n-1})z = Dz = 0$, i.e. $R^{n-1}z \in$ ker D^n. ∎

COROLLARY 2.4.1. *Suppose that all assumptions of Proposition 2.4.4 are satisfied. Then all D-monomials of degree n are eigenvectors of the operator D^n corresponding to the eigenvalue $\lambda = 0$.*

In particular, for $n = 1$, D-monomials are exponentials corresponding to the eigenvalue $\lambda = 0$.

THEOREM 2.4.1. *Suppose that X is a linear space over an algebraically closed field $\mathscr{F}$. Suppose that $D \in R(X)$. Write*

$$Q(t, s) = \sum_{k=0}^{N} q_k t^k s^{N-k} = \prod_{m=0}^{n} (t - t_m s)^{r_m},$$

where $q_0, \ldots, q_N \in \mathscr{F}, q_N = 1, t_m \neq t_j$ for $m \neq j, t_m \neq 0$ $(j, m = 1, 2, \ldots \ldots, n), r_1 + \ldots + r_n = N$ and

$$Q(t) = Q(t, 1), \quad P(t) = t^M Q(t), \tag{2.4.4}$$

where M is a non-negative integer. If there exists an $R \in \mathscr{R}_D \cap V(X)$ then $P(D) = D^M Q(D) = Q(D)D^M \in R(X)$ and

$$R^0 = R^{N+M}[Q(I, R)]^{-1} = [Q(I, R)]^{-1} R^{N+M} \in \mathscr{R}_D \tag{2.4.5}$$

$$(R_3 x)(t) = \int_a^t x(s)\,ds - (1-d)\int_a^b x(s)\,ds,$$

$$(R_4 x)(t) = \int_a^t x(s)\,ds - \frac{1}{b-a}\int_a^b (s-a)x(s)\,ds,$$

respectively;

(ii) Prove that the corresponding monomials are of the following form (where $c \in \mathbf{R}$, $n \in \mathbf{N}$ are arbitrary):

$$R_1^n z = c(t-a)^n/n!,$$

$$R_2^n z = c\left[\frac{\left(t-\dfrac{a+b}{2}\right)^n}{n!} - \right.$$

$$\left. - \sum_{k=2}^{n} \frac{1}{(k+1)!}\left(\frac{b-a}{2}\right)\frac{1+(-1)^k}{2}\frac{(t-a)^{n-k}}{(n-k)!} \right],$$

$$R_3^n z = c\left\{ \frac{1}{n!}[t-ds-(1-d)b]^n - \right.$$

$$\left. - \sum_{k=2}^{n} \frac{(-1)^k da + (1-d)b^k}{k!}\frac{(t-a)^{n-k}}{(n-k)!} \right\},$$

$$R_4^n z = c\left\{ \frac{1}{n!}\left(t-\frac{a+b}{2}\right)^n - \right.$$

$$\left. - \sum_{k=2}^{n} \frac{2k+1}{(k+1)!}[(-1)^k - 1]\frac{(b-a)^k}{2}\frac{(t-a)^n}{n!} \right\},$$

respectively.

EXERCISE 2.4.2. Let $X = C[a,b]$, $a > 0$, $D = t\dfrac{d}{dt}$ and let the operators F_1, F_2 be defined as $(F_1 x)(t) = x(a)$, $(F_2 x)(t) = x(b)$ for $x \in X$.

(i) Prove that F_1, F_2 are initial operators for D corresponding to the right inverses

$$(R_1 x)(t) = \int_a^t s^{-1} x(s)\,ds, \quad (R_2 x)(t) = -\int_t^b s^{-1} x(s)\,ds.$$

(ii) Prove that corresponding monomials are of the form

$$R_1^n z = \frac{c}{n!}\left(\ln \frac{t}{a}\right)^n, \quad R_2^n z = \frac{c}{n!}\left(\ln \frac{b}{t}\right)^n$$

where $c \in \mathbf{R}$ and $n \in \mathbf{N}$ are arbitrary.

EXERCISE 2.4.3. Suppose that all assumptions of Exercise 2.2.9 are satisfied. Prove that all monomials are of the form

$$R^m z = c\{1 + (n-1)^{m+1}\}, \quad \text{where } c \in \mathbf{R}, m \in \mathbf{N}.$$

2.5. REMARKS ON LEFT INVERTIBLE OPERATORS

An operator $\varDelta \in L_0(X)$ is said to be *left invertible* if there is an operator $L \in L(X)$ such that

$$L\varDelta = I. \tag{2.5.1}$$

Denote by $\varLambda(X)$ the set of all left invertible operators belonging to $L_0(X)$ and by $\mathscr{L}_\varDelta$ the set of all left inverses of a $\varDelta \in \varLambda(X)$.

By a simple induction we obtain

$$L^n\varDelta^n = I \ (n = 1, 2, ...) \quad \text{for } \varDelta \in \varLambda(X), \ L \in \mathscr{L}_\varDelta. \tag{2.5.2}$$

PROPERTY 2.5.1. *If $\varDelta \in \varLambda(X)$ then* $\ker \varDelta^n = \{0\}$ *for* $n = 1, 2, ...$

Proof. Indeed, let $n = 1$ and let $z \in \ker \varDelta$. Then $\varDelta z = 0$, and by (2.5.1) $z = L\varDelta z = 0$. Let $n \geqslant 1$ be arbitrary. If $z \in \ker \varDelta^n$ then $\varDelta^n z = 0$, hence $\varDelta(\varDelta^{n-1}z) = 0$ which implies that $\varDelta^{n-1}z \in \ker \varDelta^n$. By the same argument, after $n-1$ steps, we obtain $z = 0$. The arbitrariness of $z \in \ker \varDelta^n$ implies our property. ∎

Write for a $\varDelta \in \varLambda(X)$, $L \in \mathscr{L}_\varDelta$

$$G = I - \varDelta L \quad \text{on dom } L. \tag{2.5.3}$$

The operator G is a projector

$$G^2 = G. \tag{2.5.4}$$

Indeed, $G^2 = (I - \varDelta L)(I - \varDelta L) = I - \varDelta L - \varDelta L + \varDelta(L\varDelta)L = I - 2\varDelta L + \varDelta L = I - \varDelta L = G$ on dom L.

We also have

$$LG = 0 \text{ on dom } L \quad \text{and} \quad G\varDelta = 0. \tag{2.5.5}$$

Indeed, by definition, $LG = L(I-\varDelta L) = L-L\varDelta L = L-L = 0$ on dom L and $G\varDelta = (I-\varDelta L)\varDelta = \varDelta-\varDelta L\varDelta = \varDelta-\varDelta = 0$.

We shall see that the projector $G = I-\varDelta L$, defined for a left invertible operator and corresponding to a left inverse L of $\varDelta$ has, in a sense, similar properties as an initial operator F for a right invertible operator D corresponding to a right inverse R of D. Thus we shall call the projector G a *co-initial operator* for a left invertible operator $\varDelta$, corresponding to a left inverse L of $\varDelta$. Observe also that $G \neq 0$ if and only if the operator $\varDelta$ is left invertible, but not right invertible, i.e. not invertible.

PROPERTY 2.5.2. *If G is a co-initial operator for a $\varDelta \in \varLambda(X)$ corresponding to an $L \in \mathscr{L}_\varDelta$ then*

$$\{x \in \text{dom } L: Gx = x\} = \ker L.$$

Proof. Indeed, if $x \in \ker L$ then $Lx = 0$. Hence $Gx = (I-\varDelta L)x = x-\varDelta Lx = x$. Conversely, if $x \in \text{dom } L$ and $Gx = x$ then $x = Gx = (I-\varDelta L)x = x-\varDelta Lx$ which implies $Lx = (L\varDelta)Lx = (\varDelta Lx) = 0$. Thus $x \in \ker L$. ∎

PROPERTY 2.5.3. *If G is a co-initial operator for a $\varDelta \in \varLambda(X)$ corresponding to an $L \in \mathscr{L}_\varDelta$ then $\ker G = \varDelta X$.*

Proof. Indeed, if $x \in \varDelta X$ then $x = \varDelta u$ for a $u \in X$. Hence the second formula of (2.5.5) implies that $Gx = G\varDelta u = 0$. Thus $x \in \ker G$. Conversely, if $x \in \ker G$ then $Gx = 0$ and $x-\varDelta Lx = (I-\varDelta L)x = Gx = 0$. Writing $u = Lx$ we conclude that $x = \varDelta Lx = \varDelta u$. Hence $x \in \varDelta X$. ∎

PROPERTY 2.5.4. *Suppose that G is a co-initial operator for a $\varDelta \in \varLambda(X)$ corresponding to an $L \in \mathscr{L}_\varDelta$. If $x \in \varDelta^n X$ for an arbitrarily fixed n ($n = 1, 2, ...$) then $GL^k x = 0$ for $k = 0, 1, ..., n-1$.*

Proof. Indeed, since $x \in \varDelta^n X$, we have $x = \varDelta^n u$ for a $u \in X$. Formulae

(2.5.2) and (2.5.5) together imply that for $k = 0, 1, \ldots, n-1$ is $GL^k x = GL^k \varDelta^n u = GL^k \varDelta^k \varDelta^{n-k} u = G\varDelta^{n-k} u = 0$. ∎

THEOREM 2.5.1 (Taylor Formula for left invertible operators). *Suppose that G is a co-initial operator for an operator $\varDelta \in \Lambda(X)$, corresponding to a left inverse L of $\varDelta$. Then for an arbitrarily fixed $N \in \mathbf{N}$ the following identity holds*

$$I = \sum_{k=0}^{N-1} \varDelta^k GL^k + \varDelta^N L^N \quad on\ \mathrm{dom}\ L^n. \qquad (2.5.6)$$

Proof (by induction). For $N = 1$, the definition of the operator G implies that $I = I - \varDelta L + \varDelta L = G + \varDelta L$ on dom L. Suppose Formula (2.5.6) to be true for an arbitrarily fixed $N \geqslant 1$. Then, by our assumption, on dom L^{N+1} we have:

$$\varDelta^{N+1} L^{N+1} = \varDelta(\varDelta^N L^N) L = \varDelta \left(I - \sum_{k=0}^{N-1} \varDelta^k GL^k \right) L$$

$$= \varDelta L - \sum_{k=0}^{N-1} \varDelta^{k+1} GL^{k+1} = I - G - \sum_{m=1}^{N} \varDelta^m GL^m$$

$$= I - \sum_{m=0}^{N} \varDelta^m GL^m$$

which proves the required formula for $N+1$. ∎

COROLLARY 2.5.1. *Suppose that $\varDelta \in \Lambda(X)$ and $y \in \varDelta^n X$ for an arbitrarily fixed n ($n = 1, 2, \ldots$). Then $\varDelta^n x = y$ if and only if there exists an $L \in \mathscr{L}_{\lrcorner}$ such that $x = L^n y$.*

Proof. Let $y \in \varDelta^n X$ and let G be a co-initial operator for $\varDelta$ corresponding to an $L \in \mathscr{L}_{\varDelta}$. Property 2.5.4 implies that $GL^k y = 0$ for $k = 0, 1, \ldots$ $\ldots, n-1$. If $x = L^n y$ then, by Formula (2.5.6), we have

$$\varDelta^n x = \varDelta^n L^n y = \left(I - \sum_{k=0}^{n-1} \varDelta^k GL^k \right) y = y - \sum_{k=0}^{n-1} \varDelta^k GL^k y = y.$$

Conversely, if $\varDelta^n x = y$ then Formula (2.5.2) implies $x = L^n \varDelta^n x = L^n y$. ∎

Properties 2.5.2 and 2.5.3 characterize co-initial operators. Namely, we have

THEOREM 2.5.2. *Suppose that* $\Delta \in \Lambda(X)$ *and* $L \in \mathscr{L}_{\Lambda}$. *If an operator* $G \in L(X)$ *satisfies all the conditions*

$$G^2 = G, \quad \{x \in \text{dom } L \colon Gx = x\} = \ker L, \quad \ker G = \Delta X,$$

then $G = I - \Delta L$ *on* dom L.

Proof. Let $x \in$ dom L be arbitrary. Since G is a projector, we can uniquely write $x - x_1 + x_2$, where $x_1 = Gx$ and $x_2 = (I - G)x$. Since $Gx_1 = G^2x = Gx = x_1$, hence by our assumption $x_1 \in \ker L$, which implies $Lx_1 = 0$ and

$$Gx_1 = x_1 = x_1 - \Delta Lx_1. \tag{2.5.7}$$

Since $Gx_2 = G(I - G)x = Gx - G^2x = Gx - Gx = 0$, we have $x \in \ker G$. Hence by our assumption $x_2 \in \Delta X$. This implies that $x_2 = \Delta u$ for a $u \in X$. We therefore conclude

$$x_2 - \Delta Lx_2 = \Delta u - \Delta u = 0 = Gx_2. \tag{2.5.8}$$

Formulae (2.5.7) and (2.5.8) together imply that

$$Gx = Gx_1 + Gx_2 = (I - \Delta L)x_1 + (I - \Delta L)x_2 = (I - \Delta L)x.$$

The arbitrariness of $x \in$ dom L implies $G = I - \Delta L$. ∎

THEOREM 2.5.3. *Let* $\Xi \subset X'$ *be an arbitrary conjugate space* (*cf. Section* 1.5). *Suppose that* $D \in R(X)$, F *is an initial operator for* D *corresponding to an* $R \in \mathscr{R}_D$ *and* $D, F, R \in L(X, \Xi)$. *Then the conjugate operators have the following properties*: $D' \in \Lambda(X)$ *and* F' *is a co-initial operator for* D' *corresponding to a left inverse* R'.

Proof. By our assumptions

$$DR = I, F^2 = F, FR = 0, DF = 0 \text{ and } F = I - RD \text{ on dom } D.$$

Since for arbitrary $A, B \in L(X, \Xi)$, such that the superposition AB is well-defined, we have $(AB)' = B'A'$, we find

$$R'D' = I, \quad F'^2 = F', \quad R'F' = 0, \quad F'D' = 0$$
$$\text{and } F' = I - D'R' \quad \text{on dom } R' = \Xi.$$

This and Formulae (2.5.1), (2.5.2), (2.5.3), (2.5.4) and (2.5.5) together imply that $D' \in \Lambda(X)$ and F' is a co-initial operator corresponding to an $R' \in \mathscr{L}_{D'}$. ∎

Similarly we prove the following:

THEOREM 2.5.4. *Let $\Xi \subset X'$ be an arbitrary conjugate space. Suppose that $\Lambda \in \Lambda(X)$, G is a co-initial operator corresponding to an $L \in \mathscr{L}_\Lambda$ and $\Lambda, G, L \in L(X, \Xi)$. Then the conjugate operators have the following properties: $\Lambda' \in R(X)$ and G' is an initial operator for Λ' corresponding to the right inverse R'.*

We therefore have a full duality between right and left invertible operators.

COROLLARY 2.5.2. *Let $\Xi \subset X'$ be arbitrary conjugate space. If $A \in L(X, \Xi)$ is invertible then A' is also invertible. Conversely, if A' is invertible then A is invertible.*

Examples and Exercises

EXAMPLE 2.5.1. Let $X = C[0, a]$. Consider an operator defined by means of the equality

$$(Ax)(t) = \int_0^t (t-s+1)x(s)\,ds + \frac{1}{2}(t^2+t)x(0) \quad \text{for } x \in X.$$

$$(2.5.9)$$

We shall show that A is left invertible but not right invertible. Indeed, if we write: $(Rx)(t) = \int_0^t x(s)\,ds$, we find

$$(DAx)(t) = -\int_0^t x(s)\,ds + x(t) = [(I-R)x](t),$$

where $D = \mathrm{d}/\mathrm{d}t$,

which implies that the operator $A_0 = (I-R)^{-1}D$ is a left inverse of A because the operator $(I-R)^{-1}$ is well-defined (cf. Formula (2.2.41)) and $A_0 A = (I-R)^{-1}DA = (I-R)^{-1}(I-R) = I$. Put now $x(t) \equiv t$. Then $(Dx)(t) \equiv 1$. Formula (2.2.41) implies that

$$[(I-R)^{-1}Dx](t) = 1 - \int_0^t e^{t-s}ds = 1-1+e^t = e^t,$$

$$(AA_0 x)(t) = [A(I-R)^{-1}Dx](t)$$

$$= \int_0^t (t-s+1)e^s ds + \tfrac{1}{2}(t^2+t)e^0$$

$$= [(t-s+1)e^s]_0^t - \int_0^t -e^s ds + \tfrac{1}{2}(t^2+t)$$

$$= e^t - (t+1) + e^t - 1 + \tfrac{1}{2}(t^2+t)$$

$$= 2e^t + \tfrac{1}{2}(t^2-t) - 2 \neq t = x(t),$$

which proves that A is not right invertible.

EXERCISE 2.5.1. Suppose that $\varDelta \in \varLambda(X)$ and G_0 is a co-initial operator corresponding to an $L_0 \in \mathscr{L}_\varDelta$. Prove that:
   (i) Every left inverse of $\varDelta$ is of the form

$$L = L_0 + AG, \quad \text{where } A \in L(X), \text{ dom } A = G \text{ dom } L_0;$$

   (ii) Every co-initial operator is of the form

$$G = (I - \varDelta A)G_0, \quad \text{where } A \in L(X), \text{ dom } A = G \text{ dom } L_0;$$

   (iii) If G is a projection such that ker $G = \varDelta X$ then G is a co-initial operator corresponding to an $L = L_0 - L_0 G \in \mathscr{L}_\varDelta$;
   (iv) If $L_1, L_2 \in \mathscr{L}_\varDelta$ are commutative then $L_2 = L_1$;
   (v) If two co-initial operators G_1, G_2 are commutative, then $G_2 = G_1$.

EXERCISE 2.5.2. Suppose that $\varDelta \in \varLambda(X)$ and that $G_{\gamma_n}^1$ are co-initial operators for $\varDelta$ corresponding to $L_{\gamma_n} \in \mathscr{L}_\varDelta$, where $\{\gamma_n\} \subset \varGamma$, $\mathscr{L}_\varDelta = \{L_\gamma\}_{\gamma \in \varGamma}$. Prove that for arbitrary $N = 1, 2, \ldots$ the following identity holds on dom L^{N+1}:

$$I = G_{\gamma_0} + \sum_{k=1}^{N-1} \varDelta^k G_{\gamma_k} L_{\gamma_0} \ldots L_{\gamma_{k-1}} + \varDelta^N L_{\gamma_0} \ldots L_{\gamma_{n-1}}. \tag{2.5.10}$$

This formula corresponds to the Taylor–Gontcharov formula for right invertible operators.

EXERCISE 2.5.3. Suppose that $\varXi \subset X'$ is an arbitrary conjugate space, $D \in R(X)$, $F_1, F_2 \neq F_1$ are initial operators for D corresponding to $R_1, R_2 \in \mathscr{R}_D$, respectively, and $D, F_1, F_2, R_1, R_2 \in L(X, \varXi)$. Find conjugate operators for an initial operator $F_3 = dF_1 + (1-d)F_2$, where $d \in \mathbf{R}$ is arbitrarily fixed, and for the corresponding right inverse R_3.

Chapter 3

General solution of equations with right invertible operators

3.1. EQUATIONS OF ORDER ONE WITH NON-COMMUTATIVE COEFFICIENTS

To begin with, we shall consider the simplest equation with a right invertible operator:

$$Dx = y, \quad y \in X, \quad D \in R(X). \tag{3.1.1}$$

PROPOSITION 3.1.1. *Suppose that* $D \in R(X)$ *and* $\dim \ker D \neq 0$. *Then every solution of Equation* (3.1.1) *is of the form*

$$x = Ry + z, \tag{3.1.2}$$

where $z \in \ker D$ *is arbitrary and* $R \in \mathcal{R}_D$ *is arbitrarily fixed.*

Proof. Indeed, Proposition 2.1.4 (Formula (2.1.3)) implies that every element $x \in \operatorname{dom} D$ should be of the form $x = Ru + z$ for a $u \in X$ and $z \in \ker D$. We are looking for $x \in \operatorname{dom} D$ satisfying Equation (3.1.1). Then we find $u = DRu = D(x-z) = Dx - Dz = y$. Observe that by the same Proposition 2.1.4 we conclude that a change of the right inverse R implies only a change of the constant (which is arbitrarily fixed) in Formula (3.1.2). ∎

We shall consider now more general cases.

PROPOSITION 3.1.2. *Suppose that* $D \in R(X)$, $\dim \ker D \neq 0$, *F is an initial operator for D corresponding to an* $R \in \mathcal{R}_D$ *and* $A \in L_0(X)$. *Then the following identities hold*:

$$D - A = D(I - RA) \quad \text{on } \operatorname{dom} D, \tag{3.1.3}$$

$$(D - A)R = I - AR \quad \text{on } X, \tag{3.1.4}$$

$$D - A = (I - AR)D - AF \quad \text{on } \operatorname{dom} D. \tag{3.1.5}$$

Proof. Indeed, on dom D we have $D-A = D-DRA = D(I-RA)$. Since dom $R = X$ we have $(D-A)R = DR-AR = I-AR$ on X. Since $F = I-RD$ on dom D we find $D-A = D-A(RD+F) = D-ARD - -AF = (I-AR)D-AF$. ∎

PROPOSITION 3.1.3. *Suppose that all assumptions of Proposition 3.1.2 are satisfied. Then* $y \in (D-A)$ dom D *if and only if there exists a z* $\in$ ker D *such that*

$$Ry+z \in (I-RA) \text{ dom } D. \tag{3.1.6}$$

Proof. Indeed, by Proposition 2.1.4, dom $D = RX \oplus \ker D$. Suppose that $y \in (D-A)$ dom $D = D(I-RA)$ dom D for Equality (3.1.3) holds. This means that there exist $u \in X$ and $z \in \ker D$ such that $y = D(I- -RA)(Ru+z)$. Proposition 3.1.1 implies that $(I-RA)(Ru+z) = Ry+ +z_1 \in$ dom D for $z_1 \in \ker D$. But $z_1 = Fz_1 = FRy+Fz_1 = F(I- -RA)(Ry+z) = F(Ry+z) = FRy+Fz = z$ for $FR = 0$. This implies that $Ry+z \in (I-RA)$ dom D, i.e. Condition (3.1.6) is satisfied. On the other hand, suppose that Condition (3.1.6) is satisfied. It means that for given $y \in X$ and $z \in \ker D$ there exist $u \in X$ and $z_1 \in \ker D$ such that $Ry+z = (I-RA)(Ru+z_1)$. Acting on both sides of this equality by operator D we find $y = D(Ry+z) = D(I-RA)(Ry+z_1) = (D- -A)(Ry+z_1) \in (D-A)$ dom D. ∎

PROPOSITION 3.1.4. *Suppose that assumptions of Proposition 3.1.2 are satisfied. Let* $y \in X$ *be arbitrarily fixed. Then the following conditions are equivalent*:

(i) $y \in (D-A)$ dom D *and* ker $D \subset (I-RA)$ dom D;
(ii) $Ry+z \in (I-RA)$ dom D *for every* $z \in \ker D$, *i.e.* $\{Ry\} \oplus \ker D$ $\subset (I-RA)$ dom D.

Proof. In order to prove (ii) → (i) put $y = 0$ and apply Condition (3.1.4). Then we find that $z \in (I-RA)$ dom D for every $z \in \ker D$, i.e. ker $D \subset (I-RA)$ dom D. Further considerations are going on the same lines as in the proof of Proposition 3.1.3. In order to prove (i) → (ii) observe that, by assumptions, $y \in (D-A)$ dom D, and ker $D \subset (I- -RA)$ dom D. Hence, by Propositions 3.1.3 and 2.1.4 for an arbitrarily fixed $z \in \ker D$ there exist $u_1, u_2 \in X$, $z_1, z_2 \in \ker D$ such that $Ry+z_1$

$= (I - RA)(Ru_1 + z_1)$ and $z = (I - RA)(Ru_2 + z_2)$, i.e. $Ry + z_1 + z = (I - RA)[R(u_1 + u_2) + z_1 + z_2] \in (I - RA)$ dom D for $u_1, u_2 \in X_1$, $z_1 + z_2 \in \ker D$. Since $z_1 + z_2 \in \ker D$ the arbitrariness of $z \in \ker D$ implies (ii), i.e. Condition (3.1.6).

Observe that

$$(I - RA) \text{dom } D \subset \text{dom } D, \tag{3.1.7}$$

provided that all assumptions of Proposition 3.1.2 are satisfied. Indeed, dom $(I - RA) = X$ for $A, R \in L_0(X)$. Since $RAx \in$ dom D for all $x \in$ dom D, we conclude that $(I - RA)x \in$ dom D. The arbitrariness of $x \in$ dom D implies Condition (3.1.7). ∎

Propositions 2.1.4, 3.1.4 and Condition (3.1.7) together imply

COROLLARY 3.1.1. *Suppose that all assumptions of Proposition 3.1.2 are satisfied. If*

$$(I - RA) \text{dom } D = \text{dom } D, \tag{3.1.8}$$

then for every $y \in X$, $z \in \ker D$ Condition (3.1.6) is satisfied, hence $y \in (D - A)$ dom D and $\ker D \subset (I - RA)$ dom D.

DEFINITION 3.1.1. Suppose that all assumptions of Proposition 3.1.2 are satisfied. The operators $I - RA$ and $I - AR$ are said to be *resolving operators* for the operator $D - A$. If either the operator $I - RA$ or the operator $I - AR$ is invertible then the equation

$$(D - A)x = y, \quad y \in (D - A) \text{dom } D, \tag{3.1.9}$$

is said to be *well-determined*. The operators $(I - RA)^{-1}$, $(I - AR)^{-1}$, respectively, are said to be *resolvent operators* for Equation (3.1.9). If neither $I - RA$ nor $I - AR$ is invertible then Equation (3.1.9) is said to be *ill-determined* (cf. Pogorzelec, 1983).

We shall consider ill-determined Equation (3.1.9) in the case when resolving operators are either left or right invertible. The theory of ill-determined equations in these cases is essentially based on results of A. Pogorzelec obtained in her Ph. dissertation, 1983.

THEOREM 3.1.1. *Suppose that $A \in L_0(X)$, $D \in R(X)$, dim $\ker D \neq 0$ and F is an initial operator for D corresponding to an $R \in \mathcal{R}_D$.*

1. *If $I - RA \in R(X)$ and* $\dim \ker (I - RA) \neq 0$ *then*
(i) $\ker(I - RA) \subset \operatorname{dom} D$;
(ii) *for all* $R_A \in \mathscr{R}_{I - RA}, y \in (D - A) \operatorname{dom} D, z \in \ker D$ *we have*

$$R_A(Ry + z) \in \operatorname{dom} D;$$

(iii) *all solutions of Equation* (3.1.9) *are of the form*

$$x = R_A(Ry + z) + \tilde{z},$$

where $z \in \ker D, \tilde{z} \in \ker(I - RA)$ *are arbitrary,*   (3.1.10)

$\quad Fx = z \quad$ *and* $\quad \dim \ker (D - A) = \dim \ker D + \dim \ker (I - RA).$
2. *If* $I - RA \in \Lambda(X)$ *then*
(i) $FL_A u = Fu$ *for every* $L_A \in \mathscr{L}_{I - RA}$ *and* $u \in (I - RA)X$;
(ii) $L_A(Ry + z) \in \operatorname{dom} D$ *for all* $y \in (D - A) \operatorname{dom} D, \ z \in \ker D, \ L_A \in \mathscr{L}_{I - RA}$;
(iii) *all solutions of Equation* (3.1.9) *are of the form*

$$x = L_A(Ry + z), \quad \text{where } z \in \ker D \text{ is arbitrary,}   (3.1.11)$$

$\quad Fx = z \quad$ *and* $\quad \dim \ker (D - A) = \dim L_A \ker D.$

3. *If* $I - RA$ *is invertible then all solutions of Equation* (3.1.9) *are of the form*

$$x = (I - RA)^{-1}(Ry + z), \quad \text{where } z \in \ker D \text{ is arbitrary,} \quad (3.1.12)$$

$\quad Fx = z \quad$ *and* $\quad \dim \ker (D - A) = \dim \ker D.$

Proof. Suppose that $I - RA \in R(X)$ and $R_A \in \mathscr{R}_{I - AR}$ is arbitrarily fixed. By definition, we have $R_A \in L_0(X)$ and $R_A X \subset \operatorname{dom}(I - RA)$. Formula (3.1.3) implies that $D - A = D(I - RA)$ as a superposition of right invertible operators is again a right invertible operator. Suppose that $\tilde{z} \in \ker(I - RA)$ is arbitrarily fixed. Then, by definition, $\tilde{z} = RA\tilde{z}$, i.e. $\tilde{z} \in \operatorname{dom} D$, which implies $\ker(I - RA) \subset \operatorname{dom} D$.

Suppose now that $x \in \operatorname{dom} D$ and $u = R_A x$. Then $x = (I - RA)R_A x = u - RAu$, i.e. $u = x + RAu$. Hence $u \in \operatorname{dom} D$. Proposition 2.1.4 implies that $R_A(Ry + z) \in \operatorname{dom} D$ for every $y \in (D - A) \operatorname{dom} D$ and $z \in \ker D$. Since Equation (3.1.9) can be rewritten as $D(I - RA)x = y$, Proposition 3.1.1 implies that $(I - RA)x = Ry + z$, where $z \in \ker D$ is arbitrary. But $I - RA$ also is right invertible. Hence the same Proposition

3.1.1 implies that $x = R_A(Ry+z)+\tilde{z}$, where $\tilde{z} \in \ker(I-RA)$ is arbitrary and $R_A \in \mathcal{R}_{I-RA}$.

Moreover, since $FR = 0$ and $\tilde{z} = RA\tilde{z}$, then $F\tilde{z} = 0$ and $Fx = Fx - FRAx = F(I-RA)x = F(I-RA)R_A(Ry+z) = F(Ry+z) = FRy+Fz = z$ for $Fz = z$ and $(I-RA)R_A = I$.

Put now $y = 0$. Then all solutions of the homogeneous equation $(D-A)x = 0$ are of the form $x = R_A z+\tilde{z}$, where $z \in \ker D$ and $\tilde{z} \in \ker(I-RA)$ are arbitrary. Since $R_A \in \mathcal{R}_{I-RA}$, we have $\ker R_A = \{0\}$. Hence $\dim \ker(D-A) = \dim \ker D + \dim \ker(I-RA)$.

Suppose now that $I-RA \in \Lambda(X)$. Then $(I-RA)X \subset \operatorname{dom} L_A$ for all $L_A \in \mathcal{L}_{I-RA}$, and $L_A(I-RA) = I$. Let $u \in (I-RA)X$. Then $x = L_A(I-RA)x = L_A u$, for an $x \in X$ and an arbitrarily fixed $L_A \in \mathcal{L}_{I-RA}$.

Since $FR = 0$ and $F(I-RA) = F$ on X, we find $F(I-RA)L_A = FL_A$ on $(I-RA)X$. Let G be a co-initial operator for $I-RA$ corresponding to the left inverse L_A. Then $I-G = (I-RA)L_A$ and $FL_A = F(I-RA)L_A = F(I-G)$ on $(I-RA)X$. Since $u \in (I-RA)X$, we find $FL_A u = F(I-G)u = Fu - FGu$. Property 2.5.3 implies that $\ker G = (I-RA)X$. Hence $Gu = 0$ and $FL_A u = Fu$.

In order to prove that $L_A(Ry+z) \in \operatorname{dom} D$ for $y \in (D-A)\operatorname{dom} D$, $z \in \ker D$ observe that by Proposition 3.1.3 we have $Ry+z \in (I-RA)\operatorname{dom} D$. Hence $L_A(Ry+z) \in L_A(I-RA)\operatorname{dom} D = \operatorname{dom} D$.

Write Equation (3.1.9) in the form $Du = y$, where $u = (I-RA)x$. Proposition 3.1.1 implies that $u = Ry+z$, where $z \in \ker D$ is arbitrary. We therefore conclude that $x = L_A(I-RA)x = L_A u = L_A(Ry+z)$. Since $FL_A = F$ on $(I-RA)X$ we have $Fx = FL_A u = Fu = F(Ry+z) = Fz = z$. Since all solutions of the homogeneous equation $(D-A)x = 0$ are of the form $L_A z$ we find $\dim \ker(D-A) = \dim L_A \ker D$.

Points 1 and 2 immediately imply Point 3. Indeed, since in this case the operator $I-RA$ is invertible, we have $I-RA \in R(X) \cap \Lambda(X)$, $\ker(I-RA) = \{0\}$ and $L_A = R_A = (I-RA)^{-1}$. ∎

COROLLARY 3.1.2. *Suppose that all assumptions of Theorem 3.1.1 are satisfied and $I-RA \in R(X)$. Let $R_A \in \mathcal{R}_{I-RA}$ and $z \in \ker D$ be arbitrarily fixed. Then the set*

$$Z(R_A) = \{x = R_A(Ry+z)+\tilde{z} : \tilde{z} \in \ker(I-RA)\} \qquad (3.1.13)$$

of all solutions of Equation (3.1.9) *is independent of the choice of* R_A, *i.e. if* $R_A \neq R'_A \in \mathscr{R}_{I-RA}$ *then* $Z(R'_A) = Z(R_A)$.

Proof. Indeed, Proposition 2.1.2 implies that for arbitrary $u \in X$ and $R_A, R'_A \in \mathscr{R}_{I-RA}$ we have $(R_A - R'_A)u \in \ker(I - RA)$. Let $x \in Z(R_A)$ and $x' \in Z(R'_A)$ be arbitrary, i.e. $x = R_A(Ry + z) + \tilde{z}$, $x' = R'_A(Ry + z) + \tilde{z}'$, where $\tilde{z}$, $\tilde{z}' \in \ker(I - RA)$. Then $x - x' = (R_A - R'_A)(Ry + z) + \tilde{z} - \tilde{z}'$ $= \tilde{z}_0 \in \ker(I - RA)$, i.e. $x = x' + \tilde{z}_0 = R'_A(Ry + z) + z' + \tilde{z}_0 \in Z(R'_A)$. ∎

COROLLARY 3.1.3. *Suppose that all assumptions of Theorem* 3.1.1 *are satisfied and* $I - RA \in \Lambda(X)$. *Let* $L_A \in \mathscr{L}_{I-RA}$ *and* $z \in \ker D$ *be arbitrarily fixed. Then solutions of Equation* (3.1.9) *are independent of the choice of* L_A, *i.e. if* $L_A \neq L'_A \in \mathscr{L}_{I-RA}$ *then* $L_A(Ry + z) = L'_A(Ry + z)$.

Proof. Indeed, $L_A = L'_A + BG$ on $\operatorname{dom} L_A$, where $B \in L_0(X)$ and G is a co-initial operator for $I - RA$ corresponding to L_A (cf. Exercise 2.5.1). We have proved that $u = Ry + z \in (I - RA)X = \ker G$. Therefore $G(Ry + z) = 0$ and $L_A(Ry + z) = L'_A(Ry + z)$. ∎

LEMMA 3.1.1. *Suppose that* $D \in R(X)$, $\dim \ker D \neq 0$ *and* $R \in \mathscr{R}_D$. *Then for an arbitrary positive integer* k *we have*

$$\ker D^k = \ker D \oplus \dots \oplus R^{k-1}\ker D, \qquad (3.1.14)$$

$$\operatorname{dom} D^k = R^k X \oplus \ker D^k. \qquad (3.1.15)$$

Proof. Formula (3.1.15) is an immediate consequence of Corollary 2.2.3 Formula (3.1.14) will be proved by induction. For $k = 1$ this formula is. satisfied. Suppose Formula (3.1.14) to be true for an arbitrary fixed $k \geqslant 1$. It means that every element of the form $u = \sum_{j=0}^{k} R^j z_j$, where $z_0, \dots, z_k \in \ker D$, belongs to $\ker D^k$ (cf. also Proposition 2.4.3). Consider the equation $D^{k+1}v = 0$. We have $D^k(Dv) = 0$, i.e. $Dv = \sum_{j=0}^{k} R^j z_j$. Proposition 3.1.1 implies that $v = R \sum_{j=0}^{k} R^j z_j + z_{k+1}$, where $z_{k+1} \in \ker D$ is arbitrary, i.e. $v = \sum_{j=0}^{k} R^{j+1}z_j + z_{k+1} = z_{k+1} + Rz_0 + \dots + R^{k+1}z_k$

$\in \ker D^k$. The arbitrariness of $z_0, \ldots, z_{k+1}$ and Proposition 2.4.2 together imply Formula (3.1.14). ∎

The following theorems show that for y satisfying stronger conditions, i.e. "more regular" in a sense, we find "more regular" solutions of Equation (3.1.9).

THEOREM 3.1.2. *Suppose that all assumptions of Theorem 3.1.1 are satisfied. Let $I - RA \in R(X)$ and let F_A be an initial operator for $I - RA \in R(X)$ corresponding to an $R_A \subset \mathcal{R}_{I-RA}$. Suppose that $W \subset (D-A) \operatorname{dom} D^k$ for an arbitrarily fixed $k \geqslant 1$. If $\ker D \subset (I-RA) \operatorname{dom} D^k$ then*

$$Z(R_A) \subset (I-F_A) \operatorname{dom} D^k \oplus F_A \operatorname{dom} D, \qquad (3.1.16)$$

where $Z(R_A)$ is defined by Formula (3.1.13).

Proof. Suppose that $y \in W$, $z \in \ker D$ and $\tilde{z} \in \ker(I-RA)$ are arbitrarily fixed. If $y \in (D-A) \operatorname{dom} D^k = D(I-RA) \operatorname{dom} D^k$ then $Rw \in RD(I-RA) \operatorname{dom} D^k$ and conversely. Since $RX \cap \ker D = \{0\}$ and $RD = I - F$ then

$$Ry + z \in (I-F)(I-RA) \operatorname{dom} D^k \oplus \ker D. \qquad (3.1.17)$$

Lemma 3.1.1 implies that

$$\begin{aligned} \ker D &= F \ker D \\ &= F(\ker D \oplus R \ker D \oplus \ldots \oplus R^{k-1} \ker D + R^k X) \\ &= F \operatorname{dom} D^k = F(I-RA) \operatorname{dom} D^k \end{aligned}$$

for $FR = 0$. Since $\ker D \subset (I-RA) \operatorname{dom} D^k$, we conclude that F transforms the space $(I-RA) \operatorname{dom} D^k$ into itself. This, and Formula (3.1.17) together, imply that $Ry + z \in (I-RA) \operatorname{dom} D^k$. Since, by definition, $I - F_A = R_A(I-RA)$, we conclude that $R_A(Ry+z) \in (I-RA) \operatorname{dom} D^k$ and $R_A(Ry+z) + \tilde{z} \in (I-F_A) \operatorname{dom} D^k \oplus \ker(I-RA)$ for $\tilde{z} \in \ker(I-RA)$. Point 1(i) of Theorem 3.1.1 implies that $\ker(I-RA) = F_A \ker(I-RA) \subset F_A \operatorname{dom} D$. Hence $R_A(Ry+z) + \tilde{z} \in (I-F_A) \operatorname{dom} D^k \oplus F_A \operatorname{dom} D$. ∎

THEOREM 3.1.3. *Suppose that all assumptions of Theorem 3.1.1 are satisfied. Let $I - RA \in \Lambda(X)$ and let $L_A \in \mathscr{L}_{I-RA}$. Suppose that $W \subset (D - A)$ dom D^k for an arbitrary fixed $k \geqslant 1$. If $\ker D \subset (I - RA)$ dom D^k then all solutions of Equation (3.1.9) belong to dom D^k, i.e.*

$$L_A(RW \oplus \ker D) \subset \text{dom } D^k \quad \text{for all } L_A \in \mathscr{L}_{I-RA}. \qquad (3.1.18)$$

In particular

$$L_A(Ry + z) \in \text{dom } D^k \quad \text{for all } y \in W, z \in \ker D, L_A \in \mathscr{L}_{I-RA}. \qquad (3.1.19)$$

Proof. In order to prove Formula (3.1.18) we show in the same way as in the proof of Theorem 3.1.2 that the operator F transforms the space $(I - RA)$ dom D^k into itself and $\ker D = F(I - RA)$ dom D^k, i.e.

$$RW \oplus \ker D \subset (I - F)(I - RA) \text{dom } D^k \oplus F(I - RA) \text{dom } D^k.$$

Since $(I - RA)X \subset \text{dom } L_A$, we conclude that also $(I - RA)$ dom $D^k \subset \text{dom } L_A$. We therefore obtain

$$L_A(RW \oplus \ker D) \subset L_A(I - RA) \text{dom } D^k = \text{dom } D^k. \qquad \blacksquare$$

COROLLARY 3.1.4. *Suppose that all assumptions of Theorem 3.1.1 are satisfied. Suppose that the operator $I - RA$ is invertible and $W \subset (D - A)$ dom D^k for an arbitrarily fixed $k \geqslant 1$. If $\ker D \subset (I - RA)$ dom D^k then all solutions of Equation (3.1.9) belong to dom D^k, i.e.*

$$(I - RA)^{-1}(RW \oplus \ker D) \subset \text{dom } D^k. \qquad (3.1.20)$$

In particular

$$(I - RA)^{-1}(Ry + z) \in \text{dom } D^k \quad \text{for all } y \in W, z \in \ker D. \qquad (3.1.21)$$

We shall consider now Equation (3.1.9) with resolving operators of the form $I - AR$.

THEOREM 3.1.4. *Suppose that all assumptions of Theorem 3.1.1 are satisfied. Then*

1. $(D - A)$ dom $D = (I - AR)X + A \ker D$ *(in general, this last sum is not direct).*

2. *If $I - AR \in R(X)$ and dim $\ker(I - AR) \neq 0$ then*

(i) $R \ker (I - AR) \subset \mathrm{dom}\, D$;

(ii) *for all* $R^A \in \mathscr{R}_D$, $y \in (D - A)\, \mathrm{dom}\, D$, $z \in \ker D$ *we have*
$$R[R^A(y + Az) + \tilde{z}] + z \in \mathrm{dom}\, D;$$

(iii) *all solutions of Equation* (3.1.9) *are of the form*

$$x = R[R^A(y + Az) + \tilde{z}] + z,$$
where $z \in \ker D$ *and* $\tilde{z} \in \ker (I - AR)$ *are arbitrary*,   (3.1.22)
$Fx = z$ *and* $\dim \ker (D - A) = \dim (A + I) \ker D + \dim \ker (I - AR)$.

3. *If* $I - AR \in \Lambda(X)$ *then*:

(i) *for all* $L^A \in \mathscr{L}_{I-AR}$, $y \in (D - A)\, \mathrm{dom}\, D$, $z \in \ker D$, *where* $RL^A(y + Az) + z \in \mathrm{dom}\, D$;

(ii) *all solutions of Equation* (3.1.9) *are of the form*

$$x = RL^A(y + Az) + z, \textit{ where } z \in \ker D \textit{ is arbitrary},$$ (3.1.23)
$Fx = z$ *and* $\dim \ker (D - A) = \dim (L^A A + I) \ker D$.

4. *If* $I - AR$ *is invertible then* $R(I - AR)^{-1}(y + Az) + z \in \mathrm{dom}\, D$, *all solutions of Equation* (3.1.9) *are of the form*

$$x = R(I - AR)^{-1}(y + Az) + z, \textit{where } z \in \ker D \textit{ is arbitrary}, (3.1.24)$$
$Fx = z$ *and* $\dim \ker (D - A) = \dim (A + I) \ker D$.

Proof. Since $A, R \in L_0(X)$, we have $\mathrm{dom}\,(I - AR) = X$. By Formula (3.1.4) we have $(D - A)R = I - AR$ on X. Let $x \in \mathrm{dom}\, D$ be arbitrary. Then there exist $u \in X$ and $z \in \ker D$ such that $x = Ru + z$ and $(D - A)x = (D - A)Ru + (D - A)z = (I - AR)u - Az$. The arbitrariness of x implies that $(D - A)\, \mathrm{dom}\, D = (I - AR)X + A \ker D$. This algebraic sum, in general, is non direct. Indeed, if $A = F$ and $x = Ru + z$, where $u \notin \ker D$, $z \in \ker D$ then $(D - A)x = (I - FR)u - Fz = u - z \neq 0$.

Suppose that $I - AR \in R(X)$ and $\dim \ker (I - AR) \neq 0$. Then $R \ker (I - AR) \subset RX \subset \mathrm{dom}\, D$. Write Equation (3.1.9) in the form

$$(I - AR)u - Az = y,$$
where $x = Ru + z$, $z \in \ker D$ is arbitrary.   (3.1.25)

Observe that by Point 1 we have $y + Az \in (I - AR)X$ for all $z \in \ker D$. Since $I - AR \in R(X)$, Proposition 3.1.1 implies that for $R^A \in \mathscr{R}_{I-AR}$ we have $u = R^A(y + Az) + \tilde{z}$, where $z \in \ker D$ and $\tilde{z} \in \ker (I - AR)$ are arbitrary for $(I - AR)R^A = I$ and $x = Ru + z = R[R^A(y + Az) + \tilde{z}] + z$, where $z \in \ker D$, $\tilde{z} \in \ker (I - AR)$ are arbitrary.

Since x is of the form $Rw+z$, where $w \in X$, $z \in \ker D$, we conclude that $x \in \operatorname{dom} D$ and $Fx = z$. By definition, $\ker R = \{0\}$, $\ker R^A = \{0\}$ for $R \in \mathcal{R}_D$, $R^A \in \mathcal{R}_{I-AR}$. All solutions of the homogeneous Equation (3.1.9) are of the form: $x = R(R^A Az + \tilde{z}) + z$. Therefore $\dim \ker(D-A) = \dim(A+I)\ker D + \dim \ker(I-AR)$.

Suppose now that $I-AR \in \Lambda(X)$. Then from (3.1.25) we find $u = L^A(I-AR)u = L^A(y+Az)$ and $x = Ru+z = RL^A(y+Az)+z$, where $z \in \ker D$ is arbitrary. The form of x implies that $x \in \operatorname{dom} D$ and $Fx = z$. All solutions of the homogeneous Equation (3.1.9) are of the form $RL^A Az + z$. We therefore conclude that $\dim \ker(D-A) = \dim(L^A A + I)\ker D$ for $\ker R = \{0\}$.

Finally suppose that $I-AR$ is invertible. Then Points 2 and 3 together imply Point 4. Indeed, in this case $R^A = L^A = (I-AR)^{-1}$ and $\ker(I-AR) = \{0\}$. ■

COROLLARY 3.1.5. *Suppose that all assumptions of Theorem 3.1.1 are satisfied and $I-AR \in R(X)$. Let $R^A \in \mathcal{R}_{I-AR}$ and $z \in \ker D$ be arbitrarily fixed. Then the set*

$$Z'(R^A) = \{x = R[R^A(y+Az)+\tilde{z}]+z: \ \tilde{z} \in \ker(I-AR)\}$$

$$(3.1.26)$$

of all solutions of Equation (3.1.9) is independent of the choice of R^A, i.e. if $R^A \neq \tilde{R}^A \in \mathcal{R}_{I-AR}$ then $Z'(R^A) = Z'(\tilde{R}^A)$.

Proof. Indeed, Proposition 2.1.2 implies that for arbitrary $u \in X$ and $R^A, \tilde{R}^A \in \mathcal{R}_{I-AR}$ we have $(R^A - \tilde{R}^A)u \in \ker(I-AR)$. Let $x \in Z'(R^A)$ and $x' \in Z'(\tilde{R}^A)$, i.e. $x = R[R^A(y+Az)+\tilde{z}]+z$, $x' = R[\tilde{R}^A(Ry+z)+\tilde{z}']+z$, where $\tilde{z}, \tilde{z}' \in \ker(I-AR)$. Then $x-x' = R[(R^A - \tilde{R}^A)(Ry+z)+\tilde{z}-\tilde{z}'] = R\tilde{z}''$, where $\tilde{z}'' \in \ker(I-AR)$, i.e. $x = R[R^A(Ry+z)+\tilde{z}'+\tilde{z}'']+z \in Z'(R^A)$. ■

COROLLARY 3.1.6. *Suppose that all assumptions of Theorem 3.1.1 are satisfied and $I-AR \in \Lambda(X)$. Let $L^A \in \mathcal{L}_{I-AR}$ and $z \in \ker D$ be arbitrarily fixed. Then solutions of Equation (3.1.9) are independent of the choice of L^A, i.e. if $L^A \neq \tilde{L}^A \in \mathcal{L}_{I-AR}$ then $RL^A(y+Az)+z = R\tilde{L}^A(y+Az)+z$.*

Proof. Indeed, since $y + Az = (I - AR)u \in (I - AR)X = \ker G$ (cf. Formula (3.1.25)), where G is a co-initial operator for $I - AR$, in a similar way, as in the proof of Corollary 3.1.3, we conclude that $L^A(y + Az) = \tilde{L}^A(y + Az)$. ∎

THEOREM 3.1.5. *Suppose that all assumptions of Theorem 3.1.3 are satisfied. Let $W \subset (D - A)$ dom D^k for a $k \geqslant 1$ and let $y \in W$ be arbitrarily fixed.*

1. *If $I - AR \in R(X)$, dim ker $(I - AR) \neq 0$ and $R^A \in \mathscr{R}_{I-AR}$ then solutions of Equation (3.1.9) belong to dom D^k and are of the form*

$$x = R[R^A(y + Az_0) + \tilde{z}] + z_0,$$

where $z_0 \in \ker D$, $\tilde{z} \in \ker(I - AR)$ are arbitrary, (3.1.27)

i.e.

$$R[R^A(W \oplus A \ker D)] \oplus \ker D \subset \text{dom } D^k.$$ (3.1.28)

2. *If $I - AR \in \Lambda(X)$ and $L^A \in \mathscr{L}_{I-AR}$ then solutions of Equation (3.1.9) belong to dom D^k and are of the form*

$$x = RL^A(y + Az_0) + z_0, \quad \text{where } z_0 \in \ker D \text{ is arbitrary,}$$ (3.1.29)

i.e.

$$RL^A(W \oplus A \ker D) \oplus \ker D \subset \text{dom } D^k.$$ (3.1.30)

Proof. By Lemma 3.1.1 every x belonging to the domain of D^k is of the form

$$x = R^k v + x_k,$$

where $v \in X$, $x_k = \sum_{j=0}^{k-1} R^j z_j$, $z_0, \dots, z_{k-1} \in \ker D$.

This, and Formula (3.1.4) together imply

$$\begin{aligned}
(D - A)x &= (D - A)(R^k v + x_k) \\
&= (D - A)RR^{k-1}v + (D - A)x_k \\
&= (I - AR)R^{k-1}v - Az_0 + (I - AR)\sum_{j=1}^{k-1} R^{j-1}z_j \\
&= (I - AR)\Big(R^{k-1}v + \sum_{j=1}^{k-1} R^{j-1}z_j\Big) - Az_0.
\end{aligned}$$

Equation (3.1.9) can be then rewritten as follows:

$$(I-AR)u = y+Az_0, \quad \text{where } u = R^{k-1}v+\sum_{j=1}^{k-1} z_j,$$

i.e.

$$Ru+z_0 = R^k v+\sum_{j=1}^{k-1} z_j+z_0 = R^k v+x_k = x.$$

Apply now Theorem 3.1.4. If $I-AR \in R(X)$ and $R^A \in \mathscr{R}_{I-AR}$ then $u = R^A(y+Az_0)+\tilde{z}$, where $\tilde{z} \in \ker(I-AR)$ is arbitrary and $x = Ru+ +z_0 = R[R^A(y+Az_0)+\tilde{z}]+z_0$, where $z_0 \in \ker D$ is arbitrary.

If $I-AR \in \Lambda(X)$ and $L^A \in \mathscr{L}_{I-AR}$ then $u = L^A(I-AR)u = L^A(y+ +Az_0)$, i.e. $x = RL^A(y+Az_0)+z_0$, where $z_0 \in \ker D$ is arbitrary. ∎

The following theorem shows connections between left and right invertibility of resolving operators $I-AR$ and $I-RA$.

THEOREM 3.1.6. *Suppose that* $D \in R(X)$, $R \in \mathscr{R}_D$ *and* $A \in L_0(X)$. *Then*:

(i) *If* $I-RA \in R(X)$ *and* $\ker(I-RA) \neq \{0\}$ *then* $I-AR \in R(X)$, *and* $\ker(I-AR) \neq 0$.

(ii) *Suppose that* $\ker D \subset (I-RA) \operatorname{dom} D$. *If* $I-RA \in \Lambda(X)$ *then* $I-AR \in \Lambda(X)$.

Proof. Observe that the following identity holds:
$$R(I-AR) = (I-RA)R \quad \text{on } X. \tag{3.1.31}$$

In order to prove Point (i) let us consider an arbitrarily fixed operator $R_A \in \mathscr{R}_{I-RA}$. Write $R^A = DR_A R$. We shall show that the operator R^A is well-defined and is a right inverse of $I-AR$. Indeed, we have $R_A RX \subset \operatorname{dom} D$ (cf. Point 1 (ii) of Theorem 3.1.1). Then $DR_A RX \subset X$. $= \operatorname{dom}(I-AR)$. Let F be an initial operator for D corresponding to R. Since $FR = 0$, we find $F(I-RA)R_A = FR_A$ on X and $(I-AR)DR_A R = (D - ARD)R_A R = (D - A + AF)R_A R = (D - A)R_A R + AFR_A R = D(I-RA)R_A R = DR = I$, i.e. $DR_A R \in \mathscr{R}_{I-AR}$. In order to prove that $\ker(I-AR) \neq \{0\}$ we have to show that R^A is not a left inverse of $I-AR$. Indeed, let F_A be an initial operator for $I-RA$ corresponding to its right inverse R_A. Suppose that $DF_A R = 0$ on X, i.e. $F_A RX \subset \ker D$. But $\ker(I-RA) = F_A X$. It means that for every $x \in X$ there exists $z \in \ker D$ such that $F_A Rx = z$ and $0 = (I - RA)F_A Rx = (I - RA)z$, i.e. $z = RAz$. This implies that $z \in RX \subset \operatorname{dom} D$ which contradicts Proposition 2.1.4. It means that there exists a $u \neq 0$ such that $DF_A Ru \neq 0$ and $R^A(I - AR)u = DR_A R(I - AR)u = DR_A(I - RA)Ru = D(I - F_A)Ru = DRu - DF_A Ru = u - DF_A Ru \neq u$, i.e. the operator $I - AR$ is not left invertible.

In order to prove Point (ii) let us consider an arbitrarily fixed operator $L_A \in \mathscr{L}_{I-RA}$. Define on the set $(D - A)\operatorname{dom} D$ the operator L^A as follows: $L^A = DL_A R$. Theorem 3.1.3 for $k = 1$ implies that $L_A R(D - A)\operatorname{dom} D \subset \operatorname{dom} D$, i.e. the operator L^A is well-defined. Moreover, Formula (3.1.31) implies that $(I - AR)X = DR(I - AR)X = D(I - RA)RX = (D - A)RX \operatorname{dom} D = \operatorname{dom} L^A$ and $L^A(I - AR) = DL_A R(I - AR) = DL_A(I - RA)R = DR = I$. This means that $I - AR$ is left invertible and L^A is its left inverse. ∎

The following example shows that the operator $I - AR$ is, in general, not invertible.

EXAMPLE 3.1.1 (cf. Pogorzelec, 1983). Let X be the space (s) of all sequences $\{x_n\}$, where $x_n \in \mathbf{R}$, $n \in \mathbf{N}$. Let $D\{x_n\} = \{x_{n+1}\}$, $R\{x_n\} = \{x_{n-1}\}$, where we admit $x_0 = 0$ for all $x = \{x_n\} \in X$. Let $A\{x_n\} = \{y_n\}$, where $y_1 = x_2$, $y_n = x_{n+1} - (x_n + x_{n-1})$ for $n \geq 2$. It is easy to verify that for $x \in X$

$$(I - RA)\{x_n\} = \{u_n\}, \quad \text{where } u_1 = x_1, u_2 = 0,$$
$$u_n = x_{n-1} + x_{n-2} \text{ for } n \geq 3,$$

$$(I - AR)\{x_n\} = \{v_n\}, \quad \text{where } v_1 = 0, v_2 = x_1,$$
$$v_n = x_{n-1} + x_{n-2} = u_n \text{ for } n \geq 3,$$

$$L_A\{x_n\} = \{w_n\}, \quad \text{where } w_1 = x_1, w_2 = x_3 - x_1,$$
$$w_n = x_{n+1} - w_{n-1} \text{ for } n \geq 3,$$

$$(I - AR)L^A\{x_n\} = (I - AR)DL_A R\{x_n\} = \{s_n\},$$
$$\text{where } s_1 = 0, s_n = x_n \text{ for } n \geq 2.$$

This implies that $\ker(I - RA) = \ker(I - AR) = \{0\}$, $(I - RA)X \subsetneq X$ and $(I - AR)X \subsetneq X$. Moreover, for all $x \in X$ we have $(I - AR)L^A\{x_n\} = \{s_n\} \neq \{x_n\}$, i.e. the operator L^A is not an inverse of $I - AR$.

COROLLARY 3.1.7. *Suppose that all assumptions of Theorem* 3.1.6 *are satisfied. If the operator* $I-RA$ *is invertible then the operator* $I-AR$ *is also invertible.*

OPEN QUESTION. Suppose that all assumptions of Theorem 3.1.6 are satisfied. Does the left invertibility (right invertibility, invertibility) of the operator $I-AR$ imply the left invertibility (resp. right invertibility, invertibility) of the operator $I-RA$?

We shall show now that there is a possibility of reduction of all six considered cases for resolving operators to three cases only.

LEMMA 3.1.2. *Suppose that* $D \in R(X)$, $R \in \mathcal{R}_D$ *and* $A \in L_0(X)$. *Then the following conditions are equivalent:*
  (i) $A \ker D \subset (I-AR)X$;
  (ii) $Z_1 = \{x = Ry+z: y \in (D-A) \operatorname{dom} D, z \in \ker D\}$
      $\subset (I-RA) \operatorname{dom} D$;
  (iii) $Z_2 = \{x = y+Az: y \in (D-A) \operatorname{dom} D, z \in \ker D\}$
      $\subset (I-AR)X$.

Proof. The implication (iii) → (i) is immediate if we admit $y = 0$.

(i) → (iii). Let $z \in \ker D$ and $y \in (D-A) \operatorname{dom} D$ be arbitrary. By our assumptions there exist $u, v, w \in X$ and $z_1 \in \ker D$ such that $y = (D-A)(Ru+z_1)$, $Az = (I-AR)v$ and $Az_1 = (I-AR)w$. This implies that

$$
\begin{aligned}
y+Az &= (D-A)(Ru+z_1)+(I-AR)v \\
&= (DR-AR)u+Dz-Az_1+(I-AR)v \\
&= (I-AR)u-(I-AR)w+(I-AR)v \\
&= (I-AR)(u-w+v) \in (I-AR)X.
\end{aligned}
$$

The arbitrariness of y and z implies that $Z_2 \subset (I-AR)X$.

(i) → (ii). Let $z \in \ker D$ be arbitrary. By our assumptions $Az \in (I-AR)X$. This implies that there exists a $v \in X$ such that $Az = (I-AR)v$. Hence $RAz = R(I-AR)v = (I-RA)Rv$ and $z = z-RAz+RAz = (I-RA)z+(I-RA)Rv = (I-RA)(Rv+z) \in (I-RA) \operatorname{dom} D$. The arbitrariness of $z \in \ker D$ implies that $\ker D \subset (I-RA) \operatorname{dom} D$. This, and Proposition 3.1.4, together imply that $Z_1 \subset (I-RA) \operatorname{dom} D$.

(ii) → (i). Let $y = 0$ and let $z \in \ker D$ be arbitrary. By our assumption $z \in (I-RA)$ dom D, i.e. there exist $u \in X$ and $z_1 \in \ker D$ such that $z = (I-RA)(Ru+z_1)$. It means that $z-z_1 = (I-RA)Ru+(I-RA)z_1- -z_1 = R[(I-RA)u-Az_1]$. But $z_1 = z$. Indeed, if F is an initial operator for D corresponding to R then $z = Fz = F(I-RA)(Ru+z_1) = F(Ru+ +z_1) = Fz_1 = z_1$. We therefore conclude that $R[(I-AR)u-Az] = R[(I-AR)u-Az_1] = z-z_1 = 0$, i.e. $(I-AR)u-Az = 0$ and $Az = (I-AR)u \in (I-AR)X$. The arbitrariness of $z \in \ker D$ implies $A \ker D \subset (I-AR)X$. ∎

THEOREM 3.1.7. *Suppose that* $D \in R(X)$, $R \in \mathcal{R}_D$ *and* $A \in L_0(X)$. *Write for a given* $y \subset W \subset (D-A)$ dom D *and arbitrary* $z \in \ker D$:

$$Z(R_A) = \{x = R_A(Ry+z)+\tilde{z}: \tilde{z} \in \ker(I-RA)\} \quad if$$
$$I-RA \in R(X) \text{ and } R_A \in \mathcal{R}_{I-RA} \quad (cf. \text{ Formula } (3.1.13));$$

$$Z'(R^A) = \{x = R[R^A(y+Az)+\tilde{z}]+z: \tilde{z} \in \ker(I-AR)\} \quad if$$
$$I-AR \in R(X) \text{ and } R^A \in \mathcal{R}_{I-AR} \quad (cf. \text{ Formula } (3.1.26));$$

$$Z_1(L_A) = \{x = L_A(Ry+z)\} \quad if \ I-RA \in \Lambda(X)$$
$$and \ L_A \in \mathscr{L}_{I-RA};$$

$$Z_1'(L^A) = \{x = RL^A(y+Az)+z\} \quad if \ I-AR \in \Lambda(X)$$
$$and \ L^A \in \mathscr{L}_{I-AR}.$$

Then

(i) *If* $I-RA \in R(X)$ *then* $Z(R_A) = Z'(R^A)$.

(ii) *Suppose that* $A \ker D \subset (I-AR)X$. *Then* $Z_1(L_A) = Z_1'(L^A)$.

Proof. (i) By definitions and Corollaries 3.1.2, 3.1.5, 3.1.3, 3.1.6, respectively, $Z(R_A)$, $Z'(R^A)$, $Z_1(L_A)$, $Z_1'(L^A)$ are sets of solutions of Equation (3.1.9) which are independent of the choice of R_A, R^A, L_A, L^A, respectively. Let $I-RA \in R(X)$ and let $R_A \in \mathcal{R}_{I-RA}$ be arbitrary fixed. Theorem 3.1.6 implies that $I-AR \in R(X)$ and $R^A = DR_AR \in \mathcal{R}_{I-AR}$. Let $x \in Z'(R^A)$, i.e. $x = R[R^A(y+Az)+\tilde{z}]+z$, where $\tilde{z} \in \ker(I- -AR)$. Let F be an initial operator for D corresponding to R. Since $(I-RA)R_A = I$ we find $FR_A = F$ on X and $x = R[R^A(y+Az)+\tilde{z}]+z = R[DR_AR(y+Az)+\tilde{z}]+z = (I-F)R_A(Ry+RAz)+R\tilde{z}+z = (R_A- -F)[(Ry+z)-(I-RA)z]+R\tilde{z}+z = R_A(Ry+z)-z-R_A(I-RA)z+ +F(I-RA)z+R\tilde{z}+z = R_A(Ry+z)+[I-R_A(I-RA)]z+Rz = R_A(Ry+ +z)+F_Az+R\tilde{z}$, where F_A is an initial operator for $I-RA$ corresponding to its right inverse R^A. Hence $F_A z = \tilde{z}_1 \in \ker(I - RA)$. If $\tilde{z} \in \ker(I - AR)$ then $\tilde{z} = AR\tilde{z}$ and $R\tilde{z} = RAR\tilde{z}$. Hence $(I - RA)R\tilde{z} = 0$, i.e. $\tilde{z}_2 = R\tilde{z} \in \ker(I - RA)$. We therefore conclude that $x = R_A(Ry + z) + \tilde{z}_0$, where $\tilde{z}_0 = \tilde{z}_1 + \tilde{z}_2 \in \ker(I - RA)$, i.e. $x \in Z(R_A)$. The arbitrariness of $\tilde{z} \in \ker(I - AR)$ implies that solutions of Equation (3.1.9) belonging to the sets $Z(R_A)$ and $Z'(R^A)$ are the same.

(ii) Suppose that $A \ker D \subset (I - AR)X$, $I - RA \in \Lambda(X)$. Let $L_A \in \mathscr{L}_{I-RA}$ be arbitrarily fixed. Theorem 3.1.3 implies that $I - AR \in \Lambda(X)$ and $L^A = DL_A R \in \mathscr{L}_{I-AR}$. Let $x \in Z'_1(L^A)$. Then x is of the form: $x = RL^A(y + Az) + z$. Lemma 3.1.2 implies that $y + Az \in (I - AR)X$ and $L_A(y + Az) \in L_A(I - AR)X = X$ for all $y \in (D - A)\operatorname{dom} D$ and $z \in \ker D$. Let F be an initial operator for D corresponding to R. Then $x = RL^A(y + Az) + z = RDL_A R(y + Az) + z = (I - F)L_A[(Ry + z) - (I - RA)z] + z = L_A(Ry + z) - FL_A(Ry + z) - (I - F)L_A(I - RA)z + z = L_A(Ry + z) - FL_A(Ry + z) - (I - F)z + z = L_A(Ry + z) - F(Ry + z) - z + z + z = L_A(Ry + z) - Fz + z = L_A(Ry + z) \in Z'_1(L^A)$. ∎

The following example shows that not always $A \ker D \subset (I - AR)X$.

EXAMPLE 3.1.2 (cf. Pogorzelec, 1983). Let X, D, R be defined as in Example 3.1.1. Let $A\{x_n\} = \{x_{n+1} - x_n\}$. Then $(I - RA)\{x_n\} = \{u_n\}$, where $u_1 = u_2 = x_1$, $u_n = x_{n-1}$ for $n \geqslant 3$ and $(I - AR)\{x_n\} = \{v_n\}$, where $v_1 = 0$, $v_n = x_{n-1}$ for $n \geqslant 2$. If $z \in \ker D$ then $z = \{z_n\}$, where $z_1 \neq 0$ and $z_n = 0$ for $n \geqslant 2$. Then $Az = \{-z_n\} = -z$. Moreover, $\ker(I - RA) = \ker(I - AR) = \{0\}$, $(I - RA)X \underset{\neq}{\subseteq} X$, $(I - AR)X \underset{\neq}{\subseteq} X$ and $(I - AR)X \cap A \ker D = \{0\}$. Thus $A \ker D \not\subset (I - AR)X$. Observe that here we have $A \ker D = \ker D$. This show that also $\ker D \not\subset (I - AR)X$.

Recently A. Pogorzelec (1984) considered Equation (3.1.9) with the operator A such that $\operatorname{dom} A \neq X$ in the following cases:

(i) $\operatorname{dom} D \subset \operatorname{dom} A$;

(ii) $\operatorname{dom} D \not\subset \operatorname{dom} A$ but $\operatorname{dom} D \cap \operatorname{dom} A \neq \{0\}$.

Examples and Exercises

EXAMPLE 3.1.3. Suppose that $D \in R(X)$, $\dim \ker D \neq 0$, F is an initial operator for D corresponding to a Volterra right inverse R and $F_1 \neq F$ is an initial operator for D such that $F_1 e_\lambda(z) = 0$ for some $0 \neq z \in \ker D$,

$\lambda \in \mathbf{C}$, where $e_\lambda = (I - \lambda R)^{-1}$ (cf. Theorems 2.3.3 and 2.3.4). Let $A = \lambda(I - F_1)$ with arbitrarily fixed λ. Consider the operators $D - A$ and $\hat{R} = I - \lambda R + \lambda F_1 R = I - AR$. By Formula (3.1.4) we have $(D - A)R = I - AR = \hat{R}$. Condition $F_1 e_\lambda(z) = 0$ implies that ker $\hat{R} \neq \{0\}$. Indeed, since $R \in V(X)$ we have $\hat{R} e_\lambda(z) = (I - \lambda R + \lambda F_1 R) e_\lambda(z) = (I - \lambda R) e_\lambda(z) + F_1 R e_\lambda(z) = z + F_1 [e_\lambda(z) - z] = z + F_1 e_\lambda(z) - F_1 z = z - z = 0$, i.e. $e_\lambda(z) \in$ ker $\hat{R}$.

Let $y \in X$ be arbitrarily fixed. Consider the equation $\hat{R} x = y$. Since $R \in V(X)$ we have $y = \hat{R} x = (I - \lambda R) x + \lambda z$, where $z = F_1 R x \in$ ker D and $x = -\lambda e_\lambda(z) + e_\lambda(y)$. On the other hand, if for a $z \in$ ker D we have $x = -\lambda e_\lambda(z) + e_\lambda(y)$ then $(I - \lambda R) x = -\lambda z + y$, i.e. $y = \hat{R} x = (I - \lambda R) x + \lambda F_1 R x = -\lambda z + y + \lambda F_1 R x$ and $z = F_1 R x$. Since the equation $\hat{R} x = y$ has a solution for every $y \in X$, we conclude that $\hat{R}$ is a mapping onto, i.e. $\hat{R} \in R(X)$.

EXAMPLE 3.1.4. Let X, D, R, F be defined as in Example 3.1.1. Let $A_0, A_1 \in L_0(X)$ and let $A_L \in \mathscr{L}_{A_0}$ (i.e. $A_L A_0 = I$) and dom $A_L \subset A_1 X$. Consider the operator $\hat{D} = A_0 D - A_1$. Then $A_L \hat{D} = A_L(A_0 D - A_1) = A_L A_0 D - A_L A_1 = D - A$, where the operator $A = A_L A_1$ is well-defined. We therefore have a reduction to the case considered in the present section, provided that $A = A_L A_1$ is either right or left invertible or invertible.

EXAMPLE 3.1.5. Let $X = C[0, T]$, $T > 0$ (over $\mathbf{C}$). Consider the equation

$$(D - \lambda I) = y, \quad \text{where } D = \frac{d}{dt}, y \in (D - \lambda I) \text{dom } D = X.$$

Let $R = \int_0^t$. Formula (3.1.3) implies that $D - \lambda I = D(I - \lambda R)$. But $R \in V(X)$ (cf. Formula (2.2.41)) and $[(I - \lambda R)^{-1} u](t) = u(t) + \int_0^t e^{\lambda(t-s)} u(s) ds$ for $u \in X$. Point 3 of Theorem 3.1.1 implies that $x = (I - \lambda R)^{-1}(Ry + z)$, where $z \in$ ker D is arbitrary, i.e.

$$x(t) = \int_0^t y(s) ds + c + \lambda \int_0^t e^{\lambda(t-s)} \left[\int_0^s y(\sigma) d\sigma + c \right] ds.$$

By our assumptions, $y \in C[0, T]$, hence the function $\int_0^t y(s) \, ds$ $\in C^1[0, T]$. We therefore conclude that $x \in C^1[0, T] = \text{dom } D$. If we assume that $y \in (D - \lambda I) \text{dom } D^k$ for a $k > 1$, we conclude that y $\in C^{k-1}[0, T]$, hence $\int_0^t y(s) \, ds \in C^k[0, T]$ and $x \in C^k[0, T] = \text{dom } D^k$.

EXAMPLE 3.1.6. Let $X = C[0, T]$, $T > 0$ (over $\mathbf{C}$). Consider the equation

$$(D - \lambda_k I)x = y, \quad \text{where } D = d^2/dt^2,$$

$$\lambda_k = -\mu_k^{-2}, \quad \mu_k = \frac{\pi}{t_1 - t_0}(2k + 1) \quad (k = 0, \pm 1, \pm 2, \ldots)$$

$$(3.1.32)$$

$t_0, t_1 \in [0, T]$. Write

$$(Rx)(t) = \int_{t_0}^t \left(\int_{t_1}^s x(u) \, du \right) ds.$$

It is easy to verify that $D \in R(X)$, $R \in \mathcal{R}_D$ and $\ker D = \{c_1 t + c_2 : c_1, c_2 \in \mathbf{C}\}$. Then $D - \lambda_k I = D(I - \lambda_k R)$. Moreover, the operator R has the following properties (cf. Exercise 2.3.7):

(i) if $t_1 = t_0$ then R is a Volterra operator;

(ii) if $t_1 \neq t_0$ and $\mu \in \mathbf{R}$ then $\ker(I - \mu^2 R) = \{0\}$ and $I - \mu^2 R$ is invertible;

(iii) if $t_1 \neq t_0$ then all λ_k defined by Formula (3.1.32) are eigenvalues of R and $\ker(I - \lambda_k R) = \{c_1 \cos \mu_k t + c_2 \sin \mu_k t : c_1, c_2 \in \mathbf{C}\}$. Hence $I - \lambda_k R$ is not invertible. Since $\ker(I - \lambda_k R) \neq \{0\}$, the operator $I - \lambda_k R$ is also not left invertible. However, the operator $I - \lambda_k R$ is right invertible. Indeed, the equation $(I - \lambda_k R)u = v$ has a solution for every $v \in X$, namely, $x = (I - i\mu_k R_0)^{-1}(I + i\mu_k R_0)^{-1}(v + c_1 t + c_2)$, where $c_1, c_2 \in \mathbf{C}$ are arbitrary, $R_i = \int_{t_i}^t$ $(i = 0, 1)$ are Volterra right inverses of the operator d/dt and

$$[(I \pm i\mu_k R_0)^{-1} w](t) = w(t) \mp i\mu_k \int_{t_0}^t e^{\pm i\mu_k(t-s)} w(s) \, ds,$$

for $w \in X$ (cf. Formula (2.2.41)). The proof is following: let F_0 be an

initial operator for d/dt corresponding to R_0. Then $(R_1 - R_0)u = F_0 R_1 u$
$= z \in \ker d/dt$, i.e. $[(R_1 - R_0)u](t) = c \in \mathbf{C}$. Hence $(I - \lambda_k R)u = (I +$
$+ \mu_k^2 R_0 R_1)u = (I + \mu_k^2 R_0^2)u + \mu_k^2 R_0 (R_1 - R_0)u = (I + i\mu_k R_0)(I - i\mu_k R_0)u +$
$+ \mu_k^2 R_0 c = (I + i\mu_k R_0)(I - i\mu_k R_0)u - c_1 t - c_2$ and $(I + i\mu_k R_0)(i - i\mu_k R_0)u$
$= v + c_1 t + c_2$, where c_1, c_2 are arbitrary.

EXAMPLE 3.1.7. Recall that X is a Leibniz algebra (cf. Example 2.1.7)
if X is a commutative linear ring and $D \in R(X)$ satisfies the Leibniz
condition (2.1.15):

$$D(xy) = xDy + yDx \quad \text{for } x, y \in \mathrm{dom}\, D.$$

Suppose that $D \in R(X)$, where X is a Leibniz algebra. Consider
the equation

$$Dx + px = q, \quad \text{where } p, q \in X. \tag{3.1.33}$$

Suppose that $u \neq 0$ satisfies the homogeneous equation $Du + pu = 0$,
i.e. $u \in \ker(D + p)$. We are looking for solutions of the form $x = uv$,
where $v \in \mathrm{dom}\, D$. We have the following equalities:

$$q = Dx + px = D(uv) + puv = uDv + vDu + puv$$
$$= v(Du + pu) + uDv = uDv.$$

Hence, if the element u is invertible, then we find $Dv = qu^{-1}$ and
$v = R(qu^{-1}) + z$, where R is a right inverse of D and $z \in \ker D$ is arbitrary. We therefore conclude that

$$x = uv = uR(qu^{-1}) + z, \quad \text{where } z \in \ker D \text{ is arbitrary.}$$

This method of solving Equation (3.1.33) is traditionally called the
constants variation method.

EXAMPLE 3.1.8. Consider a *linear differential equation* (of order 1):

$$x'(t) + p(t)x(t) = q(t), \tag{3.1.34}$$

where $p, q \in X = C[a, b]$. We should mention here that differential
equations with an unknown function of one variable, i.e. equations
with the operator $D = d/dt$ are called *ordinary differential equations*.
Differential equations with unknown functions of several variables are
called *partial differential equations*, since they contain partial differential
operators like $\partial/\partial t$, $\partial/\partial s$, and so on.

Recall that $X = C[a, b]$ is a Leibniz algebra with respect to the operator $D = d/dt$ (cf. Example 2.1.8). We therefore can solve Equation (3.1.34) in the same manner as Equation (3.1.33) in Example 3.1.7. In order to do it, we have to find all non-trivial solutions of the homogeneous equation $u' + pu = 0$ (for this equation is, of course, satisfied by the function $u = 0$). This last equation can be written as $d \ln u/dt = u'/u = -p$. This implies that $\ln u = -P + C_1$, where P is a primitive function for p, i.e. $P'(t) = p(t)$, and C_1 is an arbitrary constant. It is convenient to introduce a new constant $C_2 = \exp C_1$ if we write "$\exp y(t)$" instead of $e^{y(t)}$. Then $C_1 = \ln C$ and $\ln u(t) = -P(t) + \ln C_2 = \ln C_2 \exp(-P(t))$, which implies $u(t) = C_2 \exp(-P(t))$. Observe that we obtain a trivial solution $u = 0$ if we put $C_2 = 0$. To apply a solution obtained in Example 3.1.7 it is enough to know only one solution of the homogeneous equation. Hence we can put $C_2 = 1$ and we find $x = u[R(qu^{-1})] + C$, where C is an arbitrary constant and, for instance, $R = \int_a^t$, i.e. $x(t) = \exp(-P(t))[Q(t) + C] = [R(qu^{-1})](t)$, where C is an arbitrary constant and $Q(t)$ is a primitive function for the function $q(t)[u(t)]^{-1} = q(t) \exp P(t)$.

In the theory of ordinary differential equations traditionally an arbitrary primitive function and an indefinite integral are denoted by the same symbol. This does not lead to any misunderstanding if we remember the sense of the symbol used and if, whenever it is necessary, we add an arbitrary constant. Using this denotation we may write general solution of Equation (3.1.34) in the form

$$x(t) = \exp\left(-\int p(t)\,dt\right)\left[\int q(t) \exp\left(\int p(t)\,dt\right)dt + c\right]$$

or

$$x(t) = e^{-\int p(t)\,dt}\left[\int q(t) e^{\int p(t)\,dt}\,dt + c\right].$$

For instance, general solution of the equation $x' + (\cos t)x = \cos t$ is of the form

$$x(t) = (e^{\sin t} + c)e^{-\sin t} = 1 + ce^{-\sin t},$$

where c is an arbitrary constant.

EXERCISE 3.1.1. Let X be the space (s) of all sequences $x = \{x_n\}$, where $x_n \in \mathbf{R}$, $n \in \mathbf{N}$, with the addition and multiplication by scalars defined as follows: $\{x_n\} + \{y_n\} = \{x_n + y_n\}$, $\lambda\{x_n\} = \{\lambda x_n\}$ for $\{x_n\}$, $\{y_n\} \in X$, $\lambda \in \mathbf{R}$. Let $D = \{x_{n+1} - x_n\}$ (cf. Examples 2.1.4 and 2.3.3 and Exercise 2.1.9). Prove that the equation $(D - \lambda I)x = y$, $y \in X$, $\lambda \in \mathbf{R}$, has a general solution of the form $x = (I - \lambda R)^{-1}(Ry + z)$, where $z \in \ker D$ is arbitrary, R and $\ker D$ are determined in Example 2.1.4 and $(I - \lambda R)^{-1}$—in Example 2.3.3.

EXERCISE 3.1.2. Suppose that X, D, R are defined as in Exercise 3.1.1. Let $S\{x_n\} = \{x_{n-1}\}$, where we assume $x_0 = 0$ for all $x \in X$. Prove that:

(i) the operator $I - RS$ is invertible in X;

(ii) all solutions of the equation $(D - S)x = y$, $y \in X$, are of the form $x = (I - RS)^{-1}(Ry + z)$, where $z \in \ker D$ is arbitrary.

EXERCISE 3.1.3. Let $X = (s)$ (cf. Exercise 3.1.1). Let $S_l\{x_n\} = \{x_{n+1}\}$, $P\{x_n\} = \{\sum\limits_{k=1}^{n+2} x_k\}$ for $\{x_n\} \in X$. Prove that:

(i) $S_l \in R(X)$ and has a right inverse S_r defined as follows: $S_r\{x_n\} = \{x_{n-1}\}$, where $x_0 = 0$ for $x \in X$;

(ii) the operator $I - PS_r$ is right invertible;

(iii) all solutions of the equation $(S_l - P)x = y$, $y \in X$, are of the form $x = S_r[R^p(w + Pz) + u] + z$, where $R^p \in \mathcal{R}_{I-PS_r}$, $z \in \ker S_l$, $u \in \ker(I - PS_r)$ are arbitrary. Determine these solutions.

EXERCISE 3.1.4. Let X, S_l, S_r be defined as in Exercise 3.1.3. Let $A\{x_n\} = \{x_{n+1} - x_{n-1}\}$, where $x_0 = 0$ for $x \in X$. Prove that:

(i) The operator $I - AS_r$ is left invertible;

(ii) Find all $y \in X$ and $z \in \ker D$ such that $y + Az \in (I - AS_r)X$;

(iii) Prove that the equation $(S_l - A)x = y$ for y satisfying the condition of Point (ii) has all solutions of the form $x = S_r L^A(y + Az) + z$, where $L^A \in \mathcal{L}_{I-AS_r}$ and $z \in \ker D$ are arbitrary. Determine these solutions.

EXERCISE 3.1.5. Suppose that X, D, R are defined as in Exercise 3.1.1. Let $A\{x_n\} = \{x_{n+1}\}$ for $x_n \in X$. Prove that:

(i) The operator $I-RA$ is left invertible;

(ii) All solutions of the equation $(D-A)x = y$, $y \in X$, are of the form $x = L_A(Ry+z)$, where $L_A \in \mathscr{L}_{I-RA}$ and $z \in \ker D$ are arbitrary. Determine these solutions.

EXERCISE 3.1.6. Let X and S_l be defined as in Example 3.1.3. Prove that:

(i) The operator R defined as follows: $R\{x_n\} = \{y_n\}$, where $y_1 = x_1$, $y_n = x_{n-1}$ for $n \geqslant 2$ is a right inverse of S_l;

(ii) The operator $I-R$ is right invertible;

(iii) $Ry+z \in (I-RA)$ dom D for these $z = \{z_n\} \in \ker D$ for which $z_1 = -y_1$, otherwise $Ry+z \notin (I-RA)$ dom D, i.e. $(I-RA)$ dom $D \neq X$.

EXERCISE 3.1.7. Let X, S_l, R be defined as in Exercise 3.1.6. Let $A\{x_n\} = \{y_n\}$, where $y_1 = x_1$, $y_n = x_{n+1} - x_n$ for $n \geqslant 2$. Prove that:

(i) The operator $I-RA$ is left invertible;

(ii) $Ry+z \in (I-RA)$ dom D for these $z = \{z_n\} \in \ker D$ for which $z_1 = -y_1$, otherwise, $Ry+z \notin (I-RA)$ dom D.

EXERCISE 3.1.8. Let X and S_l be defined as in Exercise 3.1.3. Let $A\{x_n\} = \{u_n\}$ for $x \in X$, where $u_1 = x_2$, $u_2 = x_1$, $x_n = x_{n+1} - x_{n-1}$ for $n \geqslant 3$. Prove that:

(i) The operator R defined as follows: $R\{x_n\} = \{y_n\}$, where $y_1 = x_2$, $y_n = x_{n-1}$ for $n \geqslant 2$, $x \in X$, is a right inverse of S_l;

(ii) the operator $I-RA$ is right invertible;

(iii) $Ry+z \in (I-RA)$ dom D for all y such that $y_1 = 0$ and $\{z_n\} \in \ker D$ for which $z_1 = -y_2$, otherwise, $Ry+z \notin (I-RA)$ dom D.

EXERCISE 3.1.9. Let X, S_l, R be defined as in Exercise 3.1.8. Let $A\{x_n\} = \{y_n\}$ for $x \in X$, where $y_1 = x_2$, $y_2 = x_1 - x_2$, $y_n = 0$ for $n \geqslant 3$. Prove that:

(i) The operator $I-RA$ is right invertible;

(ii) $\ker D \subset (I-RA)$ dom $D \neq X$;

(iii) $Ry \in (I-RA)$ dom D for all y such that $y_2 \neq 0$, $y_n = 0$ otherwise;

(iv) $Ry+z \in (I-RA)$ dom D for all $z \in \ker D$.

EXERCISE 3.1.10. Suppose that X, S_l are defined as in Exercise 3.1.3. Let $A\{x_n\} = \{y_n\}$ for $x \in X$, where $y_1 = x_2$, $y_n = x_{n+1} - (x_n - x_{n-1})$ for $n \geqslant 2$. Prove that:

(i) The operator R defined as follows: $R\{x_n\} = \{y_n\}$ for $x \in X$, where $y_1 = 0$, $y_n = x_{n-1}$ for $n \geqslant 2$ is a right inverse of S_l;

(ii) The operator $I - RA$ is left invertible;

(iii) $Ry \in (I - RA)$ dom D for all $y \in X$ such that $y_1 = 0$;

(iv) ker $D \subset (I - RA)X$;

(v) $Ry + z \in (I - RA)$ dom D for all $z \in$ ker D and $y \in X$ such that $y_1 = 0$.

EXERCISE 3.1.11. Let $D \in R(X)$, $R \in \mathcal{R}_D$ and $A \in L_0(X)$. Let $I - RA$ transforms X onto X. Prove that:

(i) If $I - RA$ is invertible then $Ry + z \in (I - RA)$ dom D for all $y \in X$, $z \in$ ker D, but this condition is not necessary;

(ii) A necessary condition for the operator $I - RA$ to be invertible is that ker $D \subset (I - RA)$ dom D.

EXERCISE 3.1.12. Solve the *Bernoulli differential equation*

$$\frac{dx}{dt} + p(t)x + q(t)x^n = 0, \tag{3.1.35}$$

where $p, q \in C[a, b]$, $n \neq 1$, by introduction of a new unknown function and applications of Formula (3.1.34) in Example 3.1.8.

3.2. EQUATIONS OF HIGHER ORDER WITH NON-COMMUTATIVE COEFFICIENTS

A generalization of Formulae (3.1.3) and (3.1.4) for polynomials (of higher order) in a right invertible operator with non-commutative coefficients is the following:

THEOREM 3.2.1 (on similarities). *Suppose that $D \in R(X)$, dim ker $D \neq 0$ and $F_0, \ldots, F_{N-1}$ are initial operators for D corresponding to its right inverses $R_0, \ldots, R_{N-1}$. Consider polynomials*

$$Q(D) = \sum_{k=0}^{N} Q_k D^k \quad and \quad Q\langle D\rangle = \sum_{k=0}^{N} D^k Q_k, \tag{3.2.1}$$

where $Q_0, ..., Q_{N-1} \in L_0(X)$, $Q_N = I$. Moreover, in the case of the polynomial $Q\langle D \rangle$ we shall assume that

$$Q_k X \subset \text{dom } D^N \quad (k = 0, 1, ..., N-1). \tag{3.2.2}$$

Then the following identities hold:

$$Q(D) R_0 ... R_{N-1} = S_1 \quad \text{on } X, \tag{3.2.3}$$

$$Q(D) S_2 = S_1 D^N \quad \text{on dom } D^N, \tag{3.2.4}$$

$$Q\langle D \rangle = D^N S_3 \quad \text{on } X, \tag{3.2.5}$$

$$R_{N-1} ... R_0 Q\langle D \rangle = S_3 - S_4 \quad \text{on } X, \tag{3.2.6}$$

where

$$S_1 = I + \sum_{k=0}^{N-1} Q_k R_k ... R_{N-1},$$

$$\tag{3.2.7}$$

$$S_2 = I - F_0 - \sum_{k=1}^{N-1} R_0 ... R_{k-1} F_k D^k;$$

$$S_3 = I + \sum_{k=0}^{N-1} R_{N-1} ... R_k Q_k, \tag{3.2.8}$$

$$S_4 = \sum_{k=2}^{N-1} \left(\sum_{j=1}^{k-1} R_{N-1} ... R_{k-j} F_{k-j} D^j \right) Q_k -$$

$$- F_0 - \sum_{k=1}^{N-1} R_{N-1} ... R_k F_0 Q_k. \tag{3.2.9}$$

Proof. By our assumptions we have

$$Q(D) R_0 ... R_{N-1} = \sum_{k=0}^{N} Q_k D^k R_0 ... R_{N-1}$$

$$= Q_0 R_0 ... R_{N-1} + \sum_{k=1}^{N-1} Q_k (D^k R_0 ... R_{k-1}) R_k ... R_{N-1} +$$

$$+ D^N R_0 ... R_{N-1}$$

$$= Q_0 R_0 ... R_{N-1} + \sum_{k=1}^{N-1} Q_k R_k ... R_{N-1}$$

$$= \sum_{k=0}^{N-1} Q_k R_k ... R_{N-1} + I = S_1.$$

The Taylor–Gontcharov Formula (2.2.6) and Formula (3.2.3), together imply

$$Q(D)S_2 = Q(D)\left(I - F_0 - \sum_{k=1}^{N-1} R_0 \ldots R_k F_k D^k\right) Q(D)$$

$$= Q(D)R^N D^N = S_1 D^N \quad \text{on dom } D.$$

Furthermore, we find

$$D^N S_3 = D^N\left(I + \sum_{k=0}^{N-1} R_{N-1} \ldots R_k Q_k\right)$$

$$= D^N + \sum_{k=0}^{N-1} D^N R_{N-1} \ldots R_k Q_k$$

$$= D^N + \sum_{k=0}^{N-1} D^k Q^k = \sum_{k=0}^{N} D^k Q_k = Q\langle D\rangle$$

for $Q_N = I$. This, and Taylor–Gontcharov formula (2.2.6), together imply that

$$R_{N-1} \ldots R_0 Q\langle D\rangle = \sum_{k=0}^{N} R_{N-1} \ldots R_0 D^k Q_k$$

$$= R_{N-1} \ldots R_0 Q_0 + \sum_{k=1}^{N-1} R_{N-1} \ldots R_k (R_{k-1} \ldots R_0 D^k) Q_k +$$

$$+ R_{N-1} \ldots R_0 D^N$$

$$= R_{N-1} \ldots R_0 Q_0 + R_{N-1} \ldots R_1 (I - F_0) Q_1 +$$

$$+ \sum_{k=2}^{N-1} R_{N-1} \ldots R_k \left(I - F_0 - \sum_{j=1}^{k-1} R_{k-1} \ldots R_{k-j} F_{k-j} D^j\right) Q_k +$$

$$+ I - F_0 - \sum_{k=1}^{N-1} R_{N-1} \ldots R_{N-k} F_{N-k} D^k$$

$$= R_{N-1} \ldots R_0 Q_0 + R_{N-1} \ldots R_1 Q_1 + \sum_{k=2}^{N-1} R_{N-1} \ldots R_k Q_k -$$

$$- R_{N-1} \ldots R_1 F_0 Q_1 -$$

$$- \sum_{k=2}^{N-1} R_{N-1} \ldots R_k \left(F_0 - \sum_{j=1}^{k-1} R_{k-1} \ldots R_{k-j} F_{k-j} D^j \right) Q_k +$$

$$+ I - F_0 - \sum_{k=1}^{N-1} R_{N-1} \ldots R_{N-k} F_{N-k} D^k$$

$$= I + \sum_{k=0}^{N-1} R_{N-1} \ldots R_k Q_k - \sum_{k=1}^{N-1} R_{N-1} \ldots R_k F_0 Q_k - F_0 +$$

$$+ \sum_{k=2}^{N-1} R_{N-1} \ldots R_k \left(\sum_{j=1}^{k-1} R_{k-1} \ldots R_{k-j} F_{k-j} D^j \right) Q_k$$

$$= S_3 - F_0 - \sum_{k=1}^{N-1} R_{N-1} \ldots R_k F_0 Q_k +$$

$$+ \sum_{k=2}^{N-1} \sum_{j=1}^{k-1} \left(\sum R_{N-1} \ldots R_{k-j} F_{k-j} D^j \right) Q_k = S_3 - S_4. \qquad \blacksquare$$

COROLLARY 3.2.1 (on similarities). *Suppose that $D \in R(X)$, dim ker $D \neq 0$ and F is an initial operator for D corresponding to an $R \in \mathcal{R}_D$. Assume that $Q(D)$ and $Q\langle D \rangle$ are defined by Formula* (3.2.1) *and that Condition* (3.2.2) *is satisfied. Write*

$$Q(t, s) = \sum_{k=0}^{N} Q_k t^k s^{N-k}, \quad Q\langle t, s \rangle = \sum_{k=0}^{N} s^{N-k} t^k Q_k, \qquad (3.2.10)$$

where t, s are parameters, and

$$Q^0(s) = \sum_{k=0}^{N} Q_k s^{N-k} = Q(1, s),$$

$$\qquad (3.2.11)$$

$$Q^0\langle s \rangle = \sum_{k=0}^{N} {}' s^{N-k} Q_k = Q\langle 1, s \rangle.$$

Then the following identities hold:

$$Q(D) R^N = Q(I, R) = I + Q^0(R) \quad on \ X, \qquad (3.2.12)$$

$$Q(D)\left(I - \sum_{k=0}^{N-1} R^k F D^k\right) = Q(I, R) D^N \quad \text{on dom } D^N, \qquad (3.2.13)$$

$$Q\langle D\rangle = D^N Q\langle I, R\rangle = D^N [I + Q^0\langle R\rangle] \quad \text{on dom } D^N, \qquad (3.2.14)$$

$$R^N Q\langle D\rangle = Q\langle I, R\rangle - \sum_{k=0}^{N}\left(\sum_{j=0}^{k} R^{N-k+j} F D^j\right) Q_k \quad \text{on } X. \quad (3.2.15)$$

The proof follows immediately from Theorem 3.2.1 if we put $F_0 = \ldots = F_{N-1} = F$ and $R_0 = \ldots = R_{N-1} = R$ and we observe that, by definition,

$$Q(I, R) = I + Q^0(R), \quad Q\langle I, R\rangle = I + Q^0\langle R\rangle,$$
$$Q(D) = Q(D, I), \quad Q\langle D\rangle = Q\langle D, I\rangle. \qquad \blacksquare$$

The operators $Q(I, R)$ and $Q\langle I, R\rangle$ are said to be *resolving operators* for $Q(D)$. If either $Q(I, R)$ or $Q\langle I, R\rangle$ is invertible then the corresponding inverse is said to be *resolvent operator* for $Q(D)$.

Proposition 2.1.1, Corollary 2.2.3 and Lemma 3.1.1 together imply

PROPOSITION 3.2.1. *Suppose that* $D \in R(X)$, $\dim \ker D \neq 0$ *and* $R \in \mathcal{R}_D$. *Then all solutions of the equation*

$$D^m x = y, \quad y \in X \quad (m \geqslant 1). \qquad (3.2.16)$$

are of the form

$$x = R^m y + \sum_{k=0}^{m-1} R^k z_k, \qquad (3.2.17)$$

where $z_0, \ldots, z_{m-1} \in \ker D$ *are arbitrary*.

Proof. Indeed, if x is of the form (3.2.17) then $x \in \text{dom } D^m$. Moreover,

$$D^m x = D^m R^m y + \sum_{k=0}^{m-1} D^m R^k z_k = y + \sum_{k=0}^{m-1} D^{m-k} z_k = y. \qquad \blacksquare$$

THEOREM 3.2.2. *Suppose that* $D \in R(X)$, $\dim \ker D \neq 0$, $R \in \mathcal{R}_D$, *the polynomials* $Q\langle D\rangle$ *and* $Q\langle I, R\rangle$ *are defined by Formulae (3.2.1) and (3.2.11), respectively. Consider the equation*

$$Q\langle D\rangle x = y, \quad \text{where } y \in Q\langle I, R\rangle \text{dom } D^N \qquad (3.2.18)$$

and assume $Q\langle I, R\rangle \text{ dom } D^N \subset \text{dom } D^N$.

1. *If $Q\langle I, R\rangle \in R(X)$ and $R^Q \in \mathcal{R}_{Q\langle I, R\rangle}$ then:*
(i) *all solutions of Equation (3.2.18) are of the form*

$$x = R^Q \left(R^N y + \sum_{k=0}^{N-1} R^k z_k \right) + z, \qquad (3.2.19)$$

where $z_0, \ldots, z_{N-1} \in \ker D$, $z \in \ker Q\langle I, R\rangle$ are arbitrary;
(ii) $\dim \ker Q\langle D\rangle = N \dim \ker D + \dim \ker Q\langle I, R\rangle$;
(iii) *the sets of solutions are independent of the choice of $R^Q \in \mathcal{R}_{Q\langle I, R\rangle}$.*
2. *If $Q\langle I, R\rangle \in \Lambda(X)$ and $L^Q \in \mathcal{L}_{Q\langle I, R\rangle}$ then:*
(i) *all solutions of Equation (3.2.18) are of the form*

$$x = L^Q \left(R^N y + \sum_{k=0}^{N-1} R^k z_k \right), \qquad (3.2.20)$$

where $z_0, \ldots, z_{N-1} \in \ker D$ are arbitrary;
(ii) $\dim \ker Q\langle D\rangle = \dim L^Q \ker D^N$;
(iii) *these solutions are independent of the choice of $L^Q \in \mathcal{L}_{Q\langle I, R\rangle}$.*
3. *If $Q\langle I, R\rangle$ is invertible then all solutions of Equation (3.2.18) are of the form*

$$x = [Q\langle I, R\rangle]^{-1} \left(R^N y + \sum_{k=0}^{N-1} R^k z_k \right), \qquad (3.2.21)$$

where $z_0, \ldots, z_{N-1} \in \ker D$ are arbitrary and

$$\dim \ker Q\langle D\rangle = N \dim \ker D.$$

Proof. The identity (3.2.13) permits to rewrite Equation (3.2.18) in the equivalent form

$$D^N Q\langle I, R\rangle x = y, \qquad (3.2.22)$$

where, by our assumptions, $y = Q\langle I, R\rangle \operatorname{dom} D^N \subset \operatorname{dom} D^N$. Proposition 3.2.1 implies that

$$Q\langle I, R\rangle x = R^N y + \sum_{k=0}^{N-1} R^k z_k, \qquad (3.2.23)$$

where $z_0, \ldots, z_{N-1} \in \ker D$ are arbitrary. This means that $Q\langle I, R\rangle x \in \operatorname{dom} D^N$.

Suppose now that $Q\langle I, R\rangle \in R(X)$ and $R^Q \in \mathcal{R}_{Q\langle I, R\rangle}$ is arbitrarily fixed. Again, Proposition 3.2.1 implies that

$$x = R^Q\left(R^N y + \sum_{k=0}^{N-1} R^k z_k\right) + z,$$

where $z \in \ker Q\langle I, R\rangle$ is arbitrary,

$\dim \ker Q\langle D\rangle = \dim \ker D^N Q\langle I, R\rangle = N \dim \ker D + \dim \ker Q\langle I, R\rangle$. Corollary 3.1.5 implies that the set of solutions of Equation (3.2.18) is independent of the choice of a right inverse R^Q.

Suppose now that $Q\langle I, R\rangle \in \Lambda(X)$ and $L^Q \in \mathcal{L}_{Q\langle I, R\rangle}$ is arbitrarily fixed. Since $L^Q Q\langle I, R\rangle = I$, acting on both sides of Equation (3.2.23) by the operator L^Q we find

$$x = L^Q\left(R^N y + \sum_{k=0}^{N-1} R^k z_k\right), \quad \text{where } z_0, \ldots, z_{N-1} \text{ are arbitrary.}$$

Since all solutions of the homogeneous Equation (3.2.18) are of the form $L^Q \sum_{k=0}^{N-1} R^k z_k$, where $z_0, \ldots, z_{N-1} \in \ker D$ are arbitrary, we conclude that $\dim \ker Q\langle D\rangle = \dim L^Q \ker D^N$.

Corollary 3.1.6 implies that these solutions are independent of the choice of L^Q.

If the operator $Q\langle I, R\rangle$ is invertible, then acting on both sides of Equation (3.2.23) by the operator $[Q\langle I, R\rangle]^{-1}$ we find solutions of the form (3.2.21). The invertibility of the operator $Q\langle I, R\rangle$ implies that $\dim[Q\langle I, R\rangle]^{-1} \ker D^N = \dim \ker D^N = N \dim \ker D$. ∎

THEOREM 3.2.3. *Suppose that* $D \in R(X)$, $\dim \ker D \ne 0$, $R \in \mathcal{R}_D$, *the polynomials* $Q(D)$ *and* $Q(I, R)$ *are defined by Formulae (3.2.1) and (3.2.10), respectively. Consider the equation*

$$Q(D)x = y, \quad \text{where } y \in Q(I, R)X. \tag{3.2.24}$$

1. *If* $Q(I, R) \in R(X)$ *and* $R_Q \in \mathcal{R}_{Q(I, R)}$ *then:*
(i) *all solutions of Equation (3.2.24) are of the form*

$$x = R^N[R_Q(y - y_N) + z] + \sum_{k=0}^{N-1} R^k z_k \tag{3.2.25}$$

where $z \in \ker Q(I, R)$, $z_0, \ldots, z_{N-1} \in \ker D$ *are arbitrary and*

$$y_N = \sum_{j=0}^{N} Q_j \sum_{k=j}^{N-1} R^{k-j} z_k; \tag{3.2.26}$$

(ii) dim ker $Q(D)$ = dim ker $Q(I, R)$ + N dim ker D;

(iii) *the sets of solutions are independent of the choice of* $R_Q \in \mathscr{R}_{Q(I, R)}$.

2. *If* $Q(I, R) \in \varLambda(X)$ *and* $L_Q \in \mathscr{L}_{Q(I, R)}$ *then*:

(i) *all solutions of Equation* (3.2.24) *are of the form*

$$x = R^N L_Q(y - y_N) + \sum_{k=0}^{N-1} R^k z_k, \tag{3.2.27}$$

where $z_0, \ldots, z_{N-1} \in \ker D$ *are arbitrary and* y_N *is defined by Formula* (3.2.26);

(ii) dim ker $Q(D)$ = dim $L_Q \sum_{j=0}^{N} Q_j$ ker D^{N-j} + N dim ker D;

(iii) *the sets of solutions are independent of the choice of* L_Q.

3. *If* $Q(I, R)$ *is invertible then all solutions of Equation* (3.2.24) *are of the form*

$$x = R^N [Q(I, R)]^{-1}(y - y_N) + \sum_{k=0}^{N-1} R^k z_k, \tag{3.2.28}$$

where $z_0, \ldots, z_{N-1} \in \ker D$ *and* y_N *is defined by Formula* (3.2.26). *Moreover,* dim ker $Q(D) = N$ dim ker D.

Proof. We are looking for solutions belonging to the domain of the operator D^N, i.e. for solutions which are of the form

$$x = R^N u + \sum_{k=0}^{N-1} R^k z_k, \tag{3.2.29}$$

where $u \in X$ is to be determined and $z_0, \ldots, z_{N-1} \in \ker D$ are arbitrary. By definition $D^N x = u$. Formula (3.2.12) implies that

$$Q(D)x = Q(D)\left(R^N u + \sum_{k=0}^{N-1} R^k z_k\right)$$

$$= Q(D) R^N u + Q(D) \sum_{k=0}^{N-1} R^k z_k = Q(I, R)u + y_N,$$

where

$$y_N = Q(D) \sum_{k=0}^{N-1} R^k z_k = \sum_{j=0}^{N} Q_j D^j \sum_{k=0}^{N-1} R^k z_k$$

$$= \sum_{j=1}^{N} Q_j \left[\sum_{k=0}^{j-1} D^j R^k z_k + \sum_{k=j}^{N-1} D^j R^k z_k \right] + Q_0 \sum_{k=0}^{N-1} R^k z_k$$

$$= \sum_{j=1}^{N} Q_j \left[\sum_{k=0}^{j-1} D^{j-k} z_k + \sum_{k=j}^{N-1} R^{k-j} z_k \right] + Q_0 \sum_{k=0}^{N-1} R^k z_k$$

$$= \sum_{j=1}^{N} Q_j \sum_{k=j}^{N-1} R^{k-j} z_k + Q_0 \sum_{k=0}^{N-1} R^k z_k = \sum_{j=0}^{N} Q_j \sum_{k=j}^{N-1} R^{k-j} z_k,$$

i.e. y_N is defined by Formula (3.2.26).

We therefore can rewrite Equation (3.2.24) as follows:

$$Q(I, R)u = y - y_N. \tag{3.2.30}$$

Suppose that $Q(I, R) \in R(X)$ and $R_Q \in \mathcal{R}_{Q(I,R)}$. Proposition 3.2.1 implies that

$$u = R_Q(y - y_N) + z, \quad \text{where } z \in \ker Q(I, R) \text{ is arbitrary.}$$

This, and Formula (3.2.29), together imply Formula (3.2.15). We also have: $\dim \ker Q(D) = \dim \ker Q(I, R) + N \dim \ker D$. Corollary 3.1.2 implies that the set of solutions of Equation (3.2.24) is independent of the choice of R_Q.

Suppose now that $Q(I, R) \in \Lambda(X)$ and $L_Q \in \mathcal{L}_{Q(I,R)}$. Acting on both sides of Equation (3.2.29) by the operator L_Q we find $u = L_Q(y - y_N)$. This, and Formula (3.2.30), together imply Formula (3.2.27). We also have

$$\dim \ker Q(D)$$

$$= \dim L_Q \left\{ y_N = \sum_{j=0}^{N} Q_j \sum_{k=j}^{N-1} R^{k-j} z_k \colon z_0, \dots, z_{N-1} \in \ker D \right\} +$$

$$+ \dim \ker D^N$$

$$= \dim L_Q \sum_{j=0}^{N} Q_j \ker D^{N-j} + N \dim \ker D.$$

Observe that $\sum_{j=0}^{N} Q_j \ker D^{N-j}$ is not necessarily direct. Corollary

3.1.3 implies that the set of solutions of Equation (3.2.24) is independent of the choice of $L_Q \in \mathscr{L}_{Q(I,R)}$.

Suppose finally that the operator $Q(I, R)$ is invertible. Then $u = [Q(I, R)]^{-1}(y - y_N)$ and Formula (3.2.29) implies Formula (3.2.28). Moreover, dim ker $Q(D) = N$ dim ker D. ∎

We say that the equation

$$Q(D)x = y \quad (\text{respectively}: Q\langle D \rangle x = y)$$

is of the order N, if $Q(D)$ (respectively: $Q\langle D \rangle$) *is a polynomial in D of the degree N.* Observe that general solutions of an equation of the order N consists of N arbitrary constants, i.e. N arbitrary elements of the kernel of the operator D.

Similarly, we say that the operator $Q(D)$ (respectively: $Q\langle D \rangle$) *is of the order N if $Q(D)$* (respectively: $Q\langle D \rangle$) *is a polynomial in D of the degree N.*

THEOREM 3.2.4. *Suppose that all assumptions of Theorem 3.2.2 are satisfied. Let M be an arbitrarily fixed positive integer. Consider the equations*:

$$D^M Q\langle D \rangle x = y, \tag{3.2.31}$$
$$\{R^M y\} \oplus \ker D \subset Q\langle I, R \rangle \operatorname{dom} D^N \subset \operatorname{dom} D^N,$$

$$Q\langle D \rangle D^M x = y, \tag{3.2.32}$$
$$y \in Q\langle I, R \rangle \operatorname{dom} D^{N+M} \subset \operatorname{dom} D^{N+M}.$$

1. *If $Q\langle I, R \rangle \in R(X)$ and $R^Q \in \mathscr{R}_{Q\langle I, R \rangle}$ then*:
(i) *all solutions of Equation (3.2.31) are of the form*

$$x = R^Q \left(R^{N+M} y + \sum_{k=0}^{N+M-1} R^k z_k \right) + z, \tag{3.2.33}$$

where $z_0, \dots, z_{N+M-1} \in \ker D$ and $z \in \ker Q\langle I, R \rangle$ are arbitrary;
(ii) *all solutions of Equation (3.2.32) are of the form*

$$x = R^M \left[R^Q \left(R^N y + \sum_{m=0}^{N-1} R^m z_m \right) + z \right] + \sum_{m=0}^{M-1} R^m z_{N+m}, \tag{3.2.34}$$

where $z_0, \dots, z_{N+M-1} \in \ker D$ and $z \in \ker Q\langle I, R \rangle$ are arbitrary;

(iii) *in both cases the sets of solutions are independent of the choice of* R^Q *and*

$$\dim \ker D^M Q\langle D\rangle = \dim \ker Q\langle D\rangle D^M$$
$$= \dim \ker Q\langle I, R\rangle + (M+N)\dim \ker D.$$

2. *If* $Q\langle I, R\rangle \in \Lambda(X)$ *and* $L^Q \in \mathscr{L}_{Q\langle I, R\rangle}$ *then*:
(i) *all solutions of Equation* (3.2.31) *are of the form*

$$x = L^Q \left(R^{N+M} y + \sum_{k=0}^{N+M-1} R^k z_k \right), \tag{3.2.35}$$

where $z_0, \ldots, z_{N+M-1} \in \ker D$ *are arbitrary*;
(ii) *all solutions of Equation* (3.2.32) *are of the form*

$$x = R^M L^Q \left(R^N y + \sum_{k=0}^{N-1} R^k z_k \right) + \sum_{m=0}^{M-1} R^m z_{N+m}, \tag{3.2.36}$$

where $z_0, \ldots, z_{N+M-1} \in \ker D$ *are arbitrary*;
(iii) *in both cases the sets of solutions are independent of the choice of* L^Q *and*

$$\dim \ker D^M Q\langle D\rangle = \dim L^Q \ker D^{M+N},$$
$$\dim \ker Q\langle D\rangle D^M = \dim L^Q \ker D^N + M \dim \ker D.$$

3. *If* $Q\langle I, R\rangle$ *is invertible then*:
(i) *all solutions of Equation* (3.2.31) *are of the form*

$$x = [Q\langle I, R\rangle]^{-1} \left(R^{N+M} y + \sum_{k=0}^{N+M-1} R^k z_k \right), \tag{3.2.37}$$

where $z_0, \ldots, z_{N+M-1} \in \ker D$ *are arbitrary*;
(ii) *all solutions of Equation* (3.2.32) *are of the form*

$$x = R^M [Q\langle I, R\rangle]^{-1} \left(R^N y + \sum_{k=0}^{N-1} R^k z_k \right) + \sum_{m=0}^{M-1} R^m z_{N+m}, \tag{3.2.38}$$

where $z_0, \ldots, z_{N+M-1} \in \ker D$ *are arbitrary*;
(iii) *in both cases*

$$\dim \ker D^M Q\langle D\rangle = \dim \ker Q\langle D\rangle D^M = (M+N)\dim \ker D.$$

Proof. Proposition 3.2.1 implies that Equation (3.2.31) can be written as follows:

$$Q\langle D\rangle x = R^M y + \sum_{m=0}^{M-1} R^m z_m, \tag{3.2.39}$$

where $z_0, \ldots, z_{M-1} \in \ker D$ are arbitrary.

Observe that the right side belongs to $\{R^M y\} \oplus \ker D^M$. Applying Theorem 3.2.2 to Equation (3.2.31) we obtain Formulae (3.2.33), (3.2.35), (3.2.37). Indeed, if $Q\langle I, R\rangle \in R(X)$ and $R^Q \in \mathcal{R}_{Q\langle I, R\rangle}$ then, by Formula (3.2.19)

$$x = R^Q \left[R^N \left(R^M y + \sum_{m=0}^{M-1} R^m z_m \right) + \sum_{m=0}^{N-1} R^m z_m \right] + z$$

$$= R^Q \left(R^{N+M} y + \sum_{m=0}^{M-1} R^{N+m} z_m + \sum_{m=0}^{N-1} R^m z_m \right) + z,$$

where $z_M, \ldots, z_{M+N-1} \in \ker D$ and $z \in \ker Q\langle I, R\rangle$ are arbitrary, i.e.

$$x = R^Q \left(R^{N+M} y + \sum_{k=0}^{N+M-1} R^k z_k \right) + z.$$

In a similar way we obtain Formulae (3.2.35) and (3.2.37).

In order to solve Equation (3.2.32) write $u = D^M x$. Then, by Proposition 3.2.1

$$x = R^M u + \sum_{m=0}^{M-1} R^m z_{N+m}, \tag{3.2.40}$$

where $z_N, \ldots, z_{N+M-1} \in \ker D$ are arbitrary. Then Equation (3.2.32) can be written as follows:

$$Q\langle D\rangle u = y. \tag{3.2.41}$$

Now we apply Theorem 3.2.2 and we obtain Formulae (3.2.34), (3.2.36) and (3.2.38). Points 1(iii), 2(iii) and 3(iii) also follow from Theorem 3.2.2.   ∎

THEOREM 3.2.5. *Suppose that all assumptions of Theorem 3.2.3 are satisfied. Let M be an arbitrarily fixed positive integer. Let y_N be defined by Formula (3.2.26). Consider the equations:*

$$D^M Q(D)x = y, \quad y \in D^M Q(I, R)X, \tag{3.2.42}$$

$$Q(D)D^M x = y, \quad y \in Q(I, R)X. \tag{3.2.43}$$

1. *If* $Q(I, R) \in R(X)$ *and* $R_Q \in \mathscr{R}_{Q(I, R)}$ *then*:
(i) *all solutions of Equation* (3.2.42) *are of the form*

$$x = R^N \left[R_Q \left(R^M y - y_N + \sum_{m=0}^{M-1} R^m z_{N+m} \right) + z \right] + \sum_{k=0}^{N-1} R^k z_k, \tag{3.2.44}$$

where $z_0, \ldots, z_{N+M-1} \in \ker D$ *and* $z \in \ker Q(I, R)$ *are arbitrary;*
(ii) *all solutions of Equation* (3.2.43) *are of the form*

$$x = R^{N+M} [R_Q(y - y_N) + z] + \sum_{k=0}^{N+M-1} R^k z_k,$$

where $z_0, \ldots, z_{N+M-1} \in \ker D$ *and* $z \in \ker Q(I, R)$ *are arbitrary;*
(iii) *in both cases the sets of solutions are independent of the choice of* R_Q *and*

$$\dim \ker D^M Q(D) = \dim \ker Q(D)D^M$$
$$= \dim \ker Q(I, R) + (M+N)\dim \ker D.$$

2. *If* $Q(I, R) \in \Lambda(X)$ *and* $L_Q \in \mathscr{L}_{Q(I, R)}$ *then*:
(i) *all solutions of Equation* (3.2.42) *are of the form*

$$x = R^N L_Q \left(R^M y + \sum_{m=0}^{M-1} R^m z_{N+m} - y_N \right) + \sum_{k=0}^{N-1} R^k z_k, \tag{3.2.45}$$

where $z_0, \ldots, z_{N+M-1} \in \ker D$ *are arbitrary;*
(ii) *all solutions of Equation* (3.2.43) *are of the form*

$$x = R^{N+M} L_Q(y - y_N) + \sum_{k=0}^{N+M-1} R^k z_k, \tag{3.2.46}$$

where $z_0, \ldots, z_{N+M-1} \in \ker D$ *are arbitrary;*
(iii) *in both cases the sets of solutions are independent of the choice of* L_Q *and*

$$\dim \ker Q^M Q(D) = \dim \ker Q(D)D^M$$

$$= \dim L_Q \sum_{j=0}^{N} Q_j \ker D^{N-j} + (M+N)\dim \ker D.$$

3. *If $Q(I, R)$ is invertible then*:

(i) *all solutions of Equation (3.2.42) are of the form*

$$x = R^N [Q(I, R)]^{-1} \left(R^M y + \sum_{m=0}^{M-1} R^m z_{N+m} - y_N \right) + \sum_{k=0}^{N-1} R^k z_k,$$

$$(3.2.47)$$

where $z_0, \ldots, z_{N+M-1} \in \ker D$ are arbitrary;

(ii) *all solutions of Equation (3.2.43) are of the form*

$$x = R^{N+M} [Q(I, R)]^{-1} (y - y_N) + \sum_{k=0}^{N+M-1} R^k z_k, \qquad (3.2.48)$$

where $z_0, \ldots, z_{N+M-1} \in \ker D$ are arbitrary;

(iii) $\dim \ker D^M Q(D) = \dim \ker Q(D) D^M = (M+N) \dim \ker D$.

Proof. Suppose that $Q(I, R) \in R(X)$ and $R_Q \in \mathscr{R}_{Q(I, R)}$. In order to solve Equation (3.2.42) write: $u = Q(D)x$.

Point 1(i) of Theorem 3.2.3 implies that

$$u = R^N [R_Q (y - y_N) + z] + \sum_{k=0}^{N-1} R^k z_k,$$

where $z_0, \ldots, z_{N-1} \in \ker D$ and $z \in \ker Q(I, R)$ are arbitrary, provided that $u \in Q(I, R)X$. On the other hand, we can write Equation (3.2.42) in the form $D^M Q(D)x = D^M u = y$. Proposition 3.2.1 implies that $u = R^M y + \sum_{m=0}^{M-1} R^m z_{N+m}$, where $z_N, \ldots, z_{N+M-1} \in \ker D$ are arbitrary. It means that $u \in \mathrm{dom}\, D$. We therefore conclude that $u \in Q(I, R)X \cap \cap \mathrm{dom}\, D^M$ and

$$x = R^N [R^Q (y - y_N) + z] + \sum_{k=0}^{N-1} R^k z_k$$

$$= R^N \left[R^Q \left(R^M - y_N + \sum_{m=0}^{M-1} R^m z_{N+m} \right) + z \right] + \sum_{k=0}^{N-1} R^k z_k.$$

Observe that $u \in Q(I, R)X$ implies that $y = D^M u \in D^M Q(I, R)X$.

In order to solve Equation (3.2.43) we write $v = D^N x$ and we apply

Theorem 3.2.3 to the equation $Q(D) v = y$, where $y \in Q(I, R)X$. Proposition 3.2.1 implies that

$$x = R^M v + \sum_{m=0}^{M} R^m z_{N+m}$$

$$= R^M \left\{ R^N [R_Q(y-y_N)+z] + \sum_{k=0}^{N-1} R^k z_k \right\} + \sum_{m=0}^{M} R^m z_{N+m}$$

$$= R^{M+N}[R_Q(y-y_N)+z] + \sum_{k=0}^{N+M-1} R^k z_k,$$

where $z_0, \ldots, z_{N+M-1} \in \ker D$ and $z \in \ker Q(I, R)$ are arbitrary.

Point 1(iii) of Theorem 3.2.3 implies that the sets of solutions are, in both cases, independent of the choice of R_Q and that formulae for dimensions of kernels of operators under consideration are satisfied.

Suppose now that $Q(I, R) \in \Lambda(X)$ and $L_Q \in \mathscr{L}_{Q(I,R)}$. Point 2(i) of Theorem 3.2.3 implies that the equation $Q(D)x = u$ has a solution $x = R^N L_Q(u-y_N) + \sum_{k=0}^{N-1} R^k z_k$, where $z_0, \ldots, z_{N-1} \in \ker D$ are arbitrary. Since $D^M u = y$, Proposition 3.2.1 implies that

$$x = R^N L_Q \left(R^M y + \sum_{m=0}^{M-1} R^m z_{N+m} - y_N \right) + \sum_{k=0}^{N-1} R^k z_k,$$

where $z_N, \ldots, z_{N+M-1} \in \ker D$ are arbitrary.

Write now $D^M x = v$. Proposition 3.2.1 implies that $x = R^M v + \sum_{m=0}^{M-1} R^m z_m$, where $z_0, \ldots, z_{M-1} \in \ker D$ are arbitrary. Point 2(i) of Theorem 3.2.3 implies that the equation $Q(D) v = y$ has a solution

$$v = R^N L_Q(y-y_N) + \sum_{k=0}^{N-1} R^k z_{M+k},$$

where $z_M, \ldots, z_{M+N-1} \in \ker D$ are arbitrary. We therefore conclude that

$$x = R^M \left[R^N L_Q(y-y_N) + \sum_{k=0}^{N-1} R^k z_{M+k} \right] + \sum_{m=0}^{M-1} R^m z_m$$

$$= R^{N+M} L_Q(y-y_N) + \sum_{k=0}^{N+M-1} R^k z_k.$$

Point 2(ii) of Theorem 3.2.3 implies the formulae for dimensions of kernels of operators under consideration. Point 2(iii) of this Theorem implies that the sets of solutions are independent of the choice of L_Q.

Suppose now that the operator $Q(I, R)$ is invertible. In a similar way as in the preceding cases we find Formulae (3.2.47) and (3.2.48) and formulae for dimensions of kernels of operators under consideration.

∎

Examples and Exercises

EXAMPLE 3.2.1. Let X be a linear space, $D \in R(X)$, $\dim \ker D \neq 0$, $R \in \mathcal{R}_D$ and let $A, B \in L_0$, $\operatorname{dom} D \subset AX$. Consider the equation

$$(DAD+B)x = y, \quad y \in X \quad (\operatorname{dom} D \supset A\operatorname{dom} D). \qquad (3.2.49)$$

This equation can be written as $D(AD+RB)x = y$ which implies that

$$(AD+RB)x = Ry+z_1, \quad z_1 \in \ker D \text{ is arbitrary.} \qquad (3.2.50)$$

Write now $u = Dx$. Then $x = Ru+z_2$, where $z_2 \in \ker D$ is arbitrary and Equation (3.2.50) may be written as follows:

$$Ry+z_1 = (AD+RB)(Ru+z_2) = (AD+RB)Ru+RBz_2$$
$$= (A+RBR)u+RBz_2,$$

i.e.

$$(A+RBR)u = Ry+z_1+RBz_2. \qquad (3.2.51)$$

Supposethat $T = A+RBR \in R(X)$ and $R_T \in \mathcal{R}_T$. Then $u = R_T(Ry+ +z_1+z_2)+z$ where $z \in \ker T = \ker(A+RBR)$ is arbitrary, and

$$x = R[R_{A+RBR}(Ry+z_1+z_2)+z]+z_2,$$

where $z_1, z_2 \in \ker D$, $z \in \ker(A+RBR)$ are arbitrary.

Suppose that $T \in \Lambda(X)$ and $L_T \in \mathcal{L}_T$. Then

$$u = L_T(Ry+z_1+RBz_2)$$

and

$$x = RL_{A+RBR}(Ry+z_1+RBz_2)+z_2,$$

where $z_1, z_2 \in \ker D$ are arbitrary.

If the operator $T = A+RBR$ is invertible then

$$x = R(A+RBR)^{-1}(Ry+z_1+RBz_2)+z_2,$$

where $z_1, z_2 \in \ker D$ are arbitrary.

EXERCISE 3.2.3. Suppose that all assumptions of Example 3.2.1 are satisfied. Let M be an arbitrarily fixed positive integer. Give conditions of solvability and formulae for solutions for the following equations:

(i) $(D^2 + AD + B)^M x = y$,
$$y \in X.$$
(ii) $(D^2 + DA + B)^M x = y$,

EXERCISE 3.2.4. Suppose that X, D, R are defined as in Example 2.1.10. Solve the equation

$$\left(\sum_{j=1}^{n} a_j \frac{\partial}{\partial t_j}\right)^M x = y,$$

where $y \in X$, $a_1, \ldots, a_n \in \mathbf{R}$ are given, M is an arbitrary fixed positive integer. As before we assume that $a_1, \ldots, a_n$ do not vanish simultaneously.

EXERCISE 3.2.5. Suppose that all assumptions of Corollary 3.2.1 are satisfied. Let M be an arbitrarily fixed positive integer.

(i) Assume that the operator $Q(I, R)$ is invertible. Prove that the operators $Q(D)$, $D^M Q(D)$, $Q(D) D^M$ are right invertible by an application of Formula (3.2.12). Find corresponding right inverses.

(ii) Assume that the operator $Q\langle I, R\rangle$ is invertible. Prove that the operators $Q\langle D\rangle$, $D^M Q\langle D\rangle$, $Q\langle D\rangle D^M$ are right invertible by an application of Formula (3.2.14). Find corresponding right inverses.

3.3. EQUATIONS WITH STATIONARY COEFFICIENTS

Suppose that X is a linear space over an arbitrary field $\mathscr{F}$ of scalars and that $D \in R(X)$, $R \in \mathscr{R}_D$. An operator $A \in L_0(X)$ is said to be *stationary* if

$$DA - AD = 0 \quad \text{on dom } D, \quad RA - AR = 0 \qquad (3.3.1)$$

(cf. Tasche, 1977, who considered equations with right invertible operators and with stationary coefficients in Banach spaces).

PROPOSITION 3.3.1. *If A, $B \in L_0(X)$ are stationary operators then $A + B$, AB, λA, for all $\lambda \in \mathscr{F}$, are stationary operators (by definition, $A + B$ and AB are well-defined).*

Proof. Indeed, by our assumptions

$$DA-AD = 0, \quad DB-BD = 0 \text{ on dom } D,$$
$$RA-AR = 0, \quad RB-BR = 0,$$

for $D \in R(X)$ and $R \in \mathcal{R}_D$. Thus we have on dom D

$$D(A+B)-(A+B)D = DA+DB-AD-BD = 0,$$
$$D(AB)-(AB)D = DAB-ADB+ADB-ABD$$
$$= (DA-AD)B+A(DB-BD) = 0,$$
$$D(\lambda A)-\lambda AD = \lambda DA-\lambda AD = \lambda AD-\lambda AD = 0$$

$$\text{for all } \lambda \in \mathcal{F}.$$

We have similar proofs for commutators with R. ∎

For given $D \in R(X)$ and $R \in \mathcal{R}_D$ we denote by $\mathbf{S}_{D,R}$ the set of all stationary operators, i.e. the set of all operators which commute with D and R. An immediate consequence of Proposition 3.3.1 is

COROLLARY 3.3.1. *If $D \in R(X)$ and $R \in \mathcal{R}_D$ then the set $\mathbf{S}_{D,R}$ is a subalgebra of the algebra $L_0(X)$.*

By an easy induction we conclude that

$$\text{if } A \in \mathbf{S}_{D,R} \text{ then } A^n \in \mathbf{S}_{D,R} \text{ for } n = 1, 2, \ldots \quad (3.3.2)$$

THEOREM 3.3.1. *Suppose that $D \in R(X)$, $\dim \ker D \neq 0$, $R \in \mathcal{R}_D$, $A \in \mathbf{S}_{D,R}$ and the operator $I-RA = I-AR$ is invertible. Write*

$$E_A = (I-AR)^{-1}. \quad (3.3.3)$$

Then

$$DE_A = AE_A+D \quad \text{on dom } D, \quad (3.3.4)$$
$$FE_A = F, \quad (3.3.5)$$

where F is an initial operator for D corresponding to R.

Proof. Observe that on the domain of the operator D we have $D-A = D-DRA = D(I-RA) = D(I-AR)$ which implies $(D-A)E_A = (D-A)(I-AR)^{-1} = D(I-AR)(I-AR)^{-1} = D$, i.e. Equality (3.3.4).

Formula (3.2.51) implies that a condition of solvability of Equation (3.2.50) is the following:

$$\{Ry\} \oplus (\ker D + RB \ker D) \subset (A + RBR)X. \qquad (3.2.52)$$

Observe that the algebraic sum in the parantheses is not necessarily direct.

EXAMPLE 3.2.2. Suppose that all assumptions of Example 3.2.1 are satisfied. Consider the equation:

$$(D^2 + AD + B)x = y, \quad y \in (I + AR + BR^2)X.$$

Here we have: $Q(D) = D^2 + AD + B$, $Q(D)R^2 = I + AR + BR^2$ $= Q(I, R)$. Now we can apply Theorem 2.3.3.

EXAMPLE 3.2.3. Suppose that all assumptions of Example 3.2.1 are satisfied. Consider the equation

$$(D^2 + DA + B)x = y, \quad y \in (I + RA + R^2 B) \text{dom } D^2$$

$(\text{dom } D \supset A \text{ dom } D^2)$.

Here $Q\langle D \rangle = D^2 + DA + B = D^2(I + RA + R^2 B) = D^2 Q\langle I, R \rangle$ and we can apply Theorem 2.3.2.

EXAMPLE 3.2.4. Suppose that $D \in R(X)$, dim ker $D \neq 0$ and there is a Volterra right inverse R of D. Then every solution of the equation

$$(D^2 + \lambda^2 I)u = 0 \quad (0 \neq \lambda \in \mathbf{R}) \qquad (3.2.53)$$

is of the form

$$u = c_\lambda(z_0) + s_\lambda(z_1), \quad \text{where } z_0, z_1 \in \ker D \text{ are arbitrary} \quad (3.2.54)$$

and $c_\lambda(z_0)$, $s_\lambda(z_1)$ are cosine and sine elements, respectively (cf. Definition 2.3.3).

Indeed, suppose that u is of the form (3.2.54). Formulae (2.3.14) imply that

$$\begin{aligned}
(D^2 + \lambda^2 I)u &= (D^2 + \lambda^2 I)[c_\lambda(z_0) + s_\lambda(z_1)] \\
&= D^2 c_\lambda(z_0) + D^2 s_\lambda(z_1) + \lambda^2 c_\lambda(z_0) + \lambda^2 s_\lambda(z_1) \\
&= D[-\lambda s_\lambda(z_0)] + D[\lambda c_\lambda(z_1)] + \lambda^2 c_\lambda(z_0) + \lambda^2 s_\lambda(z_1) \\
&= -\lambda^2 c_\lambda(z_0) - \lambda^2 s_\lambda(z_1) + \lambda^2 c_\lambda(z_0) + \lambda^2 s_\lambda(z_1) = 0.
\end{aligned}$$

Conversely, suppose that u is a solution of Equation (3.2.53). Then we have

$$0 = (D^2 + \lambda^2 I)u = D^2(I + \lambda^2 R^2)u.$$

Write: $w = (I + w^2 R^2)u$. Then $D^2 w = 0$. Proposition 3.2.1 implies that $w = \lambda R z_1 + z_0$, where $z_0, z_1 \in \ker D$ are arbitrary. Thus $(I + \lambda^2 R^2)u = w = \lambda R z_1 + z_0$. Since the operator $I + \lambda^2 R^2 = (I + \lambda iR)(I - \lambda iR)$, as a superposition of invertible operators, is again invertible, we conclude that $u = (I + \lambda^2 R^2)^{-1} w = (I + \lambda^2 R^2)^{-1}(\lambda R z_1 + z_0) = \lambda R(I + \lambda^2 R^2)^{-1} z_1 + (I + \lambda^2 R^2) z_0 = c_\lambda(z_0) + s_\lambda(z_1)$, which was to be proved.

Observe that the arbitrariness of $z_0, z_1 \in \ker D$ implies the following equality

$$\ker(D^2 + \lambda^2 I) = c_\lambda \ker D \oplus s_\lambda \ker D. \tag{3.2.55}$$

This sum is direct. Indeed, let $c_\lambda(z_1) + s_\lambda(z_2) = 0$ for some $z_1, z_2 \in \ker D$. Then $(I + \lambda^2 R^2)^{-1}(z_1 + \lambda R z_2) = 0$ which implies $z_1 + \lambda R z_2 = 0$, hence $z_1 = z_2 = 0$.

EXERCISE 3.2.1. Suppose that all assumptions of Example 3.2.4 are satisfied. Prove that all solutions of the equation

$$(D^2 - \lambda^2 I)u = 0 \quad (0 \neq \lambda \in \mathbf{R}) \tag{3.2.56}$$

are of the form

$$u = e_\lambda(z_0) + e_{-\lambda}(z_1), \tag{3.2.57}$$

where $z_0, z_1 \in \ker D$ are arbitrary and $e_\lambda(z)$ are exponentials defined by Formula (2.3.2), i.e.

$$\ker(D^2 - \lambda^2 I) = e_\lambda \ker D \oplus e_{-\lambda} \ker D. \tag{3.2.58}$$

EXERCISE 3.2.2. Suppose that all assumptions of Example 3.2.1 are satisfied. Let M and N be arbitrarily fixed positive integers. Give conditions of solvability and formulae for solutions for the following equations:

(i) $(D^N - A)x = y$,
(ii) $(D^N - A)^M x = y$,
(iii) $(D^N - A)(D^M - B) = y$,

for $y \in X$.

It follows from properties of the initial operator F that

$$FE_A = (I - RD)E_A = E_A - R(AE_A + D)$$
$$= E_A - RAE_A - RD = (I - RA)E_A - RD$$
$$= (I - AR)(I - AR)^{-1} - RD = I - RD = F. \qquad \blacksquare$$

COROLLARY 3.3.2. *Suppose that all assumptions of Theorem 3.3.1 are satisfied. Then*

$$DE_A z = AE_A z, \qquad (3.3.6)$$

$$\text{for all } z \in \ker D.$$

$$FE_A z = z, \qquad (3.3.7)$$

Proof. Indeed, let $z \in \ker D$ be arbitrary. Then $Dz = 0$ and $Fz = z$. Equality (3.3.4) implies that $DE_A z = AE_A z + Dz = AE_A z$. Equality (3.3.5) implies that $FE_A z = Fz = z$. $\qquad \blacksquare$

COROLLARY 3.3.3. *Suppose that all assumptions of Theorem 3.3.1 are satisfied. Then all solutions of the equation*

$$(D - A)z = y, \quad y \in X \qquad (3.3.8)$$

are of the form

$$x = E_A(Ry + z), \quad \text{where } z \in \ker D \text{ is arbitrary}. \qquad (3.3.9)$$

Proof. Indeed, write $u = (I - AR)x$. Then $x = E_A u$ and, by Equality (3.3.4) we find $Du = D(I - AR)x = (D - ADR)x = (D - A)z = y$, which implies $u = Ry + z$, where $z \in \ker D$ is arbitrary. Thus $x = E_A u = E_A(Ry + z)$, where $z \in \ker D$ is arbitrary. $\qquad \blacksquare$

An immediate consequence of our definitions is

COROLLARY 3.3.4. *Suppose that all assumptions of Theorem 3.3.1 are satisfied and $A = \lambda I$ for $\lambda \in \mathscr{F}$. Then*

$$E_{\lambda I} z = e_\lambda(z) \quad \text{for } z \in \ker D \quad (\lambda \in \mathscr{F}). \qquad (3.3.10)$$

Corollaries 3.3.2 and 3.3.4 show that the operator E_A defined by Equality (3.3.3) is a generalization of exponential operators defined in Section 2.3.

For equations of higher orders we have the following:

THEOREM 3.3.2. *Suppose that all assumptions of Theorem* 3.3.1 *are satisfied and N is an arbitrarily fixed positive integer. Then all solutions of the equation*

$$(D-A)^N x = y, \quad x \in X, \tag{3.3.11}$$

are of the form

$$x = E_A^N \left(R^N y + \sum_{k=0}^{N-1} R^k z_k \right), \tag{3.3.12}$$

where $z_0, \dots, z_{N-1} \in \ker D$ are arbitrary.

Proof. To begin with, we shall prove by induction the following identity:

$$D^k (I-AR)^k = (D-A)^k \quad \text{on dom } D^k \text{ for } k \in \mathbf{N}. \tag{3.3.13}$$

Indeed, for $k = 1$ we have $D(I-AR) = D-ADR = D-A$ on dom D. Suppose Formula (3.3.13) to be true for an arbitrarily fixed positive integer k. Then on the dom D^{k+1} we find

$$\begin{aligned} D^{k+1}(I-AR)^{k+1} &= DD^k(I-AR)^k(I-AR) = D(D-A)^k(I-AR) \\ &= (D-A)^k D(I-AR) = (D-A)^k (D-A) \\ &= (D-A)^{k+1}. \end{aligned}$$

We therefore conclude that Identity (3.3.13) hold for an arbitrary $k \in \mathbf{N}$. Write $u = (I-AR)^N x$. Then $x = E_A^N u$. Equalities (3.3.4) and (3.3.13) together imply that $D^N u = D^N (I-AR)^N x = (D-A)^N x = y$, which implies $u = R^N y + \sum_{k=0}^{N-1} R^k z_k$, where $z_0, \dots, z_{N-1} \in \ker D$ are arbitrary. Thus $x = E_A^N u = E_A^N (R^N y + \sum_{k=0}^{N-1} R_k z_k)$, where $z_0, \dots, z_{N-1} \in \ker D$ are arbitrary. ∎

THEOREM 3.3.3. *Suppose that* $D \in R(X)$, $\dim \ker D \neq 0$, $R \in \mathscr{R}_D$, $A \in \mathbf{S}_{D,R}$ *and the operators* $I-iAR$, $I+iAR$ *are invertible. Write*

$$C_A = \frac{1}{2}(E_{iA} + E_{-iA}); \quad S_A = \frac{1}{2i}(E_{iA} - E_{-iA}). \tag{3.3.14}$$

Then the operators C_A, S_A *have the following properties:*

$$C_A = (I+A^2 R^2)^{-1}, \quad S_A = AR(I+A^2 R^2)^{-1} = ARC_A, \tag{3.3.15}$$

$$DC_A = -AS_A + D \quad \text{on dom } D, \tag{3.3.16}$$

$$DS_A = AC_A, \tag{3.3.17}$$

$$(D^2 + A^2)C_A = D^2 \quad and \quad (D^2 + A^2)S_A = AD \quad on \ dom \ D^2. \tag{3.3.18}$$

Moreover, every solution of the equation

$$(D^2 + A^2)x = y, \quad y \in X, \tag{3.3.19}$$

is of the form

$$x = C_A R^2 y + C_A z_0 + S_A z_1, \tag{3.3.20}$$

where $z_0, z_1 \in \ker D$ are arbitrary.

Proof. By our assumptions, the operator $I + A^2 R^2 = (I - iAR)(I + iAR)$ is invertible as a superposition of invertible operators. Consequently,

$$C_A = \frac{1}{2}(E_{iA} + E_{-iA}) = \frac{1}{2}[(I - iAR)^{-1} + (I + iAR)^{-1}]$$

$$= \frac{1}{2}(I + A^2 R^2)^{-1}(I + iAR + I - iAR) = (I + A^2 R^2)^{-1},$$

$$S_A = \frac{1}{2i}(E_{iA} - E_{-iA}) = \frac{1}{2i}[(I - iAR)^{-1} - (I + iAR)^{-1}]$$

$$= \frac{1}{2i}(I + A^2 R^2)^{-1}[I + iAR - (I - iAR)]$$

$$= \frac{2iAR}{2i}(I + A^2 R^2)^{-1} = AR(I + A^2 R^2)^{-1}.$$

By the definition and Property (3.3.4) we find on the domain of the operator D:

$$DC_A = \frac{1}{2}(DE_{iA} + DE_{-iA}) = \frac{1}{2}(iAE_{iA} - iAE_{-iA} + D + D)$$

$$= \frac{1}{2}A(E_{iA} - E_{-iA}) + D = -A\frac{1}{2i}(E_{iA} - E_{-iA}) + D$$

$$= -AS_A + D.$$

The second Equality of (3.3.15) implies

$$DS_A = DARC_A = ADRC_A = AC_A.$$

From Equalities (3.3.16) and (3.3.17) we find on the domain of D^2:

$$(D^2+A^2)C_A = D(DC_A)+A^2C_A = D(-AS_A+D)+A^2C_A$$
$$= -ADS_A+D^2+A^2C_A = -A^2C_A+D^2+A^2C_A$$
$$= D^2,$$
$$(D^2+A^2)S_A = D(DS_A)+A^2S_A = D(AC_A)+A^2S_A$$
$$= ADC_A+A^2S_A = A(-AS_A+D)+A^2S_A$$
$$= -A^2S_A+AD+A^2S_A = AD.$$

Consider now the Equation (3.3.11). Write $u = (I+A^2R^2)x$. By definition $x = C_A u$. On the other hand $D^2u = D^2(I+A^2R^2)x = (D^2+A^2D^2R^2)x = (D^2+A^2)x = y$. Then $u = R^2y+z_0+Rz_0'$, where $z_0, z_0' \in \ker D$ are arbitrary. Hence $x = C_A u = C_A(R^2y+z_0+Rz_0')$. But, by our assumptions, $DA = AD$ on dom D, which implies that $A \ker D \subset \ker D$. Thus we can put instead of $z_0' \in \ker D$ a new constant $Az_1 \in \ker D$ where $z_1 \in \ker D$ is arbitrary. By our assumptions, $AR = RA$, $ARC_A = C_A AR$. Then from the second Equality of (3.3.15)

$$x = C_A(R^2y+z_0+Rz_0') = C_AR^2y+C_Az_0+C_ARAz_1$$
$$= C_AR^2y+C_Az_0+ARC_Az_1 = C_AR^2y+C_Az_0+S_Az_1,$$

where $z_0, z_1 \in \ker D$ are arbitrary, which was to be proved. ∎

Observe that the operators C_A, S_A defined by Formulae (3.3.14) are generalizations of cosine and sine operators introduced in Section 2.3. In particular, we have

$$C_{\lambda I} = c_\lambda, \quad S_{\lambda I} = s_\lambda \quad \text{for every scalar } \lambda, \tag{3.3.21}$$

provided that R is a Volterra operator.

THEOREM 3.3.4. *Suppose that*:

(i) $D_1, \ldots, D_n \in R(X)$ *are mutually commutative*;

(ii) $R_k \in \mathcal{R}_{D_k}$ *and* $D_j R_k = R_k D_j$ *on* dom D_j *for* $j \neq k$ $(j, k = 1, 2, \ldots$
$\ldots, n)$;

(iii) $A_1, \ldots, A_n \in \mathbf{S} = \bigcap_{k=1}^{n} \mathbf{S}_{D_k, R_k}$ *are mutually commutative*.

Write

$$D = \sum_{k=1}^{n} A_k D_k, \quad \text{dom } D = \bigcap_{k=1}^{n} \text{dom } D_k, \quad Z_0 = \bigcap_{k=1}^{n} \ker D^k.$$

$$\tag{3.3.22}$$

Then every element of the form

$$z = \sum_{j=1}^{n} \left[\sum_{k=1}^{n} (A_j R_k - A_k R_j) z_j + z_j' \right], \tag{3.3.23}$$

where $z_j, z_j' \in Z_0$ *are arbitrary* $(j = 1, \ldots, n)$, *is a solution of the equation*

$$\left(\sum_{k=1}^{n} A_k D_k \right) x = 0. \tag{3.3.24}$$

Proof. Observe that $z \in Z_0$ if and only if $D_k z = 0$, i.e. $z \in \ker D_k$, for $k = 1, 2, \ldots, n$. Suppose that z is of the form (3.3.23), where $z_j, z_j' \in Z_0$. We have

$$Dz_j' = \sum_{k=1}^{n} A_k D_k z_j' = 0, \quad Dz_j = \sum_{k=1}^{n} A_k D_k z_j = 0$$
$$(j = 1, 2, \ldots, n).$$

By our assumptions $DA_j = A_j D$ on dom D and $D_m R_k = R_k D_m$ on dom D_m for $k \neq m$ $(j, k, m = 1, 2, \ldots, n)$. Thus we find

$$Dz = \sum_{j=1}^{n} \left[\sum_{k=1}^{n} D(A_j R_k - A_k R_j) z_j + Dz_j' \right]$$

$$= \sum_{j=1}^{n} \sum_{k=1}^{n} D(A_j R_k - A_k R_j) z_j$$

$$= \sum_{j=1}^{n} \sum_{k=1}^{n} \sum_{m=1}^{n} (A_m D_m A_j R_k - A_m D_m A_k R_j) z_j$$

$$= \sum_{j=1}^{n} \sum_{k=1}^{n} \sum_{m=1}^{n} (A_m A_j D_m R_k - A_m A_k D_m R_j) z_j$$

$$= \sum_{j=1}^{n} \sum_{k=1}^{n} \left[\sum_{m=1, m \neq k}^{n} (A_m A_j D_m R_k z_j + A_k A_j D_k R_k z_j) - \right.$$

$$\left. - \sum_{m=1, m \neq k}^{n} (A_m A_k D_m R_j z_j + A_j A_k D_j R_j z_j) \right]$$

$$= \sum_{j=1}^{n} \sum_{k=1}^{n} \Big[\sum_{m=1, m\neq k}^{n} A_m A_j R_k D_m z_j -$$

$$- \sum_{m=1, m\neq j}^{n} A_m A_k R_j D_m z_j \Big] + \sum_{j=1}^{n} \sum_{k=1}^{n} (A_k A_j z_j - A_j A_k z_j)$$

$$= \sum_{j=1}^{n} \sum_{k=1}^{n} (A_k A_j - A_j A_k) z_j = 0. \qquad \blacksquare$$

PROPOSITION 3.3.2. *Suppose that* $D \in R(X)$, dim ker $D \neq 0$, F *is an initial operator for* D *corresponding to an* $R \in \mathcal{R}_D$ *and* $A \in L_0(X)$ *is stationary. If* $0 \neq x \in \ker(I - AR)$ *then* $x \in \ker(D - A)$. *If* $0 \neq x \in \ker(D - A)$ *and* $Fx = 0$ *then* $x \in \ker(I - AR)$.

Proof. Indeed, if $0 \neq x \in \ker(I - AR)$ we find $0 = D(I - AR)x = (D - DAR)x = (D - ADR)x = (D - A)x$, which implies $x \in \ker(D - A)$. If $0 \neq x \in \ker(D - A)$ and $Fx = 0$ then $x \in \text{dom } D$ and $x = x - Fx = (I - F)x = RDx = RAx = ARx$, which implies $x = \ker(I - AR)$. $\blacksquare$

THEOREM 3.3.5. *Suppose that* $D \in R(X)$, $R \in \mathcal{R}_D \cap V(X)$ *and* $A \in L_0(X)$ *is stationary and algebraic with single roots* $t_1, \ldots, t_n$ *only. Then there exists the operator* $E_A = (I - AR)^{-1} = (I - RA)^{-1} = \sum_{j=1}^{n} (I - t_j R)^{-1} P_j$,

where P_j *are disjoint projectors giving a partition of unity*

$$P_j = \prod_{m=1, m\neq j}^{n} \frac{A - t_j I}{t_m - t_j} \qquad (j = 1, 2, \ldots, n)$$

(cf. Theorem 1.4.1). Moreover, if $t_1, \ldots, t_n \neq 0$ *then* $E_A \in V(X)$.

Proof. By our assumptions that A is algebraic and Theorem 1.4.1 we have $X = \bigoplus_{j=1}^{n} X_j$, where $X_j = P_j X$ and $Ax_j = t_j x_j$ for $x_j \in X_j$ $(j = 1, 2, \ldots, n)$. Since A is stationary, we conclude that $RX_j \subset X_j$, $D(X_j \cap \text{dom } D) \subset X_j \cap \text{dom } D$ $(j = 1, 2, \ldots, n)$. The projectors P_j as polynomials in A with scalar coefficients, are all again stationary. Thus we have: $I -$

$-AR = I - RA = \sum_{j=1}^{n} (I - t_j R) P_j$ since $R \in V(X)$, we also have $R|_{X_j}$

$\in V(X_j)$ for $j = 1, 2, \ldots, n$. We therefore conclude that there exists

the operator $E_A = (I - AR)^{-1}$ and $E_A = \sum_{j=1}^{n} (I - t_j R)^{-1} P_j$.

In order to prove that $E_A \in V(X)$ observe that for an arbitrarily
fixed scalar λ we have

$$I - \lambda E_A = I - \lambda (I - AR)^{-1} = (I - AR)^{-1}[(1 - \lambda)I - AR]$$

$$= (I - AR)^{-1} \sum_{j=1}^{n} [(1 - \lambda)I - t_j R] P_j.$$

Let $\lambda \neq 1$. Then

$$I - \lambda E_A = (1 - \lambda)^{-1}(I - AR)^{-1} \sum_{j=1}^{n} \left[I - \frac{t_j}{1 - \lambda} R \right] P_j,$$

which implies that $I - \lambda E_A$ is invertible. Suppose now that $\lambda = 1$.
Then $I - E_A = -(I - AR)AR$. Consider the equation $(I - E_A)u = v$
for an arbitrarily fixed $v \in X$. This equation is equivalent to ARu
$= -(I - AR)v$, which implies $Au = DRAu = DARu = -D(I - AR)v$. If

$t_1, \ldots, t_n \neq 0$ then A is invertible, $A^{-1} = \sum_{j=1}^{n} t_j^{-1} P_j$ and we obtain

a unique solution of the equation under consideration. Therefore even
for $\lambda = 1$ the operator $I - \lambda E_A$ is invertible. This proves that $E_A \in V(X)$.

∎

Consider now the case of multiple roots.

THEOREM 3.3.6. *Suppose that* $D \in R(X)$, $R \in \mathscr{R}_D \cap V(X)$ *and* $A \in L_0(X)$
is stationary and algebraic with the characteristic polynomial $P(t)$

$= \prod_{j=1}^{n} (t - t_j)^{r_j}$. *Then* $\ker(I - AR) = \ker(I - RA) = \{0\}$.

Proof. By our assumption that A is algebraic, we have $X = \bigoplus_{j=1}^{n} X_j$,

where $X_j = P_j X$, P_j are disjoint projectors giving a partition of unity
and $(A - t_j I)^{r_j} x_j = 0$ for $x_j \in X_j$ $(j = 1, 2, \ldots, n)$ (cf. Theorem 1.4.1).

Acting on both sides of these equalities by the operators R^{r_j}, respectively, we obtain

$$(AR - t_j R)^{r_j} x_j = 0 \quad (j = 1, 2, ..., n). \tag{3.3.25}$$

If $x_j \in \ker(I - AR)$ then $x_j - ARx_j = 0$. This, and Equalities (3.3.25), together imply that $(I - t_j R)^{r_j} x_j' = 0$ for $j = 1, 2, ..., n$. Since $R \in V(X)$, we conclude that $x_j = 0$ for $j = 1, 2, ..., n$ and, consequently, $x = x_1 + \, + ... + x_n = 0$. This proves that $\ker(I - AR) = \{0\}$. ∎

COROLLARY 3.3.5. *Suppose that all assumptions of Theorem 3.3.6 are satisfied and the operator $I - AR$ is a mapping onto X. Then there exists the operator $E_A = (I - AR)^{-1}$. If, moreover, $t_1, ..., t_n \neq 0$ then $E_A \in V(X)$.*

Proof. By our assumptions and Theorem 3.3.6 $\dim \ker(I - AR) = 0$. This, and the assumption that $I - AR$ is a mapping onto, together imply that this operator is invertible. Write: $E_A = (I - AR)^{-1}$. Let λ be an arbitrarily fixed scalar. Consider the operator $I - \lambda E_A = I - \lambda(I - AR)^{-1} = (I - AR)^{-1}[(1 - \lambda)I - AR]$. Similarly as in the proof of Theorem 3.3.5, we conclude that $I - \lambda E_A$ for $\lambda \neq 1$ is invertible, as a superposition of invertible operators and $I - E_A$ is also invertible. Therefore $E_A \in V(X)$. ∎

Examples and Exercises

EXAMPLE 3.3.1. Suppose that $\Omega = [a, b] \times [c, d]$, $X = C(\Omega)$, $D = \partial/\partial t$, where $(t, s) \in \Omega$. As it has been shown in Example 2.1.2, the operator D is right invertible and has a right inverse R defined as follows:

$$(Rx)(t, s) = \int_{t_0}^{t} x(\tau, s)d\tau \quad \text{for } x \in C(\Omega), t_0 \in [a, b].$$

The space of constants is of the form

$$\ker D = \{x \in C(\Omega): x = x(t, s) = \varphi(s), \text{ where } \varphi \in C[c, d]\}.$$

Let $A \in C[c, d]$ be an arbitrary function. The operator A of multiplication by the function $A(s)$ is stationary. Indeed, for every $x \in C(\Omega)$ we find

$$[(AD-DA)x](t, s) = A(s)\frac{\partial}{\partial t} x(t, s) - \frac{\partial}{\partial t} A(s)x(t, s)$$

$$= A(s)\frac{\partial}{\partial t} x(t, s) - A(s)\frac{\partial}{\partial t} x(t, s) = 0$$

(provided that $\partial x/\partial t \in C[a, b]$),

$$[(AR-RA)x](t, s) = A(s)\int_{t_0}^{t} x(\tau, s)d\tau - \int_{t_0}^{t} A(s)x(\tau, s)d\tau$$

$$= A(s)\int_{t_0}^{t} x(\tau, s)d\tau - A(s)\int_{t_0}^{t} x(\tau, s)d\tau = 0.$$

In a similar way, as in Example 2.2.2 we can show that the operator $I-AR$ is invertible and the operator $E_A = (I-AR)^{-1}$ is of the form

$$E_A x(t, s) = x(t, s) + A(s)\int_{t_0}^{t} e^{A(s)(t-\tau)}x(\tau, s)d\tau. \tag{3.3.26}$$

Thus, by Corollary 3.3.3 every solution of the equation

$$\frac{\partial x}{\partial t} - A(s)x = y, \tag{3.3.27}$$

where $y \in C(\Omega)$, $A \in C[c, d]$ are given, is of the form

$$x = E_A(R_y + z), \quad \text{where } z \in \ker D \text{ is arbitrary,}$$

i.e.

$$x(t, s) = \int_{t_0}^{t} y(\tau, s)d\tau + z(s) +$$

$$+ A(s)\int_{t_0}^{t} e^{A(s)(t-\tau)}\left[\int_{t_0}^{\tau} y(u, s)du + z(s)\right]d\tau, \tag{3.3.28}$$

where $z \in C[c, d]$ is arbitrary.

Since in our case both the operators $I - iAR$ and $I + iAR$ are invertible (the proof is going on the same lines as for the operator $I - AR$), we find for $x \in C(\Omega)$

$$(C_A x)(t, s) = \frac{1}{2}\,[(E_{tA} + E_{-iA})x](t, s)$$

$$= x(t, s) + A(s)\int_{t_0}^{t} \frac{1}{2}[e^{iA(s)(t-\tau)} + e^{-iA(s)(t-\tau)}]x(\tau, s)\,d\tau$$

$$= x(t, s) + A(s)\int_{t_0}^{t} \cos[A(s)(t-\tau)]x(\tau, s)\,d\tau.$$

Similarly, we find for $x \in C(\Omega)$

$$(S_A x)(t, s) = x(t, s) + A(s)\int_{t_0}^{t} \sin[A(s)(t-\tau)]x(\tau, s)\,du.$$

Thus every solution of the equation

$$\frac{\partial^2 x}{\partial t^2} + A^2(s)x = y,$$

where $y \in C(\Omega)$, $A \in C[c, d]$ are given, by Theorem 3.3.3, is of the form $x = C_A R^2 y + C_A z_0 + S_A z_1$, where $z_1, z_2 \in C[c, d]$ are arbitrary, i.e. of the form

$$x(t, s) = y(t, s) +$$

$$+ A(s)\int_{t_0}^{t} \cos[A(s)(t-\tau)]\left[\int_{t_0}^{\tau} (\tau - u)y(u, s)\,du + z_0(s)\right]d\tau +$$

$$+ A(s)z_1(s)\int_{t_0}^{t} \sin[A(s)(t-\tau)]\,d\tau,$$

where $z_0, z_1 \in C[c, d]$ are arbitrary.

EXERCISE 3.3.1. Suppose that all assumptions of Theorem 3.3.1 are satisfied. Prove (by induction) that

$$D^n E_A = A^m E_A + \sum_{j=1}^{n-1} A^{n-j}D^j + D^n \quad \text{on dom } D^n \quad (n = 1, 2, \ldots).$$

EXERCISE 3.3.2. Suppose that all assumptions of Theorem 3.3.3 are satisfied. Find a formula for general solution of the equation

$$(D^2 + A^2)^n x = y, \quad \text{where } y \in X \quad (n = 1, 2, \ldots).$$

EXERCISE 3.3.3. Solve by application of Theorem 3.3.4 the following partial differential equation of order 1:

$$\sum_{k=1}^{n} a_k \frac{\partial u}{\partial t_j} = 0,$$

where $a_1, \ldots, a_n \in \mathbf{R}$ are given, $t = (t_1, \ldots, t_n) \in \Omega \subset \mathbf{R}^n$, $u \in C^1(\Omega)$.

EXERCISE 3.3.4. Suppose that X is the space of all real-valued (complex-valued) functions continuous and bounded for $t \subset \mathbf{R}$. Let $D = \mathrm{d}/\mathrm{d}t$, $(A_h x)(t) = x(t+h)$ for $h \in \mathbf{R}$ arbitrarily fixed. Prove that the operators A_h and D commute for every $h \in \mathbf{R}$. Are the operators A_h stationary for $h \neq 0$?

EXERCISE 3.3.5. Suppose that all assumptions of Theorem 3.3.5 are satisfied and either the operator $I - AR$ is a mapping onto or A has single characteristic roots only. Then the operator $E_A = (I - AR)^{-1}$ is well-defined. Prove that for all $z \in \ker D$:
  (i) $E_A z \in \ker(D - A)$;
  (ii) $FE_A z = z$.

EXERCISE 3.3.6. Suppose that all assumptions of Theorem 3.3.5 are satisfied. Prove that for all $z \in \ker D$:

  (i) $E_A z = \sum_{j=1}^{n} e_{t_j}(P_j z)$, where $E_A = (I - AR)^{-1}$, $e_\lambda = (I - \lambda R)^{-1} = E_{\lambda I}$;

  (ii) $DE_A z = |\sum_{j=1}^{n} t_j e_{t_j}(P_j z)$.

EXERCISE 3.3.7. Suppose that $\mathcal{S}$ is the so-called *Schwartz space*, i.e. the space of all complex-valued functions belonging to $C^\infty(\mathbf{R})$ which are decreasing to zero faster than any polynomial. Let $x \in \mathcal{S}$ and let

$$(\mathcal{F}x)(t) = \tilde{x}(t) = \lim_{a \to +\infty} \int_{-a}^{a} x(s)e^{-its}\mathrm{d}s.$$

This limit always exists and $\tilde{x} \in \mathcal{S}$. We shall write $(\mathcal{F}x)(t)$

$= \int\limits_{-\infty}^{\infty} x(s)e^{its}ds$. The mapping $\mathscr{F}$, which is called a *Fourier transform*, is linear and onto. The function $x = \mathscr{F}x$ is said to be the *Fourier transform of x*. Write also: $D = d/dt$ and $(Mx)(t) = itx(t)$ for $x \in \mathscr{S}$, $t \in \mathbf{R}$. It is clear that both operators, D and M, preserve the space $\mathscr{S}$. It is also well-known that the following identities hold:

$$\mathscr{F}M = -D\mathscr{F} \quad \text{and} \quad \mathscr{F}D = M\mathscr{F} \quad \text{for } x \in \mathscr{S}.$$

Prove that the operators

$$T_k = D^{2k} + M^{2k} \quad (k = 1, 2, ...)$$

commute with $\mathscr{F}$ on the space $\mathscr{S}$. Note that this property holds also for the Fourier transform in $\mathbf{R}^n$ (cf. the author, 1973).

3.4. EQUATIONS WITH SCALAR COEFFICIENTS. OPERATIONAL CALCULUS OF RIGHT INVERTIBLE OPERATORS. D-HULL

In this section we shall consider equations with scalar coefficients. We shall show that, using properties of algebraic operators, by decomposition of a given rational function into vulgar fraction, we can obtain general solutions of equations under consideration.

Suppose that X is a linear space over an arbitrary algebraically closed field $\mathscr{F}$ of scalars and $D \in R(X)$. Write

$$Q(t, s) = \sum_{k=0}^{N} q_k t^k s^{N-k} = \prod_{m=0}^{n} (t - \lambda_m s)^{r_m}, \qquad (3.4.1)$$

where $q_0, ..., q_N \in \mathscr{F}$, $q_N = 1$, $\lambda_m \neq 0$, $\lambda_m \neq \lambda_j$ for $n \neq j$ $(j, m = 1, ..., n)$, $r_1 + ... + r_n = N$ and

$$Q(t) = Q(t, 1), \quad P(t) = t^M Q(t), \qquad (3.4.2)$$

where M is a non-negative integer.

We have proved (Theorem 2.4.1) the following fact:

If there exists a right inverse R of D which is Volterra operator then the operator $P(D) = D^M Q(D) = Q(D)D^M$ is right invertible and has a right inverse

$$R^0 = R^{N+M}[Q(I, R)]^{-1} = [Q(I, R)]^{-1}R^{N+M}, \qquad (3.4.3)$$

which is a Volterra operator (cf. Formula (2.4.5)). Conversely (von Trotha, 1978*), if the operator R^0 defined by Formula (3.4.3) is a Volterra operator, then R is a Volterra operator.

THEOREM 3.4.1. *Suppose that X is a linear space over the field $\mathbf{C}$, $D \in R(X)$, $\dim \ker D \neq 0$, there exists a right inverse R of D which is a Volterra operator and the polynomials $Q(t, s)$ and $Q(t)$ are defined by Formulae (3.4.1), (3.4.2), respectively. Write*

$$W_m^\lambda = \begin{cases} (I - \lambda R)^{-m} & \text{if } \lambda \neq 0, \\ R^{m-1} & \text{if } \lambda - 0, \end{cases} \quad (m = 1, 2, \ldots), \qquad (3.4.4)$$

where $\lambda \in \mathbf{C}$ is arbitrary. We admit also that $\lambda_0 = 0$ and $r_0 = M \geqslant 0$. Then every solution of the equation

$$D^M Q(D) x = 0 \qquad (3.4.5)$$

is of the form

$$x = \sum_{j=0}^{n} \sum_{m=1}^{r} W_m^{\lambda_j} z_{jm}, \quad \text{where } z_{jm} \in \ker D \text{ are arbitrary} \quad (3.4.6)$$

and the elements $W_m^{\lambda_j} z_{jm}$ $(m = 1, \ldots, r_j; \ j = 0, 1, \ldots, n)$ are linearly independent. We admit here the following convention: if $M = 0$, i.e. in the case when the root $\lambda_0 = 0$ does not appear in the polynomial under consideration, the corresponding component in the sum (3.4.6) vanishes. Moreover,

$$\dim \ker D^M Q(D) = (M+N) \dim \ker D.$$

Proof. Observe that for $z \in \ker D$ the element $W_1^{\lambda_j} z = e_{\lambda_j}(z)$, i.e. $W_1^{\lambda_j}$ is an exponential element (cf. Formula (2.3.2)).

In particular, for $\lambda_0 = 0$ we have $W_1^{\lambda_0}(z) = z$. We shall prove that the elements $W_m^{\lambda_j} z_{jn}$ $(m = 1, \ldots, r_j; \ j = 0, 1, \ldots, n)$ are linearly independent. The operator $D^M Q(D)$ is algebraic on the space $\ker D^M Q(D)$ (cf. Section 1.4). Indeed, by definition, $D^M Q(D) x = 0$ for every $x \in \ker D^M Q(D)$. By Formulae (3.4.1) and (3.4.2) the characteristic polynomial of the operator $D^M Q(D)$ is $P(t) = t^M \prod_{m=1}^{n} (t - t_m)^{r_m} = t^M Q(t)$,

---

* Private communication, cf. Przeworska-Rolewicz, 1980a.

where $M = r_0$. Hence Theorem 1.4.1 implies that the space $\ker P(D) = \ker D^M Q(D)$ can be decomposed onto direct sum

$$\ker P(D) = X_0 \oplus X_1 \oplus \ \dots \ \oplus X_n,$$

where $X_j = \ker (D - \lambda_j I)^{r_j}$ $(j = 1, \dots, n)$.

Thus it is enough to show the linear independence of elements $W_m^{\lambda_j} z_{jm}$ $(m = 1, 2, \dots, r_j)$ corresponding to an arbitrarily fixed $\lambda_j \neq 0$ of the polynomial $P(t)$. In the case $\lambda_0 = 0$ the linear independence of elements $W_m^{\lambda_0} z_{0m} = R^{m-1} z_{0m}$ $(m = 1, 2, \dots, r_0 = M)$ follows from Proposition 2.4.2. Suppose then that $\lambda_j \neq 0$ and that elements $W_m^{\lambda_j} z_{jm}$ $(m = 1, 2, \dots, r_j)$ are linearly independent. This means that there exists complex numbers $a_1, \dots, a_{r_j}$ nonvanishing simultaneously and such that

$$0 = \sum_{m=1}^{r_j} a_m W_m^{\lambda_j} z_{jm} = \sum_{m=1}^{r_j} a_m (I - \lambda_j R)^{-m} z_{jm}.$$

Acting on both sides of this equality by the operator $(I - \lambda_j R)^{r_j}$ we obtain

$$0 = (I - \lambda_j R)^{r_j} \sum_{m=1}^{r_j} a_m W_m^{\lambda_j} z_{jm}$$

$$= \sum_{m=1}^{r_j} a_m (I - \lambda_j R)^{r_j - m} z_{jm} = \sum_{k=0}^{r_j - 1} a_{r_j - k} (I - \lambda_j R)^k z_{j, r_j - k}$$

$$= \sum_{k=0}^{r_j - 1} a_{r_j - k} \sum_{m=0}^{k} \binom{k}{m} (-1)^m \lambda_j^m R^m z_{j, r_j - k}$$

$$= \sum_{m=0}^{r_j - 1} R^m \left[\sum_{k=m}^{r_j - 1} a_{r_j - k} \binom{k}{m} (-1)^m \lambda_j^m z_{j, r_j - k} \right].$$

Writing

$$z_m = \sum_{k=m}^{r_j - 1} a_{r_j - k} \binom{k}{m} (-1)^m \lambda_j^m z_{j, r_j - k} \quad (m = 0, 1, \dots, r_j - 1),$$

we conclude that $z_m \in \ker D$ and $\sum\limits_{m=0}^{r_j-1} R^m z_m = 0$, i.e. the elements $z_0, Rz_1, \ldots, R^{r_j-1}z_{r_j-1}$ are linearly dependent. But this contradicts Proposition 2.4.2. Hence all elements $W_m^{\lambda_j} z_{jm}$ are linearly independent for $m = 1, \ldots, r_j, j = 0, 1, \ldots, n$.

We therefore obtain $(M+N)$ dim $\ker D$ elements linearly independent. To obtain the conclusion of our theorem we should prove now that each of these elements is a solution of Equation (2.4.5) because, by Point 3(iii) of Theorem 3.2.5 dim $\ker P(D) = (M+N)$ dim $\ker D$ (for the operator $Q(I, R) = \prod\limits_{m=1}^{n} (I - t_m R)^{r_m}$ is invertible as a superposition of invertible operators).

Suppose that $\lambda_0 = 0$ is a root of the multiplicity $r_0 = M \geqslant 1$. Then, by our assumptions, for an arbitrary $z \in \ker D$ and $m = 1, \ldots, r_0$ we have

$$
\begin{aligned}
D^M Q(D) W_m^{\lambda_0} z &= D^M Q(D) R^{m-1} z = Q(D) D^{r_0} R^{m-1} z \\
&= Q(D) D^{r_0 - (m-1)} D^{m-1} R^{m-1} z = Q(D) D^{r_0 + 1 - m} z = 0.
\end{aligned}
$$

Suppose now that $\lambda_j \neq 0$ is a root of the polynomial $Q(t)$ with the multiplicity $r_j \geqslant 1$. Observe that for an arbitrary $z \in \ker D$ and for $m = 1, 2, \ldots, r_j$ we have

$$
\begin{aligned}
D^M Q(D) W_m^{\lambda_j} z &= D^M \Big[\prod_{k=1}^{n} (D - \lambda_k I)^{r_k} \Big] (I - \lambda_j R)^{-m} z \\
&= D^{r_0} \Big[\prod_{k=1, k \neq j}^{n} (D - \lambda_k I)^{r_k} \Big] (D - \lambda_j I)^{r_k - m} \times \\
&\qquad\qquad \times (D - \lambda_j I)^m (I - \lambda_j R)^{-m} z \\
&= D^{r_0} \Big[\prod_{k=1, k \neq j}^{n} (D - \lambda_k I)^{r_k} \Big] (D - \lambda_k I)^{r_k - m} \times \\
&\qquad\qquad \times D^m (I - \lambda_j R)^m (I - \lambda_j R)^{-m} z \\
&= D^{r_0} \Big[\prod_{k=1, k \neq j}^{n} (D - \lambda_k I)^{r_k} \Big] (D - \lambda_k I)^{r_k - m} D^m z = 0.
\end{aligned}
$$

Thus all elements of the form $W_m^{\lambda_j} z_{jm}$, where $z_{jm} \in \ker D$ are arbitrary

$(m = 1, ..., r_j, j = 0, 1, ..., m)$ are solution of Equation (3.4.5), which was to be proved. Observe that $Q(D) D^M = D^M Q(D)$ for the polynomial $Q(D)$ has scalar coefficients.   ∎

We obtain

PROPOSITION 3.4.1. *Suppose that all assumptions of Theorem 3.4.1 are satisfied and that* $P(t) = t^M Q(t)$. *Then exponential elements* $e_\lambda(z)$, *where* $z \in \ker D$ *are arbitrary, are eigenvectors of the operator* $D^M Q(D)$ *corresponding to the eigenvalue* $\mu = P(\lambda) = \lambda^M Q(\lambda) \in \mathbf{C}$, *i.e.*

$$D^M Q(D) e_\lambda(z) = \lambda^M Q(\lambda) e_\lambda(z) \quad \text{for all } \lambda \in \mathbf{C}, z \in \ker D. \quad (3.4.7)$$

If, moreover, $\lambda = \lambda_j$ *is a root of the polynomial* $P(t)$ *then* $e_{\lambda_j}(z)$ *is a solution of Equation* (3.4.5), *i.e.*

$$D^M Q(D) e_{\lambda_j}(z) = 0 \quad \text{for } z \in \ker D \quad (j = 0, 1, ..., n). \quad (3.4.8)$$

Proof. By definition of exponential elements $D e_\lambda(z) = \lambda e_\lambda(z)$ for all $\lambda \in \mathbf{C}$, $z \in \ker D$. Thus

$$D^M Q(D) e_\lambda(z) = D^M \sum_{k=0}^{N} q_k D^k e_\lambda(z) = D^M \sum_{k=0}^{M} q_k \lambda^k e_\lambda(z)$$

$$= D^M Q(\lambda) e_\lambda(z) = Q(\lambda) D^M e_\lambda(z)$$

$$= Q(\lambda) \lambda^M e_\lambda(z) = \lambda^M Q(\lambda) e_\lambda(z).$$

If $\lambda = \lambda_j$ $(j = 1, ..., n)$ we have $Q(\lambda_j) = 0$, which implies $D^M Q(D) e_{\lambda_j}(z) = 0$. If $\lambda = \lambda_0 = 0$ and $M = r_0 \geqslant 1$ we have $D^M Q(D) e_{\lambda_0}(z) = Q(D) D^M z = 0$.   ∎

COROLLARY 3.4.1. *Suppose that all assumptions of Theorem 3.4.1 are satisfied. Then a principal space* X_j *for* $D^M Q(D)$ *corresponding to an eigenvalue* λ_j *of multiplicity* r_j *is of the form*

$$X_j = \begin{cases} \displaystyle\bigoplus_{m=0}^{r_j-1} (I - \lambda_j R)^{-m} e_{\lambda_j}(\ker D) & \text{if } j = 1, 2, ..., n, \\ \displaystyle\bigoplus_{m=0}^{r_0-1} R^m \ker D = \ker D^{r_0} & \text{if } j = 0. \end{cases} \quad (3.4.9)$$

Proof. Indeed, for arbitrary $z \in \ker D$, $\lambda_j \neq 0$ $(j = 1, ..., n)$ and k

$= 1, 2, \ldots, r_j$ we have $W_k^{\lambda_j} z = (I - \lambda_j R)^{-k} z = (I - \lambda_j R)^{-k+1}(I - \lambda_j R)^{-1} z$
$= (I - \lambda_j R)^{-k+1} e_{\lambda_j}(z)$. This implies

$$X_j = \bigoplus_{k=1}^{r_j} W_k^{\lambda_j} \ker D = \bigoplus_{k=1}^{r_j} (I - \lambda_j R)^{-k+1} e_{\lambda_j}(\ker D)$$

$$= \bigoplus_{m=0}^{r_j-1} (I - \lambda_j R)^{-m} e_{\lambda_j}(\ker D).$$

For $\lambda = 0$ we have

$$X_0 = \bigoplus_{k=1}^{r_j} W_k^0 \ker D = \bigoplus_{k=1}^{r_j} R^{k-j} \ker D$$

$$= \bigoplus_{m=0}^{r_j-1} R^m \ker D = \ker D^{r_0}. \qquad \blacksquare$$

Since the kernel of the operator $D^M Q(D)$ is well-defined by the polynomial $P(t) = t^M Q(t)$, we say that $P(t)$ is a *characteristic polynomial of $P(D)$* and its roots are *characteristic roots of $P(D)$*. We shall see that the characteristic polynomial determined also solutions of nonhomogeneous equation. Indeed, we obtain an arbitrary solution of the equation $D^M Q(D)x = y$, where $Q(D)$ is defined in Theorem 3.4.1 and $y \in X$, if we add to general solution of the homogeneous equation (3.4.5) a particular solution of the nonhomogeneous equations. For instance, Formula (3.2.28) implies that this particular solution is of the form: $R^{M+N}[Q(I, R)]^{-1} = R^M[Q(I, R)]^{-1} R^N$, since in our case $Q(t)$ is a polynomial with scalar coefficients and $Q\langle D \rangle = Q(D)$, $Q\langle I, R \rangle = Q(I, R)$. Another method is given by the following:

THEOREM 3.4.2. *Suppose that all assumptions of Theorem 3.4.1 are satisfied. Decompose the rational function $[Q(1, s)]^{-1}$ onto vulgar fractions:*

$$[Q(1, s)]^{-1} = \sum_{j=1}^{n} \sum_{m=1}^{r_j} d_{jm}(1 - \lambda_j s)^{-m}, \qquad (3.4.10)$$

where d_{jm} are well-defined scalars. Then every solution of the equation

$$D^M Q(D)x = y, \quad y \in X \qquad (3.4.11)$$

is of the form

$$x = \sum_{j=1}^{n} \sum_{m=1}^{r_j} W_k^{\lambda_j}(d_{jm} R^{N+M} y + z_{jm}) + \sum_{k=1}^{r_0} R^{k-1} z_{0k}, \qquad (3.4.12)$$

where $z_{jm} \in \ker D$ are arbitrary, scalars d_{jm} are determined by the decomposition (3.4.10) and the component $\sum\limits_{k=1}^{r_0} R^{k-1}z_{0k}$ appears only in the case when $M = r_0 \geqslant 1$.

Proof. Since all solutions of Equation (3.4.5) are determined in Theorem 3.4.1, Formula (3.2.28) implies that it is enough to show that an element of the form

$$x_0 = R^{N+M}[Q(I, R)]^{-1}y = [Q(I, R)]^{-1}R^{N+M}y$$

is a particular solution of Equation (3.4.11).

Indeed, Formula (3.2.12) implies that

$$
\begin{aligned}
D^M Q(D)x_0 &= D^M Q(D) R^{M+N}[Q(I, R)]^{-1}y \\
&= D^M Q(D) R^N [Q(I, R)]^{-1}R^M y \\
&= D^M Q(I, R)[Q(I, R)]^{-1}R^M y = D^M R^M y = y.
\end{aligned}
$$

On the other hand, from the decomposition (3.4.10) it follows that

$$
\begin{aligned}
x_0 &= [Q(I, R)]^{-1}R^{N+M}y = \sum_{j=1}^{n}\sum_{m=1}^{r_j} d_{jm}(1 - \lambda_j I)^{-m}R^{N+M}y \\
&= \sum_{j=1}^{n}\sum_{m=1}^{r_j} d_{jm} W_m^{\lambda_j} R^{N+M}y.
\end{aligned}
$$

Hence, every solution of Equation (3.4.11) is of the form

$$
\begin{aligned}
x &= x_0 + \sum_{j=0}^{n}\sum_{m=1}^{r_j} W_m^{\lambda_j} z_{jm} \\
&= \sum_{j=1}^{n}\sum_{m=1}^{r_j} d_{jm} W_m^{\lambda_j} R^{N+M}y + \sum_{j=0}^{n}\sum_{m=1}^{r_j} W_m^{\lambda_j} z_{jm} \\
&= \sum_{j=1}^{n}\sum_{m=1}^{r_j} W_m^{\lambda_j}(d_{jm} R^{N+M}y + z_{jm}) + \sum_{m=1}^{r_0} W_m^0 z_{0m} \\
&= \sum_{j=1}^{n}\sum_{m=1}^{r_j} W_m^{\lambda_j}(d_{jm} R^{N+M}y + z_{jm}) + \sum_{m=1}^{r_0} R^{m-1}z_{0m},
\end{aligned}
$$

where $z_{jm} \in \ker D$ are arbitrary $(m = 1, 2, ..., r_j, j = 0, 1, ..., n)$. ∎

We shall pass to studies of the case where the field of scalars is the field $\mathbf{R}$ of reals, i.e. the coefficients $q_0, \ldots, q_N$ of the polynomial $Q(D)$ are real. Since $\mathbf{R} \subset \mathbf{C}$, all results already proved remain true but solutions obtained can have complex coefficients. We have to transform these solutions in such a way that all coefficients will be real.

If all roots of the characteristic polynomial are real we obtain in the previous way solutions with real coefficients. Suppose now that the polynomial $Q(t)$ has complex roots. It is well-known that a polynomial with real coefficients having a complex root λ has also the conjugate number $\bar{\lambda}$ as root. (We assume here that Im $\lambda \neq 0$.) We shall apply this property. Consider two cases:

(A) The polynomial $Q(t)$ has two imaginary roots: $i\mu$ and $-i\mu$ $(0 \neq \mu \in \mathbf{R})$. Consider the space $Y = X \oplus iX$. The field of scalars for the space Y is evidently the field $\mathbf{C}$ of complexes. Write

$$\xi = x_1 + ix_2, \quad \xi^* = x_1 - ix_2 \quad \text{for every } \xi \in Y,$$

$$\text{where } x_1, x_2 \in X.$$

We shall extend the operators $D \in R(X)$ and $R \in \mathcal{R}_D \cap V(X)$ on the space Y in the following way:

$$\tilde{D}\xi = Dx_1 + iDx_2 \quad (x_1, x_2 \in \text{dom } D),$$
$$\tilde{R}\xi = Rx_1 + iRx_2 \quad (x_1, x_2 \in X).$$

This definition implies that $\tilde{D}\zeta = 0$ for a $\zeta \in Y$ if and only if $\zeta = z_1 + iz_2$, where $z_1, z_2 \in X$ and $Dz_1 = Dz_2 = 0$, i.e. $z_1, z_2 \in \ker D$. Thus $\ker \tilde{D} = \ker D \oplus i \ker D$. Observe also that $\tilde{R} \in V(Y)$. Indeed, $(I - \lambda\tilde{R})\xi = 0$, where $\xi = x_1 + ix_2$ if and only if $(I - \lambda R)x_1 = 0$ and $(I - \lambda R)x_2 = 0$. Since $R \in V(X)$, the equation $(I - \lambda\tilde{R})\xi = 0$ has only zero as a solution. These properties imply that in sequel we can denote simply the operators $\tilde{D}$ and $\tilde{R}$ by D and R, respectively, without any misunderstanding.

Similarly, as in Section 2.3 we shall write for $\lambda \in \mathbf{C}$, $\zeta, \zeta^* \in \ker D$

$$e_\lambda(\zeta) = (I - \lambda R)^{-1}\zeta, \tag{3.4.13}$$

$$c_\lambda(\zeta) = \frac{1}{2}\left[e_{\lambda i}(\zeta) + e_{-\lambda i}(\zeta^*)\right], \tag{3.4.14}$$

$$s_\lambda(\zeta) = \frac{1}{2i}\left[e_{\lambda i}(\zeta) - e_{-\lambda i}(\zeta^*)\right]. \tag{3.4.15}$$

These formulae imply the following equalities for $\lambda \in \mathbf{C}$, z_1, z_2
$\in \ker D \cap X$:

$$c_\lambda(z_1 + iz_2) = (I + \lambda^2 R^2)^{-1}(z_1 - \lambda R z_2), \qquad (3.4.16)$$

$$s_\lambda(z_1 + iz_2) = (I + \lambda^2 R^2)^{-1}(\lambda R z_1 + z_2). \qquad (3.4.17)$$

Indeed,

$$c_\lambda(z_1 + iz_2) = \frac{1}{2}\,[(e_{\lambda i}(z_1 + iz_2) + e_{-\lambda i}(z_1 - iz_2)]$$

$$= \frac{1}{2}\,[(I - \lambda iR)^{-1}(z_1 + iz_2) + (I + \lambda iR)^{-1}(z_1 - iz_2)]$$

$$= \frac{1}{2}\,(I - \lambda iR)^{-1}(I + \lambda iR)^{-1}[(I + \lambda iR)(z_1 + iz_2) +$$

$$+ (I - \lambda iR)(z_1 - iz_2)]$$

$$= \frac{1}{2}\,(I + \lambda^2 R^2)^{-1}[z_1 + iz_2 + \lambda iR z_1 + \lambda i^2 R z_2 + z_1 - iz_2 -$$

$$- \lambda iR z_1 + \lambda i^2 R z_2]$$

$$= \frac{1}{2}\,(I + \lambda^2 R^2)^{-1}(2z_1 - 2\lambda R z_2) = (I + \lambda^2 R^2)^{-1}(z_1 - \lambda R z_2),$$

$$s_\lambda(z_1 + iz_2) = \frac{1}{2i}\,[e_{\lambda i}(z_1 + iz_2) - e_{-\lambda i}(z_1 - iz_2)]$$

$$= \frac{1}{2i}\,[(I - \lambda iR)^{-1}(z_1 + iz_2) - (I + \lambda iR)^{-1}(z_1 - iz_2)]$$

$$= \frac{1}{2i}\,(I - \lambda iR)^{-1}(I + \lambda iR)^{-1}[(I + \lambda iR)(z_1 + z_2) -$$

$$- (I - \lambda iR)(z_1 - iz_2)]$$

$$= \frac{1}{2i}\,(I + \lambda^2 R^2)^{-1}[z_1 + iz_2 + \lambda iR z_1 + \lambda i^2 R z_2 - z_1 +$$

$$+ iz_2 + \lambda iR z_1 - \lambda i^2 R z_2]$$

$$= \frac{1}{2i}\,(I + \lambda^2 R^2)^{-1}(2\lambda iR z_1 + 2iz_2) = (I + \lambda^2 R^2)^{-1}(\lambda R z_1 + z_2).$$

In particular, if we put $\zeta = \zeta^*$, i.e. $z_2 = 0$, then

$$c_\lambda(z) = (I + \lambda^2 R^2)^{-1}z, \quad s_\lambda(z) = \lambda R(I + \lambda^2 R^2)^{-1}z, \qquad (3.4.18)$$

where $z \in \ker D \cap X$, $\lambda \in \mathbf{C}$ (cf. Formulae (2.3.6)). We therefore conclude that to two linearly independent solutions $e_{i\mu}(\zeta)$ and $e_{-i\mu}(\zeta)$ in the complex space Y there correspond two linearly independent solutions $c_\mu(z_1)$ and $s_\mu(z_2)$ in the real space X, where

$$z_1 = \frac{\zeta + \zeta^*}{2}, \quad z_2 = \frac{\zeta - \zeta^*}{2i} \in \ker D \cap X.$$

If the roots $i\mu$, $-i\mu$ are multiple, we construct the corresponding solutions by application of operators W_m^λ determined by Formulae (3.4.4) in a similar way as in complex case.

(B) The polynomial $Q(t)$ has two complex conjugate roots: $\lambda = \mu + + i\nu$, $\bar{\lambda} = \mu - i\nu$, where $0 \neq \mu \in \mathbf{R}$, $0 \neq \nu \in \mathbf{R}$. This case cannot be solved without additional assumptions (cf. Theorem 6.2.6).

Now we shall present results of von Trotha, 1981, which permit to characterize the set of solutions of Equation (3.4.11) provided that y belongs to a subspace of X.

DEFINITION 3.4.1. Let X be a linear space over an algebraically closed field $\mathscr{F}$ of scalars. Let $D \in R(X)$ with $\dim \ker D > 0$ and let $R \in \mathscr{R}_D$. The space X together with the operators D and R is called a *D–R space.* Sometimes, we shall write briefly: (X, D, R) instead of a D–R space X. A subspace $U \subset X$ is called a *D–R subspace of X* if there is a subspace U_1 such that

$$U_1 \subset \operatorname{dom} D, \quad RU \subset U_1 \subset U, \quad DU_1 \subset U. \tag{3.4.19}$$

Write: $(U, D|_{U_1}, R|_U)$. This triple is again a D–R space. If, in addition, $U_1 = U$ then U is said to be a *D–R-invariant subspace of X.*

DEFINITION 3.4.2. Let X, D, R be as in Definition 3.4.1. Denote by $\mathscr{F}[t]$ the ring of all polynomials in t over $\mathscr{F}$, i.e. the ring of all polynomials with coefficients belonging to $\mathscr{F}$. Let U be a D-invariant subspace of X, i.e. such subspace U that $DU \subset U$. The set

$$H_D(U) = \{x: \underset{P \in \mathscr{F}[t]}{\exists}\ P(D)x \in U\} \tag{3.4.20}$$

is said to be a *D-hull of U.*

PROPOSITION 3.4.2. *Suppose that X is a linear space over an algebraically closed field $\mathscr{F}$ of scalars, $D \in R(X)$, $\dim \ker D \neq 0$ and $R \in \mathscr{R}_D$. Let*

U be a subspace of X. Then the D-hull of U has the following properties:

(i) $U \subset H_D(U)$;

(ii) $H_D(U)$ *is a D–R invariant subspace*;

(iii) *the operator* $I - tR$ *is invertible on* $H_D(U)$ *for all* $t \in \mathscr{F}$, *i.e.* $R \in V(H_D(U))$. *Hence* $H_D(U)$ *is invariant with respect to operators of the form* $Q(R) = R^m(a_0 I + \dots + a_n R^n)^{-1}$, $a_0, \dots, a_n \in \mathscr{F}$, $a_0 \neq 0$, $n, m \in \mathbb{N}$.

Proof. (i) follows from the fact that $DU \subset U$ implies $U \subset H_D(U)$.

(ii) Suppose that $x \in H_D(U)$. Then $P(D)x \in U$, hence $P(D)(ax) \in U$ for $a \in \mathscr{F}$, $P \in \mathscr{F}[t]$, which implies $ax \in H_D(U)$ for $a \in \mathscr{F}$. Observe that $x_1, x_2 \in H_D(U)$ if and only if there exist $P_1, P_2 \in \mathscr{F}[t]$ such that $P_1(D)x_1, P_2(D)x_2 \in U$. Hence $P_1(D)P_2(D)(x_1 + x_2) = P_2(D)P_1(D)x_1 + P_1(D)P_2(D)x_2 \in U$, which implies $x_1 + x_2 \in H_D(U)$, because U, as D-invariant is $P(D)$-invariant for any polynomial P in D with scalar coefficients. Thus $H_D(U)$ is a subspace of X. Let $x \in H_D(U)$. Then $P(D)x \in U$ which implies $DP(D)Rx = P(D)DRx = P(D)x \in U$ and $Rx \in H_D(U)$.

In order to prove (iii) recall that $D - tI = D(I - tR)$ for $t \in \mathscr{F}$, hence $(D - tI)(I - tR)^{-1} = D$ on dom D for all $t \in \mathscr{F}$. If $x \in H_D(U)$ then $P(D)x \in U$ for a $P \in \mathscr{F}[t]$ and $P(D)(D - tI)(I - tR)^{-1}x \in U$ implies $(I - tR)^{-1}x \in H_D(U)$. We therefore conclude that the mapping $Q(R)$ of $H_D(U)$ into itself is one-to-one for any $Q \in \mathscr{F}[t]$ such that $Q(0) \neq 0$. ∎

The next proposition justifies the name *D*-hull.

PROPOSITION 3.4.3 (Hull operations). *Suppose that all assumptions of Proposition 3.4.2 are satisfied and U, V are D–R-invariant subspaces of X. The following hull operations hold:*

(i) $U \subset V$ *implies* $H_D(U) \subset H_D(V)$. *In particular*, $H_D(0) \subset H_D(V)$;

(ii) $H_D(H_D(U)) = H_D(U)$;

(iii) $H_D(U \cap V) = H_D(U) \cap H_D(V)$. *Hence*

$$H_D\left(\bigcap_{i \in \mathscr{J}} V_i\right) = \bigcap_{i \in \mathscr{J}} (H_D(V_i));$$

(iv) *Write*

$$S = \bigcup_{i=1}^{\infty} \ker D^i. \tag{3.4.21}$$

Then $\{0\} \neq H_D(0) = H_D(S) \neq S$;

(v) $H_D(0) = \bigcup_{P \in \mathscr{F}[t]} \ker P(D)$;

(vi) $H_D(U+V) = H_D(U) + H_D(V)$.

Proof. (i) Let $x \in H_D(U)$. Then $P(D)x \in U \subset V$ for a $P \in \mathscr{F}[t]$ which implies $x \in H_D(V)$.

(ii) Let $x \in H_D\big(H_D(U)\big)$. Then $P(D)x \in H_D(U)$ for a $P \in \mathscr{F}[t]$, which implies $Q(D)\big(P(D)x\big) \in U$ for a $Q \in \mathscr{F}[t]$ and $x \in H_D(U)$. It means that $H_D\big(H_D(U)\big) \subset H_D(U)$. The inclusion $H_D(U) \subset H_D\big(H_D(U)\big)$ follows from Point (i) of Proposition 3.4.2.

(iii) If $x \in H_D(U \cap V)$ then $P(D)x \in U \cap V$ for a $P \in \mathscr{F}[t]$, hence $x \in H_D(U) \cap H_D(V)$. If $x \in H_D(U) \cap H_D(V)$ then $P(D)x \in U$ and $Q(D)x \in V$ for some $P, Q \in \mathscr{F}[t]$, which implies $P(D)Q(D)x \in U$ and $P(D)Q(D)x \in V$, i.e. $P(D)Q(D)x \in U \cap V$. Hence $x \in H_D(U \cap V)$.

(iv) Since $\{0\} \subset S$, Point (i) of our proposition implies $H_D(0) \subset H_D(S)$. On the other hand if $x \in H_D(S)$ then $P(D)x = s \in S$ for $P \in \mathscr{F}[t]$, which implies that there exists an $n \in \mathbf{N}$ such that $D^n P(D)x = D^n s = 0$. But this means that $x \in H_D(0)$. Since $\ker D \neq \{0\}$, we conclude that $H_D(0) \neq \{0\}$. Suppose now that $S = H_D(S)$. Theorem 3.4.1 implies that $\ker(D-tI) = (I-tR)^{-1} \ker D \neq \{0\}$ for all $t \in \mathscr{F}$ for $\ker(D-tI) = \{e_t(z) = (I-tR)^{-1}z : z \in \ker D, t \in \mathscr{F}\}$. By definition, if $0 \neq x \in \ker(D-tI)$ then $x \in H_D(S)$. This implies that for all $n \in \mathbf{N}$ we have $t^n x = D^n x \neq 0$, which contradicts to our assumption.

(v) Theorem 3.4.1 implies that

$$\ker P(D) = [P(I, R)]^{-1} \ker D^n, \quad \text{where } n = \text{degree } P.$$

This and Point (iii) of Proposition 3.4.2 together imply that $\ker P(D) \in H_D\big(\ker P(D)\big) = H_D(\ker D^n) \subset H_D(S) = H_D(0)$ for any $P \in \mathscr{F}[t]$. Hence $H_D(0) \subset \bigcup_{P \in \mathscr{F}[t]} \ker P(D) \subset H_D(0)$.

(vi) Suppose now that $P(D)x = u + v \in U + V$ for a $P \in \mathscr{F}[t]$ (where $u \in U$, $v \in V$). Theorem 3.4.2 implies that $x = u_1 + v_1 + z_n$, where $u_1 + v_1 \in [P(I, R)]^{-1} R^n (u+v)$, $z_n \in \ker P(D)$, $n = \text{degree } P(t)$. But, by definition and Point (v) of our proposition, $u_1 \in U \subset H_D(U)$, $v_1 \in V \subset H_D(V)$ and $z_n \in \ker P(D) \subset H_D(0) \subset H_D(U) + H_D(V)$. Hence x

$\in H_D(U) \cap H_D(V)$, which implies $x \in H_D(U) + H_D(V)$. On the other hand, if $x_1 \in H_D(U)$, $x_2 \in H_D(V)$ then $P_1(D)x_1 \in U$, $P_2(D)x_2 \in V$ for some $P_1, P_2 \in \mathscr{F}[t]$. Then $P_1(D)P_2(D)x_1 \in U$, $P_1(D)P_2(D)x_2 \in V$, which implies $P_1(D)P_2(D)(x_1 + x_2) \in U + V$ and $x_1 + x_2 \in H_D(U+V)$.   ∎

PROPOSITION 3.4.4. *Suppose that all assumptions of Proposition 3.4.2 are satisfied. Let* $Q(t) = \sum\limits_{i=0}^{n} q_i t^i$ *and let* $P(t) = \sum\limits_{i=0}^{m} p_i t^i$, *where* $q_0 \neq 0, \ldots$
$\ldots, q_n, p_0, \ldots, p_m$ *are scalars. Write*

$$R_\infty = \bigcap_{i=0}^{\infty} R^i X, \quad D_\infty = \bigcap_{i=0}^{\infty} \mathrm{dom}\, D^i. \qquad (3.4.22)$$

Then:

(i) $Q(R)$ *maps* $R^i X$ *into itself for all* $i \in \mathbf{N}$;

(ii) *The following equalities hold*

$$R_\infty = \bigcap_{i=0}^{\infty} R^i D_\infty, \qquad (3.4.23)$$

$$R_\infty = \{ x \in X : R^n D^n x = x \text{ for all } n \in \mathbf{N} \}; \qquad (3.4.24)$$

(iii) R_∞ *is the largest* D–R *invariant subspace, where* D *and* R *are mutually inverse*;

(iv) $R^n Q(R)$ *maps* R_∞ *into itself for all* $n \in \mathbf{N}$;

(v) $P(D)$ *maps* R_∞ *into itself.*

Proof. Since R is a Volterra right inverse, the polynomial $Q(R)$ is invertible on R_∞. Moreover, $Q(R)$ commutes with R^i for all $i \in \mathbf{N}$, which proves (i).

In order to prove Formula (3.4.24) observe that $R^i D^i 0 = 0$ for all $i \in \mathbf{N}$, hence the set $\{ x \in X : R^n D^n x = x \text{ for all } n \in \mathbf{N} \}$ is not empty. If $x = R^i D^i x$ for all $i \in \mathbf{N}$ then $x \in R^i X$ for all $i \in \mathbf{N}$, hence $x \in R_\infty$. If $x \in R_\infty$ then $x = R^i x_i$, where $x_i \in X$ for all $i \in \mathbf{N}$. Hence $R^i D^i x = R^i D^i (R^i x_i) = R^i x_i = x$ for all $i \in \mathbf{N}$.

In order to prove Formula (3.4.23) observe that $\bigcap\limits_{i=0}^{\infty} R^i D_\infty \subset \bigcap\limits_{i=0}^{\infty} R^i X$ $= R_\infty$.

By Formula (3.4.23), if $x \in R_\infty$ then $x = R^i D^i x$ for all $i \in \mathbf{N}$ which implies $x \in D_\infty$. Hence $x = R^i D^i x \in R^i D^i(D_\infty) = R^i D_\infty$ for all $i \in \mathbf{N}$. We therefore conclude that $R_\infty \subset \bigcap_{i=0}^{\infty} R^i D_\infty$.

Formulae (3.4.22), (3.4.23) and (3.4.24) together imply that

$$R_\infty = \bigcap_{i=1}^{\infty} R^i X,$$

$$DR_\infty = D \bigcap_{i=1}^{\infty} R^i X \subset \bigcap_{i=1}^{\infty} D(R^i X) = \bigcap_{i=1}^{\infty} R^{i-1} X$$

$$= \bigcap_{j=0}^{\infty} R^j X = R_\infty,$$

$$RR_\infty = R \bigcap_{i=0}^{\infty} R^i X = \bigcap_{i=0}^{\infty} R^{i+1} X = R_\infty.$$

Hence R_∞ is a *D–R* invariant subspace.

We have

$$R_\infty \cap \ker D = \{0\}. \tag{3.4.25}$$

Indeed, if $x \in R_\infty \cap \ker D$ then $x = R^i D^i x$ for all $i \in \mathbf{N}$ and $Dx = 0$, which implies $x = 0$. This means that $RD = DR = I$ on R_∞. Let U be a *D–R* invariant subspace on which D and R are mutually inverse. Then $R^n D^n u = u$ for all $n \in \mathbf{N}$ and $u \in U$. The arbitrariness of $u \in U$ implies $U \subset R_\infty$.

In order to prove (iv) it is enough to observe that

$$Q(R) R_\infty = Q(R) \bigcap_{i=0}^{\infty} R^i X = \bigcap_{i=0}^{\infty} Q(R) R^i X = \bigcap_{i=0}^{\infty} R^i X = R_\infty$$

and to apply Point (iii) of our proposition.

In order to prove (v) observe that $P(D) = D^n(a_n I + a_{n-1} R + \ldots + a_i R^{n-i})$, where a_i is the first coefficient of $P(D)$ different from zero. This, and Points (i), (ii) of our proposition, together imply that $P(D)$ maps R_∞ into itself. ∎

COROLLARY 3.4.2. *Suppose that all assumptions of Proposition 3.4.2 are satisfied.*

(i) *Let U be a subspace of X. Then $R_\infty \cap U = \{0\}$ if and only if $R_\infty \cap$*
$\cap H_D(U) = \{0\}$;

(ii) $H_D(R_\infty) = R_\infty \oplus H_D(0)$.

Proof. (i) If $x \in R_\infty \cap H_D(U)$ then $x \in R_\infty$ and $x \in H_D(U)$. This and
Point (v) of Proposition 3.4.4 together imply that for any polynomial
$P(t)$ with scalar coefficients we find $P(D)x \in R_\infty$ and $P(D)x \in H_D(U)$.
Therefore, if $P(D)x = 0$ then by Point (v) of Proposition 3.4.4, we find
$R_\infty \cap H_D(U) = \{0\}$.

(ii) Propositions 3.4.2 and 3.4.3 together imply that $R_\infty + H_D(0)$
$\subset H_D(R_\infty)$. If $x \in H_D(R_\infty)$ then, by Theorem 3.4.2 $q = P(D)x$ implies
$x = [P(I, R)]^{-1} R^n q + z_n$, where $z_n \in \ker P(D)$. Point (iv) of Proposition
3.4.4 and Point (iv) of Proposition 3.4.3 together imply that $R_\infty \cap \{0\}$
$= \{0\}$ and $R_\infty \cap H_D(0) = \{0\}$. ∎

DEFINITION 3.4.3. Suppose that all assumptions of Definition 3.4.1 are
satisfied. A subspace $U \subset X$ is said to be *D-closed* if

$$P(D)x \in U \quad \text{if and only if } x \in U, \ P \in \mathscr{F}[t]. \tag{3.4.26}$$

PROPOSITION 3.4.5. *Suppose that all assumptions of Proposition 3.4.2 are
satisfied.*

(i) *A subspace $U \subset X$ is D-closed if and only if $H_D(U) = U$, i.e. the
D-hull is D-closed;*

(ii) *The intersection of D-closed subspaces is again D-closed.*

Proof. Since $U \subset H_D(U)$, Formula (3.4.26) implies that for an $x \in H_D(U)$
we have $P(D)x \in U$ for $P \in \mathscr{F}[t]$ and $x \in U$. On the other hand, $P(D)x$
$\in U$ for a $P \in \mathscr{F}[t]$ implies $x \in H_D(U) = U$.

Let $V = \bigcap_{i \in \mathscr{I}} V_i$, $V_i \subset X$ for $i \in \mathscr{I}$. Then $P(D)x \in V$ for a $P \in \mathscr{F}[t]$
if and only if $P(D)x \in V_i$ for all $i \in \mathscr{I}$. Formula (3.4.26) implies
that this is equivalent to $x \in V_i$ for $i \in \mathscr{I}$ which holds if and only if
$x \in \bigcap_{i \in \mathscr{I}} V_i$. ∎

THEOREM 3.4.3 (von Trotha, 1981). *Suppose that all assumptions of Proposition 3.4.2 are satisfied. Let U be a subspace of X and let $P \in \mathscr{F}[t]$.*

Then $H_D(U)$ is the smallest extension space of U, where every equation $P(D)x = y$ with $y \in H_D(U)$ has a solution and all solutions belong to $H_D(U)$.

Proof. Theorem 3.4.2 implies that the equation $P(D)x = y$ has a solution for every $y \in X$. Hence $P(D)x = y \in H_D(U)$ implies $x \in H_D(H_D(U)) = H_D(U)$. Let $V \supset U$ be a D-invariant extension subspace. Again Theorem 3.4.2 implies that the equation $P(D)x = y$ has solutions x. If $P(D)x = y \in V$ then, by our assumptions, $x \in V$. Formula (3.4.26) implies that V is D-closed, i.e. $H_D(V) = V$.

Observe now that for all D-invariant subspaces $V \supset U$ we have

$$\bigcap_{U \subset V} V = H_D(V) \subset H_D(U),$$

and by Point (ii) of Proposition 3.4.5

$$U \subset \bigcap_{U \subset V} V = H_D(V)$$

implies

$$H_D(U) \subset H_D\!\left(\bigcap_{U \subset V} V\right) \subset \bigcap_{U \subset V} H_D(V) = V,$$

which was to be proved.   ■

Examples and Exercises

EXAMPLE 3.4.1. Suppose that $X = C[a, b]$ and that $D = \mathrm{d}/\mathrm{d}t$, $(Rx)(t) = \int_{t_0}^{t} x(s)\,\mathrm{d}s$, for $x \in C[a, b]$, $t_0 \in [a, b]$. Example 2.3.1 shows that in the space $C[a, b]$ every exponential element is of the form $y(t) = Ce^{\lambda t}$, where $\lambda \in \mathbf{C}$, C is an arbitrary constant. (We recall that in this example there is a one-to-one correspondence between scalars and constants, as follows from Example 2.1.1.)

Formula (2.2.41) implies that for any scalar λ

$$[(I - \lambda R)^{-1}y](t) = y(t) + \lambda \int_{t_0}^{t} y(s)\, e^{\lambda(t-s)}ds$$

$$= Ce^{\lambda t} + \lambda \int_{t_0}^{t} Ce^{\lambda s}\, e^{\lambda(t-s)}ds$$

$$= Ce^{\lambda t} + Ce^{\lambda t}\lambda \int_{t_0}^{t} ds = Ce^{\lambda t}[1 + \lambda(t - t_0)].$$

By an easy induction we prove that for any scalar λ

$$[(I - \lambda R)^{-m}y](t) = Ce^{\lambda t} \sum_{k=0}^{m} \frac{\lambda^k(t - t_0)^k}{k!} \qquad (m = 1, 2, ...).$$

This, and Corollary 3.4.1, together imply that every element of a principal space of the operator $Q(D)$ corresponding to the eigenvalue λ_j of the multiplicity r_j is of the form

$$x(t) = e^{\lambda_j t} \sum_{m=0}^{r_j-1} c_m \sum_{k=0}^{m} \frac{\lambda_j^k(t - t_0)^k}{k!},$$

where $c_0, ..., c_{r_j-1}$ are arbitrary constants.
In particular, if we put

$$C_i = \sum_{m=i}^{r_j-1} c_m \sum_{\mu=0}^{m-i} \lambda_j^{\mu+i}\, \frac{(-1)^\mu}{i!\mu!}\, t_0^\mu \qquad (i = 1, 2, ..., r_j-1)$$

we find

$$x(t) = e^{\lambda_j t} \sum_{m=0}^{r_j-1} c_m \sum_{k=0}^{m} \frac{\lambda_j^k(t - t_0)^k}{k!}$$

$$= e^{\lambda_j t} \sum_{m=0}^{r_j-1} c_m \sum_{k=0}^{m} \frac{\lambda_j^k}{k!} \sum_{i=0}^{k} \binom{k}{i} t^i (-1)^{k-i} t_0^{k-i}$$

$$= e^{\lambda_j t} \sum_{m=0}^{r_j-1} c_m \sum_{i=0}^{m} t^i \sum_{k=i}^{m} \frac{\lambda_j^k}{k!} \binom{k}{i} (-1)^{k-i} t_0^{k-i}$$

$$= e^{\lambda_j t} \sum_{i=0}^{r_j-1} t^i \sum_{m=i}^{r_j-1} c_m \sum_{k=i}^{m} \frac{\lambda_j^k}{k!} \frac{k!}{i!(k-i)!} (-1)^{k-i} t_0^{k-i}$$

$$= e^{\lambda_j t} \sum_{i=0}^{r_j-1} t^i \sum_{m=i}^{r_j-1} c_m \sum_{\mu=0}^{m-i} \lambda_j^{\mu+i} \frac{(-1)^\mu}{i!\mu!} t_0^\mu = e^{\lambda_j t} \sum_{i=0}^{r_j-1} C_i t^i.$$

Finally, we conclude that all elements of a principal space corresponding to an eigenvalue λ_j with the multiplicity r_j are of the form

$$x(t) = e^{\lambda_j t} \sum_{i=0}^{r_j-1} C_i t^i, \qquad (3.4.27)$$

where $C_0, \ldots, C_{r_j-1}$ are arbitrary constants. In particular, if $\lambda_j = 0$ then elements of the corresponding principal spaces are of the form $x(t) = \sum_{i=0}^{r_j-1} C_i t^i$, where C_i are arbitrary constants. Thus the general solution of the equation

$$\sum_{k=0}^{N} p_k x^{(k)} = 0, \quad \text{where } p_0, \ldots, p_N \in \mathbf{C}, p_N = 1, \qquad (3.4.28)$$

is of the form

$$x(t) = \sum_{j=0}^{m} e^{\lambda_j t} \sum_{i=0}^{r_j-1} C_{ji} t^i, \qquad (3.4.29)$$

where C_{ji} are arbitrary constants, $\lambda_0, \ldots, \lambda_n$ are roots of the characteristic polynomial

$$P(t) = \sum_{k=0}^{N} p_k t^k = t^{r_0} \prod_{j=1}^{n} (t - \lambda_j)^{r_j} \quad (r_0 + \ldots + r_n = N).$$

If all coefficients of the polynomial $P(t)$ are real and characteristic roots λ_j are real we obtain the same formula for solutions of Equation (3.4.28). If all coefficients of the characteristic polynomial $P(t)$ are real, but there exist two complex conjugate roots $\lambda_j = \mu + i\nu$, $\bar{\lambda}_j = \mu - i\nu$, we have to find two linearly independent real solutions.

We find a solution of the non-homogeneous equation

$$\sum_{k=0}^{N} p_k x^{(k)} = y, \quad y \in C[a, b] \tag{3.4.30}$$

if we add to the general solution of the homogeneous equation (3.4.28) an arbitrary particular solution of Equation (3.4.30), for instance (cf. Theorem 3.4.2), a function

$$\left[\prod_{m=1}^{n} (I - \lambda_m R)^{r_m - 1}\right] R^N y = \sum_{j=1}^{n} \sum_{m=0}^{r_i - 1} d_{jm} W_m^{\lambda j} R^N y.$$

In some cases we can find a particular solution of the non-homogeneous equation (3.4.11) in another simpler way, as the following example shows.

EXAMPLE 3.4.2 (*Method of indetermined coefficients*). Suppose that X is a linear space over an algebraically closed field of scalars $\mathscr{F}$, $D \in R(X)$, dim ker $D \neq 0$ and $R \in \mathscr{R}_D \cap V(X)$. Let $P \in \mathscr{F}[t]$ be arbitrarily fixed. Consider the equation

$$P(D)x = y. \tag{3.4.31}$$

Suppose that y is either of the form

$$y = \sum_{m=0}^{M} e_{\mu_m}(z_m), \quad P(\mu_m) \neq 0 \quad (m = 1, 2, \dots, M), \tag{3.4.32}$$

where $\mu_i \neq \mu_j$ for $i \neq j$, $z_1, \dots, z_m \in \ker D$ are arbitrarily fixed, or of the form

$$y = c_\mu(z_1) + s_\mu(z_2), \quad P(i\mu) \neq 0, \quad P(-i\mu) \neq 0, \tag{3.4.33}$$

where $z_1, z_2 \in \ker D$ are arbitrarily fixed.

Observe that every element y either of the form (3.4.32) or of the form (3.4.33) belongs to

$$\bigcup_{Q \in \mathscr{F}[t]} \ker Q(D) = H_D(S) = H_D(0)$$

(cf. Proposition 3.4.3). Therefore, by the von Trotha Theorem 3.4.3 every solution of Equation (3.4.31) also belongs to $H_D(0)$.

Suppose now that y is of the form (3.4.32). Since $P(\mu_m) \neq 0$ for

$m = 0, 1, \ldots, M$, we conclude that $P(D)y \neq 0$, i.e. $y \notin \ker P(D)$. We are looking for solutions of Equation (3.4.31) which are of the form

$$x = \sum_{m=0}^{M} e_{\mu_m}(\zeta_m),$$

where $\zeta_0, \ldots, \zeta_M \in \ker D$ are to be determined. By properties of exponential elements we find:

$$\sum_{m=0}^{M} e_{\mu_m}(z_m) = y = P(D)x = P(D) \sum_{m=0}^{M} e_{\mu_m}(\zeta_m)$$

$$= \sum_{m=0}^{M} P(D)e_{\mu_m}(\zeta_m) = \sum_{m=0}^{M} P(\mu_m)e_{\mu_m}(\zeta_m)$$

$$= \sum_{m=0}^{M} e_{\mu_m} P(\mu_m)\zeta_m.$$

The linear independence of exponential elements corresponding to eigenvalues $\mu_0, \ldots, \mu_M$ (different by our assumptions) implies that $P(\mu_m)\zeta_m = z_m$, i.e. $\zeta_m = z_m/P(\mu_m)$ for $m = 0, 1, \ldots, M$. Therefore

$$x = \sum_{m=0}^{M} \frac{1}{P(\mu_m)}\, e_{\mu_m}(z_m)$$

is a solution of Equation (3.4.31) with y of the form (3.4.32).

Suppose now that y is of the form (3.4.33). We are looking for solutions of Equation (3.4.31) which are of the form

$$x = c_\mu(\zeta_1) + s_\mu(\zeta_2),$$

where $\zeta_1, \zeta_2 \in \ker D$ are to be determined. But, by definition

$$c_\mu = \frac{1}{2}\,(e_{\mu i} + e_{-\mu i}), \qquad s_\mu = \frac{1}{2i}\,(e_{\mu i} - e_{-\mu i})$$

and, by our assumption $P(\mu i) \neq 0$, $P(-\mu i) \neq 0$. Therefore we obtain in a similar way, as in the preceding case, that $y \notin \ker P(D)$ and

$$\frac{1}{2}z_1 + \frac{1}{2i}z_2 = P(\mu i)\left(\frac{1}{2}\zeta_1 + \frac{1}{2i}\zeta_2\right),$$

$$\frac{1}{2}z_1 - \frac{1}{2i}z_2 = P(-\mu i)\left(\frac{1}{2}\zeta_1 - \frac{1}{2i}\zeta_2\right).$$

(3.4.34)

The system of two algebraic equations with two unknowns ζ_1, ζ_2 has the determinant

$$\Delta = \frac{1}{2i} P(\mu i) P(-\mu i) \neq 0$$

by our assumptions. Thus the system has a unique solution of the form

$$\zeta_1 = \frac{1}{2i} \left\{ \left[\frac{1}{P(\mu i)} + \frac{1}{P(-\mu i)} \right] z_1 - i \left[\frac{1}{P(\mu i)} - \frac{1}{P(-\mu i)} \right] z_2 \right\},$$

$$\zeta_2 = \frac{1}{2} \left\{ \left[\frac{1}{P(\mu i)} - \frac{1}{P(-\mu i)} \right] z_1 - i \left[\frac{1}{P(\mu i)} + \frac{1}{P(-\mu i)} \right] z_2 \right\}$$

which implies

$$x = \frac{1}{2i} c_\mu (p_+ z_1 - i p_- z_2) + \frac{1}{2} s_\mu (p_- z_1 - i p_+ z_2),$$

where $p_\pm = 1/P(\mu i) \pm 1/P(-\mu i)$.

The same method can be used also in the case where $y = \sum_{m=0}^{M} R^m z_m$, $z_0, \ldots, z_M \in \ker D$ are arbitrarily fixed.

EXERCISE 3.4.1. Find a solution of Equation (3.4.30) by an application of the method of indetermined coefficients presented in Example 3.4.2 if the given function y is not a solution of the homogeneous Equation (3.4.28) and is of the form:

(i) $y = \sum_{m=0}^{M} a_k e^{\mu_k t}$, $\mu_k \neq \mu_j$ for $k \neq j$;

(ii) $y = a\cos\mu t + b\sin\mu t$;

(iii) $y = \sum_{k=0}^{M} a_k t^k$;

(iv) y is a product of functions of the form (i), (ii) and (iii).

EXERCISE 3.4.2. Suppose that y is a solution of the homogeneous Equation (3.4.28) which is of the form either (i) or (ii) or (iii) described in Exercise 3.4.1. Prove that the corresponding solution of Equation (3.4.30) can be found by multiplication of solutions determined in Exercise 3.4.1 by a polynomial of a degree dependent on the multiplicity of characteristic roots.

EXERCISE 3.4.3. Let X, D, R be determined, as in Example 3.4.1.

(i) Solve a linear homogeneous equation of order 2 with scalar coefficients

$$ax'' + bx' + cx = 0 \quad (a \neq 0)$$

in the case:

(i1) $a, b, c \in \mathbf{C}$,   (i2) $a, b, c \in \mathbf{R}$.

(ii) Solve the following linear homogeneous equations of higher order in complex and real case:

(ii1) $x''' + x = 0$,   (ii2) $x^{(4)} - 2x'' + x = 0$,
(ii3) $x^{(4)} - x = 0$,   (ii4) $x^{(4)} + x = 0$,
(ii5) $x^{(6)} - x'' = 0$,   (ii6) $x^{(6)} - x = 0$.

(iii) Determine the decomposition onto vulgar fraction of the operator $[Q(I, R)]^{-1}$ if $Q(I, R)$ is of the form:

$$I + R^3, \quad I - 2R^2 + R^4, \quad I - R^4, \quad I + R^4, \quad I - R^6.$$

(iv) Solve the following equations:

(iv1) $x''' + x = \sin t$,   (iv2) $x^{(4)} - 2x'' + x = 3e^{2t}$,
(iv3) $x^{(4)} - x = t^2 + 1$,   (iv4) $x^{(4)} + x = t \sin t + \cos t$,
(iv5) $x^{(6)} - x'' = \cos t + 1$,   (iv6) $x^{(6)} - x = e^t + t + 1$.

EXERCISE 3.4.4. Solve the *Euler differential equation*

$$at^2 x'' + btx' + cx = y \quad (\text{where } a, b, c \in \mathbf{R}, a \neq 0)$$

by substitution in the homogeneous equation $x = t^\lambda$, where λ is to be determined. Explain, why the method used is analogous to that for equations with scalar coefficients.

EXERCISE 3.4.5. Suppose that X, D, R are defined, as in Example 2.3.2.

(i) Determine operators W_m^λ.

(ii) Solve a linear equation with scalar coefficients of order 2.

EXERCISE 3.4.6. Suppose that X, D, R are defined as in Example 2.3.3.

(i) Determine operators W_m^λ.

(ii) Solve a linear equation with scalar coefficients of order two in complex and real case.

EXERCISE 3.4.7. Suppose that X is a linear space over $\mathbf{C}$, $D \in R(X)$, dim ker $D \neq 0$, $R \in \mathcal{R}_D \cap V(X)$ and A is an algebraic operator with single roots $t_1, \ldots, t_n$ only which commute with D. Prove that the equation $(D-A)x = y$, $y \in X$, is equivalent to a system of n independent equations with scalar coefficients $(D-t_j I)x_j = y_j$ $(j = 1, 2, \ldots, n)$, where $x_j = P_j x$, $y_j = P_j y$, $P_1, \ldots, P_n$ are projectors determined by A giving a partion of unit. Find a general solution of the equation

$$(D-A)x = y.$$

EXERCISE 3.4.8. Extend results of Exercise 3.4.7 for an algebraic operator with multiple roots.

EXERCISE 3.4.9. Suppose that all assumptions of Proposition 3.4.1 are satisfied. Let S, R_∞ and D_∞ be defined by Formulae (3.4.21) and (3.4.22), respectively. S is said to be the *space of finite elements* of X. R_∞ is said to be the *space of singular elements* of X. Prove that:

(i) D_∞ is the largest D–R invariant subspace of X and contains S and R_∞ (cf. also Proposition 3.4.6 on p. 218).

(ii) S is the smallest D–R invariant subspace of X containing ker D.

(iii) $S = \bigcup_{i=0}^{\infty} \ker D \oplus \ldots \oplus R^i \ker D^i$.

(iv) $S = \{x \in X : \exists_{n \in \mathbb{N}} R^n D^n x = 0\}$.

(v) $S \cap R_\infty = \{0\}$ and $S \oplus R_\infty$ is a D–R invariant subspace of X.

(vi) If $R_\infty \neq \{0\}$ then dim $R_\infty = +\infty$ (cf. von Trotha, 1981).

EXERCISE 3.4.10. Suppose that $X = C(R)$, $D = \mathrm{d}/\mathrm{d}t$, $R = \int_0^t$. Give an interpretation of results of Exercise 3.4.9 in that case.

3.5. SYSTEMS OF EQUATIONS WITH SCALAR COEFFICIENTS

Suppose that $D \in R(X)$, dim ker $D \neq 0$ and $R \in \mathcal{R}_D$. Consider a system of linear equations with scalar coefficients:

$$Dx_j - \sum_{k=1}^{M} a_{jk} x_k = y_j, \quad y_j \in X \quad (j = 1, 2, \ldots, M), \qquad (3.5.1)$$

where a_{jk} are scalars (belonging to the field under consideration). Write

$$\tilde{X} = X^M, \quad x = (x_1, \ldots, x_M), y = (y_1, \ldots, y_M) \in \tilde{X}, \qquad (3.5.2)$$

where $x_1, \ldots, x_M, y_1, \ldots, y_M \in X$,

$$\tilde{A} = (a_{jk})_{j,k=1,2,\ldots,M}; \quad \tilde{I} = (\delta_{jk}I)_{j,k=1,2,\ldots,M},$$
$$\tilde{D} = (\delta_{jk}D)_{j,k=1,2,\ldots,M}, \quad \tilde{R} = (\delta_{jk}R)_{j,k=1,2,\ldots,M}, \qquad (3.5.3)$$

where δ_{jk} denotes the *Kronecker symbol*, i.e. $\delta_{jk} = 1$ for $j = k$ and 0 otherwise.

Since the square matrices $\tilde{I}, \tilde{D}, \tilde{R}$ of dimension M are diagonal and have all term on principal diagonal equal, we conclude that

$$\tilde{D}\tilde{R} = I \text{ on } \tilde{X}, \quad \text{thus } \tilde{D} \in R(\tilde{X}), \tilde{R} \in \mathcal{R}_{\tilde{D}}, \qquad (3.5.4)$$

$$\tilde{A} \in L_0(\tilde{X}), \quad \tilde{D}\tilde{A} - \tilde{A}\tilde{D} = 0 \text{ on dom } \tilde{D}$$
$$\text{and } \tilde{R}\tilde{A} - \tilde{A}\tilde{R} = 0 \text{ on } \tilde{X}. \qquad (3.5.5)$$

Moreover, if $R \in V(X)$ then $\tilde{R} \in V(\tilde{X})$ and

$$(\tilde{I} - \lambda\tilde{R})^{-1} = (\delta_{jk}(I - \lambda R)^{-1})_{j,k=1,2,\ldots,M} \qquad (3.5.6)$$

for all scalars λ.

Indeed, if $R \in V(X)$ then the operator $(I - \lambda R)^{-1}$ exists for every scalar λ. Since diagonal matrices commute, we conclude that the operator $\tilde{I} - \lambda\tilde{R}$ is invertible for every scalar λ. Hence $\tilde{R}$ is a Volterra operator.

Using the denotations (3.5.2), (3.5.3) we can rewrite the system (3.5.1) as follows:

$$(\tilde{D} - \tilde{A})x = y, \quad y \in \tilde{X}. \qquad (3.5.7)$$

To solve this equation we shall use the following*

LEMMA 3.5.1. *Suppose that*:

(i) $\tilde{X}$ *is a linear space over* **C**;

(ii) $\tilde{D} \in R(\tilde{X})$, dim ker $\tilde{D} \neq 0$ *and* $\tilde{R} \in \mathcal{R}_{\tilde{D}} \cap V(\tilde{X})$;

(iii) $\tilde{A} \in L_0(\tilde{X})$ *is an algebraic stationary operator having the characteristic polynomial*

---

* Lemma 3.5.1 is a modification of a theorem obtained by B. Mażbic–Kulma, 1971 (cf. also Przeworska-Rolewicz, 1979, p. 120).

$$P(t) = \sum_{k=0}^{N} p_k t^k = t^{r_0} Q(t), \quad where\ p_0, \ldots, p_{N-1} \in C, \quad (3.5.8)$$

$$Q(t) = \prod_{j=1}^{n} (t - \lambda_j)^{r_j}, \quad r_0 \geqslant 0, \quad r_0 + \ldots + r_n = N \leqslant M$$

$$(\lambda_j \neq 0\ for\ j = 1, \ldots, n). \quad (3.5.9)$$

Write

$$Q(t, s) = \prod_{j=1}^{n} (t - \lambda_j s)^{r_j}, \quad (3.5.10)$$

$$\tilde{W}_m^{\lambda} = \begin{cases} (\tilde{I} - \lambda \tilde{R})^{-m} & for\ \lambda \neq 0, \\ \tilde{R}^{m-1} & for\ \lambda = 0, \end{cases} \quad (m = 1, 2, \ldots) \quad (3.5.11)$$

$$\tilde{A}_1 = \sum_{k=1}^{N} p_k \sum_{m=0}^{k-1} \tilde{A}^{k-1-m} \tilde{D}^m. \quad (3.5.12)$$

If $y \in \mathrm{dom}\ \tilde{D}^{N-1}$ then every solution of Equation (3.5.7) belonging to dom $\tilde{D}^N$ is of the form

$$x = \sum_{j=1}^{n} \sum_{m=1}^{r_j} \tilde{W}_m^{\lambda_j} (d_{jm} \tilde{R}^N \tilde{A}_1 y + z_{jm}) + \sum_{m=1}^{r_0} \tilde{R}^{m-1} z_{0m}, \quad (3.5.13)$$

where scalars d_{jm} are determined by the decomposition (3.4.10) of the rational function $[Q(1, s)]^{-1}$ onto vulgar fractions and $z_{jm} \in \ker \tilde{D}$ are principal vectors of the operator $\tilde{A}$ corresponding to a characteristic root λ_j, i.e. these constants satisfies equations

$$(\tilde{A} - \lambda_j \tilde{I})^m z_{jm} = 0, \quad z_{jm} \in \ker \tilde{D}, \quad (3.5.14)$$

$m = 1, 2, \ldots, r_j; \quad j = 0, 1, 2, \ldots, n.$

We assume that the component $\sum_{m=1}^{r_0-1} \tilde{R}^{m-1} z_{0m}$ appears only if $r_0 \geqslant 1$, i.e. if the polynomial $P(t)$ has a root $\lambda_0 = 0$.

Proof. Observe that the degree N of the characteristic polynomial of the matrix $\tilde{A}$ is not greater than the dimension M of $\tilde{A}$. Observe also that the operators $\tilde{D}^m$ and $\tilde{R}^m$ are represented by diagonal matrices of dimension M. To begin with, we shall show that Equations (3.5.14) are satisfied if and only if

$$(\tilde{A} - \lambda_j \tilde{I})^m \tilde{W}_m^{\lambda_j} z_{jm} = 0 \quad \text{for } z_{jm} \in \ker \tilde{D},$$

$$m = 1, ..., r_j; \, j = 0, 1, 2, ..., n. \qquad (3.5.15)$$

Indeed, since the operators $\tilde{A}$ and $\tilde{R}$ commute and the operators $\tilde{I} - \lambda_j \tilde{R}$ are invertible, Equations (3.5.14) for $\lambda_j \neq 0$ imply that

$$(\tilde{A} - \lambda_j \tilde{I})^m \tilde{W}_m^{\lambda_j} z_{jm} = (\tilde{A} - \lambda_j \tilde{I})^m (\tilde{I} - \lambda_j \tilde{R})^{-m} z_{jm}$$

$$= (\tilde{I} - \lambda_j \tilde{R})^{-m} (\tilde{A} - \lambda_j \tilde{I})^m z_{jm} = 0.$$

If $\lambda_j = 0$, we find

$$(\tilde{A} - \lambda_j \tilde{I})^m \tilde{W}_m^{\lambda_j} z_{jm} = (\tilde{A} - \lambda_j \tilde{I})^m \tilde{R}^m z_{jm} = \tilde{R}^m (\tilde{A} - \lambda_j \tilde{I})^m z_{jm} = 0.$$

Conversely, if $(\tilde{A} - \lambda_j \tilde{I})^m \tilde{W}_m^{\lambda_j} z_{jm} = 0$ and $\lambda_j \neq 0$, then

$$(\tilde{A} - \lambda_j \tilde{I})^m z_{jm} = (\tilde{A} - \lambda_j \tilde{I})^m (\tilde{I} - \tilde{\lambda}_j \tilde{R})^m (\tilde{I} - \lambda_j \tilde{R})^{-m} z_{jm}$$

$$= (\tilde{I} - \lambda_j \tilde{R})^m (\tilde{A} - \lambda_j \tilde{I})^m \tilde{W}_m^{\lambda_j} z_{jm} = 0.$$

If $\lambda_j = 0$ then we have

$$(\tilde{A} - \lambda_j \tilde{I})^m z_{jm} = \tilde{D}^{m-1} \tilde{R}^{m-1} (\tilde{A} - \lambda_j \tilde{I})^m z_{jm}$$

$$= \tilde{D}^{m-1} (\tilde{A} - \lambda_j \tilde{I})^m \tilde{R}^{m-1} z_{jm} = \tilde{D}^{m-1} (\tilde{A} - \lambda_j \tilde{I})^m \tilde{W}_m^{\lambda_j} z_{jm} = 0.$$

If there exists a solution $x \in \operatorname{dom} \tilde{D}$ of Equation (3.5.7) then, by our assumption that $\tilde{A}$ is stationary, we have $\tilde{D}\tilde{A} = \tilde{A}\tilde{D}$ on $\operatorname{dom} \tilde{D}$ and $\tilde{D}^2 x = \tilde{D}(\tilde{D}x) = \tilde{D}(\tilde{A}x + y) = \tilde{D}\tilde{A}x + \tilde{D}y = \tilde{A}\tilde{D}x + \tilde{D}y = \tilde{A}(\tilde{A}x + y) + \tilde{D}y = \tilde{A}^2 x + \tilde{A}y + \tilde{D}y.$

By an easy induction we conclude that for an arbitrary $N \geqslant 1$

$$\tilde{D}^k x = \tilde{A}^k x + \sum_{m=0}^{k-1} \tilde{A}^{k-1-m} \tilde{D}^m y \quad \text{for } k = 1, ..., N-1. \qquad (3.5.16)$$

Indeed, suppose that Formula (3.5.16) to be true. Then

$$\tilde{D}^{k+1} x = \tilde{D}\tilde{A}^k x + \sum_{m=0}^{k-1} \tilde{D}\tilde{A}^{k-1-m} \tilde{D}^m y$$

$$= \tilde{A}^k \tilde{D}x + \sum_{m=0}^{k-1} \tilde{A}^{k-1-m} \tilde{D}^{m+1} y$$

$$= \tilde{A}^k(\tilde{A}x+y)+\sum_{j=1}^{k} \tilde{A}^{k-j}\tilde{D}^j y$$

$$= \tilde{A}^{k+1}x+\tilde{A}^k y+\sum_{j=1}^{k} \tilde{A}^{k-j}\tilde{D}^j y = \tilde{A}^{k+1}x+\sum_{j=0}^{k} \tilde{A}^{k-j}\tilde{D}^j y,$$

which was to be shown.

Since $P(t)$ is the characteristic polynomial of the operator $\tilde{A}$, we have $P(\tilde{A}) = 0$. This, and Formula (3.5.16), together imply that

$$P(\tilde{D})x = \sum_{k=0}^{N} p_k \tilde{D}^k x = \sum_{k=0}^{N} p_k\left(\tilde{A}^k x+\sum_{m=0}^{k-1} \tilde{A}^{k-1-m}\tilde{D}^m y\right)$$

$$= \sum_{k=0}^{N} p_k \tilde{A}^k x+\sum_{k=0}^{N} p_k \sum_{m=0}^{k-1} \tilde{A}^{k-1-m}\tilde{D}^m y$$

$$= P(\tilde{A})x+\tilde{A}_1 y = \tilde{A}_1 y,$$

where $\tilde{A}_1$ is defined by Formula (3.5.12).

We therefore obtain the following equation:

$$P(\tilde{D})x = \tilde{A}_1 y. \tag{3.5.17}$$

Consider the corresponding homogeneous equation

$$P(\tilde{D})x = 0. \tag{3.5.18}$$

Theorem 3.4.1 implies that every solution of this equation can be presented in a unique way as the sum of elements $W_m^{\lambda_j}z_{jm}$, where $z_{jm} \in \ker \tilde{D}$ are arbitrary. On the other hand these solutions should also satisfy the equation $P(\tilde{A})x = 0$, which implies Condition (3.5.15). But we have shown, that these conditions are equivalent to Conditions (3.5.14). We therefore conclude that every solution of Equation (3.5.17) is of the form

$$\sum_{j=1}^{n}\sum_{m=1}^{r_j} \tilde{W}_m^{\lambda_j}z_{jm}+\sum_{m=1}^{r_0} \tilde{R}^{m-1}z_{0m},$$

where z_{jm} satisfy the Conditions (3.5.14). This, and Theorem 3.4.2 together imply that every solution of Equation (3.5.18), i.e. of Equation (3.5.7), is of the required form (3.5.13). ∎

THEOREM 3.5.1. *Suppose that X is a linear space over $\mathbf{C}$, $D \in R(X)$, dim ker $D \neq 0$ and $R \in \mathcal{R}_D \cap V(X)$. If $y_1, \ldots, y_M \in$ dom D^{N-1}, then every solution $(x_1, \ldots, x_M)$ of Equation (3.5.1) is of the form*

$$x_k = \sum_{j=1}^{n} \sum_{m=1}^{r_j} W_m^{\lambda_j}(d_{jm} R^N \eta_j + z_{jmk}) + \sum_{m=1}^{r_0} R^{m-1} z_{0mk}$$

$$(k = 1, \ldots, M), \qquad (3.5.19)$$

where $(\eta_1, \ldots, \eta_M) = \tilde{A}_1 y$, $y = (y_1, \ldots, y_M)$.

$$\tilde{A}_1 = \sum_{k=1}^{N} p_k \sum_{m=0}^{k-1} \tilde{A}^{k-1-m} \tilde{D}^m; \; \tilde{A}, \tilde{D} \text{ are determined by Formulae (3.5.3),}$$

$$P(t) = \sum_{k=0}^{N} p_k t^k = t^{r_0} Q(t), \quad p_0, \ldots, p_{N-1} \in \mathbf{C}, \quad p_N = 1,$$

$$Q(t, s) = \prod_{j=1}^{n} (t - \lambda_j s)^{r_j}, \quad r_0 \geqslant 0, \quad r_0 + \ldots + r_n = N \leqslant M$$

$$(\lambda_j \neq 0 \text{ for } j = 1, \ldots, n),$$

$Q(t, 1) = Q(t)$ *is the characteristic polynomial of the matrix of coefficients of Equation (3.5.7); scalars d_{jm} are determined by the decomposition of the rational function $[Q(1, s)]^{-1}$ onto vulgar fractions (3.4.10); vectors $z_{jm} = (z_{jm1}, \ldots, z_{jmM})$, where $z_{jmk} \in$ ker D satisfy Conditions (3.5.14).*

We assume that the components $\sum_{m=1}^{r_0} R^{m-1} z_{0mk}$ appear only if $r_0 \geqslant 1$, i.e. if the polynomial $P(t)$ has a root $\lambda_0 = 0$.

The proof is immediate if we introduce the denotations (3.5.2) and (3.5.3). Indeed, we find that the scalar matrix $\tilde{A}$ commutes with the operators $\tilde{D}$ and $\tilde{R}$, hence is stationary. Moreover, by Cayley–Hamilton Theorem (Corollary 1.4.2) $\tilde{A}$ is an algebraic operator. Since $\tilde{R}$ is a Volterra operator, we conclude that all assumptions of Lemma 3.5.1 are satisfied. Since $\tilde{W}_m^{\lambda}$ are diagonal matrices, we obtain the required formulae (3.5.19). ■

COROLLARY 3.5.1. *Suppose that all assumptions of Theorem 3.5.1 are satisfied and that the polynomial $P(t)$ has single roots only, i.e. $r_j = 1$*

for $j = 0, 1, ..., n = N \leqslant M$. Then every solution $(x_1, ..., x_M)$ of Equation (3.5.1) is of the form

$$x_k = \sum_{j=1}^{N} d_{j0}(I - \lambda_j R)^{-1} R^N \eta_j + \sum_{j=0}^{N} e_{\lambda_j}(z_{jk})$$

$$(k = 1, ..., M), \qquad (3.5.20)$$

where the constants z_{jk} satisfy the conditions:

$$\lambda_m z_{jm} - \sum_{k=1}^{M} a_{jk} z_{mk} = 0 \quad (j = 1, ..., M, m = 1, ..., N) \quad (3.5.21)$$

(d_{j0} and η_j are determined in Theorem 3.5.1).

Proof. Indeed, if we require the system $(u_1, ..., u_M)$, where u_J $= \sum_{m=0}^{N} e_{\lambda_m}(z_{mJ})$, $z_{mJ} \in \ker D$, to be a solution of the homogeneous equation (3.5.1) we find

$$0 = Du_J - \sum_{k=1}^{M} a_{jk} u_k$$

$$= \sum_{m=1}^{N} De_{\lambda_m}(z_{mJ}) - \sum_{k=1}^{M} a_{jk} \sum_{m=1}^{N} e_{\lambda_m}(z_{mk})$$

$$= \sum_{m=1}^{N} \left[\lambda_m e_{\lambda_m}(z_{mJ}) - \sum_{k=1}^{M} a_{jk} e_{\lambda_m}(z_{mk}) \right]$$

$$= \sum_{m=1}^{N} e_{\lambda_m} \left(\lambda_m z_{mj} - \sum_{k=1}^{M} a_{jk} z_{mk} \right) \quad (j = 1, ..., M).$$

The linear independence of exponentials for different characteristic roots and the assumption that dim $\ker D \neq 0$ together imply Condition (3.5.21). ∎

Writing $z_m = (z_{m1}, ..., z_{mM})$ for $m = 1, ..., N$ we obtain Condition (3.5.21) in the matrix form:

$$(\lambda_m \tilde{I} - \tilde{A}) z_m = 0 \quad (m = 1, ..., N). \qquad (3.5.22)$$

In a similar way we can also solve the following system of equations with scalar coefficients:

$$\sum_{k=1}^{M} (a_{jk}D + b_{jk}I)x_k = y_j, \quad y_j \in X \quad (j = 1, \ldots, M) \tag{3.5.23}$$

writing

$$\tilde{B} = (b_{jk})_{j,k=1,\ldots,M} \tag{3.5.24}$$

and applying the denotation (3.5.2), (3.5.3) and the following*

LEMMA 3.5.2. *Suppose that* $\tilde{X}$ *is a linear space over* **C**, $\tilde{D} \in R(\tilde{X})$, $\tilde{R} \in \mathcal{R}_{\tilde{D}} \cap$ $\cap V(\tilde{X})$, dim ker $\tilde{D} \neq 0$, *the operators* $\tilde{A}, \tilde{B} \in L_0(\tilde{X})$ *are stationary. Suppose, moreover, that the operator* $\tilde{A}$ *is invertible and the operator* $\tilde{A}^{-1}\tilde{B}$ *is algebraic. If* $\tilde{A}^{-1}y \in \text{dom } \tilde{D}^{N-1}$ *then every solution of the equation*

$$(\tilde{A}\tilde{D} + \tilde{B})x = y \tag{3.5.25}$$

is of the form

$$x = \sum_{j=1}^{n} \sum_{m=1}^{r_j} \tilde{W}_m^{\lambda_j}(d_{jm}\tilde{R}^N\tilde{A}_1\tilde{A}^{-1}y + z_{jm}) + \sum_{m=1}^{r_0} \tilde{R}^{m-1}z_{0m}, \tag{3.5.26}$$

where $\tilde{W}_m^{\lambda_j}, \tilde{A}_1, d_{jm}$ *are defined in Lemma 3.5.1, the constants* z_{jm} *satisfy the conditions*

$$(-\tilde{A}^{-1}\tilde{B} - \lambda_j\tilde{I})^m z_{jm} = 0 \quad (m = 1, 2, \ldots, r_j, j = 0, 1, \ldots, n) \tag{3.5.27}$$

and $\lambda_0, \ldots, \lambda_n$ *are roots of the characteristic polynomial* $P(t)$ *of the operator* $-\tilde{A}^{-1}\tilde{B}$.

Proof. Since, by our assumptions, $\tilde{A}\tilde{D} - \tilde{D}\tilde{A} = 0$, $\tilde{B}\tilde{D} - \tilde{D}\tilde{B} = 0$ on dom $\tilde{D}$, we find

$$\tilde{A}^{-1}\tilde{D} - \tilde{D}\tilde{A}^{-1} = \tilde{A}^{-1}(\tilde{D} - \tilde{A}\tilde{D}\tilde{A}^{-1}) = \tilde{A}^{-1}(\tilde{D}\tilde{A} - \tilde{A}\tilde{D})\tilde{A}^{-1} = 0$$

$$\text{on dom } \tilde{D},$$

---

* Lemma 3.5.2 is a modification of a theorem obtained by B. Mażbic–Kulma, 1971 (cf. also Przeworska-Rolewicz, 1979, p. 125).

Thus on dom $\tilde{D}$

$$(\tilde{A}^{-1}\tilde{B})\tilde{D} - \tilde{D}(\tilde{A}^{-1}\tilde{B}) = \tilde{A}^{-1}\tilde{B}\tilde{D} - \tilde{A}^{-1}\tilde{D}\tilde{B} = \tilde{A}^{-1}(\tilde{B}\tilde{D} - \tilde{D}\tilde{B}) = 0.$$

Hence the algebraic operator $\tilde{A}^{-1}\tilde{B}$ commutes with $\tilde{D}$ on dom $\tilde{D}$. In a similar way we show that $\tilde{A}^{-1}\tilde{B}$ commutes with $\tilde{R}$. Acting on both sides of Equation (3.5.25) by the operator $\tilde{A}^{-1}$ we obtain an equivalent equation

$$(\tilde{D} + \tilde{A}^{-1}\tilde{B})x = \tilde{A}^{-1}y.$$

Applying now to that equation Lemma 3.5.1 we obtain the required Formula (3.5.26). ∎

We can solve systems of equations with arbitrary coefficients belonging to $L(X)$ by reduction to one equation (in a matrix form) in the space $\tilde{X} = X^M$ and by application of Theorem 3.1.1.

Examples and Exercises

EXAMPLE 3.5.1. We shall solve the following system of linear differential equations with scalar coefficients

$$\frac{dx}{dt} = y + t, \quad \frac{dy}{dt} = x + 1. \tag{3.5.28}$$

Consider the homogeneous system

$$\frac{dx}{dt} = y, \quad \frac{dy}{dt} = x. \tag{3.5.29}$$

The matrix of coefficients has the form $\tilde{A} = \begin{bmatrix} 0 & 1 \\ 1 & 0 \end{bmatrix}$. We are looking for solutions of the system (3.5.29) which are of the form $x = \alpha e^{\lambda t}$, $y = \beta e^{\lambda t}$. By substitution of these functions of the system (3.5.29) we obtain the equations

$$\lambda \alpha e^{\lambda t} = \beta e^{\lambda t}, \quad \lambda \beta e^{\lambda t} = \alpha e^{\lambda t}.$$

Dividing both sides of the last equations by $e^{\lambda t}$ we obtain a system of two equations with two unknowns α and β

$$-\lambda \alpha + \beta = 0, \quad \alpha - \lambda \beta = 0. \tag{3.5.30}$$

The determinant of this system is

$$\det(\tilde{A} - \lambda \tilde{I}) = \begin{vmatrix} -\lambda & 1 \\ 1 & -\lambda \end{vmatrix} = \lambda^2 - 1 = (\lambda+1)(\lambda-1).$$

The system (3.5.30) has a non-trivial solution if and only if $\det(\tilde{A} - \lambda \tilde{I}) = 0$, i.e. for the numbers $\lambda_1 = 1$ and $\lambda_2 = -1$.

Suppose that $\lambda = \lambda_1 = 1$. Since the rank of the matrix $\tilde{A}$ is equal 1, we can put $\alpha = C_1$, where C_1 is an arbitrary constant. Then, from the first of Equations (3.5.30) we obtain $\beta = \lambda \alpha = \lambda_1 C_1 = C_1$. Thus a solution, corresponding to the characteristic root $\lambda_1 = 1$, is $x_1 = C_1 e^t$, $y_1 = C_1 e^t$.

Suppose now that $\lambda = \lambda_2 = -1$. As before, we put $\alpha = C_2$, where C_2 is an arbitrary constant, and we find $\beta = \lambda \alpha = \lambda_2 C_2 = -C_2$. Thus a solution, corresponding to the characteristic root $\lambda_2 = -1$, is $x_2 = C_2 e^{-t}$, $y_2 = -C_2 e^{-t}$.

Hence the general solution of the system (3.5.29) is

$$\begin{aligned} x &= x_1 + x_2 = C_1 e^t + C_2 e^{-t}, \\ y &= y_1 + y_2 = C_1 e^t - C_2 e^{-t}, \end{aligned} \tag{3.5.31}$$

where C_1, C_2 are arbitrary constants.

System (3.5.28) can also be solved in another way. Namely, we differentiate the first of Equations (3.5.28) and we reject the unknown y using the second of Equations (3.5.28). Then we find

$$\frac{d^2 x}{dt^2} = \frac{dx}{dt} + 1 = x + 1 + 1 = x + 2. \tag{3.5.32}$$

We therefore obtain an equation of order 2 with one unknown function x: $x'' - x = 2$.

It is easy to check that this equation has a general solution of the form $x = C_1 e^t + C_2 e^{-t} - 2$, where C_1, C_2 are arbitrary constants (i.e. x is of the form (3.5.31)). Thus $y' = x + 1 = C_1 e^t + C_2 e^{-t} - 2 + 1 = C_1 e^t + C_2 e^{-t} - 1$, which implies that $y = C_1 e^t - C_2 e^{-t} - t + C$, where C is a constant. If we require that y satisfies the first of Equations (3.5.28), we determine this constant. Namely

$$\begin{aligned} C_1 e^t - C_2 e^{-t} &= x' = y + t = C_1 e^t - C_2 e^{-t} - t + C + t \\ &= C_1 e^t - C_2 e^{-t} + C, \end{aligned} \tag{3.5.33}$$

which implies $C = 0$ and $y = C_1 e^t - C_2 e^{-t} - t$, as in Formulae (3.5.31).

This last method could be more convenient in the case of multiple characteristic roots.

EXAMPLE 3.5.2. Consider the following system of linear differential equations with scalar coefficients:

$$x' = y+z,$$
$$y' = x,$$ (3.5.34)
$$z' = -x.$$

In our case

$$\tilde{A} = \begin{bmatrix} 0 & 1 & 1 \\ 1 & 0 & 0 \\ -1 & 0 & 0 \end{bmatrix},$$

hence

$$\det(\tilde{A} - \lambda\tilde{I}) = \begin{vmatrix} -\lambda & 1 & 1 \\ & -\lambda & 0 \\ -1 & 0 & -\lambda \end{vmatrix} = -\lambda^3.$$

Thus the characteristic polynomial has one root $\lambda = 0$ with the multiplicity 3.

By differentiation of the first of Equations (3.5.34): $x'' = y'+z'$. Then, from the second and the third of Equations (3.5.34) we find: $x'' = y'+z' = x-x = 0$. Thus $x = C_0+C_1 t$, where C_0 and C_1 are arbitrary constants. Hence $y' = x = C_0+C_1 t$, which implies $y = C_0 t + (C_1/2)t^2+C_2$, where C_2 is an arbitrary constant. From the first of Equations (3.5.34) we find $z = x'-y = C_1-C_0 t-(C_1/2)t^2-C_2 = C_1-C_2-C_0 t-(C_1/2)t^2$. Thus the general solution of the system (3.5.34) $x = C_0+C_1 t$, $y = C_2+C_0 t+(C_1/2)t^2$, $z = C_1-C_2-C_0 t-(C_1/2)t$ consists of 3 arbitrary constants.

Writing $u = (x, y, z)$ and $u' = (x', y', z')$ we obtain the system (3.5.34) in the matrix form: $u' = \tilde{A}u$. The solution of this system is then of the form:

$$u = (C_0, C_2, C_1-C_2)+(C_1,C_0, -C_0)t+\left(0, \frac{C_1}{2}, -\frac{C_1}{2}\right)t^2$$

where C_0, C_1, C_2 are arbitrary constants.

EXERCISE 3.5.1. Determine values of the parameter $p \in \mathbf{C}$ for which the following system of linear differential equations: $x' = py$, $y' = (p + 1)z$, $z' = (p+2)x$:
(i) has single characteristic roots,
(ii) has multiple characteristic roots.
Determine solutions of this system in both cases.

EXERCISE 3.5.2. Solve the following system of linear differential equations:

$$x' + y' = -x + y + z,$$

$$y' + z' = x - y + z + 1,$$

$$x' + z' = x + y - z + t.$$

EXERCISE 3.5.3. Prove that every equation of order N with coefficients belonging to $L(X)$ can be written as a system of N equations of order 1. Is the converse theorem true?

EXERCISE 3.5.4. Suppose that X is a linear space over $\mathbf{C}$, $D \in R(X)$, dim ker $D \neq 0$. Solve a system of M equations of order 1 with scalar coefficients:

$$Dx_j - \sum_{k=1}^{M} a_{jk}x_k = y \quad (j = 1, ..., M), \tag{3.5.35}$$

where the matrix $\tilde{A} = (a_{jk})_{j, k = 1, 2, ..., M}$ is:
(i) an involution of order N: $\tilde{A}^N = I$,
(ii) a nilpotent operator of order N: $\tilde{A}^N = 0$ but $\tilde{A}^{N-1} \neq 0$.

EXERCISE 3.5.5. Solve the system of differential equations with real scalar coefficients:

$$\frac{dx_j}{dt} = \sum_{k=1}^{N} a_{jk}x_k + y_j, \quad y_j \in C[a, b] \quad (j = 1, 2, ..., N)$$

by an application of the Jordan Theorem 1.4.2.

3.6. GENERAL SOLUTION OF EQUATIONS WITH LEFT INVERTIBLE OPERATORS

If we consider equations with left invertible operators, the situation becomes much simpler. Indeed, if Δ is a left invertible operator then $\ker \Delta = \{0\}$ for $n = 1, 2, ...$ In a similar way, as in Section 3.2, we obtain

THEOREM 3.6.1. *Suppose that X is a linear space over an arbitrary field $\mathscr{F}$ of scalars, $\Delta \in \Lambda(X)$ and $L \in \mathscr{L}_\Delta$. Write*

$$Q(t, s) = \sum_{k=0}^{N} Q_k t^k s^{N-k}, \quad \text{where } Q_0, ..., Q_{N-1} \in L(X), \quad Q_N = I$$

(3.6.1)

and

$$Q(t) = Q(t, 1).$$ (3.6.2)

If the operator $Q(I, L)$ is invertible, then the equation

$$Q(\Delta)x = y, \quad \text{where } [Q(I, L)]^{-1}y \in \Delta^N X$$ (3.6.3)

has a unique solution

$$x = L^N[Q(I, L)]^{-1}y.$$ (3.6.4)

Proof. Formula (3.6.2) implies that

$$Q(\Delta) = \sum_{k=0}^{N} Q_k \Delta^k = \sum_{k=0}^{N} Q_k(L^{N-k}\Delta^{N-k})\Delta^k$$

$$= \left(\sum_{k=0}^{N} Q_k L^{N-k}\right)\Delta^N = Q(I, L)\Delta^N.$$

By our assumptions, Equation (3.6.3) is then equivalent to the equation

$$\Delta^N x = [Q(I, L)]^{-1}y.$$ (3.6.5)

Since $[Q(I, L)]^{-1}y \in \Delta^N x$, Corollary 2.5.1 implies that $x = L^N[Q(I, L)]^{-1}y$. Evidently, this solution is unique, because the homogeneous equation has only zero as a solution. ∎

COROLLARY 3.6.1. *Suppose that all assumptions of Theorem 3.6.1 are satisfied. If the operator $Q(I, L)$ is invertible and $y \in \Delta^M x$, $M \geqslant 0$, $[Q(I, L)]^{-1}L^M y \in \Delta^N X$ then the equation*

$$\Delta^M Q(\Delta)x = y \tag{3.6.6}$$

has a unique solution

$$x = L^N [Q(I, L)]^{-1} L^M y. \tag{3.6.7}$$

Proof. Write $u = Q(\Delta)x$. Then we have an equation

$$\Delta^M u = y, \quad \text{where } y \in \Delta^M X.$$

Corollary 2.4.1 implies that $u = L^M y$. Thus we have a new equation

$$Q(\Delta)x = L^M y, \quad \text{where } [Q(I, L)]^{-1}L^M y \in \Delta^N X.$$

By Theorem 3.6.1 we conclude that this equation has a unique solution which is of the form (3.6.7). ∎

COROLLARY 3.6.2. *Suppose that all assumptions of Theorem 3.6.1 are satisfied. If the operator $Q(I, L)$ is invertible and $[Q(I, L)]^{-1}y \in \Delta^N X$, $L^N[Q(I, L)]^{-1}y \in \Delta^M X$, $M \geqslant 0$, then the equation*

$$Q(\Delta)\Delta^M x = y \tag{3.6.8}$$

has a unique solution which is of the form

$$x = L^{M+N}[Q(I, L)]^{-1}y. \tag{3.6.9}$$

Proof. Write $u = \Delta^M x$. Then we have an equation

$$Q(\Delta)u = y, \quad \text{where } [Q(I, L)]^{-1}y \in \Delta^N X.$$

Theorem 3.6.1 implies that this equations has a unique solution $u = L^N[Q(I, L)]^{-1}y$. Thus we obtain a new equation

$$\Delta^M x = L^N[Q(I, L)]^{-1}y, \quad \text{where } L^N[Q(I, L)]^{-1}y \in \Delta^M X.$$

Corollary 2.4.1 implies that

$$x = L^M L^N[Q(I, L)]^{-1}y = L^{M+N}[Q(I, L)]^{-1}y,$$

which was to be proved. ∎

We can formulate results of this section also in another way. Namely, Theorem 2.5.1 and Corollary 2.5.1 together imply that $u \in \Delta^m X$ if and only if $GL^k u = 0$ for $k = 0, 1, \ldots, m-1$, where G is a co-initial operator for Δ corresponding to $L \in \mathcal{L}_\Delta$.

COROLLARY 3.6.3. *Suppose that all assumptions of Theorem 3.6.1 are satisfied and, moreover, G is a co-initial operator for Δ corresponding to L. If the operator $Q(I, L)$ is invertible and*

$$GL^k y = 0 \quad (k = 0, 1, \ldots, M-1, M \geqslant 0),$$
$$GL^j [Q(I, L)]^{-1} L^M y = 0 \quad (j = 0, 1, \ldots, N-1)$$

then equation (3.6.3) has a unique solution, which is of the form (3.6.7).

COROLLARY 3.6.4. *Suppose that all assumptions of Theorem 3.6.1 are satisfied and, moreover, G is a co-initial operator for Δ corresponding to L. If the operator $Q(I, L)$ is invertible and*

$$GL^k [Q(I, L)]^{-1} y = 0 \quad (k = 0, 1, \ldots, N-1),$$
$$GL^{j+N} [Q(I, L)]^{-1} y = 0 \quad (j = 0, 1, \ldots, N-1),$$

then Equation (3.6.7) has a unique solution which is of the form (3.6.9).

COROLLARY 3.6.5. *Suppose that all assumptions of Theorem 3.6.1 are satisfied. Suppose, moreover, that $Q_k = q_k I$, where $q_k \in \mathcal{F}$ ($k = 0, 1, \ldots \ldots, N-1$), $\mathcal{F}$ is an algebraically closed field and that L is a Volterra operator. If $\lambda_k \neq 0$ ($k = 1, \ldots, n$), then $Q(I, L) = \prod\limits_{k=1}^{n} (I - \lambda_k L)^{r_k}$ ($r_1 + + \ldots + r_n = N$) is invertible as a superposition of invertible operators and a unique solution of Equation (3.6.3) $Q(\Delta)x = y$, where $[Q(I, L)]^{-1} y \in \Delta^N X$, can be written in the form*

$$x = \sum_{k=1}^{n} \sum_{j=1}^{r_j} d_{kj}(I - \lambda_k L)^{-j} L^N y, \tag{3.6.10}$$

where the scalars d_{kj} are determined by the decomposition of the rational function $[Q(1, s)]^{-1}$ onto vulgar fractions:

$$[Q(1, s)]^{-1} = \sum_{k=1}^{n} \sum_{j=1}^{r_j} d_{kj}(1 - \lambda_k s)^{-j}. \tag{3.6.11}$$

Proof. Indeed, $L^N[Q(I, L)]^{-1} = [Q(I, L)]^{-1}L^N$, because the polynomial $Q(t)$ has scalar coefficients.   ∎

Since in the case of scalar coefficients of the polynomial $Q(t)$ we have $\Delta^M Q(\Delta) = Q(\Delta)\Delta^M$, Corollary 3.6.2 immediately implies

COROLLARY 3.6.6. *Suppose that all assumptions of Corollary 3.6.5 are satisfied. Then the equation*

$$\Delta^M Q(\Delta)x = y \quad (Q(\Delta)\Delta^M x = y), \tag{3.6.12}$$

where $[Q(I, L)]^{-1}y \in \Delta^{M+N}X$, *has a unique solution*

$$x = \sum_{k=1}^{n}\sum_{j=1}^{r_j} d_{kj}(I - \lambda_k L)^{-1}L^{M+N}y, \tag{3.6.13}$$

where the scalars d_{kj} *are determined by the decomposition* (3.6.11).

Examples and Exercises

EXAMPLE 3.6.1. Suppose that X, A, A_0 are defined, as in Example 2.5.1. Then $A \in \Lambda(X)$ and $A_0 \in \mathscr{L}_A$. Consider the equation

$$(A - \lambda I)x = 1. \tag{3.6.14}$$

Since $A_0 = (I - R)^{-1}\dfrac{d}{dt}$ and $y(t) \equiv 1$, we have $A_0 y = 0$. Thus $x = A_0 Ax = A_0(Ax - \lambda x) + \lambda A_0 x = A_0 y = 0$, which implies the equality

$$(I - \lambda A_0)x = 0. \tag{3.6.15}$$

But, by definition,

$$0 = (I - \lambda A_0)x = [I - \lambda(I - R)^{-1}D]x = (I - R)^{-1}[I - R - \lambda D]x$$

and we obtain an equation: $x - Rx - \lambda Dx = 0$. It is easy to check that for every $0 \neq \lambda \in \mathbf{R}$ there exists a $\mu \in \mathbf{R}$, namely $\mu = (-1 \pm \sqrt{1 + 4\lambda})/2\lambda$, such that the function $e^{\mu t}$ is a non-trivial solution of Equation (3.6.15). Thus the operator $I - \lambda A_0$ is not-invertible, which means that conditions of solvability of Equation (3.6.14) are not satisfied.

EXERCISE 3.6.1. Suppose that $\Delta \in \Lambda(X)$, $L \in \mathscr{L}_\Delta$ and the polynomial $Q(t, s)$ defined as in Theorem 3.6.1 has scalar coefficients. Does the following theorem hold: If the left inverse $L^{M+N}[Q(I, L)]^{-1}$ of the operator $\Delta^M Q(\Delta)$ is a Volterra operator then L is a Volterra operator (cf. Theorem 2.4.1)?

EXERCISE 3.6.2. Suppose that all assumptions of Theorem 3.6.1 are satisfied and G is a co-initial operator for Δ corresponding to the left inverse L.

  1. Find conditions of solvability of the equation

$$Q(\Delta)x = y \tag{3.6.16}$$

and formulae for solutions if the operator $Q(I, L)$ is:
  (i) left invertible,
  (ii) right invertible.
  2. Generalize Corollaries 3.6.1–3.6.6 for the case, when the operator $Q(I, L)$ is:
  (i) left invertible,
  (ii) right invertible
(cf. Sections 3.1 and 3.2).

Added in proof:

PROPOSITION 3.4.6. *Suppose that all assumptions of Proposition 3.4.2 are satisfied and that F is an initial operator for D corresponding to R. Then*

$$R_\infty = \{x \in D_\infty : FD^n x = 0 \ \text{for all } n \in \{0\} \cup \mathbf{N}\}$$

where R_∞ is defined by Formula (3.4.22) (cf. the author, 1986).

Chapter 4

Initial and boundary value problems

4.1. WELL-POSED AND ILL-POSED INITIAL VALUE PROBLEMS

Suppose that X is a linear space over a field $\mathscr{F}$ of scalars and that $D \in R(X)$. Write, as before

$$Q(D) = \sum_{k=0}^{N} Q_k D^k, \quad \text{where} \quad Q_0, \dots, Q_{N-1} \in L_0(X), \quad Q_N = I. \tag{4.1.1}$$

Let F be an operator for D. An *initial value problem for the operator* $Q(D)$ is the following: Find all solutions of equation

$$Q(D)x = y, \quad y \in X, \tag{4.1.2}$$

satisfying the *initial conditions*:

$$FD^k x = y_k, \quad y_k \in \ker D \quad (k = 0, 1, \dots, N-1). \tag{4.1.3}$$

We say that the initial value problem (4.1.2)–(4.1.3) is *well-posed* if this problem has a unique solution for every $y \in X, y_0, \dots, y_{N-1} \in \ker D$. If the problem (4.1.2)–(4.1.3) is well-posed then the linearity of the operators $D, Q(D), F$ implies that the corresponding homogeneous problem

$$Q(D)x = 0, \quad FD^k x = 0 \quad (k = 0, 1, \dots, N-1) \tag{4.1.4}$$

has only zero as a solution.

We say that the initial value problem (4.1.2)–(4.1.3) is *ill-posed* if either there exists $y \in X, y_0, \dots, y_{N-1} \in \ker D$ such that this problem has no solutions or the homogeneous problem (4.1.4) has at least one non-trivial solution.

LEMMA 4.1.1. *Suppose that $D \in R(X)$, dim $\ker D \neq 0$, and that F is an*

219

initial operator for D corresponding to a right inverse R. Suppose, moreover, that the operator $Q(D)$ is defined by Formula (4.1.1). Write, as before,

$$Q(t, s) = \sum_{k=0}^{N} Q_k t^k s^{N-k}. \tag{4.1.5}$$

Then

$$Q(D) = Q(I, R)D^N + \sum_{m=0}^{N-1} \left(\sum_{k=0}^{m} Q_k R^{m-k} \right) FD^m \quad \text{on dom } D^N. \tag{4.1.6}$$

Proof. Since $DF = 0$ and $Q_N = I$, from the Taylor Formula (2.2.7) we find on the domain of the operator D^N:

$$Q(D) = \sum_{k=0}^{N-1} Q_k D^k + D^N = \sum_{k=0}^{N-1} Q_k D^k \left(\sum_{m=0}^{N-1} R^m FD^m + R^N D^N \right) + D^N$$

$$= \sum_{k=0}^{N-1} Q_k D^k R^N D^N + D^N + \sum_{k=0}^{N-1} Q_k D^k \sum_{m=0}^{N-1} R^m FD^m$$

$$= \sum_{k=0}^{N-1} Q_k R^{N-k} D^N + D^N + \sum_{k=0}^{N-1} Q_k \sum_{m=0}^{N-1} D^k R^m FD^m$$

$$= \left(\sum_{k=0}^{N} Q_k R^{N-k} \right) D^N + Q_0 \sum_{m=1}^{N-1} R^m FD^m +$$

$$+ \sum_{k=1}^{N-1} Q_k \left[\sum_{m=0}^{k-1} D^{k-m} FD^m + \sum_{m=k}^{N-1} R^{m-k} FD^m \right]$$

$$= Q(I, R)D^N + \sum_{k=0}^{N-1} Q_k \left(\sum_{m=k}^{N-1} R^{m-k} FD^m \right)$$

$$= Q(I, R)D^N_{,} + \sum_{m=0}^{N-1} \left(\sum_{k=0}^{m} Q_k R^{m-k} \right) FD^m,$$

which was to be proved. ∎

THEOREM 4.1.1. *Suppose that $D \in R(X)$, dim ker $D \neq 0$, and that F is* *an initial operator for D corresponding to an $R \in \mathcal{R}_D$. Then the initial value problem for the operator D^N ($N \geqslant 1$), i.e. the problem*

$$D^N x = y, \quad y \in X, \tag{4.1.7}$$

$$FD^k x = y_k, \quad y_k \in \ker D \quad (k = 0, 1, ..., N-1) \tag{4.1.8}$$

is well-posed and its unique solution is of the form

$$x = R^N y + \sum_{k=0}^{N-1} R^k y_k. \tag{4.1.9}$$

Proof. Proposition 3.2.1 implies that every solution of Equation (4.1.7) is of the form

$$x = R^N y + \sum_{k=0}^{N-1} R^k z_k, \quad \text{where } z_0, ..., z_{N-1} \in \ker D \text{ are arbitrary.}$$

Since $FR = 0$ and $Dz_k = 0$ $(k = 0, 1, ..., N-1)$ we have for $m = 0, 1, ..., N-1$

$$y_m = FD^m x = FD^m \left[R^N y + \sum_{k=0}^{N-1} R^k z_k \right] = FR^{N-m} y + \sum_{k=0}^{N-1} FD^m R^k z_k$$

$$= \sum_{k=0}^{m-1} FD^{m-k} z_k + \sum_{k=m+1}^{N-1} FR^{k-m} z_k + Fz_m = Fz_m = z_m.$$

Hence $z_m = y_m$ $(m = 0, 1, ..., N-1)$, which implies the required formula (4.1.9). ∎

THEOREM 4.1.2. *Suppose that $D \in R(X)$, $\dim \ker D \neq 0$, and that F is an initial operator for D corresponding to an $R \in \mathcal{R}_D$. Suppose, moreover, that $Q(D)$ and $Q(t, s)$ are defined by Formulae (4.1.1) and (4.1.5), respectively. If the operator $Q(I, R)$ is invertible then the initial value problem* (4.1.2)–(4.1.3) *for the operator $Q(D)$ is well-posed and its unique solution is of the form*

$$x = R^N [Q(I, R)]^{-1} \left[y - \sum_{m=0}^{N-1} \left(\sum_{k=0}^{m} Q_k R^{m-k} \right) y_m \right] + \sum_{k=0}^{N-1} R^k y_k.$$

$$\tag{4.1.10}$$

Proof. Lemma 4.1.1 implies that Equation (4.1.2) is equivalent to the equation

$$Q(I, R)D^N x + \sum_{m=0}^{N-1} \left(\sum_{k=0}^{m} Q_k R^{m-k} \right) FD^m x = y. \qquad (4.1.11)$$

From Conditions (4.1.3) we have

$$FD^m x = y_m \quad (m = 0, 1, ..., N-1).$$

We therefore can rewrite Equation (4.1.11) in the form

$$Q(I, R)D^N x = y - \sum_{m=0}^{N-1} \left(\sum_{k=0}^{m} Q_k R^{m-k} \right) y_m.$$

Since the operator $Q(I, R)$ is invertible, we can rewrite this last equation in the form

$$D^N x = [Q(I, R)]^{-1} y_N, \quad \text{where } y_N = y - \sum_{m=0}^{N-1} \left(\sum_{k=0}^{m} Q_k R^{m-k} \right) y_m.$$

$$(4.1.12)$$

$$FD^m x = y_m \quad (m = 0, 1, ..., N-1).$$

Theorem 4.1.1 implies that the initial value problem (4.1.12) for the operator D^N is well-posed and its unique solution is of the form

$$x = R^N [Q(I, R)]^{-1} y_N + \sum_{k=0}^{N-1} R^k y_k.$$

Thus a solution, we are looking for, is of the form

$$x = R^N [Q(I, R)]^{-1} \left[y - \sum_{m=0}^{N-1} \left(\sum_{k=0}^{m} Q_k R^{m-k} \right) y_m \right] + \sum_{k=0}^{N-1} R^k y_k.$$

This solution is unique, because for $y = 0$, $y_0 = ... = y_{N-1} = 0$ we obtain $x = 0$. This is which was to be proved. ∎

COROLLARY 4.1.1. *Suppose that all assumptions of Theorem 4.1.2 are satisfied. If the operator $Q(I, R)$ is invertible then the initial value problem*

$$Q(D)D^M x = y, \quad y \in X \quad (M \geqslant 0), \qquad (4.1.13)$$

$$FD^k x = y_k, \quad y_k \in \ker D \quad (k = 0, 1, ..., N+M-1) \qquad (4.1.14)$$

is well-posed and its unique solution is of the form

$$x = R^{N+M}[Q(I, R)]^{-1}\left[y - \sum_{m=0}^{N-1}\left(\sum_{k=0}^{m} Q_m R^{m-k}\right) y_{M+m}\right] +$$

$$+ \sum_{m=0}^{N+M-1} R^m y_m. \qquad (4.1.15)$$

Proof. Write $u = D^M x$. Then we have $FD^k u = FD^{M+k}x = y_{k+M}$ for $k = 0, 1, ..., N-1$. We therefore can rewrite the problem (4.1.13)–(4.1.14) as follows:

$$Q(D)u = y, \quad FD^k u = y_{k+M} \quad (k = 0, 1, ..., N-1).$$

According to Theorem 4.1.1 this initial value problem is well-posed and its unique solution is

$$u = R^N[Q(I, R)]^{-1}\left[y - \sum_{m=0}^{N-1}\left(\sum_{k=0}^{m} Q_m R^{m-k} y_{M+m}\right)\right] + \sum_{k=0}^{N-1} R^k y_{k+M}.$$

$$(4.1.15')$$

Having already determined $u = D^M x$ we consider the following initial value problem:

$$D^M x = u, \quad FD^k x = y_k \quad (k = 0, 1, ..., M-1).$$

Theorem 4.1.1 implies that this problem is also well-posed and its unique solution is of the form

$$x = R^M u + \sum_{m=0}^{M-1} R^m y_m.$$

This and Formula (4.1.15') together imply that the problem (4.1.13)–(4.1.14) is well-posed and its unique solution is

$$x = R^M u + \sum_{m=0}^{M-1} R^m y_m$$

$$= R^M \left\{ R^N [Q(I, R)]^{-1} \left[y - \sum_{m=0}^{N-1} \left(\sum_{k=0}^{m} Q_m R^{m-k} \right) y_{k+M} \right] + \right.$$

$$\left. + \sum_{k=0}^{N-1} R^k y_{k+M} \right\} + \sum_{m=0}^{M-1} R^m y_m$$

$$= R^{N+M} [Q(I, R)]^{-1} \left[y - \sum_{m=0}^{N-1} \left(\sum_{k=0}^{m} Q_m R^{m-k} \right) y_{M+m} \right] +$$

$$+ \sum_{k=0}^{N-1} R^{k+M} y_{k+M} + \sum_{m=0}^{M-1} R^m y_m$$

$$= R^{N+M} [Q(I, R)]^{-1} \left[y - \sum_{m=0}^{N-1} \left(\sum_{k=0}^{m} Q_m R^{m-k} \right) y_{M+m} \right] +$$

$$+ \sum_{m=0}^{N+M-1} R^m y_m$$

which was to be proved.   ∎

Observe, that in Section 3.2 we have obtained similar results about general solutions of equations with the operators $D^M Q(D)$ and $Q(D) D^M$. In the case of initial value problems the situation becomes slightly different. Namely, we have only

COROLLARY 4.1.2. *Suppose that all assumptions of Theorem* 4.1.2 *are satisfied. Suppose, moreover, that the coefficients* $Q_0, \ldots, Q_{N-1}$ *are stationary operators (in the sense used in Section* 3.3*), i.e.*

$$Q_k D - D Q_k = 0, \quad Q_k R - R Q_k = 0 \quad (k = 0, 1, \ldots, N-1).$$

If the operator $Q(I, R)$ *is invertible then the initial value problem*

$$D^M Q(D) x = y, \quad y \in X \quad (M \geqslant 0) \tag{4.1.16}$$

with Conditions (4.1.14) *is well-posed and its unique solution is of the form* (4.1.15).

Indeed, by our assumptions, $D^M Q(D) = Q(D) D^M$ $(M \geqslant 0)$.   ∎

COROLLARY 4.1.3. *Suppose that all assumptions of Theorem 4.1.2 are satisfied. Suppose, moreover, that R is a Volterra operator and that the coefficients $Q_k = q_k I$, where q_k are scalars $(k = 0, 1, 2, ..., N-1)$. Then the initial value problem (4.1.13)–(4.1.14) (or (4.1.16)–(4.1.14)) is well-posed and its unique solution is of the form*

$$x = [Q(I, R)]^{-1} \left[R^{N+M} y - \sum_{m=0}^{N-1} \left(\sum_{k=0}^{m} q_m R^{N+M+m-k} \right) y_m \right] +$$

$$+ \sum_{k=0}^{N-1} R^{M+k} y_k + \sum_{k=0}^{M-1} R^k y_{k+M}. \qquad (4.1.17)$$

The proof is an immediate consequence of Corollary 4.1.2 and of the fact that, by Theorem 4.1.2, the operator $Q(I, R)$ is invertible. ∎

In the same way we obtain

COROLLARY 4.1.4. *Suppose that all assumptions of Theorems 4.1.2 and 3.4.2 are satisfied. Then the initial value problem (4.1.13)–(4.1.14) (or (4.1.16)–(4.1.14)) is well-posed and its unique solution is of the form*

$$x = \sum_{j=1}^{n} \sum_{m=1}^{r_j} d_{jm} W_{m^j}^{\lambda} \left[R^{N+M} y - \sum_{m=0}^{N-1} \left(\sum_{k=0}^{m} q_k R^{N+M+m-k} \right) y_m \right] +$$

$$+ \sum_{k=0}^{N-1} R^{M+k} y_k + \sum_{k=0}^{M-1} R^k y_{k+M}. \qquad (4.1.18)$$

Corollaries 4.1.3 and 4.1.4 show that an initial value problem for an operator $D^M Q(D)$ is always well-posed, provided that the coefficients of this operator are scalars and that the initial operator F corresponds to a Volterra right inverse.

Till this moment we have studied the case when the operator $Q(I, R)$ was invertible, i.e. the number $\lambda = -1$ was not an eigenvalue of the operator $Q^0(R) = \sum_{k=0}^{N-1} Q_k R^{N-k}$, because $Q(I, R) = I + Q^0(R)$. Examine now the case when $\lambda = -1$ is an eigenvalue of the operator $Q^0(R)$.

THEOREM 4.1.3. *Suppose that all assumptions of Theorem 4.1.2 are satisfied. Write*

$$Q^0(R) = \sum_{k=0}^{N-1} Q_k R^{N-k}, \quad i.e. \quad Q(I, R) = I + Q^0(R). \quad (4.1.19)$$

If $\lambda = -1$ is an eigenvalue of the operator Q^0 then the initial value problem (4.1.2)–(4.1.3) is ill-posed. However this problem has solutions under the following necessary and sufficient condition:

$$y_N \in [I + Q^0(R)]X, \quad where \; y_N = y - \sum_{m=0}^{N-1}\left(\sum_{k=0}^{m} Q_m R^{m-k}\right) y_m.$$

$$(4.1.20)$$

If this condition is satisfied then solutions of the problem (4.1.2)–(4.1.3) exist and are of the form

$$x = R^N(x_{-1} + w) + \sum_{k=0}^{N-1} R^k y_k, \quad (4.1.21)$$

where x_{-1} is an arbitrary eigenvector of the operator $Q^0(R)$ corresponding to the eigenvalue -1 and w is an element of the inverse-image of the element y_N by the transformation $I + Q^0(R)$.

Proof. In a similar way as Formula (4.1.12) in Theorem 4.1.2 we obtain the following equation:

$$Q(I, R) D^N x = y_N, \quad \text{where } y_N \text{ is defined by (4.1.20)},$$

i.e.

$$[I + Q^0(R)] D^N x = y_N. \quad (4.1.22)$$

The necessity and sufficiency of Condition (4.1.20) is obvious. Suppose, that this condition is satisfied. It means that there exists a $w \in X$ such that $[I + Q^0(R)]w = y_N$. Write $u = D^N x$. Then Equation (4.1.22) can be rewritten as follows

$$[I + Q^0(R)]u = [I + Q^0(R)]w, \quad \text{i.e. } [I + Q^0(R)](u - w) = 0.$$

Since $\lambda = -1$ is an eigenvalue of the operator $Q^0(R)$, we conclude that there exists an $x_{-1} \neq 0$ such that $u - w = x_{-1}$, namely x_{-1} is an arbitrary eigenvector of $Q^0(R)$ corresponding to the eigenvalue -1.

Thus we have $D^N x = x_{-1} + w$. This equation together with the initial conditions (4.1.3) has a unique solution

$$x = R^N(x_{-1} + w) + \sum_{k=0}^{N-1} R^k y_k \qquad (4.1.23)$$

for arbitrarily fixed x_{-1} and w. This is, which was to be proved. However, the problem under consideration is ill-posed, because the corresponding homogeneous problem (i.e. with $y = y_0 = \ldots = y_{N-1} = 0$) has a non-trivial solution x_{-1}. ∎

Examples and Exercises

EXAMPLE 4.1.1. Consider the following problem for ordinary differential equations:

$$x'' + \lambda x' = 6t \quad \text{for } 0 \leqslant t \leqslant T, \qquad (4.1.24)$$

$$x(0) = x_0, \quad x' = x_1, \quad x_0, x_1 \in \mathbf{R}. \qquad (4.1.25)$$

It is an initial value problem for the operator $D = d/dt$ in the space $C(0, T)$ with the initial operator $(Fx)(t) = x(0)$ corresponding to the Volterra right inverse $R = \int_0^t$ and with $Q(D) = D^2 + \lambda D = D^2(I + \lambda R)$.

Since the operator R is invertible for every $\lambda \in \mathbf{R}$, Theorem 4.1.2 implies that the problem (4.1.24)–(4.1.25) is well-posed and its unique solution is of the form

$$x = (I + \lambda R)^{-1}(R^2 y + Rx_0 + x_1).$$

But $Rx_0 = \int_0^t x_0 \, ds = x_0 t$ and

$$(R^2 y)(t) = \int_0^t \int_0^s 6u \, du = \int_0^t 3s^2 \, ds = t^3.$$

This and Formula (2.2.41) together imply that for $\lambda \neq 0$

$$x(t) = (I + \lambda R)^{-1}(R^2 y + Rx_0 + x_1)(t) = (I + \lambda R)^{-1}(t^3 + x_0 t + x_1)$$

$$= t^3 + x_0 t + x_1 - \lambda \int_0^t e^{-\lambda(t-s)}(s^3 + x_0 s + x_1) \, ds$$

$$= t^3 + x_0 t + x_1 + \frac{3}{\lambda^3}[(\lambda^2 - 2\lambda - 2) - 2e^{-\lambda t}] +$$

$$+ \frac{x_0}{\lambda}(1 - \lambda t - e^{-\lambda t}) + x_1(e^{-\lambda t} - 1)$$

$$= t^3 + \frac{1}{\lambda^3}(\lambda^3 x_1 - \lambda^2 x_0 - 6)e^{-\lambda t} + \frac{3}{\lambda^3}(\lambda^2 - 2\lambda - 2 + \lambda^2 x_0).$$

For $\lambda = 0$ we find $x(t) = (R^2 y + R x_0 + x_1)(t) = t^3 + x_0 t + x_1$.

EXAMPLE 4.1.2. Consider the following problem for ordinary differential equations:

$$x'' = e^t, \tag{4.1.26}$$

$$\int_0^1 x(s)\,ds = x_0, \quad x(1) - x(0) = x_1, \quad x_0, x_1 \in \mathbf{R}. \tag{4.1.27}$$

Since $x(1) - x(0) = \int_0^1 x'(s)\,ds$, this problem is an initial value problem for the operator $D = d/dt$ in the space $C[0, T]$, $T \geqslant 0$, with the initial operator $(Fx)(t) = \int_0^1 x(s)\,ds$ corresponding to a right inverse $(Rx)(t) = \int_0^t x(s)\,ds + \int_0^1 (s-1)x(s)\,ds$, where $x \in C[0, T]$ (cf. Example 2.2.3). Indeed, the problem (4.1.26)–(4.1.27) can be rewritten as follows:

$$D^2 x = e^t, \quad Fx = x_0, \quad FDx = x_1.$$

Theorem 4.1.1 implies that this problem is well-posed and its unique solution is of the form

$$x = R^2(e^t) + x_0 + R x_1, \quad \text{i.e.}$$

$$x(t) = \int_0^t \left[\int_0^s e^u du + \int_0^1 (u-1)e^u du\right] ds + \int_0^1 (s-1)\left[\int_0^s e^u du + \right.$$

$$\left. + \int_0^1 (u-1)e^u du\right] ds + x_0 + \int_0^t x_1\,ds + \int_0^1 (s-1)x_1\,ds$$

$$= e^t + (x_1 + 1 - e)t + x_0 - \frac{x_1}{2} - \frac{5}{2}e - \frac{1}{2}.$$

EXERCISE 4.1.1. Suppose that all assumptions of Theorem 4.1.2 are satisfied. Examine the initial value problems
   (i) (4.1.2)–(4.1.3);
   (ii) (4.1.13)–(4.1.14);
   (iii) (4.1.16)–(4.1.14)
(with stationary coefficients) if the operator $Q(I, R)$ is either left invertible or right invertible, but not invertible. Determine conditions which guarantee that these problems are well-posed.

EXERCISE 4.1.2. Suppose that the operator $Q\langle D \rangle$ is defined by Formula (3.2.10) and has stationary coefficients. Examine the initial value problems:
   (i) $D^n Q\langle D \rangle x = y$, $FD^k x = y_k \in \ker D$ $(k = 0, 1, 2, ..., M+N,$
       $M \geqslant 0)$;
   (ii) $Q\langle D \rangle D^M x = y$, $FD^k x = y_k \in \ker D$ $(k = 0, 1, 2, ..., M+N,$
       $M \geqslant 0)$
if the operator $Q\langle I, R \rangle$ is either left invertible or right invertible or invertible.

EXERCISE 4.1.3. Suppose that $D \in R(X)$, $\dim \ker D \neq 0$, F is an initial operator for D corresponding to a Volterra right inverse R. Prove that
   (i) the initial value problem

$$\sum_{j=1}^{k} \lambda^{j-1} D x_j = y_k,$$
$$\qquad\qquad\qquad\qquad (j, k = 1, ..., M)$$
$$F x_j = x_j^0, \quad \text{where } x_j^0 \in \ker D$$

can be reduced to an initial value problem in the space X^M (cf. Section 3.5).
   (ii) The obtained initial value problem is well-posed. Determine a unique solution of this problem.

EXERCISE 4.1.4. Prove that the following problem

$$\frac{dx_k}{dt} = \sum_{j=1}^{M} a_{jk} x_j + y_k, \quad x_k(0) = a_k, \quad a_k \in \mathbf{R} \quad (k = 1, ..., M)$$

has a unique solution. Determine this solution.

EXERCISE 4.1.5. Prove that the following problem

$$x^{(n)}(t) = a(t), \quad a \in C[0, T], \quad n > 1,$$

$$\int_0^T x(s)\,ds = x_0, \quad x^{(k)}(T) - x^{(k)}(0) = x_k, \quad x_k \in \mathbf{R}$$

$$(k = 1, \dots, n-1)$$

is well-posed. Determine its unique solution (cf. Example 4.1.2).

EXERCISE 4.1.6. Prove that the following problem

$$x^{(n)}(t) = p(t), \quad p \in C[a, b], \quad n > 1,$$
$$\alpha x^{(k)}(a) + (1-\alpha)x^{(k)}(b) = x_k, \quad x_k \in \mathbf{R}, \quad 0 < \alpha < 1$$

is an initial value problem for the operator $D^n = d^n/dt^n$, which is well-posed. Determine its unique solution (cf. Example 2.2.4).

EXERCISE 4.1.7. Suppose that X, D, R are defined as in Example 2.1.4. In a similar way, as in Example 4.1.1, we can find general solution of difference equations with scalar coefficients. Recall that an initial operator for the operator D is defined by means of the equality $F\{x_n\} = \{z_n\}$, where $z_n = x_1$ $(n = 1, 2, \dots)$. This means that $FDx = FD\{x_n\} = F\{x_{n+1} - x_n\} = \{z_n'\}$, where $z_n' = x_2 - x_1$ $(n = 1, 2, \dots)$. Prove that the following problem

$$\begin{aligned} x_{n+2} + ax_{n+1} + x_n &= y_n, \\ x_1 = c_1, \quad x_2 &= c_2, \end{aligned} \qquad a, y_n, c_1, c_2 \in \mathbf{R} \quad (n = 1, 2, \dots)$$

for the operator $D^2 + (a-1)D + aI$ is an initial value problem, which is well-posed. Determine its unique solution.

EXERCISE 4.1.8. Suppose that X, D, R, F are determined, as in Exercise 4.1.7. Prove that the following problem:

$$\begin{aligned} x_{n+N} - x_n &= y_n \quad (n = 1, 2, \dots), \\ x_k = c_k, \quad c_k &\in \mathbf{R} \quad (k = 1, \dots, N-1), \quad N > 1 \end{aligned}$$

is an initial value problem, which is well-posed. Determine its unique solution.

EXERCISE 4.1.9. Suppose that X, D, R, F are determined, as in Exercise 4.1.7. Prove that the following problem:

$$x_{n+2} + a^n x_n = y_n, \quad a \neq 0, \quad a \neq 1 \quad (n = 1, 2, \ldots),$$

$$x_1 = c_1, \quad x_2 = c_2, \quad c_1, c_2 \in \mathbf{R}$$

is well-posed. Determine its unique solution.

4.2. HYPERBOLIC EQUATIONS

To begin with, consider forced vibrations of a string described by the equation

$$\frac{\partial^2 u}{\partial \xi^2} - \frac{1}{c^2} \frac{\partial^2 u}{\partial \tau^2} = v \quad (c > 0) \tag{4.2.1}$$

with the initial conditions

$$u(\xi, \tau_0) = v_0(\xi), \quad u'_\tau(\xi, \tau_0) = v_1(\xi), \tag{4.2.2}$$

where $0 \leqslant \tau_0 \leqslant T$, $0 \leqslant \xi \leqslant l$. Moreover, $v_0, v_1 \in C[0, l]$, $v \in C(\Omega_0)$, where $\Omega_0 = [0, l] \times [0, T]$, are given functions.

Equation (4.2.1) can be reduced to an equation with a right invertible operator by the substitution

$$s = \xi - c\tau, \quad t = \xi + c\tau, \quad \text{i.e. } \xi = \frac{t+s}{2}, \quad \tau = \frac{t-s}{2c}, \tag{4.2.3}$$

$$x(t, s) = u\left(\frac{t+s}{2}, \frac{t-s}{2c}\right) = u(\xi, \tau),$$

$$y(t, s) = \frac{1}{4} v\left(\frac{t+s}{2}, \frac{t-s}{2c}\right) = \frac{1}{4} v(\xi, \tau), \tag{4.2.4}$$

$$y_0(t) = v_0(t - c\tau_0), \quad y_1(t) = \frac{1}{4c} v_1(t - c\tau_0).$$

Indeed, if we are looking for solutions of Equation (4.2.1) belonging to $C^2(\Omega_0) \subset C(\Omega_0)$, the continuity of the second mixed derivatives implies their equality: $u''_{\xi\tau} = u''_{\tau\xi}$ in Ω_0.

By Formulae (4.2.3), (4.2.4) we have

$$\frac{\partial x}{\partial t} = \frac{\partial u}{\partial \xi} \frac{\partial \xi}{\partial t} + \frac{\partial u}{\partial \tau} \frac{\partial \tau}{\partial t} = \frac{1}{2} \frac{\partial u}{\partial \xi} + \frac{1}{2c} \frac{\partial u}{\partial \tau},$$

$$\frac{\partial x}{\partial s} = \frac{\partial u}{\partial \xi} \frac{\partial \xi}{\partial s} + \frac{\partial u}{\partial \tau} \frac{\partial \tau}{\partial s} = \frac{1}{2} \frac{\partial u}{\partial \xi} - \frac{1}{2c} \frac{\partial u}{\partial \tau},$$

$$\frac{\partial^2 x}{\partial s \partial t} = \frac{\partial}{\partial s} \left(\frac{1}{2} \frac{\partial u}{\partial \xi} + \frac{1}{2c} \frac{\partial u}{\partial \tau} \right)$$

$$= \frac{1}{2} \frac{\partial^2 u}{\partial \xi^2} \frac{\partial \xi}{\partial s} + \frac{1}{2} \frac{\partial^2 u}{\partial \tau \partial \xi} \frac{\partial \tau}{\partial s} + \frac{1}{2c} \frac{\partial^2 u}{\partial \xi \partial \tau} \frac{\partial \xi}{\partial s} + \frac{1}{2c} \frac{\partial^2 u}{\partial \tau^2} \frac{\partial \tau}{\partial s}$$

$$= \frac{1}{4} \frac{\partial^2 u}{\partial \xi^2} - \frac{1}{4c} \frac{\partial^2 u}{\partial \tau \partial \xi} + \frac{1}{4c} \frac{\partial^2 u}{\partial \xi \partial \tau} - \frac{1}{4c^2} \frac{\partial^2 u}{\partial \tau^2}$$

$$= \frac{1}{4} \left(\frac{\partial^2 u}{\partial \xi^2} - \frac{1}{c^2} \frac{\partial^2 u}{\partial \tau^2} \right) = \frac{1}{4} v = y.$$

We therefore obtain the equation

$$\frac{\partial^2 x}{\partial s \partial t} = y, \tag{4.2.5}$$

with the conditions

$$x(t, g(t)) = y_0(t), \quad x'_t(t, g(t)) - x'_s(t, g(t)) = y_1(t), \tag{4.2.6}$$

where

$$g(t) = t - 2c\tau_0 \tag{4.2.7}$$

is a continuously differentiable function, such that $g'(t) \equiv 1 > 0$. Thus g is an invertible function.

Write

$$D = \frac{\partial^2}{\partial s \partial t}. \tag{4.2.8}$$

The domain of the operator D is $C^2(\Omega)$, where, for instance, $\Omega = [0, l] \times [0, l]$. The operator D is right invertible in the space $C(\Omega)$. Indeed, if we write

$$(R_0 x)(t, s) = \int_0^t \int_0^s x(p, q) dq dp, \quad x \in C(\Omega), \tag{4.2.9}$$

then we have for $(t, s) \in \Omega$

$$(DR_0 x)(t, s) = \frac{\partial^2}{\partial s \partial t} \int_0^t \int_0^s x(p, q) dq dp = \frac{\partial}{\partial s} \int_0^s x(t, q) dq$$

$$= x(t, s).$$

Thus we have $DR_0 = I$ on $C(\Omega)$. Moreover,

$$\ker D = \{z = z(t, s) = f_1(t) + f_2(s) \colon f_1, f_2 \in C^1[0, l]\}. \quad (4.2.10)$$

Indeed, if $z(t, s) = f_1(t) + f_2(s)$, where $f_1, f_2 \in C^1[0, l]$, then

$$\frac{\partial^2}{\partial s \partial t} z(t, s) = \frac{\partial^2}{\partial s \partial t} [f_1(t) + f_2(s)] = \frac{\partial}{\partial s} f_1'(t) + \frac{\partial}{\partial t} f_2'(s) = 0.$$

By definition (4.2.7), the invertible function g belongs to the space $C^1[0, l]$. Define an operator F by means of the equality

$$(Fx)(t, s) = \frac{1}{2} \Big\{ x\big(t, g(t)\big) + x\big(g^{-1}(s), s\big) +$$

$$+ \int_{g^{-1}(s)}^{t} [x_t'(p, g(p)) - x_s'(p, g(p))] dp \Big\}$$

$$\text{for } x \in C(\Omega). \quad (4.2.11)$$

The operator F is a projection onto the space of constants $\ker D$. Indeed, if $z(t, s) = f_1(t) + f_2(s)$, where $f_1, f_2 \in C^1[0, l]$, we find

$$\frac{\partial}{\partial t} z(t, s) = f_1'(t), \qquad \frac{\partial}{\partial s} z(t, s) = f_2'(s).$$

Since $g'(t) = 1$ for $0 \leqslant t \leqslant l$, we obtain

$$(Fz)(t, s) = \frac{1}{2} \Big\{ f_1(t) + f_2(g(s)) + f_1(g^{-1}(s)) + f_2(s) +$$

$$+ \int_{g^{-1}(s)}^{t} [f_1'(p) - f_2'(g(p))] dp \Big\}$$

$$= \frac{1}{2} \Big\{ f_1(t) + f_2(g(s)) + f_1(g^{-1}(s)) +$$

$$+ \int_{g^{-1}(s)}^{t} f_1'(p) dp + f_2(s) - \int_{g^{-1}(s)}^{t} f_2'(g(p)) g'(p) dp \Big\}$$

$$= \frac{1}{2} \Big\{ f_1(t) + f_2(g(s)) + f_1(g^{-1}(s)) + f_2(s) + f_1(t) -$$

$$- f_1(g^{-1}(s)) - f_2(g(t)) + f_2(g(g^{-1}(s))) \Big\}$$

$$= \frac{1}{2} [2f_1(t) + f_2(s) + f_2(s)] = z(t, s).$$

Hence we have $Fz = z$ for an arbitrary $z \in \ker D$. Since for an arbitrary $x \in C^1(\Omega)$ we have $Fx = z \in \ker D$, we find $F^2x = Fz = z = Fx$, which proves that F is a projection onto $\ker D$.

Theorem 2.2.4 implies that F is an initial operator for D corresponding to the inverse $R = R_0 - FR_0$. It means that for all $x \in C(\Omega)$

$$
(Rx)(t, s) = \int_0^t \int_0^s x(p, q)\,dq\,dp - \frac{1}{2}\Bigg\{\int_0^t \int_0^{g(t)} x(p, q)\,dq\,dp +
$$

$$
+ \int_0^{g^{-1}(s)} \int_0^s x(p, q)\,dq\,dp +
$$

$$
+ \int_{g^{-1}(s)}^t \Bigg[\int_0^{g(p)} x(p, q)\,dq - \int_0^p x(q, g(p))\,dq\Bigg]\,dp\Bigg\}.
$$

$$
\text{(4.2.12)}
$$

Equation (4.2.5) together with Conditions (4.2.6) can be rewritten as the following initial value problem for the right invertible operator:

$$
Dx = y, \qquad Fx = x_0, \qquad\qquad\qquad (4.2.13)
$$

where $y \in C(\Omega)$ and, by Formula (4.2.11)

$$
x_0(t, s) = \frac{1}{2}\Bigg[y_0(t) + y_0(g^{-1}(s)) + \int_{g^{-1}(s)}^t y_1(p)\,dp\Bigg], \qquad (4.2.14)
$$

where $x_0 \in \ker D$.

Theorem 4.1.1 implies that Problem (4.2.13) is well-posed and its unique solution is of the form $x = Ry + x_0$, i.e. since $g(t) = t - 2c\tau_0$, $g^{-1}(t) = t + 2c\tau_0$, we have

$$
x(t, s) = \int_0^t \int_0^s y(p, q)\,dq\,dp - \frac{1}{2}\Bigg\{\int_0^t \int_0^{g(t)} y(p, q)\,dq\,dp +
$$

$$
+ \int_0^{g^{-1}(s)} \int_0^s y(p, q)\,dq\,dp +
$$

$$
+ \int_{g^{-1}(s)}^t \Bigg[\int_0^{g(p)} y(p, q)\,dq - \int_0^p y(q, g(p))\,dq\Bigg]\,dp\Bigg\} +
$$

$$+\frac{1}{2}\left[y_0(t)+y_0\left(g^{-1}(s)\right)+\int_{g^{-1}(s)}^{t}y_1(p)\,dp\right]$$

$$=\int_0^t\int_0^s y(p,q)\,dq\,dp-\frac{1}{2}\left\{\int_0^t\int_0^{t-2c\tau_0}y(p,q)\,dq\,dp+\right.$$

$$+\int_0^{s+2c\tau_0}\int_0^s y(p,q)\,dq\,dp+\int_{s+2c\tau_0}^t\left[\int_0^{p-2c\tau_0}y(p,q)\,dq-\right.$$

$$\left.-\int_0^p y(q,p-2c\tau_0)\,dq\right]dp\right\}+$$

$$+\frac{1}{2}\left[y_0(t)+y_0(s+2c\tau_0)+\int_{s+2c\tau_0}^t y_1(p)\,dp\right].$$

Thus a unique solution of the problem (4.2.1)–(4.2.2), we are looking for, is of the form

$$u(\xi,\tau)=u\left(\frac{t+s}{2},\frac{t-s}{2c}\right)=x(t,s)=x(\xi+c\tau,\xi-c\tau).$$

(4.2.15)

We shall consider now some problems for hyperbolic equations of a more general form than (4.2.5) (i.e. also (4.2.1)). The first one is

4.2.1. Darboux problem. Suppose that $X = C(\Omega)$, where $\Omega = \{(t,s): a \leqslant t, s \leqslant b\}$. We are looking for solutions of the equation

$$\frac{\partial^2 x}{\partial t\,\partial s}=A(t,s)\frac{\partial x}{\partial t}+B(t,s)\frac{\partial x}{\partial s}+C(t,s)x+y,$$

(4.2.16)

where $A, B, C, y \in C(\Omega)$, satisfying the conditions:

$$x(t_0,s)=y_1(s),\qquad x(t,s_0)=y_2(t),$$

(4.2.17)

where $(t_0,s_0)\in\Omega$ is arbitrarily fixed, $y_1, y_2 \in C^1[a,b]$, and a *coincidence condition* is satisfied

$$y_1(s_0)=y_2(t_0).$$

(4.2.18)

The problem (4.2.16)–(4.2.18) is called the *Darboux problem* for Equation (4.2.16) (see Fig. 4.1).

Write

$$D = \frac{\partial^2}{\partial t \partial s}, \quad H = A\frac{\partial}{\partial t} + B\frac{\partial}{\partial s} + CI. \qquad (4.2.19)$$

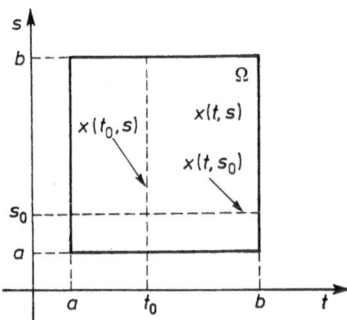

Fig. 4.1. Darboux problem

Observe that dom $H = C^1(\Omega)$ and $HC^1(\Omega) \subset C(\Omega)$. As in the preceding case we show that the operator D defined on $C^2(\Omega)$ is right invertible and has a right inverse of the form

$$(R_0 x)(t, s) = \int_{t_0}^{t} \int_{s_0}^{s} x(p, q) \, dq \, dp \quad \text{for } x \in C(\Omega). \qquad (4.2.20)$$

One can prove (cf. Section 8.4) that R_0 is a Volterra operator. As before, we conclude also that

$$\ker D = \{z = z(t, s) = f_1(t) + f_2(s) : f_1, f_2 \in C^1[a, b]\}. \qquad (4.2.21)$$

An initial operator F_0 corresponding to the right inverse R_0 has in our case an obvious form:

$$(F_0 x)(t, s) = x(t_0, s) + x(t, s_0) - x(t_0, s_0) \quad \text{for } x \in C(\Omega). \qquad (4.2.22)$$

Indeed, by definition, F_0 is a projection onto the space of constants $\ker D$ and $F_0 R_0 = 0$. We therefore can rewrite the problem (4.2.16)–(4.2.18) as an initial value problem for the operator D:

$$(D - H)x = y, \quad F_0 x = y_0, \qquad (4.2.23)$$

where $y \in C(\Omega)$ and

$$y_0(t, s) = y_1(s) + y_2(t) - y_1(t_0). \tag{4.2.24}$$

Theorem 4.1.2 implies that the problem (4.2.23) is well-posed, provided that under our assumptions the operator $I - HR_0$ is invertible (cf. Section 8.4). Thus the problem (4.2.23) has a unique solution which is of the form

$$x = R_0(I - HR_0)^{-1}(y - Hy_0) + y_0,$$

where H, R_0 and y_0 are defined by Formulae (4.2.19), (4.2.20), (4.2.24).
But

$$(Hy_0)(t, s) = A(t, s)\frac{\partial}{\partial t}[y_1(s) + y_2(t) - y_1(t_0)] +$$

$$+ B(t, s)\frac{\partial}{\partial s}[y_1(s) + y_2(t) - y_1(t_0)] +$$

$$+ C(t, s)[y_1(s) + y_2(t) - y_1(t_0)]$$
$$= A(t, s)y_2'(t) + B(t, s)y_1'(s) +$$
$$+ C(t, s)[y_1(s) + y_2(t) - y(t_0)].$$

Thus the solution of our problem is

$$x = R_0(I - HR_0)^{-1}(y - Ay_2' - By_1' - Cy_0) + y_0, \tag{4.2.25}$$

where H, R_0 and y_0 are defined by Formulae (4.2.19), (4.2.20), (4.2.24).

We should mention that this unique solution of the problem (4.2.23) can be obtained also in another way. Namely, since $(D - H)x = (D - DR_0 H)x = D(I - R_0 H)x$, we find

$$(I - R_0 H)x = R_0 y + z, \quad \text{where } z \in \ker D.$$

But $\quad y_0 = F_0 x = F_0 x - F_0 R_0 Hx = F_0(I - R_0 H)x = F_0(R_0 y + z)$
$= F_0 R_0 y + F_0 z = z$, which implies

$$x = (I - R_0 H)^{-1}(R_0 y + y_0), \tag{4.2.26}$$

provided that the operator $I - R_0 H$ is invertible.

However, the invertibility of the operator $I - HR_0$ does not imply in general case that the operator $I - R_0 H$ is invertible. This implication holds only if $HR_0 = R_0 H$, i.e. if the coefficients A, B, C of the operator H are scalars (cf. Section 3.1).

4.2.2. *Cauchy problem.* Suppose now that $X = C(\Omega)$, where Ω $= \{(t, s): 0 \leqslant t \leqslant a, 0 \leqslant s \leqslant b\}$. We are looking for solutions of Equation (4.2.16), where $A, B, C, y \in C(\Omega)$, satisfying the conditions

$$x(t, g(t)) = \sigma(t), \quad x'_t(t, g(t)) = \omega(t), \qquad (4.2.27)$$

where $g \in C^1[0, a]$, $g'(t) > 0$ for $0 \leqslant t \leqslant a$, $g(0) = 0$, $g(a) = b$, moreover, $\sigma, \omega \in C^1[0, a]$ (see Fig. 4.2). By our assumptions, the function g is invertible.

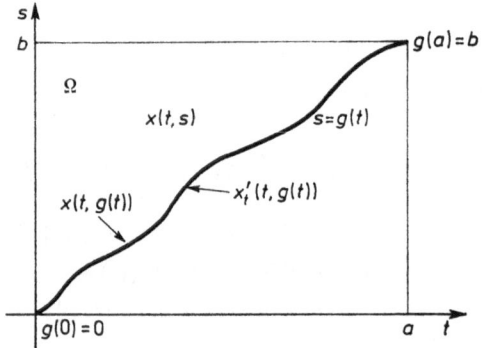

Fig. 4.2. Cauchy problem

The operators D and H are defined, as before, by Formulae (4.2.19) and the space of constants $\ker D$ by Formula (4.2.21), where f_1, f_2 $\in C^1[0, a]$.

Write

$$(Fx)(t, s) = x(g^{-1}(s), s) + \int_{g^{-1}(s)}^{t} x'_t(p, g(p)) dp \quad \text{for } x \in C^1(\Omega).$$

$$(4.2.28)$$

The operator F is a projection onto the space of constants. Indeed, suppose that $z \in \ker D$ is arbitrary. Then $z(t, s) = f_1(t) + f_2(s)$, where $f_1, f_2 \in C^1[0, a]$, $\dfrac{\partial}{\partial t} z(t, s) = f'_1(t)$ and

$$(Fz)(t, s) = f_1(g^{-1}(s)) + f_2(s) + \int_{g^{-1}(s)}^{t} f'_1(p) dp$$

$$= f_1\big(g^{-1}(s)\big)+f_2(s)+f_1(t)-f_1\big(g^{-1}(s)\big)$$
$$= f_1(t)+f_2(s) = z(t, s).$$

Thus $Fz = z$ for $z \in \ker D$. It is easy to see that for an arbitrary $x \in C(\Omega)$ we have $Fx = z \in \ker D$. Hence $F^2x = Fz = z = Fx$, which proves that F is a projection onto $\ker D$. Theorem 2.2.4 implies that F is an initial operator for D corresponding to a right inverses $R = R_0 - -FR_0$, where we can define R_0 by Formula (4.2.20). We therefore obtain for $x \in C(\Omega)$

$$(Rx)(t, s) = \int_0^t\int_0^s x(p, q)\,dq\,dp - \left\{\int_0^{g^{-1}(s)}\int_0^s x(p, q)\,dq\,dp + \right.$$
$$\left. + \int_{g^{-1}(s)}^t\left[\frac{d}{dt}\int_0^t\int_0^s x(p, q)\,dq\,dp\Big|_{\substack{t=p\\s=g(p)}}\right]dp\right\}$$
$$= \int_0^t\int_0^s x(p, q)\,dq\,dp - \int_0^{g^{-1}(s)}\int_0^s x(p, q)\,dq\,dp -$$
$$- \int_{g^{-1}(s)}^t\left[\int_0^s x(t, q)\,dq\right]_{\substack{t=p\\s=g(p)}}dp$$
$$= \int_{g^{-1}(s)}^t\int_0^s x(p, q)\,dq\,dp - \int_{g^{-1}(s)}^t\int_0^{g^{-1}(p)} x(p, q)\,dq\,dp$$
$$= \int_{g^{-1}(s)}^t\int_{g(p)}^s x(p, q)\,dq\,dp.$$

Finally

$$(Rx)(t, s) = \int_{g^{-1}(s)}^t\int_{g(p)}^s x(p, q)\,dq\,dp \quad \text{for } x \in C(\Omega). \tag{4.2.29}$$

We have to solve the following initial value problem

$$(D-H)x = y, \quad Fx = y_0, \tag{4.2.30}$$

where $y \in C(\Omega)$ and, by Conditions (4.2.27) and Formula (4.2.28)

$$y_0(t, s) = \sigma\big(g^{-1}(s)\big) + \int_{g^{-1}(s)}^{t} \omega(p)\,dp. \qquad (4.2.31)$$

If we prove that the operator $I - HR$ is invertible then we conclude by Theorem 2.2.4 that the problem (4.2.30) is well-posed* and its unique solution is of the form

$$x = R(I - HR)^{-1}(y - Hy_0) + y_0,$$

where H, R, y_0 are defined by Formula (4.2.19), (4.2.29), (4.2.31), respectively.

4.2.3. Picard problem. Suppose that $X = C(\Omega)$, where Ω is defined, as in Cauchy problem. We are looking for solutions of Equation (4.2.16), where $A, B, C, y \in C(\Omega)$, satisfying the conditions

$$x(t, 0) = \sigma(t), \quad x(g^{-1}(s), s) = \omega(s), \qquad (4.2.32)$$

where $g \in C^1[0, a]$, $g'(t) > 0$ for $0 \leqslant t \leqslant a$, $g(0) = 0$, $g(a) = b$, $\sigma, \omega \in C^1[0, a]$ and the *coincidence condition* is satisfied

$$\sigma(0) = \omega(0) \qquad (4.2.33)$$

(because $g^{-1}(0) = 0$ (see Fig. 4.3)).

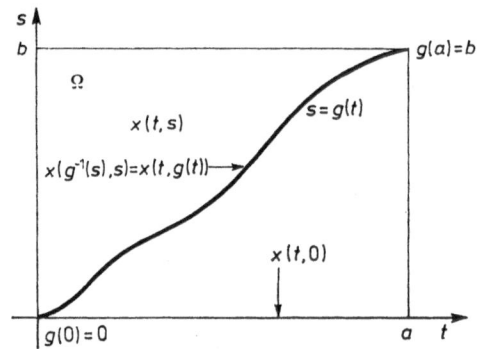

Fig. 4.3. Picard problem

---

* It is well-known that the Cauchy problem is well-posed (cf., for instance, Pogorzelski, 1966). However, the problem of invertibility of the operator $I - HR$ is out of the main interest of this book and is too complicated to be considered here.

The operators D and H are defined, as before, by Formulae (4.2.19) and the space of constants $\ker D$ by Formula (4.2.21), where $f_1, f_2 \in C^1[0, a]$.

Write

$$(Fx)(t, s) = x\big(g^{-1}(s), s\big) - x\big(g^{-1}(s), 0\big) + x(t, 0)$$

$$\text{for } x \in C(\Omega). \quad (4.2.34)$$

The operator F is a projection onto the space of constants $\ker D$. Indeed, suppose that $z \in \ker D$ is arbitrary. Then $z(t, s) = f_1(t) + f_2(t)$, where $f_1, f_2 \in C^1[0, a]$ and

$$(Fz)(t, s) = f_1\big(g^{-1}(s)\big) + f_2(s) - f_1\big(g^{-1}(s)\big) - f_2(0) + f_1(t) + f_2(0)$$

$$= f_1(t) + f_2(s) = z(t, s),$$

i.e. $Fz = z$ for $z \in \ker D$. Further arguments are going on the same line as in the Cauchy problem. Thus, by Theorem 2.2.4 we conclude that F is an initial operator for D corresponding to the right inverse $R = R_0 - FR_0$ and R_0 is defined by Formula (4.2.20). We therefore obtain for $x \in C(\Omega)$

$$(Rx)(t, s) = \int_0^t \int_0^s x(p, q)\,dq\,dp - \left\{ \int_0^{g^{-1}(s)} \int_0^s x(p, q)\,dq\,dp - \right.$$

$$- \int_0^{g^{-1}(s)} \int_0^0 x(p, q)\,dq\,dp + \left. \int_0^t \int_0^0 x(p, q)\,dq\,dp \right\}$$

$$= \int_{g(s)}^t \int_0^s x(p, q)\,dq\,dp,$$

i.e.

$$(Rx)(t, s) = \int_{g^{-1}(s)}^t \int_0^s x(p, q)\,dq\,dp \quad \text{for } x \in C(\Omega). \quad (4.2.35)$$

We have to solve the following initial value problem:

$$(D - H)x = y, \quad Fx = y_0, \quad (4.2.36)$$

where $y \in C(\Omega)$, and by Conditions (4.2.32), (4.2.33) and Formula (4.2.34)

$$y_0(t, s) = \omega(s) - \sigma\big(g^{-1}(s)\big) + \sigma(t). \quad (4.2.37)$$

If we prove that the operator $I-HR$ is invertible (cf. the footnote on p. 240) then we conclude that, by Theorem 2.2.4, the problem (4.2.36) is well-posed and its unique solution is of the form

$$x = R(I-HR)^{-1}(y-Hy_0)+y_0,$$

where H, R, y_0 are defined by Formulae (4.2.19), (4.2.35), (4.2.37), respectively.

4.2.4. Generalized Cauchy problem. Suppose that $X = C(\Omega)$, where Ω is defined, as in Cauchy problem. We are looking for solutions of Equation (4.2.16), where $A, B, C, y \in C(\Omega)$, satisfying the conditions

$$x\big(t, g_0(t)\big) = \sigma(t), \quad x_t'\big(t, g_1(t)\big) = \omega(t), \qquad (4.2.38)$$

where $g_0, g_1, \sigma, \omega \in C^1[0, a]$, $g_0(t) > 0$, $g_1'(t) > 0$ for $0 \leqslant t \leqslant a$, $g_0(0) = g_1(0) = 0$, $g_0(a) = g_1(a) = b$ (see Fig. 4.4).

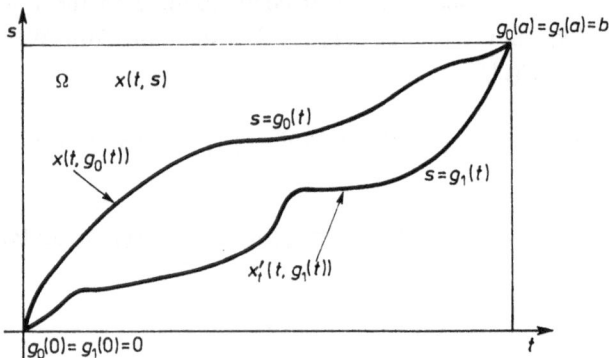

Fig. 4.4. Generalized Cauchy problem

The operators D and H are defined, as before, by Formulae (4.2.19) and the space of constants $\ker D$ by Formula (4.2.21).

Write

$$(Fx)(t, s)$$

$$= x\big(g_0^{-1}(s), s\big)+ \int\limits_{g_0^{-1}(s)} x_t'\big(p, g_1(p)\big)\mathrm{d}p \quad \text{for } x \in C^1(\Omega).$$

$$(4.2.38')$$

The operator F is a projection onto the space of constants $\ker D$. Indeed, suppose that $z \in \ker D$ is arbitrary. Then $z(t, s) = f_1(t) + f_2(s)$, where $f_1, f_2 \in C^1[0, a]$ and $z'_t(t, s) = f'_1(t)$. Thus

$$(Fz)(t, s) = f_1\left(g_0^{-1}(s)\right) + f_2(s) + \int\limits_{g_0^{-1}(s)}^{t} f'_1(p)\,dp$$

$$= f_1\left(g_0^{-1}(s)\right) + f_2(s) + f_1(t) - f_1\left(g_0^{-1}(s)\right)$$

$$= f_1(t) + f_2(s) = z(t, s)$$

and $Fz = z$ for $z \in \ker D$. Further arguments are going on the same line as in the Cauchy problem. Thus, by Theorem 2.2.4 we conclude that F is an initial operator for D corresponding to the right inverse $R = R_0 - FR_0$ and R_0 is defined by Formula (4.2.20), where we put $t_0 = s_0 = 0$. We therefore obtain for $x \in C(\Omega)$

$$(Rx)(t, s) = \int\limits_{0}^{t}\int\limits_{0}^{s} x(p, q)\,dq\,dp - \int\limits_{0}^{g_0^{-1}(s)}\int\limits_{0}^{s} x(p, q)\,dq\,dp -$$

$$- \int\limits_{g_0^{-1}(s)}^{t} \left[\frac{d}{dt}\int\limits_{0}^{t}\int\limits_{0}^{s} x(p, q)\,dq\,dp\right]_{\substack{t=p \\ s=g_1(p)}} dp$$

$$= \int\limits_{0}^{t}\int\limits_{0}^{s} x(p, q)\,dq\,dp - \int\limits_{0}^{g_0^{-1}(s)}\int\limits_{0}^{s} x(p, q)\,dq\,dp -$$

$$- \int\limits_{g_0^{-1}(s)}^{t} \left[\int\limits_{0}^{s} x(t, q)\,dq\right]_{\substack{t=p \\ s=g_1(p)}} dp$$

$$= \int\limits_{0}^{t}\int\limits_{0}^{s} x(p, q)\,dq\,dp + \int\limits_{g_0^{-1}(s)}^{0}\int\limits_{0}^{s} x(p, q)\,dq\,dp -$$

$$- \int\limits_{g_0^{-1}(s)}^{t}\int\limits_{0}^{g_1(p)} x(p, q)\,dq\,dp$$

$$= \int\limits_{g_0^{-1}(s)}^{t} \int\limits_0^s x(p, q)\,dq\,dp + \int\limits_{g_0^{-1}(s)}^{t} \int\limits_{g_1(s)}^s x(p, q)\,dq\,dp$$

$$= \int\limits_{g_0^{-1}(s)}^{t} \int\limits_{g_1(p)}^s x(p, q)\,dq\,dp,$$

$$R(x)(t, s) = \int\limits_{g_0^{-1}(s)}^{t} \int\limits_{g_1(s)}^s x(p, q)\,dq\,dp \quad \text{for } x \in C(\Omega). \qquad (4.2.39)$$

We have to solve the following initial value problem:

$$(D-H)x = y, \quad Fx = y_0,$$

where $y \in C(\Omega)$ and, by Conditions (4.2.38) and Formula (4.2.38′)

$$y_0(t, s) = \sigma\big(g_0^{-1}(s)\big) + \int\limits_{g_0^{-1}(s)}^{t} \omega(p)\,dp. \qquad (4.2.40)$$

If we prove that the operator $I - HR$ is invertible (cf. the footnote on p. 240) then we conclude that, by Theorem 4.2.2, the problem (4.2.40) is well-posed and its unique solution is of the form

$$x = R(I - HR)^{-1}(y - Hy_0) + y_0, \qquad (4.2.41)$$

where H, R, y_0 are defined by Formulae (4.2.19), (4.2.39), (4.2.41).

If $g_1 = g_0$ then the problem under consideration is the Cauchy Problem 4.2.2. If $g_1 \neq g_0$ then the problem of invertibility of the operator $I - HR$ is, in general, very difficult and needs some more studies.

Till this moment we have considered a generalization (4.2.16) of Equation (4.2.1) and problems connected with Equation (4.2.16). Observe that the coefficient c in Equation (4.2.1) is a scalar. We shall show that the method used above can be also applied in the case of a variable coefficient. It is enough to consider an operator of the form

$$D = \frac{\partial^2}{\partial \xi^2} - \frac{1}{[b(\tau)]^2} \frac{\partial^2}{\partial \tau^2}, \qquad (4.2.42)$$

where $b \in C^1[0, T]$, $b(\tau) \neq 0$ for $0 \leqslant \tau \leqslant T$ and the function

$$a(\tau) = \frac{1}{\tau} \int\limits_0^{\tau} b(\sigma)\,d\sigma \qquad (4.2.43)$$

is continuous for $0 \leqslant \tau \leqslant T$. Observe that, by our assumptions,

$$a'\tau + a = b, \qquad a''\tau + 2a' = b', \qquad (4.2.44)$$

which means that $a \in C^2[0, T]$. Indeed,

$$a'\tau + a = \left[-\frac{1}{\tau} \int_0^\tau b(\sigma) d\sigma + \frac{1}{\tau} b(\tau) \right] \tau + \frac{1}{\tau} \int_0^\tau b(\sigma) d\sigma$$

$$= -\frac{1}{\tau} \int_0^\tau b(\sigma) d\sigma + \frac{1}{\tau} \int_0^\tau b(\sigma) d\sigma + b(\tau) = b(\tau).$$

By differentiation of the first of Equalities (4.2.44) we find

$$a''\tau + 2a' = b'.$$

Write now

$$s = \xi - a(\tau)\tau, \qquad t = \xi + a(\tau)\tau. \qquad (4.2.45)$$

The first of Equalities (4.2.44) implies that

$$\frac{\partial t}{\partial \tau} = b, \qquad \frac{\partial s}{\partial \tau} = -b. \qquad (4.2.46)$$

Indeed,

$$\frac{\partial t}{\partial \tau} = a'\tau + a = b, \qquad \frac{\partial s}{\partial \tau} = -(a'\tau + a) = -b.$$

In a similar way, as for Equation (4.2.1), we find for any $u \in C^2(\Omega)$

$$\frac{\partial u}{\partial \xi} = \frac{\partial u}{\partial s} \frac{\partial s}{\partial \xi} + \frac{\partial u}{\partial t} \frac{\partial t}{\partial \xi} = u_s' + u_t',$$

$$\frac{\partial^2 u}{\partial \xi^2} = \frac{\partial}{\partial \xi}(u_s' + u_t') = u_{ss}'' \frac{\partial s}{\partial \xi} + u_{st}'' \frac{\partial t}{\partial \xi} + u_{ts}'' \frac{\partial s}{\partial \xi} + u_{tt}'' \frac{\partial t}{\partial \xi}$$

$$= u_{ss}'' + 2u_{st}'' + u_{tt}'',$$

$$\frac{\partial u}{\partial \tau} = u_s' \frac{\partial s}{\partial \tau} + u_t' \frac{\partial t}{\partial \tau} = (a'\tau + a)(u_t' - u_s') = b(u_t' - u_s'),$$

$$\frac{\partial^2 u}{\partial \tau^2} = b'(u_t' - u_s') + b\left(u_{tt}'' \frac{\partial t}{\partial \tau} + u_{ts}'' \frac{\partial s}{\partial \tau} - u_{st}'' \frac{\partial t}{\partial \tau} - u_{ss}'' \frac{\partial s}{\partial \tau} \right)$$

$$= b'(u_t' - u_s') + b^2(u_{tt}'' - 2u_{ts}'' + u_{ss}'').$$

These equalities together imply that

$$Du = \frac{\partial^2 u}{\partial \xi^2} - \frac{1}{b^2} \frac{\partial^2 u}{\partial \tau^2}$$

$$= u_{ss}'' + 2u_{st}'' + u_{tt}'' - \frac{1}{b^2} [b^2(u_{tt}'' - 2u_{ts}'' + u_{ss}'') + b'(u_t' - u_s')]$$

$$= u_{ss}'' + 2u_{st}'' + u_{tt}'' - u_{tt}'' + 2u_{ts}'' - u_{ss}'' - \frac{b'}{b^2}(u_t' - u_s')$$

$$= 4u_{ts}'' - \frac{b'}{b^2}(u_t' - u_s') = 4\frac{\partial^2 u}{\partial t \partial s} - \frac{b'}{b^2}\left(\frac{\partial u}{\partial t} - \frac{\partial u}{\partial s}\right).$$

Write

$$D_0 = \frac{\partial^2}{\partial t \partial s}, \qquad G = \frac{1}{4}\frac{b'}{b^2}\left(\frac{\partial}{\partial t} - \frac{\partial}{\partial s}\right). \qquad (4.2.47)$$

Then

$$D = 4(D_0 + G). \qquad (4.2.48)$$

The operator D_0 is right invertible and has a right inverse R_0 defined by means of Equality (4.2.9). We can prove that the operator $I + GR_0$ is invertible (cf., for instance, Pogorzelski, 1966). Then the operator

$$R = \frac{1}{4} R_0 (I + GR_0)^{-1}$$

is a right inverse of the operator D defined by (4.2.48). Indeed,

$$DR = 4(D_0 + G)\frac{1}{4} R_0(I + GR_0)^{-1} = (D_0 R_0 + GR_0)(I + GR_0)^{-1}$$

$$= (I + GR_0)(I + GR_0)^{-1} = I.$$

In similar way we can consider the case, when the coefficient of the operator D defined by Formula (4.2.42) depends on the variable ξ: $b_1 = b_1(\xi)$ (cf. Example 4.2.1).

By different application of Theorem 2.2.5 we can consider various problems analogous to the problems (4.2.1)–(4.2.4) for one-dimensional *multi-waves operators*, i.e. for the operators of the form

$$D = D_1 \ldots D_m, \qquad \text{where } D_j = \frac{\partial^2}{\partial \xi^2} - \frac{1}{c_j^2}\frac{\partial^2}{\partial \tau^2}$$

$$(j = 1, 2, \ldots, m), \qquad (4.2.49)$$

where the coefficients $c_1, \ldots, c_m$ are either scalar or variable (cf. the author, 1976).

Examples and Exercises

EXAMPLE 4.2.1. Consider the space $C(\Omega)$, where, as before, $\Omega = [0, l] \times \times [0, T]$ and the operator

$$D_1 = \frac{\partial^2}{\partial \xi^2} - \frac{1}{b_1^2} \frac{\partial^2}{\partial \tau^2}, \quad \text{where } b_1 = b_1(\xi), \qquad (4.2.50)$$

where $b_1 \in C^1[0, l]$, $b_1(\xi) \neq 0$ for $0 \leqslant \xi \leqslant l$. Write $b = 1/b_1$. Suppose that the function

$$a(\xi) = \frac{1}{\xi} \int\limits_0^\xi b(s) \, ds = -\int\limits_0^\xi \frac{ds}{b_1(s)}$$

is continuous. Write also

$$s = \tau - a(\xi)\xi, \quad t = \tau + a(\xi)\xi.$$

Then, in a similar way, as for a coefficient dependent on the variable τ, we obtain

$$D_1 = \frac{\partial^2}{\partial \xi^2} - \frac{1}{b_1^2} \frac{\partial^2}{\partial \tau^2} = -\frac{1}{b_1^2} \left(\frac{\partial^2}{\partial \tau^2} - b_1^2 \frac{\partial^2}{\partial \xi^2} \right)$$

$$= -b^2 \left(\frac{\partial^2}{\partial \tau^2} - \frac{1}{b^2} \frac{\partial^2}{\partial \xi^2} \right) = -4b^2 \frac{\partial^2}{\partial t \, \partial s} - 4 \frac{b'}{b^2} \left(\frac{\partial}{\partial t} - \frac{\partial}{\partial s} \right)$$

$$= -\frac{4}{b_1^2} \frac{\partial^2}{\partial t \, \partial s} - 4b_1' \left(\frac{\partial}{\partial t} - \frac{\partial}{\partial s} \right).$$

The operator $D_0 = \partial^2/\partial t \, \partial s$ is right invertible and has a right inverse R_0 defined by Formula (4.2.9). Write now

$$G_1 = \frac{1}{4} b_1' b_1^2 \left(\frac{\partial}{\partial t} - \frac{\partial}{\partial s} \right).$$

We can show that the operator $I + G_1 R_0$ is invertible (cf., for instance, Pogorzelski, 1966). Thus the operator

$$R_1 = -\frac{1}{4} R_0 (I - G_1 R_0)^{-1} b_1^2$$

is a right inverse of the operator D_1. Indeed,

$$D_1 R_1 = -\frac{4}{b_1^2}(D_0+G_1)\left(-\frac{1}{4}R_0\right)(I+G_1 R_0)^{-1}b_1^2$$

$$= \frac{1}{b_1^2}(D_0 R_0+G_1 R_0)(I+G_1 R_0)^{-1}b_1^2$$

$$= \frac{1}{b_1^2}(I+G_1 R_0)(I+G_1 R_0)^{-1}b_1^2 = \frac{1}{b_1^2}b_1^2 I = I.$$

EXERCISE 4.2.1. Solve the problem (4.2.1)–(4.2.2) when $\tau_0 = 0$.

EXERCISE 4.2.2. A *generalized Darboux–Picard problem* is to find solutions of Equation (4.2.16) satisfying the following conditions:

$$x\left(g_0^{-1}(s), s\right)-x\left(g_0^{-1}(s), 0\right) = \sigma_0(s),$$

$$x\left(t, g_1(t)\right)-x\left(0, g_1(t)\right) = \sigma_1(t),$$
$$(4.2.51)$$

where $\sigma_0 \in C^1[0, b]$, $\sigma_1 \in C^1[0, a]$, $g_0, g_1 \in C^1[0, a]$ are given, moreover, $g_0'(t) > 0$, $g_1'(t) > 0$ for $0 \leqslant t \leqslant a$ and

$$g_0(0) = g_1(0) = 0, \qquad g_0(a) = g_1(a) = b, \qquad \sigma_0(0) = \sigma_1(0) = 0.$$

Show that an operator F defined by means of the equality

$$(Fx)(t, s) = x\left(g_0^{-1}(s), s\right)-x\left(g_0^{-1}(s), 0\right)+$$

$$+x\left(t, g_1(t)\right)-x\left(0, g_1(t)\right)-x(0, 0) \qquad (4.2.52)$$

for $x \in C(\Omega)$ is an initial operator for the operator $D = \partial^2/\partial t\,\partial s$ corresponding to a right inverse

$$(Rx)(t, s) = \int_0^t \int_{g_1(t)}^s x(p, q)\,dq\,dp - \int_0^{g_0^{-1}(t)} \int_0^s x(p, q)\,dq\,dp.$$

Solve the problem (4.2.16)–(4.2.51) in the case $H = 0$.

EXERCISE 4.2.3. Solve the so-called *equation of transmission line*

$$\frac{\partial^2}{\partial \xi^2} - \frac{1}{c^2}\frac{\partial^2 u}{\partial \tau^2} - k\frac{\partial u}{\partial \tau} = 0 \qquad (4.2.53)$$

with the conditions

$$u(\xi, 0) = f_0(\xi), \qquad u_\tau'(\xi, 0) = f_1(\xi), \qquad (4.2.54)$$

where $f_0, f_1 \in C^1(\Omega_0)$, by the substitution (4.2.31).

EXERCISE 4.2.4. Solve problems (4.2.1)–(4.2.4) in the case $H = 0$.

EXERCISE 4.2.5. Solve a multi-wave equation ($m = 2$)

$$\left(\frac{\partial^2}{\partial\xi^2} - \frac{1}{c_1^2}\frac{\partial^2}{\partial\tau^2}\right)\left(\frac{\partial^2}{\partial\xi^2} - \frac{1}{c_2^2}\frac{\partial^2}{\partial\tau^2}\right)u = v, \quad v \in C(\Omega_0)$$

(4.2.55)

with the conditions

$$u^{(k)}(\xi, 0) = v_k, \quad v_k \in C^k(\Omega_0) \quad (k = 0, 1, 2, 3). \qquad (4.2.56)$$

EXERCISE 4.2.6. Find the general solution of the following equation:

$$\frac{\partial^2 u}{\partial\xi^2} - \frac{1}{\sin^2\tau}\frac{\partial^2 u}{\partial\tau^2} = v, \quad v \in C(\Omega_0). \qquad (4.2.57)$$

EXERCISE 4.2.7. Find the general solution of the following equation:

$$\frac{\partial^2 u}{\partial\xi^2} - \frac{1}{\xi^2}\frac{\partial^2 u}{\partial\tau^2} = v, \quad v \in C(\Omega_0) \qquad (4.2.58)$$

(apply Example 4.2.1).

EXERCISE 4.2.8. Solve Equation (4.2.57) with the conditions
  (i) $u(\xi, 0) = v_0(\xi)$, $u_\xi'(\xi, 0) = v_1(\xi)$,
  (ii) $u(\xi, 0) = v_0(\xi)$, $u(0, \tau) = v_1(\tau)$,
where $v_0, v_1 \in C^1(0, l)$ are given functions.

EXERCISE 4.2.9. Suppose that $t = (t_1, \ldots, t_q) \in \mathbf{R}^q$. Write

$$D^k = \frac{\partial^{k_1 + \ldots + k_q}}{\partial t_1^{k_1} \ldots \partial t_q^{k_q}},$$

where $k_1, \ldots, k_n$ are non-negative integers. The vector $k = (k_1, \ldots, k_q)$ is called a *multi-index*.

If $k_1 = \ldots = k_q = \alpha$, we shall write $k = (\alpha)_q$. Suppose that $k' = (k_1', \ldots, k_q')$ is also a multi-index. We shall write $k \leqslant k'$ if and only if $k_j \leqslant k_j'$ for $j = 1, \ldots, q$, $k = k'$ if and only if $k_j = k_j'$ for $j = 1, \ldots, q$ and $|k| = k_1 + \ldots + k_q$.

Consider the following differential operator

$$\hat{D} = \sum_{0 \leqslant |k| \leqslant n} a_k D^k,$$

where $a_n(t) \equiv 1$, $a_k \in C(\Omega)$, $\Omega = \{t: 0 \leqslant t_j \leqslant T_j \ (j = 1, \ldots, q)\}$.

Prove that the operator $\hat{D}$ is right invertible in the space $C(\Omega)$ and find a right inverse for $\hat{D}$ and a corresponding initial operator. In particular, consider the case $a_k(t) \equiv 0$ for $0 \leqslant |k| \leqslant n-1$.

4.3. ELLIPTIC AND POLYHARMONIC EQUATIONS

Suppose, we are given a domain $\Omega \subset \mathbf{R}^n$ $(n \geqslant 2)$. A *derivative of a function* $x \in C^1(\Omega)$ *in the direction of a line P (directional derivative)* is defined as

$$\frac{\mathrm{d}x}{\mathrm{d}P} = \sum_{j=1}^{n} \alpha_j \frac{\partial x}{\partial t_j}, \quad \text{where } t = (t_1, \ldots, t_n) \in \Omega \qquad (4.3.1)$$

and $\alpha_1, \ldots, \alpha_n$ denote the directional cosines of the line P.

Suppose that the domain Ω is bounded by a closed surface S which has at each point a tangent hyperplane. Then at each point $p \in S$ there exists a line n_p orthogonal to the tangent hyperplane. If for an arbitrary $s \in \Omega \cap n_p$ the vector $p-s$ is oriented as the line n_p, we say that n_p is an *inner normal*.

Suppose that a domain $\Omega \subset \mathbf{R}^n$ is bounded by a closed surface S of *Liapunov type*, i.e. a surface S satisfying the following conditions:

(i) The surface S has a tangent hyperplane at every point and the angle v_{pq} between two inner normals n_p and n_q at two arbitrary points $p, q \in S$ satisfies the inequality:

$$|v_{pq}| \leqslant C|p-q|^{\varkappa} \quad (0 < \varkappa \leqslant 1), \qquad (4.3.2)$$

where $|p-q|$ denotes the Euclidean distance of points p and q and the known constants $C > 0$, $\varkappa$ do not depend on the position of points p, q.

(ii) There exists a number $\delta > 0$ small enough such that a sphere of a center in arbitrary point $p \in S$ and the radius δ cuts out a part S_δ of the surface S such that an arbitrary line parallel to the inner normal n_p intersects S_δ at most at one point (Fig. 4.5).

We point out that Condition (4.3.2) is stronger than the assumption about continuity of the tangent hyperplane.

Denote by $\overline{\Omega}$ the closure of a bounded domain Ω, i.e. the set $\Omega \cup S$, where $S = \partial\Omega$ is the boundary of the domain Ω.

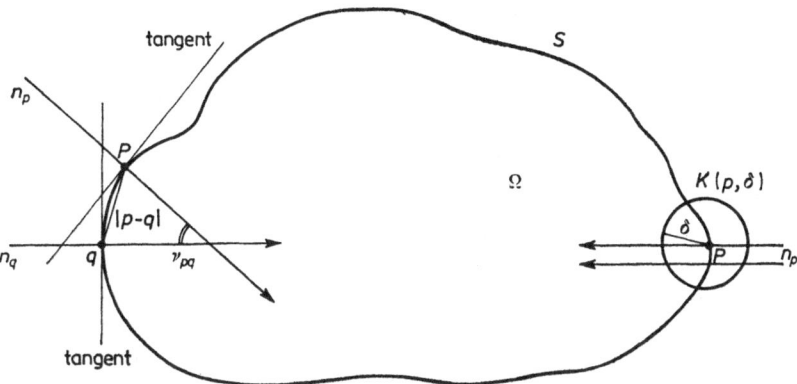

Fig. 4.5. One-dimensional Liapunov surface

Write

$$\Delta = \sum_{j=1}^{n} \frac{\partial^2}{\partial t_j^2}, \quad \text{where } t = (t_1, \ldots, t_n) \in \mathbf{R}^n. \tag{4.3.3}$$

The operator Δ is called *Laplace operator* or *Laplacian*.

A function $x \in C^2(\Omega)$ is said to be *harmonic in a domain* Ω if x satisfies *Laplace equation*

$$\Delta x = 0 \quad \text{in } \Omega. \tag{4.3.4}$$

One can prove (cf. for instance, Saks and Zygmund, 1965) that harmonic functions have the following properties:

(i) A value of a harmonic function in the center of a ball is equal to the intermediate value of this function on the sphere bounded this ball;

(ii) A harmonic function bounded and continuous in a bounded domain $\overline{\Omega}$ reaches its supremum and infimum at the boundary of this domain;

(iii) A harmonic function bounded and continuous in Ω and vanishing on the boundary of Ω is identically equal zero in Ω.

A function $x \in C^{2M}(\Omega)$ is said to be *polyharmonic* (*M-harmonic*) in a domain Ω if x satisfies the so-called *polyharmonic* (*M-harmonic*) *equation*

$$\Delta^M x = 0 \quad \text{in } \Omega \quad (M \geqslant 2). \tag{4.3.5}$$

In particular, if $M = 2$ then such functions and equations are called *biharmonic*.

We shall prove now that the Laplace operator Δ is right invertible in a suitably chosen space.

Suppose that a domain $\Omega \subset \mathbf{R}^n$ ($n \geqslant 2$) is bounded by a surface of Liapunov type. A function $G_1(t, s)$ defined for $t, s \in \Omega$, $t \neq s$, is called *Green function of the first kind for the domain Ω with respect to the Laplacian Δ* if

$$G_1(t, s) = \begin{cases} \dfrac{1}{|t-s|^{n-2}} - g(t, s) & \text{for } n \geqslant 3, \\ \ln|t-s| - g(t, s) & \text{for } n = 2, \end{cases} \qquad (4.3.6)$$

where $\Delta_t g = \displaystyle\sum_{j=1}^{n} \dfrac{\partial^2 g}{\partial t_j^2} = 0$ even if $t \to s$ (i.e. g is a harmonic function in Ω with respect to t) and

$$\lim_{t \to p \in S} G_1(t, s) = 0 \quad \text{for arbitrarily fixed } s \in \Omega. \qquad (4.3.7)$$

It is easy to check that for every $s \in \Omega$

$$\Delta_t \ln|t-s| = 0 \quad \text{for } n = 2, \qquad \Delta_t \frac{1}{|t-s|^{n-2}} = 0 \quad \text{for } n \geqslant 3.$$

Thus, by definition, $\Delta_t G_1 = 0$ in Ω for every $s \in \Omega$, i.e. $G_1(t, s)$ is a harmonic function in Ω with respect to t for every $s \in \Omega$. Moreover, from this definition follows that

$$g(t, s) = \int_S \frac{\psi(p, s)}{|t-p|^{n-2}} \, dS_p \quad \text{if } n \geqslant 3,$$

$$g(t, s) = \int_S \psi(p, s)\ln|t-p| \, dS_p \quad \text{if } n = 2, \qquad (4.3.8)$$

i.e. $g(t, s)$ is the so-called *potential of surface distribution* with the density $\psi(p, s) > 0$ for $p \in S$ and such that $\int_S \psi(p, s)dS_p = 1$ (cf., for instance, Pogorzelski, 1966).

A very important property of the Green function of the first kind is its symmetry. Namely we have

$$G_1(s, t) = G_1(t, s) \quad \text{for } t, s \in \Omega. \qquad (4.3.9)$$

Thus we also have $g(s, t) = g(t, s)$.

A function $G_2(t, s)$ defined for $t, s \in \Omega$, $t \neq s$, is called *Green function of the second kind for the domain Ω with respect to the Laplacian Δ* if G_2 satisfies Condition (4.3.6) and moreover, if t is convergent to an arbitrary point $p \in S$ and $s \in \Omega$ then the derivative of G_2 in the direction of the inner normal at the point p is convergent to a limit satisfying the following condition:

$$\lim_{t \to p \in S} \frac{\mathrm{d}G_2}{\mathrm{d}n_p} + a(p) G_2(p, s) = 0 \quad \text{for all } s \in \Omega, \qquad (4.3.10)$$

where $a(p) \leqslant 0$ for $p \in S$. We point out that the assumption $a(p) \leqslant 0$ is a sufficient condition for the existence of a Green function of the second type. From this assumption also follows that $g(t, s)$ is of the form (4.3.8) and that Condition (4.3.9) is satisfied (cf., for instance Pogorzelski, 1966).

THEOREM 4.3.1. *Suppose that a domain $\Omega \subset \mathbf{R}^n$ ($n \geqslant 2$) is bounded by a closed surface S of Liapunov type. Then the Laplace operator Δ is right invertible in the space $C(\Omega)$, a right inverse of Δ is defined by means of the equality*

$$(R_1 x)(t) = \lambda_n \int_\Omega G_1(t, s) x(s) \mathrm{d}\Omega_s \quad \text{for } x \in C(\Omega), \qquad (4.3.11)$$

where G_1 is the Green function of the first kind for the domain Ω with respect to Δ and

$$\lambda_n = \begin{cases} 1/2\pi & \text{if } n = 2, \\ \dfrac{(n-2)\Gamma(n/2)}{2\pi^{n/2}} & \text{if } n \geqslant 3, \end{cases} \qquad (4.3.12)$$

$\Gamma(t)$ *denotes the so-called Euler Γ-function, i.e. $\Gamma(t) = \int\limits_0^\infty e^s s^{t-1} \mathrm{d}s$ (the function $\Gamma(t)$ is a generalization of the function $n!$ defined only for non-negative integers n, because $\Gamma(n+1) = n!$ for $n = 0, 1, 2, \ldots$).*

The space of constants for the operator Δ is, by definition

$$\ker \Delta = \{z \in C(\overline{\Omega}) : \Delta z = 0 \text{ in } \Omega\}, \qquad (4.3.13)$$

i.e. $\ker \Delta$ is the space of all functions continuous in the closure $\overline{\Omega}$ of the domain Ω and harmonic in Ω.

An initial operator F_1 corresponding to R_1 is defined as follows:

if $x \in C(\overline{\Omega})$ then $(F_1 x)(t) = z(t)$,

where $z \in \ker \Delta$ and $x|_S = z|_S$,   (4.3.14)

i.e. to every function x continuous in the closure $\overline{\Omega}$ of Ω there corresponds a function z continuous in closure of Ω, harmonic in Ω and such that

$$x(p) = z(p) \quad for \ p \in S.$$ (4.3.15)

Proof. Consider the *Poisson equation*

$$\Delta x = y,$$ (4.3.16)

where $y \in C(\overline{\Omega})$ is arbitrarily fixed, with a *condition of Dirichlet type*

$$x|_S = 0 \quad (i.e. \ x(p) = 0 \ for \ p \in S),$$ (4.3.17)

i.e. so-called *Dirichlet problem* for the Laplacian (with a homogeneous boundary condition). It is well-known (cf. for instance, Pogorzelski, 1966) that a unique solution of the problem (4.3.16)–(4.3.17) is of the form $x = R_1 y$, where the operator R_1 is defined by Formula (4.3.11). Thus $y = \Delta x = \Delta R_1 y$. The arbitrariness of y implies that $\Delta R_1 = I$ on $C(\overline{\Omega})$. Hence Δ is right invertible in the space $C(\overline{\Omega})$ and has a right inverse R_1. We have to show that the operator F_1 defined by (4.3.14) is an initial operator for Δ corresponding to R_1. But, by definition, $F_1 x \in \ker \Delta \subset C^2(\overline{\Omega})$ and $F_1^2 = F_1$. Since $F_1 z = z$ for $z \in \ker \Delta$, we conclude that F_1 is a projection onto the space of constants $\ker \Delta$. Condition (4.3.17) implies that $F_1 R_1 = 0$, which was to be proved. ∎

This theorem has several consequences. For instance, Theorem 4.3.1 immediately implies

COROLLARY 4.3.1. *If $\Omega \subset \mathbf{R}^n$ is a domain bounded by a closed surface of Liapunov type, $Q, \varphi, y \in X = C(\overline{\Omega})$, and -1 is a regular value of the operator $Q R_1$, where R_1 is defined by Formula (4.3.11), then the Helmholtz equation*

$$\Delta x + Q(t) x = y \quad in \ \Omega$$ (4.3.18)

with the condition of Dirichlet type

$$x(p) = \varphi(p) \quad for \ p \in S = \partial \Omega$$ (4.3.19)

has a unique solution which is of the form

$$x = R_1(I - QR_1)^{-1}(y - Q\varphi) + \varphi. \qquad (4.3.20)$$

Proof. Indeed, if we write Condition (4.3.19) in the form $F_1 x = z$ where F_1 is defined by (4.3.14) and $z \in \ker \varDelta$, moreover, $z|_S = \varphi|_S$ then the problem (4.3.18)–(4.3.19) becomes an *initial value problem* for the operator $\varDelta + Q$. We point out that under our assumptions the number $\lambda = -1$ is a regular value of the operator QR_1 which is a well-known fact in the theory of partial differential equations (cf., for instance, Pogorzelski, 1966). ∎

Observe that, for the same reasons, which were used in the Darboux Problem 4.2.1, if -1 is a regular value of the operator $R_1 Q$ then a unique solution of the problem (4.3.18)–(4.3.19) can be written as follows:

$$x = (I + R_1 Q)^{-1}(R_1 y + \varphi). \qquad (4.3.21)$$

COROLLARY 4.3.2. *If $\Omega \subset \mathbf{R}^n$ is bounded by a closed surface S of Liapunov type, $a_0, \ldots, a_n$, $y \in X = C(\bar{\Omega})$, $\varphi \in C^1(\Omega)$ and -1 is a regular value of the operator AR_1, where R_1 is defined by Formula (4.3.11) and A by means of equality*

$$(Ax)(t) = \sum_{k=1}^{n} a_k(t)\frac{\partial x}{\partial t_k} + a_0(t)x$$

$$\text{for } x \in C^1(\bar{\Omega}), \ t = (t_1, \ldots, t_n) \in \bar{\Omega}, \qquad (4.3.22)$$

then the full elliptic equation

$$\varDelta x + \sum_{k=1}^{n} a_k(t)\frac{\partial x}{\partial t_k} + a_0 x = y \quad \text{in } \Omega \qquad (4.3.23)$$

with Dirichlet Condition (4.3.19) has a unique solution which is of the form

$$x = (I + AR_1)^{-1}(R_1 y + A\varphi) + \varphi. \qquad (4.3.24)$$

For the proof it is enough to observe that, by our assumptions $\varphi \in C^1(\bar{\Omega}) = \operatorname{dom} A$.

COROLLARY 4.3.3. *If $\Omega \subset \mathbf{R}^n$ is a domain bounded by a closed surface S of Liapunov type, $a_0, \ldots, a_n, y, \varphi \in C(\bar{\Omega})$, and -1 is a regular value of the operator $R_1 A$, where R_1 and A are defined by means of Equalities (4.3.11) and (4.3.22), respectively, then the problem (4.3.23)–(4.3.19) has a unique solution which is of the form*

$$x = (I + R_1 A)^{-1}(R_1 y + \varphi).$$ (4.3.25)

COROLLARY 4.3.4. *If $\Omega \subset \mathbf{R}^n$ is a domain bounded by a closed surface S of Liapunov type, $Q_N = I$, operators Q_k are defined by the equalities*

$$(Q_k x)(t) = \sum_{m=1}^{n} Q_{mk}(t) \frac{\partial x}{\partial t_m} Q_{0k}(t) x$$

$$(k = 0, 1, \ldots, N-1), \quad (4.3.26)$$

for $x \in C^1(\bar{\Omega})$, $Q_{mk}, y \in C(\bar{\Omega})$, $\varphi_k \in C^1(\bar{\Omega})$ $(m = 0, 1, \ldots, N-1)$ and -1 is a regular value of the operator $\sum_{k=0}^{N-1} Q_k R_1^{N-1} = Q(I, R_1) - I = Q^0(R_1)$, then the generalized polyharmonic equation

$$\Delta^N x + \sum_{k=0}^{N-1} Q_k \Delta^k x = y \quad \text{in } \Omega$$ (4.3.27)

with the conditions

$$(\Delta^k x)(p) = \varphi_k(p) \quad \text{for } p \in S \quad (k = 0, 1, \ldots, N-1)$$ (4.3.28)

has a unique solution which is of the form

$$x = R_1^N [Q(I, R_1)]^{-1} \left[y - \sum_{m=0}^{N-1} Q_m \left(\sum_{k=0}^{m} R_1^{m-k} \varphi_k \right) \right] + \sum_{k=0}^{N-1} R_1^k \varphi_k.$$

(4.3.29)

For the proof it is enough to observe that a derivative of a harmonic function is again a harmonic function, for the Laplacian commutes with the differential operators $\partial/\partial t_k$ $(k = 1, \ldots, n)$. Thus the introduced differential operators are well-defined in the spaces under consideration.

COROLLARY 4.3.5. *If $\Omega \subset \mathbf{R}^n$ is a domain bounded by a closed surface S of Liapunov type, $a, \varphi_0, \ldots, \varphi_{N-1}, y \in X = C(\bar{\Omega})$ and -1 is a regular* *value of the operator* $(-1)^M a R_1$, *where* R_1 *is defined by Formula* (4.3.11), *then the equation*

$$(-1)^M \Delta^M x + ax = y \qquad (4.3.30)$$

with Conditions (4.3.28) *has a unique solution which is of the form*

$$x = R_1^M [I + (-1)^M a R_1]^{-1} [y + (-1)^{M+1} a \varphi_0] + \sum_{k=0}^{N-1} R_1^k \varphi_k.$$

$$(4.3.31)$$

Observe that Theorem 4.3.1 and Corollaries 4.3.1–4.3.5 concern boundary value problems in classical sense for polynomials in Laplace operator, but their solutions are obtained by application of solutions of *initial value problems* for right invertible operators. This is a non-trivial fact which should be pointed out.

The Green function of the second kind (cf. Formula (4.3.10)) permits us to consider other problems by a reduction to initial value problems for right invertible operator. Namely, we have

THEOREM 4.3.2. *Suppose that a domain* $\Omega \subset \mathbf{R}^n$ *is bounded by a closed surface S of Liapunov type. Then the Laplace operator is right invertible in the space*

$$Y = \left\{ x \in C^1(\bar{\Omega}): \int_S \frac{dx(p)}{dn_p} \, dS_p = 0 \right\},$$

ker Δ *is defined by Formula* (4.3.13), *a right inverse of* Δ *is defined as follows*

$$(R_2 x)(t) = \lambda_n \int_\Omega G_2(t, s) x(s) \, d\Omega_s \quad \text{for } x \in Y, \qquad (4.3.32)$$

where G_2 *is the Green function of the second kind for the domain* Ω *with respect to the Laplacian* Δ *with* $a(p) \equiv 0$, λ_n *is defined by Formula* (4.3.12) *and the initial operator* F_2 *corresponding to* R_2 *is defined as follows:*

$$(F_2 x)(t) = z(t) \text{ for } x \in X, \text{ where } z \in \ker \Delta \text{ and}$$

$$\lim_{t \to p} \frac{dx(t)}{dn_p} = \lim_{t \to p} \frac{dz(t)}{dn_p} \quad \text{for } p \in S. \qquad (4.3.33)$$

Proof. Consider the Laplace equation $\Delta x = 0$ with a *condition of Neumann type*

$$\lim_{t \to p} \frac{dx(t)}{dn_p} = y_0(p) \quad \text{for } p \in S, \tag{4.3.34}$$

i.e. the so-called *Neumann problem* for the Laplace operator. It is well-known that this problem has a unique solution if and only if

$$\int_S y_0(p)\,dS_p = 0. \tag{4.3.35}$$

Indeed, let z be a function harmonic inside Ω and such that $z(p) = y_0(p)$ for $p \in S$. Then, by *Green Formula*

$$\int_\Omega x(t)\,d\Omega_t = \int_S \frac{dx(p)}{dn_p}\,dS_p \quad \text{for } x \in C^1(\bar{\Omega}) \tag{4.3.36}$$

we obtain for every function z, harmonic in Ω and satisfying Condition (4.3.34)

$$\int_S y_0(p)\,dS_p = \int_S \frac{dz(p)}{dn_p}\,dS_p = \int_\Omega (\Delta z)(t)\,d\Omega_t = 0. \tag{4.3.37}$$

Consider now Poisson Equation (4.3.16), with Neumann Condition (4.3.34). By our assumptions and Green Formula (4.3.36), if x is a solution of this problem then

$$\int_\Omega y(t)\,d\Omega_t = \int_\Omega (\Delta x)(t)\,d\Omega_t = \int_S \frac{dx(p)}{dn_p}\,dS_p = \int_S y_0(p)\,dS_p,$$

which means that $y \in Y$. Put $y_0 = 0$. By definition of the Green function of the second kind we find, if $x = R_2 y$, where $y \in Y$, then $\Delta x = \Delta R_2 y = y$, which proves that R_2 is a right inverse of Δ in Y. Moreover, $F_2 R_2 = 0$, for $\int_\Omega y(t)\,d\Omega_t = \int_S y_0(p)\,dS_p = 0$ if $y_0 = 0$. Arguing as in the proof of Theorem 4.3.1, we conclude that F_2 is a projection of Y onto the space of constants in Y, i.e. onto the set

$$Y \cap \ker \Delta = \left\{ z \in C^1(\bar{\Omega}): \Delta z = 0 \text{ in } \Omega \text{ and } \int_S z(p)\,dS_p = 0 \right\}.$$

Having already proved Theorem 4.3.2 we can obtain similar Corollaries as these following from Theorem 4.3.1, for instance

COROLLARY 4.3.6. *If* $\Omega \subset \mathbf{R}^n$ *is a domain bounded by a closed surface S of Liapunov type,* $Q_N = I$, *operators* Q_k *are defined by Equalities* (4.3.26), $Q_{mk} \in C(\bar{\Omega})$, $y \in Y$, $y_k \in Y \cap \ker \Delta$ $(m = 0, ..., n;\ k = 0, 1, ..., N-1)$ *and* -1 *is a regular value of the operator*

$$Q^0(R_2) = Q(I, R_2) - I = \sum_{k=0}^{N-1} Q_k R_2^{N-k},$$

where R_2 *is defined by Formula* (4.3.32), *y is defined in Theorem* 4.3.2, *then the generalized polyharmonic equation* (4.3.27) *with the conditions*

$$\lim_{t \to p} \frac{\mathrm{d}}{\mathrm{d}n_p} (\Delta^k x)(t) = y_k(p) \quad for\ p \in S \quad (k = 0, 1, ..., N-1)$$

(4.3.38)

has a unique solution which is of the form

$$x = R_2^N[Q(I, R_2)]^{-1} \left[y - \sum_{m=0}^{N-1} Q_m \sum_{k=0}^{m} R_2^{m-k} y_k \right] + \sum_{k=0}^{N-1} R_2^k y_k.$$

(4.3.39)

In a similar way we can examine the so-called *outer Dirichlet and Neumann problems,* i.e. problems for an unbounded domain $\Omega^* = \mathbf{R}^n \backslash \Omega$, where Ω is a domain bounded by a closed surface S of Liapunov type in spaces of functions continuous in Ω^*, and vanishing of infinity.

Using the Green function of the second kind we also obtain some results for mixed problems. Namely, in a similar way, as Theorems 4.3.1 and 4.3.2 we obtain

THEOREM 4.3.3. *Suppose that a domain* $\Omega \subset \mathbf{R}^n$ *is bounded by a closed surface S of Liapunov type. Suppose, moreover, that* $a \in C(S)$ *and* $a(p) \leqslant 0$ *for all* $p \in S$. *Then the Laplace operator is right invertible in the space*

$$Y_a = \left\{ x \in C^1(\bar{\Omega}): \lim_{t \to p} \frac{\mathrm{d}x(t)}{\mathrm{d}n_p} + a(p)x(p) = \lim_{t \to p} \frac{\mathrm{d}z(t)}{\mathrm{d}n_p} + \right.$$

$$\left. + a(p)z(t), p \in S, z \in \ker \Delta \right\},$$

$\ker \Delta$ *is defined by Formula* (4.3.13), *a right inverse of* Δ *is defined as follows:*

$$(R_3 x)(t) = \lambda_n \int_{\Omega} G_2(t, s) x(s) \mathrm{d}\Omega_s \quad for\ y \in Y_a,$$

(4.3.40)

where G_2 is the Green function of the second kind for the domain Ω with respect to the Laplacian Δ, λ_n is defined by Formula (4.3.12) and the initial operator F_3 corresponding to R_3 is defined as follows:

$$(F_3 x)(t) = z(t) \quad \text{for } x \in Y_a, \quad \text{where } z \in \ker \Delta \quad \text{and}$$

$$\lim_{t \to p} \frac{dx(t)}{dn_p} + a(p)x(p) = \lim_{t \to p} \frac{dz(t)}{dn_p} + a(p)z(p) \quad \text{for } p \in S.$$

$$(4.3.41)$$

It follows from our preceding considerations that in a similar way we could study various problems for more general differential operators of elliptic type.

Examples and Exercises

EXAMPLE 4.3.1. Consider Poisson Equation (4.3.16) with the condition

$$x(p) + a(p)(\Delta x)(p) = x_0(p) \quad \text{for } p \in S, \qquad (4.3.42)$$

where $x_0 \in C(\bar{\Omega})$ is a given function harmonic inside the domain Ω bounded by a closed surface S of Liapunov type, $a \in C(\bar{\Omega})$ is also given. Theorem 4.3.1 implies that Condition (4.3.42) can be written in the form

$$F_1(I + A\Delta)x = x_0, \quad \text{where } A \in L_0(X), \quad X = C(\bar{\Omega})$$

and F_1 is an initial operator for Δ defined by Formula (4.3.14) corresponding to a right inverse R_1 defined by Formula (4.3.11). Thus Theorem 2.2.5 implies that the operator $F = F_1(I + A\Delta)$ is an initial operator for Δ corresponding to the right inverse $R = R_1 + F_1 A$. We therefore conclude that the problem (4.3.16)–(4.3.41) is well-posed and its unique solution is of the form

$$x = Ry + x_0 = R_1 y + F_1 Ay + x_0.$$

Consider now Poisson Equation (4.3.16) with the condition

$$x(p) + (\Delta x)(b(p)) = x_0(p) \quad \text{for } p \in S, \qquad (4.3.43)$$

where x_0, Ω, S are as before, $b \in C(\bar{\Omega})$ and $b(S) \subset S$. Theorem 4.3.1 implies that Condition (4.3.43) can be written as

$$F_1(I + B\Delta)x = x_0, \quad \text{where } (Bx)(p) = x(b(p)),$$

X, F_1, R_1 are as before. Thus Theorem 2.2.5 implies that the operator $F_0 = F_1(I + B\Delta)$ is an initial operator for Δ corresponding to the right inverse $R_0 = R_1 + F_1 B$. We therefore conclude that the problem (4.3.16)–(4.3.43) is well-posed and its unique solution is of the form

$$x = Ry + x_0 = R_1 y + F_1 By + y_0 .$$

EXERCISE 4.3.1. Solve the harmonic equation

$$\Delta^2 x = y, \quad y \in C(\Omega), \tag{4.3.44}$$

where Ω is a domain bounded by a closed surface S of Liapunov type, with conditions:
   (i) of Dirichlet type,
   (ii) of Neumann type,
   (iii) of mixed type.

EXERCISE 4.3.2. Solve the equation

$$\Delta^2 x + bx = y, \quad b, y \in C(\bar{\Omega}), \tag{4.3.45}$$

where Ω is as in Exercise 4.3.1, with conditions:
   (i) of Dirichlet type,
   (ii) of Neumann type,
   (iii) of mixed type.

EXERCISE 4.3.3. Consider Poisson Equation (4.3.16) with conditions of Neumann type modified as in Example 4.3.1. Formulate conditions of solvability of such problems.

EXERCISE 4.3.4. Solve the *Tricomi equation*

$$\eta \frac{\partial^2 x}{\partial t^2} + \frac{\partial^2 x}{\partial s^2} = 0 \quad \text{in } \Omega,$$

where $\Omega = \{(t, s): t^2 + s^2 \leqslant 1, t > 0, s > 0\}$, $\eta > 0$ is a given number, with a condition of Neumann type.

EXERCISE 4.3.5. Solve the *Sjöstrand problem*, i.e. the equation

$$\frac{\partial}{\partial x} [u + a(y)u'_x + b(y)u'_y] = 0, \tag{4.3.46}$$

with the conditions

$$u(0, y) = g(y), \tag{4.3.47}$$

$$\left[\frac{\partial u}{\partial r} (r, \vartheta) + ku(r, \vartheta) \right]_{r=1} = f(\vartheta), \tag{4.3.48}$$

$r^2 = x^2 + y^2 \leqslant 1$, $-\pi \leqslant \vartheta \leqslant \pi$, where a, b, g, f are given continuous functions (cf. von Wolfersdorf, 1970, 1970a).

EXERCISE 4.3.6. Prove that the operator A_α of rotation through an angle α commute with the Laplace operator Δ in Ω. Is A_α a stationary operator for an arbitrary $\alpha \in [0, 2\pi)$? (cf. Section 3.3).

4.4. DIFFERENTIAL EQUATIONS WITH DELAYED AND ADVANCED ARGUMENT

In this section we shall study some initial value problems for ordinary linear differential equations with deviating argument which can be solved by means of right invertible operators.

To begin with, we shall prove

LEMMA 4.4.1. *Suppose that*

(i) *X is a linear space over a field $\mathscr{F}$ and $D \in R(X)$;*

(ii) *F is an initial operator for D corresponding to a right inverse R of D;*

(iii) *$P_0 \in L_0(X)$ is a projection such that both superpositions $P_0 D$ and DP_0 are well-defined and, moreover*

$$DP_0 = P_0 D \quad \text{on dom } D. \tag{4.4.1}$$

Write

$$P_1 = I - P_0, \quad X_0 = P_0 X, \quad X_1 = P_1 X; \tag{4.4.2}$$

(iv) *Write (as before)*

$$Q(D) = \sum_{k=0}^{N} Q_k D^k, \quad \text{where } Q_0, \ldots, Q_{N-1} \in L(X), \quad Q_N = 1, \tag{4.4.3}$$

$$Q(I, R) = I + Q^0(R), \quad \text{where } Q^0(R) = \sum_{k=0}^{N-1} Q_k R^{N-k}. \tag{4.4.4}$$

If the operator $I + P_1 Q^0(R)$ (which maps X into X_1) is invertible then the following problem:

$$P_1 Q(D)x = y, \quad y \in X_1, \tag{4.4.5}$$

$$P_0 D^k x = D^k x_0, \quad x_0 \in X_0 \cap \mathrm{dom}\, D^{N-1} \quad (k = 0, 1, ..., N-1)$$
$$(4.4.6)$$

with the conditions

$$FP_0 D^k x = FP_1 D^k x \quad (k = 0, 1, ..., N-1) \tag{4.4.7}$$

has a unique solution

$$x = x_0 + R^N [I + P_1 Q^0(R)]^{-1} \left(y - P_1 \sum_{m=0}^{N-1} Q_m D^m x_0 - \right.$$

$$\left. - \sum_{m=0}^{N-1} Q_m \sum_{k=0}^{m} R^{k-m} FD^k x_0 \right) + \sum_{k=0}^{N-1} R^k FD^k x_0. \tag{4.4.8}$$

Proof. By definition $P_1 = I - P_0$ is a projection operator and, moreover, by our assumption $P_1 D = (I - P_0)D = D(I - P_0) = DP_1$ on dom D. Observe that for an arbitrary $x \in X$ and a positive integer N we have

$$P_0 D^k x = x_k, \text{ where } x_k \in X_0, \text{ implies } x_k = D^k x_0$$
$$(k = 0, 1, ..., N-1). \tag{4.4.9}$$

Indeed, by definition we have $P_0 x = x_0$ and $x_k = P_0 D^k x = D^k P_0 x = D^k x_0$.

Observe that from Property (4.4.9) it follows that the imposed conditions (4.4.7) are of sufficiently general form. Formulae (4.4.6) and (4.4.7) together imply that

$$FD^k P_1 x = FP_1 D^k x = FP_0 D^k x = FD^k x_0$$
$$(k = 0, 1, ..., N-1). \tag{4.4.10}$$

Conditions (4.4.6) imply

$$P_1 Q(D) P_1 x = P_1 \sum_{m=0}^{N} Q_m D^m P_1 x$$

$$= P_1 D^N P_1 x + P_1 \sum_{m=0}^{N-1} Q_m D^m P_1 x$$

$$= P_1^2 D^N x + P_1 \sum_{m=0}^{N-1} Q_m D^m (I - P_0) x$$

$$= P_1 D^N x + P_1 \sum_{m=0}^{N-1} Q_m D^m x - P_1 \sum_{m=0}^{N-1} Q_m D^m P_0 x$$

$$= P_1 D^N x + P_1 \sum_{m=0}^{N-1} Q_m D^m x - P_1 \sum_{m=0}^{N-1} Q_m P_0 D^m x$$

$$= P_1 Q(D) x - P_1 \sum_{m=0}^{N-1} Q_m D^m x_0 = y - P_1 \sum_{m=0}^{N-1} Q_m D^m x_0 .$$

Write

$$x^0 = P_1 x, \quad y^0 = y - P_1 \sum_{m=0}^{N-1} Q_m D^m x_0,$$

$$Q_m^0 = P_1 Q_m \quad (m = 0, 1, ..., N-1),$$

$$Q_N^0 = I, \quad Q_0(D) = \sum_{m=0}^{N} Q_m^0 D^m.$$

Using these notations we obtain the equation

$$Q_0(D) x^0 = y^0, \quad \text{where } x^0, y^0 \in X_1 \tag{4.4.11}$$

together with the initial conditions

$$FD^k x^0 = FD^k x_0 \quad (k = 0, 1, ..., N-1). \tag{4.4.12}$$

These last conditions follow immediately from Formulae (4.4.10). Observe that the operator

$$I + \sum_{m=0}^{N-1} Q_m^0 R^{N-m} = I + \sum_{m=0}^{N-1} P_1 Q_m R^{N-m}$$

$$= I + P_1 \sum_{m=0}^{N-1} Q_m R^{N-m} = P_1 Q^0(R)$$

is invertible by our assumption in the space X_1. Thus Theorem 4.1.2 implies that the initial value problem (4.4.11)–(4.4.12) has a unique solution

$$x^0 = R^N [I + P_1 Q^0(R)]^{-1} \left(y^0 - \sum_{m=0}^{N-1} Q_m \sum_{k=0}^{m} R^{k-m} FD^k x_0 \right) +$$

$$+ \sum_{k=0}^{N-1} R^k FD^k x_0 .$$

Hence the problem (4.4.5)–(4.4.6)–(4.4.7) has a unique solution of the form

$$x = P_0 x + P_1 x = x_0 + x^0$$

$$= x_0 + R^N [I + P_1 Q^0(R)]^{-1} \left(y - P_1 \sum_{m=0}^{N-1} Q_m D^m x_0 - \right.$$

$$\left. - \sum_{m=0}^{N-1} Q_m \sum_{k=0}^{m} R^{k-m} FD^k x_0 \right) + \sum_{k=0}^{N-1} R^k FD^k x_0,$$

which was to be proved.   ■

Now consider the following differential equation with *delayed argument*

$$x^{(N)}(t) + \sum_{k=0}^{N-1} \sum_{j=0}^{M} a_{kj}(t) x^k(h_j(t)) = y(t), \quad t_0 < t \leqslant T, \quad (4.4.13)$$

where

$$h_j(t) \leqslant h_0(t) \equiv t \quad (j = 1, 2, ..., M) \quad \text{for } t_0 \leqslant t \leqslant T$$
$$(4.4.14)$$

with the initial conditions

$$x^{(k)}(t) = \varphi^{(k)}(t) \quad \text{on the initial set } E_{t_0} \quad (k = 0, 1, ..., N-1)$$
$$(4.4.15)$$

and

$$x^{(k)}(t_0 + 0) = \varphi^{(k)}(t_0) = 0 \quad (k = 0, 1, ..., N-1) \quad (4.4.16)$$

(cf. Fig. 4.6), where

$$E_{t_0} = \{t_0\} \cup \{h_1(t), ..., h_M(t) : h_j(t) \leqslant t_0 \text{ for } t_0 \leqslant t \leqslant T$$
$$(j = 1, ..., M)\}. \quad (4.4.17)$$

Write

$$X = \{x \in C(E_{t_0} \cup [t_0, T]) : x(t_0) = 0\}, \quad (4.4.18)$$

$$P_0 x = x|_{E_{t_0}}, \quad P_1 x = x|_{[t_0, T]}, \quad X_0 = P_0 X, \quad X_1 = P_1 X.$$

The projection operators P_0 and P_1 are well-defined on the space X. We shall assume that

$$y, a_{kj} \in X_1, \quad h_j \in C[t_0, T]$$
$$(k = 0, 1, ..., N-1; \quad j = 0, 1, ..., M) \quad (4.4.19)$$

$$\varphi, \varphi', ..., \varphi^{(N-1)} \in X_0.$$

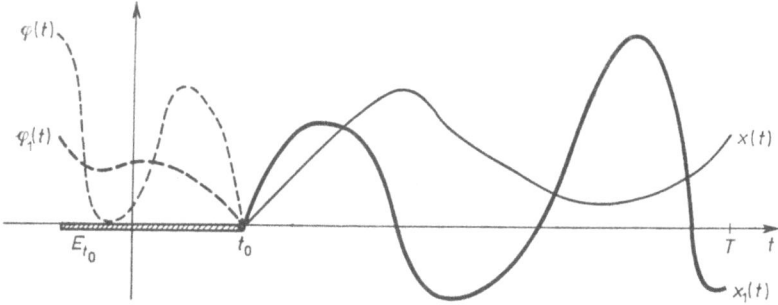

Fig. 4.6. Solutions of Equation (4.4.13)

These assumptions immediately imply that

$$y(t_0) = 0, \quad a_{kj}(t_0) = 0, \quad \varphi^{(k)}(t_0) = 0 \quad (k = 0, 1, ..., N-1;$$
$$j = 0, 1, ..., M). \quad (4.4.20)$$

We also write

$$(Q_k x)(t) = \begin{cases} \displaystyle\sum_{j=0}^{M} a_{kj}(t) x\big(h_j(t)\big) & \text{for } t_0 \leqslant t \leqslant T \\ & (k = 0, 1, ..., N-1), \\ 0 & \text{for } t \in E_{t_0}, \end{cases} \quad (4.4.21)$$

$$(Q_N x)(t) = \begin{cases} x(t) & \text{for } t_0 \leqslant t \leqslant T, \\ 0 & \text{for } t \in E_{t_0}, \end{cases}$$

$$D = \frac{\mathrm{d}}{\mathrm{d}t}, \quad (Rx)(t) = \int_{t_0}^{t} x(s)\,\mathrm{d}s, \quad (Fx)(t) = x(t_0) = 0,$$

$$(4.4.22)$$

for all $x \in X$,

$$Q^0(R) = \sum_{k=0}^{N-1} Q_k R^{N-k}. \quad (4.4.22a)$$

By definition, $P_0 Q_k = 0$, hence

$$P_1 Q_k = Q_k \quad (k = 0, 1, ..., N-1) \quad \text{and}$$
$$I + P_1 Q^0(R) = I + Q^0(R). \quad (4.4.23)$$

By usual estimations we can prove that the operator $I+Q^0(R)$ is invertible in the space X_1 (cf. Section 8.4). We therefore obtain the following

THEOREM 4.4.1. *If Conditions (4.4.19) are satisfied, then the initial value problem (4.4.13)–(4.4.14)–(4.4.15) has a unique solution*

$$x = \varphi + R^N[I+Q^0(R)]^{-1}y \qquad (4.4.24)$$

in the space X_1, where R and $Q^0(R)$ are defined by Formulae (4.4.21), (4.4.22), (4.4.22a).

Proof. Observe that the space X, the operators $D, R, F, Q_0, ..., Q_N$,
$Q(D) = \sum_{k=0}^{N} Q_k D^k, Q^0(R), P_0, P_1$ satisfy all conditions of Lemma 4.4.1.
Put $x_0 = \varphi$. By definition of the initial function φ and of the operators Q_k (Formulae (4.4.21)) we have $\sum_{m=0}^{N-1} Q_m D^m x_0 = \sum_{m=0}^{N-1} Q_m D^m \varphi = 0$ for
$k = 0, 1, ..., N-1$. Lemma 4.4.1 implies that our problem has a unique solution which is of the required form. ∎

By an obvious change of variables (i.e. $t \to -t$) we obtain

THEOREM 4.4.2. *The differential equation with advanced argument $x^{(N)}(t) +$*
$+ \sum_{k=0}^{N-1} \sum_{j=0}^{M} a_{kj}(t)x^{(k)}(h_j(t)) = y(t), \ T \leqslant t < t_0,$ *where* $t \equiv h_0(t) \leqslant h_j(t)$
$(j = 1, ..., M)$ *for $T \leqslant t \leqslant t_0$ with the conditions $x^{(k)}(t) = \varphi^{(k)}(t)$ on the set $E_{t_0}^- \ (k = 0, 1, ..., N-1)$ defined by*

$$E_{t_0}^- = \{t_0\} \cup \{h_1(t), ..., h_M(t): h_j(t) \geqslant t_0 \text{ for } T \leqslant t \leqslant t_0$$
$$(j = 1, 2, ..., M)\},$$
$$x^{(k)}(t_0 - 0) = \varphi^{(k)}(t_0) = 0 \quad (k = 0, 1, ..., N-1)$$

has a unique solution

$$x = \varphi + R^N[I+Q_-^0(R)]^{-1}y$$

in the space $X^- = \{x \in C([T, t_0] \cup E_{t_0}^-): x(t_0) = 0\}$, where:

(i) y, $a_{kj} \in X_{\bar{1}}$, $h_j \in C[T, t_0]$ $(k = 0, 1, ..., N-1; j = 0, 1, ..., M)$ $\varphi, \varphi', ..., \varphi^{(N-1)} \in X_{\bar{0}}$,

(ii) $X_{\bar{0}} = P_{\bar{0}} X^-$, $X_{\bar{1}} = P_{\bar{1}} X^-$, $P_{\bar{0}} x = x|_{E_{\bar{t}_0}}$, $P_{\bar{1}} x = x|_{[T, t_0]}$ *for* $x \in X^-$, *D, R, F are defined as in Theorem 4.4.1,*

(iii) $(Q_{\bar{k}} x)(t)$

$$= \begin{cases} \displaystyle\sum_{j=0}^{M} a_{kj}(t) x\big(h_j(t)\big) & \text{for } T \leqslant t \leqslant t_0, \\ 0 & \text{for } t \in E_{t_0}^-, \end{cases} \qquad (k = 0, 1, ..., N-1)$$

$$(Q_{\bar{k}} x)(t) = \begin{cases} x(t) & \text{for } T \leqslant t \leqslant t_0, \\ 0 & \text{for } t \in E_{t_0}^- \end{cases}$$

and

$$Q_{-}^{0}(R) = \sum_{k=0}^{N-1} Q_{\bar{k}} R^{N-k}.$$

Consider now a *differential-difference equation* with delayed argument

$$x^{(N)}(t) + \sum_{k=0}^{N-1} \sum_{j=0}^{M_k} a_{kj}(t) x^{(k)}\big(t - h_{kj}(t)\big) = y(t) \qquad (4.4.25)$$

for $t \geqslant 0$, where $h_{kj}(t)$ are given continuous increasing functions defined for $t \geqslant 0$ and such that $h_{kj}(t) \leqslant t$ for $t > 0$, $h_{kj}(0) = 0$, $h_{k0}(t) \equiv 0$ $(j = 0, 1, ..., M_k; k = 0, 1, ..., N-1)$.

Suppose that the given functions belong to the space X of all piecewise continuous functions defined for $t \geqslant 0$.

Since $t - h_{kj}(t) \geqslant 0$ for all $t \geqslant 0$ $(j = 0, 1, ..., M_k; k = 0, 1, ..., N-1)$, we conclude that in our case the initial set E_0 contains the point 0 only. Since no initial function is given, we only can assume that the values of the unknown function and its derivatives up to and including the order $N-1$ are given at 0:

$$x^{(k)}(0) = x_k, \text{ where } x_k \text{ are arbitrarily fixed constants}$$
$$(k = 0, 1, ..., N-1). \qquad (4.4.26)$$

We point out that in the case $E_{t_0} = \{t_0\}$ the method of integration "step by step" (cf. Bellman and Cooke, 1963, § 3.2) of Equation (4.4.25) cannot be used.

Put

$$D = \frac{d}{dt}, \quad (Rx)(t) = \int_0^t x(s)ds, \quad (Fx)(t) = x(0) \quad \text{for } x \in X$$

and

$$(Q_k)(t) = \sum_{j=0}^{M_k} a_{kj}(t)x(t - h_{kj}(t)) \quad (k = 0, 1, ..., N-1)$$

$$\text{for } x \in X, Q_N = I,$$

$$Q(D) = \sum_{k=0}^{N} Q_k D^k, \quad Q^0(R) = \sum_{k=0}^{N-1} Q_k R^{N-k}.$$

In a similar way as in Theorem 4.4.1 we conclude, that the operator $I + Q^0(R)$ is invertible in the space X. Thus Theorem 4.4.2 implies immediately the following

THEOREM 4.4.3. *If the given functions satisfy the above conditions, then the initial value problem* (4.4.25)–(4.4.26) *has a unique solution*

$$x = R^N [I + Q^0(R)]^{-1} \left(y - \sum_{m=0}^{N-1} Q_m \sum_{k=0}^{m} R^{k-m} x_k \right) + \sum_{k=0}^{N-1} R^k x_k.$$

All the results obtained here (cf. Przeworska-Rolewicz, 1974) are true for vector valued functions (if the coefficients are square matrices satisfying the above conditions) and for some partial differential operators.

Examples and Exercises

EXAMPLE 4.4.1. Consider the equation with delayed argument:

$$x'(t) + ax(t^2 - 1) = y(t) \quad \text{for } 0 < t \leqslant 1, \tag{4.4.27}$$

where $a > 0$, $y \in C[0, 1]$ are given. This equation has the delayed argument for $t^2 - 1 \leqslant t$ if $0 < t \leqslant 1$.

We have here $t_0 = 0$, $T = 1$. Thus the initial set $E_{t_0} = E_0$ is of the form

$$E_0 = \{0\} \cup \{t^2 - 1 : t^2 - 1 \leqslant 0 \text{ for } 0 \leqslant t \leqslant 1\}$$
$$= \{0\} \cup [-1, 0] = [-1, 0].$$

We therefore can admit the following initial condition

$$x(t) = \varphi(t), \text{ where } \varphi \in C[-1, 0] \text{ is arbitrarily fixed.}$$

Since for every $u \in C[0, 1]$ we have

$$[I + Q^0(R)u](t) = u(t) + a \int_0^{t^2 - 1} u(s) \, ds,$$

then the solution of our problem is reduced to solving the following integral equation:

$$u(t) + a \int_0^{t^2 - 1} u(s) \, ds = v(t),$$

where $v \in C[0, 1]$ is given. This equation has a unique solution which is of the form

$$u(t) = v(t) + \int_0^t K(t, s) v(s) \, ds,$$

where $K(t, s)$ is a well-determined function (cf. Section 8.4). Thus

$$\{[I + Q^0(R)]^{-1} y\}(t) = y(t) + \int_0^t K(t, s) y(s) \, ds$$

and the solution, we are looking for, is of the form

$$x(t) = \varphi(t) + \int_0^t y(s) \, ds + \int_0^t \left[\int_0^s K(s, u) y(u) \right] ds.$$

EXERCISE 4.4.1. Consider the equation

$$x'(t) = x(t^2) + 1 \quad \text{for } 0 < t \leqslant 1.$$

Determine the initial set E_0 and solve the initial value problem for this equation with the condition that $x(t) = a$ for $t \in E_0$, $a > 0$.

EXERCISE 4.4.2. Consider the equation

$$x''(t) = x(t^2) + x'(t^4) + 1 \quad \text{for } 0 < t \leqslant 1.$$

Determine the initial set E_0 and solve the initial value problem for this equation with the conditions

$$x(t) = a_0, \quad x'(t) = a_1 \quad \text{for } t \in E_0, \quad a_0, a_1 > 0.$$

EXERCISE 4.4.3. Consider the equation

$$x'(t) = x(t) + x(-t) + 1 \quad \text{for } -1 \leqslant t < 0$$

with the condition $x(t) = 1$ for $0 \leqslant t \leqslant 1$.

EXERCISE 4.4.4. Solve the equation

$$x^{(N)}(t) + \sum_{k=0}^{N-1} \sum_{j=1}^{M} t^{j+k} x^{(k)} \left(\frac{j}{M} t \right) = y(t) \quad \text{for } t \geqslant 0.$$

4.5. EQUATIONS WITH INVOLUTIONS OF ORDER n

In this section we shall consider equations with right invertible operators and involutions. The results obtained will be applied to *functional-differential equations with involutions of order n*.

THEOREM 4.5.1. *Suppose that X is a linear space over a field $\mathscr{F}$ (where either $\mathscr{F} = \mathbf{C}$ or $\mathscr{F} = \mathbf{R}$), $D \in R(X)$, F is an initial operator for D corresponding to a right inverse R of D, $S \in L_0(X)$ is an involution of order n, i.e. $S^n = I$ (cf. Section 1.4) and there exists an operator $A \in L_0(X)$ such that*

$$DS = ASD \quad \text{on } \mathrm{dom}\, D. \tag{4.5.1}$$

Then the operator A is invertible and

$$A^{-1} = S(AS)^{n-1}. \tag{4.5.2}$$

Moreover, $D_1 = A^{-1}D = S(AS)^{n-1}D \in R(X)$ and F is an initial operator for D_1 corresponding to the right inverse $R_1 = RA \in \mathscr{R}_{D_1}$.

Proof. To begin with, we shall prove that

$$DS^k = (AS)^k D \quad \text{on dom } D \quad (k = 1, 2, ...). \tag{4.5.3}$$

Indeed, for $k = 1$ Equality (4.5.3) is satisfied by our assumption (4.5.1). Suppose that Equality to be true for an arbitrary positive integer k. Then

$$DS^{k+1} = (DS^k)S = [(AS)^k D]S = (AS)^k(DS) = (AS)^k(ASD)$$
$$= (AS)^{k+1}D \quad \text{on the dom } D$$

which proves (4.5.3).

Equality (4.5.3) implies that

$$(AS)^k = DS^k R \quad \text{for } k = 1, 2, ... \tag{4.5.4}$$

Indeed, for an arbitrary positive integer k we find $(AS)^k = (AS)^k DR = DS^k R$.

Since, by our assumption, $S^n = I$, we conclude that the operator $A_1 = AS$ is also an involution of order n, i.e.

$$(AS)^n = I. \tag{4.5.5}$$

Indeed, Formula (4.5.4) implies that $(AS)^n = DS^n R = DR = I$. We therefore have

$$DS^n = (AS)^n D = D \quad \text{on dom } D. \tag{4.5.6}$$

Write $U = S(AS)^{n-1}$. We find

$$AU = AS(AS)^{n-1} = (AS)^n = I,$$
$$UA = S(AS)^{n-1}A = S(AS)^{n-1}AS^n = S(AS)^{n-1}(AS)S^{n-1}$$
$$= S(AS)^n S^{n-1} = S \cdot S^{n-1} = S^n = I$$

which implies that the operator A is invertible and $A^{-1} = U = S(AS)^{n-1}$.

Write now: $D_1 = A^{-1}D = S(AS)^{-1}D$, $R_1 = RA$. We find: $D_1 R_1 = A^{-1}DRA = A^{-1}A = I$, which proves that $D_1 \in R(X)$ and $R_1 \in \mathcal{R}_{D_1}$. Moreover, by Theorem 2.2.1, an initial operator for D corresponding to R_1 is $F_1 = I - R_1 D_1 = I - RAA^{-1}D = I - RD = F$ which was to be proved. ∎

THEOREM 4.5.2. *Suppose that all assumptions of Theorem 4.5.1 are satisfied and that*

$$Q(D, S) = \sum_{k=0}^{N} Q_k(S) D^k, \quad \text{where } Q_N = I, \quad Q_k(S) = \sum_{j=0}^{n-1} q_{jk} S^j,$$

$$(4.5.7)$$

$$q_{jk} \in \mathscr{F} \quad (k = 0, 1, ..., N-1; j = 0, 1, ..., n-1).$$

If the operator $I+Q^0(R)$, where

$$Q^0(R) = \sum_{k=0}^{N-1} Q_k(AS) R^{N-k} \qquad (4.5.8)$$

is invertible then the initial value problem

$$Q(D, S)x = y, \quad y \in \text{dom } D, \qquad (4.5.9)$$

$$FD^k x = y_k, \quad y_k \in \text{ker } D \quad (k = 0, 1, ..., N) \qquad (4.5.10)$$

is well-posed and its unique solution is of the form

$$x = R^{N+1}[I+Q^0(R)]^{-1}\left[Dy - DQ(D, S)\sum_{k=0}^{N} R^k y_k\right] + \sum_{k=0}^{N} R^k y_k.$$

$$(4.5.11)$$

Proof. Formula (4.5.4) and our assumptions together imply that

$$DQ(D, S) R^N = \sum_{k=0}^{N} DQ_k(S) D^k R^N = \sum_{k=0}^{N} DQ_k(S) R^{N-k}$$

$$= \sum_{k=0}^{N-1}\sum_{j=0}^{n-1} q_{jk} DS^j R^{N-k} + DQ_N D^N R^N$$

$$= \sum_{k=0}^{N-1}\left[\sum_{j=0}^{n-1} q_{jk}(AS)^j\right] R^{N-k-1} + D$$

$$= D + \sum_{k=0}^{N-1} Q_k(AS) R^{N-k-1}.$$

The Taylor Formula (2.2.7) implies that every solution of the problem (4.5.9)–(4.5.10) satisfies also the equality:

$$Dy = DQ(D, S)x = DQ(D, S)\left[\sum_{k=0}^{N} R^k FD^k + R^{N+1}D^{N+1}\right]x$$

$$= DQ(D, S) \left(\sum_{k=0}^{N} R^k y_k + R^{N+1} x \right)$$

$$= DQ(D, S) \sum_{k=0}^{N} R^k y_k + DQ(D, S) R^{N+1} D^{N+1} x.$$

Write

$$y_{N+1} = Dy - DQ(D, S) \sum_{k=0}^{N} R^k y_k. \tag{4.5.12}$$

Then

$$y_{N+1} = DQ(D, S) R^{N+1} D^{N+1} x = [DQ(D, S) R^N] R D^{N+1} x$$

$$= \left[D + \sum_{k=0}^{N-1} Q_k(AS) R^{N-k-1} \right] R D^{N+1} x$$

$$= DR D^{N+1} x + \left(\sum_{k=0}^{N-1} Q_k(AS) R^{N-k} \right) D^{N+1} x$$

$$= D^{N+1} x + Q^0(R) D = [I + Q^0(R)] D^{N+1} x.$$

Since, by our assumptions, the operator $I + Q^0(R)$ is invertible we obtain $D^{N+1} x = [I + Q^0(R)]^{-1} y_{N+1}$, where y_{N+1} is defined by Formula (4.5.12). This, and Conditions (4.5.10) together imply that

$$x = R^{N+1} [I + Q^0(R)]^{-1} y_{N+1} + \sum_{k=0}^{N} R^k y_k$$

$$= R^{N+1} [I + Q^0(R)]^{-1} \left[Dy - DQ(D, S) \sum_{k=0}^{N} R^k y_k \right] + \sum_{k=0}^{N} R^k y_k.$$

This is a solution, we were looking for. Since the corresponding homogeneous problem has only zero as a solution, we conclude that this solution is unique, i.e. our problem is well-posed. ■

REMARK 4.5.1. If the value $z = FD^N x$ is not given a priori then the problem (4.5.9)–(4.5.10) is not well-posed. However, if all assumptions of Theorem 4.5.2 are satisfied, then this problem has solution of the form

$$x = R^{N+1}[I+Q^0(R)]^{-1}\left[Dy-DQ(D,S)\left(\sum_{k=0}^{N-1} R^k y_k+R^N z\right)\right]+$$

$$+\sum_{k=0}^{N-1} R^k y_k+R^N z, \quad \text{where } z \in \ker D \text{ is arbitrary.} \quad (4.5.13)$$

We therefore obtain a family of solutions.

COROLLARY 4.5.1. *Suppose that all assumptions of Theorem 4.5.2 are satisfied and the operator $I+Q^0(R)$, where $Q^0(R)$ is defined by Formula (4.5.8), is invertible. Then every solution of Equation (4.5.9) is of the form*

$$x = R^{N+1}[I+Q^0(R)]^{-1}\left[Dy-DQ(D,S)\sum_{k=0}^{N} R^k z_k\right]+\sum_{k=0}^{N} R^k z_k,$$

$$(4.5.14)$$

where $z_0, ..., z_N \in \ker D$ are arbitrary.

Theorem 4.5.2 and Corollary 4.1.1 together imply

COROLLARY 4.5.2. *Suppose that all assumptions of Theorem 4.5.2 are satisfied and the operator $I+Q^0(R)$, where $Q^0(R)$ is defined by Formula (4.5.8), is invertible. Then the following initial value problem*

$$Q(D,S)D^M x = y, \quad y \in \mathrm{dom}\, D \quad (M \geqslant 0), \quad (4.5.15)$$
$$FD^k x = y_k, \quad y_k \in \ker D \quad (k = 0,1,...,N+M) \quad (4.5.16)$$

is well-posed and its unique solution is of the form

$$x = R^{N+M+1}[I+Q^0(R)]^{-1}\left[Dy-DQ(D,S)\sum_{k=0}^{N} y_k\right]+\sum_{k=0}^{N+M} R^k y_k.$$

$$(4.5.17)$$

COROLLARY 4.5.3. *Suppose that all assumptions of Theorem 4.5.2 are satisfied and the operator $I+Q^0(R)$, where $Q^0(R)$ is defined by Formula (4.5.8), is invertible. Then the general solution of Equation (4.5.15) is of the form*

$$x = R^{N+M+1}[I+Q^0(R)]^{-1}\left[Dy-DQ(D,S)\sum_{k=0}^{N} z_k\right]+\sum_{k=0}^{N+M} R^k z_k,$$

$$(4.5.18)$$

where $z_0, ..., z_{N+M} \in \ker D$ are arbitrary.

THEOREM 4.5.3. *Let $D \in R(X)$ and let F be an initial operator for D corresponding to a right inverse R of D. Suppose we are given an involution $S \in L_0(X)$ such that*

$$DS = ASD, \quad \text{where } A \in L_0(X) \text{ on dom } D. \tag{4.5.19}$$

Then the operator A is invertible and every solution of the equation

$$Q(D)x = Sx+y, \quad y \in \text{dom } D^{N-1}, \quad Q(D) = \sum_{k=0}^{N} Q_k D^k,$$

$$\tag{4.5.20}$$

$$Q_N = I, \quad Q_0, ..., Q_{N-1} \in L(X)$$

satisfies the equation

$$Hx = P(D_1)y+Sy, \tag{4.5.21}$$

where

$$H = P(D_1)Q(D)-I, \quad D_1 = A^{-1}D \in R(X),$$

$$P(D_1) = \sum_{k=0}^{N-1} P_k D_1^k, \quad P_k = SQ_k S \quad (k = 0, 1, ..., N-1)$$

with the conditions

$$FD^k SQ(D)x = FD^k x+FD^k Sy \quad (k = 0, 1, ..., N-1). \tag{4.5.22}$$

Proof. In a similar way, as in Theorem 4.5.1, we prove that the operator A is invertible and $A^{-1} = SAS$. By a simple induction we obtain the following formulae:

$$SA^k = A^{-k}S, \quad SA^{-k} = A^k S, \tag{4.5.23}$$

$$SD_1^k = D^k S, \quad SD^k = D_1^k S \quad \text{on dom } D^k \tag{4.5.24}$$

$(k = 1, 2, ...)$.

By definition and Formulae (4.5.24)

$$P(D_1)S = SQ(D), \quad \text{hence } P(D_1) = SQ(D)S \quad \text{on dom } D^N.$$

$$\tag{4.5.25}$$

Observe that the operator $D_1 = A^{-1}D \in R(X)$. Indeed, writing $R_1 = RA$, we find $D_1 R_1 = A^{-1}DRA = A^{-1}A = I$. An initial operator for D_1 corresponding to R_1 is $F_1 = I-R_1 D_1 = I-RAA^{-1}D = I-$

$-RD = F$. Since x is a solution of Equation (4.5.20), we have $Sx = Q(D)x - y$ and by Formula (4.5.25) we have

$$x = S^2x = D[Q(D)x - y] = SQ(D)x - Sy = P(D_1)Sx - Sy$$
$$= P(D_1)[Q(D)x - y] - Sy = P(D_1)Q(D)x - P(D_1)y - Sy,$$

which implies

$$Hx = P(D_1)Q(D)x - x = P(D_1)y + Sy.$$

Hence x satisfies Equation (4.5.21). Moreover,

$$FD^kSQ(D)x = FD^kS^2x + FD^kSy = FD^kx + FD^kSy$$
$$(k = 0, 1, ..., N-1),$$

i.e. x satisfies also Conditions (4.5.22).   ∎

The converse statement is, in general, not true, as the following example shows:

EXAMPLE 4.5.1. Consider the following differential equation with *reflection*:

$$x'(t) + a(t)x(t) = x(-t), \quad \text{where } a(t) = (1 - e^{2t})/(1 + e^{2t}).$$
$$(4.5.26)$$

In our case we have $D = d/dt$, $(Fx)(t) = x(0)$, $(Sx)(t) = x(-t)$, $DS = -SD$, hence $A = -I$, $y = 0$, $Q(t) = t + a$. Conditions (4.5.22) are of the form:

$$x'(0) = x(0) \qquad (4.5.27)$$

because $a(0) = 0$. Moreover, $H = (-D + a)(D + a) - I = -D^2 - aD + Da + (1 - a^2)I = 0$. Hence Equation (4.5.21) is of the form

$$x'' - ax' + (ax)' + (1 - a^2)x = 0,$$

i.e. $x'' + (1 + a' - a^2)x = 0$. It is easy to check that $1 + a' - a^2 = 0$. Thus we finally obtain $x'' = 0$. Every solution of this equation satisfying Condition (4.5.27) is of the form $x(t) = C(t + 1)$, where C is an arbitrary constant. But

$$x'(t) + a(t)x(t) - x(-t) = C + a(t)C(t + 1) - C(-t + 1)$$
$$= 2C(t - e^{2t})/(1 + e^{2t}) \equiv 0$$

if and only if $C = 0$. This means that Equation (4.5.26) has only zero as a solution. Thus there exist solutions of Equation (4.5.21) satisfying Conditions (4.5.22) which do not satisfy Equation (4.5.20). ∎

REMARK 4.5.2. If the polynomial $Q(D)$ has coefficients commutative with S, i.e. if $SQ_k = Q_k S$ for $k = 0, 1, ..., N-1$, then $P_k = SQ_k S = Q_k S^2 = Q_k$, hence $P(D_1) = Q(D_1)$. In particular the last equality holds if $Q_k = q_k I$, where q_k are scalars $(k = 0, 1, ..., N-1)$.

THEOREM 4.5.4. *Suppose that all assumptions of Theorem 4.5.3 are satisfied. Moreover, suppose that*

$$SQ_k = Q_k S \quad (k = 0, 1, ..., N-1), \quad SF = FS \qquad (4.5.28)$$

and that the operators $I+Q^0(R)$, $I+Q_1^0(R_1)$, $I-H^0$ are invertible, where we write

$$Q^0(R) = \sum_{k=0}^{N-1} Q_k R^{N-k},$$

$$Q_1^0(R_1) = \sum_{k=0}^{N-1} Q_k R_1^{N-k} = \sum_{k=0}^{N-1} Q_k (RA)^{N-k}, \qquad (4.5.29)$$

$$H^0 = R^N [I+Q^0(R)]^{-1} (RA)^N [I+Q_1^0(R_1)]^{-1}.$$

Then x is a solution of Equation (4.5.20) if and only if x satisfies Equation (4.5.21) and Conditions (4.5.22), where $P(D_1) = Q(D_1) = Q(A^{-1}D)$.

Proof. Observe that, by our assumptions and Remark 4.5.2, we have $P(D_1) = Q(D_1)$, which implies $R = Q(D_1)Q(D)-I$. The necessity has been proved by Theorem 4.5.3.

Sufficiency. Suppose that x is a solution of (4.5.21)–(4.5.22). Put $u = Q(D)x-Sx-y$. Then, by Formula (4.5.25), since $Q(D_1)Q(D)x = x+Q(D_1)y+Sy$, we find

$$\begin{aligned}
Q(D_1)u &= Q(D_1)Q(D)x-Q(D_1)Sx-Q(D_1)y \\
&= Q(D_1)Q(D)x-SQ(D)x-Q(D_1)y \\
&= x-SQ(D)x+Sy = -S[Q(D)x-Sx-y] = -Su.
\end{aligned}$$

Write $v = Su$. Then we have $Q(D)\, v = Q(D)\, Su = SQ(D_1)u = -S^2 u$
$= -u = -Sv$.   Hence $Hv = Q(D_1)Q(D)\, v - v = Q(D_1)(-Sv) - v$
$= -SQ(D)\, v - v = -S(-Sv) - v = S^2 v - v = 0$. Thus v satisfies the equation

$$Hv = 0, \quad \text{where } H = Q(D_1)Q(D) - I. \tag{4.5.30}$$

Since $FS = SF$, Conditions (4.5.22) and Formula (4.5.25) together imply

$$FD^k v = 0 \quad \text{for } k = 0, 1, \dots, N-1. \tag{4.5.31}$$

Indeed,

$$
\begin{aligned}
FD^k v &= FD^k Su = FD^k SQ(D)x - Sx - y \\
&= FD^k SQ(D)x - FD^k S^2 x - FD^k Sy \\
&= FD^k SQ(D)x - FD^k x - FD^k Sy = 0.
\end{aligned}
$$

Write $w = Q(D)\, v$, Formulae (4.5.31) and Equality $Q(D)\, v = -Sv$ together imply

$$FD_1^k w = 0 \quad \text{for } k = 0, 1, \dots, N-1. \tag{4.5.32}$$

Indeed, for $k = 0, 1, \dots, N-1$ we have $FD_1^k w = FD_1^k Q(D)v$
$= FD_1^k(-Sv) = -FD_1^k Sv = -FSD^k v = -SFD^k v = 0$. From Equality (4.5.30) we conclude that

$$Q(D_1)w = -Sv. \tag{4.5.33}$$

Indeed, $Q(D_1)w + Sw = Q(D_1)Q(D)\, v + SQ(D)\, v = Q(D_1)Q(D)\, v +$
$+ S(-Sv) = Q(D_1)Q(D)\, v - S^2 v = Q(D_1)Q(D)\, v - v = Hv = 0$. Since by our assumptions the operator $I + Q^0(R)$ is invertible, Formulae (4.5.32)–(4.5.33) and Theorem 4.1.2 together imply

$$w = R_1^N[I + Q_1^0(R_1)]^{-1}v. \tag{4.5.34}$$

Indeed,

$$w = R_1^N[I + Q_1^0(R_1)]^{-1}\left(-Sw - \sum_{j=0}^{N-1} Q_j \sum_{k=0}^{j} R_1^{k-j}FD_1^k w\right) +$$

$$+ \sum_{k=0}^{N-1} R_1^k FD_1^k w$$

$$= -R_1^N[I + Q_1^0(R_1)]^{-1}Sw = -R_1^N[I + Q_1^0(R_1)]^{-1}SQ(D)v$$

$$= -R_1^N[I+Q_1^0(R_1)]^{-1}S(-Sv) = R_1^N[I+Q_1^0(R_1)]^{-1}S^2v$$
$$= R_1^N[I+Q_1^0(R_1)]^{-1}v$$

because, by Theorem 4.5.1, $F_1 = F$.

Observe now that, by definition, $Q(D) v = w$, and that the operator $I+Q^0(R)$ is invertible by our assumption. Hence in the same way, as above, we conclude from Formulae (4.5.34), (4.5.31) and Theorem 4.1.2 that

$$v = R^N[I+Q^0(R)]^{-1}\left(w - \sum_{j=0}^{N-1} Q_j \sum_{k=0}^{j} R^{k-j}FD^kv\right)$$
$$= R^N[I+Q^0(R)]^{-1}w = R^N[I+Q^0(R)]^{-1}R_1^N[I+Q_1^0(R_1)]^{-1}v$$
$$= H^0v.$$

Thus $(I-H^0) v = 0$. Since the operator $I-H^0$ is invertible by our assumption, we have $v = 0$ and $Q(D)x-Sx-y = u = Sv = 0$, which proves that x is a solution of Equation (4.5.20).   ∎

Examples and Exercises

EXAMPLE 4.5.2. Consider the following ordinary functional-differential equation of *Carleman type*

$$x''(t)+Q_1(t)x'(t)+Q_0(t)x(t) = x(g(t))+y(t), \qquad (4.5.35)$$

where we assume that: (i) $g(t) \not\equiv t$ is a continuously differentiable function mapping the interval (a, b) (where we may have $a = -\infty$, $b = +\infty$) onto itself and satisfying the so-called *Carleman condition*: $g(g(t)) = t$ for all $t \in (a, b)$; (ii) Q_0, Q_1 are real-valued continuously differentiable functions determined on (a, b) and such that $Q_k(g(t)) = Q_k(t)$ for $k = 0, 1$ and $t \in (a, b)$; (iii) y is a real-valued continuously differentiable function determined on (a, b).

It follows from our assumptions, that $g'(t) \neq 0$ for all $t \in (a, b)$ and that there exists a unique fixed-point of g, i.e. such a point $c \in (a, b)$ that $g(c) = c$ (cf. Przeworska-Rolewicz, 1973, Chapter VIII, Proposition 1.3). We put $D = d/dt$, $(Fx)(t) = x(c)$, $(Rx)(t) = \int_c^t x(s)ds$, $(Sx)(t) = x(g(t))$. We conclude that F is an initial operator for D corresponding to the right inverse R of D and that S is an involution. Moreover, by our assumptions,

$$(SFx)(t) = x(g(c)) = (FSx)(t),$$
$$(Ax)(t) = g'(t)x(t), \quad \text{hence } A \text{ is invertible,}$$
$$(SQ_k x)(t) = Q_k(g(t))x(g(t)) = Q_k(t)x(g(t))$$
$$= (Q_k Sx)(t) \quad (k = 0, 1).$$

We have also

$$Q^0(R)x(t) = (Q_0 R^2 x + Q_1 Rx)(t) = \int_c^t M(t, s)x(s)\,ds,$$

where $M(t, s) = Q_0(t-s) + Q_1(t)$ is a continuous function for $a \leqslant t$, $s \leqslant b$;

$$[Q_1^0(RA)x](t) = [Q_0(RA)^2 x + Q_1(RA)x](t) = \int_c^t M_1(t, s)x(s)\,ds,$$

where $M_1(t, s) = Q_0(t)[g(t) - g(s)] + Q_1(t)g'(t)$ is a continuous function for $a \leqslant t$, $s \leqslant b$. The operators $I + Q^0(R)$, $I + Q_1^0(RA)$ are invertible in the space $C[a_1, b_1]$, where $a < a_1 < c < b_1 < b$ are arbitrarily fixed (cf. Section 8.4, also Pogorzelski, 1966). Moreover we conclude that the operator

$$I - H^0 = I - R^2[I + Q^0(R)]^{-1}(RA)^2[I + Q_1^0(RA)]^{-1}$$

is also invertible in the space $X = C[a_1, b_1]$. All conditions of Theorem 4.5.4 are satisfied. Then $x \in X$ is a solution of Equation (4.5.34) if and only if (4.5.21)–(4.5.22) hold. But in our case

$$H = \left[(g')^{-1}\frac{d}{dt}(g')^{-1}\frac{d}{dt} + (g')^{-1}Q_1\frac{d}{dt} + Q_0 \right] \times$$
$$\times \left(\frac{d^2}{dt^2} + Q_1\frac{d}{dt} + Q_0 \right)$$
$$= (g')^{-2}\left\{ \frac{d^2}{dt^2} + [g'Q_1 - (g')^{-1}g'']\frac{d}{dt} + g'^2 Q_0 \right\} \times$$
$$\times \left(\frac{d^2}{dt^2} + Q_1\frac{d}{dt} + Q_0 \right).$$

Conditions (4.5.22) are of the form

$$0 = x''(g(c)) + Q_1(g(c))x'(g(c)) + Q_0(g(c))x(g(c)) -$$
$$\qquad\qquad\qquad\qquad -x(c) - y(g(c))$$
$$= x''(c) + Q_1(c)x'(c) + [Q_0(c) - 1]x(c) - y(c),$$

$$0 = \left\{ \frac{d}{dt} [x''(g(t)) + Q_1(g(t))x'(g(t)) + Q_0(g(t))x(g(t)) - \right.$$
$$\left. - x(t) - y(g(t))]_{t=c} \right\}$$
$$= g'(c)\{x'''(c) + Q_1(c)x''(c) +$$
$$+ [Q_1'(c) + Q_0(c) - (g'(c))^{-1}]x'(c) +$$
$$+ (g'(c))^{-1}Q_0'(c)x(c) - y'(c)\}.$$

For instance, if we put $g(t) = -t$, $Q_1(t) \equiv 0$, $Q_0(t) \equiv \lambda$, we have $g'(t) \equiv -1 \neq 0$, $c = 0$ and we conclude that every solution of the equation $x'' + \lambda x = x(-t) + y(t)$ is a solution of the equation

$$x^{(4)}(t) + 2\lambda x''(t) + \lambda^2 x(t) = y''(t) + y(t) + y(-t)$$

satisfying the conditions

$$x''(0) + (\lambda - 1)x(0) = y(0); \quad x''(0) + (\lambda + 1)x'(0) = y'(0).$$

EXAMPLE 4.5.3. Consider a partial functional-differential equation

$$\frac{\partial^2 x(t, s)}{\partial t \, \partial s} + P(t, s)x(t, s) = x(s, t) + y(t, s), \tag{4.5.36}$$

where we assume that $P, y \in C^1(\Omega)$, $\Omega = [a, b] \times [a, b]$ and, moreover, $P(s, t) = P(t, s)$ in Ω. We are looking for solutions of (4.5.36) belonging to $C^2(\Omega)$. We put $D = \partial^2/\partial t \, \partial s$, $(Fx)(t, s) = x(s, s) + \int_s^t x_t'(u, u) \, du$,

$(Rx)(t, s) = \int_s^t [\int_u^s x(u, w) dw] \, du$ and $(Sx)(t, s) = x(s, t)$. We conclude that F is an initial operator for D corresponding to the right inverse R of D (cf. Cauchy problem 4.2.2 for $g(t) \equiv t$) and that S is an involution. Moreover, since $(SDx)(t, s) = (DSx)(t, s)$, we have $A = I$. Also $SF = FS$ and $SP = PS$. Since $A = I$, in our case $I + Q_1^0(R_1) = I + Q^0(R) = I + RP$ is obviously invertible and $H^0 = I - [R(I + RP)^{-1}]^2$ is also invertible. Therefore all conditions of Theorem 4.5.4 are satisfied. Observe that $H = [Q(D)]^2 - I = (D + P)^2 - I = D^2 + DP + PD + P^2 - I$. Hence x is a solution of Equation (4.5.36) if x satisfies the equation

$$x_{ttss}^{(4)} + 2P x_{ts}'' + P_t' x_s' + (2P_{ts}'' + P^2 - 1)x = w, \tag{4.5.37}$$

where $w(t, s) = y''_{ts}(t, s) + P(t, s) y(t, s) + y(s, t)$, with the condition

$$x''_{ts}(s, s) + P(s, s) x(s, s) - y(s, s) + \int_s^t [x'''_{tts}(u, u) +$$

$$+ P(u, u) x'_t(u, u) + P'_t(u, u) x(u, u) - y'_t(u, u)] du = 0.$$

General solution of Equation (4.5.37) can be easily obtained by application of Corollary 4.5.1 (where $N = n = 2$).

EXERCISE 4.5.1. Solve the following functional-differential equation of Carleman type

$$\frac{\partial^2 x(t, s)}{\partial t \, \partial s} + q x(t, s) = x\big(g(t), h(s)\big) + y(t, s) \quad \text{in } \Omega, \qquad (4.5.38)$$

where $\Omega = [0, a] \times [0, a]$, $a > 0$, $q \in \mathbf{R}$, $y \in C(\Omega)$, $g, h \in C^1[0, a]$, moreover, $g(t) \not\equiv t, h(t) \not\equiv t$ and

$$g\big(g(t)\big) \equiv t, \quad h\big(h(t)\big) \equiv t \quad \text{for } 0 \leqslant t \leqslant a \qquad (4.5.39)$$

and t_0, s_0 are fixed points of functions g, h, respectively, i.e. $g(t_0) = t_0$, $h(s_0) = s_0$. Here we put for $x \in C(\Omega)$

$$D = \frac{\partial^2}{\partial t \, \partial s}, \quad (Rx)(t, s) = \int_{t_0}^t \int_{s_0}^s x(p, q) \, dq \, dp,$$

$$(Fx)(t, s) = x(t, s_0) - x(t_0, s) - x(t_0, s_0),$$

$$(Sx)(t, s) = x\big(g(t), h(s)\big).$$

We conclude that F is an initial operator for D corresponding to R (cf. the Darboux problem 4.2.1), $S^2 = I$, $DS = ASD$, where $(Ax)(t, s) = g'(t) h'(s) x(t, s)$ for $x \in C(\Omega)$ and $SF = FS$.

EXERCISE 4.5.2. Suppose that all assumptions of Exercise 4.5.1 are satisfied and, moreover, $h = g$. Solve Equation (4.5.38).

EXERCISE 4.5.3. Solve the equation

$$\frac{\partial^2 x(t, s)}{\partial t^2} = a^2 \frac{\partial^2 x(t, s)}{\partial s^2} + b^2 \frac{\partial^2 x(-t, s)}{\partial s^2}$$

(cf. Section 4.2, also Viner, 1970).

EXERCISE 4.5.4. Solve the following functional-differential equation of Carleman type:

$$x'(t) = x\big(g(t)\big) + y(t), \tag{4.5.40}$$

where $g \in C^1(a, b)$ (where maybe $a = -\infty$, $b = +\infty$), $g(t) \not\equiv t$, but $g\big(g(t)\big) \equiv t$ for $a < t < b$, $y \in C(a, b)$. In particular solve the following equations:

(i) $x'(t) = x\left(\dfrac{1}{t}\right)$, $0 < t < \infty$;

(ii) $x'(t) = x\big(g(t)\big)$, where

$$g(t) = \begin{cases} -at & \text{for } t > 0, \\ 0 & \text{for } t = 0, \text{ where } a > 0 \text{ is arbitrarily fixed}, \\ -\dfrac{1}{a}t & \text{for } t < 0 \end{cases}$$

(cf. Przeworska-Rolewicz, 1973, Chapter VIII);

(iii) $x'(t) = x(c-t)$, where $c \in \mathbf{R}$ is arbitrarily fixed, $t \in \mathbf{R}$;

(iv) $x'(t) = x\big(g(t)\big)$ for $0 < t < \infty$, where

$$g(t) = \begin{cases} t^{-1/k} & \text{for } t \geqslant 1, \\ t^{-k} & \text{for } 0 < t < 1, \end{cases}$$

k is an arbitrarily fixed positive integer (cf. Przeworska-Rolewicz, 1973, Chapter VIII).

EXERCISE 4.5.5. Solve the following functional-differential equation of Carleman type:

$$x'(t) = \sum_{k=0}^{n-1} a_k x(\varepsilon^k t) + y(t),$$

where $a_0, \dots, a_{n-1} \in \mathbf{C}$, $\varepsilon = e^{2\pi i/n}$, y is a complex-valued continuous function for $t \in \mathbf{C}$, $|t| \leqslant R$, $R > 0$.

EXERCISE 4.5.6. Solve the following functional-differential equation

$$x''(t) = a_0 x(t) + a_1 x(it) + a_2 x(-it),$$

where $a_0, a_1 \in \mathbf{C}$, $t \in \mathbf{R}$.

EXERCISE 4.5.7. Suppose that $Q(D, S)$ is defined by means of Formula (4.5.7). Prove that

$$DQ(D, S)\left(\sum_{k=0}^{N} R^k z_k\right) = \sum_{k=0}^{N-1} Q_k \sum_{m=k+1}^{N} R^{m+k+1} z_m$$

for arbitrary $z_0, \ldots, z_N \in \ker D$. Apply this formula in Theorem 4.5.2 and in Corollaries 4.5.1 and 4.5.2.

4.6. WELL-POSED AND ILL-POSED BOUNDARY VALUE PROBLEMS

Let X be a linear space (over a field $\mathscr{F}$) and let $D \in R(X)$ with $\ker D \neq \{0\}$. Let $F_0, \ldots, F_{N-1}$ be initial operators for D corresponding to $R_0, \ldots, R_{N-1} \in \mathscr{R}_D$, respectively. Consider equation

$$Q(D)x = y, \quad y \in X, \quad Q(D) = \sum_{k=0}^{N} Q_k D^k, \quad Q_k \in L_0(X),$$

$$Q_N = I, \quad N \geqslant 2. \quad (4.6.1)$$

A *boundary value problem* (BVP) for the operator $Q(D)$ is to find all solutions of Equation (4.6.1) which satisfy *boundary conditions*

$$F_k x = y_k, \quad y_k \in \ker D \quad (k = 0, 1, \ldots, N-1). \quad (4.6.2)$$

A *first mixed boundary value problem* (FMBVP) for the operator $Q(D)$ is to find all solutions of Equation (4.6.1) satisfying *mixed boundary conditions*

$$F_k D^k x = y_k, \quad y_k \in \ker D \quad (k = 0, 1, \ldots, N-1). \quad (4.6.3)$$

A *second mixed boundary value problem* (SMBVP) for the operator $Q(D)$ is to find all solutions of Equation (4.6.1) satisfying *mixed boundary conditions*

$$\begin{aligned} F_k x &= y_k & & (k = 0, 1, \ldots, M < N-1), \\ F_k D^k x &= y_k & y_k \in \ker D \quad & (k = M+1, \ldots, N-1). \end{aligned}$$

$$(4.6.4)$$

A *third mixed boundary value problem* (TMBVP) for the operator $Q(D)$ is to find all solutions of Equation (4.6.1) satisfying *mixed boundary conditions*

$$\left(F_{0k} + \sum_{j=1}^{N-1} P_{jk} F_{jk} D^j\right) x = y_k, \quad y_k \in \ker D$$

$$(k = 0, 1, ..., N-1), \quad (4.6.5)$$

where the operator P_{jk} $(j, k = 1, ..., N-1)$ maps $\ker D$ into itself, F_{jk} are initial operators for D.

So that we have four different problems for the operator $Q(D)$. Each of these problems is *well-posed* if it has a unique solution for every $y \in X$, $y_0, ..., y_{N-1} \in \ker D$. This implies that the corresponding homogeneous problems have only zero as a solution.

We say that each of these problems is *ill-posed* if either there exist $y \in X$, $y_0, ..., y_{N-1} \in \ker D$ such that the problem under consideration has no solutions or the corresponding homogeneous problem has at least one non-trivial solution.

THEOREM 4.6.1. *Suppose that $D \in R(X)$, dim $\ker D \neq 0$, F_{jk} are initial operators for D, $P_{jk} \in L(X)$ maps $\ker D$ into itself $(j, k = 0, 1, 2, ..., N-1)$ and $Q(D)$ is defined by Formula (4.6.1). Then $TMBVP$ (4.6.1)– (4.6.5) can be reduced to a BVP of the form (4.6.1)–(4.6.2).*

Proof. Theorem 2.2.6 implies, that each of operators $F_k = F_{0k} +$
$$+ \sum_{j=1}^{N-1} P_{jk} F_{jk} D^j \ (k = 0, 1, ..., N-1)$$ is an initial operator for D, hence the conditions (4.6.5) can be rewritten as $F_k x = y_k$ $(k = 0, 1, ..., N-1)$, i.e. in the form of the boundary condition (4.6.2). ∎

The Taylor–Gontcharov Formula implies

THEOREM 4.6.2. *Suppose that $D \in R(X)$, dim $\ker D \neq 0$, $F_0, ..., F_{N-1}$ are initial operators for D corresponding to the right inverse $R_0, ..., R_{N-1}$ of D, respectively, and the operator $Q(D)$ is defined by Formula (4.6.1). Write*

$$Q^0 = \sum_{m=0}^{N-1} Q_m R_m \ ... \ R_{N-1}. \tag{4.6.6}$$

If the operator $I+Q^0$ is invertible then FMBVP (4.6.1)–(4.6.3) *is well-posed and its unique solution is*

$$x = R_0 \ldots R_{N-1}(I+Q^0)^{-1}y_N + y_0 + \sum_{k=1}^{N-1} R_0 \ldots R_{k-1}y_k, \qquad (4.6.7)$$

$$y_N = y - Q_0 y_0 \sum_{m=1}^{N-1} Q_m \left(\sum_{k=m+1}^{N-1} R_m \ldots R_{k-1}y_k + y_m \right). \qquad (4.6.8)$$

Proof. The Taylor–Gontcharov Formula (2.2.6) implies that Equation (4.6.1) can be rewritten as follows:

$$y = Q(D)x = Q(D)\left[F_0 x + \sum_{k=1}^{N-1} R_0 \ldots R_{k-1}F_k D^k x + \right.$$

$$\left. + R_0 \ldots R_{N-1}D^N x \right]$$

$$= \sum_{m=0}^{N} Q_m D^m \left[y_0 + \sum_{k=1}^{N-1} R_0 \ldots R_{k-1}y_k + R_0 \ldots R_{N-1}D^N x \right]$$

$$= Q_0 y_0 + \sum_{m=1}^{N-1} Q_m \left(\sum_{k=1}^{N-1} D^m R_0 \ldots R_{k-1}y_k + \right.$$

$$\left. + D^m R_0 \ldots R_{N-1}D^N x \right) + D^N R_0 \ldots R_{N-1}D^N x +$$

$$+ Q_0 R_0 \ldots R_{N-1}D^N x$$

$$= Q_0 y_0 + \sum_{m=1}^{N-1} Q_m \left(\sum_{k=1}^{m} D^{m-k}y_k + \sum_{k=m+1}^{N-1} R_m \ldots R_{k-1}y_k + \right.$$

$$\left. + R_m \ldots R_{N-1}D^N x \right) + D^N x + Q_0 R_0 \ldots R_{N-1}D^N x$$

$$= Q_0 y_0 + \sum_{m=1}^{N-1} Q_m \left(y_m + \sum_{k=m+1}^{N-1} R_m \ldots R_{k-1}y_k \right) +$$

$$+ \left(\sum_{m=1}^{N-1} Q_m R_m \ldots R_{N-1} \right) D^N x + D^N x + Q_0 R_0 \ldots R_{N-1}D^N x$$

$$= y - y_N + (I+Q^0)D^N x,$$

for the denotations (4.6.6), (4.6.8) is used. We obtain then an equivalent equation

$$(I+Q^0)D^N x = y_N. \tag{4.6.9}$$

Since, by our assumptions, the operator $I+Q^0$ is invertible, Equation (4.6.9) is equivalent to the equation

$$D^N x = (I+Q^0)^{-1} y_N, \tag{4.6.10}$$

which has solutions of the form

$$x = R_0 \dots R_{N-1}(I+Q^0)^{-1} y_N + \sum_{k=0}^{N-1} R_0 \dots R_{k-1} z_k, \tag{4.6.11}$$

$z_0, \dots, z_{N-1} \in \ker D$.

We have now to determine elements $z_0, \dots, z_{N-1} \in \ker D$. Acting on both sides of (4.6.11) by the operators $I, D, \dots, D^{N-1}$ we find for $j = 0, 1, \dots, N-1$

$$D^j x = D^j R_0 \dots R_{N-1}(I+Q^0)^{-1} y_N + \sum_{k=0}^{N-1} D^j R_0 \dots R_{k-1} z_k$$

$$= R_j \dots R_{N-1}(I+Q^0)^{-1} y_N + \sum_{k=0}^{j} D^{j-k} z_k +$$

$$+ \sum_{k=j+1}^{N-1} R_j \dots R_{k-1} z_k$$

$$= R_j \dots R_{N-1}(I+Q^0)^{-1} y_N + z_j + \sum_{k=j+1}^{N-1} R_j \dots R_{k-1} z_k.$$

Hence Conditions (4.6.3) imply the following equalities:

$$y_j = F_j D^j x = F_j R_j \dots R_{N-1}(I+Q^0)^{-1} y_N + F_j z_j +$$

$$+ \sum_{k=j+1}^{N-1} F_j R_j \dots R_{k-1} z_k = z_j$$

for $j = 0, 1, \dots, N-1$, because $F_j R_j = 0$ and $F z_j = z_j$.

We therefore conclude that x is of the form (4.6.7) and that x is a unique solution of the problem (4.6.1)–(4.6.3), because for $y = 0$, $y_0 = \dots = y_{N-1} = 0$ we obtain $x = 0$.   ∎

Observe that for $F_0 = \ldots = F_{N-1} = F$ we obtain $R_0 = \ldots = R_{N-1} = R$ and Theorem 4.1.2, i.e. solutions of an initial value problem for the operator $Q(D)$.

From the proof of Theorem 4.6.2 in the case $Q_0 = \ldots = Q_{N-1} = 0$ we have $Q^0 = 0$ and we obtain

COROLLARY 4.6.1. *Suppose that $D \in R(X)$, $\dim \ker D \neq 0$, $F_0, \ldots, F_{N-1}$ are initial operators for D corresponding to the right inverses $R_0, \ldots, R_{N-1}$ of D, respectively. Then FMBVP for the operator D^N, i.e. the problem*

$$D^N x = y, \quad y \in X, \tag{4.6.12}$$

$$F_k D^k x = y_k, \quad y_k \in \ker D \quad (k = 0, 1, \ldots, N-1) \tag{4.6.13}$$

is well-posed and its unique solution is of the form

$$x = R_0 \ldots R_{N-1} y + y_0 + \sum_{k=1}^{N-1} R_0 \ldots R_{k-1} y_k. \tag{4.6.14}$$

In a similar way, as Corollaries 4.1.1 and 4.1.2 we obtain

COROLLARY 4.6.2. *Suppose that $D \in R(X)$, $\dim \ker D \neq 0$, $F_0, \ldots$ $\ldots, F_{N+M-1}$ are initial operators for D corresponding to the right inverses $R_0, \ldots, R_{N+M-1}$ of D, respectively, and the operators $Q(D)$ and Q^0 are defined by Formulae (4.6.1) and (4.6.6), respectively. If the operator $I+Q^0$ is invertible then FMBVP for the operator $Q(D)D^M$, i.e. the problem*

$$Q(D) D^M x = y, \quad y \in X \quad (M \geqslant 0), \tag{4.6.15}$$

$$F_k D^k x = y_k, \quad y_k \in \ker D \quad (k = 0, 1, \ldots, N+M-1) \tag{4.6.16}$$

is well-posed and its unique solution is of the form

$$x = R_0 \ldots R_{M+N-1}(I+Q^0)^{-1} y_{N+M}^0 + y_0 + \sum_{k=1}^{M+N-1} R_0 \ldots R_{k-1} y_k,$$

$$\tag{4.6.17}$$

where

$$y_{N+M}^0 = y - Q_0 y_M - \sum_{m=1}^{N-1} Q_m \left(\sum_{k=m+1}^{N-1} R_m \ldots R_{k-1} y_{M+k} + y_{M+m} \right).$$

$$\tag{4.6.18}$$

Proof. Write $u = D^M x$. Then $F_{k+M} D^k u = F_{k+M} D^{k+M} x = y_{k+M}$ for $k = 0, 1, \ldots, N-1$.

By our assumptions, Theorem 4.1.2 implies that FMBVP

$$Q(D)u = y, \quad F_{k+M} D^k u = y_{k+M} \quad (k = 0, 1, \ldots, N-1)$$

is well-posed and its unique solution is

$$u = R_M \ldots R_{M+N-1}(I+Q^0)^{-1} y^0_{N+M} + y_M +$$
$$+ \sum_{k=1}^{N-1} R_M \ldots R_{M+k-1} y_{M+k}, \qquad (4.6.19)$$

where y^0_{N+M} is defined by Formula (4.6.18).

On the other hand, Corollary 4.6.1 implies that FMBVP

$$D^M x = u, \quad F_k D^k x = y_k \quad (k = 0, 1, \ldots, M-1)$$

is also well-posed and its unique solution is

$$x = R_0 \ldots R_{M+1} u + y_0 + \sum_{k=1}^{M-1} R_0 \ldots R_{k-1} y_k.$$

This and Formula (4.6.19) together imply that the problem (4.6.12)–(4.6.13) is well-posed and its unique solution is

$$x = R_0 \ldots R_{M-1} u + y_0 + \sum_{k=1}^{M-1} R_0 \ldots R_{k-1} y_k$$

$$= R_0 \ldots R_{M-1} R_M \ldots R_{M+N-1}(I+Q^0)^{-1} y^{0}_{N+M} + y_M +$$
$$+ \sum_{k=1}^{N-1} R_M \ldots R_{M+k-1} y_{M+k} + y_0 + \sum_{k=1}^{M-1} R_0 \ldots R_{k-1} y_k$$

$$= R_0 \ldots R_{M+N-1}(I+Q^0)^{-1} y^0_{N+M} + y_0 + \sum_{k=1}^{M-1} R_0 \ldots R_{k-1} y_k +$$
$$+ R_0 \ldots R_{M-1} y_M + \sum_{k=1}^{N-1} R_0 \ldots R_{M-1} R_M \ldots R_{M+k-1} y_{M+k}$$

$$= R_0 \ldots R_{M+N-1}(I+Q^0)^{-1} y^0_{N+M} + y_0 + \sum_{k=1}^{M+N-1} R_0 \ldots R_{k+1} y_k,$$

which was to be proved. ∎

An immediate consequence is

COROLLARY 4.6.3. *Suppose that all assumptions of Corollary 4.6.2 are satisfied and that operators $Q_0, \ldots, Q_{N-1}$ commute with D. Consider the equation*

$$D^M Q(D)x = y, \quad y \in X. \tag{4.6.20}$$

Then FMBVP (4.6.20)–(4.6.16) is well-posed and its unique solution is of the form (4.6.17).

Till this moment we have considered the case when the operator $I+Q^0$ is invertible. In this case -1 is not an eigenvalue of the operator Q^0. Consider now the case when -1 is an eigenvalue of the operator Q^0. We have the following

THEOREM 4.6.3. *Suppose that all assumptions of Theorem 4.6.2 are satisfied. If -1 is an eigenvalue of the operator Q^0 then FMBVP (4.6.1)–(4.6.3) is ill-posed. However, this problem has solutions if and only if*

$$y_N \in (I+Q^0)X, \tag{4.6.21}$$

where y_N is defined by Formula (4.6.8).

If this condition is satisfied then solutions exist and are of the form

$$x = R_0 \ldots R_{N-1} w + x_{-1} + y_0 + \sum_{k=1}^{N-1} R_0 \ldots R_{k-1} y_k,$$

where x_{-1} is an eigenvector of Q^0 corresponding to the eigenvalue -1 and w is an element of the inverse-image of the element y_N by the mapping $I+Q^0$.

Proof. In the same way, as in the proof of Theorem 4.6.2, we obtain Equation (4.6.9). Further the proof is going on the same lines, as the proof of Theorem 4.1.3, by application of Corollary 4.6.1.   ∎

We therefore have solved in a full generality the first mixed boundary value problem and we have proved that the third mixed boundary value problem can be reduced to a boundary value problem. We still have to study boundary value problems and the second boundary value problems. These problems are much more difficult and we can solve by our methods some particular case only.

DEFINITION 4.6.1. A system of initial operators $\{F_0, \ldots, F_{N-1}\}$ for an operator $D \in R(X)$ such that dim ker $D \neq 0$ is said to be *admissible*, if there exists a right inverse R of D such that for every $y_0, \ldots, y_{N-1} \in \ker D$ there exist $z_0, \ldots, z_{N-1} \in \ker D$ satisfying the following conditions:

$$\sum_{k=0}^{N-1} F_j R^k z_k = y_k \quad (j = 0, 1, \ldots, N-1), \tag{4.6.22}$$

$$y_0 = y_1 = \ldots = y_{N-1} = 0 \text{ implies } z_0 = z_1 = \ldots = z_{N-1} = 0. \tag{4.6.23}$$

In other words, the system $\{F_0, \ldots, F_{N-1}\}$ is *admissible* if the elements $z_0, \ldots, z_{N-1}$ satisfying property (4.6.22) are uniquely determined. So that the correspondence ζ_N between the system $\{y_0, \ldots, y_{N-1}\}$ and $\{z_0, \ldots, z_{N-1}\}$ is linear and one-to-one.

In the sequel we shall write $\zeta_N \{y_0, \ldots, y_{N-1}\} = \{z_0, \ldots, z_{N-1}\}$.

The mapping ζ_N, in general, depends on properties of the model considered, and as far as it is known, it cannot be described in the general case. *Added in proof*: Form of ζ_N is already known (1987).

EXAMPLE 4.6.1. Suppose that $X = C[0, 1]$, $D = d/dt$, $0 \leqslant t_0 < t_1 < \ldots$ $\ldots < t_{N-1} = 1$. Write $(F_j x)(t) = x(t_j)$ $(j = 0, 1, \ldots, N-1)$, $(Rx)(t) = \int_0^t x(s) \, ds$. We shall show that the system $\{F_0, \ldots, F_{N-1}\}$ is admissible. Indeed, since $y_0, \ldots, y_{N-1}, z_0, \ldots, z_{N-1}$ are constant functions in our case, we have for $k = 1, \ldots, N-1$

$$R^k z_k = \int_0^t \frac{(t-s)^{k-1}}{(k-1)!} z_k \, ds = z_k \frac{t^k}{k!} \quad \text{and} \quad R^0 z_0 = z_0,$$

hence $F_j R^k z_k = z_k t^k / k!$ $(j, k = 0, 1, \ldots, N-1)$.

Thus we obtain a system of N algebraic equations with N unknowns:

$$\sum_{k=0}^{N-1} \frac{t_j^k}{k!} z_k = y_j \quad (j = 0, 1, \ldots, N-1). \tag{4.6.24}$$

The determinant of this system is

$$\Delta = \det \left[\frac{t_j^k}{k!} \right]_{j,\,k=0,\,1,\,\ldots,\,N-1}$$

$$= \left[\prod_{k=1}^{N-1} \frac{1}{k!} \right] V(t_0, \ldots, t_{N-1}),$$

where $V(t_0, \ldots, t_{N-1})$ is the Vandermonde determinant of N different numbers $t_0, \ldots, t_{N-1}$. Hence $\Delta \neq 0$ and the system (4.6.24) has a unique solution for every $y_0, \ldots, y_{N-1} \in \ker D$. This implies that the system $\{F_0, \ldots, F_{N-1}\}$ is admissible.   ∎

THEOREM 4.6.4. *Suppose that* $F_0, \ldots, F_{N-1}$ *are initial operators for an operator* $D \in R(X)$ *such that* $\dim \ker D \neq 0$. *Then the following conditions are equivalent*:

(i) *The system* $\{F_0, \ldots, F_{N-1}\}$ *is admissible*;

(ii) *The boundary value problem*

$$D^N x = 0, \tag{4.6.25}$$

$$F_j x = y_j, \quad y_j \in \ker D \quad (j = 0, 1, \ldots, N-1) \tag{4.6.26}$$

has a unique solution for every $y_0, \ldots, y_{N-1} \in \ker D$;

(iii) *For every system of elements* $\{y_0, \ldots, y_{N-1}\} \subset \ker D$ *there exists a unique system of elements* $\{\eta_0, \ldots, \eta_{N-1}\} \subset \ker D$ *such that*

$$\eta_0 + \sum_{k=1}^{N-1} F_j R_0 \ldots R_{k-1} \eta_k = y_j \quad (j = 0, 1, \ldots, N-1), \tag{4.6.27}$$

where $R_0, \ldots, R_{N-1}$ *are right inverses of* D *corresponding to* $F_0, \ldots, F_{N-1}$, *respectively. In the sequel we shall write* $\mathfrak{z}_N(y_0, \ldots, y_{N-1}) = (\eta_0, \ldots, \eta_{N-1})$.

Proof. (i) → (ii). Suppose that $\{F_0, \ldots, F_{N-1}\}$ is an admissible system of initial operators. Write $z = \sum_{k=0}^{N-1} R^k z_k$, where $\{z_0, \ldots, z_{N-1}\} = \zeta_N\{y_0, \ldots, y_{N-1}\}$. It is easy to see that $D^N z = 0$. By definition, $F_j z = \sum_{k=0}^{N-1} F_j R^k z_k = y_j$ $(j = 0, 1, \ldots, N-1)$ and $\{z_0, \ldots$

..., z_{N-1}} is uniquely determined for a fixed system of elements {y_0, ..., y_{N-1}}. This proves the condition (ii).

(ii) → (i). Every solution of the equation (4.6.25) is of the form $x = \sum_{k=0}^{N-1} R^k z_k$, where $z_0, ..., z_{N-1} \in \ker D$ are arbitrary. Our assumptions, that the system (4.6.25)–(4.6.26) has a unique solution for every $y_0, ..., y_{N-1} \in \ker D$, and the condition (4.6.26) which now is of the form

$$ F_j z = \sum_{k=0}^{N-1} R^k z_k = y_j \quad (j = 0, 1, ..., N-1) $$

together imply that there exist uniquely determined $z_0^j, ..., z_{N-1} \in \ker D$ satisfying this condition. Hence both conditions (4.6.22)–(4.6.23) are satisfied, which implies that the system {$F_0, ..., F_{N-1}$} is admissible.

(ii) → (iii). The proof is analogous to that of the implication (ii) → (i). We only write $x = z_0 + \sum_{k=1}^{N-1} R_0 ... R_{k-1} z_k$.

(iii) → (ii). Suppose that (iii) is satisfied. Write $\eta = \eta_0 + \sum_{k=1}^{N-1} R_0 ... ... R_{N-1} \eta_k$. Then $D^N \eta = 0$ and, moreover, from the condition (4.6.27) we find

$$ F_j \eta = F_j \eta_0 + \sum_{k=1}^{N-1} F_j R_0 ... R_{k-1} \eta_k $$

$$ = \eta_0 + \sum_{k=1}^{N-1} F_j R_0 ... R_{k-1} \eta_k = y_j $$

for $j = 0, 1, ..., N-1$ and the system {$\eta_0, ..., \eta_{N-1}$} is unique for every fixed $y_0, ..., y_{N-1} \in \ker D$. We therefore have proved that (i) → (ii) → (iii), hence (i) → (iii). ∎

We recall (Proposition 2.2.2) that for every $z \in \ker D \neq \{0\}$ there exists an $x \in X$ such that $F_\beta R_\alpha x = z$, where F_β, F_α, $\alpha \neq \beta$, are initial operators corresponding to R_β, R_α, respectively.

We say that a family $\{R_\gamma\}_{\gamma \in \Gamma_0 C \Gamma} \subset \mathcal{R}_D$ has *the property* (c) *if for* every $z \in \ker D$ and $\alpha, \beta \in \Gamma_0$ there is a scalar $d_{\alpha\beta}$ such that

$$F_\beta R_\alpha z = d_{\alpha\beta} z, \quad d_{\alpha\beta} \neq 0 \quad \text{for } \alpha \neq \beta, \qquad (4.6.28)$$

where $\{F_\gamma\}_{\gamma \in \Gamma_0 C \Gamma}$ is the induced family of initial operators.

Just from the definition and properties of definite integrals for right invertible operators we conclude that

$$d_{\alpha\beta} = 0 \quad \text{if and only if } \alpha = \beta, \qquad (4.6.29)$$

$$(\alpha, \beta, \gamma \in \Gamma_0).$$

$$d_{\beta\alpha} = -d_{\alpha\beta}, \quad d_{\alpha\gamma} + d_{\gamma\beta} = d_{\alpha\beta} \qquad (4.6.30)$$

PROPOSITION 4.6.1. *Suppose that* $D \in R(X)$, $\dim \ker D \neq 0$ *and* F_α, F_β $(\alpha \neq \beta)$ *are initial operators for* D *corresponding to* $R_\alpha, R_\beta \in R_D$. *If the pair* $\{R_\alpha, R_\beta\}$ *has the property* (c) *then* $\{F_\alpha, F_\beta\}$ *is an admissible pair and*

$$\mathfrak{z}_2\{y_\alpha, y_\beta\} = \left\{ y_\alpha, \frac{1}{d_{\alpha\beta}} (y_\beta - y_\alpha) \right\} \quad \text{for every } y_\alpha, y_\beta \in \ker D.$$

Proof. By our assumptions the system

$$z_\alpha + F_\alpha R_\alpha z_\beta = y_\alpha, \quad z_\alpha + F_\beta R_\alpha z_\beta = y_\beta \quad (\alpha \neq \beta)$$

can be written as

$$z_\alpha = y_\alpha, \quad z_\alpha + d_{\alpha\beta} z_\beta = y_\beta$$

which implies that this system has a unique solution $z_\alpha = y_\alpha$, $z_\beta = (y_\beta - y_\alpha)/d_{\alpha\beta}$ for every $z_\alpha, z_\beta \in \ker D$. Therefore the pair $\{F_\alpha, F_\beta\}$ is admissible (cf. Theorem 4.6.4, Point (ii)). ∎

PROPOSITION 4.6.2. *If the system* $\{F_0, \ldots, F_{N-1}\}$ *of initial operators for a* $D \in R(X)$ *with* $\dim \ker D \neq 0$ *is admissible then BVP* (4.6.1)–(4.6.2) *for the operator* D^N $(N \geq 2)$ *is well-posed and its unique solution is of the form*

$$x = R_0 \ldots R_{N-1} y + z_0 + \sum_{k=1}^{N-1} R_0 \ldots R_{k-1} z_k, \qquad (4.6.31)$$

where

$$\{z_0, \ldots, z_{N-1}\} = \mathfrak{z}_N \{y_0 - F_1 R_0 \ldots R_{N-1} y, \ldots, y_{N-1} -$$
$$- F_{N-1} R_0 \ldots R_{N-1} y\}$$

and $\mathfrak{z}_N$ *is defined in Theorem 4.6.4.*

Proof. Write $u = x - R_0 \ldots R_{N-1}y$. Then $D^N u = D^N x - D^N R_0 \ldots R_{N-1}y$
$= y - y = 0$. Moreover, $F_j u = F_j x - F_j R_0 \ldots R_{N-1}y = y_j - F_j R_0 \ldots$
$\ldots R_{N-1}y$ $(j = 0, 1, \ldots, N-1)$. Hence $F_0 u = y_0$, $F_j u = y_j - F_j R_0 \ldots$
$\ldots R_{N-1}y_j$ for $j = 1, \ldots, N-1$. Conditions (ii) and (iii) of Theorem
4.6.4 together imply that $u = z_0 + \sum\limits_{k=1}^{N-1} R_0 \ldots R_{k-1}z_k$, where $z_0, \ldots, z_{N-1}$
are defined as above and then u is a unique element having the required
properties. Therefore BVP (4.6.1)–(4.6.2) is for the operator D^N
well-posed and has a unique solution of the form (4.6.31). ∎

THEOREM 4.6.5. *Suppose that the system* $\{F_0, \ldots, F_M\}$ *of initial operators
for a* $D \in R(X)$, dim ker $D \neq 0$, *is admissible. Then SMBVP* (4.6.1)–
(4.6.4) *for the operator* D^N $(N \geqslant 2)$ *is*

$$x = y_0 + \sum_{k=1}^{M} R_0 \ldots R_k z_k + \sum_{k=M+1}^{N-1} R_0 \ldots R_{k-1}y_k + R_0 \ldots R_{N-1}y$$

$$(4.6.32)$$

where

$$\{z_0, \ldots, z_M\} = \mathfrak{Z}_{M+1}\{y_0^*, \ldots, y_M^*\}, \quad y_0^* = y_0, \qquad (4.6.33)$$

$$y_j^* = y_j - y_0 - \sum_{k=M+1}^{N-1} F_j R_0 \ldots R_{k-1}y_k - R_0 \ldots R_{N-1}y \qquad (4.6.34)$$

$(j = 1, \ldots, M)$, $\mathfrak{Z}_{M+1}$ *is defined in Theorem* 4.6.4.

Proof. Write $z_k = F_k D^k x$ $(k = 1, \ldots, M)$. From Taylor Formula (2.2.7)
we have

$$x = F_0 x + \sum_{k=1}^{N-1} R_0 \ldots R_{k-1}F_k D^k x + R_0 \ldots R_{N-1}D^N x$$

$$= y_0 + \sum_{k=1}^{M} R_0 \ldots R_{k-1}z_k + \sum_{k=M+1}^{N-1} R_0 \ldots R_{k-1}y_k + R_0 \ldots R_{N-1}y$$

$$(4.6.35)$$

since by our assumption

$$D^N x = y, \quad F_0 x = y_0, \quad F_j D^j x = y_j \quad \text{for } j = M+1, \ldots, N-1.$$

Acting on both sides of Equation (4.6.35) by operators F_j $(j = 1, ..., M)$ we have

$$y_j = F_j x = F_j y_0 + \sum_{k=1}^{M} F_j R_0 \ldots R_{k-1} z_k +$$

$$+ \sum_{k=M+1}^{N-1} F_j R_0 \ldots R_{k-1} y_k + F_j R_0 \ldots R_{N-1} y.$$

Hence, applying the denotation (4.6.34) we obtain $M+1$ equations with $M+1$ unknowns

$$z_0 + \sum_{k=1}^{M} F_j R_0 \ldots R_{k-1} z_k = y_j^* \quad (j = 0, 1, ..., M). \qquad (4.6.36)$$

Since, by our assumption, the system $\{F_0, ..., F_M\}$ is admissible, the system (4.6.36) has the unique solution of the form (4.6.33) which gives the required formula (4.6.32), because $F_0 x = y_0$. ∎

Observe that we have solutions of SMBVP and TMBVP in these cases, when we are able to solve explicitly the corresponding BVP (Theorems 4.6.1, 4.6.2, 4.6.4, 4.6.5 and Proposition 4.6.2).

Consider now a mixed boundary value problem which can be reduced to an initial value problem.

THEOREM 4.6.6. *Suppose that* $D \in R(X)$, $F_0, ..., F_M$ *are initial operators for* D *corresponding to* $R_1, ..., R_M \in \mathcal{R}_D$, *respectively, and that the operators* $A_1, ..., A_M \in L_0(X)$ *are invertible. Moreover, suppose that all superpositions* $A_M \ldots A_1$, $A_j D$, $D A_j$, $A_j^{-1} R_j$ $(j = 1, ..., M)$ *are well-determined. Write*

$$D^0 = D A_M D \ldots A_1 D; \quad R^0 = R_0 A_1^{-1} \ldots R_{M-1} A_M^{-1} R_M. \qquad (4.6.37)$$

Consider the following mixed boundary value problem:

$$(D^0 + \lambda I) x = y, \quad y \in X, \quad \lambda \text{ is a scalar}, \qquad (4.6.38)$$

$$F_0 x = x_0, \; F_1 A_1 Dx = x_1, \; ..., \; F_M A_M \ldots A_1 Dx = x_M. \qquad (4.6.39)$$

(i) *If $\lambda \neq 0$ and $-1/\lambda$ is a regular value of the operator R^0, then the problem* (4.6.38)–(4.6.39) *is well-posed and its unique solution is of the form*

$$x = (I + \lambda R^0)^{-1}(R^0 y + y_0), \qquad (4.6.40)$$

where

$$y_0 = R_0 A_1^{-1} \dots R_{M-1} A_M^{-1} x_M + \dots + R_0 A_1^{-1} x_1 + x_0; \qquad (4.6.41)$$

(ii) *If $\lambda = 0$, then the problem* (4.6.38)–(4.6.39) *is well-posed and its unique solution is of the form*

$$x = R^0 y + y_0, \quad \text{where } y_0 \text{ is defined by Formula } (4.6.41);$$

(iii) *If $\lambda \neq 0$ and $-1/\lambda$ is an eigenvalue of the operator R^0, then the problem* (4.6.38)–(4.6.39) *is ill-posed. However, this problem has solutions if and only if*

$$R^0 y + y_0 \in (I + \lambda R^0)X, \text{ where } y_0 \text{ is defined by Formula } (4.6.41).$$

If this condition is satisfied, then solutions of problem (4.6.38)–(4.6.39) *exist and are of the form*

$$x = R^0 w + y_0 + x_{-1/\lambda}, \qquad (4.6.42)$$

where $x_{-1/\lambda}$ is an arbitrary eigenvalue vector of the operator R^0 corresponding to the eigenvalue $-1/\lambda$ and w is an element of the inverse-image of the element $R^0 y + y_0$, by the transformation $I + \lambda R^0$.

Proof. Observe that, according to Formulae (4.6.37) we have

$$\begin{aligned} D^0 R^0 &= (DA_M D \dots A_1 D)(R_0 A_1^{-1} \dots R_{M-1} A_M^{-1} R_M) \\ &= DA_M D \dots A_1 DR_0 A_1^{-1} \dots R_{M-1} A_M^{-1} R_M \\ &= DA_M D \dots A_1 A_1^{-1} \dots A_M^{-1} R_M \\ &= \dots = DA_M A_M^{-1} R_M = DR_M = I. \end{aligned}$$

Thus $D^0 \in R(X)$ and $R^0 \in \mathcal{R}_{D^0}$. An initial operator F^0 for D^0 corresponding to R^0 is of the form

$$\begin{aligned} F^0 = F_0 + R_0 A_1^{-1} F_1 A_1 D + \dots + \\ + R_0 A_1^{-1} \dots R_{M-1} A_M^{-1} F_M A_M D \dots A_1 D. \end{aligned} \qquad (4.6.43)$$

Indeed, by definition we have

$$F^0 = I - R^0 D^0$$
$$= I - (R_0 A_1^{-1} \dots R_{M-1} A_M^{-1} \dots R_M)(DA_M D \dots A_1 D)$$
$$= I - R_0 A_1^{-1} \dots R_{M-1} A_M^{-1}(R_M D) A_M \dots A_1 D$$
$$= I - R_0 A_1^{-1} \dots R_{M-1} A_M^{-1}(I - F_M) A_M \dots A_1 D$$
$$= I - R_0 A_1^{-1} \dots R_{M-1} A_M^{-1} A_M \dots A_1 D +$$
$$+ R_0 A_1^{-1} \dots R_{M-1} A_M^{-1} F_M A_M \dots A_1 D = \dots$$
$$= I - R_0 D + R_0 A_1^{-1} F_1 A_1 D + \dots +$$
$$+ R_0 A_1^{-1} \dots R_{M-1} A_M^{-1} F_M A_M \dots A_1 D$$
$$= F_0 + R_0 A_1^{-1} F_1 A_1 D + \dots$$
$$\dots + R_0 A_1^{-1} \dots R_{M-1} A_M^{-1} F_M A_M \dots A_1 D.$$

Conditions (4.6.39) and Formulae (4.6.43), (4.6.41) together imply that

$$F^0 x = F_0 x + R_0 A_1^{-1} F_1 A_1 Dx + \dots +$$
$$+ R_0 A_1^{-1} \dots R_M A_M^{-1} F_M A_M \dots A_1 Dx$$
$$= x_0 + R_0 A_1^{-1} x_1 + \dots + R_0 A_1^{-1} \dots R_{M-1} A_M^{-1} x_M + y_0.$$

We therefore can rewrite problem (4.6.38)–(4.6.39) as follows:

$$(D^0 + \lambda I)x = y, \quad F^0 x = y_0,$$

where $y_0 \in \ker D^0$ is determined by Formula (4.6.41). This implies the conclusions of our theorem (cf. also Section 4.1). ∎

It is possible to show that under some additional condition ill-posed problems of the considered type have also solutions.

THEOREM 4.6.7. *Suppose that X is a linear space over the reals, $D \in R(X)$, $\dim \ker D \neq 0$, F_0 and F_1 are initial operators for D corresponding to $R_0, R_1 \in \mathscr{R}_D$ respectively and that $R_0 \in V(X)$. If there is a scalar $\lambda \in \mathbf{R}$ such that*

$$F_1 s_\lambda(z) = 0 \quad \text{for } z \in \ker D, \quad z \neq 0, \tag{4.6.44}$$

where the operator s_λ is defined by Formula (2.3.9), then the boundary value problem for the operator $D^2 + \lambda^2 I$ with the initial operators F_0, F_1 is ill-posed, i.e. the homogeneous problem

$$(D^2 + \lambda^2 I)x = 0, \quad F_0 x = 0, \quad F_1 x = 0 \qquad (4.6.45)$$

has a non-trivial solution

$$x = s_\lambda(z), \quad \text{where } z \neq 0, \quad z \in \ker D. \qquad (4.6.46)$$

Proof. By our assumption dim ker $D = 0$. Suppose that the scalar λ satisfies Condition (4.6.44). Observe that a general solution of the equation $(D^2 + \lambda^2 I)x = 0$ is of the form:

$$x = c_\lambda(z_0) + s_\lambda(z_1), \quad \text{where } z_0, z_1 \in \ker D \text{ are arbitrary}$$
$$(4.6.47)$$

(cf. Example 3.2.4).

We are looking for z_0, z_1 such that the conditions $F_0 x = 0$, $F_1 x = 0$ are satisfied. Then

$$F_0 c_\lambda(z_0) = z_0, \quad F_0 s_\lambda(z_1) = 0. \qquad (4.6.48)$$

Indeed, since $F_0 R_0 = 0$, we have

$$F_0 s_\lambda(z_1) = \lambda F_0 R_0 (I + \lambda^2 R_0^2)^{-1} z_1 = 0.$$

Moreover, Formula (2.3.14) implies that

$$\begin{aligned}
F_0 c_\lambda(z_0) &= (I - R_0 D) c_\lambda(z_0) = c_\lambda(z_0) - R_0 D c_\lambda(z_0) \\
&= c_\lambda(z_0) - R_0(-\lambda) s_\lambda(z_0) \\
&= (I + \lambda^2 R_0^2)^{-1} z_0 + \lambda R_0 \lambda R_0 (I + \lambda^2 R_0^2)^{-1} z_0 \\
&= (I + \lambda^2 R_0^2)^{-1}(I + \lambda^2 R_0^2) z_0 = z_0.
\end{aligned}$$

Formulae (4.6.47), (4.6.48) together imply that

$$0 = F_0 x = F_0 c_\lambda(z_0) + F_0 s_\lambda(z_1) = z_0, \quad \text{hence } z_0 = 0.$$

Thus $x = s_\lambda(z_1)$. Condition (4.6.44) implies that for every $z_1 \in \ker D$ we have $F_1 x = F_1 s_\lambda(z_1) = 0$. Hence elements $s_\lambda(z_1)$ for $z_1 \in \ker D$, $z_1 \neq 0$, are non-trivial solutions of the problem (4.6.45) which was to be proved. ∎

Tasche, 1977a, also 1978, has considered Equation (4.6.1) with the following conditions:

$$G_k x = z_k, \quad z_k \in \ker D \quad (k = 1, ..., N), \qquad (4.6.49)$$

where the operators $G_1, ..., G_N$ transforming dom D^N onto ker D satisfy the following assumptions:

(i) $Q(D)x = 0$, $G_k x = 0$ $(k = 1, ..., N)$ implies $x = 0$;

(ii) For every $z \in \ker D$ and $k = 1, 2, ..., N$ there exists $x \in \ker Q(D)$ such that $G_j x = \delta_{jk} z$ $(j = 1, ..., N)$.

He solved boundary value problem of type (4.6.1)–(4.6.49) under assumptions that for given operators $Q(D)$, $G_1, ..., G_N$ there exists a *normal system*, i.e. operators $A_1, ..., A_N$, such that:

(i) $A_1 z_1 + ... + A_N z_N = 0$ for $z_1, ..., z_N \in \ker D$ implies $z_1 = ... = z_N = 0$;

(ii) for every $x \in \ker Q(D)$ there exists $z_1, ..., z_N \in \ker D$ such that $x = A_1 z_1 + ... + A_N z_N$;

(iii) $G_k A_j = \delta_{jk} I$ on $\ker D$ $(j, k = 1, ..., N)$.

Examples and Exercises

EXAMPLE 4.6.2. Consider the *equation of a harmonic oscillator*

$$x'' + \lambda^2 x = 0, \quad \text{where } 0 \neq \lambda \in \mathbf{R} \text{ is a parameter} \qquad (4.6.50)$$

with homogeneous boundary conditions:

$$x(a) = 0, \quad x(b) = 0 \quad (b \neq a). \qquad (4.6.51)$$

We shall show that this problem has infinitely many non-trivial solutions, i.e. is ill-posed. Indeed, general solution of Equation (4.6.50) is of the form $x(t) = c_1 \cos \lambda t + c_2 \sin \lambda t$, where c_1, c_2 are arbitrary constants. If we require Conditions (4.6.51) to be satisfied, we obtain the following system of equations:

$$c_1 \cos \lambda a + c_2 \sin \lambda a = 0, \quad c_1 \cos \lambda b + c_2 \sin \lambda b = 0. \qquad (4.6.52)$$

This system has a non-trivial solution (c_1, c_2) if and only if the determinant

$$\Delta = \begin{vmatrix} \cos \lambda a & \sin \lambda a \\ \cos \lambda b & \sin \lambda b \end{vmatrix} = \cos \lambda a \sin \lambda b - \cos \lambda b \sin \lambda a = \sin \lambda (b - a)$$

is equal to zero. But

$$\sin \lambda (b - a) = 0 \quad \text{if and only if} \quad \lambda (b - a) = k\pi$$

$(k = 0, \pm 1, \pm 2, ...)$.

Thus we have obtained infinitely many values of the parameter λ for which the determinant vanishes:

$$\lambda_k = \frac{\pi}{b-a} k \quad (k = 0, \pm 1, \pm 2, ...).$$

We therefore conclude that for $\lambda = \lambda_k$ the problem (4.6.50)–(4.6.51) is ill-posed. Since for $\lambda = \lambda_k$ we have from the first equation of (4.6.52)

$$c_2/c_1 = \cot \lambda_k a \quad (\text{if } \sin \lambda_k a \neq 0)$$

we put $c_1 = \alpha$ where $\alpha \in \mathbf{R}$ is arbitrary, $c_2 = \alpha \cot \lambda_k a$ and we obtain the corresponding non-trivial solution:

$$\begin{aligned}
x_k(t) &= -\alpha \cos \lambda_k t + \alpha \cot \lambda_k a \sin \lambda_k t \\
&= \alpha \frac{\sin \lambda_k a \cos \lambda_k t + \cos \lambda_k a \sin \lambda_k t}{\cot \lambda_k a} \\
&= \alpha \tan \lambda_k a \sin \lambda_k (t+a) \quad (k = 0, \pm 1, \pm 2, ...).
\end{aligned}$$

In a similar way we find solutions if $\cos \lambda_k a \neq 0$ but $\sin \lambda_k a = 0$.

EXAMPLE 4.6.3. Consider for Equation (4.6.50) FMBVP with conditions:

$$x(a) = 0, \quad x'(b) = 0 \quad (b \neq a). \tag{4.6.53}$$

Let us require that a solution of Equation (4.6.50) satisfies Conditions (4.6.53). Then we obtain the following system of equations:

$$c_1 \cos \lambda a + c_2 \sin \lambda a = 0,$$
$$-\lambda c_1 \sin \lambda b + \lambda c_2 \cos \lambda b = 0.$$

This system has a non-trivial solution if and only if the determinant

$$\begin{aligned}
\Delta &= \begin{vmatrix} \cos \lambda a & \sin \lambda a \\ -\lambda \sin \lambda b & \lambda \cos \lambda b \end{vmatrix} \\
&= \lambda \cos \lambda a \cos \lambda b + \lambda \sin \lambda a \sin \lambda b = \lambda \cos \lambda (b-a)
\end{aligned}$$

is equal zero. But, by our assumptions $\lambda \neq 0$ and $\cos \lambda (b-a) = 0$ if and only if $\lambda(b-a) = \pm \frac{\pi}{2} + k\pi$ $(k = 0, \pm 1, \pm 2, ...)$, i.e. if

$$\lambda = \lambda_k, \text{ where } \lambda_k = \frac{1}{2} \frac{\pi}{b-a} (2k \pm 1) \quad (k = 0, \pm 1, \pm 2, ...).$$

We therefore conclude that for $\lambda = \lambda_k$ the problem (4.6.50)–(4.6.53) is ill-posed. Non-trivial solutions corresponding to the values λ_k can be found in a similar way, as in the preceding example.

EXAMPLE 4.6.4. Consider the equation

$$\frac{d}{dt} A(t) \frac{d}{dt} x(t) + \lambda x(t) = y(t), \tag{4.6.54}$$

where $A, y \in C[a, b]$, $A(t) > 0$ for $a \leqslant t \leqslant b \neq a$, $\lambda \in \mathbf{R}$ is a parameter with the mixed boundary conditions

$$x(a) = x_a, \quad A(b)x'(b) = x_b, \quad \text{where } x_a, x_b \in \mathbf{R} \text{ are given.} \tag{4.6.55}$$

The problem (4.6.54)–(4.6.55) could be solved by application of Theorem 4.6.6. In particular, if $\lambda = 0$ then this problem is well-posed and its unique solution is

$$x(t) = \int\limits_a^t \frac{1}{A(s)} \left[\int\limits_b^s y(u)\,du \right] ds + x_a + x_b \int\limits_a^t \frac{ds}{A(s)}.$$

EXERCISE 4.6.1. Suppose that $X = C[a, b]$, $D = d/dt$. Give examples of operators P_{jk} satisfying assumptions of TMBVP (4.6.1)–(4.6.5) for $N = 2$.

EXERCISE 4.6.2. Determine values $t_0, t_1 \in [a, b]$ for which FMBVP $x'' - \lambda x = y$, $x(t_0) = 0$, $x'(t_1) = 1$ ($t_1 \neq t_0$), $y \in C[a, b]$, is well-posed.

EXERCISE 4.6.3. Is FMBVP

$$x^{(N)} = y, \quad x^{(k)}\left(\frac{k}{N}\right) = a_k, \quad a_k \in \mathbf{R} \quad (k = 0, 1, \dots, N-1)$$

well-posed? If yes, determine its solution.

EXERCISE 4.6.4. Applying the results of Example 4.6.1 solve the so-called *Nicoletti boundary value problem* (*multipoints boundary value problem*) for the equation $x^{(N)} = y$, i.e. determine solutions of this equation satisfying boundary conditions: $x(t_k) = a_k$ ($k = 0, 1, \dots, N-1$), where $a \leqslant t_0 < t_1 < \dots < t_{N-1} \leqslant b$; $y \in C[a, b]$.

EXERCISE 4.6.5. Applying the results of Example 4.6.1 solve SMBVP: $x^{(N)} = y$, $x(t_0) = a_0$, $x(t_1) = a_1$, $x^{(k)}(t_k) = a_k$ ($k = 2, ..., N-1$), where $N \geqslant 3$, $a_0, ..., a_{N-1}$, $t_0, ..., t_{N-1}$, y are as in Exercise 4.6.4.

EXERCISE 4.6.6. Solve SMBVP for the so-called *spline equation* $(ay'')'' + \lambda^2 y'' = 0$, $y(0) = y'(0) = y''(l) = y'''(0) = 0$, where $\lambda \in \mathbf{R}$ is a parameter, in the case:
   (i) the coefficient $a \neq 0$ is scalar,
   (ii) the coefficient $a \in C[0, 1]$, $a(t) > 0$ for $0 \leqslant t \leqslant l$.

EXERCISE 4.6.7. Suppose that $X = C[a, b]$, $D = d/dt$. Does the family $\{R_\gamma\}_{\gamma \in \Gamma_0}$ possess the property (c)? We recall that in this case $\Gamma_0 = [a, b]$,

$$(R_\gamma x)(t) = \int_\gamma^t x(s)ds \text{ for } x \in C[a, b] \text{ (cf. Example 2.1.1)}.$$

EXERCISE 4.6.8. Suppose that X, D, R are defined, as in Example 2.1.4. Does the operator R possess the property (c)?

EXERCISE 4.6.9. Give an example of a pair of initial operators which are not admissible.

EXERCISE 4.6.10. Give an example of a right invertible operator D such that there exists its right inverse which has no property (c).

EXERCISE 4.6.11. Suppose that X, D, $\{R_\gamma\}_{\gamma \in \Gamma_0}$ are defined as in Exercise 4.6.7. Apply Theorem 4.6.6 in the cases when:
   (i) $M = 2$, $A_1 = A_2 = I$, $\lambda = \mu^2$, $\mu \in \mathbf{R}$;
   (ii) $M = 2$, $A_1 = A_2 = I$, $\lambda = -\mu^2$, $\mu \in \mathbf{R}$.

Does the equality $I - \dfrac{1}{d_{\alpha\beta}} F_\alpha R_\beta = 0$ hold for some $\alpha, \beta \in \Gamma_0$?

If yes, boundary value problems with the initial operators F_α, F_β are well-posed. Give examples of values of the parameter λ for which the operator $I + \lambda R^0$ is not invertible.

EXERCISE 4.6.12. Solve BVP for the operator $(d^2/dt^2) + I$ with the following conditions:
   (i) $x(0) = 0$, $x\left(\dfrac{\pi}{4}\right) = 0$;

(ii) $x(0) = 0$, $x\left(\dfrac{\pi}{2}\right) = 0$;

(iii) $x(0) = 0$, $x(\pi) = 0$.

Which of these problems is well-posed? For ill-posed problems determine eigenspaces.

EXERCISE 4.6.13. Suppose that all assumptions of Theorem 4.6.2 are satisfied and the system $\{F_N, \dots, F_{N+M-1}\}$ of initial operators for the operator D is admissible. Is SMBVP for the operators $Q(D)D^M$ and $D^M Q(D)$ well-posed? If yes, determine solutions.

EXERCISE 4.6.14. Is the following FMBVP

$$x'' + \lambda^2 x = y, \quad \int\limits_a^b x(s)\,ds = y_0, \quad x'(c) = y_1,$$

where $y \in C[a, b]$, $a \leqslant c \leqslant b$, $y_0, y_1 \in \mathbf{R}$ are given, $\lambda \in \mathbf{R}$, well-posed? (Example 2.2.3). Determine solutions, if they exist.

EXERCISE 4.6.15. Is the following problem $x'' + \lambda^2 x = y$, $\int\limits_a^b x(s)\,ds = y_0$, $x(b) - x(a) = y$, where y, y_0, y_1, λ are as in Exercise 4.6.14, well-posed? Observe that $x(b) - x(a) = \int\limits_a^b x'(s)\,ds$ and apply Example 2.2.3.

EXERCISE 4.6.16. Solve FMBVP $\Delta^2 x = y$, $x|_s = y_0$, $\dfrac{dx}{dn_p}\bigg|_s = y_1$ in the space Y determined in Theorem 4.3.2 (cf. also Example 4.3.1).

EXERCISE 4.6.17. Let $D \in R(X)$. Suppose that the family $\{R_\gamma\}_{\gamma \in \Gamma_0} \subset \{R_\gamma\}_{\gamma \in \Gamma} = \mathcal{R}_D$ has the property (c). Is a system of initial operators $\{F_\gamma\}_{\gamma \in \Gamma_0}$ for every $\Gamma_0 = \{\gamma_1, \dots, \gamma_N\} \subset \Gamma$ admissible?

EXERCISE 4.6.18. Generalize Proposition 4.6.1 for an arbitrary system of $n > 2$ right inverses having the property (c).

Chapter 5

Periodic operators and elements. Shift operators. Shift invariant spaces

5.1. PERIODIC OPERATORS AND ELEMENTS

Suppose that X is a linear space over C and that $D \in R(X)$. An operator $S \in L_0(X)$ is said to be *D-invariant* if

$$SD = DS \quad \text{on dom } D. \tag{5.1.1}$$

An operator $A \in L_0(X)$ is said to be *S-periodic* if there exists a *D*-invariant operator S such that

$$AS = SA. \tag{5.1.2}$$

PROPERTY 5.1.1. *Suppose that $D \in R(X)$ and S is D-invariant. Then for an arbitrary positive integer m:*
   (i) *the operator S^m is D-invariant;*
   (ii) *if A is S-periodic then A is S^m-periodic.*

Indeed, by an easy induction we show that Equality (5.1.1) implies

$$S^m D = D S^m \quad \text{on dom } D \quad (m = 1, 2, \ldots) \tag{5.1.1'}$$

and that Equality (5.1.2) implies

$$A S^m = S^m A \quad (m = 1, 2, \ldots). \tag{5.1.2'}$$

Suppose we are given an arbitrarily fixed positive integer N and a *D*-invariant operator S. Write

$$X_{S,N} = \{x \in X : S^N x = x\}. \tag{5.1.3}$$

If $\dim X_{S,N} > 0$ then every element $x \in X_{S,N}$ will be called an S^N-*periodic element*. The definition of the space $X_{S,N}$ implies that the operator S is an involution of order N on the space $X_{S,N}$.*

---

* cf. Muhamadev, 1967, who defined a *p*-periodic operator acting in a Banach space E as follows: a bounded linear operator U acting in E is said to be *p-periodic* if $U^p = I$ and $U^i x = x$ for $0 < i < p$ implies $x = 0$.

Thus we have (cf. Section 1.4)

$$X_{S,N} = \bigoplus_{j=1}^{N} X_j, \quad \text{where } X_j = P_j X_{S,N} \quad \text{and}$$

$$P_j = \frac{1}{N} \sum_{k=0}^{N-1} \varepsilon^{-kj} S^k \quad (j = 1, ..., N), \quad \varepsilon = e^{2\pi i/N}. \quad (5.1.4)$$

PROPOSITION 5.1.1. *Suppose that S is a D-invariant operator and* $\dim X_{S,N}$ > 0 *for* $N > 1$. *If* $x_j \in X_j$ $(j = 1, 2, ..., N)$ *then* $Dx_j \in X_j$, *where the spaces* X_j *are defined by Formulae* (5.1.4).

Proof. Indeed, $x_j \in X_j$ if and only if $Sx_j = \varepsilon^j x_j$ (cf. Section 1.4). Thus $SDx_j = DSx_j = D(\varepsilon^j x_j) = \varepsilon^j Dx_j$ by Condition (5.1.1), which implies $Dx_j \in X_j$ $(j = 1, ..., N)$. ■

Proposition 5.1.1 implies that each of spaces $X_1, ..., X_N$ is invariant under the operator D. Hence the operator D is invariant with respect to the decomposition (5.1.4).

PROPOSITION 5.1.2. *Suppose that all assumptions of Proposition 5.1.1 are satisfied. Then* $z \in \ker D$ *implies* $Sz \in \ker D$ *and* $S^N z \in \ker D$ *for all* $N \in \mathbf{N}$.
  In other words: *a D-invariant operator preserves the space of constants.*

Proof. Indeed, if $z \in \ker D$ then $Dz = 0$. Hence Condition (5.1.1) implies that $DSz = SDz = 0$. Thus $Sz \in \ker D$. By an easy induction we prove that $S^N z \in \ker D$ for all $N \in \mathbf{N}$. ■

THEOREM 5.1.1. *Suppose that* $D \in R(X)$, $\dim \ker D \neq 0$ *and* $R \in \mathscr{R}_D \cap$ $\cap V(X)$. *If an operator* $S \in L_0(X)$ *is D-invariant then for every* $\lambda \in \mathbf{R}$, $z \in \ker D$ *there exists* $z_1 \in \ker D$ *such that*

$$Se_\lambda(z) = e_\lambda(z_1), \quad \text{where } e_\lambda = (I - \lambda R)^{-1}, \quad (5.1.5)$$

i.e. the operator S preserves the eigenspaces of the operator D. One can say also that S preserves all exponentials up to a constant.

Proof. By definition $e_\lambda = (I - \lambda R)^{-1}$. Thus

$$(D - \lambda I)e_\lambda = (D - \lambda I)(I - \lambda R)^{-1} = D(I - \lambda R)(I - \lambda R)^{-1} = D.$$

Hence $De_\lambda = \lambda e_\lambda + D$ on dom D. Condition (5.1.1) implies that DSe_λ $= SDe_\lambda = \lambda Se_\lambda + SD = \lambda Se_\lambda + DS$ for all scalars λ. Proposition 5.1.2 implies that for all $\lambda \in \mathbf{C}$ and for arbitrary $z \in \ker D$ we have $DSe_\lambda(z) = \lambda Se_\lambda(z) + DSz = \lambda Se_\lambda(z) + Dz = \lambda Se_\lambda(z)$. We therefore con-clude that the element $Se_\lambda(z)$ is also an eigenvector of D corresponding to the eigenvalue λ. Thus there exists a constant z_1 such that $Se_\lambda(z)$ $= e_\lambda(z_1)$, which was to be proved. ∎

COROLLARY 5.1.1. *Suppose that all assumptions of Theorem 5.1.1 are satisfied and F is an initial operator for D corresponding to R. Then the constant z_1 in Formula (5.1.5) is defined by means of the equality*

$$z_1 = FSe_\lambda(z).\tag{5.1.6}$$

THEOREM 5.1.2. *Suppose that $D \in R(X)$, $S \in L_0(X)$ is a D-invariant operator, the operators $Q_{km} \in L_0(X)$ are S-periodic $(m = 0, 1, \dots, M;$ $k = 0, 1, \dots, N-1)$. Write*

$$Q_m(S) = \sum_{k=0}^{N-1} Q_{km} S^k, \quad Q(D, S) = \sum_{m=0}^{M} Q_m(S) D^{m+M_1} \, (M_1 \geq 0).$$
$$\tag{5.1.7}$$

If $\dim X_{S,N} > 0$ for an $N > 1$ then

$$Q(D, S) = \sum_{j=1}^{N} Q(D, \varepsilon^j) P_j \quad \text{on } X_{S,N},\tag{5.1.8}$$

where the operators P_j are defined by decomposition (5.1.4).

Proof. Since the operators D and P_j $(j = 1, 2, \dots, N)$ commute, the decomposition (5.1.4) implies that for $j = 1, 2, \dots, N$

$$Q(D, S)P_j = \sum_{m=0}^{M} Q_m(S) D^{m+M_1} P_j = \sum_{m=0}^{M} D^{m+M_1} Q_m(S) P_j$$

$$= \sum_{m=0}^{M} D^{m+M_1} \sum_{k=0}^{N-1} Q_{km} S^k P_j = \sum_{m=0}^{M} D^{m+M_1} \sum_{k=0}^{N-1} \varepsilon^{jk} Q_{km} P_j$$

$$= \sum_{m=0}^{M} D^{m+M_1} Q_m(\varepsilon^j) P_j = \sum_{m=0}^{M} Q_m(\varepsilon^j) D^{m+M_1} P_j = Q(D, \varepsilon^j) P_j.$$

Since $\sum_{j=1}^{N} P_j = I$, we finally obtain

$$Q(D, S) = Q(D, S) \sum_{j=1}^{N} P_j = \sum_{j=1}^{N} Q(D, S) P_j = \sum_{j=1}^{N} Q(D, \varepsilon^j) P_j.$$

∎

COROLLARY 5.1.2. *Suppose that all assumptions of Theorem 5.1.2 are satisfied. Then the equation*

$$Q(D, S)x = y, \quad y \in X_{S,N} \tag{5.1.9}$$

is equivalent in the space $X_{S,N}$ to N independent equations

$$Q(D, \varepsilon^j)x_j = y_j, \quad \text{where} \quad x_j = P_j x, \quad y_j = P_j y \in X_j$$
$$(j = 1, ..., N) \tag{5.1.10}$$

and the operators P_j and spaces X_j are defined by decomposition (5.1.4).

Proof. The definition of the operators P_j and elements x_j, y_j $(j = 1, ..., N)$ implies that $x = x_1 + ... + x_N$, $y = y_1 + ... + y_N$ and that this decomposition is uniquely determined. Since the operators Q_{km} are S-periodic, we have $SQ_{km}x_j = Q_{km}Sx_j = \varepsilon^j Q_{km}x_j$ for $j = 1, ..., N$, $k = 0, 1, ..., N-1$, $m = 0, 1, ..., M$. Thus $Q_{km}X_j \subset X_j$. This implies that the operators $Q(D, \varepsilon^j)$ preserve all spaces X_j. Indeed, if $x_j \in X_j$ then $Q(D, \varepsilon^j)x_j = \sum_{m=0}^{M} \sum_{k=0}^{N-1} Q_{km}D^{m+M_1}x_j \in X_j$ $(j = 1, ..., N)$ by Proposition 5.1.1. Hence Theorem 5.1.1 implies that

$$Q(D, S)x = Q(D, S) \sum_{j=1}^{N} P_j x = \sum_{j=1}^{N} Q(D, S) P_j x$$

$$= \sum_{j=1}^{N} Q(D, \varepsilon^j) P_j x = \sum_{j=1}^{N} Q(D, \varepsilon^j) x_j.$$

Since $y = \sum_{j=1}^{N} P_j y = \sum_{j=1}^{N} y_j$ and since the space $X_{S,N}$ is a direct sum of spaces X_j $(j = 1, ..., N)$ and, moreover, $x_j, y_j \in X_j$ $(j = 1, ..., N)$, we conclude that the equation (5.1.9) is equivalent to the system (5.1.10).

∎

COROLLARY 5.1.3. *Suppose that all assumptions of Theorem 5.1.2 are satisfied. If each of equations (5.1.10) has a solution* $x_j \in X_j$ *(j = 1, ..., N) then the equation (5.1.9) has a solution* $x = x_1 + ... + x_N$ *which is* S^N-*periodic. Conversely if the equation (5.1.9) has an* S^N-*periodic solution* x, *then j-th equation (5.1.10) has a solution* $x_j = P_j x \in X_j$, *where the operator* P_j *is defined by decomposition (5.1.4) (j = 1, ..., N).*

Proof. If Equation (5.1.9) has an S^N-periodic solution x then by definition $x \in X_{S,N}$. Hence $S^N x = x$ and $x = x_1 + ... + x_N$, where $x_j = P_j x \in X_j$ $(j = 1, ..., N)$. Corollary 5.1.2 implies that the equation (5.1.9) is equivalent to the system (5.1.10). Thus the j-th equation (5.1.10) has a solution $x_j \in X_j$. Conversely, suppose that the j-th equation (5.1.10) has a solution $x_j \in X_j$ $(j = 1, ..., N)$. Since $P_j^2 = P_j$, $P_j \neq P_k$ for $j \neq k$ and $x_j = P_j x$ for an $x \in X$ $(j, k = 1, ..., N)$, writting $x = x_1 + $

$$+ ... + x_N \text{ we find } P_j x = P_j \sum_{k=1}^{N} x_k = \sum_{k=1}^{N} P_j P_k x = P_j x = x_j (j = 1, ...$$

$..., N)$. Corollary 5.1.2 implies that $x \in X_{S,N}$ and that x is a solution of Equation (5.1.9). ∎

COROLLARY 5.1.4. *Suppose that all assumptions of Theorem 5.1.2 are satisfied and that* $Q_M(S) = I$. *Write*

$$Q_j^0(R) = \sum_{m=0}^{M-1} Q_m(\varepsilon^j) R^{M_1-m} \quad \text{for } R \in \mathcal{R}_D \quad (j = 1, ..., N).$$

$$(5.1.11)$$

If there exists an operator $R \in \mathcal{R}_D$ *such that* $[I + Q_j^0(R)]X_j \subset X_j$ *and the operators* $I + Q_j^0(R)$ *are invertible (j = 1, ..., N) then the general solution of Equation (5.1.9) is of the form:*

$$x = \sum_{j=1}^{N} x_j \in X_{S,N},$$

$$(5.1.12)$$

where

$$x_j = R^{M+M_1}[I + Q_j^0(R)]^{-1} y_{M+M_1,j} + \sum_{k=0}^{M+M_1-1} R^k z_{jk},$$

$$(5.1.12')$$

$$y_{M+M_1, j} = y_j - \sum_{m=0}^{M-1} Q_m(\varepsilon^j) \sum_{k=m}^{M-1} R^{k-m} z_{jm}, \quad z_{jk} \in \ker D$$

$(j = 1, ..., N, k = 0, 1, ..., M+M_1-1)$.

Proof. The proof is an immediate consequence of the fact that the equation (5.1.9) is equivalent to the system (5.1.10) and that, by our assumptions, the j-th equation (5.1.10) has a solution $x_j \in X_j$. ∎

COROLLARY 5.1.5. *Suppose that all assumptions of Corollary 5.1.4 are satisfied and that the operator D is invertible on the space $X_{S, N}$. Then the Equation (5.1.9) has a unique solution*

$$x = \sum_{j=1}^{N} x_j, \quad \text{where } x_j = R^{M+M_1-1}[I+Q_j^0(R)]^{-1} y_j \qquad (5.1.13)$$

$(j = 1, ..., N)$, $R = D^{-1}$.

Proof. The proof follows immediately from Corollary 5.1.4 and from the fact that in our case $\ker D = \{0\}$. Hence all constants z_{jk} appearing in Formulae (5.1.12') vanish. ∎

In a similar way as Theorem 5.1.2 and Corollaries 5.1.1–5.1.5 we can consider the operator

$$Q(D, S) = D^{M_1} \sum_{m=0}^{M} Q_m(S) D^m. \qquad (5.1.14)$$

In the case, when the coefficients Q_{km} of the polynomials $Q_m(S)$ commute with D and S (for instance, are scalars) one can apply the above results directly.

Examples and Exercises

EXAMPLE 5.1.1. Kahane, 1969, has proved that every d/dt-invariant linear operator mapping each of the spaces $C[a, b]$ and $C^1[a, b]$ into itself (and bounded, cf. p. 480) is a scalar multiple of the identity.

EXAMPLE 5.1.2. A function $x \in C(\mathbf{R})$ is said to be *p-periodic* if $x(t+p) = x(t)$ for all $t \in \mathbf{R}$. Write $(Sx)(t) = x(t+p)$ for $x \in C(\mathbf{R})$, $t \in \mathbf{R}$. The translation S is a $\dfrac{d}{dt}$ -invariant operator, since we have $S\dfrac{d}{dt} = \dfrac{d}{dt} S$

on $C^1(\mathbf{R}) = \operatorname{dom} \dfrac{d}{dt}$. It is easy to verify that the operator A of multiplication by a p-periodic function $a(t)$ is an S-periodic operator. Indeed, for all $x \in C(\mathbf{R})$, $t \in \mathbf{R}$, we find $(ASx - SAx)(t) = a(t)x(t+p) - (ax)(t+p)$ $= a(t)x(t) - a(t+p)x(t+p) = a(t)x(t) - a(t)x(t) = 0$.

EXERCISE 5.1.1. Prove that the operator S defined as $(Sx)(t) = x(rt)$, where $x \in C[a, T]$, $a > 0$, $r > 0$, $t \in [a, T]$, is $t\dfrac{d}{dt}$ -invariant.

EXERCISE 5.1.2. Suppose that all assumptions of Theorem 5.1.2 are satisfied and the operators $Q_j^0(R)$ are defined by Formulae (5.1.11). Generalize Colorrary 5.1.4 for the case when:
   (i) the operators are either left or right invertible,
   (ii) $Q(D, S)$ is defined by Formula (5.1.14).

EXERCISE 5.1.3. Suppose that $X = (s)$ and:
   (i) $D\{x_n\} = \{x_{n+1}\}$,
   (ii) $D\{x_n\} = \{x_{n+1} - x_n\}$
for $x = \{x_n\} \in X$. Give in both cases examples of D-invariant operators and periodic elements.

5.2. R-SHIFTS AND D-SHIFTS

In the sequel we shall write $A(\mathbf{R})$ instead of one of the sets: $\mathbf{R}, \mathbf{R}^+$ $= \{r \in \mathbf{R}: r \geqslant 0\}$.

DEFINITION 5.2.1. Suppose that $D \in R(X)$, $\dim \ker D \neq 0$, and F is an initial operator for D corresponding to an $R \in \mathcal{R}_D$. Define a family $S_{A(\mathbf{R})} = \{S_h\}_{h \in A(\mathbf{R})} \subset L_0(X)$ of linear operators in the following way:
$$S_0 = I, \tag{5.2.1}$$

$$S_h R^k F = \sum_{j=0}^{k} \frac{h^{k-j}}{(k-j)!} R^j F \quad \text{for all } h \in A(\mathbf{R}), \tag{5.2.2}$$

$k \in \{0\} \cup \mathbf{N}$.

We say that the family $S_{A(\mathbf{R})}$ *determines R-shifts for D and that S_h* is an *R-shift* on *h* for an arbitrary $h \in A(\mathbf{R})$.

In particular, for $k = 0$ we have

$$S_h F = F \quad \text{for all } h \in A(\mathbf{R}). \tag{5.2.3}$$

PROPOSITION 5.2.1. *Suppose that $D \in R(X)$, dim ker $D \neq 0$ and F is an initial operator for D corresponding to an $R \in \mathcal{R}_D$. Suppose, we are given a family $S_{A(\mathbf{R})} = \{S_h\}_{h \in A(\mathbf{R})}$ of R-shifts. Then for all $h \in A(\mathbf{R})$, $z \in$ ker D*

$$S_h R^k z = \sum_{j=0}^{k} \frac{h^{k-j}}{(k-j)!} R^j z \quad (k = 0, 1, 2, \ldots). \tag{5.2.4}$$

In particular

$$S_h z = z \quad \text{for all } z \in \ker D, \quad h \in A(\mathbf{R}), \tag{5.2.5}$$

i.e. *R-shifts preserve constants.*

Proof. Let $z \in$ ker D be arbitrarily fixed. Then there exists an $x \in X$ such that $Fx = z$, for F is a mapping onto. Thus we have for all $h \in A(\mathbf{R})$

$$S_h R^k z = S_h R^k F x = \sum_{j=0}^{k} \frac{h^{k-j}}{(k-j)!} R^j F x$$

$$= \sum_{j=0}^{k} \frac{h^{k-j}}{(k-j)!} R^j z.$$

Putting $k = 0$ we obtain Formula (5.2.5). ∎

THEOREM 5.2.1. *Suppose that $D \in R(X)$, dim ker $D \neq 0$ and F is an initial operator for D corresponding to an $R \in \mathcal{R}_D$. Suppose that we are given a family $S_{A(\mathbf{R})} = \{S_h\}_{h \in A(\mathbf{R})}$ of R-shifts. Then*

(i) *If $A(\mathbf{R}) = \mathbf{R}^+$ then $S_{A(\mathbf{R})}$ is a commutative semigroup,*

　　if $A(\mathbf{R}) = \mathbf{R}$ then $S_{A(\mathbf{R})}$ is an Abelian group
with respect to the superposition of operators, i.e.

$$S_{h_1} S_{h_2} = S_{h_2} S_{h_1} = S_{h_1 + h_2} \quad \text{for all } h_1, h_2 \in A(\mathbf{R});$$

(ii) *For all $h \in A(\mathbf{R})$ the operators S_h are D-invariant and uniquely determined on the set $P(R)$ of D-polynomials.*

Proof. (i) Suppose that $A(\mathbf{R}) = \mathbf{R}^+$. Then for arbitrary $h_1, h_2 \in \mathbf{R}^+$ Definition 5.2.1 implies that

$$S_{h_1+h_2} R^k F = \sum_{j=0}^{k} \frac{1}{(k-j)!} (h_1+h_2)^{k-j} R^j F$$

$$= \sum_{m=0}^{k} \frac{1}{m!} (h_1+h_2)^m R^{k-m} F$$

$$= \sum_{m=0}^{k} \frac{1}{m!} \sum_{j=0}^{m} \binom{m}{j} h_1^{m-j} h_2^j R^{k-m} F$$

$$= \sum_{j=0}^{k} \left(\sum_{m=j}^{k} \frac{1}{(m-j)!} h_1^{m-j} R^{k-m} z \right) \frac{1}{j!} h_2^j$$

$$= \sum_{j=0}^{k} \left(\sum_{n=0}^{k-j} \frac{1}{n!} h_1^n R^{k-n-j} F \right) \frac{1}{j!} h_2^j$$

$$= \sum_{j=0}^{k} \frac{1}{j!} h_2^j S_{h_1} R^{k-j} F = S_{h_1} \sum_{j=0}^{k} \frac{1}{j!} h_2^j R^{k-j} F$$

$$= S_{h_1} S_{h_2} R^k F.$$

Since we have

$$S_{h_1} S_{h_2} R^k F = S_{h_2+h_1} R^k F = S_{h_1+h_2} R^k F = S_{h_1} S_{h_2} R^k F,$$

we conclude that $S_{\mathbf{R}^+}$ is a commutative semigroup.

By the same arguments we conclude that $S_{\mathbf{R}}$ is an Abelian group.

(ii) Suppose that $0 \neq h \in A(\mathbf{R})$, $k \in \mathbf{N}$, $z \in \ker D$ are arbitrary fixed. Since $Dz = 0$, Proposition 5.2.1 implies that

$$(DS_h - S_h D) R^k z = DS_h R^k z - S_h DR^k z$$

$$= D \sum_{j=0}^{k} \frac{h^{k-j}}{(k-j)!} R^j z - S_h R^{k-1} z$$

$$= \frac{1}{k!} Dz + \sum_{j=1}^{k} \frac{h^{k-j}}{(k-j)!} R^{j-1} z - \sum_{m=0}^{k-1} \frac{h^{k-1-m}}{(k-1-m)!} R^m z$$

$$= \sum_{m=0}^{k-1} \left[\frac{h^{k-1-m}}{(k-1-m)!} R^m z - \frac{h^{k-1-m}}{(k-1-m)!} R^m z \right] = 0.$$

If $k = 0$, we have $(DS_h - S_h D)z = DS_h z - S_h Dz = Dz = 0$. This implies that S_h for $h \in A(\mathbf{R})$, are *D*-invariant on the set $P(R)$. Definition 5.2.1 and Proposition 5.2.1 together imply that S_h are uniquely determined on the set $P(R)$. ∎

COROLLARY 5.2.1. *Suppose that all assumptions of Theorem 5.2.1 are satisfied. Then*

$$S_h^n = S_{nh} \quad \text{for all } h \in A(\mathbf{R}), \quad n \in \mathbf{N} \cup \{0\}. \tag{5.2.6}$$

This is an immediate consequence of Point (i) in Theorem 5.2.1. ∎

PROPOSITION 5.2.2. *Suppose that $D \in R(X)$, dim ker $D \neq 0$, F is an initial operator for D corresponding to $R \in \mathcal{R}_D$ and $S_{A(\mathbf{R})} = \{S_h\}_{h \in A(\mathbf{R})}$ is a family of R-shifts. Then*

$$FS_h RFx = hFx \quad \text{for all } x \in X, \quad h \in A(\mathbf{R}). \tag{5.2.7}$$

Proof. By definition, $F^2 = F$ and $FR = 0$. Suppose that $x \in X$ and $h \in A(\mathbf{R})$ are arbitrarily fixed. Let $z = Fx$. Then $z \in \ker D$. Proposition 5.2.1 and Definition 5.2.1 together imply that $S_h RFx = S_h Rz = Rz + hz$. Hence $FS_h RFx = F(RFx + hFx) = FRFx + hF^2 x = hFx$. ∎

THEOREM 5.2.2. *Suppose that $D \in R(X)$, F is an initial operator for D corresponding to an $R \in \mathcal{R}_D$ and $S_{A(\mathbf{R})} = \{S_h\}_{h \in A(\mathbf{R})}$ is a family of D-invariant R-shifts. Write*

$$F_n = FS_h^n, \quad D_n = D + \frac{1}{h} F_n \quad \text{for all } n \in \mathbf{N}, \quad h \in A(\mathbf{R}). \tag{5.2.8}$$

Then $D_n \in R(X)$ and F_n is an initial operator for D and D_n, corresponding to the right inverse

$$R_n = R - F_n R = R - FS_h^n R \quad (n = 0, 1, 2, \ldots). \tag{5.2.9}$$

Moreover, if we consider the space of S_h^n-periodic elements

$$X_{S_h, n} = \{x \in X: S_h^n x = x\}, \quad \dim X_{S_h, n} > 0 \quad (n \in \mathbf{N}),$$

then by Formula (5.2.6) we have $S_h^n = S_{nh}$, so that

$$X_{S_h, n} = X_{S_{nh}} = \{x \in X: S_h^n x = x\} \tag{5.2.10}$$

and

$$F_n x = Fx \quad \text{for all } x \in X_{S_{nh}} \quad (n \in \mathbf{N}). \tag{5.2.11}$$

Proof. Let $n \in \mathbf{N}$ and $h \in A(\mathbf{R})$ be arbitrarily fixed. Then F_n is a projection. Indeed, Formula (5.2.3) implies that $F_n^2 = FS_h^n FS_h^n = F^2 S_h^n = FS_h^n = F_n$. Moreover, F_n is a projection onto the kernel D. Indeed, our assumptions and Formula (5.2.5) together imply that $F_n z = FS_h^n z = FS_{nh} z = Fz = z$ for all $z \in \ker D$. Therefore F_n is an initial operator for D corresponding to the right inverse $R_n = R - F_n R = R - FS_{nh} R$ (Theorem 2.2.4). The operator $D_n = D + F_n/h$ is also right invertible, because we have $F_n R_n = 0$ (by definition) and $DF_n = DFS_{nh} = 0$, which implies $D_n R_n = (D + F_n/h) R_n = DR_n + F_n/h R_n = DR_n = D(R - F_n R) = DR - DF_n R = DR = I$. Thus $R_n \in \mathcal{R}_{D_n}$ and F_n is an initial operator for D_n corresponding to the right inverse R_n.

Suppose now that $x \in X_{S_{nh}}$, i.e. $S_h^n x = x$. Then $F_n x = FS_h^n x = Fx$, i.e. Equality (5.2.11) holds for $n \in \mathbf{N}$.

By definition of the operators D_n, F_n, R_n, we have the following identities:

$$D_n R_n = I \text{ on } X_{S_{nh}}, \quad R_n D_n x = I - Fx \text{ on } X_{S_{nh}} \cap \operatorname{dom} D. \tag{5.2.12}$$

∎

THEOREM 5.2.3. *Suppose that all assumptions of Theorem 5.2.2 are satisfied and that $D_n, F_n, R_n, X_{S_{nh}}$ are defined by Formulae (5.2.8)–(5.2.10) respectively. Write*

$$X_{S_{nh}}^{(m)} = X_{S_{nh}} \cap \operatorname{dom} D^m, \quad \hat{X}_{S_{nh}} = \{x \in X_{S_{nh}}: F_n Rx = 0\}$$
$$(n, m = 1, 2, \ldots). \tag{5.2.13}$$

Then the operator D_n maps $X_{S_{nh}}^{(1)}$ into $X_{S_{nh}}$, however the operator R_n maps $\hat{X}_{S_{nh}}$ onto $X_{S_{nh}}^{(1)}$ and the following equalities hold

$$D_n x = Dx - \frac{1}{h} Fx; \quad D_n R_n = I, \quad R_n D_n x = (I - F)x$$
$$\text{for } x \in X_{S_{nh}}^{(1)} \quad (n = 1, 2, \ldots). \tag{5.2.14}$$

Proof. Suppose that $n \in \mathbb{N}$, $h \in A(\mathbf{R})$ and $x \in X_{S_{nh}}^{(1)}$ are arbitrarily fixed. Then $x \in \operatorname{dom} D$, $S_h^n x = x$ and $y = D_n x = Dx + F_n x/h = Dx + FS_h^n x = Dx + Fx/h$. Since, by our assumptions, $S_h D = DS_h$ on the domain of D, we have $S_h^n D = DS_h^n$ on $\operatorname{dom} D$. This and Formula (5.2.3) together imply that

$$S_h^n y = S_h^n \left(Dx + \frac{1}{h} Fx \right) = DS^n x + \frac{1}{h} S_h^n Fx = Dx + \frac{1}{h} Fx = y,$$

i.e. $y \in X_{S_{nh}}$. The arbitrariness of $x \in X_{S_{nh}}^{(1)}$ implies that $D_n X_{S_{nh}}^{(1)} \subset X_{S_{nh}}$.

Suppose now that $x \in \hat{X}_{S_{nh}}$ is arbitrarily fixed. Then $S_h^n x = x$ and $FS_h Rx = F_n Rx = 0$. Put $y = R_n x = Rx - FS_h Rx = Rx$. Since $D_n R_n = I$, we have $x = D_n R_n x = D_n y$. Thus $y \in \operatorname{dom} D$. Furthermore, $Fy = FR_n x = FRx - RF_n Rx = FRx = 0$ and $S_h^n y = S_h^n Rx$, i.e. $F_n y = FS_h^n y = FS_h^n Rx = F_n Rx$. Then $(I - F_n)y = y - F_n y = Rx - F_n Rx = R_n x = y$, which implies that $F_n y = 0$. But $S_h^n x = S_h^n Dy = DS_h^n y$. Then $S_h^n y = Rx + FS_h^n y = Rx + F_n y = Rx = y$, which implies that $y \in X_{S_{nh}}^{(1)} = X_{nh} \cap \operatorname{dom} D$. The arbitrariness of $x \in \hat{X}_{S_{nh}}$ implies that $R_n \hat{X}_{nh} \subset X_{S_{nh}}^{(1)}$. We have obtained also $F_n x = FS_h^n x = Fx$ and $D_n x = Dx + F_n x/h = Dx + Fx/h$ and $R_n D_n x = (I - F_n)x = (I - F)x$ for $x \in X_{S_{nh}}^{(1)}$. ∎

EXAMPLE 5.2.1. Suppose that $X = C[0, T]$, $D = \mathrm{d}/\mathrm{d}t$, $(Rx)(t) = \int_0^t x(s)\,\mathrm{d}s$, $(Fx)(t) = x(0)$, $(Sx)(t) = x(t + \omega)$ for $x \in X$. Consider the space of all ω-periodic functions, i.e. the space X_{S_h} where $h = \omega$, for $(Sx)(t) = x(t + \omega) = x(t)$. In a similar way, as it was shown in Corollary 5.2.1, we conclude that S_h is an R-shift on $h = \omega$. It is easy to check that S_h is also a D-invariant R-shift. Examine the condition $FS_n R = 0$ which appears in Theorem 5.2.3. We have

$$(FS_h Rx)(t) = F\left[S_h \int_0^t x(s)\,\mathrm{d}s \right] = F\left[\int_0^{t+\omega} x(s)\,\mathrm{d}s \right]$$

$$= \left[\int_0^{t+\omega} x(s)\,\mathrm{d}s \right]_{t=0} = \int_0^\omega x(s)\,\mathrm{d}s.$$

We conclude that the condition $FS_h Rx = 0$ in our case means that a definite integral on the period should vanish for some $x \in X_{S_h}$.

Observe that, by Formula (5.2.6), it is enough to consider in our further considerations only the case $n = 1$.

PROPOSITION 5.2.3. *Suppose that all assumptions of Theorem 5.2.2 are satisfied. Suppose that $z \neq 0$ is an arbitrary constant. If $h \neq 0$ then $FS_h Rz \neq 0$.*

Proof. Indeed, if S_h is an R-shift on h then $FS_h Rz = F(Rz + hz) = FRz + + hFz = hz \neq 0$. ∎

Example 5.2.1 and Proposition 5.2.3 together show that the right inverses R_n introduced by means of Equality (5.2.14) are not the best ones for periodic problems. However, we have the following

THEOREM 5.2.4. *Suppose that $D \in R(X)$, dim ker $D \neq 0$, F is an initial operator for D corresponding to an $R \in \mathcal{R}_D$, and $S_{A(R)} = \{S_h\}_{h \in A(R)}$ is a family of R-shifts. Suppose that operators D_1 and F_1 are defined by Formulae (5.2.8). Write*

$$R_1^0 = R + F_1 R - \frac{1}{h} RF_1 R = \left(I + FS_h - \frac{1}{h} RFS_h\right)R$$

$$\text{for } h \in A(\mathbf{R}). \quad (5.2.15)$$

Then the operator R_1^0 maps the space X_{S_h} onto the space $X_{S_h}^{(1)}$ and

$$D_1 R_1^0 = I \text{ on } X_{S_h}, \quad R_1^0 D_1 = I \text{ on } X_{S_h}^{(1)}, \quad (5.2.16)$$

i.e. the operator D_1 is invertible and its right inverse is $D_1^{-1} = R_1^0$. Moreover, an initial operator F_1^0 corresponding to R_1^0 vanishes on the space X_{S_h}.

Proof. Suppose that $h \in A(\mathbf{R})$ and $x \in X_{S_h}$ are arbitrarily fixed. Then $S_h x = x$, $F_1 x = FS_h x = x$. Write $z = F_1 Rx = FS_h Rx \in \ker D$. Since $DR = I$, $FR = 0$, we find

$$D_1 R_1^0 x = \left(D + \frac{1}{h} F_1\right)\left(Rx + F_1 R - \frac{1}{h} RF_1 R\right)x$$

$$= \left(D + \frac{1}{h} F_1\right)\left(Rx + F_1 Rx - \frac{1}{h} RF_1 R\right)x$$

$$= \left(D + \frac{1}{h} F_1\right)\left(Rx + z - \frac{1}{h} Rz\right)$$

$$= DRx + Dz - \frac{1}{h} DRz + \frac{1}{h} F_1 Rx + \frac{1}{h} F_1 z - \frac{1}{h^2} F_1 Rz$$

$$= x - \frac{1}{h} z + \frac{1}{h} z + \frac{1}{h} FS_h z - \frac{1}{h^2} FS_h Rz$$

$$= x + \frac{1}{h} Fz - \frac{1}{h} F(Rz + hz)$$

$$= x + \frac{1}{h} z - \frac{1}{h} FRz - \frac{1}{h} Fz = x + \frac{1}{h} z - \frac{1}{h} z = x.$$

The arbitrariness of $x \in X_{S_h}$ implies the first of Identities (5.2.16). Suppose now that $x \in X_{S_h}^{(1)}$ is arbitrarily fixed. Then $x \in \text{dom } D$ and $S_h x = x$, $F_1 x = FS_h x = Fx$. By Proposition 5.2.2 $FS_h RFx = hFx$ and

$$R_1^0 D_1 x = \left(R + F_1 R - \frac{1}{h} RF_1 R\right)\left(D + \frac{1}{h} F_1\right) x$$

$$= \left(I + FS_h - \frac{1}{h} FRS_h\right) R \left(D + \frac{1}{h} FS_h\right) x$$

$$= \left(I + FS_h - \frac{1}{h} RFS_h\right)\left(RDx + \frac{1}{h} RFS_h x\right)$$

$$= \left(I + FS_h - \frac{1}{h} RFS_h\right)\left(x - Fx + \frac{1}{h} RFx\right)$$

$$= x - Fx + \frac{1}{h} RFx + FS_h x - FS_h Fx + \frac{1}{h} FS_h RFx -$$

$$- \frac{1}{h} RFS_h x + \frac{1}{h} RFS_h Fx - \frac{1}{h^2} RFS_h RFx$$

$$= x - Fx + \frac{1}{h} RFx + Fx - F^2 x + \frac{1}{h} hFx -$$

$$- \frac{1}{h} RFx + \frac{1}{h} RF^2 x - \frac{1}{h^2} hRFx$$

$$= x + \frac{1}{h} RFx - Fx + Fx - \frac{1}{h} RFx + \frac{1}{h} RFx - \frac{1}{h} RFx$$

$$= x.$$

The arbitrariness of $x \in X_{S_h}^{(1)}$ implies the second of Identities (5.2.16). Write $y = R_1^0 x$. Then $y \in \mathrm{dom}\, D$. Indeed, since $DF = 0$, we find

$$Dy = D\left(R + F_1 R - \frac{1}{h} RF_1 R\right)x$$

$$= DRx + DFS_h Rx - \frac{1}{h} DRFS_h Rx = x - \frac{1}{h} FS_h Rx.$$

Moreover,

$$S_h y = S_h \left(I + F_1 - \frac{1}{h} RF_1\right) Rx$$

$$= S_h Rx + S_h FS_h Rx - \frac{1}{h} S_h RFS_h Rx$$

$$= S_h Rx + FS_h Rx - \frac{1}{h}(RFS_h Rx + hFS_h Rx)$$

$$= S_h Rx + FS_h Rx - \frac{1}{h} RFS_h Rx - FS_h Rx$$

$$= S_h Rx - \frac{1}{h} RFS_h x.$$

Hence

$$u = y - S_h y = R_1^0 x - \left(S_h Rx - \frac{1}{h} RFS_h Rx\right)$$

$$= Rx + FS_h Rx - \frac{1}{h} RFS_h Rx - S_h Rx + \frac{1}{h} RFS_h Rx$$

$$= Rx - S_h Rx + FS_h Rx$$

and

$$Du = D(y - S_h y) = DRx - DS_h Rx + DFS_h Rx$$

$$= x - S_h DRx = x - S_h x = 0,$$

which implies that $u \in \ker D$. On the other hand, $Fu = FRx - FS_h Rx + F^2 S_h Rx = -FS_h Rx + FS_h Rx = 0$. We therefore conclude that $u = Fu = 0$, i.e. $S_h y = y$ and $y \in X_{S_h}^{(1)}$. The arbitrariness of $x \in X_{S_h}$ implies that $R_1^0 X_{S_h} \subset X_{S_h}^{(1)}$. An initial operator F_1^0 corresponding to R_1^0 is defined on $X_{S_h}^{(1)}$ by the equality $F_1^0 = I - R_1^0 D_1 = I - I = 0$. ∎

THEOREM 5.2.5. *Suppose that $D \in R(X)$, dim ker $D \neq 0$, F_0 is an initial operator for an $R_0 \in \mathcal{R}_D$ and $S_{A(\mathbf{R})} = \{S_h\}_{h \in \mathbf{R}}$ is a family of R_0-shifts. Suppose that $F \in L_0(X)$ is a projection onto the space of constants and that there exists a real $d \neq 0$ such that*

$$FR_0^n z = \frac{d^n}{n!} z \quad \text{for all } z \in \ker D \quad (n = 0, 1, 2, \ldots). \quad (5.2.17)$$

Then F is an initial operator for D corresponding to the right inverse

$$R = R_0 - FR_0 \qquad (5.2.18)$$

and $\{S_h\}_{h \in \mathbf{R}}$ is a family of R-shifts.

Proof. Our assumptions and Theorem 2.2.4 together imply that F is an initial operator for D corresponding to a right inverse R defined by means of Equality (5.2.18). We shall prove that

$$R^n z = S_{-d} R_0^n z \quad \text{for all } z \in \ker D \quad (n = 0, 1, 2, \ldots). \quad (5.2.19)$$

Proof (by induction). Suppose that $z \in \ker D$ is arbitrarily fixed. If $n = 0$ then $R^0 z = z = S_{-d} z = S_{-d} R_0^0 z$. If $n = 1$ then $Rz = (R_0 - -FR_0)z = R_0 z - FR_0 z = R_0 z - dz = S_{-d} z$. Suppose Formula (5.2.19) to be true for an arbitrarily fixed $n \geqslant 1$. Then

$$R^{n+1} z = R(R^n z) = R(S_{-d} R_0^n z) = (R_0 - FR_0) S_{-d} R_0^n z$$

$$= (R_0 - FR_0) \sum_{j=0}^{n} \frac{(-d)^{n-j}}{(n-j)!} R_0^j z$$

$$= \sum_{j=0}^{n} \frac{(-d)^{n-j}}{(n-j)!} (R_0^{j+1} z - FR_0^{j+1} z)$$

$$= \sum_{j=0}^{n} \frac{(-d)^{n-j}}{(n-j)!} \left(R_0^{j+1} z - \frac{d^{j+1}}{(j+1)!} z \right)$$

$$= \sum_{j=0}^{n} \frac{(-d)^{n-j}}{(n-j)!} R_0^{j+1} z - \sum_{j=0}^{n} \frac{(-1)^{n-j} d^{n+1}}{(n-j)!(j+1)!} z$$

$$= \sum_{m=1}^{n+1} \frac{(-d)^{n+1-m}}{(n+1-m)!} R_0^m z + \frac{d^{n+1}}{(n+1)!} \left[\sum_{j=0}^{n} \frac{(-1)^{n-j}(n+1)!}{(n-j)!(j+1)!} \right] z$$

$$= \sum_{m=0}^{n+1} \frac{(-d)^{n+1-m}}{(n+1-m)!} R_0^m z - \frac{(-d)^{n+1}}{(n+1)!} z +$$

$$+ \frac{d^{n+1}}{(n+1)!} \sum_{j=0}^{n} \binom{n+1}{j+1} (-1)^{n-j} z$$

$$= S_{-d} R_0^{n+1} z - \frac{d^{n+1}}{(n+1)!} \left[\frac{(-1)^{n+1}}{(n+1)!} + \right.$$

$$\left. + \sum_{m=1}^{n+1} \binom{n+1}{m} (-1)^{n+1-m} \right] z$$

$$= S_{-d} R_0^{n+1} z - \frac{d^{n+1}}{(n+1)!} \left[\sum_{m=0}^{n+1} \binom{n+1}{m} (-1)^{n+1-m} \right] z$$

$$= S_{-d} R_0^{n+1} z,$$

since the sum $\sum_{m=0}^{n} \binom{n+1}{m} (-1)^{n+1-m}$ vanishes. This proves Formula (5.2.19) for all non-negative integers. Having already proved this Formula, we conclude from Point (i) of Theorem 5.2.1 that for arbitrary $z \in \ker D$, $h \in \mathbf{R}$ and non-negative integer n we have

$$S_h R^n z = S_h S_{-d} R_0^n z = S_{-d} S_h R_0^n z = S_{-d} \sum_{j=0}^{n} \frac{h^{n-j}}{(n-j)!} R_0^j z$$

$$= \sum_{j=0}^{n} \frac{h^{n-j}}{(n-j)!} S_{-d} R_0^j z = \sum_{j=0}^{n} \frac{h^{n-j}}{(n-j)!} R^j z,$$

which proves that S_h are R-shifts. ∎

We shall determine now another class of shifts for right invertible operators by means of exponential operator e_λ defined by Formula (2.3.2).

DEFINITION 5.2.2. Suppose that $D \in R(X)$, dim ker $D \neq 0$ and F is an initial operator for D corresponding to an $R \in \mathscr{R}_D \cap V(X)$. Define a family $S_{A(\mathbf{R})} = \{S_h\}_{h \in A(\mathbf{R})} \subset L_0(X)$ in the following way:

$$S_0 = I, \tag{5.2.20}$$

$$S_h e_\lambda F = e^{\lambda h} e_\lambda F \quad \text{for all } h \in A(\mathbf{R}), \quad \lambda \in \mathbf{C}. \tag{5.2.21}$$

We say that the family $S_{A(\mathbf{R})}$ *determine D-shifts* and that S_h is a *D-shift* on h for an arbitrary $h \in A(\mathbf{R})$.

In particular, for $\lambda = 0$ we have

$$S_h F = F \quad \text{for all } h \in A(\mathbf{R}). \tag{5.2.22}$$

PROPOSITION 5.2.4. *Suppose that all assumptions of Definition 5.2.2 are satisfied and that* $S_{A(\mathbf{R})} = \{S_h\}_{h \in A(\mathbf{R})}$ *is a family of D-shifts. Then*

$$S_h e_\lambda(z) = e^{\lambda h} e_\lambda(z) \quad \text{for all } h \in A(\mathbf{R}), \quad z \in \ker D. \tag{5.2.23}$$

In particular,

$$S_h z = z \quad \text{for all } z \in \ker D, \quad h \in A(\mathbf{R}), \tag{5.2.24}$$

i.e. D-shifts preserve constants.

Proof. Let $z \in \ker D$ be arbitrarily fixed. Then there is an $x \in X$ such that $Fx = z$. Thus we have $S_h e_\lambda(z) = S_h e_\lambda(Fx) = S_h e_\lambda Fx = e^{\lambda h} e_\lambda Fx = e^{\lambda h} e_\lambda(z)$ for all $h \in A(\mathbf{R})$. Putting $\lambda = 0$ we obtain Formula (5.2.24). ∎

THEOREM 5.2.6. *Suppose that $D \in R(X)$, dim ker $D \neq 0$ and F is an initial operator for D corresponding to an $R \in \mathscr{R}_D \cap V(X)$. Suppose that we are given a family* $S_{A(\mathbf{R})} = \{S_h\}_{h \in A(\mathbf{R})}$ *of D-shifts. Then:*

(i) *If $A(\mathbf{R}) = \mathbf{R}^+$ then $S_{A(\mathbf{R})}$ is a commutative semigroup, if $A(\mathbf{R}) = \mathbf{R}$ then $S_{A(\mathbf{R})}$ is an Abelian group with respect to the superposition of operators, i.e.*

$$S_{h_1} S_{h_2} = S_{h_2} S_{h_1} = S_{h_1 + h_2} \quad \text{for all } h_1, h_2 \in A(\mathbf{R}); \tag{5.2.25}$$

(ii) *For all $h \in A(\mathbf{R})$ the operators are D-invariant and uniquely determined on the set $E(R)$ of all exponentials.*

Proof. (i) Suppose that $A(\mathbf{R}) = \mathbf{R}^+$. For arbitrary $h_1, h_2 \in A(\mathbf{R}), \lambda \in \mathbf{C}$ and $z \in \ker D$ we find

$$S_{h_1} S_{h_2} e_\lambda(z) = S_{h_1} e^{\lambda h_2} e_\lambda(z) = e^{\lambda h_2} S_{h_1} e_\lambda(z)$$
$$= e^{\lambda h_2} e^{\lambda h_1} e_\lambda(z) = e^{\lambda(h_1 + h_2)} e_\lambda(z) = S_{h_1 + h_2} e_\lambda(z).$$

We therefore conclude that a superposition of D-shifts on h_1 and h_2 is again a D-shift on $h_1 + h_2$. The commutativity of the operators S_{h_1} and S_{h_2} follows from the commutativity of the addition in $A(\mathbf{R})$. Thus $S_{\mathbf{R}^+}$ is a commutative semigroup. If $0 \neq h \in \mathbf{R}$ then the D-shift S_{-h} is an inverse operator for D-shift S_h, for $S_h S_{-h} = S_{-h} S_h = S_{h-h} = S_0 = I$. Then $S_{\mathbf{R}}$ is an Abelian group.

(ii) Suppose now that $0 \neq h \in A(\mathbf{R})$, $\lambda \in \mathbf{C}$, $z \in \ker D$ are arbitrarily fixed. Since $De_\lambda(z) = \lambda e_\lambda(z)$, Proposition 5.2.4 implies that

$$(DS_h - S_h D)e_\lambda(z) = D[e^{\lambda h}e_\lambda(z)] - S_h[\lambda e_\lambda(z)]$$
$$= e^{\lambda h}e_\lambda(z) - e^{\lambda h}e_\lambda(z) = 0.$$

Hence S_h are D-invariant on the set $E(R)$. Definition 5.2.2 and Proposition 5.2.4 together imply that S_h are uniquely determined on the set $E(R)$. ∎

COROLLARY 5.2.2. *Suppose that all assumptions of Theorem 5.2.6 are satisfied. Then*

$$S_h^n = S_{nh} \quad \text{for all } h \in A(\mathbf{R}), \quad n \in \mathbf{N} \cup \{0\}. \tag{5.2.26}$$

This is an immediate consequence of Point (i) in Theorem 5.2.6.

Observe that a D-shift on h is an operator such that the constant z_1 appearing in Theorem 2.3.3 is equal $e^{\lambda h}z$ for all $z \in \ker D$ and $\lambda \in \mathbf{C}$: $z_1 = e^{\lambda h}z$. Hence we have

PROPOSITION 5.2.5. *If S_h is a D-shift on $h \in A(\mathbf{R})$ and F is an initial operator corresponding to the operator R then*

$$FS_h e_\lambda(z) = e^{\lambda h}z \quad \text{for all } \lambda \in \mathbf{C}, \quad z \in \ker D.$$

Proposition 5.2.5 does not imply that $\dim \ker D = 1$. Namely, we have

PROPOSITION 5.2.6. *There exist a linear space X and an operator $D \in R(X)$ and a D-invariant D-shift S_h on $h \in A(\mathbf{R})$ that $\dim \ker D = n$, where n is an arbitrarily fixed positive integer.*

Proof. Indeed, if $D \in R(X)$, $\dim \ker D = 1$ and S_h is a D-invariant D-shift on h, we can consider the space $Y = X^n$ ($n = 1, 2, ...$). Write:

$x = (x_1, ..., x_n) \in Y$, $\hat{D}x = (Dx_1, ..., Dx_n)$, $\hat{S}_h x = (S_h x_1, ..., S_h x_n)$, $\hat{R}x = (Rx_1, ..., Rx_n)$. It is easy to verify that $\hat{D}$ is a right invertible operator in the space Y, $\hat{R} \in \mathcal{R}_{\hat{D}} \cap V(Y)$, dim ker $\hat{D} \neq 0$ and that $\hat{S}_h \hat{D} = \hat{D} \hat{S}_h$ on dom $\hat{D}$. Write $\hat{e}_\lambda = (I - \hat{R})^{-1}$. Then $\hat{S}_h \hat{e}_\lambda(z) = (S_h e_\lambda(z_1), ..., S_h e_\lambda(z_n)) = (e^{\lambda h} z_1, ..., e^{\lambda h} z_n) = e^{\lambda h}(z_1, ..., z_n) = e^{\lambda h}z$ for all $\lambda \in \mathbb{C}$ and $z = (z_1, ..., z_n) \in$ ker $\hat{D} = ($ker $D)^n$. Thus the operator $\hat{S}_h$ is a $\hat{D}$-shift on h. On the other hand, it is well-known that there exist an operator D such that dim ker $D = 1$, namely the operator $D = d/dt$. A D-invariant D-shift for this operator is a usual shift on h: $(Sx)(t) = x(t+h)$. ∎

PROPOSITION 5.2.7. *Suppose that all assumptions of Theorem 5.2.6 are satisfied. Then for all $h \in A(\mathbb{R})$, $z \in$ ker D:*

(i) *exponential elements $e_\lambda(z)$ are S_h-periodic for $\lambda = 2\pi i k/h$, where k is an arbitrary integer;*

(ii) *cosine and sine elements $c_\lambda(z)$ and $s_\lambda(z)$ are S_h-periodic for $\lambda = 2\pi k/h$, where k is an arbitrary integer.*

Proof. (i) Suppose that $\lambda = 2\pi i k/h$, k is an arbitrary integer. Since S_h is a D-shift on h, we find

$$S_h e_\lambda(z) = e^{\lambda h} e_\lambda(z) = e^{(2\pi i k/h)h} e_\lambda(z)$$
$$= (e^{2\pi i})^k e_\lambda(z) = e_\lambda(z).$$

(ii) Suppose that $\lambda = 2\pi k/h$, where k is an arbitrary integer. We find

$$S_h c_\lambda(z) = S_h \frac{1}{2}(e_{\lambda i} + e_{-\lambda i}) = \frac{1}{2}[S_h e_{\lambda i}(z) + S_h e_{-\lambda i}(z)]$$

$$= \frac{1}{2}[e^{(2\pi i k/h)h} e_{\lambda i}(z) + e^{(-2\pi i k/h)h} e_{-\lambda i}(z)]$$

$$= \frac{1}{2}[(e^{2\pi i})^k e_{\lambda i}(z) + (e^{-2\pi i})^k e_{-\lambda i}(z)]$$

$$= \frac{1}{2}[e_{\lambda i}(z) + e_{-\lambda i}(z)] = c_\lambda(z).$$

A similar proof for elements $s_\lambda(z)$. ∎

PROPOSITION 5.2.8. *Suppose that $D \in R(X)$, dim ker $D \neq 0$, F is an initial operator for D corresponding to an $R \in \mathcal{R}_D \cap V(X)$ and that $S_{A(\mathbb{R})}$*

$= \{S_h\}_{h \in A(\mathbf{R})}$ *is a family of D-shifts. Then for all* $0 \neq h \in A(\mathbf{R})$, $z \in \ker D$

$$FS_h \, Re_\lambda(z) \neq 0 \quad \text{for } \lambda \neq 2\pi i k/h, \tag{5.2.27}$$

k is an arbitrary integer.

Proof. Let $0 \neq h \in A(\mathbf{R})$ and $\lambda \neq 2\pi i k/h$ (*k* is an arbitrary integer) be arbitrarily fixed. Then

$$FS_h \, Re_\lambda(z) = FS_h \frac{1}{\lambda} [e_\lambda(z) - z] = \frac{1}{\lambda} [FS_h e_\lambda(z) - FS_h z]$$

$$= \frac{1}{\lambda} \{F[e^{\lambda h} e_\lambda(z)] - Fz\} = \frac{1}{\lambda} [e^{\lambda h} Fe_\lambda(z) - z]$$

$$= \frac{1}{\lambda} (e^{\lambda h} z - z) = \frac{1}{\lambda} (e^{\lambda h} - 1) z \neq 0$$

by our assumptions.   ∎

THEOREM 5.2.7. *Suppose that* $D \in R(X)$, $\dim \ker D \neq 0$, $R \in \mathscr{R}_D$, $A \in S_{D, R}$ *(i.e. is stationary) and* $S_h \in L_0(X)$ *is an R-shift on* $0 \neq h \in \mathbf{R}$. *Then A is* S_h-*periodic on the set* $P(R)$ *of all D-polynomials*:

$$AS_h = S_h A \quad \text{on } P(R). \tag{5.2.28}$$

Proof. Since $DA = AD$ we have $A \ker D \subset \ker D$, which implies that $Az \in \ker D$ for every $z \in \ker D$. Since S_h is an *R*-shift on *h*, we have for $k = 0, 1, 2, \ldots$ and $z \in \ker D$

$$AS_h R^k z = A \sum_{j=0}^{k} \frac{h^{k-j}}{(k-j)!} R^j z = \sum_{j=0}^{k} \frac{h^{k-j}}{(k-j)!} AR^j z$$

$$= \sum_{j=0}^{k} \frac{h^{k-j}}{(k-j)!} R^j Az = S_h R^k Az = S_h AR^k z$$

and

$$(AS_h - S_h A) R^k z = 0 \quad \text{for all } z \in \ker D \text{ and } k = 0, 1, 2, \ldots \ \blacksquare$$

COROLLARY 5.2.3. *Suppose that all assumptions of Theorem 5.2.7 are satisfied. Then*:

(i) *If the operator* $I-AR$ *is invertible then* $E_A = (I-AR)^{-1}$ *is* S_h-*periodic on the set* $P(R)$:

$$S_h E_A = E_A S_h \quad \text{on } P(R); \tag{5.2.29}$$

(ii) *If the operators* $I-iAR$, $I+iAR$ *are invertible then the operators* C_A, S_A *are* S_h-*periodic on the set* $P(R)$:

$$S_h C_A = C_A S_h, \quad S_h S_A = S_A S_h \quad \text{on } P(R). \tag{5.2.30}$$

Proof. Formula (5.2.29) and our assumptions together imply that $S_h(I-AR) = S_h - S_h AR = S_h - AS_h R = S_h - ARS_h = (I-AR)S_h$ on $P(R)$. Then $S_h E_A = S_h(I-AR)^{-1} = (I-AR)^{-1}S_h = E_A S_h$ on $P(R)$ and $S_h C_A = \dfrac{1}{2} S_h(E_{iA}+E_{-iA}) = \dfrac{1}{2}(E_{iA}+E_{-iA})S_h = C_A S_h$ on $P(R)$. A similar proof for the operator S_A. ∎

COROLLARY 5.2.4. *Suppose that all assumptions of Theorem 5.2.7 are satisfied and that the operators* D_1 *and* R_1^0 *are defined by Formulae* (5.2.8), (5.2.15):

$$D_1 = D - \frac{1}{h} FS_h, \quad R_1^0 = \left(I + FS_h - \frac{1}{h} RFS_h\right)S_h, \tag{5.2.31}$$

where F *is an initial operator for* D *corresponding to* R. *Then*

$$D_1 A = AD_1, \quad R_1^0 A = AR_1^0 \quad \text{on } P(R). \tag{5.2.32}$$

Proof. Our assumptions and Corollary 5.2.3 together imply that on $P(R)$

$$D_1 A = \left(D - \frac{1}{h} FS_h\right)A = DA - \frac{1}{h} FS_h A = AD - \frac{1}{h} FAS_h$$

$$= AD - \frac{1}{h} AFS_h = A\left(D - \frac{1}{h} FS_h\right) = AD_1,$$

$$R_1^0 A = \left(I + FS_h - \frac{1}{h} RFS_h\right)RA = \left(I + FS_h - \frac{1}{h} RFS_h\right)AR$$

$$= \left(A + AFS_h - A\frac{1}{h} RFS_h\right)R = A\left(I + FS_h - \frac{1}{h} RFS_h\right)R$$

$$= AR_1^0. \quad ∎$$

Examples and Exercises

EXAMPLE 5.2.2. Let $C(\mathbf{R})$ be the space of all functions continuous on real line. Let $D = \mathrm{d}/\mathrm{d}t$, $R = \int_0^t$, $(Fx)(t) = x(0)$, $(S_h x)(t) = x(t+h)$ for all $x \in X$, $h \in \mathbf{R}$. Then S_h are R-shifts acting in X, since all assumptions of Theorem 5.2.1 are satisfied.

EXAMPLE 5.2.3. Let X, D, R, F be defined as in Example 5.2.2. Suppose, we are given a continuous function $g\colon \mathbf{R} \to \mathbf{R}$ such that $g(t) \not\equiv t$ and $g'(t) > 0$ for $t \in \mathbf{R}$.* Write: $(Ax)(t) = x(g(t))$ for $x \in X$. Observe that by our assumptions the operator $A \in L_0(X)$ is invertible and $(A^{-1}y)(t) = y(g^{-1}(t))$ for $y \in X$, where g^{-1} denotes the inverse function for g. It is easy to see that $\hat{D} = A^{-1}DA \in R(X)$, $\hat{R} = A^{-1}RA \in \mathscr{R}_{\hat{D}}$, $\hat{F} = A^{-1}FA$, $\ker \hat{D} = \{A^{-1}z\colon z \in \ker D\} = A^{-1}\ker D$. Indeed, $\hat{D}\hat{R} = DAA^{-1}R = DR = I$, $\hat{F} = I - \hat{R}\hat{D} = I - A^{-1}RDA = A^{-1}(I - RD)A = A^{-1}FA$ and $\hat{D}\hat{z} = 0$ if and only if $\hat{z} = A^{-1}z$, where $z \in \ker D$. By our definitions we have

$$(\hat{R}x)(t) = \int_0^{g^{-1}(t)} x(g(s))\,\mathrm{d}s \quad \text{for } x \in X, \tag{5.2.33}$$

$$\hat{z}(t) = z(g^{-1}(t)) \equiv c \in \mathbf{R} \quad \text{for } z \in \ker D. \tag{5.2.34}$$

It is well-known that

$$(R^k z)(t) = c\,\frac{t^k}{k!} \quad \text{for } z(t) \equiv c \quad (k = 0, 1, 2, \ldots).$$

Thus, by Formula (5.2.8)

$$(\hat{R}^k \hat{z})(t) = c\,\frac{[g^{-1}(t)]^k}{k!} \quad \text{for } \hat{z}(t) \equiv c \quad (k = 0, 1, 2, \ldots).$$
$$\tag{5.2.35}$$

Write

$$\gamma_h(t) = g(g^{-1}(t)+h) \quad \text{for } h \in \mathbf{R}, \tag{5.2.36}$$

$$(S_h x)(t) = x(\gamma_h(t)) \quad \text{for } h \in \mathbf{R}, x \in X. \tag{5.2.37}$$

---

* We can consider in the same manner the space $C(\mathbf{R}^+)$ of all functions continuous on the half-axis $\mathbf{R}^+$ and a continuous function $g\colon \mathbf{R}^+ \to \mathbf{R}^+$, such that $g'(t) > 0$ for $t \in \mathbf{R}^+$.

We shall prove that $\hat{S}_{\mathbf{R}} = \{\hat{S}_h\}_{h \in \mathbf{R}}$ is a family of $\hat{R}$-shifts. Indeed, let $k \in \mathbf{N} \cup \{0\}$ and $c \in \mathbf{R}$ be arbitrary and let $z(t) \equiv c$. Then we have

$$\sum_{j=0}^{k} \frac{h^{k-j}}{(k-j)!} (\hat{R}^j \hat{z})(t)$$

$$= \sum_{j=0}^{k} \frac{h^{k-j}}{(k-j)!} c \frac{[g^{-1}(t)]^j}{j!} = \frac{c}{k!} \sum_{j=0}^{k} \frac{k!}{j!(k-j)!} h^{k-j} [g^{-1}(t)]^j$$

$$= \frac{c}{k!} \sum_{j=0}^{k} \binom{k}{j} h^{k-j} [g^{-1}(t)]^j = \frac{c}{k!} [g^{-1}(t) + h]^k$$

$$= \frac{c}{k!} [g^{-1}(\gamma_h(t))]^k = (\hat{R}^k \hat{z})(u)|_{u = \gamma_h(t)} = (\hat{S}_h \hat{R}^k \hat{z})(t)$$

which proves that $\hat{S}_h$ are $\hat{R}$-shifts $(h \in \mathbf{R})$. On the other hand, we have

$$(A^{-1} S_h R^k z)(t) = c \frac{[g^{-1}(t) + h]^k}{k!} \quad \text{for } z \in \ker D$$

which implies

$$A^{-1} S_h R^k F = \hat{S}_h \hat{R}^k A^{-1} F \quad \text{for } h \in \mathbf{R} \quad (k = 0, 1, 2, \ldots).$$

We should point out that the function γ_h defined by means of Formula (5.2.36) is a continuous solution of the so-called *translation equation* (cf. Moszner, 1973, 1973a, Moszner and Tabor, 1976).

EXAMPLE 5.2.4. Suppose X, D, R, F, S_h are defined as in Example 5.2.1. We shall show that

$$(R_1^0 x)(t) = \int_0^t x(s)\,ds + \left(\frac{t}{\omega} + 1\right) \int_0^\omega x(s)\,ds \tag{5.2.38}$$

for ω-periodic $x \in X$.

Indeed, we have shown in Example 5.2.1 that $(FS_h Rx)(t) = \int_0^\omega x(s)\,ds$, where $h = \omega$. Hence

$$(RFS_h Rx) = \int_0^t \left[\int_0^\omega x(s)\,ds\right] dt = \left(\int_0^\omega x(s)\,ds\right)\left(\int_0^t dt\right) = t \int_0^\omega x(s)\,ds$$

and

$$(R_1^0 x)(t) = \left(Rx + FS_h Rx + \frac{1}{h} RFS_h Rx\right)(t)$$

$$= \int_0^t x(s)\,ds + \int_0^\omega x(s)\,ds + \frac{t}{\omega} \int_0^\omega x(s)\,ds$$

$$= \int_0^t x(s)\,ds + \left(\frac{t}{\omega} + 1\right) \int_0^\omega x(s)\,ds$$

because we have for ω-periodic functions $x(t+\omega) = x(t)$.

EXAMPLE 5.2.5. An R-shift, in general, is not an R^n-shift ($n = 2, 3, 4, ...$). Indeed, let $D \in R(X)$, $R \in \mathscr{R}_D$ and let S_h be an R-shift on $h \neq 0$. Write $\hat{D} = D^2$. Then $\hat{z} \in \ker \hat{D}$ if and only if $z = Rz_1 + z_0$, where $z_1, z_0 \in \ker D$ are arbitrary. Thus we have

$$S_h \hat{z} = S_h(Rz_1 + z_0) = Rz_1 + hz_1 + z_0 = z + hz_1 \neq z \quad \text{if } z_1 \neq 0.$$

EXERCISE 5.2.1. Suppose that $X_1, ..., X_n$ are linear spaces over $\mathbf{C}$, $D_j \in R(X_j)$, $\dim \ker D_j \neq 0$, $R_j \in \mathscr{R}_{D_j}$ ($j = 1, 2, ..., n$). Suppose, moreover, that $S_h^{(j)}$ is an R_j-shift on $h \in A(\mathbf{R})$ ($j = 1, 2, ..., n$). Write

$$X = X_1 \times ... \times X_n, \quad x = (x_1, ..., x_n) \in X, \tag{5.2.39}$$
$$Dx = (D_1 x_1, ..., D_n x_n), \quad Rx = (R_1 x_1, ..., R_n x_n), \tag{5.2.40}$$
$$S_h x = (S_h^{(1)} x_1, ..., S_h^{(n)} x_n) \quad \text{for } x \in X, \tag{5.2.41}$$
$$z = (z_1, ..., z_n) \in \ker D = \ker D_1 \times ... \times \ker D_n. \tag{5.2.42}$$

Prove that $D \in R(X)$, $R \in \mathscr{R}_D$, $\dim \ker D = \sum_{j=1}^{n} \dim \ker D_j \neq 0$, $\sum_{k=0}^{n} R^k z$ are D-polynomials and S_h are R-shifts on h.

EXERCISE 5.2.2. Suppose that X_j, D_j, R_j, $S_h^{(j)}$, X are defined, as in Exercise 5.2.1 and that $X_1 = ... = X_n$. Write

$$D = [\delta_{jk} D_j]_{j, k=1, ..., n}, \quad R = [\delta_{jk} R_j]_{j, k=1, ..., n}. \tag{5.2.43}$$

Prove that $D \in R(X)$, $R \in \mathscr{R}_D$, $\dim \ker D \neq 0$ and the operator $S_h = [\delta_{jk} S_k^{(j)}]_{j, k=1, ..., n}$ is an R-shift on h.

EXAMPLE 5.2.6. Suppose that $X = C[0, \infty)$ and $D = t\dfrac{\mathrm{d}}{\mathrm{d}t}$. The operator D is right invertible in X and has a Volterra right inverse R of the form:

$$(Rx)(t) = \int_a^t x(s)s^{-1}\mathrm{d}s, \quad a > 0.$$ It is easy to verify that exponential elements for the operator D are of the form Ct^λ, where C is an arbitrary real number, λ is an arbitrary complex number. Define the operators S_h by means of the equality $(S_h x)(t) = x(rt)$, where $r > 0$. Hence S_h is an operator of multiplication of argument by a number $r > 0$. Write: $h = \ln r$. Since $S_h(Ct^\lambda) = C(rt)^\lambda = Cr^\lambda t^\lambda = C(e^{\lambda h})t^\lambda = Ce^{\lambda h}t^\lambda = e^{\lambda h}Ct^\lambda$,

we find that S_h are $t\dfrac{\mathrm{d}}{\mathrm{d}t}$ shifts on $h = \ln r$. The operator S_h is also an R-shift on h. Indeed, $z \in \ker D$ if and only if $z(t) \equiv c$, where $c \in \mathbf{R}$. By an easy induction we can prove that

$$(R^k z)(t) = c\,\frac{(\ln t - \ln a)^k}{k!} \quad (k = 0, 1, 2, \ldots).$$

Thus we have for all $c \in \mathbf{R}$, $k = 0, 1, 2, \ldots$

$$(S_h R^k z)(t) = S_h \frac{c}{k!}\,(\ln t - \ln a)^k$$

$$= \frac{c}{k!}\,[\ln(rt) - \ln a]^k = \frac{c}{k!}\,(\ln t - \ln a + \ln r)^k$$

$$= \frac{c}{k!}\sum_{j=0}^{k}\binom{k}{j}(\ln r)^{k-j}(\ln t - \ln a)^j$$

$$= c\sum_{j=0}^{k}\frac{1}{k!}\,\frac{k!}{(k-j)!\,j!}\,(\ln r)^{k-j}(\ln t - \ln a)^j$$

$$= \sum_{j=0}^{k}\frac{h^{k-j}}{(k-j)!}\,\frac{c}{j!}\,(\ln t - \ln a)^j$$

$$= \sum_{j=0}^{k}\frac{h^{k-j}}{(k-j)!}\,(R^j z)(t)$$

which proves that S_h is an R-shift on $h = \ln r$.

EXAMPLE 5.2.7. A D-shift S_h $(h \in A(\mathbf{R}))$ is not, in general, a D^n-shift $(n \in \mathbf{N})$. Indeed, let $\hat{D} = D^2$ and let S_h be a D-shift on $h \neq 0$. Let $\lambda = 0$. Then $\hat{z} \in \ker \hat{D}$ if and only if $\hat{z} = Rz_1 + z_0$, where $z_0, z_1 \in \ker D$ are arbitrary (Compare Remark 4.2.1). We have $S_h \hat{D} = S_h D^2 = D^2 S_h = \hat{D} S_h$ on dom $\hat{D}$. But $S_h \hat{z} = S_h Rz_1 + S_h z_0 = S_h Rz_1 + z_0$ and $u = \hat{z} - S_h z = Rz_1 + z_0 - (S_h Rz_1 + z_0) = Rz_1 - S_h Rz_1$. Hence $Du = DRz_1 - DS_h Rz_1 = z_1 - S_h DRz_1 = z_1 - S_h z_1 = z_1 - z_1 = 0$, which implies that $u \in \ker D$ and $S_h Rz_1 = Rz_1 - u$. Therefore $u = -FS_h Rz_1$ and, in general, $u \neq 0$, hence $S_h \hat{z} \neq \hat{z}$, which contradicts to our assumption that S_h is a $\hat{D}$-shift.

EXERCISE 5.2.3. Suppose that $X_1, \ldots, X_n$ are linear spaces over $\mathbf{C}$, $D_j \in R(X_j)$, $\dim \ker D_j \neq 0$, $R_j \in \mathcal{R}_{D_j} \cap V(X_j)$ $(j = 1, 2, \ldots, n)$. Write $e_\lambda^{(j)} = (I - \lambda R_j)^{-1}$ for $\lambda \in \mathbf{C}$ $(j = 1, 2, \ldots, n)$. Suppose, moreover, that operators S_j^h are D_j-shift on $h \in A(\mathbf{R})$, respectively, i.e. $D_j S_j^h = S_j^h D_j$ on dom D_j $(j = 1, 2, \ldots, n)$ and $S_j^h e_\lambda^{(j)}(z_j) = e^{\lambda h_j} e_\lambda^{(j)}(z_j)$ for $\lambda \in \mathbf{C}$, $z_j \in \ker D$ $(j = 1, 2, \ldots, n)$.

Let X, D, R will be defined by Formulae (5.2.39), (5.2.40) and let $S^h x = (S_1^h x_1, \ldots, S_n^h x_n)$ for $x \in X$, $e_\lambda(z) = (e^{(1)}(z_1), \ldots, e^{(n)}(z_n))$ for $z = (z_1, \ldots, z_n) \in \ker D = \ker D_1 \times \ldots \times \ker D_n$. Prove that:

(i) $D \in R(X)$, $R \in \mathcal{R}_D \cap V(X)$, $\dim \ker D = \sum_{j=1}^{n} \dim \ker D_j \neq 0$ and that $e_\lambda(z)$ are exponentials for D;

(ii) $S^h D = DS^h$ on dom D and S^h is a D-shift on $h = h_1 + \ldots + h_n$.

EXERCISE 5.2.4. Suppose that $X_j, X, D_j, S_j^h, e_\lambda^{(j)}(j = 1, \ldots, n)$ are defined as in Exercise 5.2.3, $X_1 = X_2 = \ldots = X_n$ and D, R are defined by Formulae (5.2.43). Write

$$S^h = [\delta_{jk} S_j^h]_{j, k = 1, 2, \ldots, n}, \quad e_\lambda = [\delta_{jk} e_\lambda^{(j)}]_{j, k = 1, 2, \ldots, n}.$$

Prove that S^h is a D-shift on $h = h_1 + \ldots + h_n$.

EXERCISE 5.2.5. Suppose that all assumptions of Example 5.2.3 are satisfied. Prove that the equation

$$(I - S_h)x = y, \quad \text{where } y \in \ker (I - S_h) \tag{5.2.44}$$

has a solution of the form

$$x(t) = \frac{1}{h} g^{-1}(t) y(t).$$ (5.2.45)

EXERCISE 5.2.6. Suppose that $X = (s)$ over $\mathbf{C}$. Suppose that we are given a sequence $\{\lambda_n\} \subset \mathbf{C}$ such that $|\lambda_n| > 0$ for all $n \in \mathbf{N}$. Write

$$\mu_n = \lambda_1 \dots \lambda_n, \qquad \delta = \{\delta_{1,n}\},$$

where

$$\delta_{1,n} = \begin{cases} 1 & \text{for } n = 1 \\ 0 & \text{otherwise} \end{cases} \quad (n \in \mathbf{N}).$$

Define for all $x \in X$ the operators D, R in the following way:

$$D\{x_n\} = \{\lambda_n x_{n+1}\}, \qquad R\{x_n\} = \{\lambda_n^{-1} x_{n-1}\},$$

where $x_0 = 0$ for all $x \in X$.

Prove that:

(i) $D \in R(X)$, the operator F defined by the equality $F\{x_n\} = x_1 \delta$ is an initial operator for D corresponding to its right inverse R;

(ii) $\ker D = \{z = c\delta : c \in \mathbf{C}\}$;

(iii) $R^k z = c\lambda_{k+1}^{-k} \{\delta_{1,n-k}\}$ for $k \in \mathbf{N}$, $z \in \ker D$ (we admit: $\delta_{1,n-k} = 0$ for $n \le k$);

(iv) R-shifts are defined by the equality

$$S_h R^k z = c \sum_{j=0}^{k} \frac{h^{k-j}}{(k-j)!} \lambda_{j+1}^{-j} \{\delta_{1,n-j}\}$$

for $k \in \mathbf{N}$, $z \in \ker D$, $c \in \mathbf{C}$, $h \in \mathbf{R}$;

(v) exponentials are of the form: $e_\lambda(z) = c\{\lambda^n \mu_n^{-1}\}$ for $z \in \ker D$, $c, \lambda \in \mathbf{C}$, and D-shifts are of the form: $S_h e_\lambda(z) = ce^{\lambda h}\{\lambda^n \mu_n^{-1}\}$, $h \in \mathbf{R}$;

(vi) if S_h is a D-shift then exponentials are S_h-periodic for $\lambda = 2\pi i k/h$, k is an arbitrary integer.

5.3. EXISTENCE OF PERIODIC SOLUTIONS

In this section we shall look for periodic solutions of equations with periodic coefficients.

THEOREM 5.3.1. *Suppose that $D \in R(X)$, dim ker $D \neq 0$, F is an initial operator for D corresponding to an $R \in \mathcal{R}_D$ and $S_h \in L_0(X)$ is an R-shift on h. Then the equation*

$$Dx = y, \quad y \in X_{S_h} \qquad \qquad (5.3.1)$$

with the initial condition

$$Fx = x_0, \quad x_0 \in \ker D \qquad \qquad (5.3.2)$$

has a unique S_h-periodic solution which is of the form

$$x = \left(I + FS_h - \frac{1}{h} RFS_h \right) Ry - x_0.$$

Proof. Let D_1, F_1, R_1^0 be defined by Formulae (5.2.8) and (5.2.15). We are looking for S_h-periodic solutions of the problem (5.3.1)–(5.3.2), i.e. for such x that $S_h x = x$ and $F_1 x = Fx = x_0$. Since $x_0 \in \ker D$ and S_h is an R-shift on h, Theorem 5.2.4 implies that

$$
\begin{aligned}
x = R_1^0 D_1 x &= R_1^0 \left(Dx + \frac{1}{h} F_1 x \right) = R_1^0 \left(y + \frac{1}{h} Fx \right) \\
&= R_1^0 \left(y + \frac{1}{h} x_0 \right) = \left(I + F_1 S_h - \frac{1}{h} RF_1 S_h \right) R \left(y + \frac{1}{h} x_0 \right) \\
&= \left(I + FS_h - \frac{1}{h} RFS_h \right) Ry + \frac{1}{h} \left(Rx_0 + FS_h Rx_0 - \right. \\
&\qquad\qquad\qquad\qquad\qquad\qquad\qquad \left. - \frac{1}{h} RFS_h Rx_0 \right) \\
&= \left(I + FS_h - \frac{1}{h} RFS_h \right) Ry - \frac{1}{h} \left[Rx_0 + F(Rx_0 + hx_0) - \right. \\
&\qquad\qquad\qquad\qquad\qquad\qquad\qquad \left. - \frac{1}{h} RF(Rx_0 + hx_0) \right] \\
&= \left(I + FS_h + \frac{1}{h} RFS_h \right) Ry - \frac{1}{h} \left(Rx_0 + FRx_0 + hFx_0 - \right. \\
&\qquad\qquad\qquad\qquad\qquad\qquad\qquad \left. - \frac{1}{h} RFRx_0 - RFx_0 \right)
\end{aligned}
$$

$$= \left(I+FS_h+\frac{1}{h}\,RFS_h\right)Ry-\frac{1}{h}\,(Rx_0+hx_0-Rx_0)$$

$$= \left(I+FS_h+\frac{1}{h}\,RFS_h\right)Ry-x_0. \qquad\blacksquare$$

An immediate consequence of this theorem is

COROLLARY 5.3.1. *Suppose that all assumptions of Theorem 5.3.1 are satisfied. Then all S_h-periodic solutions of Equation (5.3.1) are of the form*

$$x = \left(I-FS_h-\frac{1}{h}\,RFS_h\right)Ry+z, \quad \text{where } z \in \ker D \text{ is arbitrary.}$$

COROLLARY 5.3.2. *Suppose that all assumptions of Theorem 5.3.1 are satisfied. Then for an arbitrary positive integer n the equation*

$$D^n x = y, \quad y \in X_{S_h} \tag{5.3.3}$$

with the initial conditions

$$FD^k x = x_k, \quad x_k \in \ker D \quad (k = 0, 1, \ldots, n-1) \tag{5.3.4}$$

has a unique S_h-periodic solution which is of the form

$$x = \left[\left(I+FS_h-\frac{1}{h}\,RFS_h\right)R\right]^n y -$$
$$- \sum_{k=0}^{n-1}\left[\left(I+FS_h-\frac{1}{h}\,RFS_h\right)R\right]^k x_k. \tag{5.3.5}$$

Proof (by induction). For $n = 1$ our statement is just Theorem 5.3.1. Suppose that for an arbitrary fixed n the equation (5.3.3) together with the conditions (5.3.4) has a unique S_h-periodic solution which is of the form (5.3.5). Write, as before, $R_1^0 = \left(I+FS_h-\frac{1}{h}\,RFS_h\right)R$ and consider the equation

$$D^{n+1} x = y, \quad y \in X_{S_h} \tag{5.3.6}$$

with the initial conditions

$$FD^k x = x_k, \quad x_k \in \ker D \quad (k = 0, 1, \ldots, n). \tag{5.3.7}$$

Write $u = D^n x$ and consider the problem

$$Du = y, \quad Fu = x_n. \tag{5.3.8}$$

Theorem 5.3.1 implies that the problem (5.3.8) has a unique S_h-periodic solution which is of the form

$$u = R_1^0 y = -x_n. \tag{5.3.9}$$

Now consider the problem

$$D^n x = u, \quad FD^k x = x_k \quad (k = 0, 1, \ldots, n-1).$$

By our inductive assumption this problem has a unique S_h-periodic solution which is of the form

$$x = (R_1^0)^n u - \sum_{k=0}^{n-1} (R_1^0)^k x_k.$$

This implies that a unique S_h-periodic solution of the problem (5.3.6)–(5.3.7), we are looking for, is of the form

$$x = (R_1^0)^n u - \sum_{k=0}^{n-1} (R_1^0)^k x_k = (R_1^0)^n (R_0^1 y - x_n) - \sum_{k=0}^{n-1} (R_1^0)^k x_k$$

$$= (R_1^0)^{n+1} y - \sum_{k=0}^{n} (R_1^0)^k x_k = \left[\left(I + FS_h - \frac{1}{h} RFS_h\right) R\right]^{n+1} y -$$

$$- \sum_{k=0}^{n} \left[\left(I + FS_h - \frac{1}{h} RFS_h\right) R\right]^k x_k$$

which was to be proved. ∎

COROLLARY 5.3.3. *Suppose that all assumptions of Theorem 5.3.1 are satisfied. Then for an arbitrary positive integer n all S_h-periodic solutions of Equation (5.3.3) are of the form*

$$x = \left[\left(I + FS_h - \frac{1}{h} RFS_h\right) R\right]^n y + \sum_{k=0}^{n-1} \left[\left(I + FS_h - \frac{1}{h} RFS_h\right) R\right]^k z_k,$$

$$\tag{5.3.10}$$

where $z_k \in \ker D$ are arbitrary $(k = 0, 1, \ldots, n-1)$.

THEOREM 5.3.2. *Suppose that all assumptions of Theorem 5.3.1 are satis-fied. Suppose, moreover, that $\tilde{D} \in R(X)$, $\dim \ker \tilde{D} \neq 0$, $\tilde{F}$ is an initial operator for $\tilde{D}$ corresponding to an $\tilde{R} \in \mathscr{R}_{\tilde{D}}$ and $\tilde{S}_h \in L_0(X)$ is an $\tilde{R}$-shift. Write*

$$\tilde{R}_1^0 = \left(I - \tilde{F}\tilde{S}_h - \frac{1}{h} \tilde{R}\tilde{F}\tilde{S}_h \right) \tilde{R}. \tag{5.3.11}$$

If $y \in X_{\tilde{S}_h}$, $\tilde{R}_1^0 y \in X_{S_h}$ then the equation

$$\tilde{D}Dx = y \tag{5.3.12}$$

together with the initial conditions

$$Fx = x_0, \quad \tilde{F}Dx = \tilde{x}_0, \quad \text{where } x_0 \in \ker D, \ \tilde{x}_0 \in \ker \tilde{D} \cap X_{S_h}. \tag{5.3.13}$$

has a unique S_h-periodic solution

$$x = R_1^0 \tilde{R}_1^0 y - R_1^0 \tilde{x}_0 - x_0. \tag{5.3.14}$$

Proof. Since $\tilde{x}_0 \in \ker \tilde{D} \cap X_{S_h}$, we have $\tilde{D}\tilde{x}_0 = 0$ and $S_h \tilde{x}_0 = \tilde{x}_0$. Write: $u = Dx$. Theorem 5.3.1 implies that the equation $\tilde{D}u = y$ with the condition $\tilde{F}u = \tilde{x}_0$ has a unique $\tilde{S}_h$-periodic solution $u = \tilde{R}_1^0 y - \tilde{x}_0$. Since $Dx = u$, we have an initial value problem: $Dx = \tilde{R}_1^0 y - \tilde{x}_0$, $Fx = x_0$, where, by our assumptions, $\tilde{R}_1^0 y - \tilde{x}_0 \in X_{S_h}$. Thus the last problem has a unique S_h-periodic solution

$$x = R_1^0(\tilde{R}_1^0 y - \tilde{x}_0) - x_0 = R_1^0 \tilde{R}_1 y - R_1^0 \tilde{x}_0 - x_0. \qquad \blacksquare$$

An immediate consequence is

COROLLARY 5.3.4. *Suppose that all assumptions of Theorem 5.3.2 are satisfied. If $y \in X_{\tilde{S}_h}$, $\tilde{R}_1^0 y \in X_{S_h}$ then all S_h-periodic solutions of Equation* (5.3.12) *are of the form*

$$x = R_1^0 \tilde{R}_1^0 y + R_1^0 z_1 + z_0, \tag{5.3.15}$$

where $z_0 \in \ker D$, $z_1 \in \ker \tilde{D} \cap X_{S_h}$ are arbitrary.

COROLLARY 5.3.5. *Suppose that all assumptions of Theorem 5.3.2 are satisfied. Suppose, moreover, that D and $\tilde{D}$ are commutative and y, $R_1^0 y$,*

$\tilde{R}_1^0 y \in X_{S_h} \cap X_{\tilde{S}_h}$. *Then Equation* (5.3.12) *has solutions of the form* (5.3.15) *which are S_h-periodic and $\tilde{S}_h$-periodic simultaneously.*

It is an immediate consequence of Corollary 5.3.3, if we apply the equality $D\tilde{D} = \tilde{D}D$ on dom $D \cap$ dom $\tilde{D}$ and change the role of D and $\tilde{D}$ in the proof of Theorem 5.3.2.

PROPOSITION 5.3.1. *Suppose that $D \in R(X)$, $\dim \ker D \neq 0$, F is an initial operator for D corresponding to an $R \in \mathcal{R}_D$, $S_h \in L_0(X)$ is an R-shift on $h \neq 0$ and D_1 is defined by Formula* (5.3.8). *Then*

$$D_1^n F = h^{-n} F \quad on \ X_{S_h} \quad (n = 0, 1, 2, ...). \tag{5.3.16}$$

Proof (by induction). For $n = 0$ Formula (5.3.16) becomes trivial. For $n = 1$, since $DF = 0$ and $F^2 = F$, $FS_h = F$ on X_{S_h} we have by definition: $D_1 F = \left(D + \dfrac{1}{h} F\right)F = \dfrac{1}{h} F^2 = h^{-1} F$.

Suppose Formula (5.3.16) to be true for an arbitrarily positive integer n. Then $D_1^{n+1} F = D_1(D_1^n F) = D_1 h^{-n} F = h^{-n} D_1 F = h^{-n} h^{-1} F = h^{-(n+1)} F$, which was to be proved. ∎

PROPOSITION 5.3.2. *Suppose that all assumptions of Proposition 5.3.1 are satisfied. Then*

$$D^n = D_1^n + F_n^h \quad on \ X_{S_h}^{(n)} = X_{S_h} \cap \operatorname{dom} D^n \quad (n = 0, 1, 2, ...), \tag{5.3.17}$$

where

$$F_n^h = -\sum_{j=0}^{n-1} h^{-n+j} FD^j, \quad F_0^h = 0. \tag{5.3.18}$$

Proof (by induction). For $n = 0$ Formula (5.3.17) becomes trivial. For $n = 1$ the definition of D_1 and the equality $F_1 = FS_h = F$ on X_{S_h} together imply that on X_{S_h} we have $D = D_1 - \dfrac{1}{h} F_1 = D_1 - \dfrac{1}{h} F = D_1 + F_1^h$. Suppose Formula (5.3.17) to be true for an arbitrarily fixed positive integer n. Proposition 5.3.1 implies that on $X_{S_h}^{(n+1)}$ we have (by our induction assumption):

$$D^{n+1} = D^n D = (D_1^n + F_n^h)D = D_1^n(D_1 - h^{-1}F) - F_n^h D$$

$$= D_1^n(D_1 - h^{-1}F) - \sum_{j=0}^{n-1} h^{-n+j}FD^j D$$

$$= D_1^{n+1} - h^{-1}D_1^n F - \sum_{j=0}^{n-1} h^{-n+1}FD^{j+1}$$

$$= D_1^{n+1} - h^{-1}h^{-n}F - \sum_{k=1}^{n} h^{-(n+1)+k}FD^k$$

$$= D_1^{n+1} - h^{-(n+1)}F - \sum_{k=1}^{n} h^{-(n+1)+k}FD^k$$

$$= D_1^n - \sum_{k=0}^{n} h^{-(n+1)+k}FD^k = D_1^{n+1} + F_{n+1}^h$$

which was to be proved. ∎

An immediate consequence of Proposition 5.3.2 is

COROLLARY 5.3.6. *Suppose that all assumptions of Proposition 5.3.1 are satisfied. Then*

$$D^n x = D_1^n x + \tilde{x}_n \quad \text{for all } x \in X_{S_h}^{(n+1)} = X_{S_h} \cap \mathrm{dom}\, D^{n+1}, \quad (5.3.19)$$

where

$$\tilde{x}_n = -\sum_{j=0}^{n-1} h^{-n+j}FD^j x, \quad \tilde{x}_0 = 0. \qquad (5.3.20)$$

THEOREM 5.3.3. *Suppose that $D \in R(X)$, $\dim \ker D \neq 0$, F is an initial operator for D corresponding to an $R \in \mathcal{R}_D$, $S_h \in L_0(X)$ is an R-shift on $h \neq 0$ and D_1, F_1, R_1^0 are defined by Formulae (5.2.8), (5.2.15). Write, as before*

$$Q(D) = \sum_{k=0}^{N} Q_k D^k, \quad \text{where } Q_k \in L_0(X), \; S_h Q_k = Q_k S_h,$$

$$(k = 0, 1, ..., N-1), Q_N = I, \qquad (5.3.21)$$

$$Q(t, s) = \sum_{k=0}^{N} Q_k t^k s^{N-k}. \qquad (5.3.22)$$

If the operator $Q(I, R_1^0)$ is invertible in the space $X_{S_h}^{(N)}$ then the equation

$$Q(D)x = y, \quad y \in X_{S_h} \tag{5.3.23}$$

together with the initial conditions

$$FD^k x = x_k, \quad x_k \in \ker D \quad (k = 0, 1, 2, \ldots, N-1) \tag{5.3.24}$$

has a unique S_h-periodic solution

$$x = (R_1^0)^N [Q(I, R_1^0)]^{-1}(y + y_0), \tag{5.3.25}$$

where

$$y_0 = -\sum_{k=1}^{N} Q_k \sum_{j=0}^{k-1} h^{-k+1} x_j \in X_{S_h}. \tag{5.3.26}$$

Proof. Observe that coefficients Q_k, as S_h-periodic operators, preserve the space X_{S_h}. Corollary 5.3.6 implies that

$$y = Q(D)x = \sum_{k=0}^{N} Q_k D^k x = \sum_{k=0}^{N} Q_k(D_1^k x + \tilde{x}_k)$$

$$= \sum_{k=0}^{N} Q_k D_1^k - \sum_{k=1}^{N} Q_k \sum_{j=0}^{k-1} h^{-k+j} FD^j x$$

$$= Q(D_1) - \sum_{k=1}^{N} Q_k \sum_{j=0}^{k-1} h^{-k+j} x_j = Q(D_1)x - y_0$$

because $\tilde{x}_0 = 0$. Observe that $y_0 \in X_{S_h}$. Indeed, since all constants $x_j \in X_{S_h}$ and operators Q_k are S_h-periodic, we find

$$S_h y_0 = -\sum_{k=1}^{N} S_h Q_k \sum_{j=0}^{k-1} h^{-k+j} x_j$$

$$= -\sum_{k=1}^{N} Q_k \sum_{j=0}^{k-1} h^{-k+j} S_h x_j$$

$$= -\sum_{k=1}^{N} Q_k \sum_{j=0}^{k-1} h^{-k+j} x_j = y_0.$$

We have obtained the equation

$$Q(D_1)x = y + y_0, \quad \text{where } y + y_0 \in X_{S_h}. \tag{5.3.27}$$

Put $u = D_1^N x$. Then $x = (R_1^0)^N u$, because $D_1 R_1^0 = R_1^0 D_1 = I$ on X_{S_h}. Hence

$$y + y_0 = Q(D_1)x = Q(D_1)(R_1^0)^N u = \sum_{k=0}^{N} Q_k D_1^k (R_1^0)^N$$

$$= \sum_{k=0}^{N} Q_k (R_1^0)^{N-k} u = Q(I, R_1^0)u.$$

But the operator $Q(I, R_1^0)$ is invertible in the space X_{S_h} which implies that $u = [Q(I, R_1^0)]^{-1}(y+y_0)$ and $x = (R_1^0)^N u = (R_1^0)^N [Q(I, R_1^0)]^{-1}(y+y_0)$. This is a unique S_h-periodic solution of the problem (5.3.23), (5.3.24), we were looking for.   ∎

An immediate consequence of Theorem 5.3.3 and Corollary 5.3.6 is

COROLLARY 5.3.7. *Suppose that all assumptions of Theorem 5.3.3 are satisfied and that the operator $Q(I, R_1^0)$ is invertible. Then all S_h-periodic solutions of Equation (5.3.23) are of the form*

$$x = (R_1^0)^N [Q(I, R_1^0)]^{-1} \left(y + \sum_{k=1}^{N} Q_k z_k \right), \tag{5.3.28}$$

where $z_1, \ldots, z_N \in \ker D$ are arbitrary.

COROLLARY 5.3.8. *Suppose that all assumptions of Theorem 5.3.3 are satisfied and that the operator $Q(I, R_1^0)$ is invertible. Then all S_h-periodic solutions of the equation*

$$D^M Q(D)x = y, \quad y \in X_{S_h} \quad (M \geqslant 0), \tag{5.3.29}$$

are of the form

$$x = (R_1^0)^N [Q(I, R_1^0)]^{-1} \left[(R_1^0)^M y + \sum_{k=0}^{M-1} (R_1^0)^k z_{N+k+1} + \sum_{k=1}^{N} Q_k z_k \right], \tag{5.3.30}$$

where $z_1, \ldots, z_{N+M} \in \ker D$ are arbitrary.

Proof. Put $u = Q(D)x$. Then we have the equation $D^M u = y$. Corollary 5.3.2 implies

$$u = (R_1^0)^M y + \sum_{k=0}^{M-1} (R_1^0)^k z_{N+k},$$

where $z_{N+k} \in \ker D$ are arbitrary.

Having already determined u, we apply Corollary 5.3.7 and we obtain

$$x = (R_1^0)^N [Q(I, R_1^0)]^{-1} \left(u + \sum_{k=1}^{N} Q_k z_k \right)$$

$$= (R_1^0)^N [Q(I, R_1^0)]^{-1} \left[(R_1^0)^M y + \sum_{k=0}^{M-1} (R_1^0)^k z_{N+k+1} + \sum_{k=1}^{N} Q_k z_k \right],$$

where $z_1, \ldots, z_{N+M} \in \ker D$ are arbitrary.   ∎

COROLLARY 5.3.9. *Suppose that all assumptions of Theorem 5.3.3 are satisfied and that the operator $Q(I, R_1^0)$ is invertible. Then all S_h-periodic solutions of the equation*

$$Q(D) D^M x = y, \quad y \in X_{S_h} \quad (M \geqslant 0) \tag{5.3.31}$$

are of the form

$$x = (R_1^0)^{M+N} [Q(I, R_1^0)]^{-1} \left(y + \sum_{k=1}^{N} Q_k z_k \right) + \sum_{k=0}^{M-1} (R_1^0)^k z_{N+k+1}, \tag{5.3.32}$$

where $z_1, \ldots, z_{N+M} \in \ker D$ are arbitrary.

Proof. Put $u = D^M x$. Corollary 5.3.7 implies that all S_h-periodic solutions of the equation $Q(D)u = y$ are of the form

$$u = (R_1^0)^N [Q(I, R_1^0)]^{-1} \left(y + \sum_{k=1}^{N} Q_k z_k \right),$$

where $z_1, \ldots, z_N \in \ker D$ are arbitrary.

Corollary 5.3.2 implies then that

$$x = (R_1^0)^M u + \sum_{k=0}^{M-1} (R_1^0)^k z_{N+k+1}$$

$$= (R_1^0)^M \left\{ (R_1^0)^N [Q(I, R_1^0)]^{-1} \left(y + \sum_{k=1}^{N} Q_k z_k \right) \right\} + \sum_{k=0}^{M-1} (R^0)^k z_{N+k+1}$$

$$= (R_1^0)^{M+N} [Q(I, R_1^0)]^{-1} \left(y + \sum_{k=1}^{N} Q_k z_k \right) + \sum_{k=0}^{M-1} (R_1^0)^k z_{N+k+1},$$

where $z_1, \ldots, z_{N+M} \in \ker D$ are arbitrary.   ∎

THEOREM 5.3.4. *Suppose that $D \in R(X)$, $\dim \ker D \neq 0$, F is an initial operator for D corresponding to an $R \in \mathcal{R}_D$, $S_h \in L_0(X)$ is an R-shift on $h \neq 0$ and D_1, R_1^0 are defined by Formulae (5.2.14), (5.2.15), respectively. If for a scalar $\lambda \neq 0$ the operator $I + \lambda R_1^0$ is invertible in the space X_{S_h} then*

(i) *An initial value problem*

$$(D + \lambda I) x = y, \quad y \in X_{S_h}, \tag{5.3.33}$$

$$Fx = x_0, \quad x_0 \in \ker D \tag{5.3.34}$$

has a unique S_h-periodic solution

$$x = (I + \lambda R_1^0)^{-1} \left(R_1^0 y + \frac{1}{h} x_0 \right)$$

$$= \Big(I + \lambda R + \lambda F S_h R -$$

$$- \frac{\lambda}{h} R F S_h R \Big)^{-1} \left(R + F S_h R - \frac{1}{h} R F S_h R \right) \left(y + \frac{1}{h} x_0 \right);$$

$$\tag{5.3.35}$$

(ii) *All S_h-periodic solutions of Equation (5.3.33) are of the form*

$$x = \Big(I + \lambda R + \lambda F S_h R - \frac{\lambda}{h} R F S_h R \Big)^{-1} \left(R + F S_h R - \right.$$

$$\left. - \frac{1}{h} R F S_h R \right) \left(y + \frac{1}{h} z \right), \tag{5.3.36}$$

where $z \in \ker D$ is arbitrary.

Proof. Recall that $D = D_1 - \frac{1}{h} F$ on X_{S_h}. Theorem 5.2.3 implies that

$D_i R_1^0 = I$ on X_{S_h}. Thus the problem (5.3.33), (5.3.34) can be rewritten as follows:

$$y = (D + \lambda I)x = Dx + \lambda x = D_1 x - \frac{1}{h} Fx + \lambda x$$

$$= D_1 x + \lambda x - \frac{1}{h} x_0 = (D_1 + \lambda I)x - \frac{1}{h} x_0$$

$$= D_1(I + \lambda R_1^0)x - \frac{1}{h} x_0,$$

i.e. we obtain an equation

$$D_1(I + \lambda R_1^0)x = y + \frac{1}{h} x_0.$$

Since D_1 is invertible , we find

$$(I + \lambda R_1^0)x = R_1^0 \left(y + \frac{1}{h} x_0 \right).$$

The invertibility of the operator $I + \lambda R_1^0$ implies Formula (5.3.35). Hence x determined by this formula is a unique S_h-periodic solution, we were looking for. An immediate consequence of this fact is that all S_h-periodic solutions of Equation (5.3.33) are of the form (5.3.36). ∎

COROLLARY 5.3.10. *Suppose that all assumptions of Theorem 5.2.7 are satisfied, the operators D_1 and R_1^0 are defined by Formulae (5.2.31) and*

the operators $I - AR$ and $I - A(R + R_1^0) = I - A \left(2I + FS_h - \frac{1}{h} RFS_h \right) R$

are invertible in the space X_{S_h}. Then the equation

$$Dx = Ax + y, \quad y \in X_{S_h} \tag{5.3.37}$$

has a unique S_h-periodic solution satisfying the condition

$$Fx = x_0 \tag{5.3.38}$$

which is of the form

$$x = (I - AR - AR_1^0)^{-1} \left(R_1^0 y + \frac{1}{h} x_0 \right). \tag{5.3.39}$$

Proof. Write: $u = (I - AR)x$. Then $Fu = Fx - FARx = Fx - AFRx$
$= Fx = x_0$. Since $x = (I - AR)^{-1}u = E_A u$, we find

$$u = R_1^0 D_1 u = R_1^0 \left(Du + \frac{1}{h} Fu \right)$$

$$= R_1^0 (D - A) E_A u - A E_A u + \frac{1}{h} x_0$$

$$= R_1^0 (D - A) x - A E_A u + \frac{1}{h} x_0 = R_1^0 \left(y + \frac{1}{h} x_0 - A E_A u \right).$$

Hence $(I - R_1^0 A E_A) u = R_1^0 \left(y + \frac{1}{h} x_0 \right)$. But

$$I - R_1^0 A E_A = I - A R_1^0 (I - AR)^{-1} = (I - AR - A R_1^0)(I - AR)^{-1}$$
$$= (I - AR - A R_1^0) E_A$$

and this operator is invertible as a superposition of invertible operators. Thus

$$u = (I - R_1^0 A E_A)^{-1} R_1^0 \left(y + \frac{1}{h} x_0 \right)$$

$$= E_A^{-1} (I - AR - A R_1^0)^{-1} \left(R_1^0 y + \frac{1}{h} x_0 \right)$$

and

$$x = E_A u = (I - AR - A R_1^0)^{-1} \left(R_1^0 y + \frac{1}{h} x_0 \right).$$

This solution is S_h-periodic because we were dealing in the space X_{S_h} only.

Exercises

EXERCISE 5.3.1. Generalize Theorem 5.3.3 and Corollaries 5.3.7, 5.3.8, 5.3.9 for the case when:
(i) the operator $Q(I, R_1^0)$ is either right or left invertible;
(ii) we consider instead of the operator $Q(D)$ the operator $Q\langle D \rangle$ defined by Formula (3.2.1) and the corresponding operator $Q\langle I, R_1^0 \rangle$ defined by Formula (3.2.10) is either invertible or left invertible or right invertible.

EXERCISE 5.3.2. Generalize Theorem 5.3.4 for the case when the operator $I + \lambda R$ is either left or right invertible.

EXERCISE 5.3.3. Generalize Corollary 5.3.10 for the case when the operators $I - AR$ and $I - A(R + R_1^0)$ are either left or right invertible.

EXERCISE 5.3.4. Suppose that all assumptions of Theorem 5.3.4 are satisfied. Write:

$$Q(D) = \sum_{k=0}^{N} q_k D^k, \quad \text{where } q_0, \ldots, q_{N-1} \in \mathbf{C}, \; q_N = 1,$$

$$Q(t, s) = \sum_{k=0}^{N} q_k t^k s^{N-k}.$$

(i) Prove that the operator $Q(I, R_1^0)$ has an inverse mapping $X_{S_h}^{(n)}$ into X_{S_h}, provided that $R_0^1 \in V(X_{S_h})$;

(ii) Find all S_h-periodic solutions of the equation
$$D^M Q(D) x = Q(D) D^M x = y, \quad y \in X_{S_h}, \quad M \geqslant 0,$$
provided that the operator $Q(I, R_1^0)$ is invertible.

5.4. CANONICAL MAPPING

We shall introduce in this section a mapping induced by a family of shifts. Properties of this mapping for some more specified spaces and operators will permit to us to prove results of a great importance.

DEFINITION 5.4.1. Suppose that X is a linear space over $\mathbf{C}$, $D \in R(X)$, dim ker $D \neq 0$, F is an initial operator for D corresponding to an $R \in \mathscr{R}_D$ and either $S_{A(\mathbf{R})} = \{S_h\}_{h \in A(\mathbf{R})}$ is a family of R-shifts or $R \in V(X)$ and $S_{A(\mathbf{R})}$ is a family of D-shifts. A mapping $\varkappa$ of X is said to be *canonical for* $S_{A(\mathbf{R})}$ if
$$\varkappa x = \{F S_h x\}_{h \in A(\mathbf{R})} \quad \text{for } x \in X. \tag{5.4.1}$$

It is clear that $\varkappa$ maps X into the set of functions defined on $A(\mathbf{R})$ with values in ker D.

Write
$$(\varkappa x)(h) = x^\wedge(h), \quad \text{where } x^\wedge(h) = F S_h x \text{ for all } x \in X, \; h \in A(\mathbf{R}). \tag{5.4.2}$$

An immediate consequence of this definition is:

PROPOSITION 5.4.1. *Suppose that all assumptions of Definition 5.4.1 are satisfied. Then the functions* $x^{\wedge}$ *of real variable* $h \in A(\mathbf{R})$ *and with values in* ker D *defined by Formula* (5.4.2) *have the following properties*:

(i) *For all* $x, y \in X$, $\lambda, \mu \in \mathbf{C}$

$$(\lambda x + \mu y)^{\wedge} = \lambda x^{\wedge} + \mu y^{\wedge};$$

(ii) $x^{\wedge}(0) = Fx$ *for all* $x \in X$;

(iii) *If* $x, y \in X$ *and* $x = y$ *then* $x^{\wedge} = y^{\wedge}$;

(iv) *If* $x \in X$ *is* S_h-periodic for an $h \in A(\mathbf{R})$ *then* $x^{\wedge}(nh) = x^{\wedge}(0)$ *for all* $n \in \mathbf{N}$.

THEOREM 5.4.1. *Suppose that* $D \in R(X)$, dim ker $D \neq 0$, F *is an initial operator for* D *corresponding to an* $R \in \mathcal{R}_D$ *and* $S_{A(\mathbf{R})}$ *is the family of R-shifts. Then the induced canonical mapping* $\varkappa$ *has the following properties*:

(i) $\varkappa$ *is linear*;

(ii) *The restriction of* $\varkappa$ *to the set* $P(R)$ *of all D-polynomials is an isomorphism*:

$$\ker \varkappa|_{P(R)} = \{0\}; \tag{5.4.3}$$

(iii) $\varkappa$ *separates points of* $P(R)$, *i.e.*

$$\varkappa x = \varkappa y \text{ if and only if } x = y \text{ for } x, y \in P(R). \tag{5.4.4}$$

Proof. The linearity of $\varkappa$ is an immediate consequence of Formula (5.4.2) and Point (i) of Proposition 5.4.1. In order to prove Points (ii) and (iii) of our theorem we have to show that

$$x^{\wedge} = y^{\wedge} \text{ if and only if } x = y \text{ for } x, y \in P(R). \tag{5.4.5}$$

Observe that by Point (iii) of Proposition 5.4.1, if $x = y$ then $x^{\wedge} = y^{\wedge}$. Hence we have only to prove that $x \neq y$ implies $x^{\wedge} \neq y^{\wedge}$ for $x, y \in P(R)$. Recall that elements $R^k z_1$ and $R^m z_2$, where $z_1, z_2 \in$ ker D, are linearly independent for $m \neq k$, even if $z_1 = z_2$ (cf. Propositions 2.4.3 and 2.4.4). Write: $x = R^k z_1, y = R^k z_2$, $z_1 \neq z_2$, $z_1, z_2 \in$ ker D, $k \in \mathbf{N} \cup \{0\}$ is arbitrarily fixed. Suppose that $x^{\wedge} = y^{\wedge}$. Since $FR = 0$, we have for all $h \in A(\mathbf{R})$

$$0 = x^{\wedge}(h) - y^{\wedge}(h) = FS_h x - FS_h y = FS_h R^k z_1 - FS_h R^k z_2$$

$$= FS_h R^k(z_1 - z_2) = F \sum_{j=0}^{k} \frac{h^{k-j}}{(k-j)!} R^j(z_1 - z_2)$$

$$= \sum_{j=0}^{k} \frac{h^{k-j}}{(k-j)!} FR^j(z_1 - z_2) = \frac{h^k}{k!} F(z_1 - z_2) = \frac{h^k}{k!}(z_1 - z_2),$$

which implies $z_1 = z_2$. But this contradicts our assumption that $z_1 \neq z_2$. Thus $x^\wedge \neq y^\wedge$ and Formula (5.4.5) is already proved. Hence $\varkappa x = \{x^\wedge(h)\}_{h \in A(R)} = \{y^\wedge(h)\}_{h \in A(R)} = \varkappa y$ if and only if $x = y$ for $x, y \in P(R)$. In particular, if $x \in P(R)$, we conclude that $\varkappa x = 0$ if and only if $x = 0$, i.e. Formula (5.4.3) is satisfied. ∎

THEOREM 5.4.2. *Suppose that* $D \in R(X)$, $\dim \ker D \neq 0$, F *is an initial operator corresponding to an* $R \in \mathscr{R}_D \cap V(X)$ *and* $S_{A(R)}$ *is the family of D-shifts. Then the induced canonical mapping* $\varkappa$ *has the following properties*:

(i) $\varkappa$ *is linear*;

(ii) *The restriction of* $\varkappa$ *to the set* $E(R)$ *of all exponentials is an isomorphism*:

$$\ker \varkappa|_{E(R)} = \{0\}; \tag{5.4.6}$$

(iii) $\varkappa$ *separates points of* $E(R)$, *i.e.*

$$\varkappa x = \varkappa y \quad \text{if and only if } x = y \text{ for } x, y \in E(R). \tag{5.4.7}$$

Proof. The linearity of $\varkappa$ is an immediate consequence of Formula (5.4.2) and Point (i) of Proposition 5.4.1. In order to prove Points (ii) and (iii) of our theorem we have to show that

$$x^\wedge = y^\wedge \quad \text{if and only if } x = y \text{ for } x, y \in E(R). \tag{5.4.8}$$

By Point (i) of Proposition 5.4.1 if $x = y$ then $x^\wedge = y^\wedge$. Hence it is enough to prove that $x \neq y$ implies $x^\wedge \neq y^\wedge$. Recall that exponential elements $e_\lambda(z_1)$, $e_\mu(z_2)$, where $z_1, z_2 \in \ker D$, are linearly independent for $\lambda \neq \mu$, even in the case, when $z_1 = z_2$ (cf. Proposition 2.3.4). Consider elements $x = e_\lambda(z)$ and $y = e_\mu(z)$, where $0 \neq z \in \ker D$ and

$$\lambda \neq \mu + 2\pi i k/h, \quad \text{where } k \text{ is an arbitrary integer.} \tag{5.4.9}$$

Suppose that $x^\wedge = y^\wedge$. Then for all $h \in A(\mathbf{R})$ we have

$$0 = x^\wedge(h) - y^\wedge(h) = FS_h x - FS_h y = FS_h[e_\lambda(z) - e_\mu(z)]$$
$$= F[e^{\lambda h} e_\lambda(z) - e^{\mu h} e_\mu(z)] = e^{\lambda h} Fe_\lambda(z) - e^{-\mu h} Fe_\mu(z)$$
$$= e^{\lambda h} z - e^{\mu h} z = (e^{\lambda h} - e^{\mu h}) z = e^{\lambda h}(1 - e^{(\mu - \lambda)h}) z$$

which implies $e^{(\mu - \lambda)h} = 1$ for $z \neq 0$ and $e^{\lambda h} \neq 0$. Hence $\mu - \lambda = 2\pi i k/h$, where k is an integer, which contradicts our assumption (5.4.7). Therefore $x^\wedge = y^\wedge$ and Formula (5.4.6) is already proved. Further the proof is going on the same lines as the proof of Theorem 5.4.1. ∎

Examples and Exercises

EXAMPLE 5.4.1. Suppose that X is the space $C(\mathbf{R})$ of all functions continuous on real line, $D = \mathrm{d}/\mathrm{d}t$, $(Ry)(t) = \int_0^t x(s)\,\mathrm{d}s$, $(Fx)(t) = x(0)$, $(S_h x)(t) = x(t+h)$ for all $x \in X$, $h \in \mathbf{R}$. Then $\{S_h\}_{h \in \mathbf{R}}$ is a family of R-shifts and D-shifts simultaneously. Write

$$\hat{D} = D^2, \quad \hat{R} = R^2. \tag{5.4.10}$$

Then $\hat{D} \in R(X)$, $\hat{R}$ is a right inverse for $\hat{D}$, an initial operator $\hat{F}$ corresponding to $\hat{R}$ is of the form: $\hat{F} = F + RFD$ and

$$\ker \hat{D} = \{\hat{z} = z_0 + Rz_1 : z_0, z_1 \in \ker D\}$$
$$= \{c_0 + c_1 t : c_0, c_1 \in \mathbf{R}\}.$$

It is easy to check that all exponential elements for the operator $\hat{D}$ are of the form:

$$\hat{e}_\lambda(\hat{z}) = c_0 \operatorname{ch} \sqrt{\lambda}\, t + \frac{c_1}{\sqrt{\lambda}} \operatorname{sh} \sqrt{\lambda}\, t$$
$$= \frac{1}{2}\left[\left(c_0 + \frac{c_1}{\sqrt{\lambda}}\right) e^{\sqrt{\lambda} t} + \left(c_0 - \frac{c_1}{\sqrt{\lambda}}\right) e^{-\sqrt{\lambda} t}\right], \quad \lambda \in \mathbf{C} \tag{5.4.11}$$

and monomials are of the form:

$$\hat{R}^k \hat{z} = c_0 \frac{t^{2k}}{(2k)!} + c_1 \frac{t^{2k+1}}{(2k+1)!}, \quad k \in \mathbf{N}, c_0, c_1 \in \mathbf{R}. \tag{5.4.12}$$

Observe that S_h are neither $\hat{R}$-shifts nor $\hat{D}$-shifts. Define $\{\hat{S}_h\}_{h \in \mathbf{R}}$ as $\hat{R}$-shifts. Then for all $k \in \mathbf{N} \cup \{0\}$, $h \in \mathbf{R}$ we have

$$\hat{S}_h \hat{R}^k \hat{F} = \sum_{j=0}^{k} \frac{h^{k-j}}{(k-j)!} \hat{R}^j \hat{F} = \sum_{j=0}^{k} \frac{h^{k-j}}{(k-j)!} R^{2j}(F+RFD)$$

$$= \sum_{j=0}^{k} \frac{h^{k-j}}{(k-j)!} (R^{2j}+R^{2j+1}FD).$$

Hence for all $k \in \mathbf{N} \cup \{0\}$, $\hat{z} = z_0 + Rz_1$, $z_0, z_1 \in \ker D, h \in \mathbf{R}$ we find

$$\hat{S}_h \hat{R}^k \hat{z} = \sum_{j=0}^{k} \frac{h^{k-j}}{(k-j)!} (R^{2j}z_0 + R^{2j+1}z_1)$$

$$= \sum_{j=0}^{k} \frac{h^{k-j}}{(k-j)!} \left(c_0 \frac{t^{2j}}{(2j)!} + c_1 \frac{t^{2j+1}}{(2j+1)!} \right).$$

Thus $\hat{S}_h$ are well-determined $\hat{R}$-shifts. Since the operator $\hat{R} = R^2$ is a Volterra operator (for the operator $I - \lambda\hat{R} = I - \lambda R^2 = (I - \sqrt{\lambda} R)(I + \sqrt{\lambda} R)$ is invertible for all $\lambda \in \mathbf{C}$ as a superposition of invertible operators), we may suppose that $\hat{S}_h$ are also $\hat{D}$-shifts, i.e. the following equality holds:

$$\hat{S}_h \hat{e}_\lambda(\hat{z}) = e^{\lambda h} \hat{e}_\lambda(\hat{z}) \quad \text{for all } h \in \mathbf{R}, \lambda \in \mathbf{C}, \hat{z} \in \ker \hat{D}. \quad (5.4.13)$$

We shall prove Formula (5.4.13) later, using stronger properties of operators under consideration (cf. Section 8.2).

Let x be a monomial, i.e. a function of the form (5.4.12). Then for all $h, t \in \mathbf{R}$ we find, according with Formula (5.4.1), $Fx = 0$ if $k \neq 0$ and

$$[x^{\hat{}}(h)](t) = (\hat{F}\hat{S}_h x)(t)$$

$$= \hat{F} \sum_{j=1}^{k} \frac{h^{k-j}}{(k-j)!} \left(c_0 \frac{t^{2j}}{(2j)!} + c_1 \frac{t^{2j+1}}{(2j+1)!} \right)$$

$$= \sum_{j=0}^{k} \frac{h^{k-j}}{(k-j)!} (F+RFD) \left(c_0 \frac{t^{2j}}{(2j)!} + c_1 \frac{t^{2j+1}}{(2j+1)!} \right)$$

$$= \frac{h^k}{k!}(c_0+c_1 t)+\sum_{j=1}^{k}\frac{h^{k-j}}{(k-j)!}\,RF\left(c_0\frac{t^{2j-1}}{(2j-1)!}+c_1\frac{t^{2j}}{(2j)!}\right)$$

$$= \frac{h^k}{k!}(c_0+c_1 t).$$

Therefore, if x is of the form (5.4.12) then

$$\varkappa x = \{x^\wedge(h)\}_{h\in\mathbf{R}} = \frac{1}{k!}(c_0+c_1 t)\{h^k\}_{h\in\mathbf{R}}.$$

EXERCISE 5.4.1. Determine canonical mapping for shifts defined in Examples 5.2.3, 5.2.6 and Exercises 5.2.1–5.2.4, 5.2.6.

5.5. BOUNDARY VALUE PROBLEMS FOR STATIONARY LINEAR SYSTEMS WITH SHIFTS

Suppose that $D \in R(X)$, dim ker $D \neq 0$, $F, F_1 \neq F$ are initial operators for D corresponding to right inverses R, R_1, respectively and $\{S_h\}_{h\in\mathbf{R}}$ is the family of D-invariants R-shifts (D-shifts, respectively). Let $A, B \in L_0(X)$ be stationary operators. Suppose that the operator $I-AR = I-RA$ is invertible and write: $E_A = (I-AR)^{-1}$. By our assumptions, A and B commute with D and R, hence with F, F_1 and S_h for $h \in \mathbf{R}$. Recall that the following identities hold:

$$(D-A)E_A R = I, \tag{5.5.1}$$
$$E_A R(D-A)+E_A F = I \quad \text{on dom } D$$

(cf. Section 3.3).

Consider the system

$$Dx = Ax+BS_{-h}x+y, \quad y \in X \tag{5.5.2}$$

with the boundary conditions:

$$Fx = x_0, \quad F_1 x = x_1, \quad x_0, x_1 \in \ker D. \tag{5.5.3}$$

The system (5.5.2) can be rewritten as follows:

$$(D-A)x = BS_{-h}x+y. \tag{5.5.4}$$

Applying (5.5.1) and the first of Conditions (5.5.3) we find

$$x = E_A R(D-A)x + E_A Fx = E_A R(BS_{-h}x+y) + E_A x_0.$$

From the second of Conditions (5.5.3) we have

$$x_1 = F_1 x = F_1 E_A R(BS_{-h}x+y) + x_0. \qquad (5.5.5)$$

Denote by $F_1^{\{-1)}(E)$ the *inverse image* of a set $E \subset \ker D$, i.e.

$$F_1^{\{-1)}(E) = \{y \in \operatorname{dom} D: F_1 y \in E\}. \qquad (5.5.6)$$

In particular

$$F_1^{(-1)}(x_1) = \{y \in \operatorname{dom} D: F_1 y = x_1\}. \qquad (5.5.7)$$

If D is right invertible, but not left invertible, i.e. $\ker D \neq \{0\}$, then the sets $F_1^{\{-1)}(E)$ are non-trivial. Then from (5.5.3) we find

$$E_A R(BS_{-h}x+y) + x_0 = y_0, \quad \text{where } y_0 \in F_1^{(-1)}(x_1),$$

i.e.

$$R(BS_{-h}x+y) + x_0 = (I-AR)y_0, \quad \text{where } y_0 \in F_1^{(-1)}(x_1).$$

Since $y_0 \in \operatorname{dom} D$ and $x_0 \in \ker D$, acting on both sides of the last equality by the operator DS_{-h} we obtain

$$Bx+y = S_h D(I-AR)y_0 = S_h(D-A)y_0,$$
$$\text{where } y_0 \in F_1^{(-1)}(x_1),$$

i.e.

$$Bx = S_h(D-A)y_0 - y, \quad \text{where } y_0 \in F_1^{(-1)}(x_1). \qquad (5.5.8)$$

An immediate consequence of this equality, our assumptions and the first of Conditions (5.5.3) is the following

THEOREM 5.5.1. *Suppose that B is left invertible (invertible) and B_L is a left inverse (inverse of B). If there exists a $y_0 \in F_1^{(-1)}(x_1)$ such that*

$$FB_L S_h(D-A)y_0 - y = x_0 \qquad (5.5.9)$$

then there exists a solution of the boundary value problem (5.5.2)–(5.5.3) and is of the form

$$x = B_L S_h(D-A)y_0 - y. \qquad (5.5.10)$$

If B is neither left invertible nor invertible we can find solutions of the problem (5.5.2)–(5.5.3) by the following

THEOREM 5.5.2. *Denote by* $B^{(-1)}(E)$ *and* $F^{(-1)}(E)$ *the inverse images of a set* $E \subset \text{dom } D$. *If the set*

$$E_0 = B^{(-1)}[S_h(D-A)y_0 - y] \cap F^{(-1)}(x_0) \qquad \text{where } y_0 \in F_1^{\{-1\}}(x_1)$$

$$(5.5.11)$$

is non-empty, then any element $x \in E_0$ *is a solution of the boundary value problem* (5.5.2)–(5.5.3).

Proof. The proof follows from the fact that any element of the set E_0 satisfies (5.5.8), hence also (5.5.2) and the boundary conditions (5.5.3). ∎

Suppose now that U is a linear space over the same field of scalars and that the operator C is defined on the space U and has its range in X. Consider the system:

$$Dx = Ax + BS_{-h}x + Cu \tag{5.5.12}$$

with the boundary conditions (5.5.3).

THEOREM 5.5.3. *Suppose that* B *is left invertible (invertible) and* B_L *is a left inverse of* B *(inverse). If there exists a* $u \in U$ *which satisfies the equation*

$$B_L S_h Cu = y_2 + Fy_1 - x_0, \tag{5.5.13}$$

where

$$y_0 \in F_1^{\{-1\}}(x_1), \qquad y_1 = B_L S_h(D-A)y_0,$$
$$y_2 \in F_1^{\{-1\}}[(F_1 - F)y_1 + x_0 - x_1] \tag{5.5.14}$$

then all solutions of the boundary value problem are of the form

$$x = (I-F)B_L S_h(D-A)y_0 - y_2 + x_0, \tag{5.5.15}$$

where y_0, y_1, y_2 *are determined by* (5.5.14) *and the set* $F_1^{\{-1\}}(x_1)$ *plays a role of a family of parameters.*

Proof. In a similar way, as Equality (5.5.8), we conclude that Equation (5.5.12) can be reduced to an equation of the form:

$$Bx + S_h Cu = S_h(D-A)y_0, \qquad y_0 \in F_1^{\{-1\}}(x_1). \tag{5.5.16}$$

By our assumptions, acting on both sides of this equation by the operator B_L we find

$$x = B_L S_h (D-A) y_0 - B_L S_h Cu = y_1 - B_L S_h Cu, \qquad (5.5.17)$$

where $y_0 \in F_1^{(-1)}(x_1)$. Write $v = B_L S_h Cu$. Then $x = y_1 - v$ and from (5.5.3) we find

$$x_0 = Fx = Fy_1 - Fv, \quad x_1 = F_1 x = F_1 y_1 - F_1 v,$$

i.e. we have conditions of solvability:

$$Fv = x_3, \quad F_1 v = x_4, \qquad (5.5.18)$$

where $x_3 = Fy_1 - x_0 \in \ker D$, $x_4 = F_1 y_1 - x_1 \in \ker D$, $v = B_L S_h Cu$.

The elements x_3, x_4 are given (up to y_0). There exists a $w \in X$ such that $F_1 Rw = x_4 - x_3 \in \ker D$. This implies that $v = Rw + x_3 \in \operatorname{dom} D$ and $Dv = w$, $Fv = FRw + Fx_3 = x_3$, $F_1 v = F_1 Rw + x_3 = x_4 - x_3 + x_3 = x_4$. Write:

$$y_3 = x_4 - x_3 = (F_1 - F) y_1 + x_0 - x_1 = (F_1 - F) y_1, \qquad (5.5.19)$$

where y_1 is determined by (5.5.14). Then

$$Rw = v - x_3 \in F_1^{(-1)}(y_3),$$

i.e. $v = y_2 + x_3$, where $y_2 \in F_1^{(-1)}(y_3)$. Finally we get $x = y_1 - v = y_1 - y_2 - x_3 = y_1 - y_2 - Fy_1 + x_0 = (I-F) y_1 - y_2 + x_0$, where y_0, y_1, y_2 are determined by Formulae (5.5.14). Observe that the set $F_1^{(-1)}(x_1)$ plays a role of the family of parameters for the problem (5.5.12)–(5.5.3) and that

$$B_L S_h Cu = v = y_2 + x_3 = y_2 + Fy_1 - x_0. \qquad (5.5.20)$$

∎

COROLLARY 5.5.1. *Suppose that all assumptions of Theorem 5.5.3 are satisfied, C is left invertible (invertible) and C_L is a left inverse (inverse) of C. Then*

$$u = C_L(D-A) y_0 + BFy_1 - S_{-h} B(y_1 - y_2 + x_0), \qquad (5.5.21)$$

where y_0, y_1, y_2 are defined by Formula (5.5.14).

Proof. Since B is stationary we have $BS_h = S_h B$. Also $S_h z = z$ for all

$z \in \ker D$. This and Formulae (5.5.12), (5.5.17) and (5.5.19), together imply

$$
\begin{aligned}
Cu &= (D-A)y_0 - S_{-h}Bx = (D-A)y_0 - S_{-h}B(y_1 - B_L S_h Cu) \\
&= (D-A)y_0 - S_{-h}B(y_1 - y_2 + x_0 + Fy_1) \\
&= (D-A)y_0 - S_{-h}B(y_1 - y_2 + x_0) + BS_{-h}Fy_1 \\
&= (D-A)y_0 + BFy_1 - S_{-h}B(y_1 - y_2 + x_0).
\end{aligned}
$$

Hence

$$
u = C_L Cu = C_L[(D-A)y_0 + BFy_1 - S_{-h}B(y_1 - y_2 + x_0)]. \quad \blacksquare
$$

If B is neither left invertible nor invertible then the corresponding results can be obtained if we apply Theorem 5.5.2 instead of Theorem 5.5.1 (cf. the proof of Theorem 5.5.3 which is based on the proof of Theorem 5.5.1).

COROLLARY 5.5.2. *Suppose that all assumptions of Theorem 5.5.3 are satisfied. Consider the system*

$$
Dx = Ax + BS_{-h}x + S_{-h}Cu, \quad u \in U \tag{5.5.22}
$$

together with the boundary conditions (5.5.3). *If there exists a $u \in U$ which satisfies the equation*

$$
B_L S_h Cu = y_2 + Fy_1 - x_0, \tag{5.5.23}
$$

where y_0, y_1, y_2 are defined by Formula (5.5.14), *then all solutions of the boundary value problem* (5.5.22)–(5.5.3) *are of the form* (5.5.15). *If, in addition, C is left invertible (invertible) and C_L is a left inverse (inverse) of C then*

$$
u = C_L[S_h(D-A)y_0 - B(y_1 - y_2 + x_0 - Fy_1)]. \tag{5.5.24}
$$

Proof. The proof is going on the same lines as the proofs of Theorem 5.5.3 and Corollary 5.5.1 if we put $S_{-h}Cu$ instead of Cu. $\quad \blacksquare$

Observe that in all considerations dimensions of the spaces X and U did not play any role.

EXAMPLE 5.5.1. Let $X = U = C(\mathbf{R} \to \mathbf{R}^n)$, i.e. X and U are spaces of all continuous n-dimensional vector functions of a real variable. Let $a, b \in \mathbf{R}$ be arbitrarily fixed and $b \neq a$. Write:

$$
D = \frac{d}{dt}, \quad R = \int_a^t, \quad (Fx)(t) = x(a), \quad R_1 = \int_b^t,
$$

$$(F_1 x)(t) = x(b), \quad (S_{-h}x)(t) = x(t-h) \quad \text{for } x \in X, h \in \mathbf{R}.$$

Let A, B, C be square $n \times n$ scalar matrices. By definition, A, B, C are stationary. In particular, $S_{-h}C = CS_{-h}$ $(h \in \mathbf{R})$. Suppose also that B is invertible. Consider the system

$$x'(t) = Ax(t) + Bx(t-h) + Cu(t-h) \tag{5.5.25}$$

with the boundary conditions

$$x(a) = x_0, \quad x(b) = x_1 \quad (a, b \in \mathbf{R}, b \neq a). \tag{5.5.26}$$

This system satisfies all the assumptions of Theorem 5.5.3 and Corollary 5.5.2. Let $y_0 \in X$ be an arbitrary function such that $y_0(b) = x_1$. Then $y_1(t) = [B^{-1}S_h(D-A)y_0](t) = B^{-1}[y_0'(t+h) - Ay_0(t+h)]$. Let $y_2 \in X$ be an arbitrary function such that $y_1(b) - y_1(a) + x_0 - x_1 = y_2(b)$, i.e.

$$y_2(b) = B^{-1}[y_0'(b+h) - Ay_0(b+h) - y_0'(a+h) + Ay_0(a+h)] + \\ + x_0 - x_1,$$

$$x(t) = [(I-F)S_{-h}B^{-1}(D-A)y_0 - y_2 + x_0](t)$$
$$= B^{-1}[y_0'(t+h) - Ay_0(t+h) - y_0(a+h) + Ay_0(a+h) - \\ - y_2(t) + x_0].$$

EXAMPLE 5.5.2. Let $X, U, D, R, F, \{S_h\}_{h \in \mathbf{R}}, A, B, C$ be defined as in Example 5.5.1. Let

$$(F_1 x)(t) = \frac{1}{b-a} \int_a^b x(s) \, ds \quad \text{for } x \in X.$$

The corresponding right inverse is of the form:

$$(R_1 x)(t) = \int_a^t x(s) \, ds + \frac{1}{b-a} \int_a^b (b-s)x(s) \, ds \quad \text{for } x \in X.$$

Consider the system (5.5.25) with the non-local conditions:

$$x(a) = x_0, \quad \frac{1}{b-a} \int_a^b x(s) \, ds = x_1. \tag{5.5.27}$$

For instance: Given $x_1(t) = \varphi(t)$ for $t \in [a-h, h]$. Again, let y_0 be an arbitrary function such that $\int_a^b y_0(t) \, dt = (b-a)x_1$. Further considerations are going on the same lines.

EXERCISE 5.5.1. Let X, U, D, R, F, $\{S_h\}_{h \in \mathbf{R}}$, A, B, C be defined as in Example 5.5.1. Consider the following boundary condition instead of the second of Conditions (5.5.26) for Equation (5.5.25):

(i) $(F_1 x)(t) = x(a) + dx'(b)$ for $x \in X$ where $d \neq 0$ is a given real;

(ii) $(F_1 x)(t) = dx(a) + \dfrac{1-d}{b-a} \displaystyle\int\limits_a^b x(s)\,ds$ for $x \in X, 0 \neq d \in \mathbf{R}$

and find conditions of solvability of the corresponding boundary value problems.

EXERCISE 5.5.2. Suppose that all assumptions of Example 5.2.3 are satisfied and $(Fx)(t) = x(0)$, $(F_1 x)(t) = x(T)$, $T > 0$. Solve the boundary value problem (5.5.2)–(5.5.3) with scalar $n \times n$ matrices A, B.

EXERCISE 5.5.3. Suppose that $X = (s)$, $D\{x_n\} = \{x_{n+1} - x_n\}$, $F\{x_n\} = x_1\{e_n\}$, where $e_n = 1$ for $n \in \mathbf{N}$, $F_1\{x_n\} = x_2\{e_n\}$. Solve the boundary value problem (5.5.2)–(5.5.3) with scalar coefficients $A, B \in \mathbf{R}$.

Added in proof:

EXAMPLE 5.2.8. Suppose that all assumptions of Proposition 5.2.7 are satisfied. Let $x = a_1 e_{\lambda_1}(z) + \ldots + a_n e_{\lambda_n}(z)$, where $z \in \ker D$, $a_1, \ldots, a_n$ are scalars. Write:

$$a^x(h) = a_1 \exp(\lambda_1 h) + \ldots + a_n \exp(\lambda_n h).$$

Then

$$\varkappa x = \{a^x(h)\}_{h \in \mathbf{R}} \cdot z.$$

In particular,

$$F_h c_\lambda(z) = z \cdot \cos \lambda h, \qquad F_h s_\lambda(z) = z \cdot \sin \lambda h, \qquad \text{where } F_h = F S_h.$$

Moreover, if $\lambda = 2\pi k / h$, where k is an arbitrary integer, then

$$F_0 s_\lambda(z) = F_{h/2}\, s_\lambda(z) = F_{h/4}\, c_\lambda(z) = 0;$$
$$F_0 c_\lambda(z) = -F_{h/2}\, c_\lambda(z) = F_{h/4}\, s_\lambda(z) = z.$$

Similarly, if S_h are R-shifts and $x = a_0 z + a_1 Rz + \ldots + a_n R^n z$, $a^x(h) = a_0 + a_1 h + \ldots + a_n h^n$ then $\varkappa x = \{a^x(h)\}_{h \in \mathbf{R}} \cdot z$.

Chapter 6

D-algebras

Let X be a commutative algebra (i.e. a commutative linear ring) and let $D \in R(X)$. X is said to be a *D-algebra* if dim ker $D > 0$ and the following condition is satisfied

$$x \in \text{dom } D \text{ and } y \in \text{dom } D \quad \text{implies } xy \in \text{dom } D, \quad (6.1.1)$$

i.e. dom D is a subalgebra of X.

Here and in the sequel we shall assume that X is a D-algebra. Write

$$f_D(x, y) = D(xy) - c_D(xDy + yDx) \quad (6.1.2)$$

for all $x, y \in \text{dom } D$, where

(i) c_D is a scalar dependent on D only;

(ii) $f_D: \text{dom } D \times \text{dom } D \to \text{dom } D$ is a *bilinear* and *symmetric* mapping, i.e. linear in each variable and

$$f_D(y, x) = f_D(x, y) \quad \text{for all } x, y \in \text{dom } D. \quad (6.1.3)$$

Using the denotation (6.1.2) we can write

$$D(xy) = c_D(xDy + yDx) + f_D(x, y) \quad \text{for } x, y \in \text{dom } D. \quad (6.1.4)$$

The bilinear operator f_D is said to be a *non-Leibniz component.* Non-Leibniz components for powers of D are determined by a recursion formula, as it is shown by the following

THEOREM 6.1.1. *Suppose that X is a D-algebra. Then for all $k \in \mathbb{N}$, $x, y \in \text{dom } D^k$ such that $f_D(x, y) \in \text{dom } D^k$ we have $xy \in \text{dom } D^k$ and*

$$D^k(xy) = c_D^k(xD^ky + yD^kx) + f_D^{(k)}(x, y), \quad (6.1.5)$$

where we denote

$$f_D^{(0)} = 0, \quad f_D^{(1)} = f_D$$

and for $k = 2, 3, \ldots,$

$$f_D^{(k)}(x, y) = c_D^k[(Dx)(D^{k-1}y) + (D^{k-1}x)(Dy)] +$$
$$+ c_D^{k-1}[f_D(x, D^{k-1}y) + f_D(D^{k-1}x, y)] +$$
$$+ Df_D^{(k-1)}(x, y) \qquad (6.1.6)$$

or, in another (equivalent) form:

$$f_D^{(k)}(x, y) = c_D^k[(Dx)(D^{k-1}y) + (D^{k-1}x)(Dy)] +$$
$$+ c_D[f_D^{(k-1)}(x, Dy) + f_D^{(k-1)}(Dx, y)] +$$
$$+ D^{k-1}f_D(x, y) \qquad (6.1.6')$$

and $f_D^{(k)}$ *are bilinear symmetric mappings of* dom $D^k \times$ dom D^k *into* dom D^k $(k = 1, 2, \ldots)$.

Proof. Let $2 \leqslant N \in \mathbf{N}$ be arbitrarily fixed. Suppose Formula (6.1.5) to be true for $k = 1, 2, \ldots, N-1$. By (6.1.1), if $x, y \in$ dom D^N then $D^{N-1}x, D^{N-1}y \in$ dom D. Since X is a D-algebra, we conclude that also $xD^{N-1}y, yD^{N-1}x \in$ dom D and

$$D^N(xy) = D[D^{N-1}(xy)]$$
$$= D[c_D^{N-1}(xD^{N-1}y + yD^{N-1}x) + f_D^{(N-1)}(x, y)]$$
$$= c_D^N[xD^Ny + (Dx)(D^{N-1}y) + yD^Nx + (Dy)(D^{N-1}x)] +$$
$$+ c_D^{N-1}[f_D^{(N-1)}(x, D^{N-1}y) + f_D^{(N-1)}(y, D^{N-1}x)] + Df_D^{(N-1)}(x, y)$$
$$= c_D^N(xD^Ny + yD^Nx) + c_D^N[(Dx)(D^{N-1}y) + (D^{N-1}x)(Dy)] +$$
$$+ c_D^N[f_D^{(N-1)}(x, D^{N-1}y) + f_D^{(N-1)}(D^{N-1}x, y)] + Df_D^{(N-1)}(x, y)$$
$$= c_D^N(xD^Ny + yD^Nx) + f_D^{(N)}(x, y).$$

A similar proof for Formula (6.1.6').

Observe that by definition $f_D^{(k)}$ is a symmetric bilinear mapping of dom $D^k \times$ dom D^k into dom D^k $(k = 1, 2, \ldots)$. ∎

For superpositions of right invertible operators we have

COROLLARY 6.1.1. *Suppose that* $D_1, D_2 \in R(X)$, $D = D_2D_1$ *and* X *is a D-algebra. Then the operator D satisfies Condition* (6.1.2) *with*

$$c_D = c_{D_1}c_{D_2},$$
$$f_D(x, y) = f_{D_2}(x, y) + D_2f_{D_1}(x, y) + \qquad (6.1.7)$$
$$+ c_{D_1}c_{D_2}[(D_1x)(D_2y) + (D_2x)(D_1y)]$$

for $x, y \in$ dom $D = $ dom $D_1 \cap$ dom D_2.

Proof. Indeed, we have for all $x, y \in \text{dom } D = \text{dom } D_2 D_1$

$$
\begin{aligned}
D(xy) &= D_2[D_1(xy)] = D_2[c_{D_1}(xD_1 y + yD_1 x) + f_{D_1}(x, y)] \\
&= c_{D_1}[c_{D_2}(xD_2 D_1 y + yD_2 D_1 x) + (D_1 x)(D_2 y) + \\
&\quad + (D_1 y)(D_2 x)] + f_{D_2}(x, y) + D_2 f_{D_1}(x, y) \\
&= c_{D_1} c_{D_2}(xDy + yDx) + c_{D_1} c_{D_2}[(D_1 x)(D_2 y) + \\
&\quad + (D_2 x)(D_1 y)] + f_{D_2}(x, y) + D_2 f_{D_1}(x, y). \qquad \blacksquare
\end{aligned}
$$

COROLLARY 6.1.2. *Suppose that* X *is a D-algebra and that* f_D, c_D *are defined as in Formula* (6.1.2). *Write* $\hat{D} = pD$ *where* $p \neq 0$ *is an arbitrarily fixed scalar. Then* X *is an pD-algebra and*

$$
c_{pD} = c_D, \quad f_{pD} = pf_D, \quad f_{pD}^{(k)} = p^k f_D^{(k)} \quad \text{for } k \in \mathbf{N},
$$

where $f_D^{(k)}$ *are defined by Formulae* (6.1.6) *(or* (6.1.6')).

Proof. It is easy to verify that the operator pD is right invertible and that dom $(pD) = $ dom D which implies that X is a pD-algebra. Since we have

$$
\begin{aligned}
pD(xy) &= pc_D(xDy + yDx) + pf_D(x, y) \\
&= c_D[x(pD)y + y(pD)x] + pf_D(x, y)
\end{aligned}
$$

we conclude that $c_{pD} = c_D$ and $f_{pD} = pf_D$. By an easy induction we find $f_{pD}^{(k)} = p^k f_D^{(k)}$ for all $k \in \mathbf{N}$. $\qquad \blacksquare$

Without loss of generality we can assume here and in the sequel that $c_D \neq 0$. Namely, we have

PROPOSITION 6.1.1 (cf. Dudek, 1978). *Suppose that* X *is a D-algebra. Then* D *satisfies the condition*

$$
D(xy) = xDy + yDx + d(Dx)(Dy) \tag{6.1.8}
$$

for $x, y \in \text{dom } D$, *where* $d \neq 0$ *is an arbitrarily fixed scalar, if and only if*

$$
\hat{D} = I + dD \in R(X) \text{ is multiplicative,} \tag{6.1.9}
$$

i.e.

$$
\hat{D}(xy) = (\hat{D}x)(\hat{D}y) \quad \text{for } x, y \in \text{dom } D. \tag{6.1.10}
$$

Proof. Indeed, suppose that D satisfies (6.1.8) and $\hat{D}$ is of the form (6.1.9). Then for $x, y \in \text{dom } D$ we find

$$
\begin{aligned}
\hat{D}(xy) &= (I+dD)(xy) = xy+d[(xDy+yDx)+d(Dx)(Dy)] \\
&= d(xDy+yDx)+xy+d^2(Dx)(Dy) \\
&= (x+dDx)(y+dDy) = (\hat{D}x)(\hat{D}y).
\end{aligned}
$$

It means that

$$
c_{\hat{D}} = 0 \quad \text{and} \quad f_{\hat{D}}(x, y) = (Dx)(Dy). \tag{6.1.11}
$$

Conversely, suppose that $\hat{D} \in R(X)$ is multiplicative and consider the operator $D = \dfrac{1}{d}(\hat{D}-I)$. Then $\hat{D} = I+dD$ and for all $x, y \in \text{dom } \hat{D}$ we have

$$
\begin{aligned}
D(xy) &= \frac{1}{d}(\hat{D}-I)(xy) = \frac{1}{d}\hat{D}(xy)-\frac{1}{d}xy \\
&= \frac{1}{d}(\hat{D}x)(\hat{D}y)-\frac{1}{d}xy = \frac{1}{d}(x+dDx)(y+dDy)-\frac{1}{d}xy \\
&= xDy+yDx+d(Dx)(Dy),
\end{aligned}
$$

i.e.

$$
c_D = 1, \quad f_D(x, y) = d(Dx)(Dy). \tag{6.1.11'}
$$

∎

Formula (6.1.4) is not accidentally chosen, as well as the form of a non-Leibniz component. Namely, we have the following

THEOREM 6.1.2 (cf. Targonski, 1967). *Let X be a commutative algebra and let $A \in L(X)$ satisfy the so-called Bourlet-condition:* [*]

$$
A(xy) = f(x, y, Ax, Ay) \quad \text{for } x, y \in \text{dom } A. \tag{6.1.12}
$$

If f is an entire function then f is necessarily of the form:

$$
f(x, y, Ax, Ay) = \alpha(Ax)(Ay)+\beta(xAy+yAx)+\gamma xy, \tag{6.1.13}
$$

where α, β, γ are given scalars. If X has a unit $e \in \text{dom } A$ and no zero divisors, then A is either a multiple of the identity or belongs to one of the following three types:

(H) $A_0(xy) = \alpha(A_0 x)(A_0 y)$, *where* $A_0 = A-\mu I$;

[*] cf. C. Bourlet, *C.R. Acad. Sci. Paris*, **124** (1827), 1431–1433.

(M) $A(xy) = \dfrac{1}{2}(xAy+yAx);$

(D) $A_1(xy) = xA_1y+yA_1x$, where $A_1 = A+\gamma I$.

An operator of the type (M) *is a multiplication by a fixed element* Ae, i.e. $Ax = xAe$ *for all* $x \in$ dom D. *An operator of the type* (D) *always annihilates the unit*, i.e. $A_1e = 0$. *Indeed*, $A_1e = A_1e^2 = eA_1e+$ $+eA_1e = 2A_1e$, *which implies* $A_1e = 0$.

The proof can be found in the Targonski's book, 1967.

If X is a D-algebra over reals and we want to pass to D-algebras over complexes then we shall extend it in a standard way. Namely, write $Y = X+iX$. The set Y is again a commutative algebra with the addition and multiplication by scalars defined by Formulae (1.2.7) and (1.2.8), respectively, and with the multiplication of elements defined as follows:

$$(a+ib)(c+id) = (ac-bd)+i(ad+bc), \qquad (6.1.14)$$

for $a, b, c, d \in Y$.

Let $A = L(X)$ and let $u = x+iy$, where $x, y \in$ dom A. Write

$$Au = Ax+iAy. \qquad (6.1.15)$$

By this definition $A \in L(Y)$. Indeed, $A(\lambda u) = \lambda Au$ for all $\lambda \in \mathbf{C}$, $u \in Y$. Observe that $\overline{Au} = A\bar{u}$, where we write $\bar{u} = \overline{x+iy} = x-iy$. Indeed, we have $\overline{Au} = \overline{Ax+iAy} = Ax-iAy = A(x-iy) = A\bar{u}$.

It is easy to verify that $Y = X+iX$ is a D-algebra over $\mathbf{C}$, provided that X is a D-algebra over $\mathbf{R}$. Initial operators and right inverses are extended according with Formula (6.1.15). Write Formula (6.1.2) in the complex case: For all $u, v \in$ dom D (in Y) we have

$$f_D(u, v) = D(uv)-c_D(uDv+vDu),$$

where c_D is a coefficient defined by Formula (6.1.2) in the D-algebra X. By this definition, since $c_D \in \mathbf{R}$, we have

$$\overline{f_D(\bar{u}, \bar{v})} = f_D(u, v) = f_D(v, u) \quad \text{for all } u, v \in \text{dom } D. \quad (6.1.16)$$

Indeed,

$$\overline{f_D(\bar{u}, \bar{v})} = \overline{D(\bar{u}\bar{v})}-\overline{c_D(\bar{u}D\bar{v}+\bar{v}D\bar{u})}$$
$$= \overline{D(\bar{u}\bar{v})}-c_D(\overline{u D\bar{v}}+\overline{v D\bar{u}})$$

$$= D(uv) - c_D(uDv + vDu)$$
$$= f_D(u, v) = f_D(v, u).$$

This implies that (cf. Corollary 6.1.2)

$$c_{\lambda D} = \lambda c_D \quad (\lambda \in \mathbf{R}), \tag{6.1.17}$$
$$f_{\lambda D}(u, v) = \lambda f_D(u, v) \quad \text{for all } u, v \in \text{dom } D \quad \text{(in } X + iX).$$

PROPOSITION 6.1.2. *Let X be a D-algebra with the unit $e \in \text{dom } D$ and let c_D, f_D be defined by Formula (6.1.2). Then $f_D(e, e) = 0$ if and only if either $c_D = 1/2$ or $e \in \ker D$, i.e. $De = 0$.*

Proof. Since $e \in \text{dom } D$ and $e^2 = e$, we have $De = De^2 = c_D(eDe + eDe) + f_D(e, e) = 2c_D De + f_D(e, e)$, which implies

$$f_D(e, e) = (1 - 2c_D)De. \tag{6.1.18}$$

If $f_D(e, e) = 0$ then either $c_D = 1/2$ or $De = 0$. If $c_D = 1/2$ then $f_D(e, e) = 0$. If $e \in \ker D$ then $De = 0$ and also $f_D(e, e) = 0$ (cf. Theorem 6.1.2). ∎

We shall assume here and in the sequel that a unit in a D-algebra (if exists) belongs to the domain D.

Immediate consequences of this proposition are the following:

COROLLARY 6.1.3. *Suppose that all assumptions of Proposition 6.1.2 are satisfied. If $c_D \neq 1/2$ then $De = \gamma e$ for a scalar γ if and only if $f_D(e, e) = \gamma(1 - 2c_D)^{-1}e$.*

COROLLARY 6.1.4. *Suppose that all assumptions of Proposition 6.1.2 are satisfied and that $c_D \neq 1/2$. Then $e \in \ker D$ if and only if $f_D(e, e) = 0$.*

In other words: *a unit is not necessarily a constant.*

COROLLARY 6.1.5. *Suppose that X is a D-algebra and $z \in \ker D$. Then $z^2 \in \ker D$ if and only if $f_D(z, z) = 0$.*

Proof. Indeed, we have $Dz^2 = c_D(zDz + zDz) + f_D(z, z) = f_D(z, z)$. ∎

This means that *a square of a constant is not necessarily a constant.*

THEOREM 6.1.3. *If X is a D-algebra, $x \in \mathrm{dom}\, D$ and $n \geqslant 2$ is an arbitrary positive integer then $x^n \in \mathrm{dom}\, D$ and*

$$Dx^n = d_n x^{n-1} Dx + \sum_{j=0}^{n-2} c_D^j x^j f_D(x, x^{n-1-j}), \qquad (6.1.19)$$

where

$$d_1 = 1; \quad d_2 = 2c_D; \quad d_n = 2c_D^{n-1} + \sum_{j=1}^{n-2} c_D^j \quad \text{for } n \geqslant 2.$$

$$(6.1.20)$$

Proof (by induction). For $n = 1$, Equality (6.1.19) is trivially satisfied with $d_1 = 1$. For $n = 2$ and $x \in \mathrm{dom}\, D^2$ we have $Dx^2 = c_D(xDx + {}+xDx) + f_D(x, x) = 2c_D xDx + f_D(x, x)$, i.e. $d_2 = 2c_D$. Suppose Formula (6.1.19) to be true for an arbitrary $n \geqslant 2$. Then, by our induction assumption, we find

$$Dx^{n+1} = D(xx^n) = c_D(xDx^n + x^n Dx) + f_D(x, x^n)$$

$$= c_D x \left[d_n x^{n-1} Dx + \sum_{j=0}^{n-1} c_D^j x^j f_D(x, x^{n-1-j}) \right] +$$

$$+ c_D x^n Dx + f_D(x, x^n)$$

$$= c_D x^n Dx + c_D \left(2c_D^{n-1} + \sum_{j=1}^{n-2} c_D^j \right) x^n Dx +$$

$$+ \sum_{j=0}^{n-1} c_D^{j+1} x^{j+1} f_D(x, x^{n-1-j}) + f_D(x, x^n)$$

$$= \left[c_D + 2c_D^n + \sum_{j=2}^{n-1} c_D^j \right] x^n Dx + \sum_{j=1}^{n} c_D^j x^j f_D(x, x^{n-j}) + f_D(x, x^n)$$

$$= \left[2c_D^n + \sum_{j=1}^{n-1} c_D^j \right] x^n Dx + \sum_{j=0}^{n} c_D^j x^j f_D(x, x^{n-j})$$

$$= d_{n+1} x^n Dx + \sum_{j=0}^{n} c_D^j f_D(x, x^{n-1}),$$

which was to be proved.   ∎

COROLLARY 6.1.6. *If X is a D-algebra then for all $x \in \mathrm{dom}\ D^n$, $z \in \ker D$, $n \in \mathbf{N}$*

$$D^n(xz) = c_D^n z D^n x + f_D^{(n)}(z, D^n x), \qquad (6.1.21)$$

$$Dz^n = \sum_{j=0}^{n-2} c_D^j z^j f_D(z, z^{n-1-j}). \qquad (6.1.22)$$

Proof. Indeed, since $Dz = 0$, we have $D^n(xz) = c_D^n(xD^n z + zD^n x) + f_D^{(n)}(z, D^n x) = c_D^n z D^n x + f_D^{(n)}(z, Dx)$. Formula (6.1.19) implies (6.1.22). ∎

EXAMPLE 6.1.1. A D-algebra X is called a *Leibniz D-algebra* (briefly: *L-algebra*) if D satisfies the *Leibniz condition* (2.1.15):

$$D(xy) = xDy + yDx \quad \text{for } x, y \in \mathrm{dom}\ D.$$

Here we have $c_D = 1$ and $f_D = 0$. Hence, if X has a unit e then by Proposition 6.1.2 $De = 0$, i.e. e is a constant. Formula (6.1.6) for Leibniz D-algebras is of the form:

$$f_D^{(n)}(x, y) = \sum_{k=1}^{n-1} \binom{n}{k} (D^k x)(D^{n-k} y) \quad \text{for } x, y \in \mathrm{dom}\ D^n, \ n \geqslant 2.$$
$$(6.1.23)$$

This is a generalization of the well-known *Leibniz formula*.

EXAMPLE 6.1.1a. Let $X = C[a, b]$ with the usual multiplication of functions. Then X is a Leibniz $\dfrac{\mathrm{d}}{\mathrm{d}t}$-algebra. This algebra has a unit, namely the function $e(t) \equiv 1$ and zero divisors. Indeed, let $a \leqslant t_0 \leqslant b$ and $0 \neq x \in C[a, b]$ be arbitrarily fixed and such that $x(t_0) = 0$. Write:

$$x_1(t) = \begin{cases} x(t) & \text{if } a \leqslant t \leqslant t_0, \\ 0 & \text{if } t_0 \leqslant t \leqslant b, \end{cases}$$

$$x_2(t) = \begin{cases} 0 & \text{if } a \leqslant t \leqslant t_0, \\ x(t) & \text{if } t_0 \leqslant t \leqslant b. \end{cases}$$

The functions x_1 and x_2 do not vanish simultaneously, however, $x_1 x_2 = 0$ for $a \leqslant t \leqslant b$.

EXAMPLE 6.1.1b. Let $X = C(\Omega)$, where $\Omega = \{(t, s)\colon a \leqslant t \leqslant b;\ c \leqslant s \leqslant d\}$

with usual multiplication of functions. Let $D_1 = \partial/\partial t$, $D_2 = \partial/\partial s$. Then X is a Leibniz D_1-algebra and simultaneously a Leibniz D_2-algebra.

EXAMPLE 6.1.2. A *D*-algebra X is called a *quasi-Leibniz D-algebra* (briefly: *QL-algebra*) if D satisfies the following condition:

$$D(xy) = xDy + yDx + d(Dx)(Dy) \quad \text{for } x, y \in \text{dom } D, \quad (6.1.23')$$

where *d* is a scalar dependent on D only. If $d = 0$ we obtain an *L*-algebra. Here we have $c_D = 1$ and $f_D(x, y) = d(Dx)(Dy)$. By induction we prove that

$$f_D^{(1)}(x, y) = d(Dx)(Dy) \quad \text{for } x, y \in \text{dom } D,$$

$$f_D^{(n)}(x, y) = \sum_{k=1}^{n-1} \binom{n}{k} d^k \sum_{j=0}^{n-k} \binom{n-k}{j} (D^{k+j}x)(D^{n-j}y) \tag{6.1.24}$$

for $x, y \in \text{dom } D^n$, $n \geqslant 2$.

EXAMPLE 6.1.2a. Let X be the space (s) of all real sequences $(x_n)_{n \in \mathbf{N}}$ with the usual multiplication by coordinates, i.e.

$$xy = \{x_n\}\{y_n\} = \{x_n y_n\} \quad \text{for } x = \{x_n\}, \ y = \{y_n\} \in X.$$

Let $Dx = \{x_{n+1} - x_n\}$ for $x \in X$. Then $D \in R(X)$ (cf. Example 2.1.4). It is easy to verify that X with this multiplication is a *D*-algebra with unit $e = \{e_n\}$, where $e_n = 1$ for all $n \in \mathbf{N}$. Moreover, for $x, y \in X$

$$
\begin{aligned}
D(xy) &= \{x_{n+1} y_{n+1} - x_n y_n\} \\
&= \{(x_{n+1} - x_n)(y_{n+1} - y_n) + (x_{n+1} - x_n) y_n + (y_{n+1} - y_n) x_n\} \\
&= \{x_{n+1} - x_n\}\{y_{n+1} - y_n\} + \{y_n\}\{x_{n+1} - x_n\} + \\
&\quad + \{x_n\}\{y_{n+1} - y_n\} \\
&= xDy + yDx + (Dx)(Dy).
\end{aligned}
$$

Therefore X is a *QL*-algebra with $d = 1$. This algebra has zero divisors. Indeed, if $x = \{x_n\}$, when $x_{2n} = 0$, $x_{2n-1} \neq 0$ for $n = 1, 2, \ldots$ and $y = \{y_n\}$, where $y_{2n} \neq 0$, $y_{2n-1} = 0$ for $n = 1, 2, \ldots$ are arbitrarily fixed, then $x \neq 0$, $y \neq 0$, but $xy = \{x_n\}\{y_n\} = \{x_n y_n\} = 0$.

EXAMPLE 6.1.2b. Let X be a Leibniz *D*-algebra. Then X is not a Leibniz D^2-algebra. Indeed, for all $x, y \in \text{dom } D^2$ we have $D^2(xy) = xD^2y +$

$+yD^2x+2(Dx)(Dy)$. Hence $c_D = 1$, but $f_D(x, y) = 2(Dx)(Dy)$. However, X is a QL-algebra with $d = 2$. In general, for an arbitrary $n \geqslant 2$ a D^n-algebra with $f_{D^n} = f_D^{(n)}$, where $f_D^{(n)}$ is defined by Formula (6.1.23).

EXAMPLE 6.1.3. A D-algebra X is called a *Duhamel D-algebra* (briefly: *Dh-algebra*) if D satisfies the following condition for all $x, y \in \text{dom } D$:

$$D(xy) = xDy+f_1(x)y+f_2(y)x+g_1(x)Dy+g_2(y)Dx, \quad (6.1.24')$$

where f_1, f_2, g_1, g_2 are linear functionals dependent on D only. Since X is a commutative algebra, we have also

$$D(xy) = D(yx) = yDx+f_1(y)x+f_2(x)y+g_1(y)Dx+g_2(x)Dy.$$
$$(6.1.24'')$$

Formulae (6.1.24') and (6.1.24'') imply that for $x, y \in \text{dom } D$ we have

$$D(xy) = \frac{1}{2}(xDy+yDx)+f_D(x, y), \quad (6.1.25)$$

where

$$f_D(x, y) = \frac{1}{2}\{[f_1(x)+f_2(y)]y+[f_1(y)+f_2(x)]x+$$
$$+ [g_1(x)+g_2(x)]Dy+[g_1(y)+g_2(y)]Dx\}.$$

We have $c_D = 1/2$ (cf. Proposition 6.1.2 and Corollaries 6.1.3 and 6.1.4).

A *Dh*-algebra is said to be a *Dh₀-algebra* or *simple Duhamel D-algebra* if $f_1 = f_2 = g_1 = g_2 = 0$. This means that in a Dh_0-algebra we have $c_D = 1/2$ and $f_D = 0$. We therefore conclude that in Dh_0-algebras we have

$$D^n(xy) = \frac{1}{2^n}(xD^ny+yD^nx)+f_D^{(n)}(x, y) \quad (6.1.26)$$

for all $x, y \in \text{dom } D^n, n \in \mathbf{N}$, where

$$f_D^{(n)}(x, y) = \frac{1}{2^n}\sum_{k=1}^{n-1}\binom{n}{k}(D^kx)(D^{n-k}y) \quad \text{for } n \geqslant 2.$$

Suppose that X is a simple Duhamel D-algebra and $R \in \mathscr{R}_D \cap V(X)$.

Then either dim ker $D = 0$ or X has zero divisors. Indeed, assume that X has no zero divisors and dim ker $D > 0$, i.e. there exists a $z \in$ ker D such that $z \neq 0$. Let $\lambda \neq 0$ be an arbitrarily chosen scalar. Write $x = e_\lambda(z) = (I - \lambda R)^{-1}z$. By our assumption $x \neq 0$ and $Dx = \lambda x$. Moreover, since $f_1 = f_2 = g_1 = g_2 = 0$, Formulae (6.1.24′) and (6.1.24″) together imply that

$$D(zx) = zDx = \lambda zx; \quad D(zx) = D(xz) = xDz = 0$$

which implies $zx = 0$. But $z \neq 0$, $x \neq 0$ and X has no zero divisors, which contradicts our assumptions. Thus either ker $D = \{0\}$ or X has zero divisors.

EXAMPLE 6.1.3a. Let X be the space (s) of all real sequences $\{x_n\}_{n \in \mathbb{N}}$ with the *convolution* as multiplication, i.e.

$$x * y = \{x_n\} * \{y_n\} = \left\{ \sum_{k=1}^{n} x_k y_{n+1-k} \right\} \tag{6.1.27}$$

for $x = \{x_n\}$, $y = \{y_n\} \in X$.

It is easy to verify that X, with this multiplication, is a commutative algebra with unit $e = \{e_n\}$, where $e_1 = 1$, $e_n = 0$ for $n \geqslant 2$. Let $Dx = \{x_{n+1} - x_n\}$ for $x \in X$. Then $D \in R(X)$ (cf. Example 2.1.4) and

$$D(x * y) = \left\{ \sum_{k=1}^{n+1} x_k y_{n+2-k} - \sum_{k=1}^{n} x_k y_{n+1-k} \right\}$$

$$= \left\{ \sum_{k=1}^{n} x_k (y_{n+2-k} - y_{n+1-k}) \right\} + \{x_{n+1}\} y_1$$

$$= x * Dy + y_1 \{x_{n+1}\}.$$

Since X is a commutative algebra in the same way we find

$$D(x * y) = D(y * x) = y * Dx + x_1 \{y_{n+1}\}.$$

Hence for $x, y \in X$ we have

$$D(x * y) = \tfrac{1}{2}(x * Dy + y * Dx) + \tfrac{1}{2}[x_1 \{y_{n+1}\} + y_1 \{x_{n+1}\}] \tag{6.1.28}$$

and

$$f_D(x, y) = \tfrac{1}{2}[x_1 \{y_{n+1}\} + y_1 \{x_{n+1}\}],$$

but

$$\tfrac{1}{2}[x_1\{y_{n+1}\}+y_1\{x_{n+1}\}]$$
$$= \tfrac{1}{2}x_1[\{y_{n+1}-y_n\}+\{y_n\}]+\tfrac{1}{2}y_1[\{x_{n+1}-x_n\}+\{x_n\}]$$
$$= \tfrac{1}{2}[x_1(Dy+y)+y_1(Dx+x)].$$

Hence X is a Duhamel algebra with $f_1(x) = g_1(x) = f_2(x) = g_2(x)$ $= \tfrac{1}{2}x_1$ for $x \in X$.

The algebra X has no zero divisors. Indeed, suppose that $x = \{x_n\}$, $y = \{y_n\} \in X$, $x_1 \neq 0$ and $x*y = 0$, i.e.

$$\sum_{k=1}^{n} x_k y_{n+1-k} = 0 \quad \text{for } n = 1, 2, \ldots$$

For $n = 1$ we have $x_1 y_1 = 0$. Since $x_1 \neq 0$, we find $y_1 = 0$. Suppose that for an arbitrarily fixed $n \geqslant 2$ we have $y_1 = y_2 = \ldots = y_{n-1} = 0$. Then

$$0 = \sum_{k=1}^{n} x_k y_{n+1-k} = x_1 y_n \quad \text{which implies } y_n = 0.$$

We therefore have proved by induction that $y_n = 0$ for all $n \in \mathbf{N}$. Hence $y = \{y_n\} = 0$. Since $x_1 \neq 0$, we have $x \neq 0$. Thus $x \neq 0$ and $x \overset{=}{*} y = 0$ implies $y = 0$, which was to be proved.

EXAMPLE 6.1.3b. Consider the space $C[0, T]$, the operator $D = \mathrm{d}/\mathrm{d}t$ and its right inverse defined by the equality $(Rx)(t) = \int_0^t x(s)\mathrm{d}s$ for $x \in C[0, T]$. An initial operator F corresponding to R is $(Fx)(t) = x(0)$ for $x \in C[0, T]$.

Define now a *convolution* of functions belonging to $C[0, T]$ by means of the equality

$$(x*y)(t) = \int_0^t x(s)y(t-s)\mathrm{d}s \quad \text{for } x, y \in C[0, T]. \qquad (6.1.29)$$

We shall show that $C[0, T]$ with the multiplication defined as a convolution (6.1.29) is a commutative algebra without unit and with divisors of zero.

To begin with, we prove that this multiplication is associative, commutative and distributive with respect to the addition, i.e. for all $x, y, z \in C[0, T]$ we have

$$x*(y*z) = (x*y)*z, \quad y*x = x*y, \quad x*(y+z) = x*y+x*z.$$

Indeed,

$$(y*x)(t) = \int_0^t y(s)x(t-s)\,ds = -\int_t^0 y(t-u)x(u)\,du$$

$$= \int_0^t x(u)y(t-u)\,du = (x*y)(t),$$

$$[x*(y+z)](t) = \int_0^t x(s)[y(t-s)+z(t-s)]\,ds$$

$$= \int_0^t x(s)y(t-s)\,ds + \int_0^t x(s)z(t-s)\,ds$$

$$= (x*y)(t)+(x*z)(t),$$

$$[(x*y)*z](t) = \int_0^t \left[\int_0^s x(u)y(s-u)\,du\right] z(t-s)\,ds$$

$$= \int_0^t \left[\int_u^t y(s-u)z(t-s)\,ds\right] x(u)\,du$$

$$= \int_0^t x(u) \left[\int_0^{t-u} y(v)z(t-u-v)\,dv\right] du$$

$$= \int_0^t x(u)[(y*z)(t-u)]\,du = [x*(y*z)](t).$$

Thus $C[0, T]$ is a commutative algebra. This algebra has divisors of zero, namely a function $x_0 \in C[0, T]$ and such that $x_0(t) = 0$ for $0 \leqslant t \leqslant T/2$. Indeed,

$$(x_0*x_0)(t) = \int_0^t x_0(s)x_0(t-s)\,ds = 0 \quad \text{for } 0 \leqslant t \leqslant T$$

because the first component of the integrand function vanishes for $0 \leqslant t \leqslant T/2$ and the second one vanishes if $T/2 \leqslant t \leqslant T$ for $0 \leqslant t - -s \leqslant t-T/2 \leqslant T/2$.

We prove now the so-called *Duhamel formula*

$$\frac{d}{dt}(x*y)(t) = \left(x*\frac{d}{dt}y\right)(t) + x(t)y(0) - x(0)y(t) \qquad (6.1.30)$$

for all $x, y \in C^1[0, T]$. Indeed, since in our case dom $D = C^1[0, T]$, we find

$$\frac{d}{dt}(x*y)(t) = \frac{d}{dt}\int_0^t x(s)y(t-s)\,ds$$

$$= [x(s)y(t-s)]_0^t + \int_0^t x(s)\frac{d}{dt}y(t-s)\,ds$$

$$= x(t)y(0) - x(0)y(t) + \left(x*\frac{d}{dt}y\right)(t).$$

Thus $C[0, T]$ with the convolution (6.1.29) as multiplication is a *Dh*-algebra such that $f_1(x) = x(0), f_2(x) = -x(0)$ and $g_1 = g_2 = 0$ for $x \in X$.

Suppose now that the algebra $C[0, T]$ has a unit e. Observe that for every $x, y \in C[0, T]$ we have $(x*y)(0) = 0$. In particular, $e(0) = (e*e)(0) = 0$. Let $x \in C^1[0, T]$ be an arbitrary function such that $x(0) \neq 0$. Formula (6.1.30) implies that

$$\frac{d}{dt}x(t) = \frac{d}{dt}(x*e)(t) = \left(e*\frac{d}{dt}x\right)(t) + e(0)x(t) - x(0)e(t)$$

$$= \frac{d}{dt}x(t) - x(0)e(t).$$

We therefore conclude that $-x(0)e(t) \equiv 0$, hence $e(t) \equiv 0$, which contradicts the required property of a unit: $x*e = e*x = x$ for all $x \in X$.

Consider now the set

$$X_0 = \{x \in C[0, T]: x(0) = 0\}. \qquad (6.1.31)$$

It is easy to verify that X_0 is a subalgebra of $X = C[0, T]$ with the convolution as multiplication. Moreover, X_0 is a *Dh*-algebra such that $f_1 = f_2 = g_1 = g_2 = 0$, because the definition (6.1.31) and Formula (6.1.30) together imply

$$\frac{d}{dt}(x*y) = x*\frac{d}{dt}y = y*\frac{d}{dt}x$$

for all $x \in X_0 \cap \operatorname{dom} D = \{x \in C^1[0, T]: x(0) = 0\}$.

Hence X_0 is a simple Duhamel *D*-algebra.

EXAMPLE 6.1.4. Let X, D_1, D_2 be defined as in Example 6.1.1a and let $D = D_1 D_2 = \partial^2/\partial t\,\partial s$. Then X is a *D*-algebra and for all $x, y \in \operatorname{dom} D$ we have

$$
\begin{aligned}
D(xy) &= xDy + yDx + f_D(x, y), \\
f_D(x, y) &= x_t' y_s' + x_s' y_t' = (D_1 x)(D_2 y) + (D_2 x)(D_1 y)
\end{aligned}
\qquad (6.1.32)
$$

(cf. Section 4.2).

EXAMPLE 6.1.5. Let $\Omega \subset \mathbf{R}^n$ be a domain with the boundary S of Liapunov type and let $X = C(\bar{\Omega})$ with usual multiplication of functions. Let D be the Laplace operator $\Delta = \sum_{i=1}^{n} \partial^2/\partial t_i^2$ (cf. Section 4.3). Then X is a *D*-algebra and for all $x, y \in C^2(\bar{\Omega})$ we have

$$\Delta(xy) = x\Delta y + y\Delta x + f_D(x, y), \quad \text{where } c_D = 1,$$

$$f_D(x, y) = 2\sum_{i=1}^{n}\left(\frac{\partial x}{\partial t_i}\right)\left(\frac{\partial y}{\partial t_i}\right).$$

(6.1.33)

EXAMPLE 6.1.6. Suppose that $X = C[a, b]$ (where $-\infty \leqslant a, b \leqslant +\infty$) with usual multiplication of functions and that $t_0, t_1 \in (a, b)$ are arbitrarily fixed. Define the operator D as follows

$$(Dx)(t) = \frac{x(t) - x(t_0)}{t - t_0} \quad \text{for all } x \in X, \ t \in [a, b]. \qquad (6.1.34)$$

The domain of D consists of all functions continuous in $[a, b]$ and differentiable at the point t_0. The kernel of D consists of all constant functions. Define the operator R as follows:

$$(Rx)(t) = (t - t_0)x(t) - (t_1 - t_0)x(t_1). \qquad (6.1.35)$$

It is easy to verify that $DR = I$ on X. Hence $D \in R(X)$. An initial operator F corresponding to R is defined by the equality: $(Rx)(t) = x(t_1)$. Moreover, X is a D-algebra with unit e, where e is a constant function equal to 1, and for all $x, y \in \text{dom } D$ we have

$$D(xy) = yDx + (F_0 x)Dy, \quad \text{where } (F_0 x)(t) = x(t_0). \quad (6.1.36)$$

Since multiplication in X is commutative, we also have

$$D(xy) = D(yx) = xFy + (F_0 y)(Dx). \quad (6.1.37)$$

Formulae (6.1.36) and (6.1.37) together imply that

$$D(xy) = \tfrac{1}{2}(xDy + yDx) + f_D(x, y),$$
$$f_D(x, y) = (F_0 x)Dy + (F_0 y)Dx, \quad (F_0 x)(t) = x(t_0). \quad (6.1.38)$$

EXAMPLE 6.1.7. Let $X = (s)$ with multiplication by coordinates (cf. Example 6.1.2a). Let $p = \{p_n\} \subset \mathbf{R}$ be arbitrarily fixed. Write

$$Sx = \{x_{n+1}\}, \quad D_p x = Sx - px \quad \text{for } x = \{x_n\} \in X. \quad (6.1.39)$$

The operator D_p is defined on the whole space X, the kernel of D_p consists of all sequences $c\{p_0 p_1 \ldots p_{n-1}\}$, where $p_0 = 1$, $c \in \mathbf{R}$. The operator D_p is right invertible and a right inverse for D_p can be defined as follows: for a given $x \in X$ we have

$$y = R_p x \quad \text{if } y_1 = 0, y_2 = y_1, y_n = \sum_{k=1}^{n-1} p_{n-1} \ldots p_{k+1} x_k + x_{n-1}$$

for $n \geqslant 3$. The corresponding initial operator is given by the equality $F_p x = x_1 \{p_0 p_1 \ldots p_{n-1}\}$. Observe that X is a D_p-algebra with unit $e = \{e_n\}$, where $e_n = 1$ for $n \in \mathbf{N}$. It is easy to verify that D_p satisfies the following condition for all $x, y \in X$:

$$D_p(xy) = xD_p y + yD_p x + f_{D_p}(x, y),$$
$$(6.1.40)$$
$$f_{D_p}(x, y) = (p-1)(xD_p y + yD_p x) + (D_p x)(D_p y) + (p^2 - p)xy.$$

Here we have $c_{D_p} = 1$. Observe that in the case $p_n = 1$ for all $n \in \mathbf{N}$ we obtain Condition (6.1.23) with $d = 1$ (cf. also Example 6.1.2a), i.e. a quasi-Leibniz D-algebra.

Let $q = \{q_n\} \subset \mathbf{R}$ be also arbitrarily fixed. Consider the operators D_p and D_q. It is easy to verify that

$$D_p + D_q = 2D_{(p+q)/2} \quad \text{or} \quad D_{p+q} = \tfrac{1}{2}(D_{2p} + D_{2q}), \quad (6.1.41)$$
$$D_p D_q = D_{p+q}^2 + (p+q)D_{p+q} + pqI. \quad (6.1.42)$$

EXAMPLE 6.1.7a. Let X, D_p, S be defined as in Example 6.1.7. Suppose that $p = \{p_n\}$ is a constant sequence, i.e. $p_n = c_D \neq 0$ for all $n \in \mathbb{N}$, where c_D is a real dependent on D only. In this case it is more convenient to rewrite Formula (6.1.40) in the form: $D_p = D$ and

$$D(xy) = c_D(xDy+yDx)+f_D(x, y), \tag{6.1.43}$$
$$f_D(x, y) = (Dx)(Dy)+(c_D^2-c_D)xy \quad \text{for } x, y \in \text{dom } D.$$

Indeed, Formula (6.1.2) implies that for $p_n = c_D$ $(n \in \mathbb{N})$

$$\begin{aligned}
D(xy) &= xDy+yDx+(c_D-1)(xDy+yDx)+ \\
&\quad +(Dx)(Dy)+(c_D^2-c_D)xy \\
&= c_D(xDy+yDx)+(Dx)(Dy)+(c_D^2-c_D)xy.
\end{aligned}$$

Observe that the coefficient c_D appearing in (6.1.43) can be arbitrary real.

If $p_n = 0$ for all $n \in \mathbb{N}$ then we have $p = 0$, $D_0 = S$ and $S(xy) = (Sx)(Sy)$. Therefore in this case we have $c_D = 0$ and $f_S(x, y) = (Sx)(Sy)$. Observe that the operator S defined by (6.1.39) is right invertible. A right inverse of S can be defined as follows: $R\{x_n\} = \{x_{n-1}\}$, where we admit $x_0 = 0$ for all $x \in X$.

In order to study properties of D-algebras with respect to initial operators we have to introduce some new definitions.

Suppose that X is a commutative algebra. An operator $A \in L(X)$ is said to be *averaging* if

$$A(xAy) = (Ax)(Ay) \quad \text{for } x, y \in \text{dom } A, \tag{6.1.44}$$

$A \in L(X)$ is said to be a *Reynolds operator* if

$$A(xy) = (Ax)(Ay)+A[(x-Ax)(y-Ay)] \quad \text{for } x, y \in \text{dom } A \tag{6.1.45}$$

(cf. Rota, 1964).

Suppose that X is a D-algebra and F is an initial operator for D. Then F is said to be *almost averaging* if

$$F(zx) = zFx \quad \text{for all } x \in X, z \in \text{ker } D. \tag{6.1.46}$$

The last property is very useful and has been used in several applications.

By definition, any multiplicative initial operator is almost averaging. Indeed, since $Fz = z$ for $z \in \ker D$, we find that $F(zx) = (Fz)(Fx)$ $= zFx$ for $x \in X$.

The converse statement is not true, as several examples show.

THEOREM 6.1.4. *Suppose that X is a D-algebra and F is an initial operator for D. Then the following conditions are equivalent*:
(i) *F is almost averaging*,
(ii) *F is averaging*.

Proof. Suppose that F is averaging and $x \in X$, $z \in \ker D$ are arbitrary· Since F is a projection onto $\ker D$, there exists $y \in X$ such that $Fy = z$. The condition (6.1.44) implies

$$F(zx) = F(xFy) = (Fx)(Fy) = zFx,$$

i.e. F is almost averaging. Conversely, suppose that F is almost averaging and x, y are arbitrary. Since $z = Fy \in \ker D$ and condition (6.1.46) holds, we have

$$F(zFy) = F(zx) = zFx = (Fx)(Fy),$$

i.e. F is averaging.   ∎

THEOREM 6.1.5. *Suppose that X is a D-algebra with unit $e \in \ker D$ and F is an almost averaging initial operator. Then*
(i) *F is a Reynolds operator*;
(ii) *We have*
$$Fz^n = z^n \quad \text{for all } z \in \ker D, n \in \mathbf{N};   \tag{6.1.47}$$
(iii) *Any power of a constant is again a constant*;
(iv) *We have*
$$f_D(z, z) = 0 \quad \text{for all } z \in \ker D.   \tag{6.1.48}$$

Proof. By our assumption $Fe = e$. By Theorem 6.1.4, F is averaging. An averaging operator in a commutative algebra with unit such that $Fe = e$ is a Reynolds operator (cf. Rota, 1964). Since F is a Reynolds operator in a commutative algebra with unit (over a field of characteristic zero), the following identity holds:

$$nF[x(Fx)^{n-1}] = (n-1)F[(Fx)^n] + (Fx)^n \quad \text{for all } x \in X, n \in \mathbf{N}$$
$$\tag{6.1.49}$$

(cf. also Rota, 1964). But $z = Fz \in \ker D$ and F is almost averaging. Then Identity (6.1.49) can be rewritten as follows:

$$nF(xz^{n-1}) = (n-1)Fz^n + z^n \quad \text{for } n \in \mathbb{N}, \text{ where } z = Fx.$$

For $n = 2$ we have

$$2z^2 = 2zFx = 2F(xz) = Fz^2 + z^2, \tag{6.1.50}$$

which implies that $Fz^2 = z^2$ and z^2 is a constant for $Fz^2 \in \ker D$. Suppose that $Fz^k = z^k$ for an arbitrary fixed $2 \leqslant k \leqslant \mathbb{N}$. Then $z_k = Fz^k \in \ker D$. Since F is averaging and (6.1.50) holds, we find $(k+1)z^{k+1} = (k+1)z^2Fx = (k+1)F(xz^k) = kFz^{k+1} + z^{k+1}$ which implies $Fz^{k+1} = z^{k+1}$. Therefore $Fz^n = z^n$ for all $n \in \mathbb{N}$. Since $Fz^n \in \ker D$ for $n \in \mathbb{N}$, we conclude that any power of a constant is again a constant. Since $z^2 \in \ker D$ for any $z \in \ker D$, we conclude that

$$0 = Dz^2 = 2c_D z Dz + f_D(z, z) = f_D(z, z),$$

i.e. the non-Leibniz component f_D vanishes on constants. ∎

A *D*-algebra X is said to be *almost Leibniz* if

$$f_D(x, z) = 0 \quad \text{for all } x \in \text{dom } D, \, z \in \ker D. \tag{6.1.51}$$

The following *D*-algebras are almost Leibniz:
(i) Leibniz *D*-algebras, since we have

$$D(xy) = xDy + yDx \quad \text{and } f_D = 0 \quad \text{for } x, y \in \text{dom } D;$$

(ii) Quasi-Leibniz *D*-algebras, since we have

$$D(xy) = xDy + yDx + d(Dx)(Dy) \quad \text{for } x, y \in \text{dom } D$$

(where $d \neq 0$ is a given scalar dependent on D only). Hence $f_D(x, z) = d(Dx)(Dz) = 0$ for $z \in \ker D$;
(iii) Simple Duhamel *D*-algebras, since we have

$$D(xy) = xDy \quad \text{and } f_D = 0 \quad \text{for } x, y \in \text{dom } D.$$

Indeed, the commutativity of multiplication in X implies that $D(xy) = yDx$. We therefore conclude that $D(xy) = \frac{1}{2}(xDy + yDx)$ for $x, y \in \text{dom } D$, which implies $f_D = 0$;
(iv) Suppose that X is a Leibniz *D*-algebra. Then for an arbitrary positive integer $n > 1$, we have $f_D^{(n)}(x, z) = 0$ if $z \in \ker D$. Indeed, the Leibniz condition implies that for $x, y \in \text{dom } D^n$

$$D^n(xy) = \sum_{k=0}^{n} \binom{n}{k} (D^k x)(D^{n-k}y)$$

$$= xD^n y + yD^n x + \sum_{k=1}^{n-1} \binom{n}{k} (D^k x)(D^{n-k}y)$$

(cf. Formula (6.1.23)). Hence

$$f_D^{(n)}(x, z) = \sum_{k=1}^{n-1} \binom{n}{k}(D^k x)(D^{n-k}z) = 0 \quad \text{for } x \in \text{dom } D^n, z \in \ker D$$

(cf. also Exercise 2.1.6).

THEOREM 6.1.6. *Suppose that X is an almost Leibniz D-algebra. Then*

$$f_D^{(n)}(x, z) = 0 \quad \text{for all } x \in \text{dom } D, z \in \ker D \quad (n \in \mathbf{N}). \quad (6.1.52)$$

Proof (by induction). For $n = 1$ Formula (6.1.52) follows just from the definition (6.1.51). Suppose Formula (6.1.52) to be true for an arbitrarily fixed $n \geqslant 1$. Then, by Formula (6.1.6) we find for $x \in \text{dom } D^n$, $z \in \ker D$

$$f_D^{(n+1)}(x, z) = c_D^n[(Dx)(D^n z) + (Dz)(D^n x)] +$$
$$+ c_D^{n-1}[f_D(x, D^n z) + f_D(z, D^n x)] + Df_D^{(n)}(x, z) = 0. \quad \blacksquare$$

COROLLARY 6.1.7. *Suppose that X is an almost Leibniz D-algebra. If $z_1, z_2 \in \ker D$ then $z_1 z_2 \in \ker D$.*

Proof. Indeed, if $z_1, z_2 \in \ker D$ then $D(z_1 z_2) = c_D(z_1 Dz_2 + z_2 Dz_1) + f_D(z_1, z_2) = 0$. $\blacksquare$

COROLLARY 6.1.8. *Suppose that X is an almost Leibniz D-algebra. Then*

$$D^n(xz) = c_D^n z D^n x \quad \text{for all } x \in \text{dom } D^n, z \in \ker D \ (n \in \mathbf{N}),$$
$$(6.1.53)$$

and

$$z \in \ker D \quad \text{implies } z^n \in \ker D \ (n \in \mathbf{N}), \quad (6.1.54)$$

i.e. a power of a constant is again a constant. If X has a unit e then

$$z D^n e = 0 \quad \text{for all } z \in \ker D, n \in \mathbf{N}. \quad (6.1.55)$$

Proof. Indeed, Formulae (6.1.52) and (6.1.19) together imply (6.1.53). Formulae (6.1.52) and (6.1.51) together imply that $Dz^n = 0$. Hence $z^n \in \ker D$ for $n \in N$ and z^n is a constant. Formulae (6.1.22) and (6.1.52) together imply (6.1.55).   ∎

COROLLARY 6.1.9. *If X is an almost Leibniz D-algebra with unit e, and constants are not zero divisors, then $e \in \ker D$, i.e. the unit is a constant. Moreover, $c_D = 1$.*

COROLLARY 6.1.10. *Suppose that X is an almost Leibniz D-algebra and $P(t) = \sum_{k=1}^{n} p_k t^k$, where $p_1, \ldots, p_n$ are scalars. Then $P(z) \in \ker D$ for an arbitrary $z \in \ker D$, i.e. a polynomial (with scalar coefficients) in a constant is again a constant.*

COROLLARY 6.1.11. *Suppose that X is an almost Leibniz D-algebra and F is an initial operator for D. Then*

$$F(zFx) = zFx \quad \text{for all } z \in \ker D, x \in X. \tag{6.1.56}$$

Proof. Indeed, for an arbitrarily fixed $x \in X$ we have $z_1 = Fx \in \ker D$. This, and Corollary 6.1.6 together imply that $F(zFx) = F(zz_1) = zz_1 = zFx$.   ∎

COROLLARY 6.1.12. *Suppose that X is an almost Leibniz D-algebra and F is a Reynolds initial operator for D. Then F is averaging.*

Proof. Indeed, if F is a Reynolds operator then for all $z \in \ker D$, $x \in X$ we find $z = Fz$ and $F(xz) = (Fx)(Fz) + F[(x - Fx)(z - Fx)] = (Fx)(Fz) = zFx$, i.e. F is almost averaging. Theorem 6.1.4 implies that F is averaging.

Recall that in Theorem 6.1.5 we have proved that every almost averaging operator in a D-algebra is a Reynolds operator.

Exercises

EXERCISE 6.1.1. Prove that the space $X_a = \{x \in C[0, T]: x(t) = 0$ for $0 \leqslant t \leqslant a\}$, $a \in R$ is arbitrarily fixed, with the multiplication defined as a convolution (6.1.29) and with D defined by means of the equality: $(Dx)(t) = tx(t)$ for $x \in X_a$ is a Leibniz D-algebra without unit and with zero divisors.

EXERCISE 6.1.2. Suppose that all assumptions of Exercise 2.2.3 are satisfied. Prove that the initial operator F for $D = \mathrm{d}/\mathrm{d}t$ defined by Formula (2.2.42) is almost averaging. In particular, F defined by (2.2.42') is almost averaging (we assume that X is a D-algebra with the usual multiplication of functions).

EXERCISE 6.1.3. Prove that initial operators for $D = \mathrm{d}/\mathrm{d}t$ defined by Formulae (2.2.44), (2.2.45), (2.2.46) are almost averaging. (We assume that X is a D-algebra with the usual multiplication of functions.)

EXERCISE 6.1.4. Prove that initial operators for the difference operator defined in Exercises 2.2.9, 2.2.10, 2.2.11 are almost averaging (we assume that X is a D-algebra with the multiplication by coordinates).

EXERCISE 6.1.5. Consider D-algebras defined in Examples 6.1.3a, 6.1.3b, 6.1.6, 6.1.7. Indicate (if they exist) multiplicative and almost averaging initial operators.

EXERCISE 6.1.6. Let X be a commutative algebra. An operator $A \in L_0(X)$ is said to be a *Bourlet operator* if it satisfies Condition (6.1.12). Prove that A is a Bourlet operator if and only if there is a function g such that

$$Ax^2 = g(x, Ax) \quad \text{for } x \in X \tag{6.1.57}$$

(we have: $g(x, y) = f(x, x, y, y)$).

EXERCISE 6.1.7. Let X be a commutative algebra with unit e. Let f be a symmetric bilinear mapping satisfying the condition proposed by R. Ger:*

$$f(x, yz) - f(xy, z) = zf(x, y) - xf(y, z) \quad \text{for } x, y, z \in X. \tag{6.1.58}$$

---

* Private communication.

Write $Ax = -\frac{1}{2}f(x, e)$ for $x \in X$. Prove that A is a linear operator satisfying the condition $A(xy) = \frac{1}{2}(xAy + yAx) + f(x, y)$ for $x, y \in X$, i.e. if $A \in R(X)$ then X is an A-algebra and $f_A = f$, $c_A = \frac{1}{2}$. It means that a non-Leibniz component f determines a corresponding right invertible operator. Prove that $e \in \ker A$.

EXERCISE 6.1.8. Let X be a commutative algebra A with unit e. Let f be a symmetric bilinear mapping satisfying the condition

$$f(x, yz) + f(xy, z) + f(xz, y) = zf(x, y) + xf(y, z) + yf(x, z)$$

$$(6.1.59)$$

for $x, y, z \in X$.

Write $Ax = f(x, e)$ for $x \in X$. Prove that A is a linear operator satisfying the condition $A(xy) = xAy + yAx + f(x, y)$, i.e. if $A \in R(X)$ then X is an A-algebra and $c_A = 1$, $f_A = -f$.

EXERCISE 6.1.9. Let X be a commutative algebra with unit e. Let f be a symmetric bilinear mapping satisfying the condition

$$f(x, yz) + f(y, xz) + f(xy, z) = 2[zf(x, y) + xf(y, z) + yf(x, z)]$$

$$(6.1.60)$$

for $x, y, z \in X$.

Write $Ax = f(x, e)$ for $x \in X$. Prove that $A(xy) = 2(xAy + yAx)$ i.e. if A is right invertible then X is an A-algebra and $c_A = 2$, $f_A = 0$. Prove that $e \in \ker A$.

EXERCISE 6.1.10. Suppose that X is QL-algebra, i.e. the operator $D \in R(X)$ satisfies Condition (6.1.23′). Suppose that X has a unit e. Prove that:
   (1) The following condition is satisfied by the element De:

$$(De)(e + dDe) = 0;$$

$$(6.1.61)$$

   (2) There are 3 types of QL-algebras with unit:
   (i) $e \in \ker D$,
   (ii) $De = -e/d$, i.e. De is a scalar multiple of the unit,
   (iii) $e \notin \ker D$ and $De \neq -e/d$, however Condition (6.1.61) is satisfied. If neither De nor $e + dDe$ are zero divisors then a QL-algebra of the type (iii) can be decomposed onto QL-algebras of the types (i) and (ii) (cf. Dudek, 1981a).

EXERCISE 6.1.11. Suppose that X is a commutative algebra with unit e and A is a Reynolds operator, i.e. A satisfies Condition (6.1.45). Prove that:

(i) $g^2 + Ag^2 = 2Ag$, where $g = Ae$;

(ii) If $e \in \ker A$ then $A^2 = 0$;

(iii) If $e \in \ker A$ and A is a projection then $A = 0$.

6.2. INTEGRATION. EXPONENTIALS. TRIGONOMETRIC IDENTITY

In this section we shall study properties of right inverses and exponentials in D-algebras.

THEOREM 6.2.1 (*Generalized integration by parts formula**). *Let X be a D-algebra and let F be an initial operator for D corresponding to an $R \in \mathscr{R}_D$. Then for all $x \in \mathrm{dom}\, D$, $y \in X$ and for every positive integer n*

$$R^n(xy) = c_D^{-n}xR^ny - \sum_{j=1}^{n} c_D^{-j}R^{n-j}\{c_D R[(Dx)R^j y] +$$
$$+ F(xR^j y) + Rf_D(x, R^j y)\}. \tag{6.2.1}$$

Proof. Let $x, y \in \mathrm{dom}\, D$. Write: $u = Ry$. Then $u \in \mathrm{dom}\, D$ and $xDu = c_D^{-1}(xu) - uDx - c_D^{-1}f_D(x, u)$. Since $RD = I - F$ on dom D and $y = Du$, acting on both sides of this equality by the operator R we find

$$\begin{aligned}
R(xy) = R(xDu) &= c_D^{-1}RD(xu) - R(uRx) - c_D^{-1}Rf_D(x, u) \\
&= c_D^{-1}[xu - F(xu) - c_D R(uDx) - Rf_D(x, u)] \\
&= c_D^{-1}[xRy - F(xRy)] - R[(Dx)(Ry)] - c_D^{-1}Rf_D(x, Ry)
\end{aligned}$$

which proves Formula (6.2.1) for $n = 1$. Suppose now Formula (6.2.1) to be true for an arbitrarily fixed $n \geqslant 1$. Then

$$R^{n+1}(xy) = R[R^n(xy)]$$

$$= c_D^{-n}R(xR^ny) - \sum_{j=1}^{n} c_D^{-j}R^{n+1-j}\{c_D R[(Dx)R^j y] +$$
$$+ F(xR^j y) + Rf_D(x, R^j y)\}$$

---

* This formula and Formulae (6.2.2), (6.2.4)–(6.2.7) in the Leibniz case have been proved by H. von Trotha (cf. Przeworska-Rolewicz and von Trotha, 1981).

$$= c_D^{-(n+1)}xR^{n+1}y - c_D^{-n}R[(Dx)(R^{n+1}y)] - c_D^{-(n+1)}F(xR^{n+1}y) -$$
$$- c_D^{-(n+1)}Rf_D(x, R^{n+1}y) -$$
$$- \sum_{j=1}^{n} c_D^{-j}R^{n+1-j}\{c_DR[(Dx)R^jy] + F(xR^jy) + Rf_D(x, R^jy)\}$$

$$= c_D^{-(n+1)}xR^{n+1}y - \sum_{j=1}^{n+1} c_D^{-j}R^{n+1-j}\{c_DR[(Dx)(R^jy)] +$$
$$+ F(xR^jy) + Rf_D(x, R^jy)\}$$

which proves Formula (6.2.1) for $n+1$.   ∎

COROLLARY 6.2.1. *Suppose that all assumptions of Theorem 6.2.1 are satisfied. Then the following formula of integration by parts holds:*

$$R(xDy) = c_D^{-1}[xy - F(xy) - Rf_D(x, y)] - R(yDx), \qquad (6.2.2)$$

for all $x, y \in$ dom D.

   If $F_1 \neq F$ is also an initial operator for D we obtain a formula of integration by parts for definite integrals:

$$F_1 R(xDy) = c_D^{-1}[(F_1 - F)(xy) - F_1 Rf_D(x, y)] - F_1 R(yDx),$$
$$(6.2.3)$$

for $x, y \in$ dom D.

Proof. Indeed, Formula (6.2.2) follows from (6.2.1) if we put $n = 1$ and Dy instead of y. Acting on both sides of (6.2.2) by the operator F_1 and applying the property: $F_1 F = F$, we obtain immediately Formula (6.2.3) (cf. also Example 2.1.8 concerning the Leibniz case, when we have $f_D = 0$).   ∎

COROLLARY 6.2.2. *Suppose that all assumptions of Theorem 6.2.1 are satisfied.*
   (i) *If $z \in$ ker D, $x \in$ dom D then*

$$R^n(zx) = c_D^{-n}zR^nx - \sum_{j=1}^{n} c_D^{-j}R^{n-j}[F(zR^jx) + Rf_D(z, R^jx)];$$

$$(6.2.4)$$

(ii) *If X has a unit e, $z \in \ker D$ and $x \in \mathrm{dom}\, D$ then*

$$R^n x = c_D^{-n} x R^n e - \sum_{j=1}^{n} c_D^{-j} R^{n-j} \{ c_D R[(Dx) R^j e] +$$

$$+ F(x R^j e) + f_D(x, R^j e) \}, \qquad (6.2.5)$$

$$R^n z = c_D^{-n} z R^n e - \sum_{j=1}^{n} c_D^{-j} R^{n-j} [F(z R^j e) + R f_D(z, R^j e)]; \qquad (6.2.6)$$

(iii) *If X has a unit e then the remainder in the Taylor Formula is of the form*

$$R^n D^n x = c_D^{-n} (D^n x) R^n e -$$

$$- \sum_{j=1}^{n} c_D^{-j} R^{n-j} \{ c_D R[(D^{n+1} x) R^j e] +$$

$$+ F[(D^n x) R^j y] + y + R f_D(D^n x, R^j e) \}, \qquad (6.2.7)$$

provided that $x \in \mathrm{dom}\, D^{n+1}$;

(iv) *If X has a unit e and F is multiplicative then for $x \in \mathrm{dom}\, D$, $z \in \ker D$*

$$R^n x = c_D^{-n} x R^n e - \sum_{j=1}^{n} c_D^{-j} R^{n-j} \{ c_D R[(De)(R^j e)] + R f_D(x, R^j e) \}, \qquad (6.2.8)$$

$$R^n z = c_D^{-n} z R^n e - \sum_{j=1}^{n} c_D^{-j} R^{n+1-j} f_D(z, R^j e). \qquad (6.2.9)$$

Proof. In order to prove Formula (6.2.4) observe that $Dz = 0$ and apply Formula (6.2.1) for $y = z$. If X has a unit e then applying Formula (6.2.1) for $y = e$ we obtain (6.2.5) for $xe = x$. Applying Formula (6.2.5) for $x = z \in \ker D$ we obtain (6.2.6). Write $w = D^n x$, $y = e$. Then $Dw = D^{n+1} x$. Applying Formula (6.2.5) to the element $w \in \mathrm{dom}\, D$, we obtain the required Formula (6.2.7).

We have $FR = 0$ by definition. If F is multiplicative, then $F(x R^j e) = (Fx)(FR^j e) = 0$ for $j = 1, \ldots, n$. This and Formula (6.2.5) together imply (6.2.8). Applying (6.2.8) for $z \in \ker D$, we obtain (6.2.9). ∎

Formulae (6.2.1)–(6.2.9) are, as a matter of fact, formulae of integration in D-algebras.

PROPOSITION 6.2.1. *Suppose that X is an almost Leibniz D-algebra and F is an almost averaging initial operator for D corresponding to an $R \in \mathscr{R}_D$. Then*

$$F(zR^n x) = 0 \quad \text{for } x \in X, z \in \ker D, n \in \mathbf{N}, \tag{6.2.10}$$

$$R^n(zx) = c_D^{-n} z R^n x. \tag{6.2.11}$$

Proof. Indeed, since F is almost averaging and $FR = 0$, we find $F(zR^n x) = (Fz)(FR^n x) = zFR^n x = 0$ for $x \in X$, $z \in \ker D$ and $n \in \mathbf{N}$. Since by our assumption $f_D(z, x) = 0$ for $z \in \ker D$, $x \in X$, Formula (6.2.9) immediately imply (6.2.11). ∎

THEOREM 6.2.2 (*Integration of unit formula*). *Suppose that X is an almost Leibniz D-algebra with unit e such that constants are not zero divisors. Suppose that F is an initial operator for D corresponding to a right inverse R. Then*

$$R^n e = \frac{(Re)^n}{d(n)} - \sum_{k=2}^{n} \frac{1}{d(k)} R^{n-k} F[(Re)^k] \quad (n = 1, 2, \ldots),$$

$$\tag{6.2.12}$$

where we assume that

$$d(n) = d_1 \ldots d_n \neq 0 \quad (n = 1, 2, \ldots) \tag{6.2.13}$$

and $d_1, \ldots, d_n$ are defined by Formulae (6.1.20):

$$d_1 = 1, \quad d_2 = 2c_D, \quad d_n = 2c_D^{n-1} + \sum_{j=1}^{n-2} c_D^j \quad \text{for } n \geqslant 3.$$

$$\tag{6.2.14}$$

Proof. Observe that by our assumption $d_1 \neq 0$, $d_2 \neq 0$ and

$$d_{n+1} = c_D(d_n + 1), \quad d(n+1) = d_{n+1} d(n), \quad d(n) \neq 0$$
$$(n = 1, 2, \ldots).$$

Our assumptions and Corollary 6.1.9 together imply that $De = 0$. Then e is a constant. Write: $g = Re$. Then $Dg = DRe = e$. Theorem

6.1.3 and our assumptions together imply that $Dg^n = d_n g^{n-1} Dg$ $= d_n g^{n-1} e = d_n g^{n-1}$ $(n = 1, 2, ...)$. Hence

$$g^n - Fg^n = (I-F)g^n = RDg^n = d_n Rg^{n-1}$$

$$(n = 2, 3, ...). \qquad (6.2.15)$$

Observe that $d(2) R^2 e = d_2 Rg = g^2 - d_1 Fg^2$. Suppose that for an arbitrarily fixed $n \geqslant 2$ we have

$$R^n e = \frac{g^n}{d(n)} - \sum_{k=2}^{n} \frac{1}{d(k)} R^{n-k} Fg^k. \qquad (6.2.16)$$

Then

$$R^{n+1} e = R(R^n e) = \frac{1}{d(n)} Rg^n - \sum_{k=2}^{n} \frac{1}{d(k)} R^{n+1-k} Fg^k$$

$$= \frac{1}{d_{n+1} d(n)} (g^{n+1} - Fg^{n+1}) - \sum_{k=2}^{n} \frac{1}{d(k)} R^{n+1-k} Fg^k$$

$$= \frac{1}{d(n+1)} g^{n+1} - \frac{1}{d(n+1)} Fg^{n+1} - \sum_{k=2}^{n} \frac{1}{d(k)} R^{n+1-k} Fg^k$$

$$= \frac{g^{n+1}}{d(n+1)} - \sum_{k=2}^{n+1} \frac{1}{d(k)} R^{n+1-k} Fg^k$$

which was to be proved.   ∎

Note that in the Leibniz* case $c_D = 1$. Hence $d_n = n$ and

$$d(n) = n! \quad (n = 1, 2, ...). \qquad (6.2.17)$$

PROPOSITION 6.2.2. *Suppose that X is a D-algebra and* dim ker $D \neq 0$. *Suppose that $z_1, z_2 \in$ ker D. Then $z_1 z_2 \in$ ker D if and only if either $f_D = 0$ or X is an almost Leibniz D-algebra.*

Proof. Let $z_1, z_2 \in$ ker D. Then $Dz_1 = Dz_2 = 0$ and

$$D(z_1 z_2) = c_D(z_1 Dz_2 + z_2 Dz_1) + f_D(z_1 z_2).$$

---

* cf. Przeworska-Rolewicz and von Trotha, 1981. The present proof is simpler, even in the Leibniz case.

Then $D(z_1 z_2) = 0$ if and only if $f_D(z_1, z_2) = 0$. This condition is satisfied if either $f_D = 0$ or $f_D(x, z) = 0$ for all $x \in \mathrm{dom}\, D$, $z \in \ker D$, i.e. if X is an almost Leibniz D-algebra (cf. also Corollary 6.1.7). ∎

COROLLARY 6.2.3. *If X is either a Leibniz D-algebra or a quasi-Leibniz D-algebra or a simple Duhamel D-algebra then $z_1 z_2 \in \ker D$ for all $z_1, z_2 \in \ker D$.*

Proof. Indeed, all these D-algebras are almost Leibniz D-algebras. ∎

PROPOSITION 6.2.3. *Suppose that X is a D-algebra, $\dim \ker D \neq 0$ and $R \in \mathscr{R}_D \cap V(X)$. Then*

$$[c_\lambda(z)]^2 + [s_\lambda(z)]^2 = e_{\lambda i}(z) e_{-\lambda i}(z) \quad \text{for all } \lambda \in \mathbf{R},\, z \in \ker D,$$
$$(6.2.18)$$

where $e_\lambda, c_\lambda, s_\lambda$ are defined by Formulae (2.3.4), (2.3.9), respectively.

Proof. By definition, for arbitrary $\lambda \in \mathbf{R}$ and $z \in \ker D$ we have

$$
\begin{aligned}
[c_\lambda(z)]^2 + [s_\lambda(z)]^2 &= \frac{1}{4}\,[e_{\lambda i}(z) + e_{-\lambda i}(z)]^2 + \frac{1}{4 i^2}\,[e_{\lambda i}(z) - e_{-\lambda i}(z)]^2 \\
&= \frac{1}{4}\,\{[e_{\lambda i}(z)]^2 + 2 e_{\lambda i}(z) e_{-\lambda i}(z) + [e_{-\lambda i}(z)]^2\} - \\
&\quad - \frac{1}{4}\,\{[e_{\lambda i}(z)]^2 - 2 e_{\lambda i}(z) e_{-\lambda i}(z) + [e_{-\lambda i}(z)]^2\} \\
&= e_{\lambda i}(z) e_{-\lambda i}(z). \qquad\qquad ∎
\end{aligned}
$$

PROPOSITION 6.2.4. *Suppose that all assumptions of Proposition 6.2.3 are satisfied. Then for all $\lambda, \mu \in \mathbf{C}$, $z_1, z_2 \in \ker D$ there exist a $\nu \in \mathbf{C}$ and a $z_3 \in \ker D$ such that*

$$e_\lambda(z_1) e_\mu(z_2) = e_\nu(z_3),\qquad\qquad (6.2.19)$$

where

$$
\nu = \begin{cases}
\lambda + \mu & \text{if } c_D = 1, f_D = 0, \\
c_D(\lambda + \mu) & \text{if } c_D \neq 1, f_D = 0, \\
\lambda + \mu + d\lambda\mu & \text{if } c_D = 1, f_D(x, y) = d(Dx)(Dy) \\
& \quad \text{for } x, y \in \mathrm{dom}\, D.
\end{cases}
\qquad (6.2.20)
$$

$\neq 0$, $R \in \mathscr{R}_D \cap V(X)$ and $S \in L_0(X)$ is a D-invariant operator. An element $x \in X$ is said to be *exponential-periodic* if

$$x = e_\lambda(z)v, \quad \text{where } \lambda \text{ is a scalar, } z \in \ker D \text{ and } Sv = v,$$
$$(6.2.29)$$

i.e. v is an S-periodic element.

A linear combination of such elements will be also called an *exponential-periodic element*.

The following question arises: We have assumed that S is a D-invariant operator, i.e.

$$SD = DS \quad \text{on dom } D. \tag{6.2.30}$$

Does Equality (6.2.30) hold for exponential-periodic elements in some D-algebras? The answer is given by the following

THEOREM 6.2.3. *Let X be a D-algebra over $\mathbf{C}$, $\dim \ker D \neq 0$ and $R \in \mathscr{R}_D \cap V(X)$. Suppose that $S \in L_0(X)$ is D-invariant, multiplicative and has the following property: there exists $0 \neq r \in \mathbf{R}$ such that*

$$Se_\lambda(z) = e^{\lambda r} e_\lambda(z) \quad \text{for all } \lambda \in \mathbf{C}, z \in \ker D.* \tag{6.2.31}$$

Write

$$X^0 = \{x = e_\lambda(z)v: \lambda \in \mathbf{C}, z \in \ker D, v \in \text{dom } D, Sv = v\}. \tag{6.2.32}$$

Then Equality (6.2.30) holds on X^0 if and only if

$$\text{either } f_D(e_\lambda(z), v) = \gamma_\lambda e_\lambda(z) Dv \quad \text{for } 0 \neq \gamma_\lambda \in \mathbf{C}, \lambda \neq 0$$
$$\text{or } f_D = 0. \tag{6.2.33}$$

Proof. Let $x \in X^0$ be arbitrarily fixed. By our assumptions, if we write $u = e_\lambda(z)$ then $x = uv \in \text{dom } D$ and

$$SDv = DSv = Dv,$$
$$SDu = SDe_\lambda(z) = \lambda Se_\lambda(z) = \lambda e^{\lambda r} e_\lambda(z) = \lambda e^{\lambda r} u. \tag{6.2.33'}$$

Hence

$$SDx = SD(uv) = S[c_D(uDv + vDu)] + Sf_D(u, v)$$

---

* Observe that (6.2.31) implies $Sz = z$ for all $z \in \ker D$. In particular, Property (6.2.31) is satisfied, if S is a D-shift.

$$= c_D[S(uDv) + S(vDu)] + Sf_D(u, v)$$
$$= c_D[(Su)(SDv) + (Sv)(SDu)] + Sf_D(u, v)$$
$$= c_D(e^{\lambda r}uDv + v\lambda e^{\lambda r}u) + Sf_D(u, v)$$
$$= c_D e^{\lambda} u(Dv + \lambda v) + Sf_D(u, v).$$

On the other hand

$$DSx = DS(uv) = D[(Su)(Sv)] = D[e^{\lambda r}(uv)] = e^{\lambda r}D(uv)$$
$$= e^{\lambda r}[c_D(uDv + vDu) + f_D(u, v)]$$
$$= e^{\lambda r}[c_D(uDv + \lambda uv) + f_D(u, v)]$$
$$= e^{\lambda r}c_D u(\lambda v + Dv) + e^{\lambda r}f_D(u, v).$$

This implies that $SDx = DSx$ for $x \in X^0$ if and only if $Sf_D(u, v)$ $= e^{\lambda r}f_D(u, v)$. If $f_D = 0$ then this last equality is trivially satisfied. If $f_D \neq 0$, according to the assumed Property (6.2.31), this equality implies that f_D is of the form (6.2.33). ∎

COROLLARY 6.2.9. *Suppose that all assumptions of Theorem 6.2.3 are satisfied and either $f_D = 0$ (in particular X is a Leibniz D-algebra or simple Duhamel D-algebra) or X is a quasi-Leibniz D-algebra. Then Equality (6.2.30) holds for exponential-periodic elements.*

Proof. Indeed, suppose that $f_D \neq 0$. Then X is a quasi-Leibniz D-algebra, i.e. $f_D(e_\lambda(z), v) = d[De_\lambda(z)](Dv) = \lambda de_\lambda(z)Dv$, where $d \neq 0$. Hence γ_λ $= \lambda d \neq 0$ for $\lambda \neq 0$. ∎

PROPOSITION 6.2.5. *Suppose that X is a D-algebra over $\mathbf{C}$, dim ker $D \neq 0$, $R \in \mathcal{R}_D \cap V(X)$ and exponentials are not zero divisors. Suppose, moreover, that S_h is a multiplicative D-shift on $h \in \mathbf{R}$, dim $X_{S_{Nh}} > 0$ (where $X_{S_{Nh}}$ is defined by Formula (5.2.10)). Then $x \in X_j$ (cf. Formula (5.1.4)) if and only if*

$$u = e_{-\lambda}(z)v, \quad \text{where } Sv = v, \lambda = \frac{2\pi i}{Nh}j \quad (j = 1, 2, ..., N),$$
$$\lambda \in \mathbf{C}, \quad z \in \ker D, \quad (6.2.34)$$

i.e. a general form of S_{nh}-periodic elements is

$$x = \sum_{j=1}^{N} e_{2\pi ij/Nh}(z_j)v_j, \quad \text{where } z_j \in \ker D, S_{Nh}v_j = v_j$$

$$(j = 1, 2, ..., N). \quad (6.2.35)$$

In the first case X is a Leibniz D-algebra, in the second one if $c_D = \frac{1}{2}$ then X is a simple Duhamel D-algebra, in the third one, X is a quasi-Leibniz D-algebra.

Proof. For all $z_1, z_2 \in \ker D$, $\lambda, \mu \in \mathbf{C}$ we have

$$D[e_\lambda(z_1)e_\mu(z_2)]$$
$$= c_D[e_\lambda(z_1)De_\mu(z_2) + e_\mu(z_2)De_\lambda(z_1)] + f_D(e_\lambda(z_1), e_\mu(z_2))$$
$$= c_D[\mu e_\lambda(z_1)e_\mu(z_2) + \lambda e_\mu(z_2)e_\lambda(z_1)] + f_D(e_\lambda(z_1), e_\mu(z_2))$$
$$= c_D(\lambda + \mu)e_\lambda(z_1)e_\mu(z_2) + f_D(e_\lambda(z_1), e_\mu(z_2)).$$

If $f_D = 0$ then we find

$$D[e_\lambda(z_1)e_\mu(z_2)] = c_D(\lambda + \mu)e_\lambda(z_1)e_\mu(z_2), \qquad (6.2.21)$$

which implies that the element $e_\lambda(z_1)e_\mu(z_2)$ is an exponential corresponding to the eigenvalue $v = c_D(\lambda + \mu)$, i.e. is of the form $e_v(z_3)$, where $z_3 \in \ker D$. If $f_D(x, y) = d(Dx)(Dy)$ for $x, y \in \mathrm{dom}\ D$ then

$$f_D(e_\lambda(z_1), e_\mu(z_2)) = d[De_\lambda(z_1)][De_\mu(z_2)] = d\lambda\mu e_\lambda(z_1)e_\mu(z_2)$$

which implies

$$D[e_\lambda(z_1)e_\mu(z_2)] = (\lambda + \mu + d\lambda\mu)e_\lambda(z_1)e_\mu(z_2) \qquad (6.2.22)$$

for $c_D = 1$. Therefore the element $e_\lambda(z_1)e_\mu(z_2)$ is an exponential corresponding to the eigenvalue $v = \lambda + \mu + d\lambda\mu$, i.e. is of the form $e_v(z_3)$, where $z_3 \in \ker D$. ∎

COROLLARY 6.2.4. *Suppose that all assumptions of Proposition 6.2.3 are satisfied. Then*

$$e_\lambda(\ker D)e_\mu(\ker D) \subset e_v(\ker D) \quad \text{for all } \lambda, \mu \in \mathbf{C}, \qquad (6.2.23)$$

where v is defined by Formula (6.2.20).

COROLLARY 6.2.5. *Suppose that all assumptions of Proposition 6.2.3 are satisfied. Then for all $\lambda \in \mathbf{R}$, $z \in \ker D$ there exists a $z_0 \in \ker D$ such that*

$$[c_\lambda(z)]^2 + [s_\lambda(z)]^2 = e_v(z_0), \qquad (6.2.24)$$

where

$$v = \begin{cases} 0 & \text{if } f_D = 0, \\ -d\lambda^2 & \text{if } X \text{ is a } QL\text{-algebra.} \end{cases}$$

Proof. Indeed, if we put instead of λ the number λi, where $\lambda \in \mathbf{R}$, and $\mu = -\lambda i$, then Propositions 6.2.3 and 6.2.4 together imply that $\nu = 0$ if $f_D = 0$ and $\nu = d(\lambda i)(-\lambda i) = -d\lambda^2$ if X is a quasi-Leibniz algebra. Hence Formula (6.2.24) is satisfied. ∎

COROLLARY 6.2.6. *Suppose that all assumptions of Proposition 6.2.3 are satisfied. If $f_D = 0$ then the trigonometric identity holds, i.e. for arbitrary $z \in \ker D$, $\lambda \in \mathbf{R}$ there exists a $z_0 \in \ker D$ such that*

$$[c_\lambda(z)]^2 + [s_\lambda(z)]^2 = z_0. \tag{6.2.25}$$

Proof. Indeed, since $f_D = 0$, we have $\nu = 0$ and $e_\nu(z_0) = e_0(z_0) = z_0$. Then Formula (6.2.24) implies (6.2.25). ∎

COROLLARY 6.2.7. *Suppose that all assumptions of Proposition 6.2.3 are satisfied and the initial operator F corresponding to R is multiplicative, i.e. $(Fx)(Fy) = F(xy)$ for $x, y \in X$. If $f_D = 0$ then*

$$e_\lambda(z_1)e_\mu(z_2) = e_{c_D(\lambda+\mu)}(z_1 z_2) \quad \text{for all } z_1, z_2 \in \ker D, \quad \lambda, \mu \in \mathbf{C}, \tag{6.2.26}$$

$$[c_\lambda(z)]^2 + [s_\lambda(z)]^2 = z^2 \quad \text{for all } z \in \ker D, \quad \lambda \in \mathbf{R}. \tag{6.2.27}$$

Proof. Indeed, Formula (2.3.3) and Proposition 6.2.4 together imply that for all $\lambda, \mu \in \mathbf{C}$, $z_1, z_2 \in \ker D$ we have $Fe_\lambda(z_1) = z_1$, $Fe_\mu(z_2) = z_2$. Hence $[Fe_\lambda(z_1)][Fe_\mu(z_2)] = z_1 z_2 = Fe_{c_D(\lambda+\mu)}(z_1 z_2)$. Formulae (6.2.18) and (6.2.20) together imply that for all $\lambda \in \mathbf{R}$, $z \in \ker D$ we have

$$[c_\lambda(z)]^2 + [s_\lambda(z)]^2 = e_{\lambda i}(z)e_{-\lambda i}(z) = e_{c_D(\lambda i - \lambda i)}(z^2)$$
$$= e_0(z^2) = z^2. \qquad ∎$$

COROLLARY 6.2.8. *Suppose that all assumptions of Proposition 6.2.3 are satisfied, the initial operator F corresponding to R is multiplicative and X has a unit e. If $f_D = 0$ then*

$$[c_\lambda(e)]^2 + [s_\lambda(e)]^2 = e \quad \text{for all } \lambda \in \mathbf{R}. \tag{6.2.28}$$

Indeed, $e^2 = e$, and, by Proposition 6.1.2, $e \in \ker D$ for $f_D = 0$. ∎

DEFINITION 6.2.1. Suppose that X is a *D*-algebra over $\mathbf{C}$, dim ker *D*

Proof. If $S_{Nh}v = v$ and $x_j = e_{2\pi i j/Nh}(z_j)v$, where $z_j \in \ker D$ ($j = 1, 2, \ldots$
$\ldots, N$) then

$$S_{Nh}x_j = S_{Nh}[e_{2\pi i j/Nh}(z_j)v]$$
$$= S_{Nh}[e_{2\pi i j/Nh}(z_j)]S_{Nh}v = e^{h \cdot 2\pi i j/Nh}e_{2\pi i j/Nh}(z_j)v$$
$$= e^{2\pi i j/N}e_{2\pi i j/Nh}(z_j)v = \varepsilon^j x_j,$$

where $\varepsilon = e^{2\pi i/N}$. Hence $x_j \in X_j$ and $x = \sum\limits_{j=1}^{N} x_j \in X_{S_{Nh}}$.

On the other hand, suppose that $x_j \in X_j, j = 1, 2, \ldots, N$. Then $S_{Nh}x_j = \varepsilon^j x_j$. Observe the element $u = e_{2\pi i j/Nh}(z)$, where $0 \neq z \in \ker D$ is arbitrarily fixed, belongs to X_j for $S_{Nh}u = \varepsilon^j u$. We are looking for $x_j = uv$, where $v \neq 0$ is to be determined. Then $\varepsilon^j uv = \varepsilon^j x_j = \varepsilon^j S_{Nh}(uv)$ $= (S_{Nh}u)(S_{Nh}v) = \varepsilon^j uS_{Nh}v$, i.e. $\varepsilon^j u(v - S_{Nh}v) = 0$. The arbitrariness of $z \in \ker D$ and our assumption that exponentials are not zero divisors together imply that $S_{Nh}v = v$. ∎

THEOREM 6.2.4. *Suppose that X is a D-algebra over* **C**, *dim $\ker D \neq 0$ $R \in \mathcal{R}_D \cap V(X)$, $S_h \in L_0(X)$ is a multiplicative D-shift on $0 \neq h \in \mathbf{R}$. Then S_h is an algebraic operator on the space*

$$X_{EP}(\lambda_1, \ldots, \lambda_m)$$

$$= \left\{ x = \sum_{m=1}^{M} e_{\lambda_m}(z_m)v_m \colon S_h v_m = v_m \in X, z_m \in \ker D, \lambda_m \in \mathbf{C}, \right.$$

$$\lambda_m \neq \lambda_j + 2\pi i k/h \text{ for } m \neq j \ (j, m = 1, 2, \ldots, M;$$

$$\left. k = 0, \pm 1, \pm 2, \ldots) \right\} \quad (6.2.36)$$

with the characteristic polynomial (with single roots only):

$$P(t) = \prod_{m=1}^{M} (t - t_m), \quad \text{where } t_m = e^{\lambda_m h},$$

and

$$X_{EP}(\lambda_1, \ldots, \lambda_m) = \bigoplus_{i=1}^{M} X_m, \quad \text{where } X_m = P_m X, \quad P_m = P_m(S_h),$$

$$P_m(t) = \prod_{j=1, j \neq m}^{M} \frac{t - t_j}{t_m - t_j} \quad (m = 1, 2, \ldots, M). \quad (6.2.37)$$

Proof. Suppose that $x = \sum\limits_{m=1}^{M} e_{\lambda_m}(z_m)v_m \in X_{EP}(\lambda_1, \dots, \lambda_m)$ is arbitrarily fixed. Then, by our assumptions,

$$S_h x = \sum_{m=1}^{M} S_h[e_{\lambda_m}(z_m)v_m] = \sum_{m=1}^{M} [S_h e_{\lambda_m}(z_m)](S_h v_m)$$

$$= \sum_{m=1}^{M} e^{\lambda_m h} e_{\lambda_m}(z_m)v_m = \sum_{m=1}^{M} t_m e_{\lambda_m}(z_m)v_m.$$

By an easy induction we prove that

$$S_h^k x = \sum_{m=1}^{M} t_m^k e_{\lambda_m}(z_m)v_m \quad (k = 0, 1, 2, \dots).$$

If we write $P(t) = \prod\limits_{m=1}^{M} (t - t_m) = \sum\limits_{k=1}^{M} p_k t^k$ with $p_M = 1$ then we have $P(t_m) = 0$ for $m = 1, 2, \dots, M$. Thus

$$P(S_h)x = \sum_{k=1}^{M} p_k S_h^k x = \sum_{k=1}^{M} p_k \sum_{m=1}^{M} t_m^k e_{\lambda_m}(z_m)v_m$$

$$= \sum_{m=1}^{M} \left(\sum_{k=1}^{M} p_k t_m^k\right) e_{\lambda_m}(z_m)v_m = \sum_{m=1}^{M} P(t_m) e_{\lambda_m}(z_m)v_m = 0.$$

The arbitrariness of $x \in X_{EP}(\lambda_1, \dots, \lambda_m)$ implies that $P(S_h) = 0$ on $X_{EP}(\lambda_1, \dots, \lambda_m)$. The roots of the polynomial $P(t)$ are single by our assumption that $\lambda_m \neq \lambda_j + 2\pi i k/h$ for an arbitrary integer k, what implies $t_m \neq t_j$ for $m \neq j$.

The decomposition onto direct sum of spaces $X_1, \dots, X_m$ follows from Formula (5.1.4). ∎

COROLLARY 6.2.10. *Suppose that all assumptions of Theorem 6.2.4 are satisfied. Then* $x \in X_m$ *(where X_m are defined by Formulae (6.2.37)) if and only if*

$$x = e_{\mu_m}(z)v, \quad \text{where } z \in \ker D, \ S_h v = v, \ \mu_m = \lambda_m + \frac{2\pi i}{h}j$$

$$(j = 0, \pm 1, \pm 2, \dots, m = 1, 2, \dots, M). \quad (6.2.38)$$

Proof. Indeed, suppose that $x = e_{\mu_m}(z)v$, where $z \in \ker D$, $S_h v = v$. Then

$$S_h x = S_h[e_{\mu_m}(z)v] = [S_h e_{\mu_m}(z)](S_h v) = e^{\mu_m h} e_{\mu_m}(z)v = e^{\mu_m h} x.$$

Hence $e^{\mu_m h} = t_m = e^{\lambda_m h}$ if and only if $\mu_m = \lambda_m + \dfrac{2\pi i}{h} j$, where j is an integer. Clearly, if x is of the form (6.2.38) then $x \in X_m$. ∎

In Section 3.4 we have determined real solutions of an equation with scalar coefficients in the case when the characteristic polynomial has purely imaginery roots (cf. Point (A) after Theorem 3.4.2). The case of complex roots (Point (B)) has not been solved, since it was not possible without additional assumptions. We shall study this case now.

THEOREM 6.2.5. *Suppose that all assumptions of Theorem 3.4.1 are satisfied and the polynomial $Q(t)$ has two conjugate complex roots*: $\lambda = \mu + i\nu$, $\bar{\lambda} = \mu - i\nu$ *(where $0 \neq \mu, \nu \in \mathbf{R}$). If X is a D-algebra such that $f_D = 0$ then the corresponding linearly independent solutions are exponential-periodic elements of the form*

$$x_1 = e_{\mu/c_D}(z_1)c_{\nu/c_D}(z_2), \quad x_2 = e_{\mu/c_D}(z_1)s_{\nu/c_D}(z_2), \qquad (6.2.39)$$

where $z_1, z_2 \in \ker D$ are arbitrary.

Proof. In the same way, as in Point (A) of Section 3.4, we extend the space X to a space $Y = X \oplus iX$ over complexes and all operators under consideration. Then in the space Y the equation $Q(D)x = 0$ has two linearly independent solutions: $y_1 = e_\lambda(\zeta_1)$, $y_2 = e_{\bar{\lambda}}(\zeta_2)$, $\zeta_1, \zeta_2 \in \ker D$ in Y are arbitrary. On the other hand, Formulae (6.2.20) imply that for every $z_1, z_2 \in \ker D$ there exist $z_3, z_4 \in \ker D$ such that

$$x_1 + ix_2 = e_{\mu/c_D}(z_1)c_{\nu/c_D}(z_2) + ie_{\mu/c_D}(z_1)s_{\nu/c_D}(z_2)$$

$$= e_{\mu/c_D}(z_1)\left\{\frac{1}{2}[e_{i\nu/c_D}(z_2) + e_{-i\nu/c_D}(z_2)] + \right.$$

$$\left. + \frac{1}{2i}[e_{i\nu/c_D}(z_2) - e_{-i\nu/c_D}(z_2)]\right\}$$

$$= e_{\mu/c_D}(z_1)e_{i\nu/c_D}(z_2) = e_{\mu+i\nu}(z_3) = e_\lambda(z_3)$$

and, similarly,

$$x_1 - ix_2 = e_{\mu - i\nu}(z_4) = e_{\bar{\lambda}}(z_4).$$

Hence

$$x_1 = \frac{1}{2}[e_\lambda(z_3) + e_{\bar{\lambda}}(z_4)], \quad x_2 = \frac{1}{2i}[e_\lambda(z_3) - e_{\bar{\lambda}}(z_4)]$$

are solutions of the equation $Q(D)x = 0$, we are looking for, since a linear combination of elements belonging to ker $Q(D)$ again is an element of ker $Q(D)$.   ∎

Examples and Exercises

EXAMPLE 6.2.1. Suppose that all assumptions of Example 6.1.1 are satisfied. Then X is a Leibniz D-algebra and Formula (6.2.2) of integration by parts is of the form

$$R(xDy) = xy - F(xy) - R(yDx) \quad \text{for } x, y \in \text{dom } D, \quad (6.2.40)$$

since $c_D = 1, f_D = 1$.

Suppose now that all assumptions of Example 6.1.1a are satisfied. Then $X = C[a, b]$ is a Leibniz $\frac{d}{dt}$-algebra and the formula of integration by parts for indefinite integrals is

$$\int x(t)y'(t)\,dt = xy - \int x'(t)y(t)\,dt \quad \text{for } x, y \in C^1[a, b].$$

$$(6.2.41)$$

Indeed, since $z = F(xy)$ is a constant, the definition of an indefinite (2.1.8) implies Formula (6.2.41). The formula of integration by parts for definite integrals (cf. Formula (6.2.3)) is

$$\int_a^b x(t)y'(t)\,dt = [x(t)y(t)]_a^b - \int_a^b x'(t)y(t)\,dt \quad \text{for } x, y \in C^1[a, b],$$

where $[x(t)y(t)]_a^b = x(b)y(b) - x(a)y(a)$.

EXAMPLE 6.2.2. Suppose that all assumptions of Theorem 6.2.2 are satisfied and F is a multiplicative initial operator. Then Formula (6.2.12) of integration of unit is of the form:

$$R^n e = \frac{(Re)^n}{d(n)} \quad \text{for all } n \in \mathbf{N}. \quad (6.2.42)$$

Indeed, since $FR = 0$ we find for all $k \geqslant 2$
$$F[(Re)^k] = Fg^k = (Fg)^k = (FRg)^k = 0.$$

EXAMPLE 6.2.3. Suppose that all assumptions of Example 6.1.1a are satisfied. Since $e(t) \equiv 1$ is the unit of the Leibniz $\dfrac{d}{dt}$-algebra $C[a, b]$, $(Re)(t) = \int_a^t 1 \, ds = t - a$ and the initial operator F corresponding to R is multiplicative, Formula (6.2.42) implies that the formula of integration of unit is

$$\left[\left(\int_a^t\right)^n e\right](t) = \frac{(t-a)^n}{n!} \tag{6.2.42'}$$

for $d(n) = n!$ in the Leibniz case.

EXERCISE 6.2.1. Suppose that X is a D-algebra, $\dim \ker D \neq 0$, F is a multiplicative initial operator corresponding to an $R \in \mathscr{R}_D \cap V(X)$. Prove that:

(i) If $f_D = 0$ and $c_D = 1$ then $[e_\lambda(z)]^n = e_{n\lambda}(z)$, for all $n \in \mathbf{N}$, $z \in \ker D$ and $\lambda \in \mathbf{C}$;

(ii) If X is a quasi-Leibniz D-algebra then
$$[e_\lambda(z)]^n = e_{\mu_n}(z),$$
where
$$\mu_n = \lambda\left(n + \sum_{j=1}^{n-1} d^j \lambda^j\right) \quad \text{for } n \geqslant 2, z \in \ker D, \lambda \in \mathbf{C}.$$

Find a formula for $[e_\lambda(z)]^n$ if $f_D = 0$ but $c_D \neq 0$, where $n \in \mathbf{N}$, $z \in \ker D$, $\lambda \in \mathbf{C}$ are arbitrary.

EXERCISE 6.2.2. Suppose that X is a D-algebra, $\dim \ker D \neq 0$ and $R \in \mathscr{R}_D \cap V(X)$. Prove that for all $\lambda, \mu \in \mathbf{R}$, $z \in \ker D$

$$c_\lambda(z) s_\mu(z) + s_\lambda(z) c_\mu(z) = \frac{1}{2i} [e_{\lambda i}(z) e_{\mu i}(z) - e_{-\lambda i}(z) e_{-\mu i}(z)],$$

$$c_\lambda(z) c_\mu(z) - s_\lambda(z) s_\mu(z) = \frac{1}{2} [e_{\lambda i}(z) e_{\mu i}(z) + e_{-\lambda i}(z) e_{-\mu i}(z)].$$

EXERCISE 6.2.3. Suppose that X is a D-algebra such that $f_D = 0$, dim ker $D \neq 0$ and F is a multiplicative initial operator for D corresponding to an $R \in \mathscr{R}_D \cap V(X)$. Prove that for all $\lambda, \mu \in \mathbf{R}$, $z \in \ker D$

$$c_\lambda(z) s_\mu(z) + s_\lambda(z) c_\mu(z) = s_{\lambda+\mu}(z),$$
$$c_\lambda(z) c_\mu(z) - s_\lambda(z) s_\mu(z) = c_{\lambda+\mu}(z).$$

Find the corresponding formulae if F is not multiplicative.

EXERCISE 6.2.4. Suppose that X is a D-algebra, dim ker $D \neq 0$ and $R \in \mathscr{R}_D \cap V(X)$. Let operators ch_λ and sh_λ be defined by Formulae (2.3.23). Prove that:

(i) $[\mathrm{ch}_\lambda(z)]^2 - [\mathrm{sh}_\lambda(z)]^2 = e_\lambda(z) e_{-\lambda}(z)$ for all $z \in \ker D$, $\lambda \in \mathbf{R}$;

(ii) If X is a quasi-Leibniz D-algebra then for all $\lambda \in \mathbf{R}$, $z \in \ker D$ there exists a $z_0 \in \ker D$ such that

$$[\mathrm{ch}_\lambda(z)]^2 - [\mathrm{sh}_\lambda(z)]^2 = e_{d\lambda^2}(z_0);$$

(iii) If $f_D = 0$ then for all $\lambda \in \mathbf{R}$, $z \in \ker D$ there exists $z_0 \in \ker D$ such that

$$[\mathrm{ch}_\lambda(z)]^2 - [\mathrm{sh}_\lambda(z)]^2 = z_0;$$

(iv) If $f_D = 0$ and the initial operator F corresponding to R is multiplicative then for all $\lambda \in \mathbf{R}$ and $z \in \ker D$

$$[\mathrm{ch}_\lambda(z)]^2 - [\mathrm{sh}_\lambda(z)]^2 = z^2.$$

If, moreover, X has a unit e then for all $\lambda \in \mathbf{C}$

$$[\mathrm{ch}_\lambda(e)]^2 - [\mathrm{sh}_\lambda(e)]^2 = e.$$

EXERCISE 6.2.5 (cf. Tasche, 1976). Suppose that all assumptions of Exercise 2.3.3 are satisfied. Prove that X is a D-algebra over $\mathbf{C}$ with the multiplication of function defined by a convolution

$$(x * y)(t) = \int_1^t x(t/s) y(s) s^{-1} ds \quad \text{for } x, y \in X. \tag{6.2.43}$$

EXERCISE 6.2.6 (cf. Dimovski, 1973; Tasche, 1976). Suppose that all assumptions of Exercise 2.3.4 are satisfied. Prove that X with the multiplication defined by a convolution

$$(x*y)(t)$$

$$= \frac{1}{r(t)} \int_0^t x\{u^{-1}[u(t)-u(s)]\} r\{u^{-1}[u(t)-u(s)]\} r(s) y(s) u'(s)\, ds$$

$$\text{for } x, y \in X \quad (6.2.44)$$

is a D-algebra over $\mathbf{C}$.

EXERCISE 6.2.7 (cf. Tasche, 1976). Suppose that all assumptions of Exercise 2.3.5 are satisfied. Prove that:

(i) X with the multiplication defined by a convolution
$$(x*y)(t)$$

$$= \frac{1}{4} t^2 \int_0^1 \int_0^1 x\left(t \sqrt{uv}\right) y\left[t \sqrt{(1-u)(1-v)}\right] du\, dv$$

$$\text{for } x, y \in X \quad (6.2.45)$$

is a D-algebra over $\mathbf{C}$ (cf. also Dimovski, 1966);

(ii) Powers of exponentials are of the form

$$[e_\lambda(z)]^{n+1} = \begin{cases} \dfrac{c^n}{\lambda^{n/2}} \dfrac{t^n}{2^n n!} \cdot J_n(\sqrt{\lambda}\, t) & \text{for } \lambda \neq 0, \\[2em] c^n \dfrac{t^n}{(n!)^2} & \text{for } \lambda = 0 \\[1em] & (n = 0, 1, 2, \ldots), \end{cases}$$

where c is an arbitrary constant and J_n denotes the Bessel function of order n (cf. Formula (2.3.29)).

EXERCISE 6.2.8. Suppose that all assumptions of Exercise 2.3.6 are satisfied. Prove that:

(i) X with the multiplication defined as a convolution

$$(x*y)(t, s) = \int_0^t \int_0^s x(t-u, s-v) y(u, v)\, du\, dv \quad \text{for } x, y \in X$$

$$(6.2.46)$$

is a D-algebra over $\mathbf{C}$;

(ii) Powers of exponentials are of the form

$$[e_\lambda(z)^{n+1}] = \begin{cases} C^{n+1} \dfrac{(ts)^{n/2}}{(n!)^{n/2}} J_n(\sqrt{\lambda ts}\,) & \text{for } \lambda \neq 0, \\[2ex] C^{n+1} \dfrac{1}{(n!)^2} \left(\dfrac{ts}{2}\right)^{2n} & \text{for } \lambda = 0 \\[2ex] & (n = 0, 1, 2, \ldots), \end{cases}$$

where C is an arbitrary constant and J_n denotes the Bessel function of order n (cf. Formula (2.3.29), also Tasche, 1976).

EXERCISE 6.2.9. Suppose that X is a D-algebra, $\dim \ker D \neq 0$ and x_λ are exponentials defined for all $\lambda \in \mathbf{C}$. Suppose that $f_D(x, y) = axy + b(Dx)(Dy)$ for $x, y \in \operatorname{dom} D$. Prove that:

(i) For every $\lambda, \mu \in \mathbf{C}$ there exists $\nu \in \mathbf{C}$ such that $x_\lambda x_\mu = x_\nu$, where $\nu = a + b\lambda\mu + c_D(\lambda + \mu)$;

(ii) There exists $\lambda, \mu \in \mathbf{C}$ such that $x_\lambda x_\mu = x_0 \in \ker D$. Does the trigonometric identity hold for some $\lambda \in \mathbf{C}$?

EXERCISE 6.2.10. Suppose that X is a D-algebra, $\dim \ker D \neq 0$, and x^λ are exponentials defined for all $\lambda \in \mathbf{C}$. Suppose that $f_D(x_\lambda, x_\mu) = g(\lambda, \mu)x_\lambda x_\mu$ for $\lambda, \mu \in \mathbf{C}$, where g is a given symmetric function of 2 complex variables (i.e. $g(\lambda, \mu) = g(\mu, \lambda)$). Prove that:

(i) For all $\lambda, \mu \in \mathbf{C}$ we have $x_\lambda x_\mu = x_\nu$, where $\nu = c_D(\lambda + \mu) + g(\lambda, \mu)$, in particular $x_\lambda x_{-\lambda} = g(\lambda, -\lambda)$;

(ii) For all $z \in \ker D$, $\lambda \in \mathbf{R}$ there exists $z_1 \in \ker D$ such that $e_{\lambda i}(z)e_{-\lambda i}(z) = e_{g(\lambda i, -\lambda i)}(z_1)$;

(iii) If $R \in \mathcal{R}_D \cap V(X)$ and there exists a $\lambda_0 \in \mathbf{R}$ such that $g(\lambda_0 i, -\lambda_0 i) = 0$, then the trigonometric identity holds, i.e. for all $z \in \ker D$ there exists a $z_1 \in \ker D$ such that

$$[c_{\lambda_0}(z)]^2 + [s_{\lambda_0}(z)]^2 = z_1.$$

EXERCISE 6.2.11. Suppose that $U \in L_0(X)$ is arbitrarily fixed. Write $\Delta A = AU - UA$ for every $A \in L_0(X)$. Prove that $\Delta(AB) = A(\Delta B) + (\Delta B)A$ for all $A, B \in L_0(X)$.

EXERCISE 6.2.12. Suppose that X is a non-commutative algebra over

$\mathbf{C}, D \in R(X)$, dim ker $D \neq 0$ and there exists an $R \in \mathscr{R}_D \cap V(X)$. Suppose that D satisfies the following condition:

$$D(xy+yx) = xDy+(Dx)y+(Dy)x+yDx \quad \text{for } x, y \in \text{dom } D.$$

This formula is a generalization of the Leibniz condition in the non-commutative case. Prove that for every $\lambda, \mu \in \mathbf{C}$, $z_1, z_2 \in \text{ker } D$ there exists a $z_3 \in \text{ker } D$ such that

$$e_\lambda(z_1)e_\mu(z_2)+e_\mu(z_2)e_\lambda(z_1) = e_{\lambda+\mu}(z_3),$$

(cf. Kornacki, 1986). Does the trigonometric identity hold?

EXERCISE 6.2.13. Prove the *generalized Floquet theorem*: Suppose that D is a Leibniz D-algebra, F is an initial operator for D corresponding to an $R \in \mathscr{R}_D \cap V(X)$, $S_h \in L_0(X)$ is a multiplicative D-shift and R-shift on $0 \neq h \in \mathbf{R}$ simultaneously,

$$D_1 = D+\frac{1}{h} FS_h^n, \quad R_1^0 = \left(I+FS_h-\frac{1}{h} RFS_h\right)R$$

and

$$\tilde{Q}(t, s) = \sum_{k=0}^{N} \tilde{Q}_k t^k S_h^{N-k}, \quad \tilde{Q}(t) = \tilde{Q}(t, 1),$$

where

$$\tilde{Q}_k = \sum_{m=k}^{N} \binom{m}{k} \lambda^{m-k}Q_m, \quad Q_0, ..., Q_N \in L_0(X) \text{ are } S_h\text{-periodic,}$$

$$\lambda \in \mathbf{C}.$$

If the operator $\tilde{Q}(I, R_1^0)$ is invertible in the space X_{S_h} then the equation $Q(D)x = 0$ has exponential-periodic solutions which are of the form:

$$x = e_\lambda(z)v,$$

where

$$S_h v = v, \quad \lambda = \frac{2\pi i}{h} k \quad (k = 0, \pm 1, \pm 2, ...),$$

$$v = (R_1^0)^N[\tilde{Q}(I, R_1^0)]^{-1} \sum_{k=1}^{N} \tilde{Q}_k \sum_{j=1}^{k-1} h^{j-k}z_j,$$

$z_0, \ldots, z_N \in \ker D$ are arbitrary. If, in addition, F is multiplicative and $N = 2$ then there exist two exponential-periodic solutions x, y of the equation $(D^2 + Q_1 D + Q_0)x = 0$ such that $Fx = 0$, $FDx = x_1 \neq 0$ and $Fy = y_0 \neq 0$, $FDy = 0$, where $x_1, y_0 \in \ker D$ (cf. Przeworska-Rolewicz, 1980, Theorem 8.4 and Corollary 8.5).

The Floquet theorem shows that *every linear equation with a polynomial in D and with S_h-periodic coefficients has at least one exponential-periodic solution* (provided that the operator $\tilde{Q}(I, R_1^0)$ is invertible in the space X_{S_h}). Formulate the corresponding theorem for ordinary differential equation of the order 2 in the space $C(\mathbf{R})$.

EXERCISE 6.2.14. Suppose that X is a D-algebra, $R \in \mathscr{R}_D$ and S_h is a multiplicative R-shift on $0 \neq h \in \mathbf{R}$. Write for arbitrarily fixed positive integers $M, n_0, \ldots, n_M$:

$$X_{PP} = \lim \left\{ u = \sum_{m=0}^{M} v_m R^m z_m : z_m \in \ker D, S_{n_m h} v = v \right.$$

$$(m = 0, 1, \ldots, M) \right\}. \tag{6.2.47}$$

X_{PP} is the space of *D-polynomial-periodic elements*. Prove that S_h is an algebraic operator on X_{PP} with the characteristic polynomial

$$P(t) = \prod_{m=0}^{M} (t^{n_m} - 1)^{M+1} \text{ with the roots } \varepsilon_m^j = \varepsilon^{2\pi i j/n_m} \ (m = 0, 1, \ldots, M;$$

$j = 0, 1, \ldots, n_m - 1)$, each of the multiplicity $M+1$ (cf. Przeworska-Rolewicz, 1980, Corollary 9.4; also Włodarska-Dymitruk, 1975, in the case $D = d/dt$ in the space of continuous functions on real line).

EXERCISE 6.2.15. Suppose that X is either a Leibniz D-algebra or a quasi-Leibniz D-algebra or a simple Duhamel D-algebra. Suppose that $R \in \mathscr{R}_D \cap V(X)$ and S_h is a multiplicative D-shift and R-shift on $0 \neq h \in \mathbf{R}$ simultaneously. Write for arbitrary positive integers $M, K_j, N \ (j = 0, 1, \ldots, M)$:

$$X_{PEP}(\lambda_j; K_j; N_j; M)$$

$$= \lim \left\{ x = \sum_{j=0}^{M} \left[\sum_{k=0}^{K_j} v_{jk} R^k z_{jk} \right] e_{\lambda_j}(z_j) : z_{jk}, z_j \in \ker D, \right.$$

$$v_{jk} \in X, \; S_{N_j h} v_{jk} = v_{jk}, \; \lambda_m \neq \lambda_j + 2\pi i l/h \text{ for } m \neq j$$
$$(l = 0, \pm 1, \pm 2, \ldots; j = 0, 1, \ldots, M, k = 0, 1, \ldots, K_j) \Big\}.$$

$$(6.2.48)$$

The space X_{PEP} is the space of *D-polynomial-exponential-periodic elements*. Prove that the operator S_h is algebraic on X_{PEP} with the characteristic polynomial

$$P(t) = \prod_{j=0}^{M} (t^{N_j} - t_j^{N_j})^{K_j+1}, \quad \text{where } t_j = e^{\lambda_j h} \; (j = 0, 1, \ldots, M).$$

The polynomial $P(t)$ has roots

$$t_{jm} = t_j \varepsilon_j^m, \quad \text{where } \varepsilon_j = e^{2\pi i/N_j}$$

$(m = 0, 1, \ldots, N_j - 1, j = 0, 1, \ldots, M)$ of multiplicity K_j, respectively (cf. the author, 1980, Theorem 9.4).

EXERCISE 6.2.16. Suppose that X is a D-algebra with unit e, $R \in \mathcal{R}_D$, S_h is a multiplicative R-shift (D-shift, respectively, provided that $R \in V(X)$) on $0 \neq h \in \mathbf{R}$. Prove that the equation

$$(I - S_h)x = y, \quad \text{where } y \in \ker (I - S_h) \tag{6.2.49}$$

has a general solution of the form

$$x = \frac{1}{h} y Re + x_h, \quad \text{where } x_h \in \ker (I - S_h) \text{ is arbitrary.}$$

$$(6.2.50)$$

EXERCISE 6.2.17. Suppose that X is a quasi-Leibniz D-algebra with unit $e \in \ker D$ and F is an initial operator for D corresponding to an $R \in \mathcal{R}_D$. Find a formula for $R^n e$.

6.3. CONSTANTS VARIATION METHOD. WROŃSKI THEOREMS

Suppose that X is a D-algebra. Consider a linear equation of the form:

$$Q(D)x = y, \quad y \in X, \quad Q(D) = \sum_{k=0}^{N} a_k D^k,$$

$$a_0, \ldots, a_{N-1} \in X, \quad a_N = I, \tag{6.3.1}$$

i.e. the coefficients of the polynomial $Q(D)$ are *operators of multiplication by elements of X.*

Let X be a Leibniz D-algebra. If we know at least one solution of the homogeneous equation

$$Q(D)u = 0 \tag{6.3.2}$$

then we are looking for solutions of Equation (6.3.1) which are of the form:

$$x = uv, \quad \text{where } v \in \text{dom } D^N \text{ is to be determined.} \tag{6.3.3}$$

We find

$$y = Q(D)x = \sum_{k=0}^{N} a_k D^k x = \sum_{k=0}^{N} a_k D^k (uv)$$

$$= \sum_{k=0}^{N} a_k \sum_{j=0}^{k} \binom{k}{j} (D^{k-j}u)(D^j v) = \sum_{j=0}^{N} \left[\sum_{k=0}^{N} \binom{k}{j} a_k D^{k-j} u \right] D^j v.$$

Write

$$b_j = \sum_{k=j}^{N} \binom{k}{j} a_k D^{k-j} u \quad (j = 0, 1, \ldots, N) \quad \text{and } w = Dv.$$

$$\tag{6.3.4}$$

Then $b_0, \ldots, b_N \in X$. Observe that, by our assumption,

$$b_0 = \sum_{k=0}^{N} a_k D^k u = Q(D)u = 0.$$

Hence

$$Q(D)x = \sum_{j=0}^{N} b_j D^j v = \sum_{j=1}^{N} b_j D^j v = \sum_{j=1}^{N} b_j D^{j-1} w$$

$$= \sum_{m=0}^{N-1} b_{m+1} D^m w.$$

Thus w satisfies an equation of order $N-1$:

$$\sum_{m=0}^{N-1} b_{m+1} D^m w = y, \tag{6.3.5}$$

where $b_1, \ldots, b_N$ are determined by Formulae (6.3.4). Having already known the element w, we can easily find x. Namely, since $w = Dv$, we have $v = Rw + z$, where R is a right inverse of D and $z \in \ker D$ is arbitrary. Thus $x = uv = u(Rw + z)$, where $z \in \ker D$ is arbitrary. Therefore, we are able to solve Equation (6.3.2) by a reduction of its order.

In particular, suppose that $N = 1$ and that u is a solution of the equation

$$Du + a_0 u = 0. \tag{6.3.6}$$

We are looking for solutions of the equation

$$Dx = a_0 x = y, \quad y = X \text{ of the form } x = uv,$$

$$\text{where } v \in \operatorname{dom} D. \tag{6.3.7}$$

We find

$$y = Dx + a_0 x = D(uv) + a_0 uv = uDv + vDu + a_0 uv$$
$$= v(Du + a_0 u) + uDv = uDv.$$

Hence, if $u \in X$ is an invertible element, we obtain $Dv = yu^{-1}$ and $v = R(yu^{-1}) + z$, where R is a right inverse of D and z is an arbitrary constant. This implies that $x = uv = uR(yu^{-1}) + z$, where $z \in \ker D$ is arbitrary and $R \in \mathcal{R}_D$.

This method traditionally is called *constants variation method* (cf. Example 6.3.2).

In order to determine all solutions of Equation (6.3.2) in some D-algebras we have to introduce new necessary notions. Recall that determinants in commutative algebras are defined by Formulae (1.1.18) and (1.1.19).

DEFINITION 6.3.1. Let X be a commutative algebra, let $A \in L(X)$ and let $\dim \ker A \neq 0$. Suppose that $x_1, \ldots, x_N \in \operatorname{dom} A^{N-1}$ for an arbitrarily fixed positive integer $N \geqslant 2$. The determinant

$$W_A(x_1, \ldots, x_n) = \det (A^{j-1} x_k)_{j,k=1,2,\ldots,N}$$

$$= \begin{vmatrix} x_1 & x_2 & \ldots & x_N \\ Ax_1 & Ax_2 & \ldots & Ax_N \\ \cdots\cdots\cdots\cdots\cdots\cdots\cdots\cdots \\ A^{N-1}x_1 & A^{N-1}x_2 & \ldots & A^{N-1}x_N \end{vmatrix} \tag{6.3.8}$$

is called the *Wroński A-determinant of elements* $x_1, \ldots, x_N$, briefly: an *A-Wrońskian*.

PROPOSITION 6.3.1. *Suppose that X is a commutative algebra with unit e, $A \in L(X)$, dim ker $A \neq 0$ and $x_1, \ldots, x_N \in$ dom A^{N-1}. If the A-Wrońskian $W_A(x_1, \ldots, x_N)$ is invertible then elements $x_1, \ldots, x_N$ are linearly independent.*

Proof. Suppose that $x_1, \ldots, x_N \in$ dom A^{N-1}. Consider the equality

$$c_1 x_1 + \ldots + c_N x_N = 0, \quad \text{where } c_1, \ldots, c_N \text{ are scalars.}$$

Acting on both sides of this equality by operators $A^1, A^2, \ldots, A^{N-1}$ we obtain a system of N homogeneous equations with N unknowns $c_1, \ldots, c_N$:

$$c_1 A^k x_1 + c_2 A^k x_2 + \ldots + c_N A^k x_N = 0 \quad (k = 0, 1, \ldots, N-1).$$

$$(6.3.9)$$

Since the A-Wrońskian $W_A(x_1, \ldots, x_N)$ is an invertible determinant of the system (6.3.9), we conclude that this system has only a trivial solution $c_1 = c_2 = \ldots = c_N = 0$ (cf. Cramer Formulae (1.1.21)). This implies the linear independence of elements $x_1, \ldots, x_N$. ∎

DEFINITION 6.3.2. Suppose that X is a commutative algebra, $A \in L(X)$, dim ker $A \neq 0$. Elements $x_1, \ldots, x_N \in X$ are said to be *A-linearly dependent* if there exist $z_1, \ldots, z_N \in$ ker A non-vanishing simultaneously and such that a *A-linear combination of* $x_1, \ldots, x_N$, i.e. the element

$$z_1 x_1 + \ldots + z_N x_N, \quad z_1, \ldots, z_N \in \ker A \qquad (6.3.10)$$

is equal zero. Elements $x_1, \ldots, x_N$ are said to be *A-linearly independent* if the condition $z_1 x_1 + \ldots + z_N x_N = 0$ implies $z_1 = \ldots = z_N = 0$. In other words, elements $x_1, \ldots, x_N$ are A-linearly independent if there is no $z_1, \ldots, z_N \in$ ker A non-vanishing simultaneously such that $z_1 x_1 + \ldots + z_N x_N = 0$.

PROPOSITION 6.3.2. *Suppose that X is a D-algebra with the unit $e \in$ ker D. Then the linear dependence of elements $x_1, \ldots, x_N \in X$ implies their D-linear dependence, i.e. the D-linear independence of elements $x_1, \ldots, x_N$ implies their linear independence.*

Proof. By our assumption there exist scalars $c_1, \ldots, c_N$ non-vanishing simultaneously such that

$$c_1 x_1 + \ldots + c_N x_N = 0. \qquad (6.3.11)$$

Write $z_j = c_j e$ for $j = 1, \ldots, N$. Since $De = 0$, we find $Dz_j = D(c_j e) = c_j De = 0$, which implies $z_j \in \ker D$ $(j = 1, \ldots, N)$. Equality (6.3.11) implies that

$$\sum_{j=1}^{N} z_j x_j = \sum_{j=1}^{N} c_j e x_j = \sum_{j=1}^{N} c_j x_j = 0$$

which proves the D-linear dependence of elements $x_1, \ldots, x_N$. If elements $x_1, \ldots, x_N$ are D-linearly independent then there is no $z_1, \ldots, z_n \in \ker D$ non-vanishing simultaneously such that $z_1 x_1 + \ldots + z_N x_N = 0$. In particular, scalars $c_1, \ldots, c_N$ such that $z_j = c_j e$ non-vanishing simultaneously, also do not exist. This implies the linear independence of elements $x_1, \ldots, x_N$. ∎

The converse statement is not true, as it is shown by the following

EXAMPLE 6.3.1. Suppose that $X = C[0, T], D = \mathrm{d}^2/\mathrm{d}t^2$. Then the functions $x_1(t) = t$ and $x_2(t) = 1-t$ are linearly independent. However, since $z_2(t) = -t$ and $z_1(t) = x_1(t) = 1-t$ belongs to $\ker D$ for $(-t)'' = (-1)' = 0$ and $(1-t)'' = (-1)' = 0$, we conclude that

$$(z_1 x_1 + z_2 x_2)(t) = (1-t)t - t(1-t) = 0.$$

Hence x_1 and x_2 are $\mathrm{d}^2/\mathrm{d}t^2$-linearly dependent.

In general, in any D-algebra all constants are D-linearly dependent (even if they are linearly independent). Indeed, let $z_1, z_2 \in \ker D$ be linearly independent. However, $z_1 z_2 - z_2 z_1 = 0$. Hence z_1 and z_2 are D-linearly dependent.

PROPOSITION 6.3.3. *Suppose that X is an almost Leibniz D-algebra. If the $c_D D$-Wroński $W_{c_D D}(x_1, \ldots, x_N)$ is invertible, then elements $x_1, \ldots, x_N \in \operatorname{dom} D^{N-1}$ are D-linearly independent.*

Proof. Suppose that the $c_D D$-Wroński $W_{c_D D}(x_1, \ldots, x_N)$ is invertible and the elements $x_1, \ldots, x_N$ are D-linearly dependent, i.e. there exist $z_1, \ldots, z_N \in \ker D$ non-vanishing simultaneously such that

$$z_1 x_1 + \ldots + z_N x_N = 0. \tag{6.3.12}$$

Since $Dz_j = 0$ for $j = 1, \ldots, N$, acting on both sides of Equality (6.3.12) and applying Formula (6.1.21) we find

$$0 = D^k \sum_{j=1}^{N} z_j x_j = \sum_{j=1}^{N} D^k(z_j x_j) = \sum_{j=1}^{N} c_D^k z_j D^k x_j + f^{(k)}(z_j, x_j)$$

$$= \sum_{j=1}^{N} z_j c_D^k D^k x_j \qquad (k = 1, \ldots, N-1)$$

since $f^{(k)}(z_j, x_j) = 0$ for $z_j \in \ker D$. Thus we have obtained a system of N homogeneous equations with N unknowns $z_1, \ldots, z_N$:

$$\sum_{j=1}^{N} z_j c_D^k D^k x_j = 0 \qquad (k = 0, 1, \ldots, N-1). \tag{6.3.13}$$

The determinant of this system, i.e. the $c_D D$-Wrońskian $W_{c_D D}(x_1, \ldots, x_N)$, is invertible by our assumption. Thus the Cramer Formulae (1.1.21) imply that the system (6.3.13) has only a trivial solution $z_1 = \ldots = z_n = 0$. But this contradicts our assumption that not all $z_1, \ldots, z_N = 0$ vanish simultaneously. Hence elements $x_1, \ldots, x_N$ are D-linearly independent. ∎

DEFINITION 6.3.3. Suppose that X is a D-algebra. A system $x_1, \ldots, x_N \in \operatorname{dom} D^{N-1}$ is said to be *fundamental* if and only if its $c_D D$-Wrońskian $W_{c_D D}(x_1, \ldots, x_N)$ is invertible.

Proposition 6.3.3 implies that elements of a fundamental system are D-linearly independent.

THEOREM 6.3.1*. *Suppose that X is an almost Leibniz D-algebra with $c_D = 1$ (in particular, a Leibniz D-algebra, a quasi-Leibniz D-algebra), $Q(D)$ is defined by Formulae (6.3.1) and $x_1, \ldots, x_N \in \ker D$ is a fundamental system. Then every D-linear combination of elements $x_1, \ldots, x_N$ belong also to $\ker Q(D)$, i.e. satisfies Equation (6.3.2).*

---

* Theorems 6.3.1–6.3.4 have been proved in a slightly different way by Dudek, 1978, 1980, 1980a.

Proof. By our assumption $Q(D)x_j = 0$ for $j = 1, ..., N$. Suppose that $x = \sum_{j=1}^{N} z_j x_j$, where $z_1, ..., z_N \in \ker D$ are arbitrarily fixed. Then $Dz_j = 0$, $f_D^{(k)}(z_j, x_j) = 0$ $(j = 1, ..., N, k = 0, 1, ..., N-1)$ and Formula (6.1.21) implies that

$$Q(D)x = \sum_{k=0}^{N} a_k D^k x = \sum_{k=0}^{N} a_k \left[\sum_{j=1}^{N} D^k(z_j x_j) \right]$$

$$= \sum_{j=1}^{N} a_0 z_j x_j + \sum_{k=1}^{N} a_k [z_j D^k x_j + f_D^{(k)}(z_j, x_j)]$$

$$= \sum_{j=1}^{N} \left(a_0 z_j x_j + \sum_{k=1}^{N} a_k z_j D^k x_j \right) = \sum_{j=1}^{N} z_j \left(\sum_{k=0}^{N} a_k D^k x_j \right)$$

$$= \sum_{j=1}^{N} z_j Q(D) x_j = 0. \qquad \blacksquare$$

If we know a fundamental system of N solutions of the homogeneous equation (6.3.2) of order N then we can obtain general solution of the non-homogeneous equation (6.3.1) by the *constants variation method* applied in another way. Namely, we have

THEOREM 6.3.2. *Suppose that all assumptions of Theorem 6.3.1 are satisfied. If x is of the form*

$$x = \sum_{j=1}^{N} x_j \{(-1)^{N+1} R[W_D^{-1} W_j(y)] + z_j\}, \qquad (6.3.14)$$

where $R \in \mathcal{R}_D$, $z_1, ..., z_N \in \ker D$ are arbitrary, $W_D = W_D(x_1, ..., x_N)$ and $W_j(y)$ is a determinant obtained from W_D if we put instead of the j-th column the column $(0, ..., 0, y)$, then x is a solution of Equation (6.3.1).

If, in addition, the operator

$$Q(I, R) = \sum_{k=1}^{n} a_k R^{N-k} \qquad (6.3.15)$$

is invertible then every solution of Equation (6.3.1) is of the form (6.3.14).

Proof. By our assumption $Q(D)x_j = 0$ for $j = 1, 2, ..., N$ and the D-Wrońskian $W_D = W_D(x_1, ..., x_N)$ is invertible. By Theorem 6.3.1 if x is of the form $x = z_1 x_1 + ... + z_N x_N$, where $z_1, ..., z_N \in \ker D$ are arbitrary, then x is a solution of Equation (6.3.2).

We therefore suppose that solutions of Equation (6.3.1) we are looking for, are of the form

$$x = u_1 x_1 + ... + u_N x_N, \tag{6.3.16}$$

where $u_1, ..., u_N \in \operatorname{dom} D$ are to be determined. Since $u_1, ..., u_N$ are arbitrary, we can admit some additional conditions. Namely, assume that

$$\sum_{j=1}^{N} x_j Du_j = 0,$$

$$\sum_{j=1}^{N} (D^k x_j)(Du_j) = 0 \quad (k = 1, ..., N-2), \tag{6.3.17}$$

$$f_D(D^{k-1}x_j, u_j) = 0 \quad (k = 1, ..., N-1). \tag{6.3.18}$$

Then

$$Dx = \sum_{j=1}^{N} D(x_j u_j) = \sum_{j=1}^{N} [x_j Du_j + u_j Dx_j + f_D(x_j, u_j)]$$

$$= \sum_{j=1}^{N} u_j Dx_j,$$

$$D^2 x = D(Dx) = \sum_{j=1}^{N} D(u_j Dx_j)$$

$$= \sum_{j=1}^{N} [(Du_j)(Dx_j) + u_j D^2 x_j + f_D(Dx_j, u_j)]$$

$$= \sum_{j=1}^{N} u_j D^2 x_j,$$

$$\cdots\cdots\cdots\cdots\cdots\cdots\cdots\cdots\cdots\cdots\cdots\cdots$$

$$D^{N-1}x = D(D^{N-2}x) = \sum_{j=1}^{N} D(u_j D^{N-2}x_j)$$

$$= \sum_{j=1}^{N} [(Du_j)(D^{N-2}x_j) + u_j D^{N-1}x_j + f_D(D^{N-2}x_j, u_j)]$$

$$= \sum_{j=1}^{N} u_j D^{N-1}x_j,$$

$$D^N x = D(D^{N-1}x) = \sum_{j=1}^{N} D(u_j D^{N-1}x_j)$$

$$= \sum_{j=1}^{N} [(Du_j)(D^{N-1}x_j) + u_j D^N x_j + f_D(D^{N-1}x_j, u_j)].$$

Since $a_N = I$ and $Q(D)x_j = 0$ for $j = 1, \ldots, N$ we obtain

$$Q(D)x = \sum_{k=0}^{N} a_k D^k x$$

$$= \sum_{k=0}^{N-1} a_k \sum_{j=1}^{N} u_j D^k x_j +$$

$$+ \sum_{j=1}^{N} [(Du_j)(D^{N-1}x_j) + u_j D^N x_j + f_D(D^{N-1}x_j, u_j)]$$

$$= \sum_{j=1}^{N} u_j \sum_{k=0}^{N} a_k D^k x_j + \sum_{j=1}^{N} (Du_j)(D^{N-1}x_j)$$

$$= \sum_{j=1}^{N} u_j Q(D)x_j + \sum_{j=1}^{N} (Du_j)(D^{N-1}x_j)$$

$$= \sum_{j=1}^{N} (Du_j)(D^{N-1}x_j).$$

Hence

$$\sum_{j=1}^{N} (Du_j)(D^{N-1}x_j) = y \qquad (6.3.19)$$

and together with Equations (6.3.17) we obtain a non-homogeneous system of N equations with N unknowns Du_j:

$$\sum_{j=1}^{N} (D^k x_j)(Du_j) = \delta_{N-1, k} y \quad (k = 0, 1, \ldots, N-1) \qquad (6.3.20)$$

(where $\delta_{k,j}$ is the *Kronecker symbol*, i.e. $\delta_{k,k} = 1$ and $\delta_{k,j} = 0$ for $k \neq j$). The determinant of this system is the *D*-Wrońskian $W = W_D(x_1, \ldots, x_N)$ which is, by our assumption, invertible. Applying the Cramer formulae (1.1.21) we find

$$Du_j = (-1)^{N+j} W_D^{-1} W_j(y), \qquad (6.3.21)$$

where the determinant $W_j(y)$ is obtained from the *D*-Wrońskian $W_D = W_D(x_1, \ldots, x_N)$ if we put instead of the *j*-th column the column $(0, \ldots, 0, y)$ of right side terms. Let R be a right inverse of D. Then

$$u_j = (-1)^{N+j} R[W_D^{-1} W_j(y)] + z_j \qquad (j = 1, \ldots, N),$$

where $z_1, \ldots, z_N \in \ker D$ are arbitrary. This, and Formula (6.1.16), together imply Formula (6.3.14).

Observe that by Formula (6.3.14) the homogeneous equation (6.3.2) has at least Ndim ker D linearly independent solutions. On the other hand, if the operator $Q(I, R)$ defined by Formula (6.3.15) is invertible then Theorem 3.2.5, Point 3(iii) implies that dim ker $Q(D) = N$dim ker D. This means that every solution of Equation (6.3.2) is of the form $\sum_{j=1}^{N} z_j x_j$, where $z_1, \ldots, z_N \in \ker D$ are arbitrary, provided that the operator $Q(I, R)$ is invertible. We therefore conclude that every solution of Equation (6.3.1) is of the form (6.3.14), provided that the operator $Q(I, R)$ is invertible. ∎

REMARK 6.3.1. If X is a Leibniz *D*-algebra then $f_D = 0$ and Conditions (6.3.18) are automatically satisfied. If X is a quasi-Leibniz *D*-algebra then $f_D(x, y) = d(Dx)(Dy)$ for all $x, y \in \operatorname{dom} D$ $(d \neq 0)$. Then $0 = f_D(D^{k-1}x_j, u_j) = d(D^k x_j)(Du_j)$, which implies $(D^k x_j)(Du_j) = 0$ for $j = 1, \ldots, N, k = 1, \ldots, N-1$. Hence also $\sum_{j=1}^{N} (D^k x_j)(Du_j) = 0$ for $k = 1, \ldots, N-1$, i.e. Conditions (6.3.18) imply Conditions (6.3.17).

By an easy induction we obtain

LEMMA 6.3.1. *If X is an almost Leibniz D-algebra with $c_D = 1$ then*

$$D \prod_{j=1}^{n} a_j = \sum_{k=1}^{n} (Da_k) \prod_{j=1, j\neq k}^{n} a_j + f_D\left(\prod_{j=1}^{n-1} a_j, a_n\right) + a_n A_n \qquad (6.3.22)$$

for $a_1, \ldots, a_n \in \operatorname{dom} D$, where $A_n = a_n A_{n-1}$ for $n \geqslant 3$, $A_2 = f_D(a_1, a_2)$.

In particular, if X is a quasi-Leibniz D-algebra, i.e. $f_D(x, y) = d(Dx)(Dy)$ for $x, y \in \text{dom } D$ then Formula (6.3.22) has the form

$$D \prod_{j=1}^{n} a_j = \sum_{k=1}^{n-1} (Da_k) \prod_{j=1, j \neq k}^{n} a_j + d(Da_n)\left(D \prod_{j=1}^{n-1} a_j\right) + A_n,$$

$$(6.3.23)$$

for $a_1, \ldots, a_n \in \text{dom } D$ and $n \geqslant 2$.

If X is a Leibniz D-algebra then $f_D = 0$ and Formula (6.3.22) has the form

$$D \prod_{j=1}^{n} a_j = \sum_{k=1}^{n} (Da_k) \prod_{j=1, j \neq k}^{n} a_j \qquad (6.3.24)$$

for $a_1, \ldots, a_n \in \text{dom } D$.

THEOREM 6.3.3. *Suppose that X is an almost Leibniz D-algebra with $c_D = 1$ and with a unit e. If $\{x_1, \ldots, x_N\} \subset \text{dom } D^N$ is a fundamental system then there exists an operator $Q(D)$ of the order N such that every D-linear combination of elements $x_1, \ldots, x_N$ is a solution of the equation $Q(D)x = 0$, namely*

$$Q(D) = \sum_{k=0}^{N} a_k D^k, \qquad (6.3.25)$$

where

$$a_k = (-1)^{N+k} W_D^{-1} W_k \in X \quad (k = 0, 1, \ldots, N),$$
$$\text{in particular, } a_N = e, \quad (6.3.26)$$

W_k are minor determinants obtained by cancelling in the D-Wrońskian $W_D = W_D(x_1, \ldots, x_N, x)$ the $(k+1)$-th row and the last column.

Proof. Clearly, elements $a_0, \ldots, a_{N-1}$ defined by Formulae (6.3.26) belong to D-algebra X. The determinant $W_N = W_D(x_1, \ldots, x_N)$ is invertible by our assumption that the system $\{x_1, \ldots, x_N\}$ is fundamental. Therefore, if $Q(D) = \sum_{k=0}^{N} a_k D^k$ then $a_N = (-1)^{2N} W_D^{-1} W_N = W_D^{-1} W_D = e$ and $Q(D)x = 0$, which was to be proved. ∎

COROLLARY 6.3.1. *Suppose that X is a Leibniz D-algebra with unit e. Then*

$$DW_D = (-1)^N a_{N-1} W_D, \quad \text{where } a_{N-1} = W_D^{-1} W_{N-1}, \quad (6.3.27)$$

$W_D = W_D(x_1, \ldots, x_N)$, $\{x_1, \ldots, x_N\} \subset \operatorname{dom} D^N$ *is a fundamental system,* W_{N-1} *is a minor determinant obtained by cancelling in the determinant* $W_D(x_1, \ldots, x_N, x)$ *the* $(N-1)$*-th row and the last column.*

Proof. By definition we have $a_{N-1} W_D = (-1)^{N-1+N} W_{N-1} = -W_{N-1}$. If we develop the determinant W_{N-1} with respect to the last row and if we apply Formula (6.3.24) then we find

$$a_{N-1} W_D = -W_{N-1} = -\sum_{k=1}^{N} (-1)^{N+k-1} (D^N x_k) \prod_{\substack{j=1, \, j \neq k \\ m=0,1,\ldots,N-2}}^{N} D^m x_j$$

$$= (-1)^N \sum_{k=1}^{N} D \left[(-1)^k \prod_{\substack{j=1, \, j \neq k \\ m=0,1,\ldots,N-1}}^{N} D^m x_j \right]$$

$$= (-1)^N D \sum_{k=1}^{N} (-1)^k \prod_{\substack{j=1, \, j \neq k \\ m=0,1,\ldots,N-1}}^{N} D^m x_j = (-1)^N D W_D$$

which implies Formula (6.3.27).   ∎

THEOREM 6.3.4. *Suppose that* X *is a Leibniz D-algebra with unit e and F is an initial operator for D corresponding to R. Suppose, moreover, that the operator* $I - (-1)^N a_{N-1} R$ *is invertible, where* a_{N-1} *is defined in Corollary 6.3.1. Then*

$$W_D(x_1, \ldots, x_N) = 0 \quad \text{if and only if } FW_D(x_1, \ldots, x_N) = 0,$$
$$(6.3.28)$$

i.e. the D-Wrońskian is determined by its initial value.

Proof. Formula (6.3.27) implies that $DW_D = (-1)^N a_{N-1} W_D = 0$, i.e. $0 = [D - (-1)^N a_{N-1}] W_D = D[I - (-1)^N a_{N-1} R] W_D$, which implies $[I - (-1)^N a_{N-1} R] W_D = z$, where $z \in \ker D$. But $FR = 0$, hence $FW_D = Fz = z$. Since the operator $I - (-1)^N a_{N-1} R$ is invertible, we conclude that

$$W_D = [I - (-1)^N a_{N-1} R]^{-1} FW_D. \quad (6.3.29)$$

This implies that $W_D = 0$ if and only if $FW_D = 0$.   ∎

Equality (6.3.29) is the *Liouville Formula* for *D*-Wrońskian.

Examples and Exercises

EXAMPLE 6.3.2. We shall solve an ordinary differential equation of order 1:

$$x' + px = q, \quad \text{where } p, q \in C[a, b] \tag{6.3.30}$$

by the constant variation method. In the first step we shall solve the homogeneous equation

$$x' + px = 0. \tag{6.3.31}$$

Clearly, the function $x(t) \equiv 0$ satisfies this equation, but it is a trivial solution. We are looking for non-trivial solutions of Equation (6.3.31). We can rewrite Equation (6.3.31) as follows

$$\frac{d}{dt} \ln x = \frac{x'}{x} = -p(t),$$

which implies that $\ln x = -P(t) + C_1$, where $P(t)$ is a primitive function for $p(t)$, i.e. $P'(t) = p(t)$, and C_1 is an arbitrary constant. It is convenient to write $C = e^{C_1}$. Then $C_1 = \ln C$ and $\ln x = -P(t) + \ln C = \ln C e^{-P(t)}$, which implies $x(t) = C e^{-P(t)}$. Observe that we obtain a trivial solution $x = 0$ if we put $C = 0$.

We assume now, according with our previous considerations, that C is not a constant, but a function $C = C(t)$. This function has to be chosen in such a way that the function $x(t) = C(t) e^{-P(t)}$ becomes a solution of the non-homogeneous equation (6.3.30). Since $P' = p$, we obtain

$$q = x' + px = C'e^{-P} + Ce^{-P}(-P') + pCe^{-P}$$
$$= C'e^{-P} - Ce^{-P}p + Ce^{-P}p = C'e^{-P}.$$

Thus $C' = qe^P$ and $C(t) = Q(t) + c$, where Q is a primitive function for qe^P and c is an arbitrary constant. Finally

$$x(t) = C(t)e^{-P(t)} = e^{-P(t)}[Q(t) + c], \tag{6.3.32}$$

where c is an arbitrary constant and P, Q are primitive functions for p and qe^P, respectively.

In the theory of ordinary differential equations traditionally arbitrary primitive functions are denoted by the same symbol as indefinite integrals. This does not lead to any misunderstanding if we remember the meaning of the symbol used and if we add an arbitrary constant.

Using this denotation we can rewrite general solution (6.3.32) of Equation (6.3.30) as follows:

$$x(t) = e^{-\int p(t)dt}\left[\int q(t)e^{\int p(t)dt}\,dt + c\right]$$ (6.3.33)

or

$$x(t) = \exp\left(-\int p(t)dt\right)\left[\int q(t)\exp\left(\int p(t)dt\right)dt + c\right]$$ (6.3.34)

if we write $\exp y(t)$ instead of $e^{y(t)}$.

For instance, the general solution of the equation

$$x' + \cos tx = \cos t$$

is $x(t) = (e^{\sin t} + c)e^{-\sin t} = 1 + ce^{-\sin t}$, where c is an arbitrary constant.

EXAMPLE 6.3.3. Suppose that $D \in R(X)$ and X is a quasi-Leibniz D-algebra, i.e. $D(xy) = xDy + yDx + d(Dx)(Dy)$ for all $x, y \in \text{dom } D$, where $d \neq 0$ is a given scalar.

Assume that $N = 2$. Let $x_1, x_2 \in \text{dom } D$ and let the D-Wrońskian $W_D(x_1, x_2)$ be invertible.

Suppose that x_1, x_2 are solutions of the equation

$$(D^2 + a_1 D + a_0)u = 0, \quad \text{where } a_0, a_1 \in X.$$ (6.3.35)

Then every D-linear combination of elements $x_1, x_2 \in \text{dom } D$ is a solution of Equation (6.3.35). Indeed, let $x = z_1 x_1 + z_2 x_2$, where $z_1, z_2 \in \ker D$ are arbitrary. Then, by our assumptions, since $D(z_1 x_1 + z_2 x_2) = z_1 Dx_1 + z_2 Dx_2$, we find

$$
\begin{aligned}
(D^2 + a_1 D + a_0)x &= (D^2 + a_1 D + a_0)(z_1 x_1 + z_2 x_2) \\
&= D(z_1 Dx_1 + z_2 Dx_2) + a_1(z_1 Dx_1 + z_2 Dx_2) + a_0(z_1 x_1 + z_2 x_2) \\
&= z_1 D^2 x_1 + (Dz_1)(Dx_1) + d(Dz_1)(D^2 x_1) + z_2 D^2 x_2 + \\
&\quad + (Dz_2)(Dx_2) + d(Dz_2)(D^2 x_2) + z_1 a_1 Dx_1 + z_2 a_1 Dx_2 + \\
&\quad + z_1 a_0 x_1 + z_2 a_0 x_2 \\
&= z_1(D^2 x_1 + a_1 Dx_1 + a_0 x_1) + z_2(D^2 x_2 + a_1 Dx_2 + a_0 x_2) = 0.
\end{aligned}
$$

Thus Theorem 6.3.1 holds. Suppose now that $\{x_1, x_2\}$ is a fundamental system, i.e. the D-Wrońskian $W_D(x_1, x_2)$ is invertible. We are looking for solutions of the equation

$$(D^2 + a_1 D + a_0)x = y, \quad y \in X.$$ (6.3.36)

We assume, as in Theorem 6.3.2, that these solutions are of the form $x = u_1 x_1 + u_2 x_2$, where $u_1, u_2 \in \mathrm{dom}\, D$ and we put an additional condition

$$(x_1 + dDx_1)\, Du_1 + (x_2 + dDx_2)\, Du_2 = 0. \tag{6.3.37}$$

In a similar way, as in the proof of Theorem 6.3.2, we obtain that under Condition (6.3.37)

$$(Dx_1 + dD^2 x_1)\, Du_1 + (Dx_2 + dD^2 x_2)\, Du_2 = y. \tag{6.3.38}$$

The determinant of the system (6.3.37)–(6.3.38) is

$$\begin{aligned}
\varDelta &= (x_1 + dDx_1)(Dx_2 + dD^2 x_2) - (x_2 + dDx_2)(Dx_1 + dD^2 x_1) \\
&= W_D(x_1, x_2) + d(x_1 D^2 x_2 - x_2 D^2 x_1) \\
&= W_D(x_1, x_2) + dx_1(-a_1 Dx_2 - a_0 x_2) - x_2(-a_1 Dx_1 - a_0 x_1) \\
&= W_D(x_1, x_2) - da_1 W(x_1, x_2) = (e - da_1)W_D(x_1, x_2).
\end{aligned}$$

Since $W_D(x_1, x_2)$ is invertible, we conclude (in the case when, for instance, X has a unit e and $e - da_1$ is invertible), that the system (6.3.37)–(6.3.38) has a unique solution which is of the form

$$Du_1 = -\varDelta^{-1}(x_2 + dDx_2)y, \qquad Du_2 = \varDelta^{-1}(x_1 + dDx_1)y.$$

Thus for an $R \in \mathscr{R}_D$ and for arbitrary $z_1 z_2 \in \ker D$ solutions, we are looking for, are of the form

$$\begin{aligned}
x = &-x_1 R[\varDelta^{-1}(x_2 + dDx_2)y] + x_2 R[\varDelta^{-1}(x_1 + dDx_1)] - x_1 z_1 + \\
&+ x_2 z_2,
\end{aligned}$$

where $\varDelta = (e - da_1)\, W_D(x_1, x_2)$.

Observe that

$$DW_D(x_1, x_2) = (da_0 - a_1)W_D(x_1, x_2). \tag{6.3.39}$$

In a similar way, as in the proof of Theorem 6.3.4, we conclude that

$$W_D(x_1, x_2) = [I - (da_0 - a_1)R]^{-1} F W_D(x_1, x_2) \tag{6.3.40}$$

provided that the operator $I - (da_0 - a_1)R$ is invertible and F is an initial operator for D corresponding to R.

EXERCISE 6.3.1. Suppose that X is a Leibniz D-algebra and that F is a multiplicative initial operator for D corresponding to an $R \in \mathscr{R}_D$. Prove that:

(i) $FW_D(x_1, \ldots, x_N) = \det (FD^{k-1}x_j)_{j,\, k=1,\ldots,N}$;

(ii) The initial value FW_D of the D-Wrońskian $W_D(x_1, ..., x_N)$ is invertible if and only if $W_D(x_1, ..., x_N)$ is invertible;

(iii) If all assumptions of Theorem 6.3.4 are satisfied then the initial value problem for the operator $Q(D) = \sum_{k=0}^{N} a_k D^k$, where $a_0, ..., a_{N-1}$ $\in X$, $a_N = I$, is well-posed. Determine a unique solution of this problem.

EXERCISE 6.3.2. Solve an ordinary linear differential equation of order 2: $x'' + p(t)x' + q(t)x = r(t)$, where $p, q, r \in C[a, b]$, under assumption that there are known two linearly independent solutions x_1, x_2 of the homogeneous equation $x'' + p(t)x' + q(t)x = 0$ by application of Theorem 6.3.2.

EXERCISE 6.3.3. Solve an ordinary linear differential equation of order 2: $x'' + a^2 x = b(t)$, where $a \in \mathbf{R}$, $b \in C[a, b]$, if there is known a solution $x_1(t) = \sin at$ of the homogeneous equation $x'' + a^2 x = 0$.

EXERCISE 6.3.4. Suppose that X, D, R are defined as in Example 2.1.2. Using the fact that the function $u_1(t, s) = (t+1)(s+1)$ is a solution of the homogeneous equation $\partial u_1 / \partial t - u_1 / (t+1) = 0$ solve a partial linear differential equation of order 1:

$$\frac{\partial x}{\partial t} - \frac{1}{t+1} x = \frac{1}{(t+1)(s+1)}.$$

EXERCISE 6.3.5. Suppose that X, D, R are defined as in Example 2.1.2. Solve the partial linear differential equation of order 2

$$\frac{\partial^2 x}{\partial t^2} + x = t(\sin s + \cos s)$$

by an application of Theorem 6.3.2, checking that the functions $x_1(t, s)$ $= (s+1)\cos t$, $x_2(t, s) = (s+1)\sin t$ are two linearly independent solutions of the homogeneous equation $\partial^2 x / \partial t^2 + x = 0$.

EXERCISE 6.3.6. Suppose that X, D, R are defined as in Example 2.1.4. If we define the multiplication in X as follows: $\{x_n\}\{y_n\} = \{x_n y_n\}$ for arbitrary $x = \{x_n\}$, $y = \{y_n\} \in X$ then X is a quasi-Leibniz D-algebra such that $d = 1$. Apply results of Example 6.3.3 for solving linear equations with variable coefficients of order 2, i.e. equations, which can be written in the form:

$$x_{n+2} + a_{1,n}x_{n+1} + a_{0,n}x_n = y_n \quad (n = 1, 2, \ldots),$$

where $a_0 = \{a_{0,n}\}$, $a_1 = \{a_{1,n}\}$, $y = \{y_n\}$ are given elements of X.

EXERCISE 6.3.7. Suppose that X is a Duhamel D-algebra. Show that results analogous to those obtained in Theorems 6.3.2–6.3.4 are true in X, provided that $N = 2$.

EXERCISE 6.3.8. Suppose that X, D, R are defined as in Example 2.1.4. If we define the multiplication in X as a convolution: $\{x_n\} * \{y_n\}$ $= \left\{ \sum\limits_{k=1}^{n} x_k y_{n+1-k} \right\}$ for arbitrary $x = \{x_n\}$, $y = \{y_n\} \in X$, then X becomes a Duhamel D-algebra. Apply results of Exercise 6.3.7 for solving linear equations of order 2 with coefficients belonging to X.

EXERCISE 6.3.9. Suppose that $X = C[a, b]$, $D = \mathrm{d}/\mathrm{d}t$. Solve the system of n ordinary linear differential equations of order 1

$$\frac{\mathrm{d}x_j(t)}{\mathrm{d}t} = \sum_{k=1}^{n} a_{jk}(t)x_k(t) \quad (j = 1, 2, \ldots, n),$$

where $a_{jk} \in C[a, b]$ $(j, k = 1, 2, \ldots, n)$ by a reduction to one equation of order n with variable coefficients. In particular, solve this system for $n = 2$.

EXERCISE 6.3.10. Denote by Ω the Cartesian product of n copies of the interval $[a, b]$ $(n \geqslant 2)$. Let $X = C(\Omega)$. Prove that $u \in C^1(\Omega) \subset X$ is a solution of a linear partial differential equation of order 1 with variable coefficients

$$\sum_{j=1}^{n} P_j(t) \frac{\partial x}{\partial t_j} = 0,$$

where $P_1, \ldots, P_n \in C^1(\Omega)$, $t = (t_1, \ldots, t_n) \in \Omega$ if and only if $x = \Phi(u_1, \ldots, u_n)$, where Φ is an arbitrary differentiable function of arguments $u_1, \ldots, u_n$ and $u_j(t_1, \ldots, t_n) = c_j$, c_j are arbitrary constants

$(j = 1, ..., n)$, is general solution of the following system of ordinary differential equations of order 1:

$$\frac{dt_1}{P_1} = \frac{dt_2}{P_2} = \cdots = \frac{dt_n}{P_n}.$$

EXERCISE 6.3.11. Suppose that Ω is defined as in Exercise 6.3.10 and that the functions $P_1, ..., P_n, Q$ are continuous and continuously differentiable for $t \in \Omega$ and $u \in [c, d]$. Prove that $u \in C^1(\Omega)$ is a solution of a partial differential equation of order 1:

$$\sum_{j=1}^{n} P_j(t, u) \frac{\partial u}{\partial t_j} = Q(t, u), \quad t = (t_1, ..., t_n) \in \Omega,$$

if and only if $u = \Psi(v_0, ..., v_n)$, where Ψ is an arbitrary differentiable function of its argument and $v_j(t_1, ..., t_n, u) = c_j, c_j$ are arbitrary constants $(j = 0, 1, ..., n)$, is general solution of the following system of ordinary differential equations of order 1

$$\frac{dt_1}{P_1} = \frac{dt_2}{P_2} = \cdots = \frac{dt_n}{P_n} = \frac{du}{Q}.$$

We point out that this result can be easily reduced to the results of Exercise 6.3.10.

EXAMPLE 6.3.4. We shall solve the system

$$\frac{dx}{dt} = y + t, \quad \frac{dy}{dt} = x + 1 \tag{6.3.41}$$

of ordinary differential equations with scalar coefficients considered in Example 3.5.1 by the constants variation method. We recall that the solution of the homogeneous system

$$\frac{dx}{dt} = y, \quad \frac{dy}{dt} = x \tag{6.3.42}$$

is of the form

$$x(t) = C_1 e^t + C_2 e^{-t}, \quad y(t) = C_1 e^t - C_2 e^{-t}. \tag{6.3.43}$$

Suppose that $C_1 = C_1(t), C_2 = C_2(t)$ are chosen in such a way that the system (6.3.41) is satisfied. Since $x' = (C_1' - C_1)e^t + (C_2' - C_2)e^{-t}$,

$y' = (C_1' + C_1)e^t - (C_2' - C_2)e^t$ we find

$$(C_1' + C_1)e^t + (C_2' - C_2)e^{-t} = C_1 e^t - C_2 e^{-t} + t,$$
$$(C_1' + C_1)e^t - (C_2' - C_2)e^{-t} = C_1 e^t + C_2 e^{-t} + 1.$$

After reduction we obtain the following system:

$$C_1' e^t + C_2' e^{-t} = t, \quad C_1' e^t - C_2' e^{-t} = 1. \tag{6.3.44}$$

The determinant of the system (6.3.44) is

$$\Delta = \begin{vmatrix} e^t & e^{-t} \\ e^t & -e^{-t} \end{vmatrix} = -e^t e^{-t} - e^t e^{-t} = -2 \neq 0.$$

Thus the system (6.3.44) has a unique solution

$$C_1' = -\frac{1}{2} \begin{vmatrix} t & e^{-t} \\ 1 & -e^{-t} \end{vmatrix} = \frac{1}{2}(t+1)e^{-t},$$

$$C_2' = -\frac{1}{2} \begin{vmatrix} e^t & t \\ e^t & 1 \end{vmatrix} = \frac{1}{2}(t-1)e^t.$$

By integration by parts we find

$$C_1(t) = \int \frac{1}{2}(t+1)e^{-t} dt = -\frac{1}{2}(t+2)e^{-t} + c_1,$$

$$C_2(t) = \int \frac{1}{2}(t-1)e^t dt = \frac{1}{2}(t-2)e^t + c_2,$$

where c_1, c_2 are arbitrary constants. Hence general solution of System (6.3.41) is

$$\begin{aligned}
x(t) &= C_1 e^t + C_2 e^{-t} \\
&= [-\tfrac{1}{2}(t+2)e^{-t} + c_1]e^t + [\tfrac{1}{2}(t-2)e^t + c_2]e^{-t} \\
&= c_1 e^t + c_2 e^{-t} - 2, \\
y(t) &= C_1 e^t - C_2 e^{-t} \\
&= [-\tfrac{1}{2}(t+2)e^{-t} + c_1]e^t + [-\tfrac{1}{2}(t-2)e^t + c_2]e^{-t} \\
&= c_1 e^t - c_2 e^{-t} - t,
\end{aligned}$$

where c_1, c_2 are arbitrary constants.

EXERCISE 6.3.12. Suppose that X is a Leibniz D-algebra and $R \in \mathscr{R}_D$. Solve the equation

$$[a_2 D^2 + (a_1 Rz_1 + b_1 I)D + (a_0 Rz_0 + b_2 I)]x = 0, \qquad (6.3.45)$$

where $z_0, z_1 \in \ker D$ and reals a_0, a_1, a_2, b_0, b_1 are given. Does the assumption that X has a unit e simplify the result obtained? Apply this result to an ordinary differential equation of order 2:

$$a_2 y'' + (a_1 t + b_1)y' + (a_0 t + b_0)y = 0$$

(cf. also Yosida, 1983).

6.4. FOURIER METHOD

We recall that an initial operator F for $D \in R(X)$ is said to be *almost averaging* if $F(zx) = zFx$ for all $z \in \ker D$, $x \in X$ (cf. Formula (6.1.46)).

THEOREM 6.4.1. *Suppose that*

(i) *X is an almost Leibniz D_1-algebra and D_2-algebra with $c_{D_1} = c_D$ = 1 simultaneously (over* **C**);

(ii) *F_0, F_1, F_2 are almost averaging initial operators corresponding to $R_0, R_1 \in \mathscr{R}_{D_1}$ and $R_2 \in \mathscr{R}_{D_2}$, respectively, and $R_0 \in V(X)$;*

(iii) *there is a scalar $\lambda \neq 0$ such that*

$$F_1 s_\lambda(z) = 0 \quad \text{and} \quad u = s_\lambda(z) \in \ker D_2 \quad \text{for all } z \in \ker D_1,$$
$$(6.4.1)$$

where

$$s_\lambda = \lambda R_0 (I + \lambda^2 R_0^2)^{-1}; \qquad (6.4.2)$$

(iv) *$0 \neq v \in \ker D_1$ is an eigenvector of the operator R_2 corresponding to the eigenvalue $-\lambda^{-2}$.*

Then there exists a non-trivial solution of the homogeneous equation

$$(D_2 - D_1^2)x = 0 \qquad (6.4.3)$$

with a homogeneous initial condition

$$F_2 x = 0 \qquad (6.4.4)$$

and with homogeneous boundary conditions

$$F_0 x = 0, \quad F_1 x = 0, \qquad (6.4.5)$$

which is of the form

$$x = uv, \quad where \ u = s_\lambda(z), \quad 0 \neq z \in \ker D_1, \tag{6.4.6}$$

i.e. this problem is ill-posed.

Proof. We shall show that every element of the form (6.4.6) is a non-trivial solution of the problem in question. Indeed, by our assumptions we have (even for $0 \neq z \in \ker D_1$):

$$D_1 v = 0, \quad D_2 u = D_2 s_\lambda(z) = 0,$$
$$R_2 v = -\lambda^{-2} v, \quad i.e. \ v = -\lambda^2 R_2 v,$$
$$(D_1^2 + \lambda^2 I)u = (D_1^2 + \lambda^2 I)s_\lambda(z) = 0$$

(cf. Example 3.2.4). Since $v \in \ker D_1$, $u \in \ker D_2$ and X is simultaneously an almost Leibniz D_1- and D_2-algebra, we find $f_{D_1}(u, v) = 0$, $f_{D_2}(u, v) = f_{D_2}^{(2)}(u, v) = 0$. Hence

$$(D_2 - D_1^2)x = (D_2 - D_1^2)(uv) = D_2(uv) - D_1^2(uv)$$
$$= uD_2 v + vD_2 u + f_{D_2}(u, v) -$$
$$- [uD_1^2 v + vD_1^2 u + f_{D_1}^{(2)}(u, v)]$$
$$= uD_2 v - vD_1^2 u = uD_2(-\lambda^2 R_2 v) - vD_1^2 u$$
$$= -\lambda^2 uD_2 R_2 v - vD_1^2 u = -v(\lambda^2 I + D_1^2)u = 0$$

which proves that x is a non-trivial solution of Equation (6.4.3). In order to prove that Conditions (6.4.4) and (6.4.5) are satisfied, recall that F_0, F_1, F_2 are almost averaging initial operators and $u \in \ker D_2$, $v \in \ker D_1$. Since $F_j R_j = 0$ for $j = 0, 1, 2$, we find

$$F_2 x = F_2(uv) = uF_2 v = uF_2(-\lambda^2 R_2 v) = -\lambda^2 uF_2 R_2 v = 0,$$
$$F_0 x = F_0(uv) = vF_0 u = vF_0[\lambda R_0(I + \lambda^2 R_0^2)^{-1}u]$$
$$= \lambda vF_0 R_0(I + \lambda^2 R_0^2)^{-1}u = 0,$$
$$F_1 x = F_1(uv) = vF_1 u = F_1 s_\lambda(z) = 0.$$

Hence x is a solution, we were looking for. ∎

The method described in Theorem 6.4.1 and in the following theorems is called in classical cases the *Fourier method*.

Observe that, in general, a sum and a difference (or any linear combination) of right invertible operators is not a right invertible operator. Hence the Fourier method permits to study equations with operators of these types. The fundamental idea of this method is a *separation of variables* which leads to independent problems for each of them.

THEOREM 6.4.2. *Suppose that the assumptions* (i), (ii) *and* (iv) *of Theorem* 6.4.1 *are satisfied and that* (*instead of* (iii))

(iii)′ *There exists a scalar* $\lambda \neq 0$ *such that*

$$F_1 c_\lambda(z) = 0 \quad \text{and} \quad c_\lambda(z) \in \ker D_2 \quad \text{for all } z \in \ker D_1, \quad (6.4.7)$$

where

$$c_\lambda = (I + \lambda^2 R_0^2)^{-1}. \quad (6.4.8)$$

Then there exists a non-trivial solution of Equation (6.4.3) *satisfying the initial condition* (6.4.4) *and mixed boundary conditions*:

$$F_0 x = 0, \quad F_1 D_1 x = 0,$$

which is of the form (6.4.6), *i.e. this problem is ill-posed.*

Proof. In the same way, as in the proof of Theorem (6.4.1), we prove that x is a solution of Equation (6.4.3) satisfying Condition (6.4.4) and the first of Conditions (6.4.5). Since $F_1 R_1 = 0$, $v \in \ker D_1$, $D_1 u = D_1 s_\lambda(z) = \lambda c_\lambda(z)$, we find $F_1 D_1 u = \lambda F_1 c_\lambda(z) = 0$ for all $z \in \ker D$. Hence $F_1 D_1 x = F_1 D_1(u, v) = F_1(v D_1 u) = v F_1 D_1 u = 0$ and x satisfies also the second of Conditions (6.4.5). This proves that $x \neq 0$ is a solution, we were looking for. ∎

COROLLARY 6.4.1. *Suppose that all assumptions of Theorem* 6.4.2 *are satisfied and* $F_0 = F_1$, *hence* $R_0 = R_1$. *Then there exists a non-trivial solution of Equation* (6.4.3) *with initial conditions*

$$F_1 x = 0, \quad F_1 D_1 x = 0, \quad F_2 x = 0$$

which is of the form (6.4.6).

THEOREM 6.4.3. *Suppose that the assumptions* (i) *and* (ii) *of Theorem* 6.4.1 *are satisfied and* $R_1, R_2 \in V(X)$. *If there exist a scalar* $\lambda \neq 0$ *and* $z_0, z_1 \in \ker D_1$, $z_2 \in \ker D_2$ *such that the elements*

$$u = c_\lambda(z_0) + s_\lambda(z_1) \in \ker D_2,$$
$$\text{where } c_\lambda = (I + \lambda^2 R_1^2)^{-1}, \quad s_\lambda = \lambda R_1 (I + \lambda^2 R_1^2)^{-1}, \quad (6.4.9)$$
$$v = \tilde{e}_{-\lambda^2}(z_2) \in \ker D_1, \quad \text{where } \tilde{e}_{-\lambda^2} = (I + \lambda^2 R_2)^{-1}$$

are invertible, then Equation (6.4.3) *has a solution* $x = uv$ (*which is invertible, then different than* 0).

Proof. Suppose that $x = uv$, where $u \in \ker D_2$, $v \in \ker D_1$ are invertible, is a solution of Equation (6.4.3). Then, in the same way, as in the proof of Theorem 6.4.1, we conclude that $uD_2v - vD_1^2 u = 0$, i.e. $uD_2v = vD_1^2 u$. Since u and v are invertible, we may write: $u^{-1}D_1^2 u = v^{-1}D_2v = -\lambda^2 I$. This leads to two independent equations:

$$(D_1^2 + \lambda^2)u = 0 \quad \text{and} \quad (D_2 + \lambda^2 I)v = 0,$$

which have solutions of the form (6.4.9).   ∎

THEOREM 6.4.4. *Suppose that*:
(i) X *is an almost Leibniz D_1-algebra with* $c_{D_1} = c_{D_2} = 1$;
(ii) F_0, F_1, F_2, F_3 *are almost averaging initial operators corresponding to* $R_0, R_1 \in \mathscr{R}_{D_1}$ *and* $R_2, R_3 \in \mathscr{R}_{D_2}$, *respectively*;
(iii) $R_0, R_2 \in V(X)$;
(iv) *There is a scalar* $\lambda \neq 0$ *such that*

$$F_1 s_\lambda(z) = 0, \quad F_3 s_\lambda^{\#}(\zeta) = 0,$$
$$u = s_\lambda(z) \in \ker D_2, \quad v = s_\lambda^{\#}(\zeta) \in \ker D_1$$

for all $z \in \ker D_1$, $\zeta \in \ker D_2$, *where*

$$s_\lambda = \lambda R_0(I + \lambda^2 R_0^2)^{-1}, \quad s_\lambda^{\#} = \lambda R_2(I + \lambda^2 R_2)^{-1}. \quad (6.4.10)$$

Then there exists a non-trivial solution of the homogeneous equation

$$(D_1^2 - D_2^2)x = 0 \quad (6.4.11)$$

with homogeneous boundary conditions

$$F_0 x = F_1 x = F_2 x = F_3 x = 0 \quad (6.4.12)$$

which is of the form:

$$x = uv, \quad \text{where } u = s_\lambda(z), v = s_\lambda^{\#}(\zeta),$$
$$0 \neq z \in \ker D_1, \quad 0 \neq \zeta \in \ker D_2, \quad (6.4.13$$

i.e. this problem is ill-posed.

Proof. By definition $D_1^2 u = -\lambda^2 u$, $D_2^2 u = -\lambda^2 v$. Hence, in a similar way, as in the proof of Theorem 6.4.1, we find

$$(D_1^2 - D_2^2)x = (D_1^2 - D_2^2)(uv) = vD_1^2 u - uD_2^2 v$$
$$= -\lambda^2 uv + \lambda^2 uv = 0.$$

Also $F_0 u = F_0 s_\lambda(z) = \lambda F_0 R_0(I + \lambda^2 R_0^2)^{-1}z = 0$ and $F_2 v = 0$ for similar reasons. By our assumptions in Point (iv) $F_1 u = 0$ and $F_3 v = 0$. Since $u \in \ker D_2$, $v \in \ker D_1$ and X is simultaneously an almost Leibniz D_1- and D_2-algebra, we conclude that $F_0 x = F_0(uv) = v F_0 u = 0$, $F_2 x = F_2(uv) = u F_2 v = 0$ and, for similar reasons, $F_1 x = F_3 x = 0$. ■

THEOREM 6.4.5. *Suppose that assumptions* (i)–(ii) *of Theorem* 6.4.4 *are satisfied and*

(iii) $R_0 \in V(X)$;

(iv) *There is a scalar* $\lambda \neq 0$ *such that*

$$F_1 s_\lambda(z) = 0 \quad \text{for } z \in \ker D,$$
$$\text{where } u = s_\lambda(z) \in \ker D_2, \quad s_\lambda = \lambda R_0(I + \lambda^2 R_0^2)^{-1}; \quad (6.4.14)$$

(v) $0 \neq v \in \ker D$ *is an eigenvector of the operator* $R_2 R_3$ *corresponding to the eigenvalue* $-\lambda^{-2}$.

Then there exists a non-trivial solution of Equation (6.4.11) *satisfying mixed boundary conditions*

$$F_0 x = 0, \quad F_1 x = 0, \quad F_2 x = 0, \quad F_3 D_2 x = 0 \qquad (6.4.15)$$

which is of the form

$$x = uv, \quad \text{where } u = s_\lambda(z), \quad 0 \neq z \in \ker D_1, \qquad (6.4.16)$$

i.e. this problem is ill-posed.

Proof. In a similar way, as in the proof of Theorems 6.4.1, 6.4.2, we prove that x is of the form (6.4.16) and conditions $F_0 x = 0$, $F_1 x = 0$ are satisfied. By our assumptions, $F_2 R_2 = 0$, $F_3 R_3 = 0$, $v = -\lambda^2 R_2 R_3$, $D_2 v = -\lambda^2 R_3 v$, $D_2 u = 0$. Since F_2 and F_3 are almost averaging, this implies that $F_2 x = F_2(uv) = u F_2 v = -\lambda^2 u F_2 R_2 R_3 = 0$, $F_3 D x = F_3 D_2(uv) = F_3[u D_2 v + f_{D_2}(u, v)] = F_3(u D_2 v) = u F_3 D_2 v = -\lambda^2 u F_3 R_3 v = 0$, which proves that x satisfies all the required conditions. ■

COROLLARY 6.4.2. *Suppose that all assumptions of Theorem* 6.4.4 *are satisfied and that* $F_3 = F_2$, *hence* $R_3 = R_2$. *Then there exists a non-trivial solution of Equation* (6.4.11) *with boundary conditions* $F_0 x = 0$, $F_1 x = 0$ *and with initial conditions* $F_2 x = 0$, $F_2 D_2 x = 0$.

Examples and Exercises

EXAMPLE 6.4.1. Consider the one-dimensional *heat equation*

$$\frac{\partial}{\partial t} x(t, s) - \frac{\partial^2}{\partial s^2} x(t, s) = 0 \quad \text{for } 0 \leqslant t \leqslant T, 0 \leqslant s \leqslant a$$

$$(6.4.17)$$

and the homogeneous boundary conditions:

$$x(t, 0) = 0, \quad x(t, a) = 0 \quad \text{for } 0 < t < T. \qquad (6.4.18)$$

The space under consideration is $X = C(\Omega)$, where $\Omega = \{(t, s): 0 \leqslant t \leqslant T; 0 \leqslant s \leqslant a\}$. Write:

$$D_1 = \partial/\partial s, \quad D_2 = \partial/\partial t,$$
$$(F_0 x)(t, s) = x(t, 0), \quad (F_1 x)(t, s) = x(t, a), \quad \text{for } x \in C(\Omega).$$

It is easy to verify that all assumptions of Theorem 6.4.3 are satisfied. Indeed, since $u(s) = s_\lambda(z)(s) + c_\lambda(z)(s) = C_1 \cos \lambda s + C \sin \lambda s$, $C_1, C \in \mathbf{R}$, we conclude that $u \in \ker D$, $u(0) = 0$ if and only if $C_1 = 0$ and $u(a) = 0$ if and only if $u(a) = C \sin \lambda a = 0$, i.e. if $\lambda a = \pi k$, where k is an arbitrary integer. Hence for all $\lambda_k = \pi k/a$, where k is an arbitrary integer, we have $u(a) = 0$, u is invertible, and, moreover, all solutions of the equation $(D_1 + \lambda_k^2 I)v = 0$, i.e. the functions $e^{-\lambda_k^2 t}$ are invertible. We therefore obtain infinitely many non-trivial solutions of the problem in question:

$$x_k(t, s) = C_k \sin \lambda_k s e^{-\lambda_k^2 t},$$

where $\lambda_k = \dfrac{\pi k}{a}$ $(k = 0, \pm 1, \pm 2, ...)$, $C_k \in \mathbf{R}$.

Observe that the main idea of the Fourier method is a *separation of variables* which leads to solving of independent initial and boundary value problems for ordinary differential equations.

EXERCISE 6.4.1. Generalize Theorems 6.4.1–6.4.5 for the case when $c_{D_1} \neq 1$, $c_{D_2} \neq 1$.

EXERCISE 6.4.2. Solve the heat equation (6.4.17) with boundary conditions $x(t, 0) = 0$, $x_t'(t, a) = 0$ $(0 < t < T)$, by an application of Theorem 6.4.2.

EXERCISE 6.4.3. Solve the heat equation (6.4.17) with the conditions

$x(0, s) = 0$, $x'_t(0, s) = 0$ for $0 < s < a$ and $x(t, 0) = 0$, $x'_s(t, 0)$ for $0 < t < T$ by an application of Corollary 6.4.1.

EXERCISE 6.4.4. Solve the one-dimensional *wave equation*

$$\frac{\partial^2 x}{\partial t^2} - \frac{\partial^2 x}{\partial s^2} = 0 \quad (0 \leqslant t \leqslant T, 0 \leqslant s \leqslant a)$$

with the conditions:

(i) $x(0, s) = 0$, $x(T, s) = 0$, $x(t, 0) = 0$, $x(t, a) = 0$;
(ii) $x(0, s) = 0$, $x(T, s) = 0$, $x(t, 0) = 0$, $x'_s(t, a) = 0$;
(iii) $x(0, s) = 0$, $x(T, s) = 0$, $x(t, 0) = 0$, $x'_s(t, 0) = 0$;
(iv) $x(0, s) = 0$, $x(T, s) = 0$,

$$x(t, s) = \frac{1}{a} \int_0^a x(t, u)\,du, \quad x'_s(t, 0) = 0.$$

EXERCISE 6.4.5. Solve the one-dimensional *Schrödinger equation*

$$\frac{\partial x}{\partial t} = -i\frac{\partial^2 x}{\partial s^2} \quad (0 \leqslant t \leqslant T, 0 \leqslant a \leqslant s)$$

with the conditions:

(i) $x(0, s) = 0$, $x(t, 0) = 0$, $x(t, a) = 0$;
(ii) $x(0, s) = 0$, $x(t, 0) = 0$, $x'_s(t, a) = 0$;
(iii) $x(0, s) = 0$, $x(t, 0) = 0$, $x'_s(t, 0) = 0$;

(iv) $x(t, s) = \frac{1}{T} \int_0^T x(u, s)\,du$, $x(t, 0) = 0$, $x'_s(t, 0) = 0$.

EXERCISE 6.4.6. Apply Theorems 6.4.1, 6.4.2 and Corollary 6.4.1, respectively, to the equation

$$\frac{\partial x}{\partial t} = P(D)x, \quad \text{where } D = \frac{\partial}{\partial s},$$

$P(t)$ is an arbitrary polynomial with real coefficients such that $P(0) \neq 0$.

EXERCISE 6.4.7. Apply (by a modification) the Fourier method for the *diffusion equation*

$$\frac{\partial}{\partial s}\left(d\,\frac{\partial x}{\partial s}\right) = c\,\frac{\partial x}{\partial t} \quad (0 \leqslant t \leqslant T, 0 \leqslant s \leqslant a),$$

where $c \in \mathbf{R}$ is a given coefficient and:

(i) $d \in \mathbf{R}$ is a given scalar coefficient;

(ii) $d = d(t, s) \neq 0$ is a given function (may be satisfying some additional conditions which permit to apply the Fourier method).

6.5. GREEN FORMULAE. PICONE IDENTITY.
EULER–LAGRANGE EQUATION

In this section we shall prove some theorems which generalize well-known results of the classical analysis for D-algebras.

THEOREM 6.5.1. *Let X be a D-algebra. Write*

$$Q(D) = \sum_{k=0}^{N} Q_k D^k, \quad Q^+\langle D \rangle = \sum_{k=0}^{N} (-1)^k D^k Q_k, \qquad (6.5.1)$$

where $Q_0, Q_1, \ldots, Q_N \in L_0(\operatorname{dom} D^N), N \geq 1$. Then for every $x, y \in \operatorname{dom} D^N$ the following identity holds:

$$xQ(D)y - yQ^+\langle D \rangle x = \sum_{k=0}^{N} \{(-1)^{k+1} c_D^{-k} [D^k(yQ_k x) -$$

$$- f_D^{(k)}(Q_k x, y)] + [xQ_k - (Q_k x)]D^k y + [1 + (-1)^k](Q_k x) D^k y\}.$$

$$(6.5.2)$$

Proof. Our assumptions imply that

$$xQ(D)y - yQ^+\langle D \rangle x = \sum_{k=0}^{N} [xQ_k D^k y - y(-1)^k D^k Q_k x]$$

$$= \sum_{k=0}^{N} \{(-1)^{k+1} [(Q_k x) D^k y + y D^k(Q_k x)] +$$

$$+ [xQ_k D^k y + (-1)^k (Q_k x) D^k y]\}$$

$$= \sum_{k=0}^{N} \{(-1)^{k+1} c_D^{-k} [D^k(yQ_k x) - f_D^{(k)}(Q_k x, y)] +$$

$$+ [xQ_k - (Q_k x)]D^k y + [1 + (-1)^k](Q_k x) D^k y\}. \quad \blacksquare$$

Formula (6.5.2) is the *Lagrange formula* for polynomials in D with operator coefficients.

COROLLARY 6.5.1. *Let X be a D-algebra and let $Q(D)$ and $Q^+\langle D\rangle$ be defined by Formula (6.5.1). Then for all $F_\alpha, F_\beta \in \mathscr{F}_D$, $R_\alpha \in \mathscr{R}_D$ ($\alpha \neq \beta$) and for all $x, y \in \mathrm{dom}\ D^N$ the following identity holds:*

$$F_\beta R_\alpha[xQ(D)y - yQ^+\langle D\rangle x]$$

$$= \sum_{j=0}^{N-1} (-1)^j c_D^{-(j+1)}(F_\beta - F_\alpha) D^j(yQ_{j+1}x) + F_\beta R_\alpha g_{Q(D)}(x,y) +$$

$$+ h_{Q(D)}(x,y), \quad (6.5.3)$$

where

$$g_{Q(D)}(x,y) = \sum_{k=1}^{N} \{(-1)^k c_D^{-k} f_D^{(k)}(Q_k x, y) +$$

$$+ [1 + (-1)^k](Q_k x) D^k y\} + yQ_0 x, \quad (6.5.4)$$

$$h_{Q(D)}(x,y) = \sum_{k=0}^{N} [xQ_k - (Q_k x)]D^k y. \quad (6.5.5)$$

Proof. Formulae (6.5.1), (6.5.2), (6.5.4), (6.5.5) together imply that for all $F_\alpha, F_\beta \in \mathscr{F}_D$, $R_\alpha \in \mathscr{R}_D$, $\alpha \neq \beta$ and $x, y \in \mathrm{dom}\ D^N$ we have

$$F_\beta R_\alpha[xQ(D)y - yQ^+\langle D\rangle x]$$

$$= F_\beta R_\alpha \Big\{ -yQ_0 x + \sum_{k=1}^{N} (-1)^{k+1} c_D^{-k}[D^k(yQ_k x)] +$$

$$+ \sum_{k=0}^{N} (-1)^{k+2} c_D^{-k} f_D^{(k)}(x,y) + \sum_{k=0}^{N} [xQ_k - (Q_k x)]D^k y +$$

$$+ \sum_{k=1}^{N} [1 + (-1)^k](Q_k x) D^k y \Big\}$$

$$= F_\beta R_\alpha \Big\{ \sum_{j=0}^{N-1} (-1)^{j+2} c_D^{-(j+1)}[D^{j+1}(yQ_{j+1}x)] +$$

$$+ \sum_{k=0}^{N} c_D^{-k} f_D^{(k)}(x,y) + \sum_{k=0}^{N} [xQ_k - (Q_k x)]D^k y +$$

$$+ \sum_{k=1}^{N} [1 + (-1)^k](Q_k x) D^k y + yQ_0 x \Big\}$$

$$= \sum_{j=0}^{N-1} (-1)^j c_D^{-(j+1)} F_\beta R_\alpha D[D^j(yQ_{j+1}x)] + F_\beta R_\alpha[g_{Q(D)}(x, y) +$$

$$+ h_{Q(D)}(x, y)]$$

$$= \sum_{j=0}^{N-1} (-1)^j c_D^{-(j+1)} (F_\beta - F_\alpha) D^j(yQ_{j+1}x) + F_\beta R_\alpha[g_{Q(D)}(x, y) +$$

$$+ h_{Q(D)}(x, y)]$$

since $f_D^{(0)} = 0$. ∎

Formula (6.5.3) is the *Green formula* for polynomials in D with operator coefficients. Now we shall specify coefficients. Observe that the bilinear operator appearing in this formula maps the space of constants into itself.

COROLLARY 6.5.2. *Suppose that all assumptions of Corollary 6.5.1 are satisfied. If $Q_0, ..., Q_N$ are operators of multiplication by elements of X, i.e.*

$$Q_k x = a_k x, \quad \text{where } a_k \in X \ (k = 0, 1, ..., N) \quad \text{for } x \in X$$
$$(6.5.6)$$

then

$$Q(D)x = \sum_{k=0}^{N} a_k D^k x, \quad Q^+\langle D \rangle x = \sum_{k=0}^{N} (-1)^k D^k(a_k x)$$

$$\text{for } x \in X, \quad (6.5.7)$$

$$h_{Q(D)} = 0,$$

where $h_{Q(D)}$ is defined by (6.5.5), and the Green formula is of the form

$$F_\beta R_\alpha[xQ(D)y - yQ^+\langle D \rangle x]$$

$$= \sum_{j=0}^{N-1} (-1)^j c_D^{-(j+1)} (F_\beta - F_\alpha) D^j(axy) + F_\beta R_\alpha g_{Q(D)}(x, y)$$

where

$$g_{Q(D)}(x, y) = \sum_{k=1}^{N} \left\{ (-1)^k c_D^{-k} f_D^{(k)}(a_k x, y) + \right.$$

$$\left. + [1 + (-1)^k] a_k x D^k y \right\} + a_0 xy.$$

Indeed, by our definition, for all $x, y \in \text{dom } D^N$,

$$h_{Q(D)}(x, y) = \sum_{k=0}^{N} [xQ_k - (Q_k x)]D^k y = \sum_{k=0}^{N} (xa_k - a_k x)D^k y = 0$$

and $yQ_0 x = a_0 xy$.   ∎

COROLLARY 6.5.3. *Suppose that all assumptions of Corollary 6.5.1 are satisfied. If the coefficients* $Q_0, \ldots, Q_N$ *commute with* D: $DQ_k = Q_k D$ $(k = 0, 1, \ldots, N)$, *then*

$$Q^+\langle D \rangle = Q(-D) = Q(D^+) \quad \text{where } D^+ = -D. \qquad (6.5.8)$$

Proof. Indeed, by our assumptions we have

$$Q^+\langle D \rangle = \sum_{k=0}^{N} (-1)^k D^k Q_k = \sum_{k=0}^{N} Q_k(-D)^k = Q(-D) = Q(D^+).$$

■

COROLLARY 6.5.4. *Suppose that all assumptions of Corollary 6.5.1 are satisfied. If the coefficients of* $Q(D)$ *are scalars, i.e. if* $Q_k = q_k I$, *where* $q_k \in \mathbf{R}$ $(k = 0, 1, \ldots, N)$, $q_N \neq 0$, *then*

$$Q^+\langle D \rangle = Q(-D) = Q(D^+) \quad \text{for all } x, y \in \text{dom } D^N,$$

$$g_{Q(D)}(x, y)$$
$$= \sum_{k=1}^{N} q_k \{(-1)^k c_D^{-k} f_D^{(k)}(x, y) + [1 + (-1)^k] x D^k y\} + q_0 xy,$$

$$h_{Q(D)}(x, y) = 0$$

and the Green formula is of the form

$$F_\beta R_\alpha [xQ(D)y - yQ(-D)x]$$
$$= \sum_{j=0}^{N-1} (-1)^j c_D^{-(j+1)} q_j (F_\beta - F_\alpha) D^j (xy) + F_\beta R_\alpha g_{Q(D)}(x, y).$$

$$(6.5.9)$$

Proof. Indeed, since $Q_0, \ldots, Q_N$ commute with D, Corollary 6.5.3 implies that $Q^+\langle D \rangle = Q(-D)$. Moreover, we have

$$g_{Q(D)}(x, y) = \sum_{k=1}^{N} \{(-1)^k c_D^{-k} f_D^{(k)}(q_k x, y) +$$
$$+ [1 + (-1)^k] q_k x D^k y\} + q_0 xy$$

$$= \sum_{k=1}^{N} q_k \{(-1)^k c_D^{-k} f_D^{(k)}(x, y) + [1 + (-1)^k] x D^k y\} + q_0 xy$$

since $f_D^{(k)}$ are bilinear operators. We also find

$$h_{Q(D)}(x, y) = \sum_{k=0}^{N} (xQ_k - Q_k x) D^k y = \sum_{k=0}^{N} q_k(x - x) D^k y = 0.$$

Therefore the Green formula in our case is of the form (6.5.9). ∎

COROLLARY 6.5.5. *Suppose that all assumptions of Corollary 6.5.4 are satisfied. Then the Green formula (6.5.9) can be written as follows:*

$$F_\beta R_\alpha [xQ(D)y - yQ(-D)x]$$
$$= c_D^{-1}(F_\beta - F_\alpha)[Q(-c_D^{-1}D) - (-1)^N q_N D^N](xy) + F_\beta R_\alpha g_{Q(D)}(x, y)$$
$$(6.5.10)$$

for all $x, y \in \operatorname{dom} D^N$, *where* $g_{Q(D)}$ *is as in Corollary 6.5.4.*

Proof. Indeed, Formula (6.5.8) implies that for all $x, y \in \operatorname{dom} D^N$

$$F_\beta R_\alpha [xQ(D)y - yQ(-D)x]$$
$$= \sum_{j=0}^{N-1} (-1)^j c_D^{-(j+1)} q_j (F_\beta - F_\alpha) D^j (xy) + F_\beta R_\alpha g_{Q(D)}(x, y)$$
$$= c_D^{-1}(F_\beta - F_\alpha) \sum_{j=0}^{N-1} q_j (-1)^j c_D^{-j} D^j (xy) + F_\beta R_\alpha g_{Q(D)}(x, y).$$

If $c_D = 1$, then Formula (6.5.9) is of the form:

$$F_\beta R_\alpha [xQ(D)y - yQ(-D)x]$$
$$= (F_\beta - F_\alpha)[Q(-D) - (-1)^N q_N D^N](xy) + F_\beta R_\alpha g_{Q(D)}(x, y),$$
$$(6.5.11)$$

where

$$g_{Q(D)}(x, y) = \sum_{k=1}^{N} q_k \{(-1)^k f_D^{(k)}(x, y) + [1 + (-1)^k] x D^k y\} + q_0 xy.$$

∎

COROLLARY 6.5.6. *Suppose that all assumptions of Corollary 6.5.1 are satisfied. Let* x, y *be solutions of equations*

$$Q(D)y = v, \qquad Q^+\langle D \rangle x = u \qquad\qquad (6.5.12)$$

respectively, where $u, v \in X$ *are given. Then*

$$F_\beta R_\alpha(xv - yu)$$

$$= \sum_{j=0}^{N-1} (-1)^j c_D^{-(j+1)} (F_\beta - F_\alpha) D^j(yQ_{j+1}x) +$$

$$+ F_\beta R_\alpha g_{Q(D)}(x, y) + h_{Q(D)}(x, y). \qquad (6.5.13)$$

The proof immediately follows from Formula (6.5.3).

COROLLARY 6.5.7. *Let* X *be a* D-*algebra. Let* N *be an arbitrarily fixed positive integer and let* $x, y \in \operatorname{dom} D^N$. *Then the Lagrange formula for the operator* D^N *is of the form*:

$$xD^N y - y(D^+)^N x$$

$$= (-1)^{N+1} c_D^{-N} D^N(xy) - f_D^{(N)}(x, y) + [1 + (-1)^N] xD^N y, \qquad (6.5.14)$$

where $D^+ = -D$ *and the Green formula is of the form*

$$F_\beta R_\alpha [xD^N y - y(D^+)^N x]$$

$$= (-1)^{N+1} c_D^{-N} (F_\beta - F_\alpha) D^{N-1}(xy) +$$

$$+ (-1)^N c_D^{-N} f_D^{(N)}(x, y) + [1 + (-1)^N] xD^N y, \qquad (6.5.15)$$

where $F_\alpha, F_\beta \in \mathscr{F}_D, R_\alpha \in \mathscr{R}_D, \alpha \neq \beta$.

Proof. Indeed, if we put in Theorem 6.5.1 and Corollary 6.5.1 $Q_0 = Q_1 = \ldots = Q_{N-1} = 0, Q_N = I$ then we obtain

$$Q(D) = D^N, \quad Q^+\langle D\rangle = (-1)^N D^N = (D^+)^N. \qquad \blacksquare$$

COROLLARY 6.5.8. *Suppose that* $D_1, D_2 \in R(X), D = D_1 D_2$ *and* X *is a* D-*algebra. Then the Lagrange and Green formulae for the superposition* $D = D_2 D_1$ *are of the form: for all* $x, y \in \operatorname{dom} D$

$$xD_2 D_1 y - yD_2^+ D_1^+ x$$

$$= c_{D_1}^{-1} c_{D_2}^{-1} \{D_2 D_1(xy) - f_{D_2}(x, y) - D_2 f_{D_1}(x, y) +$$

$$+ c_{D_1}^{-1} c_{D_2}^{-1} [(D_1 x)(D_2 y) + (D_2 x)(D_1 y)]\}, \qquad (6.5.16)$$

$$F_\beta R_\alpha(xD_2 D_1 y - yD_2^+ D_1^+ x)$$

$$= c_{D_1 D_2}(F_\beta - F_\alpha)(xy) - c_{D_1 D_2} F_\beta R_\alpha \{f_{D_2}(x, y) + D_2 f_{D_1}(x, y) +$$

$$+ c_{D_1} c_{D_2} [(D_1 x)(D_2 y) + (D_2 x)(D_1 y)]\}, \qquad (6.5.17)$$

where we admit $D_1^+ = -D_1, D_2^+ = -D_2, D^+ = -D$ *and*

$$F_\beta R_\alpha = (F_\beta^{(1)} R_\alpha^{(1)} + R_\beta^{(1)} F_\beta^{(2)}) R_\alpha^{(2)}, \qquad (6.5.18)$$

$F_\alpha^{(i)} F_\beta^{(i)} \in \mathscr{F}_{D_i}, R_\alpha^{(i)}, R_\beta^{(i)} \in \mathscr{R}_{D_i}$ $(i = 1, 2)$ *and* $\alpha \neq \beta$.

Proof. Indeed $D^+ = -D_2^+D_1^+ = -(-D_2)(-D_1) = -D_2D_1 = -D$.
It is well-known (cf. Section 2.2) that

$$F_\beta = F_\beta^{(1)}+R_\beta^{(1)}F_\beta^{(2)}D_1 \in \mathscr{F}_D, \quad F_\alpha = F_\alpha^{(1)}+R_\alpha^{(1)}F_\alpha^{(2)}D_1 \in \mathscr{F}_D,$$
$$R_\alpha = R_\alpha^{(1)}R_\alpha^{(2)} \in \mathscr{R}_D.$$

Hence

$$F_\beta R_\alpha = (F_\beta^{(1)}+R_\beta^{(1)}F_\beta^{(2)}D_1)R_\alpha^{(1)}R_\alpha^{(2)} = (F_\beta^{(1)}R_\alpha^{(1)}+R_\beta^{(1)}F_\beta^{(2)})R_\alpha^{(2)}\cdot$$
$$F_\beta-F_\alpha = F_\beta^{(1)}-F_\alpha^{(1)}+(R_\beta^{(1)}F_\beta^{(2)}-R_\alpha^{(1)}F_\alpha^{(2)})D_1.$$

Since $D^+ = -D$, Formulae (6.5.2) and (6.5.3) imply that for all $x, y \in \text{dom } D$

$$xDy-yD^+x = xDy+yDx = c_D^{-1}[D(xy)-f_D(x,y)],$$
$$F_\beta R_\alpha(xDy-yD^+x) = F_\beta R_\alpha c_D^{-1}[D(xy)-f_D(x,y)]$$
$$= c_D^{-1}F_\beta R_\alpha D(xy)-c_D^{-1}F_\beta R_\alpha f_D(x,y)$$
$$= c_D^{-1}(F_\beta-F_\alpha)(xy)-c_D^{-1}F_\beta R_\alpha f_D(x,y).$$

This implies the required formulae (6.5.16) and (6.5.17).

COROLLARY 6.5.9. *Suppose that $D_1, D_2 \in R(X)$, $D = D_2^q D_1^p$, $p,q \in \mathbf{N}^\iota$ are arbitrarily fixed and X is a D-algebra. Then we have $c_D = c_{D_1}^p c_{D_2}^q$,*

$$f_D(x,y) = f_{D_2}^{(q)}(x,y)+D_2^q f_{D_1}^{(p)}(x,y)+$$
$$+c_{D_1}^p c_{D_2}^q[(D_1^p x)(D_2^q y)+(D_2^q x)(D_1^p y)] \quad (6.5.19)$$

for $x,y \in \text{dom } D$,

$$D^+ = (-1)^{p+q}D, \quad (6.5.20)$$

$$F_\gamma = \sum_{k=0}^{p-1} R_\gamma^{(1)k}F_\gamma^{(1)}D_1^k+(R_\gamma^{(1)})^p \sum_{j=0}^{q-1}(R_\gamma^{(2)})^j F_\gamma^{(2)}D_2^j D_1^p$$
$$(\gamma = \alpha \text{ or } \gamma = \beta), \quad (6.5.21)$$

$$R_\alpha = (R_\alpha^{(1)})^p(R_\alpha^{(2)})^q, \quad F_\alpha^{(i)}, F_\beta^{(i)} \in \mathscr{F}_{D_i}, \quad R_\alpha^{(i)}, R_\beta^{(i)} \in \mathscr{R}_{D_i}$$
$$(i = 1,2). \quad (6.5.22)$$

The Lagrange formula is of the form

$$xDy-yD^+x = c_D^{-1}[D(xy)-f_D(x,y)].$$

The Green formula is of the form

$$F_\beta R_\alpha(xDy-yD^+x) = c_D^{-1}(F_\beta-F_\alpha)(xy)-F_\beta R_\alpha c_D^{-1}f(x,y)$$
$$(6.5.23)$$

for $x,y \in \text{dom } D$.

The proof follows from Corollary 6.5.8, Theorem 6.5.1 and Taylor Formulae for operators D_1 and D_2 (cf. Section 2.2). ∎

Immediate consequences are

COROLLARY 6.5.10. *Let $X, D, c_D, f_D, D^+, F_\alpha, F_\beta, R_\alpha$ be defined as in Corollary 6.5.9. Let*

$$Q(D) = \sum_{k=0}^{N} Q_k D^k, \quad Q^+\langle D\rangle = \sum_{k=0}^{N} (-1)^k D^k Q_k,$$

where $Q_k \in L_0(\mathrm{dom}\, D^N)$ $(k = 0, 1, ..., N)$ and Q_N is invertible. Then the Green formula (6.5.3) holds.

COROLLARY 6.5.11. *Let $X, D, c_D, f_D, F_\alpha, F_\beta, R_\alpha$ be defined as in Corollary 6.5.9 (with $p = q = 1$). Let*

$$P(D) = D_2 D_1 + AD_1 + BD_2 + C,$$
$$P^+\langle D\rangle = D_2 D_1 - D_1 A - D_2 B + C, \tag{6.5.24}$$

$A, B, C \in L_0(X)$.

Then the Green formula for all $x, y \in \mathrm{dom}\, D$ is of the form

$$F_\beta R_\alpha[xP(D)y - yP^+\langle D\rangle x]$$
$$= c_{D_1}^{-1}c_{D_2}^{-1}(F_\beta - F_\alpha)(xy) + c_{D_1}^{-1}(F_\beta^{(1)} - F_\alpha^{(1)})(yAx) +$$
$$+ c_{D_2}^{-1}(F_\beta^{(2)} - F_\alpha^{(2)})(yBx) + F_\beta R_\alpha\{c_{D_1}^{-1}c_{D_2}^{-1}[f_{D_2}(x, y) +$$
$$+ D_2 f_D(x,y)] + [(D_1 x)(D_2 y) + (D_2 x)(D_1 y)] + c_{D_1}^{-1}f_{D_1}(Ax,y) +$$
$$+ c_{D_2}^{-1}f_{D_2}(Bx, y) + [x, A]D_1 y + [x, B]D_2 y + yCx - xCy\},$$

$$\tag{6.5.25}$$

where $D = D_2 D_1$, $[x, A] = Ax - xA$, $[x, B] = Bx - xB$ for $x \in X$ and

$$F_\gamma = F_\gamma^{(1)} + R_\gamma^{(1)}F_\gamma^{(2)}D_1 \in \mathscr{F}_D = \mathscr{F}_{D_2 D_1} \quad (\gamma = \alpha, \beta),$$
$$R_\alpha = R_\alpha^{(1)}R_\alpha^{(2)} \in \mathscr{R}_D = \mathscr{R}_{D_1 D_2}, \quad F_\alpha^{(i)}, F_\beta^{(i)} \in \mathscr{F}_{D_i}, \quad R_\alpha^{(i)} \in \mathscr{R}_{D_i}$$
$$(i = 1, 2).$$

Proof. Indeed, observe that for $x, y \in \mathrm{dom}\, D$ we have

$$xP(D)y - yP^+\langle D\rangle x = xD_2 D_1 y - yD_2 D_1 x + xAD_1 y -$$
$$- yD_1 Ax + xBD_2 y - yD_2 Bx + xCy - yCx$$

$$= xDy - yDx + (Ax)D_1y - yD_1(Ax) + (Bx)(D_2y) - yD_2(Bx) +$$
$$+ xCy - yCx + [x, A]D_1y + [x, B]D_2y$$
$$= c_D^{-1}[D(x, y) - f_D(x, y)] + c_{D_1}^{-1}[D_1(yAx) - f_{D_1}(Ax, y)] +$$
$$+ c_{D_2}^{-1}[D_2(yBx) - f_{D_2}(Bx, y)] + [x, A]D_1y + [x, B]D_2y +$$
$$+ yCx - xCy.$$

Further the proof is going in a similar way as in Corollaries 6.5.8 and 6.5.9.   ∎

Till now we have considered D-algebras over reals. Now we shall pass to D-algebras over C, in the same manner, as it was done in Section 6.1. Then Formula (6.1.16) implies

$$\bar{u}Dv - \overline{vD^+u} = c_D^{-1}D(\bar{u}v) - f_D(\bar{u}, v) \quad \text{for all } u, v \in \text{dom } D.$$
$$(6.5.26)$$

Indeed, since $D^+ = -D$, we find
$$\bar{u}Dv - \overline{vD^+u} = \bar{u}Dv - vD^+\bar{u} = \bar{u}Dv + vD\bar{u}$$
$$= c_D^{-1}[D(\bar{u}v) - f_D(\bar{u}, v)].$$

We also have
$$(\lambda D)^+ = \bar{\lambda}D^+ \quad \text{for all } \lambda \in C,$$
$$(6.5.27)$$
i.e. the operator D^+ is *antilinear* in the complex case.

Indeed, by Formulae (6.5.26), (6.5.27) we have for $\hat{D} = \lambda D$
$$\bar{u}(\lambda D)v - \overline{v(\lambda D)^+u} = \bar{u}\hat{D}v - v\hat{D}^+\bar{u} = c_D^{-1}[\hat{D}(\bar{u}v) - f_{\hat{D}}(\bar{u}, v)]$$
$$= c_D^{-1}[\lambda D(\bar{u}v) - f_{\lambda D}(\bar{u}, v)] = \lambda c_D^{-1}[D(\bar{u}v) - f_D(\bar{u}, v)]$$
$$= \lambda(\bar{u}Dv - vD^+\bar{u}).$$

This implies that for all $\lambda \in C$, $u \in \text{dom } D$ we have $\bar{\lambda}D^+\bar{u} = \overline{\lambda D^+u}$ $= \overline{(\lambda D)^+u} = (\bar{\lambda}D)^+u.$
The arbitrariness of $u \in \text{dom } D$ implies (6.5.27). In particular we have
$$(iD)^+ = iD.$$
$$(6.5.28)$$

All further considerations are going in a similar way as in the real case. In particular, the Green formula (6.5.3) in the complex case is of the form
$$F_\beta R_\alpha[\bar{x}Q(D)y - \overline{y\bar{Q}^+\langle D\rangle x}]$$

$$= \sum_{j=0}^{N-1} (-1)^j c_D^{-(j+1)} (F_\beta - F_\alpha)(y Q_{j+1} \bar{x}) +$$

$$+ F_\beta R_\alpha [g_{Q(D)}(\bar{x}, y) + h_{Q(D)}(\bar{x}, y)], \qquad (6.5.29)$$

where $g_{Q(D)}$, $h_{Q(D)}$ are defined by Formulae (6.5.4), (6.5.5).

Indeed, if we put $\bar{x}$ instead of x and we shall make use of (6.1.16) we obtain

$$Q^+ \langle D \rangle \bar{x} = \sum_{k=0}^{N} (-q)^k \overline{D^k Q_k \bar{x}} = \sum_{k=0}^{N} (-1)^k D^k Q_k \bar{x}$$

which implies Formula (6.5.29).

Now we shall pass to the proof of the Picone identity in *D*-algebras.

PROPOSITION 6.5.1. *Suppose that X is a D-algebra with unit e and x $\in$ dom D is invertible. If $x^{-1} \in$ dom D then*

$$Dx^{-1} = -x^{-2}Dx + c_D^{-1} x^{-1} [De - f_D(x, x^{-1})]. \qquad (6.5.30)$$

Proof. Indeed, $De = D(x \cdot x^{-1}) = c_D(xDx^{-1} + x^{-1}Dx) + f_D(x, x^{-1})$. ∎

PROPOSITION 6.5.2. *Suppose that X is a D-algebra with unit e, x, y $\in$ dom D are invertible and $y^{-1} \in$ dom D. Then*

$$D(xy^{-1}) = c_D y^{-2}(yDx - xDy) + xy^{-1}[De - f_D(y, y^{-1})] +$$

$$+ f_D(x, y^{-1}). \qquad (6.5.31)$$

Proof. Indeed, Formula (6.5.30) implies that

$$D(xy^{-1}) = c_D(xDy^{-1} + y^{-1}Dx) + f_D(x, y^{-1})$$

$$= c_D \{x(-y^2 Dy + c_D^{-1} y^{-1}[De - f_D(y, y^{-1})]) + y^{-1}Dx\} +$$

$$+ f_D(x, y^{-1})$$

$$= c_D y^{-2}(xDy - yDx) +$$

$$+ xy^{-1}[De - f_D(y, y^{-1})] + f_D(x, y^{-1}). \qquad ∎$$

THEOREM 6.5.2. *Suppose that X is a D-algebra with unit e and that $p_1, p_2 \in$ dom D, $q_1, q_2 \in X$. If $x_i \in \ker[D(p_i D) + q_i]$ for $i = 1, 2$ and are invertible then the following identity holds*

$$D[x_1^2(p_1 x_1^{-1} Dx_1 - p_2 x_2^{-1} Dx_2)] = D[x_1 x_2^{-1}(p_1 x_2 Dx_1 - p_2 x_1 Dx_2)]$$

$$= c_D[(q_2 - q_1)x_1^2 + (p_1 - p_2)(Dx_1)^2 + p_2(Dx_1 - c_D x_1 x_2^{-1} Dx_2)^2] +$$

$$+ q_D(x_1, x_2) + r_D(x_1, x_2),$$

where

$$q_D(x_1, x_2) = (1 - c_D) c_D^2 p_2 x_1 x_2^{-1}(Dx_2)(2Dx_1) + x_1 x_2^{-1}(Dx_2),$$

$$r_D(x_1, x_2) = c_D\{[x_1^2 x_2^{-1}(De - f_D(x_2, x_2^{-1})) + f_D(x_1, x_2^{-1}) - \tag{6.5.32}$$

$$- c_D x_2^{-1} f_D(x_1, x_1)] p_2 (Dx_2)\} +$$

$$+ f_D(x_1, p_1 Dx_1) - f_D(x_1^2 x_2^{-1}, p_2 Dx_2).$$

Proof. By our assumptions we have $D(p_i Dx_i) = -q_i x_i$ for $i = 1, 2$ and (by Formula (6.5.31))

$$D[x_1^2(p_1 x_1^{-1} Dx_1 - p_2 x_2^{-1} Dx_2)] = D[x_1 x_2^{-1}(p_1 x_2 Dx_1 - p_2 x_1 Dx_2)]$$

$$= D[x_1(p_1 Dx_1)] - D[(x_1^2 x_2^{-1})(p_2 Dx_2)]$$

$$= c_D\{p_1(Dx_1)^2 + x_1 D(p_1 Dx_1) - D(x_1^2 x_2^{-1}) p_2 Dx_2 - x_1^2 x_2^{-1} D(p_2 Dx_2)\} +$$

$$+ f_D(x_1, p_1 Dx_1) - f_D(x_1^2 x_2^{-1}, p_2 Dx_2)$$

$$= c_D\{(p_1 - p_2)(Dx_1)^2 + x_1(-q_1 x_1) - (x_1^2 x_2^{-1})(-q_2 x_2) + p_2(Dx_1)^2 -$$

$$- [c_D x_2^{-2}(x_2 Dx_1^2 - x_1^2 Dx_2) + x_1^2 x_2^{-1}(De - f_D(x_2, x_2^{-1})) +$$

$$+ f_D(x_1, x_2^{-1})] p_2 Dx_2 + p_2(Dx_1)^2\} + f_D(x_1, p_1 Dx_1) - f_D(x_1^2 x_2^{-1}, p_2 Dx_2)$$

$$= c_D\{(p_1 - p_2)(Dx_1)^2 + (q_2 - q_1)x_1^2 + p_2(Dx_1)^2 -$$

$$- 2c_D^2 p_2 x_1 x_2^{-1}(Dx_1)(Dx_2) + c_D p_2 x_1^2 x_2^{-2}(Dx_2)^2 -$$

$$- c_D x_2^{-1} p_2(Dx_2) f_D(x_1, x_1) + [x_1^2 x_2^{-1}(De - f_D(x_2, x_2^{-1})) +$$

$$+ f_D(x_1, x_2^{-1})] p_2 Dx_2\} + f_D(x_1, p_1 Dx_1) - f_D(x_1 x_2^{-1}, p_2 Dx_2)$$

$$= c_D\{(p_1 - p_2)(Dx_1)^2 + (q_2 - q_1)x_1^2 + p_2(Dx_1 - c_D x_1 x_2^{-2} Dx_2)^2 +$$

$$+ 2(1 - c_D) c_D p_2 x_1 x_2^{-1}(Dx_1)(Dx_2) + (1 - c_D) c_D p_2 x_1^2 x_2^{-2}(Dx_2)^2\} +$$

$$+ r_D(x_1, x_2)$$

$$= c_D\{(p_1 - p_2)(Dx_1)^2 + (q_2 - q_1)x_1^2 + [p_2(Dx_1 - c_D x_1 x_2^{-2} Dx_2)]^2\}$$

$$+ q_D(x_1, x_2) + r_D(x_1, x_2). \qquad \blacksquare$$

An immediate consequence of this theorem is

THEOREM 6.5.3. *Suppose that all assumptions of Theorem 6.5.2 are satisfied and* F_α, F_β, $\alpha \neq \beta$, *are initial operators for* D *corresponding to* R_α, R_β $\in \mathcal{R}_D$, *respectively. The Picone identity is*

$$(F_\beta - F_\alpha)[x_1^2(p_1 x_1^{-1} Dx_1 - p_2 x_2^{-1} Dx_2)]$$

$$= c_D I_\alpha^\beta[(q_2 - q_1)x_1^2 + (p_1 - p_2)(Dx_1)^2 +$$

$$+ p_2(Dx_1 - c_D x_1 x_2^{-1} Dx_2)^2] +$$

$$+ I_\alpha^\beta[q_D(x_1, x_2) + r_D(x_1, x_2)], \tag{6.5.33}$$

where q_D *and* r_D *are defined by Formulae (6.5.32).*

COROLLARY 6.5.12. *Suppose that all assumptions of Theorem 6.5.3 are satisfied.*

(i) *If $f_D = 0$ then $r_D(x_1, x_2) = 0$ and the Picone identity is of the form*:

$$(F_\beta - F_\alpha)[x_1^2(p_1 x_1^{-1}Dx_1 - p_2 x_2^{-1}Dx_2)]$$
$$= c_D I_\alpha^\beta[(q_2 - q_1)x_1^2 + (p_1 - p_2)(Dx_1)^2 +$$
$$+ p_2(Dx_1 - c_D x_1 x_2^{-1}Dx_2)^2] + I_\alpha^\beta q_D(x_1, x_2); \qquad (6.5.34)$$

(ii) *If $c_D = 1$ then $q_D(x_1, x_2) = 0$ and the Picone identity is of the form*:

$$(F_\beta - F_\alpha)[x_1^2(p_1 x_1^{-1}Dx_2 - p_a x_2^{-1}Dx_2)]$$
$$= I_\alpha^\beta[(q_2 - q_1)x_1^2 + (p_1 - p_2)(Dx_1)^2 + p_2(Dx_1 - x_1 x_2^{-1}Dx_2)^2] +$$
$$+ I_\alpha^\beta r_D(x_1, x_2); \qquad (6.5.35)$$

(iii) *If $f_D = 0$ and $c_D = 1$ then $r_D(x_1, x_2) = q_D(x_1, x_2) = 0$ and*

$$(F_\beta - F_\alpha)[x_1^2(p_1 x_1^{-1}Dx_1 - p_2 x_2^{-1}Dx_2)]$$
$$= I_\alpha^\beta[(q_2 - q_1)x_1^2 + (p_1 - p_2)(Dx_1)^2 +$$
$$+ p_2(Dx_1 - x_1 x_2^{-1}Dx_2)^2]. \qquad (6.5.36)$$

Proof. Indeed, if $f_D = 0$ then $De = 0$, hence $r_D(x_1, x_2) = 0$. If $c_D = 1$ then $q_D(x_1, x_2) = 0$. ∎

Observe that $f_D = 0$ and $c_D = 1$ imply that X is a Leibniz D-algebra and Formula (6.5.36) has a form of the classical Picone identity (cf. Example 6.5.2).

Recall that

$$D_\infty = \bigcap_{k=0}^\infty D_k, \quad \text{where } D_0 = X \text{ and } D_k = \text{dom } D^k \text{ for } k \in \mathbb{N}$$
$$(6.5.37)$$

(cf. Formula (3.4.22)).

DEFINITION 6.5.1. Let X be a D-algebra. Let F_1, F be initial operators for D corresponding to R_1, $R \in \mathcal{R}_D$, respectively, and such that $F_1 \neq F$. Write:

$$E_m = \{u \in D_m: F_1 D^j u = F D^j u = 0 \text{ for } j = 0, 1, ..., m\}$$
$$(m = 0, 1, ..., +\infty). \qquad (6.5.38)$$

A definite integral $F_1 R$ is said to be *total on* E_m if

$$F_1 R(xh) = 0 \text{ for all } h \in E_m \text{ implies } x = 0 \qquad (6.5.39)$$

(m is arbitrarily fixed).

PROPOSITION 6.5.3. *Suppose that all assumptions of Definition 6.5.1 are satisfied. Write*:

$$E'_m = \{v \in X: F_1 R^k v = 0 \text{ for } k = 0, 1, ..., m\}$$

$$(m = 0, 1, ..., +\infty). \qquad (6.5.40)$$

If $y \in E'_m$ *then* $h_m = R^{m+1}y \in E_m$ *for all* $m \in \mathbf{N} \cup \{0\}$. *If* $y \in E'_\infty$ *then* $h_m = R^{m+1}y \in E_\infty$ *for all* $m \in \mathbf{N} \cup \{0\}$.

Proof. Indeed, if $y \in E'_m$ then $D^{m+1}h_m = D^{m+1}R^{m+1}y = y$. Hence $h_m \in \mathrm{dom}\, D^{m+1}$. Let $j = 0, 1, ..., m$ be arbitrarily fixed. Then $FD^j h_m = FD^j R^{m+1}y = FR^{m+1-j}y = 0$ and $F_1 D^j h_m = F_1 D^j R^{m+1}y = F_1 R^{m+1-j}y = 0$ for $y \in E'_m$. This means that $h_m \in E_m$. Suppose now that $y \in E'_\infty$. Then $F_1 D^j h_m = FD^j h_m = 0$ for $j = 0, 1, ..., m$, where $m \in \mathbf{N} \cup \{0\}$ is arbitrary. This implies that $h_m \in E_m$ for all $m \in \mathbf{N} \cup \{0\}$. ∎

THEOREM 6.5.4. *Suppose that all assumptions of Definition 6.5.1 are satisfied. Let* F_1, F *be multiplicative. Then*

$$F_1 R(yh + xDy) = -F_1 R[(Dx - y)h] - c_D^{-1} F_1 R f_D(x, h) \qquad (6.5.41)$$

for all $x \in \mathrm{dom}\, D$, $y \in X$ *and* $h \in E_m$ ($m = 0, 1, ..., +\infty$).

Proof. Write: $w = R^{m+1}y$. Thus $w \in \mathrm{dom}\, D^{m+1}$ and $y = D^{m+1}w$. Our assumptions and Formula (6.2.3) of integration by parts for definite integrals together imply that

$$F_1 R(yh + xDh) = F_1 R(hD^{m+1}w + xDh)$$

$$= c_D^{-1} F_1 R\{c_D[hD^{m+1}w + (D^m w)(Dh)] + f_D(h, D^m w)\} +$$

$$+ F_1 R[xDh - (D^m w)Dh - c_D^{-1} f_D(h, D^m w)]$$

$$= c_D^{-1} F_1 RD(hD^m w) + F_1 R\{[(x - D^m w)Dh] - c_D^{-1} f_D(h, D^m w)\}$$

$$= c_D^{-1}(F_1 - F)(hD^m w) + c_D^{-1}\{(F_1 - F)[(x - D^m w)h] -$$

$$- F_1 R f_D(x - D^m w, h)\} - F_1 R[h(Dx - D^{m+1}w)] -$$

$$- c_D^{-1} F_1 R f_D(D^m w, h).$$

Since the operators F_1, F are multiplicative and $h \in E_m$, we conclude that $(F_1-F)(hD^m w) = [(F_1-F)h][(F_1-F_1)(D^m w)] = 0$ and, similarly, $(F_1-F)[(x-D^m w)h] = 0$. Thus

$$
\begin{aligned}
F_1 R(yh+xDh) &= -c_D^{-1}F_1 R[f_D(x-D^m w, h)+f_D(D^m w, h)]-\\
&\quad -F_1 R[h(Dx-D^{m+1}w)]\\
&= -c_D^{-1}F_1 Rf_D(x, h)-F_1 R[(Dx-y)h],
\end{aligned}
$$

which was to be proved.   ∎

Immediate consequences of this theorem are:

COROLLARY 6.5.13. *Suppose that all assumptions of Definition 6.5.1 are satisfied and F_1, F are multiplicative. Then $F_1 R(yh+xDh) = 0$ for all $x \in \text{dom } D$, $y \in X$, $h \in E_m$ $(m = 0, 1, \ldots, +\infty)$ if and only if*

$$F_1 R[(Dx-y)h]+c_D^{-1}F_1 Rf_D(x, h) = 0. \qquad (6.5.42)$$

COROLLARY 6.5.14. *Suppose that all assumptions of Definition 6.5.1 are satisfied, F_1, F are multiplicative and the integral $F_1 R$ is total on E_m. If for given $x \in \text{dom } D$, $y \in X$ and for all $h \in E_m$ $(m = 0, 1, \ldots, +\infty)$ we have $F_1 R(yh+xDh) = 0$ then $y-Dx = 0$ if and only if $F_1 Rf_D(x, h) = 0$ for all $h \in E_m$.*

Proof. By our assumptions and Corollary 6.5.13 we have

$$F_1 R[(Dx-y)h]+c_D^{-1}F_1 Rf_D(x, h) = 0. \qquad (6.5.43)$$

If $F_1 Rf_D(x, h) = 0$ then $F_1 R[(Dx-y)h] = 0$. Since $F_1 R$ is total on E_m, we conclude that $Dx-y = 0$. Conversely, if $Dx-y = 0$ then Formula (6.5.43) implies $F_1 Rf_D(x, h) = 0$.   ∎

Equation (6.5.42) is a *generalized Euler–Lagrange equation*, well-known in the variational calculus.

Examples and Exercises

EXAMPLE 6.5.1. Suppose that $X = C(\Omega)$, where

$$\Omega = \{(t, s): 0 \leqslant t \leqslant a, 0 \leqslant s \leqslant b\}, \quad D = \frac{\partial^2}{\partial t\, \partial s},$$

$$R_0 = \int_0^t\int_0^s, \quad (F_0 x)(t, s) = x(t, 0)+x(0, s)-x(0, 0).$$

The operator $D = D_2 D_1$, where $D_2 = \partial/\partial t$, $D_1 = \partial/\partial s$, is right invertible, F_0 is an initial operator for D corresponding to its right inverse R_0. Moreover, F_0 is an initial operator induced by the classical Darboux problem for the operator D. To have Green formulae for the operator

$$P(D) = \frac{\partial^2}{\partial t\, \partial s} + A\,\frac{\partial}{\partial t} + B\,\frac{\partial}{\partial s} + C, \quad A, B, C \in X \qquad (6.5.44)$$

we have to find some more initial operators and right inverses for D and to apply Corollary 6.5.10. Suppose then that we are given functions $h, g \in C^1[0, a]$ such that $g'(t) > 0$, $h'(t) > 0$ of $0 \leqslant t \leqslant a$, $g(0) = h(0) = 0$, $g(a) = h(a) = b$.

Consider the following operators defined for $x \in C^1(\Omega)$

$$(F_1 x)(t, s) = x\big(g^{-1}(s), s\big) + \int_{g^{-1}(s)}^{t} x_t'(p, g(p))\,\mathrm{d}p,$$

$$(F_2 x)(t, s) = x\big(g^{-1}(s), s\big) - x\big(g^{-1}(s), 0\big) + x(t, 0),$$

$$(F_3 x)(t, s) = x\big(g(s), s\big) + \int_{g^{-1}(s)}^{t} x_t'(p, h(p))\,\mathrm{d}p.$$

All these operators are initial operators for D because they are projection onto $\ker D$. The operator F_1 is induced by the Cauchy problem for D, the operator F_2 is induced by the Picard problem for D, the operator F_3 is induced by the generalized Cauchy problem for D (i.e. such a problem, where we are given values $x(t, g(t))$ and $x_t'(t, h(t))$ (cf. Section 4.2).

It is easy to verify that for all $x \in X$ we have

$$(F_1 R_0 x)(t, s) = -\int_0^{g^{-1}(s)} \left[\int_0^{s} x(p, q)\,\mathrm{d}q\right]\mathrm{d}p -$$

$$- \int_{g^{-1}(s)}^{t} \left[\int_0^{g(p)} x(p, q)\,\mathrm{d}q\right]\mathrm{d}p,$$

$$(F_2 R_0 x)(t, s) = \int_0^{g^{-1}(s)} \left[\int_0^{s} x(p, q)\,\mathrm{d}q\right]\mathrm{d}p,$$

$$(F_3 R_0 x)(t, s) = \int_{g^{-1}(s)}^{0} \left[\int_{0}^{s} x(p, q) dq \right] dp -$$

$$- \int_{g^{-1}(s)}^{0} \left[\int_{0}^{h(p)} x(p, q) dq \right] dp.$$

Using these last expressions we can derive three different Green formulae for the operator $P(D)$ defined by Formula (6.5.44).

EXAMPLE 6.5.2. Suppose that x_1, x_2 are solutions of the following linear ordinary differential equations:

$$\frac{d}{dt} \left(p_i \frac{dx_i}{dt} \right) + q_i x_i = 0, \quad p_i \in C^1[a, b], \quad q_i \in C[a, b]$$

$$(i = 1, 2).$$

The following formula is well-known as the *Picone identity*:

$$\left[x_1^2 \left(p_2 \frac{1}{x_2} \frac{dx_2}{dt} - p_1 \frac{1}{x_1} \frac{dx_1}{dt} \right) \right]_a^b$$

$$= - \int_a^b (p_1 - p_2) \left(\frac{dx_1}{dt} \right)^2 dt + \int_a^b (q_1 - q_2) x_1^2 dt -$$

$$- \int_a^b p_2 \frac{1}{x_2^2} \left(x_2 \frac{dx_1}{dt} - x_1 \frac{dx_2}{dt} \right)^2 dt.$$

This formula follows immediately from Formula (6.5.36) if we admit

$$D = \frac{d}{dt}, \quad R_1 = \int_b^t, \quad (Fx)(t) = x(a), \quad (F_1 x)(t) = x(b)$$

for $x \in X = C[a, b]$.

EXAMPLE 6.5.3. Suppose that $X = C[a, b]$, $D = d/dt$, $R = \int_a^t$, $R_1 = \int_b^t$.
In this case $(Fx)(t) = x(a)$, $(F_1 x)(t) = x(b)$ for $x \in X$ and for $m = 0, 1, ..., +\infty$

$$E_m = \{u \in C^m[a, b] \colon u^{(j)}(a) = u^{(j)}(b) = 0 \ (j = 0, 1, ..., m)\}.$$

Corollary 6.5.14 in this case is the well-known, so-called *fundamental lemma of the variational calculus*. Indeed, we have $0 = F_1 R[(Dx-y)h](t)$

$$= \int_a^b [y(t)h(t)+x(t)h'(t)]dt \text{ and } f_D = 0, c_D = 1, \text{ which implies } y-x' = 0.$$

As follows from Example 6.1.5 for the Laplace operator $\varDelta$ we have

$$f_D(x,y) = 2 \sum_{j=1}^{n} (\partial x/\partial t_j)(\partial y/\partial t_j).$$ Observe that initial operators induced

by Dirichlet and Neumann boundary conditions (cf. Section 4.3) are multiplicative. Thus Corollary 6.5.14 can be also applied in this case.

EXERCISE 6.5.1. Find a form of the Green formula if X and D are defined as in Examples 6.1.2a, 6.1.3a, 6.1.3b, 6.1.4, 6.1.5, respectively.

EXERCISE 6.5.2. Find a form of the Picone identity if X and D are defined as in Examples 6.1.2a, 6.1.3a, 6.1.3b, 6.1.4, 6.1.5.

EXERCISE 6.5.3. Find a form of the Euler–Lagrange equations if X and D are defined as in Examples 6.1.2a, 6.1.3a, 6.1.3b, 6.1.4, 6.1.5. Formulate Corollary 6.5.14 in all these cases.

EXERCISE 6.5.4. Let $X = (s)$ (over $\mathbf{R}$ or $\mathbf{C}$). Let, as before, $D\{x_n\} = \{x_{n+1}-x_n\}$ for $\{x_n\} \in X$. Write: $e = \{e_n\}$ with $e_n = 1$ for $n \in \mathbf{N}$. Recall that R_1 is defined as follows: $R_1\{x_n\} = \{y_n\}$, where $y_1 = 0$,

$$y_n = \sum_{k=0}^{n-1} x_k \text{ for } n > 0,$$ is a right inverse of D. Prove that:

(i) The operators F_m defined by the equality: $F_m\{x_n\} = x_m e$ are initial operators for D corresponding to right inverses $R_m = R_1 - F_m R_1$ and $R_m\{x_n\} = \{y_{n,m}\}$, where

$$y_{n,m} = \begin{cases} -\sum_{k=1}^{m-1} x_k & \text{for } n = 1, \\ \\ \sum_{k=1}^{n-1} x_k - \sum_{k=1}^{m-1} x_k & \text{for } n \geqslant 2 \end{cases} \quad (m = 2, 3, ...);$$

(ii) The Green formula for the operator D is of the form:

$$F_m R_1(xDy+yDx) = (F_m-F_1)(xy) - F_m R_1(Dx)(Dy)$$

for $x, y \in X, m \geqslant 2$;

(iii) $F_m R_1 \{u_n\} = e \sum\limits_{k=1}^{m-1} u_k$ for $\{u_n\} \in X$;

(iv) For all $\{x_n\}, \{y_n\} \in X$ the Green formula implies the equalities:

$$\sum_{k=1}^{m-1} x_k(y_{k+1}-y_k)+y_k(x_{k+1}-x_k)$$

$$= x_m y_m - x_1 y_1 - \sum_{k=1}^{m-1} (x_{k+1}-x_k)(y_{k+1}-y_k) \quad (m \geqslant 2);$$

(v) $\tilde{F}_j = \sum\limits_{m=1}^{j} d_m F_m$ with $\sum\limits_{m=1}^{j} d_m = 1, d_m \in \mathbf{R}$ are initial operators

for D and $\tilde{F}_j \{x_n\} = e \sum\limits_{m=1}^{j} d_m x_m$, $\tilde{F}_j R_1 \{u_n\} = e \sum\limits_{m=2}^{j} d_m \sum\limits_{k=1}^{m-1} u_k$ for $j \geqslant 2$;

(vi) The Green formula with the operators $\tilde{F}_j$ yields to the equalities:

$$\sum_{m=2}^{j} d_m \sum_{k=1}^{m-1} x_k(y_{k+1}-y_k)+y_k(x_{k+1}-x_k)$$

$$= \left(\sum_{m=2}^{j} d_m x_m\right)\left(\sum_{m=2}^{j} d_m y_m\right) - x_1 y_1 -$$

$$- \sum_{m=2}^{j} d_m \sum_{k=1}^{m-1} (x_{k+1}-x_k)(y_{k+1}-y_k) \quad (j \geqslant 2).$$

Chapter 7

Perturbations and nonlinear problems

7.1. FINITE DIMENSIONAL PERTURBATIONS

This section contains results obtained by H. Nowak (1982, 1984).

Let $D \in R(X)$. A finite-dimensional perturbation $K \in L_0(X)$ is said to be *D-admissible* if $KX \subset \operatorname{dom} D$ (cf. Section 1.3). Recall that every finite-dimensional operator $K \in L_0(X)$ is of the form:

$$Kx = \sum_{j=1}^{n} f_j(x) x_j \quad \text{for all } x \in X, \tag{7.1.1}$$

where $f_1, \dots, f_n \in X'$ and $x_1, \dots, x_n \in X$ are linearly independent systems of linear functionals over X and elements of X, respectively (cf. Proposition 1.3.1, Formula (1.3.24)). We will show that finite-dimensional D-admissible perturbations preserve properties of solutions of problems under consideration.

THEOREM 7.1.1. *Suppose that* $D \in R(X)$, $\dim \ker D \neq 0$, $R \in \mathscr{R}_D$ *and* $K \in L_0(X)$ *is a D-admissible finite-dimensional perturbation. Write*:

$$\Delta_K(R) = \det \left(f_j(Rx_i) + \delta_{ij} \right)_{i,j=1,\dots,n}, \tag{7.1.2}$$

where $n = \dim K$ *and* δ_{ij} *denotes the Kronecker symbol, i.e.* $\delta_{ii} = 1$ *and* $\delta_{ij} = 0$ *for* $i \neq j$. *Consider the perturbed equation*:

$$(D+K)x = y, \quad y \in X. \tag{7.1.3}$$

(i) *If* $\Delta_K(R) \neq 0$ *then all solutions of Equation (7.1.3) are of the form*:

$$x = Ry - \sum_{j=1}^{n} a_j Rx_j + z, \tag{7.1.4}$$

*where $z \in \ker D$ is arbitrary and the coefficients $a_j = f_j(x)$ for $j = 1, 2, \ldots$
$\ldots, n$ are uniquely determined by the Cramer system*

$$a_j + \sum_{i=1}^{n} a_i f_j(Rx_i) = f_j(Ry+z) \quad (j = 1, 2, \ldots, n); \qquad (7.1.5)$$

(ii) *If $\Delta_K(R) = 0$ but the rank $\left(f_j(Rx_i) + \delta_{ij}\right)_{i,j=1,\ldots,n} = m < n$,
$X_m = \lin \{\varphi_j = f_j(Ry+z): z \in \ker D, y \in X, j = 1, \ldots, n\}$ and $\dim X_m
= m$ then all solutions of Equation (7.1.3) are of the form (7.1.4),
where $z \in \ker D$ is arbitrary, $f_j(\ker D) \subset X_m$ for $j = 1, \ldots, m$ and*

$$a_i = f_i(x) = \begin{cases} f_i(x, \alpha_0, \ldots, \alpha_{n-m-1}) & \text{for } i = 1, \ldots, m, \\ \alpha_{i-m-1} & \text{for } i = m+1, \ldots, n, \end{cases}$$
$$\qquad (7.1.6)$$

where $\alpha_0, \ldots, \alpha_{n-m-1}$ are parameters.

Proof. Since, by our assumptions, $DR = I$ and K is of the form (7.1.1)
we can rewrite Equation (7.1.3) in the following way:

$$D\left[x + \sum_{j=1}^{n} f_j(x) Rx_j - Ry\right] = 0$$

which implies

$$x + \sum_{j=1}^{n} f_j(x) Rx_j - Ry = z \in \ker D$$

or

$$x + \sum_{j=1}^{n} f_j(x) Rx_j = Ry+z, \quad \text{where } z \in \ker D \text{ is arbitrary.}$$
$$\qquad (7.1.7)$$

Acting on both sides of Equation (7.1.7) by the functionals $f_1, \ldots, f_n$
we obtain a system

$$f_j(x) + \sum_{i=1}^{n} f_i(x) f_j(Rx_i) = f_j(Ry+z) \quad (j = 1, \ldots, n) \qquad (7.1.8)$$

of algebraic linear equations with n unknowns $a_j = f_j(x)$ $(j = 1, ...,n)$, i.e. the system

$$a_j + \sum_{i=1}^{n} a_i f_j(Rx_i) = f_j(Ry+z) \quad (j = 1, ..., n). \tag{7.1.9}$$

(i) If $\Delta_K(R) \neq 0$ the system (7.1.8) (or (7.1.9)) is a Cramer system. Therefore the coefficients $a_j = f_j(x)$ $(j = 1, ..., n)$ are uniquely determined by Cramer Formulae and all solutions of Equation (7.1.3), i.e. all solutions of the system (7.1.7) are of the form (7.1.5).

(ii) If $\Delta_K(R) = 0$ then not for all possible vectors

$$f(Ry+z) = \{f_1(Ry+z), ..., f_n(Ry+z)\},$$

$$z \in \ker D \text{ is arbitrary}, \tag{7.1.10}$$

there exists a solution of the system (7.1.8) (i.e. the system (7.1.9)). However, if rank $(f_j(Rx_i) + \delta_{ij}) = m < n$ then a necessary and sufficient condition for this system to have a solution is that the vector $f(Ry+z)$ defined by Formula (7.1.10) belongs to an m-dimensional subspace X_m of X. In this case the system (7.1.8), hence also Equation (7.1.3), has infinitely many solutions of the form (7.1.4) with the coefficients a_j dependent of $n-m$ parameters $\alpha_0, ..., \alpha_{n-m-1}$ defined by Formulae (7.1.6). ∎

Immediate consequences of this theorem are

COROLLARY 7.1.1. *Suppose that all assumptions of Theorem 7.1.1 are satisfied and F is an initial operator for D corresponding to R. Consider Equation (7.1.3) together with the initial condition*

$$Fx = x_0, \quad x_0 \in \ker D. \tag{7.1.10'}$$

(i) *If $\Delta_K(R) \neq 0$ then the perturbed initial value problem (7.1.3)–(7.1.10) is well-posed and its unique solution is*

$$x = Ry - \sum_{j=1}^{n} a_j Rx_j + y_0, \tag{7.1.11}$$

where $a_1, ..., a_n$ is a uniquely determined solution of the Cramer system (7.1.9).

(ii) *If $\Delta_K(R) = 0$ the problem (7.1.3)–(7.1.10) is ill-posed. However if* rank $\left(f_j(Rx_j) + \delta_{ij}\right)_{i, j = 1, 2, \ldots, n} = m < n$ *then a necessary and sufficient condition for this problem to have a solution is that the vector* $\{f_1(Ry + y_0), \ldots, f_n(Ry + y_0) = f(Ry + y_0)\}$ *belongs to an m-dimensional subspace X_m of X. In this case the problem (7.1.3)–(7.1.10) has solutions of the form (7.1.11) which depend on $n - m$ parameters (cf. Formulae (7.1.8))*

COROLLARY 7.1.2. *Suppose that all assumptions of Theorem 7.1.1 are satisfied. If $\Delta_K \neq 0$ then K is a nullity and deficiency preserving perturbation, i.e.*

$$\alpha_{D+K} = \dim \ker(D+K) = \dim \ker D = \alpha_D,$$

$$\beta_{D+K} = \operatorname{codim} X/(D+K)\operatorname{dom} D = X/D\operatorname{dom} D = \beta_D = 0.$$

If $\Delta_K = 0$ but rank $\left(f_j(Rx_i) + \delta_{ij}\right)_{i, j = 1, \ldots, n} = m < n = \dim K$ *then*

$$\alpha_{D+K} = \dim \ker(D+K) = \dim \ker D + n - m = \alpha_D + n + m,$$

$$\beta_{D+K} = \operatorname{codim} X/(D+K)\operatorname{dom} D = \operatorname{codim} X/(D\operatorname{dom} D)$$
$$= \beta_D + n - m.$$

It means that in both cases K is an index preserving perturbation (cf. Section 1.2).

THEOREM 7.1.2. *Suppose that $D \in R(X)$, F is an initial operator for D corresponding to an $R \in \mathcal{R}_D$, S_h is an R-shift on $h \neq 0$ and K is a finite-dimensional D-admissible perturbation with S_h-periodic $x_1, \ldots, x_n$ (cf. Formula (7.1.1)). Consider the equation*

$$(D+K)x = y \quad \text{where } y \in X_{S_h}. \tag{7.1.12}$$

Let $\Delta_K(R_1^0)$ be defined by Formula (7.1.2), where

$$R_1^0 = \left(I + FS_h - \frac{1}{h} RFS_h\right) R \in \mathcal{R}_{D_1}, \quad D_1 = D + \frac{1}{h} FS_h$$

(cf. Formulae (5.2.20), (5.2.13)).

(i) *If $\Delta_K(R_1^0) \neq 0$ then all S_h-periodic solutions of Equation (7.1.12) are of the form*

$$x + R_1^0 y - \sum_{j=1}^{n} a_j R_1^0 x_j + z, \quad \text{where } z \in \ker D \text{ is arbitrary} \tag{7.1.13}$$

and the coefficients $a_j = f_j(x)$ $(j = 1, ..., n)$ are uniquely determined by the Cramer system

$$a_j + \sum_{i=1}^{n} a_i f_j(R_1^0 x_i) = f_j(R_1^0 + z) \quad (j = 1, 2, ..., n). \qquad (7.1.14)$$

If, moreover, we are given an initial condition (7.1.10) then the initial value problem (7.1.12)–(7.1.10) has a unique S_h-periodic solution

$$x = R_1^0 y - \sum_{j=1}^{n} a_j R_1^0 x_j - y_0;$$

(ii) *If $\Delta_K(R_1^0) = 0$ and rank $\big(f_j(R_1^0 y) + \delta_{ij}\big)_{i, j=1, 2, ..., n} = m < n = \dim K$ then a necessary and sufficient condition for Equation (7.1.12) to have a solution is that the vector $f(R_1^0 y + z) = \{f_1(R_1^0 y + z), ..., f_n(R_1^0 y + z)\}$ belongs to an m-dimensional subspace X_m^h of the space X_{S_h}. In this case Equation (7.1.12) has S_h-periodic solution of the form (7.1.14) dependent on $n - m$ parameters (cf. Formula (7.1.8)).*

The proof follows from Theorem 5.3.1 for the equation $Dx = y$, $y \in X_{S_h}$ and is going on similar lines as the proof of Theorem 7.1.1. ∎

Examples and Exercises

EXAMPLE 7.1.1. Consider a homogeneous *trace spline* with the one fixed end described by the equation

$$(px'')'' + \lambda^2 x'' + q \int_0^l x''(s) \, ds = 0 \quad \text{in } X = C[0, l], \qquad (7.1.15)$$

where l, p, q, x are given positive numbers, with the mixed boundary conditions:

$$x(0) = x_0^0, \quad x'(0) = x_1^0, \quad x''(l) = x_2^0, \quad x'''(0) = x_3^0. \qquad (7.1.16)$$

Write $D = \mathrm{d}/\mathrm{d}t$, $(Fx)(t) = x(0)$, $(F_1 x)(t) = x(l)$, $R = \int_0^t$, $R_1 = \int_l^t$,

$(Kx)(t) = \dfrac{q}{p} \int_0^l x(s) \, ds$ for $x \in X$. By definition, K is a one-dimensional operator with $f_1(x) = \int_0^l$ and $x_1 = q/p$. Write $v = D^2 x$. Then Equation (7.1.15) together with Conditions (7.1.16) can be rewritten as follows

$$(D^2 + \lambda^2 p^{-1} I + K)v = 0, \quad F_1 v = x_2, \quad FDv = x_3. \quad (7.1.17)$$

But the operator $D_0 = D^2 + \lambda^2 p^{-1} I$ is again right invertible for $D_0 = D^2(I + \lambda^2 p^{-1} R_1^2)$ and $R_1 \in V(X)$ and has a right inverse $R_0 = (I + \lambda^2 p^{-1} R_1^2)^{-1} R^2$. Since $D^2 + \lambda^2 p^{-1} I + K = D^2(I + \lambda^2 p^{-1} R_1^2 + R_1^2 K) = D_0 + R_1^2 K$ and the operator $K_0 = R_1^2 K$ is a finite-dimensional operator, as a superposition with a finite-dimensional operator, we write the system (7.1.17) in the form

$$(D_0 + K_0)v = 0, \quad F_0 v = v_0,$$
$$\text{where } F_0 v = F_1 v + R_1 FDv = x_2 + R_1 x_3, \quad (7.1.18)$$

hence $v_0 = x_2^0 + x_3^0(t-l)$.

The operator K_0 is of the form

$$(K_0 x)(t) = (R_1^2 Kx)(t) = \int_l^t \left[(u-l) \frac{q}{p} \int_0^l x(s)\, ds \right] du$$

$$= \frac{q}{p} \frac{(l-t)^2}{2} \int_0^l x(s)\, ds,$$

i.e. is also a one-dimensional operator with $f_1 = \dfrac{q}{p} \int_0^l$ and $x_1 = (l-t)^2/2$.

We may apply now Corollary 7.1.1. In our case

$$\Delta_K = \Delta_K(R_0) = f_1(R_0 x_1 + v_0) - 1$$

$$= \frac{q}{p} \int_0^l \left[R_0 \frac{(l-s)^2}{2} + x_2^0 + x_3^0(s-l) \right] ds + 1$$

$$= 1 + \frac{q}{p} \left\{ \int_0^l \left[R_0 \frac{(l-s)^2}{2} \right] ds + x_2^0 l + x_3^0 \frac{(t-l)^2}{2} \right\} \neq 0$$

since $\int_0^l \left[R_0 \dfrac{(l-s)^2}{2} \right] ds$ consists of a polynomial of degree $\geqslant 4$. Hence the problem (7.1.18) has a unique solution

$$v = -a_1 R_0 x_1 + v_0, \quad \text{where } a_1 = \Delta_K^{-1} f_1(v_0),$$

i.e.

$$v = x_2^0 + x_3^0 l - \Delta_K^{-1} \frac{q}{p} \left[x_2^0 l + x_3^0 \frac{(t-l)^2}{2} \right] + R_0 \frac{(l-t)^2}{2}.$$

But $D^2 x = v$ and $Fx = x_0^0$, $FDx = x_1^0$. Hence the problem (7.1.15)–(7.1.16) has a unique solution

$$x = R^2 v + R x_1^0 + x_0^0 = R^2 v + x_1^0 t + x_0^0.$$

If $x_0^0 = x_1^0 = x_2^0 = x_3^0 = 0$, i.e. the problem in question has homogeneous boundary conditions, then has a unique solution of the form

$$x = R^2 R_0 \frac{(l-t)^2}{2} = R^2 (I + \lambda^2 p^{-1} R_1^2)^{-1} R^2 \frac{(l-t)^2}{2}$$

$$= (I + \lambda^2 p^{-1} R^2)^{-1} R^4 \frac{(l-t)^2}{2}.$$

The operator $(I + \lambda^2 p^{-1} R_1^2)^{-1} = (I + \lambda \sqrt{p^{-1}} R_1)^{-1} (I - \lambda \sqrt{p^{-1}} R_1)^{-1}$ can be calculated according with Formula (2.2.41). Also we have $R^4 \dfrac{(l-t)^2}{2}$

$$= \int_0^t \frac{(t-s)^3}{3!} \cdot \frac{(l-s)^2}{2} ds.$$

EXERCISE 7.1.1. Suppose that all assumptions of Theorem 7.1.1 are satisfied. Write

$$Q(D) = \sum_{k=0}^{N} Q_k D^k, \quad Q_0, ..., Q_{N-1} \in L_0(X), \quad Q_N = I.$$

Solve the following perturbed equations:
(i) $Q(D+K)x = y$,
(ii) $(D+K)^M Q(D+K)x = y$,
(iii) $(D+K)^M Q(D+K)x$, $\quad (M \geqslant 0)$,
(iv) $Q(D+K)(D+K)^M v = y$,

using results of Section 3.2. We recall that a superposition of any operator with a finite-dimensional operator (if exists) is again a finite-dimensional operator.

EXERCISE 7.1.2. Suppose that all assumptions of Theorem 7.1.2 are satisfied and $Q(D) = \sum\limits_{k=0}^{N} Q_k D^k$, where $Q_k \in L_0(X)$ $(k = 0, 1, ..., N-1)$ are S_h-periodic and $Q_N = I$. Find S_h-periodic solutions of the perturbed equation $Q(D+K)x = y, y \in S_h$, provided that the operator $Q(I, R_1^0) = \sum\limits_{k=0}^{N} Q_k (R_1^0)^{N-k}$ is invertible in the space X_{S_h}.

EXERCISE 7.1.3. Suppose that all assumptions of Theorem 7.1.2 are satisfied, X is a Leibniz D-algebra and $R \in V(X)$. Find all exponential-periodic solutions of Equation (7.1.12), provided th at y is an exponentia periodic element.

7.2. PERTURBATIONS BY MEANS O F RIGHT INVERSES

We shall consider in this section equations with right invertible operators perturbed by polynomials in right inverses and we shall find solutions of such problems in a closed form. Note that perturbations ol a general type has been considered by Przeworska-Rolewicz (1976b, cf. also 1973b), also by H. Lausch and the author, 1986.

   All results of this section have been communicated to Przeworska-Rolewicz by H. von Trotha (1975).

PROPOSITION 7.2.1. *Suppose that* $D_1, D_2 \in R(X)$, $R \in \mathcal{R}_{D_1} \cap \mathcal{R}_{D_2}$, $\mathrm{dom}\, D_1 = \mathrm{dom}\, D_2 = D_{12} \subset X$. *Denote by* F_1, F_2 *initial operators for* D_1 *and* D_2, *respectively, corresponding to* R. *Then*

$$F_i = I - RD_i \quad on\ D_{12} \quad (i = 1, 2), \tag{7.2.1}$$

$$D_1 = D_1 F_2 + D_2 \quad on\ D_{12}, \tag{7.2.2}$$

$$I = R(D_1 - D_1 F_2) + F_2 \quad on\ D_{12}, \tag{7.2.3}$$

$$F_i F_j = F_i \quad (i, j = 1, 2) \quad on\ D_{12}, \tag{7.2.4}$$

$$F_i F_j = I \quad on\ \ker D_j \quad (i = 1, 2), \tag{7.2.5}$$

which implies that F_i is a one-to-one mapping of ker D_j *onto* ker D_i. *Moreover, x is a solution of an initial value problem*

$$D_1 x = y, \quad F_2 x = z_2, \quad \text{where } y \in X, \ z_2 \in \ker D_2 \qquad (7.2.6)$$

if and only if

$$x = R(y - D_1 z_2) + z_2. \qquad (7.2.7)$$

Proof. Observe that, by definition, we have $D_1 R = D_2 R = I$ and $D_1 F_2 + D_2 = D_1(I - RD_2) + D_2 = D_1 - D_1(RD_2) + D_2 = D_1 - (RD_2) + D_2 = D_1 - (D_1 R)D_2 + D_2 = D_1 - D_2 + D_2 = D_1$ on $D_{12} = \text{dom } D_1 = \text{dom } D_2$, which proves Formula (7.2.2). Thus on D_{12} we find $R(D_1 - D_1 F_2) + F_2 = R(D_1 F_2 + D_2 - D_1 F_2) + F_2 = RD_2 + I - RD_2 = I$, which proves (7.2.3). For $i, j = 1, 2$ we have $F_i F_j = (I - RD_i)(I - RD_j) = I - RD_i - RD_j + (RD_i)(RD_j) = I - RD_i - RD_j + R(D_i R)D_j = I - RD_i - RD_j + RD_j = I - RD_i = F_i$ on D_{12}, i.e. Formula (7.2.4) holds. Suppose now that $z_i \in \ker D_i$ $(i = 1, 2)$ are arbitrary. Then for $i = 1, 2$ we find $D_i z_i = 0$ and, by Formula (7.2.4), $F_i F_j z_i = F_i z_i = (I - RD_i)z_i = z_i - RD_i z_i = z_i$. The arbitrariness of z_i implies that $F_i F_j = I$ on $\ker D_i$ $(i, j = 1, 2)$. Formula (7.2.3) implies that every solution $x \in D_{12}$ of the problem (7.2.6) is of the form

$$x = R(D_1 - D_2 F_2) + F_2 x = R(D_1 x - D_1 z_2) + z_2$$
$$= R(y - D_1 z_2) + z_2.$$

On the other hand, the first equation of (7.2.6) has general solution of the form $x = Ry + z_1$, where $z_1 = F_1 x \in \ker D_1$. By Formula (7.2.4), since $F_i z_i = z_i$ $(i = 1, 2)$, we find $z_1 = F_1 x = F_1 F_2 x = F_1 z_2 = F_1 F_2 z_2 = z_2$. Hence $F_1 z_2 = z_2$ and $x = Ry + z_1 = Ry + z_2 = Ry + F_1 z_2 = Ry + z_2 - RD_1 z_2 = R(y - D_1 z_2) + z_2$. ∎

PROPOSITION 7.2.2. *Suppose that* $D_1 \in L_0(X)$, $D_2 \in R(X)$, $R_2 \in \mathscr{R}_{D_2}$, $\text{dom } D_1 = \text{dom } D_2 = D_{12}$ *and* $D_1 R_2 = W$, *where* $W \in L_0(X)$ *is invertible on X. Then* $D_1 \in R(X)$ *and there is a one-to-one correspondence between* $\ker D_1$ *and* $\ker D_2$. *If* F_2 *is an initial operator for* D_2 *corresponding to* R_2 *then x is a solution of the initial value problem*

$$D_1 x = y, \quad F_2 x = z_2, \quad \text{where } y \in X, \ z_2 \in \ker D_2 \qquad (7.2.8)$$

if and only if

$$x = R_2 W^{-1}(y - D_1 z_2) + z_2. \tag{7.2.9}$$

Proof. Write $R = R_2 W^{-1}$ and $D = WD_2$. Then $\operatorname{dom} D = \operatorname{dom} D_2 = D_{12}$ and $D_1 R = D_1(R_2 W^{-1}) = D_1 R_2 W^{-1} = WW^{-1} = I$, $DR = W(D_2) \times \times (R_2 W^{-1}) = W(D_2 R_2)W^{-1} = WW^{-1} = I$, which implies $R \in \mathcal{R}_{D_1} \cap \mathcal{R}_D$. Let F be an initial operator for D corresponding to R. Then on $\operatorname{dom} D = D_{12}$ we have $F = I - RD = I - (R_2 W^{-1})(WD_2) = I - R_2 W^{-1} WD_2 = I - R_2 D_2 = F_2$. If $z_2 \in \ker D_2$ then $z_2 = F_2 z_2 = F z_2$ and $D_2 = DF z_2 = 0$. Hence $z_2 \in \ker D$ which implies $\ker D_2 \subset \ker D$. A similar proof for the inclusion $\ker D \subset \ker D_2$. Hence $\ker D = \ker D_2$. Let F_1 be an initial operator for D_1 corresponding to $R_1 = R_2 W^{-1} \in \mathcal{R}_{D_1}$. Then for an arbitrary $x \in X$ we have $F(F_1 x) = F(I - R_1 D_1)x = F(I - R_2 W^{-1}D_1)x = Fx - F_2 R_2 W^{-1}D_1 x = Fx$. Let $z_1 = F_1 x$. Hence $FF_1 = F$.

Proposition 7.2.1 implies that $FF_1 = I$ on $\ker D = \ker D_2$. On the other hand, for all $z_1 \in \ker D_1$ we find $Fz_1 = FF_1 z_1 = F_1 z_1 - FR_1 D_1 z_1 = F_2 R_2 W^{-1}D_1 z_1 = F_1 z_1 = z_1$. Hence $z_1 = F_1 z_1 = F_1 F z_1$ and $F_1 F = I$ on $\ker D_1$. Since we have $FF_1 = I$ on $\ker D$ and $F_1 F = I$ on $\ker D_1$, we conclude that there is a one-to-one correspondence between elements of $\ker D_2 = \ker D$ and $\ker D_1$. Since $F = F_2$, Proposition 7.2.1 implies that x is a solution of the problem (7.2.8) if and only if $x = R(y - D_1 z_2) + z_2 = R_2 W^{-1}(y - D_1 z_2) + z_2$. ∎

THEOREM 7.2.1. *Suppose that we are given a family* $\{D_i\}_{i \in \mathbf{Z}} \subset R(X)$ *with* $\dim \ker D_i \neq 0$ *for all* $i \in \mathbf{Z}$ *(where* $\mathbf{Z}$ *denotes the set of all integers) and a family of their right inverses* $\{R_i\}_{i \in \mathbf{Z}}$, *i.e.* $D_i R_i = I$ *for* $i \in \mathbf{Z}$. *Write for* $N \geqslant 0$, $M \geqslant 1$

$$Q(D) = \sum_{i=0}^{N} Q_i D_i \dots D_1, \quad \text{where we put } D_i \dots D_1 = I \text{ if } i = 0,$$

$$\tag{7.2.10}$$

$$Q^0(R) = \sum_{i=0}^{N-1} Q_i R_{i+1} \dots R_N,$$

$$T_{i+1}(R) = \sum_{k=i+1}^{M-1} T_k R_{-k} \dots R_{-(i+1)}, \quad T_M(R) = 0,$$

$$T(R) = \sum_{i=0}^{M-1} T_i R_{-i} \dots R_0, \tag{7.2.11}$$

where $Q_i, T_j \in L_0(X)$ $(i = 0, 1, \dots, N, j = 0, 1, \dots, M-1)$, $Q_N \neq 0$, $T_{M-1}(R) \neq 0$. Define the operator W by means of the equalities

$$W = \begin{cases} Q_N + Q^0(R) + T(R) R_1 \dots R_N & \text{if } N \geqslant 1, \\ Q_0 + T(R) & \text{if } N = 0, Q_0 \neq 0, \\ T_i + T_{i+1}(R) \ (i = 0, 1, \dots, M-1) \\ \quad \text{if } N = 0, Q_0 = 0, T_0 = \dots = T_{i-1} = 0, T_i \neq 0. \end{cases} \tag{7.2.12}$$

If the operator W is invertible then
(i) For $N \geqslant 1$ all solutions of the equation

$$[Q(D) + T(R)]x = y, \quad y \in X \tag{7.2.13}$$

are of the form

$$x = R_1 \dots R_N [Q_N + Q^0(R) + T(R) R_1 \dots R_N]^{-1} y^0 + z_1 +$$

$$+ \sum_{i=1}^{N-1} R_1 \dots R_i z_{i+1}, \tag{7.2.14}$$

where

$$y^0 = y - \left\{ Q_0 z_1 + \sum_{i=0}^{N-2} Q_i \sum_{k=i+1}^{N-1} R_{i+1} \dots R_k z_{k+1} + Q_{N-1} z_N + \right.$$

$$\left. + \sum_{i=0}^{M-1} T_i \sum_{k=0}^{N-1} R_{-i} \dots R_0 \dots R_k z_{k+1} \right\},$$

$$\text{where } z_1, \dots, z_n \in \ker D \text{ are arbitrary}; \tag{7.2.15}$$

(ii) For $N = 0, Q_0 \neq 0$ Equation (7.2.13) has a unique solution

$$x = [Q_0 + T(R)]^{-1} y; \tag{7.2.16}$$

(iii) For $N = 0, Q_0 = 0$ Equation (7.2.13) has a unique solution

$$x = D_0 \dots D_{-i} [T_i + T_{i+1}(R)]^{-1} y, \tag{7.2.17}$$

provided that

$$T_0 = \ldots = T_{i-1} = 0, \quad T_i \neq 0$$
$$\text{and } [T_i + T_{i+1}(R)]^{-1} y \in R_{-i} \ldots R_0 X. \qquad (7.2.18)$$

Proof. Write

$$D_1^0 = Q(D) + T(R), \quad D_2^0 = D_N \ldots D_1, \quad R_2^0 = R_1 \ldots R_N. \qquad (7.2.19)$$

By our assumptions $D_2^0 R_2^0 = I$ on X. Observe that

$$\text{dom } D_1^0 = \text{dom } D_2^0. \qquad (7.2.20)$$

Indeed, $\text{dom} Q_i = X$ $(i = 0, 1, \ldots, N)$, $\text{dom} Q_i D_i \ldots D_1 = \text{dom} D_i \ldots$
$\ldots D_1 \subset \text{dom} D_1$ $(i = 1, \ldots, N)$ and $\text{dom} T(R) = X$, which implies
$\text{dom} D_1^0 = \bigcap\limits_{i=0}^{N} \text{dom} Q_i D_i \ldots D_1 = \text{dom} D_N \ldots D_1 = \text{dom} D_2^0 = D_{12}$. Observe also that

$$D_1^0 R_2^0 = W, \qquad (7.2.21)$$

where the operator $W \in L_0(X)$ defined by Formulae (7.2.12) is invertible.
Indeed, by our assumptions

$$Q(D) R_1 \ldots R_N = \left(\sum_{i=0}^{N} Q_0 D_i \ldots D_1 \right) R_1 \ldots R_N$$

$$= Q_0 R_1 \ldots R_N + \sum_{i=1}^{N} Q_i (D_i \ldots D_1 R_1 \ldots R_i) R_{i+1} \ldots R_N$$

$$= Q_N (D_N \ldots D_1 R_1 \ldots R_N)$$

$$= Q_0 R_1 \ldots R_N + \sum_{i=1}^{N-1} Q_i R_{i+1} \ldots R_N + Q_N$$

$$= Q_N + \sum_{i=0}^{N-1} Q_i R_{i+1} \ldots R_N = Q_N + Q^0(R).$$

Thus

$$D_1^0 R_2^0 = Q(D) + T(R) R_1 \ldots R_N = Q(D) R_1 \ldots R_N + T(R) R_1 \ldots R_N$$
$$= Q_N + Q^0(R) + T(R) R_1 \ldots R_N = W,$$

where the operator W is invertible by our assumptions.

Let F_2^0, F_i $(i = 1, 2, ..., N)$ be initial operators for D_2^0, D_i, respectively, corresponding to their right inverses R_2^0, R_i. Theorem 2.2.8 implies that

$$F_2^0 = I - R_2^0 D_2^0 = I - R_1 \ldots R_N D_N \ldots D_1$$

$$= F_1 + \sum_{i=1}^{N-1} R_1 \ldots R_i F_{i+1} D_i \ldots D_1 \qquad (7.2.22)$$

on dom $D_2^0 = $ dom $D_N \ldots D_1$.

Hence, by similar arguments, as these used in the proof of Corollary 2.2.2, we conclude that

$$\ker D_2^0 = \ker D_1 \oplus R_1 \ker D_2 \oplus \ldots \oplus R_1 \ldots R_{N-1} \ker D_N, \qquad (7.2.23)$$

which implies

$$\dim \ker D_N \ldots D_1 = \dim \ker D_2^0 = \sum_{i=1}^{N} \dim \ker D_i. \qquad (7.2.24)$$

Observe now that

$$T(R) F_2^0 = \sum_{i=0}^{M-1} T_i \sum_{k=0}^{N-1} R_{-i} \ldots R_k F_{k+1} D_k \ldots D_1. \qquad (7.2.25)$$

Indeed, Formulae (7.2.11) and (7.2.22) together imply that

$$T(R) F_2^0 = \sum_{i=0}^{M-1} T_i R_{-i} \ldots R_0 \left(F_1 + \sum_{k=1}^{N-1} R_1 \ldots R_k F_{k+1} D_k \ldots D_1 \right)$$

$$= \sum_{i=0}^{M-1} T_i \sum_{k=0}^{N-1} R_{-i} \ldots R_k F_{k+1} D_k \ldots D_1.$$

Formulae (7.2.10) and (7.2.22) together imply that

$$Q(D) F_2^0 = Q_0 F_1 + \sum_{i=1}^{N-1} Q_i \sum_{k=i+1}^{N-2} R_{i+1} \ldots R_k F_{k+1} D_k \ldots D_1 +$$

$$+ Q_{N-1} F_N D_N \ldots D_1. \qquad (7.2.26)$$

Indeed, since $D_k F_k = 0$ $(k = 1, 2, \ldots N)$ and $D_i \ldots D_1 = I$ if $i = 0$ we find

$$Q(D)F_2^0 = \left(Q_0 + \sum_{i=1}^{N-1} Q_i D_i \ldots D_1\right) \times$$

$$\times \left(F_1 + \sum_{k=1}^{N-1} R_1 \ldots R_k F_{k+1} D_k \ldots D_1\right)$$

$$= Q_0 F_1 + Q_0 \sum_{k=1}^{N-1} R_1 \ldots R_k F_{k+1} D_k \ldots D_1 +$$

$$+ \sum_{i=1}^{N-1} Q_i D_i \ldots D_1 F_1 +$$

$$+ \sum_{i=1}^{N-1} Q_i \sum_{k=1}^{N-1} D_i \ldots D_1 R_1 \ldots R_k F_{k+1} D_k \ldots D_1$$

$$= Q_0 F_1 + Q_0 \sum_{k=1}^{N-1} R_1 \ldots R_k F_{k+1} D_k \ldots D_1 +$$

$$+ \sum_{i=1}^{N-2} Q_i \left(\sum_{k=1}^{i} D_i \ldots D_{k+1} D_k \ldots D_1 +\right.$$

$$+ \left.\sum_{k=i+1}^{N-1} R_{i+1} \ldots R_k F_{k+1} D_k \ldots D_1\right) +$$

$$+ Q_{N-1} \left(\sum_{k=1}^{i} D_i \ldots D_{k+1} F_{k+1} D_k \ldots D_1 + F_N D_N \ldots D_1\right)$$

$$= Q_0 F_1 + Q_0 \sum_{k=1}^{N-1} R_1 \ldots R_k F_{k+1} D_{k+1} \ldots D_1 +$$

$$+ \sum_{i=1}^{N} Q_i \sum_{k=i+1}^{N-2} R_{i+1} \ldots R_k F_{k+1} D_k \ldots D_1 +$$

$$+ Q_{N-1} F_N D_N \ldots D_1$$

$$= Q_0 F_1 + \sum_{i=0}^{N} Q_i \sum_{k=i+1}^{N-2} R_{i+1} \ldots R_k F_{k+1} D_k \ldots D_1 +$$

$$+ Q_{N-1} F_N D_N \ldots D_1 .$$

Formulae (7.2.19)–(7.2.21) together imply that all assumptions of Proposition 7.2.2 are satisfied. We therefore conclude that all solutions of Equation (7.2.13)

$$D_1^0 x = [Q(D)x + T(R)]x = y, \quad y \in X$$

are of the form

$$x = R_2^0 W^{-1}(y - D_1^0 z_2^0 + z_2^0), \quad \text{where } z_2^0 \in \ker D_2^0 \text{ is arbitrary,} \tag{7.2.27}$$

since the operator W defined by (7.2.13) is invertible by our assumptions.

Formulae (7.2.22) and (7.2.26) together imply that $z_2^0 = z_1 +$

$$+ \sum_{i=1}^{N-1} R_1 \dots R_i z_{i+1}, \quad \text{where } z_i \in \ker D_i \text{ are arbitrary } (i = 1, \dots, N)$$

$$D_1^0 z_2^0 = Q_0 z_1 + \sum_{i=0}^{N-1} Q_i \sum_{k=i+1}^{N-2} R_{i+1} \dots R_k z_{k+1} + Q_{N-1} z_N +$$

$$+ \sum_{i=1}^{M-1} T_i \sum_{k=0}^{N-1} R_{-i} \dots R_k z_{k+1}. \tag{7.2.28}$$

We obtain the conclusion (ii) if we put $N = 0$, $Q_0 \neq 0$ and the conclusion (iii) if we put $N = 0$, $Q_0 = 0$ and we assume that Conditions (7.2.18) are satisfied. Observe that these conditions are necessary and sufficient conditions of solvability of Equation (7.2.13) in the case (iii). ∎

COROLLARY 7.2.1. *Suppose that all assumptions of Theorem 7.2.1 are satisfied. Then*

$$\dim \ker[Q(D) + T(R)] = \sum_{i=1}^{N} \dim \ker D_i. \tag{7.2.29}$$

Proof. Indeed, Proposition 7.2.2 implies that there is a one-to-one correspondence between the sets $\ker D_1^0$ and $\ker D_2^0$. Hence their dimensions are equal. This, and Formula (7.2.24), together imply Formula (7.2.29). ∎

COROLLARY 7.2.2. *Suppose that all assumptions of Theorem 7.2.1 are satisfied, $i = 1, \dots, N$, $D_1 = \dots = D_N = D$, $R_1 = R_2 = \dots = R_N = R$,*

$$Q_k = a_k I \ (k = 0, 1, \dots, N), \quad Q^0(R) = Q(I, R) \text{ and } T(R) = \sum_{k=0}^{M} b_k R^k,$$

where $a_0, \ldots, a_N, b_0, \ldots, b_M$ are scalars. Consider the equation

$$[Q(D) + T(R)]x = y, \quad y \in X. \tag{7.2.30}$$

(i) If $N \geqslant 1$ and the operator $Q(I, R) + R^N T(R)$ is invertible then all solutions of Equation (7.2.30) are of the form

$$x = [Q(I, R) + R^N T(R)]^{-1} \left[R^N y + \sum_{i=0}^{N-1} R^i z_i \right], \tag{7.2.31}$$

where $z_i \in \ker D$ are arbitrary;

(ii) If $N = 0$, $a_0 \neq 0$ and the operator $a_0 I = T(R)$ is invertible then Equation (7.2.30) has a unique solution

$$x = [a_0 I + T(R)]^{-1} y; \tag{7.2.32}$$

(iii) If $N = 0$, $b_0 = \ldots = b_{i-1} = 0$, $b_i \neq 0$ $(i = 1, \ldots, M-1)$ and the operator $D^i T(R)$ is invertible then Equation (7.2.30) has a unique solution:

$$x = D^i [D^i T(R)]^{-1} y, \quad provided\ that\ y \in R^i X. \tag{7.2.33}$$

Proof. The proof is going on the same lines as the proof of Theorem 7.2.1.

The condition $y \in R^i X$ is an obvious necessary and sufficient condition of solvability of Equation (7.2.30) in the case (iii). ∎

There is another direct proof of Corollary 7.2.2 given also by von Trotha (1975).

COROLLARY 7.2.3. *Suppose that all assumptions of Corollary 7.2.2 are satisfied and, moreover, R is a Volterra operator.*

(i) *If $N \geqslant 1$, then the operator $Q(I, R) + R^N T(R)$ is invertible;*

(ii) *If $N = 0$ and $a_0 \neq 0$ then the operator $a_0 I + T(R)$ is invertible;*

(iii) *If $N = 0$, $b_0 = \ldots = b_{i-1} = 0$, $b_i \neq 0$ $(i = 1, \ldots, M-1)$ then the operator $D^i T(R)$ is invertible.*

Thus all solutions of Equation (7.2.30) are of the form (7.2.31), (7.2.32), (7.2.33), respectively.

Proof. Indeed, in the cases (i) and (ii) the operators

$$Q(I, R) + R^N T(R) = a_N I + \sum_{k=1}^{N-1} a_k R^k + R^N T(R) \quad and\ a_0 I + T(R)$$

are polynomials in R with non-vanishing free terms a_N and a_0, respectively. Thus they can be written as superpositions of operatsıos of the form $I - \lambda R$ which are invertible for every $\lambda \in \mathcal{F}$ by our assumption that R is a Volterra operator. In the case (iii) we have, by our assumption, $M \geqslant 2$, $i = 1, \ldots, M-1$ and

$$
D^i T(R) = D^i \sum_{k=1}^{M} b_k R^k = D^i \sum_{k=1}^{M} b_k R^k = \sum_{k=1}^{M} b_k D^i R^k
$$

$$
= \sum_{k=i}^{M} b_k R^{k-i} = b_i I + \ldots + b_M R^{M-i}.
$$

Thus $D_i^i T(R)$ is a polynomial in R with a non-vanishing free term, hence is invertible.   ∎

COROLLARY 7.2.4. *If all assumptions of Corollary 7.2.2 are satisfied and if R is a Volterra operator then*

$$
\dim \ker [Q(D) + T(R)] = N \dim \ker D, \tag{7.2.34}
$$

where $N = \deg Q$.

If Q', Q'' and T', T'' satisfy such assumptions as Q and T in Corollary 7.2.2 and R is a Volterra operator then

$$
\dim \ker [Q'(D) + T'(R)][Q''(D) + T''(R)] = (N' + N'') \dim \ker D, \tag{7.2.35}
$$

where $N' = \deg Q'$ and $N'' = \deg Q''$.

Examples and Exercises

EXAMPLE 7.2.1. Consider the *differential-integral equation*:

$$
x^N(t) - \frac{1}{(M-1)!} \int_0^t (t-s)^{M-1} x(s) \, ds = y(t), \tag{7.2.36}
$$

where $y \in C[0, T]$, $N \geqslant 1$, $M \geqslant 1$. Put $D = d/dt$, $(Rx)(t) = \int_0^t x(s) \, ds$. Then by Formula (2.2.28) we have

$$
\int_0^t (t-s)^{M-1} x(s) \, ds = (M-1)! \int_0^t \frac{(t-s)^{M-1}}{(M-1)!} x(s) \, ds
$$

$$
= (M-1)! (R^M x)(t).
$$

Hence Equation (7.2.36) can be rewritten as follows:

$$(D^N - R^M)x = y.$$ (7.2.37)

Since D and R satisfy all assumptions of Corollary 7.2.3, $N \geqslant 1$, $Q(I, R) = I$, $T(R) = -R^M$, Point (i) of this corollary implies that

$$x = (I - R^N R^M)^{-1} \left(R^N y + \sum_{k=0}^{N-1} R^k z_k \right),$$

where $z_k \in \ker D$ are arbitrary, i.e.

$$x = (I - R^{N+M})^{-1} \left(R^N y + \sum_{k=0}^{N-1} R^k z_k \right),$$

where $z_0, \ldots, z_{N-1} \in \ker D$ are arbitrary.

Write $z_k(t) = c_k \in \mathbf{R}$ $(k = 0, 1, \ldots, N-1)$. Formulae (2.2.28) and (2.2.29) together imply that

$$u(t) = \left(R^N y + \sum_{k=0}^{N-1} R^k z_k \right)(t)$$

$$= \int_0^t \frac{(t-s)^{N-1}}{(N-1)!} y(s)\,ds + \sum_{k=0}^{N-1} c_k \frac{t^k}{k!} \in C[0, T]$$

and $x = (I - R^{N+M})^{-1} u$, where u is an already known function.

On the other hand

$$(I - R^{N+M})^{-1} = \left[\prod_{j=1}^{N+M} (I - \varepsilon^j R) \right]^{-1} = \prod_{j=1}^{N+M} (I - \varepsilon^j R)^{-1},$$

where $\varepsilon = e^{2\pi i/(N+M)}$ and the operators $(I - \lambda R)^{-1}$ for every λ are determined by Formula (2.2.41).

Observe that we can obtain in another way solutions of Equation (7.2.37), but under a strong assumption that $y \in C^{(M)}[0, T]$, i.e. $y \in \operatorname{dom} D^M$. Namely acting on both sides of Equation (7.2.37) by the operator D^M we find

$$(D^{M+N} - I)x = D^M y, \quad \text{i.e.} \quad x^{(M+N)} - x = y^{(M)}.$$

EXERCISE 7.2.1. Suppose that all assumptions of Corollary 7.2.3 are satisfied, $N \geqslant 1$, $a_0 \neq 0$ and F is an initial operator for D corresponding to R. Prove that an initial value problem

$$[Q(D)+T(R)]x = y, \quad y \in X,$$
$$FD^k x = y_k, \quad y_k \in \ker D \quad (k = 0, 1, ..., N-1) \tag{7.2.38}$$

is well-posed and its unique solution is of the form

$$x = [Q(I, R)+R^N T(R)]^{-1} \left(R^N y + \sum_{i=0}^{N-1} \sum_{k=0}^{i} a_{N+k-i} R^i y_k\right)$$

(von Trotha, 1975, private communication).

EXERCISE 7.2.2. Suppose that all assumptions of Corollary 7.2.3 are satisfied and that $y \in X_0 = \{x \in X : R^k D^k x = 0 \text{ for } k = 1, 2, ...\}$. Prove that the initial value problem (7.2.38) is well-posed and its unique solution $x \in X_0$ is of the form

$$x = \begin{cases} [Q(I, R)+R^N T(R)]^{-1} R^N y & \text{if } N \geqslant 1, \\ [a_0 I+T(R)]^{-1} y & \text{if } N = 0, a_0 \neq 0, \\ D^i[D^i T(R)]^{-1} y & \text{if } N = 0, a_0 = 0 \end{cases}$$

(in the last case we assume that $b_0 = ... = b_{i-1} = 0$, $b_i \neq 0$ ($i = 1, ...$ ..., $M-1$)) (von Trotha, 1975, private communication).

EXERCISE 7.2.3. Solve the following differential-integral equations, where

$$K(t, s) = \sum_{k=1}^{M} b_k \frac{(t-s)^{k-1}}{(k-1)!} \quad (b_1, ..., b_M \in R):$$

(i) $x^{(N)}(t)+\int_0^t K(t, s)x(s)ds = 0$;

(ii) $x''(t)+\lambda^2 x(t)+\int_0^t K(t, s)x(s)ds = 0$;

(iii) $x^{(N)}(t)+\lambda^N x(t)+\int_0^t K(t, s)x(s)ds = 0$;

(iv) $\sum_{k=0}^{N} a_k x^{(k)}+\int_0^t K(t, s)x(s)ds = 0$

(cf. also Section 8.4 and Example 7.2.1).

EXERCISE 7.2.4. Solve the following differential-integral equations:

(i) $x^{(N)}(t)+\sum_{k=1}^{M}b_k(t)\int_0^t\frac{(t-s)^{k-1}}{(k-1)!}x(s)ds = y(t),$

where $b_1, ..., b_M, y \in C[0, T]$ are given;

(ii) $x''(t)+\lambda^2 x(t)+\mu(t)\int_0^t x(s)ds = y(t),$

where $y, \mu \in C[0, T]$ are given (cf. also Section 8.4 and Example 7.2.1).

7.3. QUASI-LINEAR AND NONLINEAR PROBLEMS

In this section we shall consider some quasi-linear and nonlinear problems with right invertible operators to show, how one can solve problems of this type in algebraic way.

Suppose that X is a linear space and A is a nonlinear mapping of X into itself. We say that the mapping A is *invertible in the space X* if the equation

$$A(x) = y$$

has a unique solution x belonging to X for every given $y \in X$.

THEOREM 7.3.1. *Suppose that $D \in R(X)$, $\dim \ker D \neq 0$, $F_0, ..., F_{M+N-1}$ are initial operators for D corresponding to right inverses $R_0, ..., R_{M+N-1}$, respectively $(N \geqslant 1, M \geqslant 0)$. Write as before*

$$Q(D) = \sum_{k=0}^{N}Q_k D^k, \quad \text{where } Q_0, ..., Q_{N-1} \in L_0(X), Q_N = I$$

$$(7.3.1)$$

and suppose that -1 is a regular value of the operator

$$Q^0 = \sum_{k=0}^{N-1}Q_k R_k ... R_{N-1}, \qquad (7.3.2)$$

i.e. the operator $I+Q^0$ is invertible. Suppose, moreover, that A is a nonlinear mapping of X into itself. Consider the mixed boundary value problem

$$Q(D)D^M x+A(x) = y, \quad y \in X, \qquad (7.3.3)$$

$$F_k D^k x = y_k, \quad y_k \in \ker D \quad (k = 0, 1, ..., M+N-1). \quad (7.3.4)$$

Then

(i) *The problem* (7.3.3)–(7.3.4) *is equivalent to the following equation*

$$x + A^0(x) = y^0, \quad \text{where } A^0 = R_0 \ldots R_{M+N-1}(I+Q^0)^{-1}A,$$
(7.3.5)

$$y^0 = R_0 \ldots R_{M+N-1}(I+Q^0)^{-1}y_{N+M} + y_0 + \sum_{k=1}^{N-1} R_0 \ldots R_k y_k,$$
(7.3.6)

$$y_{N+M} = y - Q_0 y_M - \sum_{m=1}^{N-1} Q_m \Big(\sum_{k=m+1}^{N-1} R_m \ldots R_{k-1} y_{M+k} + y_{M+m} \Big);$$
(7.3.7)

(ii) *If the mapping* $I+A^0$, *where* A^0 *is defined by the second of Formulae* (7.3.5), *is invertible in the space* X *then the problem* (7.3.3)–(7.3.4) *is well-posed and its unique solution is*

$$x = (I+A^0)^{-1}(y^0), \quad \text{where } y^0 \text{ is defined by } (7.3.6), (7.3.7).$$
(7.3.8)

Proof. Write Equation (7.3.3) in the form

$$Q(D)D^M x = y^*, \quad \text{where } y^* = y - A(x).$$
(7.3.9)

Corollary 4.6.2 implies that the problem (7.3.9)–(7.3.4) is well-posed and its unique solution is

$$x = R_0 \ldots R_{M+N-1}(I+Q^0)^{-1}\Big[y^* - Q_0 y_M - $$

$$- \sum_{m=1}^{N-1} Q_m \Big(\sum_{k=m+1}^{N-1} R_m \ldots R_{k-1} y_{M+k} + y_{M+m} \Big) \Big] + $$

$$+ y_0 + \sum_{k=1}^{M+N-1} R_0 \ldots R_{k-1} y_k$$

$$= R_0 \ldots R_{M+N-1}(I+Q^0)^{-1}\Big[y - Ax - Q_0 y_M - $$

$$- \sum_{m=1}^{N-1} Q_m (R_m \ldots R_{k-1} y_{M+k} + y_{M+m}) \Big] + y_0 + $$

$$+ \sum_{k=1}^{M+N-1} R_0 \dots R_{k-1} y_k$$

$$= -A^0(x) + R_0 \dots R_{M+N-1}(I+Q^0)^{-1} y_{M+N} + y_0 +$$

$$+ \sum_{k=1}^{M+N-1} R_0 \dots R_{k-1} y_k = -A^0(x) + y^0$$

which was to be proved. Suppose now that the mapping $I+A^0$, where A^0 is defined by the second of Formulae (7.3.5), is invertible in the space X. Then $x = (I+A^0)^{-1}(y^0)$ and is a unique solution of the problem in question. ∎

An immediate consequence of this theorem and Corollary 4.6.3 is

COROLLARY 7.3.1. *Suppose that all assumption of Theorem 7.3.1 are satisfied, the operators* $Q_0, \dots, Q_{N-1}$ *commute with D and the mapping* $I+A^0$ *is invertible in the space X. Then the mixed boundary value problem for the equation*

$$D^M Q(D)x + A(x) = y, \quad y \in X \tag{7.3.10}$$

with Conditions (7.3.4) is well-posed and its unique solution is of the form (7.3.5).

COROLLARY 7.3.2. *Suppose that all assumptions of Theorem 7.3.1 are satisfied and, moreover,* $F_k = F, R_k = R$ *for* $k = 0, 1, \dots, M+N-1$. *If the mapping* $I+A^0$, *where* $A^0 = R^{M+N}[Q(I, R)]^{-1}A$, $Q(I, R) = I+$ $+Q^0 = \sum_{k=0}^{N} Q_k R^{N-k}$, *is invertible then the initial value problem for Equation (7.3.3) with conditions*

$$FD^k x = y_k, \quad y_k \in \ker D \quad (k = 0, 1, \dots, M+N-1) \tag{7.3.11}$$

is well-posed and its unique solution is

$$x = (I+A^0)^{-1} \left\{ R^{M+N}[Q(I, R)]^{-1} \left[y - \sum_{m=0}^{N-1} Q_m \left(\sum_{k=m+1}^{N-1} y_{m+k} + \right. \right. \right.$$

$$\left. \left. \left. + y_{M+m} \right) \right] + \sum_{k=0}^{M+N-1} R^k y^k \right\}. \tag{7.3.12}$$

COROLLARY 7.3.3. *Suppose that all assumptions of Corollary 7.3.2 are satisfied and the operators $Q_0, \ldots, Q_{N-1}$ commute with D. If the mapping $I + A^0$ is invertible then the initial value problem (7.3.10)–(7.3.11) is well-posed and its unique solution is of the form (7.3.12).*

COROLLARY 7.3.4. *Suppose that all assumptions of Theorem 7.3.1 are satisfied. If the mapping $I + A^0$ is invertible then every solution of Equation (7.3.3) is of the form*

$$x = (I + A^0)^{-1} \left\{ R_0 \ldots R_{M+N-1}(I + Q^0)^{-1} \left[y - Q_0 z_M - \right. \right.$$

$$- \sum_{m=1}^{N-1} Q_m \sum_{k=m+1}^{N-1} R_m \ldots R_{k-1} z_{M+k} + z_{M+m} \Big] + z_0 +$$

$$\left. + \sum_{k=1}^{N-1} R_0 \ldots R_{k-1} z_k \right\},$$

$$(7.3.13)$$

where $z_0, \ldots, z_{M+N-1} \in \ker D$ are arbitrary.

If -1 is an eigenvalue of the operator Q^0 defined by Formula (7.3.2), i.e. the operator $I + Q^0$ is not invertible, we can obtain conditions of solvability of the problem under consideration in a similar way, as in Section 3.2. As regards the invertibility of the mapping $I + A^0$, where A^0 is defined by the second of Formulae (7.3.5), and solvability of Equation (7.3.3) we have a similar situation. It means that different conditions of solvability of Equation (7.3.3) lead to different solutions of our problem.

Consider now purely nonlinear problems of order 1 (for simplicity only).

THEOREM 7.3.2. *Suppose that $D \in R(X)$, dim $\ker D \neq 0$, F is an initial operator for D corresponding to an $R \in \mathcal{R}_D$ and A is a nonlinear mapping of the space X into itself.*

(i) *If the mapping $I - RA$ is invertible in the space X then the initial value problem*

$$Dx - A(x) = y, \quad y \in X, \tag{7.3.14}$$

$$Fx = x_0, \quad x_0 \in \ker D \tag{7.3.15}$$

is well-posed and its unique solution is of the form

$$x = (I - RA)^{-1}(Ry + x_0), \tag{7.3.16}$$

so that all solutions of Equation (7.3.14) are of the form

$$x = (I - RA)^{-1}(Ry + z), \quad \text{where } z \in \ker D \text{ is arbitrary}, \tag{7.3.17}$$

provided that

$$x \in \{u \in \mathrm{dom}\, D : u - RA(u) \in Ry \oplus \ker D\}; \tag{7.3.18}$$

(ii) *If the mapping* $I - AR$ *is invertible in the space* X *then the problem* (7.3.14)–(7.3.15) *is well-posed and its unique solution is of the form*

$$x = R[(I - AR)^{-1}y - x_0] + x_0, \tag{7.3.19}$$

so that all solutions of Equation (7.3.14) are

$$x = R[(I - AR)^{-1}y - z] + z, \quad \text{where } z \in \ker D \text{ is arbitrary}, \tag{7.3.20}$$

provided that

$$y \in (I - AR)X. \tag{7.3.21}$$

Proof. Observe that Equation (7.3.14) can be written as follows:
$$D[x - A(x)] = y.$$

If Condition (7.3.18) is satisfied then we obtain a new equation
$$x - RA(x) = Ry + x_0. \tag{7.3.22}$$

This, and our assumptions, together imply that Equation (7.3.22) has a unique solution of the form (7.3.16) and that general solution of Equation (7.3.14) is of the form (7.3.17).

In order to prove Point (ii) put $u = Dx$. Then $x = Ry + z$, where $z \in \ker D$ is arbitrary. In particular, if Condition (7.3.15) has to be satisfied then $z = x_0$ and $x = Ry + x_0$. Equation (7.3.14) can be rewritten as follows:

$$u - A(Ry + z) = y. \tag{7.3.23}$$

But $Ry + z \in \mathrm{dom}\, D$ for all $z \in \ker D$. Hence, if Condition (7.3.21) is satisfied and the mapping $I - AR$ is invertible all solutions of Equation (7.3.14) are of the form (7.3.20) and a unique solution of the problem (7.3.14)–(7.3.15) is of the form (7.3.19). ∎

If neither $I-AR$ nor $I-RA$ is invertible then we can examine Equation (7.3.14) in a similar way, as in Section 3.1.

Examples and Exercises

EXAMPLE 7.3.1. Consider the following ordinary differential equation:

$$\frac{d^n x}{dt^n} = a(t) x^m + y, \tag{7.3.24}$$

where $a, y \in C[t_0, T]$, $n \geq 1$, $m \geq 1$ are given, with the initial conditions

$$x^{(k)}(t_0) = a_k, \quad a_k \in \mathbf{R} \quad (k = 0, 1, ..., n-1), \tag{7.3.25}$$

where $D = d/dt$, $R = \int_{t_0}^{t}$, $(Fx)(t) = x(t_0)$ for $x \in X$. Then the problem (7.3.24)–(7.3.25) can be rewritten as follows: $D^n x = ax^m + y$, $FD^k x = a_k$ $(k = 0, 1, ..., n-1)$. Here we have $Q(D) = D$. Corollary 7.3.2 implies that our problem can be reduced to the equation

$$x + R^n(ax^m) = R^n y + \sum_{k=0}^{n-1} R^k a_k, \tag{7.3.26}$$

i.e., according with Formulae (2.2.28) and (2.2.29) to the integral equation

$$x(t) + \int_{t_0}^{t} \frac{(t-s)^{n-1}}{(n-1)!} a(s) x^m(s) ds = y_1(t), \tag{7.3.27}$$

where

$$y_1(t) = \int_{t_0}^{t} \frac{(t-s)^{n-1}}{(n-1)!} y(s) ds + \sum_{k=0}^{n-1} a_k \frac{(t-t_0)^k}{k!}.$$

One can prove that the Equation (7.3.27) has a unique solution for every y_1, i.e. for every $y \in X$, which means that the mapping $I+A$, where $A(x) = R^n(ax^m)$ for $x \in X$, is invertible in the space X (cf. for instance Pogorzelski, 1966, also Section 8.4).

EXERCISE 7.3.1. Suppose that X is a linear space (over: (i) complexes, (ii) reals, $D \in R(X)$ and A is a nonlinear mapping of X into itself.

Write: $Q(D) = \sum\limits_{k=0}^{N} q_k D^k$, where $q_0, \ldots, q_N \in \mathbf{R}$, $q_n = 1$, and $Q(t)$

$= \prod\limits_{j=1}^{n} (t - t_j)^{r_j}$ $(r_1 + \ldots + r_n = N$, $t_i \neq t_j$ for $i \neq j)$. Reduce the equation

$$D^M Q(D) x + A(x) = y, \quad y \in X \tag{7.3.28}$$

to an equation without the operator D by an application of Theorem 7.3.1.

EXERCISE 7.3.2. Suppose that X, D, $Q(D)$ are defined as in Exercise 7.3.1 and F is an initial operator for D corresponding to an $R \in \mathscr{R}_D \cap V(X)$. Solve an initial value problem for Equation (7.3.28) with the conditions:

$$FD^k x = y_k, \quad y_k \in \ker D \quad (k = 0, 1, \ldots, N-1). \tag{7.3.29}$$

EXERCISE 7.3.3. Suppose that X is a D-algebra over $\mathbf{C}$ and F is an initial operator for D. Give a necessary and sufficient condition for the following initial value problem:

$$D^N x + \sum\limits_{k=0}^{M} a_k x^k = y, \quad \text{where } a_0, \ldots, a_M \in X \text{ are given}, \tag{7.3.30}$$

$$FD^k x = y_k, \quad y_k \in \ker D \quad (k = 0, 1, \ldots, N-1) \tag{7.3.31}$$

to be well-posed.

7.4. METHOD OF VARIABLES SEPARABLE

In this section we shall consider problems which can be solved by a separation of variables.

THEOREM 7.4.1. *Suppose that $D \in R(X)$, $\dim \ker D \neq 0$ and $R \in \mathscr{R}_D$. Let $\{H_x\}_{x \in X}$ be a family of mappings of the space X into itself which depend in a nonlinear way on $x \in X$. Then every solution of the equation*

$$Dx = H_x y, \quad \text{where } y \in X \tag{7.4.1}$$

is a solution of the equation

$$x - RH_x y = z, \quad \text{where } z \in \ker D \text{ is arbitrary}. \tag{7.4.2}$$

Conversely, every solution of Equation (7.4.2) *belonging to* dom D *is a solution of Equation* (7.4.1).

Proof. Suppose that $x \in$ dom D is a solution of Equation (7.4.1). Then $x = RH_x y + z$, where $z \in$ ker D, hence x is a solution of Equation (7.4.2). Conversely, suppose that $x \in$ dom D is a solution of Equation (7.4.2). Then $Dx - H_x y = D(x - RH_x y) = Dz = 0$, i.e. x satisfies Equation (7.4.1). ∎

COROLLARY 7.4.1. *Suppose that all assumptions of Theorem* 7.4.1 *are satisfied and F is an initial operator for D corresponding to R. Then every solution of the initial value problem for Equation* (7.4.1) *with the condition*

$$Fx = x_0, \quad x_0 \in \text{ker } D \tag{7.4.3}$$

is a solution of the equation

$$x - RH_x y = x_0. \tag{7.4.4}$$

Conversely, every solution belonging to dom D *of Equation* (7.4.4) *is a solution of the problem* (7.4.1)–(7.4.3).

Proof. Indeed, $x \in$ dom D is a solution of Equation (7.4.4) if and only if it is a solution of Equation (7.4.1). If x satisfies Condition (7.4.3) then $z = Fz = F(x - RH_x y) = Fx - FRH_x y = Fx = x_0$. ∎

The following theorem generalizes Theorem 7.4.1:

THEOREM 7.4.2. *Suppose that $D \in R(X)$,* dim ker $D \neq 0$, $R_0, \dots, R_{M+N-1}$
$\in \mathcal{R}_D$, $Q(D) = \sum_{k=0}^{N} Q_k D^k$, *where* $Q_0, \dots, Q_{N-1} \in L_0(X), Q_N = I$ *and the operator $I + Q^0$, where*

$$Q^0 = \sum_{k=0}^{N-1} Q_k R_{M+k} \cdots R_{M+N-1}$$

is invertible. Let $\{H_x\}_{x \in X}$ be a family of mappings of the space X into itself which depend in a nonlinear way on $x \in X$. Then every solution of the equation

$$Q(D)D^M x = H_x y, \quad y \in X \quad (M \geqslant 0) \tag{7.4.5}$$

satisfies the equation

$$x - R_0 \dots R_{M+N-1}(I+Q^0)^{-1}H_x y$$

$$= R_0 \dots R_{M+N-1}(I+Q^0)^{-1} \sum_{m=0}^{N-1} Q_m \left(\sum_{k=m+1}^{N-1} R_m \dots R_{k-1} z_{M+k} + \right.$$

$$\left. + z_{M+m} \right) + z_0 + \sum_{k=1}^{M+N-1} R_0 \dots R_{k-1} z_k, \qquad (7.4.6)$$

where $z_0, \dots, z_{N+M-1} \in \ker D$ are arbitrary.

Conversely, every solution of Equation (7.4.6) belonging to dom D^{N+M} satisfies Equation (7.4.5).

Proof. Write: $D_1 = Q(D)D^M$, $R_1 = R_0 \dots R_{M+N-1}(I+Q^0)^{-1}$. Then

$$D_1 R_1 = Q(D)D^M R_0 \dots R_{M+N-1}(I+Q^0)^{-1}$$

$$= \sum_{k=0}^{N} Q_k D^k D^M R_0 \dots R_{M+N-1}(I+Q^0)^{-1}$$

$$= \sum_{k=0}^{N} Q_k R_{M+k} \dots R_{M+N-1}(I+Q^0)^{-1}$$

$$= (I+Q^0)(I+Q^0)^{-1} = I.$$

Thus $D_1 \in R(X)$, $R_1 \in \mathcal{R}_{D_1}$. We also have

$$\ker D_1 = \left\{ z = R_0 \dots R_{M+N-1}(I+Q^0)^{-1} \left(\sum_{m=0}^{N-1} Q_m \sum_{k=m+1}^{N-1} R_m \dots \right. \right.$$

$$\left. \dots R_{k-1} z_{M+k} + z_{M+m} \right) + z_0 + \sum_{k=1}^{M+N-1} R_0 \dots R_{k-1} z_k :$$

$$\left. z_0, \dots, z_{M+N-1} \in \ker D \right\}.$$

This, and Theorem 7.4.1, together imply the conclusion of our theorem. ∎

An immediate consequence of Theorem 7.4.2 is

COROLLARY 7.4.2. *Suppose that all assumptions of Theorem 7.4.2 are satisfied and* $R_0 = \dots = R_{M+N-1} = R$. *Then every solution of Equation* (7.4.5) *satisfies the equation*

$$x - R^{M+N}[Q(I, R)]^{-1}H_x y$$

$$= R^{M+N}[Q(I, R)]^{-1}\left(\sum_{m=0}^{N-1} Q_m \sum_{k=m}^{N-1} R^{k-m}z_k\right) + \sum_{k=0}^{M+N-1} R^k z_k,$$

$$(7.4.7)$$

where $z_0, \ldots, z_{M+N-1} \in \ker D$ are arbitrary and $Q(I, R) = I + Q^0$
$= \sum_{k=1}^{N} Q_k R^{N-k}$. Conversely, every solution of Equation (7.4.7) belonging
to dom D^{M+N} is a solution of Equation (7.4.5).

THEOREM 7.4.3. Suppose that X is a D-algebra with unit e, F is an initial
operator for D corresponding to an $R \in \mathcal{R}_D$ and $\{a_x\}_{x \in X} \subset X$ is a family
of invertible elements dependent on the parameter $x \in X$. Then every
solution of the equation

$$Dx = a_x y, \quad y \in X \qquad (7.4.8)$$

is a solution of the equation

$$a_x x - a_x R(a_x y) = z, \quad \text{where } z \in \ker D \text{ is arbitrary.} \qquad (7.4.9)$$

Conversely, every solution of Equation (7.4.9) belonging to dom D
is a solution of Equation (7.4.8). If, in addition, we have an initial condition (7.4.3) $Fx = x_0$ then Equation (7.4.9) has the form

$$x - R(a_x y) = x_0. \qquad (7.4.10)$$

Proof. Equation (7.4.8) can be written in the following form: $a_x^{-1}Dx$
$= y$. Write for all $x \in X$: $D_1 = a_x^{-1}D$, $R_1 = Ra_x$. Then dom D_1
$=$ dom D, $D_1 R_1 = a_x^{-1}DRa_x = a_x^{-1}a_x = e$, i.e. $D_1 R_1 x = x$ for all
$x \in X$, which implies that $D_1 \in R(X)$, $R_1 \in \mathcal{R}_{D_1}$. We also have on dom D:
$F_1 = I - R_1 D_1 = I - Ra_x^{-1}a_x D = I - RD = F$, i.e. an initial operator
F_1 corresponding to R_1 is equal to F. Moreover, $\ker D_1 = a_x^{-1} \ker D$
for all $x \in X$. Hence we can write Equation (7.4.8) in the form: $D_1 x = y$,
which implies that $x = R_1 y + z_1$, where $z_1 \in \ker D_1$ is arbitrary. But
$z_1 = a_x^{-1}z$, where $z \in \ker D$ is arbitrary. Hence $x = R_1(a_x y) + a_x^{-1}z$,
$z \in \ker D$. Multiplying this equality by a_x we obtain Equation (7.4.9).

Applying the initial condition $Fx = x_0$ we find $a_x^{-1}z = z_1 = F_1 z_1$ $= x - R_1 y = F_1 x - F_1 R_1 y = F_1 x = Fx = x_0$, which implies $z = a_x x_0$. We therefore can write Equation (7.4.9) in the form: $a_x x - a_x R(a_x y)$ $= a_x x_0$, i.e. $a_x x - R(a_x y) = a_x x_0$. Since elements a_x are invertible, we finally obtain Equation (7.4.10). ∎

Examples and Exercises

EXAMPLE 7.4.1. Consider an ordinary differential equation

$$x'(t) = h(x)g(t), \tag{7.4.11}$$

where we assume that the functions h, g are continuous for $x, t \in \mathbf{R}$, respectively, and $h \neq 0$. Write: $D = d/dt$, $H_x u = h(x)u$ for all $x \in X$ $= C(\mathbf{R})$, $u \in X$, $y = g(t)$. Then Equation (7.4.11) can be written as $Dx = H_x y$. All assumptions of Theorem 7.4.1 (or Theorem 7.4.3) are satisfied. Thus this equation is equivalent to the equation $x - RH_x y$ $= z$, where $z \in \ker D$, i.e. to an integral equation

$$x(t) - \int_0^t h\big(x(s)\big)g(s)\,\mathrm{d}s = c, \quad \text{where } c \in \mathbf{R}. \tag{7.4.12}$$

The Bielecki theorem in Section 8.4 shows that this equation has a unique solution for every fixed c (provided that h satisfies the Lipschitz condition). However, having already proved the existence and uniqueness of a solution we can determine this solution by *method of variables separable*. Namely, denote by h_1 and g_1 primitive functions for functions $1/h$ and g, respectively. Then we can rewrite Equation (7.4.11) as follows:

$$[h(x)]^{-1}x'(t) = g(t)$$

or

$$\int \frac{\mathrm{d}x}{h(x)} = \int g(t)\,\mathrm{d}t, \quad \text{i.e. } h_1(x) = g_1(t) + C,$$

where C is an arbitrary constant. If the function h_1 is one-to-one then the function h_1^{-1} exists and we find

$$x(t) = h_1^{-1}[g_1(t) + C], \quad \text{where } C \text{ is an arbitrary constant.}$$

EXAMPLE 7.4.2. Suppose that the functions h and g are as in Example 7.4.1. Consider a differential equation of order $n \geqslant 1$

$$x^{(n)} = h(x)g(t). \tag{7.4.13}$$

Using the denotations introduced in Example 7.4.1 we can rewrite Equation (7.4.13) as follows: $D^n x = H_x y$. Theorem 7.4.1 implies that this equation is equivalent to the equation $x - R^n H_x y = \sum_{k=0}^{n-1} R^k c_k$, where c_k are arbitrary constants. The last equation can be written in the form

$$x(t) - \int_{t_0}^{t} \frac{(t-s)^{n-1}}{(n-1)!} g(s) h(x(s)) ds = \sum_{k=0}^{n-1} c_k \frac{t^k}{k!},$$

where $c_0, \dots, c_{n-1}$ are arbitrary constants.

EXERCISE 7.4.1. Reduce the following equations to equations with variables separable:

(i) $x'(t) = f(at+bx+c)$ by substitution $u = at+bx+c$;

(ii) $x'(t) = f(x/t)$ by substitution $x = tu$, $x' = u+tu'$;

(iii) $x'(t) = f\left(\dfrac{a_1 t + b_1 x + c_1}{a_2 t + b_2 x + c_2}\right)$

(a) if $\begin{vmatrix} a_1 & b_1 \\ a_2 & b_2 \end{vmatrix} \neq 0$ by substitution $t = t_0 + \tau$, $x = x_0 + \xi$, where (t_0, x_0) is the point of intersection of lines $a_1 t + b_1 x + c_1 = 0$ and $a_2 t + b_2 x + c_2 = 0$;

(b) if $\begin{vmatrix} a_1 & b_1 \\ a_2 & b_2 \end{vmatrix} = 0$ then there exists a scalar $\lambda \neq 0$ such that $a_2 = \lambda a_1$, $b_2 = \lambda b_1$. We substitute $u = a_1 t + b_1 x$ and we reduce the equation to that considered at Point (ii);

(iv) $f(t, x', x) = 0$ by substitution $x' = u$, $x'' = u'$;

(v) $f(x, x', x'') = 0$ by substitution $x' = u$, $x'' = \dfrac{du}{dt} = \dfrac{du}{dx}\dfrac{dx}{dt} = u\dfrac{du}{dx}$;

(vi) $f(t, x, x', x'') = 0$, where $f(t, x, u, v)$ is a *homogeneous function* of variables x, u, v, i.e. $f(t, \lambda x, \lambda u, \lambda v) = \lambda f(t, x, u, v)$ for $\lambda \in \mathbf{R}$, by substitution $x = e^u$.

EXERCISE 7.4.2. Formulate and prove Theorem 7.4.2 and Corollary 7.4.1 for the operator $D^M Q(D)$.

EXERCISE 7.4.3. Give necessary and sufficient conditions of solvability of a partial differential equation

$$\frac{\partial x(t, s)}{\partial t} = h\big(s, x(t, s)\big)g(t).$$

EXERCISE 7.4.4. Give necessary and sufficient conditions of solvability of a difference equation

$$x_{n+1} - x_n = a_n h(x_n) \quad (n = 1, 2, \ldots)$$

in space $X = (s)$ (over $\mathbf{R}$), where h is a given real valued function of the real variable $t \in \mathbf{R}$.

Added in proof:

An operator $\Gamma \in L(X)$ is said to be a *D-perturbation* if $\operatorname{dom} \Gamma \supset \operatorname{dom} D$ and $D + \Gamma \in R(X)$, i.e. if $D + \Gamma$ is again a right invertible operator. If Γ is a *D-perturbation* then Γ preserves the deficiency: $\beta_{D+\Gamma} = \beta_D$. Any operator of the form:

$$A = \Gamma F, \quad \text{where } F \in \mathscr{F}_D, \Gamma \subset L(X) \text{ and } \operatorname{dom} \Gamma \supset \operatorname{dom} D$$

is a *D-perturbation*. Indeed, if F correspond to an $R \in \mathscr{R}_D$ then $(D + \Gamma F)R = DR + \Gamma FR = DR = I$ (cf. Lausch and the author, 1986).

Chapter 8

Why using metric properties in algebraic analysis?

In the present chapter we shall consider metric properties of right invertible operators strictly connected with their algebraic properties.

8.1. RIGHT INVERTIBLE OPERATORS IN LINEAR METRIC SPACES

A set X is said to be a *metric space* if on $X \times X$ there is defined a non-negative function $\varrho(x, y)$, called *metric*, which satisfies the following conditions:

(i) $\varrho(x, y) = 0$ if and only if $x = y$;

(ii) $\varrho(y, x) = \varrho(x, y)$;

(iii) $\varrho(x, y) \leqslant \varrho(x, z) + \varrho(z, y)$ (triangle inequality);

for all $x, y, z \in X$.

We define the *distance* of points $x, y \in X$ as the value $\varrho(x, y)$.

Suppose that X is a metric space with the metric ϱ. A sequence $\{x_n\} \subset X$ is *convergent* to an $x_0 \in X$ if $\lim_{n \to \infty} \varrho(x_n, x_0) = 0$. The point x_0 is called the *limit* of the sequence $\{x_n\}$. We write: either $x_n \to x_0$ or $x_0 = \lim_{n \to \infty} x_n$. The limit of a sequence (if exists) is, by definition, unique.

Indeed, suppose that $x_n \to x_0$ and $x_n \to x_1$. Then the triangle inequality implies $\varrho(x_1, x_0) \leqslant \varrho(x_n, x_0) + \varrho(x_n, x_1) \to 0$. Thus $\varrho(x_1, x_0) = 0$ and $x_1 = x_0$.

A set $Y \subset X$ is said to be *closed* if Y consists of limits of all convergent sequences $\{x_n\} \subset Y$. A set $Z \subset X$ is said to be *open* if for every $x_0 \in Z$ there exists $\varepsilon > 0$ such that the set $K_\varepsilon(x_0) = \{x \in X: \varrho(x, x_0) < \varepsilon\} \subset Z$. If a set $Y \subset X$ is closed then the set $X \setminus Y$ is open. Conversely, if $Z \subset X$ is an open set then $X \setminus Z$ is a closed set. The set $K_\varepsilon(x_0)$ is said to be a *ball* in the space X with the radius ε and the center x_0. We say also that $K_\varepsilon(x_0)$ is a *neighbourhood* of the point x_0.

A sequence $\{x_n\} \subset X$ is said to be *fundamental* (or a *Cauchy sequence*) if $\{x_n\}$ satisfies the Cauchy condition, i.e. if for every $\varepsilon > 0$ there exists a positive integer N such that for all indices $n, m > N$ we have $\varrho(x_n, x_m) < \varepsilon$. It is easy to observe that a convergent sequence is fundamental.

A metric space X is said to be *complete* if every fundamental sequence $\{x_n\} \subset X$ is convergent to an $x_0 \in X$.

We say that a subset Y of a complete metric space X is *dense* in X and we write: $\overline{Y} = X$, if for every $x \in X$ and for every $\varepsilon > 0$ there exists a $y \in Y$ such that $\varrho(x, y) < \varepsilon$. By definition, X is the smallest closed set containing Y.

Suppose, we are given two metric spaces X and Y with metrics ϱ_X and ϱ_Y, respectively. A mapping f of X into Y is *continuous* if for every $x \in X$ and every $\{x_n\} \subset X$, $\lim_{n \to \infty} \varrho_X(x_n, x) = 0$ implies

$$\lim_{n \to \infty} \varrho_Y\bigl(f(x_n), f(x)\bigr) = 0.$$

Consider now a set X, which is simultaneously a metric space and a linear space (over complexes or reals). We say that X is a *linear metric space* if the operations of addition and multiplication by scalars are continuous, i.e. if x_n, y_n $(n = 1, 2, ...)$, $x, y \in X$, t_n $(n = 1, 2, ...)$, t are scalars and $x_n \to x, y_n \to y, t_n \to t$ then

$$x_n + y_n \to x + y, \quad t_n x_n \to tx.$$

Two metrics $\varrho(x, y)$ and $\varrho'(x, y)$ are said to be *equivalent* if $\varrho(x_n, x) \to 0$ if and only if $\varrho'(x_n, x) \to 0$.

A metric $\varrho(x, y)$ in a linear metric space X is said to be *invariant* if

$$\varrho(x + z, y + z) = \varrho(x, y) \quad \text{for all } x, y \in X.$$

Kakutani (1936) proved the following theorem: Let X be a linear metric space with a metric ϱ. Then there exists an invariant metric ϱ' equivalent to ϱ.

The reader can find the proof of Kakutani theorem, other properties of linear metric spaces and examples in the book of S. Rolewicz, 1985.

Suppose that X is a linear metric space with an invariant metric $\varrho(x, y)$. Write

$$\|x\| = \varrho(x, 0) \quad \text{for all } x \in X.$$

We say that $||x||$ is an *F-norm*. Properties (i)–(iii) of the metric ϱ imply the following properties of the induced *F*-norm:

(i) $||x|| = 0$ if and only if $x = 0$;

(ii) $||x|| = ||-x||$;

(iii) $||x+y|| \leqslant ||x|| + ||y||$ (triangle inequality)

(for all $x, y \in X$).

The continuity of multiplication by scalars implies that

(iv) if t_n $(n \in \mathbf{N})$, t are scalars, $\{x_n\} \subset X$, $x \in X$ and $t_n \to t$, $x_n \to x$ then $||t_n x_n - tx|| \to 0$.

Conversely, if on the space X there is defined a non-negative function $||x||$ satisfying Conditions (i)–(iv) then this function induces an invariant metric $\varrho(x, y) = ||x-y||$.

An *F*-norm is said to be *homogeneous* if $||tx|| = |t| \cdot ||x||$ for all scalars t and $x \in X$.

Every homogeneous *F*-norm is called briefly a *norm*. A linear metric space X with a homogeneous *F*-norm, i.e. with a norm, is said to be a *normed space*. A complete normed space is said to be a *Banach space*.

Suppose, we are given two linear metric spaces (over the same field of scalars) with *F*-norms $|| \ ||_X$ and $|| \ ||_Y$, respectively. A linear operator A mapping X into Y is said to be *closed* if $\{x_n\} \subset$ dom A and $\lim_{n\to\infty} x_n = x \in X$ and $\lim_{n\to\infty} Ax_n = y \in Y$ implies $Ax = y$. A is said to be *continuous* if is closed and $y = Ax$. If A is continuous and its domain is closed then is closed, but not conversely.

Let X be a linear space (over reals or complexes). A non-negative function $||x||$ defined on X is said to be a *pseudonorm* if satisfies the following conditions:

$$||x+y|| \leqslant ||x|| + ||y||; \quad ||tx|| = |t| \cdot ||x|| \quad \text{for } t > 0.$$

By these properties $||0|| = 0$.

Let X be a linear space with the pseudonorm $||x||$. Then the set $K = \{x \in X : ||x|| < 1\}$ is a convex and open set containing 0. If X is a linear space over reals and $E \subset X$ is an open convex set containing 0 then there exists a continuous pseudonorm $||x||$ such that $E = \{x \in X : ||x|| < 1\}$.

A linear metric space X is said to be *locally convex* if every open set containing zero contains an open convex set containing zero. One can prove that a linear metric space is locally convex if its metric ϱ is induced by a countable sequence $\{||x||_n\}_{n \in \mathbb{N}}$ of homogeneous F-norms, i.e.

$$\varrho(x, y) = \sum_{n=1}^{\infty} \frac{1}{2^n} \frac{||x - y||_n}{1 + ||x - y||_n} .$$

Every Banach space X is locally convex, but not conversely.

A set E in a linear metric space is said to be *precompact* if and only if for every $\varepsilon > 0$ there exists a finite ε-net, of E, i.e. a system of points $\{x_1, ..., x_n\} \subset X$ such that for every $x \in E$ there exists an $i \in \{1, ..., n\}$ satisfying the condition: $\varrho(x, x_i) < \varepsilon$. Let X and Y be linear metric spaces (over the same field of scalars). An operator T mapping X into Y is said to be *compact* if there exists an open set E containing zero such that the set TE is precompact.

If X is a Banach space, $A \in L_0(X)$ is continuous and the number $||A|| = \inf_{||x|| \leqslant 1} ||Ax||$ is finite, then $||A||$ is said to be a *norm of the operator* A. By definition, we have $||Ax|| \leqslant ||A|| \cdot ||x||$ for $x \in X$. For that reason A is also called a *bounded operator*.

In the sequel we shall use the following

THEOREM 8.1.1 (*Banach principle of contractive mappings*). *If A is a contractive mapping (briefly: contraction) of a linear space X into itself, i.e. if there is a positive number $q < 1$ such that*

$$||A(x_1) - A(x_2)|| \leqslant q||x_1 - x_2|| \quad \textit{for all } x_1, x_2 \in X, \qquad (8.1.1)$$

then the equation

$$x = A(x) + y, \quad \textit{where } y \in X, \qquad (8.1.2)$$

has a unique solution which is a limit (in norm) of a sequence of successive approximations:

$$x = \lim_{n \to \infty} x_n, \quad \textit{where } x_{n+1} = Ax_n + y \quad (n = 0, 1, 2, ...) \; (8.1.3)$$

and $x_0 \in X$ is arbitrarily fixed.

Theorem 8.1.1 holds also in an arbitrary metric space, if we assume that

$$\varrho\big(A(x_1), A(x_2)\big) \leqslant q\varrho(x_1, x_2) \quad \textit{for } x_1, x_2 \in X \quad (0 < q < 1).$$

COROLLARY 8.1.1. *Suppose that A is a contractive mapping of a Banach space X into itself with a constant $0 < q < 1$. Suppose, we are given a sequence $\{A_n\}$ of contractive mappings such that*

$$\lim_{n \to \infty} \|A_n(x) - A(x)\| = 0 \quad \text{for all } x \in X \tag{8.1.4}$$

and

$$\|A_n x_1 - A_n x_2\| \leqslant q_n \|x_1 - x_2\|, \quad \text{where } 0 < q_n \leqslant q < 1 \tag{8.1.5}$$

and $x_1, x_2 \in X$ are arbitrary. Then a unique solution x of Equation (8.1.2) is a limit (in norm) of a sequence of solutions of equations

$$u_n = A_n u_n + y_n, \quad \text{where } \lim_{n \to \infty} y_n = y, \quad y \in X, \tag{8.1.6}$$

i.e. $\lim_{n \to \infty} u_n = x.$

Proof. By our assumptions, each of Equations (8.1.6) has a unique solution u_n and

$$\begin{aligned}
\|x - u_n\| &= \|[A(x) + y] - [A_n(u_n) + y_n]\| \\
&\leqslant \|y - y_n\| + \|A(x) - A_n(x)\| + \|A_n(x) - A_n(u_n)\| \\
&\leqslant \|y - y_n\| + \|A(x) - A_n(x)\| + q_n \|x - u_n\| \\
&\leqslant \|y - y_n\| + \|A(x) - A_n(x)\| + q \|x - u_n\|.
\end{aligned}$$

Since we have $q < 1$, we conclude that

$$\|x - u_n\| \leqslant \frac{1}{1-q} \left(\|A(x) - A_n(x)\| + \|y - y_n\| \right) \to 0 \quad \text{as } n \to \infty.$$

This implies that $x = \lim_{n \to \infty} u_n$. ∎

This Corollary shows that a unique solution of Equation (8.1.2) is continuous with respect to the operator A and the given element y, i.e. small enough changes of A and y lead to small changes of this solution.

Let X be a linear space over an algebraically closed field $\mathscr{F}$. An operator $A \in L_0(X)$ is said to be *locally algebraic* if for every $x \in X$ there exists a polynomial $P_x(t)$ with coefficients in $\mathscr{F}$ such that $P_x(A)x = 0$. In sequel we shall use the following

THEOREM 8.1.2 (Kaplansky, 1957). *If X is a complete linear metric space then every continuous locally algebraic operator acting in X is algebraic.*

(For the proof see also: the author and S. Rolewicz, 1968, p. 167).

THEOREM 8.1.3 (Mikusiński, 1958; Sikorski, 1958). *Let X be a linear space over an algebraically closed field $\mathscr{F}$. Let $P(t)$ and $Q(t)$ be arbitrary polynomials with coefficients in $\mathscr{F}$. Suppose that an operator $A \in L_0(X)$ satisfies the following conditions:*

$$\alpha_{P(A)} = \alpha_A \deg P, \quad \text{where } \alpha_A = \dim \ker A \neq 0,$$
$$\alpha_{Q(A)P(A)} \leqslant \alpha_{Q(A)} + \alpha_{P(A)}, \tag{8.1.7}$$

A is locally algebraic. $\tag{8.1.8}$

Then there exists an operator $T \in L_0(X)$ such that

$$AT - TA = I. \tag{8.1.9}$$

Observe, that the operator T, which appears in Condition (8.1.9), plays the role of argument of functions. Indeed, if we put $D = \mathrm{d}/\mathrm{d}t$, $(Tx)(t) = tx(t)$ for $x \in C^1[a, b]$, we obtain

$$(DT - TD)x(t) = \frac{\mathrm{d}}{\mathrm{d}t} tx(t) - t \frac{\mathrm{d}}{\mathrm{d}t} x(t)$$

$$= x(t) + t \frac{\mathrm{d}x(t)}{\mathrm{d}t} - t \frac{\mathrm{d}x}{\mathrm{d}t} = x(t).$$

The arbitrariness of x implies $DT - TD = I$.

PROPOSITION 8.1.1. *Let X be a linear space over an algebraically closed field $\mathscr{F}$ and let an operator $T \in L_0(X)$ satisfy Condition (8.1.9). Then the operator $T_a = aI + T$ also satisfies Condition (8.1.9) for every $a \in \mathscr{F}$.*

Proof. Indeed, $ATa - TaA = A(aI + T) - (aI + T)A = aA - aA + AT - TA = AT - TA = I.$ ∎

THEOREM 8.1.4 (Przeworska-Rolewicz and S. Rolewicz, 1968, p. 71). *Suppose that X is a linear space over an algebraically closed field $\mathscr{F}$ and an operator $A \in L_0(X)$ is algebraic. Then $BA - AB \neq \lambda I$ for every $B \in L_0(X)$ and $0 \neq \lambda \in \mathscr{F}$.*

Proof. Suppose that $A \in L_0(X)$ is algebraic and there exists an operator B and a scalar $\lambda \neq 0$ such that $BA - AB = \lambda I$. Denote by $P(t)$ $= \sum_{k=0}^{N} p_k t^k, p_N = 1$, the characteristic polynomial of A. By an easy induction, we find

$$BA^n - A^n B = n\lambda A^{n-1} \quad (n = 1, 2, ...). \tag{8.1.10}$$

Indeed, suppose Equality (8.1.10) to be true for every n. Then $BA^{n+1} - A^{n+1}B = BA^{n+1} - ABA^n + ABA^n - A^{n+1}BA = (BA - AB)A^n + A(BA^n - A^n B) = \lambda I A^n + n A \lambda A^{n-1} = (n+1)\lambda A^n$.

Since $P(A) = 0$, $\lambda \neq 0$, Equality (8.1.10) implies

$$0 = BP(A) - P(A)B = \sum_{k=1}^{N} p_k (BA^k - A^k B) = \sum_{k=1}^{N} k p_k A^{k-1}$$

$$= \sum_{m=0}^{N-1} (m+1) p_{m+1} A^m.$$

We therefore have proved that there exists a polynomial $P'(t)$ of degree $N-1 = \deg P - 1$ such that $P'(A) = 0$. This is a contradiction with our assumption that $P(t)$ is a characteristic polynomial for A. ∎

An immediate consequence of Theorem 8.1.4 is

COROLLARY 8.1.2. *Suppose that X is a linear space over an algebraically closed field $\mathscr{F}$ and $A \in L_0(X)$. Let $P(t) \neq t$ be an arbitrary polynomial with coefficients in $\mathscr{F}$. Then linear operators T such that $AT - TA = I$ on $\ker P(A)$ do not exist.*

Proof. Indeed, by definition, the operator A is algebraic on the space $\ker P(A)$ for $P(A)x = 0$ if $x \in \ker P(A)$. ∎

THEOREM 8.1.5. *Suppose that X is a complete linear metric space over C, $D \in R(X)$ is locally algebraic and $\ker D \neq \{0\}$. Then D is not continuous.*

Proof. Suppose that the operator $D \in R(X)$ is locally algebraic and continuous. Then, by Kaplansky Theorem 8.1.2, D is algebraic. On the other hand, Theorem 3.4.1 implies that $\alpha_{P(D)} = \alpha_P \deg P$ for every polynomial $P(t)$ with coefficients in $\mathbf{C}$. The same theorem implies that $\alpha_{P(D)Q(D)} = \alpha_{P(D)} + \alpha_{Q(D)}$. Indeed, if

$$P(t) = \prod_{j=1}^{n} (t - t_j)^{r_i}, \quad r_1 + \ldots + r_n = N = \deg P,$$

$$t_i \neq t_j \text{ for } i \neq j,$$

$$Q(t) = \prod_{k=1}^{m} (t - \tau_k)^{\varrho_k}, \quad \varrho_1 + \ldots + \varrho_m = M = \deg Q,$$

$$\tau_i \neq \tau_j \text{ for } i \neq j,$$

then

$$P(t)Q(t) = \left[\prod_{j=1}^{n} (t - t_j)^{r_j} \right] \left[\prod_{k=1}^{m} (t - \tau_k)^{\varrho_k} \right] = \prod_{j=1}^{n+m} (t - s_j)^{\sigma_j},$$

where

$$s_j = \begin{cases} t_j & \text{for } 1 \leqslant j \leqslant n, \\ \tau_j & \text{for } n+1 \leqslant j \leqslant m, \end{cases} \quad \sigma_j = \begin{cases} r_j & \text{for } 1 \leqslant j \leqslant n, \\ \varrho_j & \text{for } n+j \leqslant j \leqslant m. \end{cases}$$

Hence $\deg PQ = M + N$ and $\alpha_{P(D)Q(D)} = (\deg PQ)\alpha_D = (M+N)\alpha_D = M\alpha_D + N\alpha_D = \alpha_{P(D)} + \alpha_{Q(D)}$.

Since D is locally algebraic, all conditions of Theorem 8.1.3 are satisfied. We therefore conclude that there is an operator $T \in L_0(X)$ such that $DT - TD = I$ on dom D. By Corollary 8.1.2 this is a contradiction with our assumption that D is algebraic. Thus D is not continuous. ∎

REMARK 8.1.1. If X is a complete linear metric space, $D \in R(X)$, ker D is closed and there is a continuous $R \in \mathcal{R}_D$, then D is closed.

PROPOSITION 8.1.2. *Let X be a Leibniz D-algebra with unit e. Then there exists an operator $T \in L_0(X)$ such that $DT - TD = I$ on dom D, namely, $Tx = gx$ for all $x \in X$, where $g = Re$ and $R \in \mathcal{R}_D$.*

Proof. Let $Tx = gx$ for all $x \in X$, where $g = Re$ and $R \in \mathcal{R}_D$. Then, by definition, for all $x \in$ dom D we have $DTx - TDx = D(gx) - gDx = D[(Re)x] - gDx = (DRe)x + (Re)Dx - gDx = ex + gDx - gDx = x$. The arbitrariness of $x \in$ dom D implies $DT - TD = I$ on dom D. ∎

This, and Theorem 8.1.5, together imply

COROLLARY 8.1.3. *Suppose that all assumptions of Proposition* 8.1.2 *are satisfied. Then D is not an algebraic operator on* dom *D.*

Note that in the proofs of Proposition 8.1.2 and Corollary 8.1.3 the assumption that X is a Leibniz D-algebra is essential.

COROLLARY 8.1.4. *Suppose that all assumptions of Proposition* 8.1.2 *are satisfied and* $Tx = gx$ *for all* $x \in X$, *where* $g = Re$. *Then*

$$T^n z = z(Re)^n \quad \text{for all } z \in \ker D, \, n \in \mathbf{N}.$$

Proof. Indeed, $Tz = zRe$ by our definition. Suppose that $T^n z = z(Re)^n$ for an arbitrarily fixed positive integer n and for all $z \in \ker D$. Then $T^{n+1}z = T(T^n z) = T[z(Re)^n] = (Re)z(Re)^n = z(Re)^{n+1}$. ∎

THEOREM 8.1.6. *Suppose that X is a complete linear metric space, $D \in R(X)$ is closed,* dim $\ker D \neq 0$, $Q(D) = \sum\limits_{k=0}^{N} Q_k D^k$, *where* $Q_0, \ldots, Q_{N-1} \in L_0(X)$, *is closed,* $P(R) \subset \ker Q(D)$ *and* $\overline{P(R)} = X$.* *Then* $Q(D) = 0$ *if and only if*

$$Q(D)R^k z = 0 \quad \text{for all } z \in \ker D.$$

Proof. Necessity is obvious. Sufficiency follows from the fact that, by our assumptions, we have $X = \overline{P(R)} \subset \overline{\ker Q(D)} = \ker Q(D)$, which implies $Q(D) = 0$. ∎

THEOREM 8.1.7. *Suppose that X is a Banach space, $D \in R(X)$,* dim $\ker D \neq 0$ *and* $R \in \mathcal{R}_D$ *is quasinilpotent, i.e.* $\lim\limits_{n \to \infty} \sqrt[n]{||R^n||} = 0$. *Then*

(i) $R \in V(X)$;
(ii) *D-shifts exist and coincide with R-shifts.*

Proof. By our assumption the series $\sum\limits_{n=0}^{\infty} \lambda^n R^n$ is convergent, hence the

---

* i.e. the set of all D-polynomials is dense in X, cf. also Remark 8.1.1 on p. 484.

operators $(I- \lambda R)^{-1} = \sum\limits_{n=0}^{\infty} \lambda^n R^n$ exist for all $\lambda \in C$. This implies that $R \in V(X)$ and exponentials $e_\lambda(z) = (I- \lambda R)^{-1}$ exist for all $\lambda \in C$, $z \in \ker D$. Hence there also exist D-shifts. Let $z \in \ker D$ be arbitrarily fixed and let $\{S_h\}_{h \in \mathbf{R}}$ be a family of R-shifts. Then, by definition, for all $h \in \mathbf{R}$, $\lambda \in C$ we have

$$S_h e_\lambda(z) = S_h(I- \lambda R)^{-1}z = S_h \sum_{n=0}^{\infty} \lambda^n R^n z = \sum_{n=0}^{\infty} \lambda^n S_h R^n z$$

$$= \sum_{n=0}^{\infty} \lambda^n \sum_{j=0}^{n} \frac{h^{n-j}}{(n-j)!} R^j z = \left(\sum_{n=0}^{\infty} \lambda^n \sum_{k=0}^{n} \frac{h^k}{k!} R^{n-k} \right) z$$

$$= \left(\sum_{k=0}^{\infty} \frac{h^k}{k!} \sum_{n=k}^{\infty} \lambda^n R^{n-k} \right) z$$

$$= \left(\sum_{k=0}^{\infty} \frac{h^k}{k!} \sum_{m=0}^{\infty} \lambda^{k+m} R^m \right) z = \left(\sum_{k=0}^{\infty} \frac{h^k}{k!} \lambda^k \sum_{m=0}^{\infty} \lambda^m R^m \right) z$$

$$= \left(\sum_{k=0}^{\infty} \frac{h^k}{k!} \lambda^k \right)(I- \lambda R)^{-1}z = e^{\lambda h} e_\lambda(z),$$

which means that R-shifts are also D-shifts. A similar proof shows that D-shifts are R-shifts.   ∎

PROPOSITION 8.1.3. *Suppose that* $D \in R(X)$, $\dim \ker D \neq 0$, $R \in \mathcal{R}_D$ *and either* $\{S_h\}_{h \in \mathbf{R}}$ *is a family of* R-shifts *or* $R \in V(X)$ *and* $\{S_h\}_{h \in \mathbf{R}}$ *is a family of* D-shifts. If $y \in X$ *and the series* $\sum\limits_{n=0}^{\infty} S_{nh} y$ *is convergent* (*in the norm*) *then the equation*

$$(I-S_h)x = y \tag{8.1.11}$$

has a unique solution $x = \sum\limits_{n=0}^{\infty} S_{nh} y$.

Proof. Indeed, $(I-S_h)x = x - S_h x = \sum\limits_{n=0}^{\infty} (S_{nh} y - S_h S_{nh} y) = \sum\limits_{n=0}^{\infty} (S_{nh} - S_{(n+1)h})y = y$ (cf. also Exercise 6.2.15).   ∎

Other properties of right invertible operators in Banach spaces were considered recently by Tasche, 1982, 1986, Delvos and Schempp, 1983, Dimovski, 1982, Grabowski, 1983. Existence, or non-existence of right inverses in specified C^∞ spaces, has been studied by Cohoon, 1969, 1970, 1971 and Trèves, 1970.

Examples and Exercises

EXAMPLE 8.1.1. Suppose that X is a Leibniz D-algebra with unit e and F is an almost averaging initial operator for D corresponding to an $R \in \mathscr{R}_D$, i.e. $F(zx) = zFx$ for all $z \in \ker D$, $x \in X$. Recall Formula (6.2.12) for integration of unit

$$R^n e = \frac{(Re)^n}{n!} - \sum_{k=2}^{n} \frac{1}{k!} R^{n-k} F[(Re)^k] \quad \text{for } n \geqslant 2.$$

Suppose, moreover, that X is a *Banach algebra* over $\mathbf{C}$, i.e. a Banach space with the property that $||xy|| \leqslant ||x|| \cdot ||y||$ for $x, y \in X$ (with respect to the given addition, multiplication and multiplication by scalars). We can admit that $||e|| = 1$.

Assume that R and F are bounded. Then:

(i) If $|\lambda| < 1/||R||$ then the series $\sum_{n=0}^{\infty} \lambda^n R^n e$ is convergent, exponentials exist and are of the form

$$e_\lambda(z) = z \sum_{n=0}^{\infty} \lambda^n R^n e \quad \text{for } |\lambda| < 1/||R||, z \in \ker D; \qquad (8.1.12)$$

(ii) If F is multiplicative then the series $\sum_{n=0}^{\infty} \lambda^n R^n e$ is convergent for all $\lambda \in \mathbf{C}$. Therefore exponentials exist for all $\lambda \in \mathbf{C}$ and are of the form (8.1.12).

Indeed, for all $n \in \mathbf{N}$ we have

$$\left\| \sum_{k=0}^{n} \lambda^k R^k e \right\| = \left\| \sum_{k=0}^{n} \lambda^k \left\{ \frac{(Re)^k}{k!} - \sum_{j=2}^{k} \frac{1}{j!} R^{k-j} F[(Re)^j] \right\} \right\|$$

$$\leqslant \sum_{k=0}^{n} |\lambda|^k \left[\frac{||R||^k}{k!} + \sum_{j=2}^{k} \frac{1}{j!} ||R||^{k-j} ||F|| \cdot ||R||^j \right]$$

$$= \sum_{k=0}^{n} |\lambda|^k \|R\|^k \left[\frac{1}{k!} + \|F\| \sum_{j=2}^{k} \frac{1}{j!} \right]$$

$$\leqslant \sum_{k=0}^{n} \frac{|\lambda|^k \|R\|^k}{k!} + \|F\| \sum_{k=0}^{n} |\lambda|^k \|R\|^k$$

$$\leqslant e^{|\lambda| \cdot \|R\|} + \|F\| \sum_{k=0}^{n} |\lambda|^k \|R\|^k$$

because

$$\sum_{j=2}^{k} \frac{1}{j!} \leqslant \sum_{j=2}^{\infty} \frac{1}{j!} = e - 2 < 1.$$

We therefore conclude that the series in question is convergent for $|\lambda| < 1/\|R\|$. Let $z \in \ker D$ and $|\lambda| < 1/\|R\|$. Then

$$\sum_{n=0}^{\infty} \lambda^n R^n z = z \sum_{n=0}^{\infty} \lambda^n R^n e = z (I - \lambda R)^{-1} e.$$

Hence exponentials are

$$e_\lambda(z) = (I - \lambda R)^{-1} z = \sum_{n=0}^{\infty} \lambda^n R^n z = z (I - \lambda R)^{-1} e$$

for $z \in \ker D$ and $|\lambda| < 1/\|R\|$.

Suppose now that F is multiplicative. Since $FR = 0$, we have $F[(Re)^k]$ $= (FRe)^k = 0$ and $R^n e = (Re)^n/n!$ for all $n \in \mathbf{N}$. This implies that for all $\lambda \in \mathbf{C}$

$$\left\| \sum_{n=0}^{\infty} \lambda^n R^n e \right\| = \left\| \sum_{n=0}^{\infty} \lambda^n \frac{(Re)^n}{n!} \right\| \leqslant \sum_{n=0}^{\infty} \frac{|\lambda|^n \|R\|^n}{n!} = e^{|\lambda| \cdot \|R\|}.$$

Then exponentials exist and are of the form (8.1.12).

EXERCISE 8.1.1. Suppose that $X = C[0, T]$, $T > 0$, $D = d/dt$, $(Fx)(t)$ $= \dfrac{1}{T} \int_0^T x(s) \, ds$ for $x \in X$. Prove that:

(i) F is not a multiplicative initial operator for D and $R \notin V(X)$;

(ii) Exponentials exist for $|\lambda| < 2/3T$.

EXERCISE 8.1.2. Suppose that X is a Leibniz D-algebra with unit e and that F is a multiplicative initial operator for D, corresponding to R. Prove that R-shifts are of the form: $S_h R^k z = z(Re+he)^k/k!$ for all $h \in \mathbf{R}, k \in \mathbf{N} \cup \{0\}, z \in \ker D$.

EXERCISE 8.1.3. Suppose that all assumptions of Example 8.1.1 are satisfied and the operator F is multiplicative. Prove that:

   (i) D-shifts exist and coincide with R-shifts;

   (ii) In particular, for all $\lambda \in \mathbf{C}, z \in \ker D$,

$$S_h e_\lambda(z) = z e^{\lambda(Re+he)}, \quad \text{where } e^{\lambda(Re+he)} = \sum_{n=0}^{\infty} \lambda^n \frac{(Re+he)^n}{n!}.$$

EXERCISE 8.1.4. Suppose that X is a simple Duhamel algebra with unit e and F is an almost averaging initial operator for D corresponding to an $R \in \mathscr{R}_D$. Prove that:

   (i) R-shifts exist;

   (ii) If X is a Banach algebra over $\mathbf{C}$ (with respect to the given addition, multiplication and multiplication by scalars), F and R are bounded, then:

   (a) The series $\sum_{n=0}^{\infty} \lambda^n R^n e$ is convergent for $|\lambda| < 1/\|R\|$;

   (b) Exponentials exist for $|\lambda| < 1/\|R\|$ and are of the form

$$e_\lambda(z) = z \sum_{n=0}^{\infty} \lambda^n R^n e, \quad \text{where } z \in \ker D;$$

   (c) $R \notin V(X)$ (even in the case when F is multiplicative);

   (d) D-shifts exist for $|\lambda| < 1/\|R\|$.

EXERCISE 8.1.5. Suppose that X is a quasi-Leibniz D-algebra with unit e satisfying the condition

$$e + dDe = 0 \quad (d \neq 0) \tag{8.1.13}$$

and F is an initial operator for D corresponding to $R \in \mathscr{R}_D$. Prove that:

   (i) If the operator $I + dR$ is invertible then the unit e is an exponential for D corresponding to the eigenvalue $\lambda = -d$;

(ii) If $R \in V(X)$ then either $d = 1$ or $d = -1$;

(iii) The following formula holds for all $n \geqslant 2$:

$$R^n e = \frac{(-1)^n}{d^n} \left[e + \sum_{k=0}^{n-1} (-1)^{k+1} d^k R^k F e \right]; \qquad (8.1.14)$$

(iv) If F is almost averaging, X is a Banach algebra (over **C** with respect to the given addition, multiplication and multiplication by scalars) and both, F and R, are bounded then for all $|\lambda| < d/\|R\|$, the series $\sum_{n=0}^{\infty} \lambda^n R^n e$ is convergent, exponentials exist and are of the form

$$e_\lambda(z) = z \sum_{n=0}^{\infty} \lambda^n R^n e, \text{ where } z \in \ker D \text{ and } D\text{-shifts exist.}$$

EXERCISE 8.1.6. Suppose that all assumptions of Proposition 8.1.3 are satisfied and the series $\sum_{n=0}^{\infty} S_{nh} y$ is convergent for all $y \in X \backslash \ker (I - S_h)$ $= X \backslash X_{S_h}$. Prove that on this set the operator $I - S_h$ is invertible and $(I - S_h)^{-1} = \sum_{n=0}^{\infty} S_{nh}$ on $X \backslash X_{S_h}$ (cf. also Exercise 6.2.15).

8.2. CANONICAL MAPPING AND SEMIGROUPS

We have introduced by Definition 5.4.1 canonical mappings $\varkappa$ for R-shifts and D-shifts. We shall study now metric properties of these mappings. Theorem 5.4.1 immediately implies the following:

THEOREM 8.2.1. *Suppose that X is a complete linear metric space, $D \in R(X)$, $\dim \ker D \neq 0$, F is a continuous initial operator for D corresponding to an $R \in \mathscr{R}_D$, $S_{A(R)}$ is a family of continuous R-shifts and the set $P(R)$ of D-polynomials is dense in $X: \overline{P(R)} = X$. Then the induced canonical mapping is an isomorphism: $\ker \varkappa = \{0\}$, so that $\varkappa$ separates points, i.e. $\varkappa x = \varkappa y$ if and only if $x = y$ for $x, y \in X$. In particular, $x^\wedge = y^\wedge$ if and only if $x = y$.*

Theorem 5.4.2 immediately implies the following:

THEOREM 8.2.2. *Suppose that X is a complete linear metric locally convex space, $D \in R(X)$, dim ker $D \neq 0$, F is a continuous initial operator for D corresponding to an $R \in \mathscr{R}_D \cap V(X)$, $S_{A(\mathbf{R})}$ is a family of continuous D-shifts and the set $E(R)$ of exponentials is dense in X: $\overline{E(R)} = X$. Then the induced canonical mapping is an isomorphism: $\ker \varkappa = \{0\}$. So that $\varkappa$ separates points, i.e. $\varkappa x = \varkappa y$ if and only if $x = y$ for $x, y \in X$. In particular, $x^\wedge = y^\wedge$ if and only if $x = y$.*

DEFINITION 8.2.1 (cf. Yosida, 1965). Suppose that X is a complete linear metric locally convex space. A semigroup of continuous operators $\{S_t\}_{t \in A(\mathbf{R})} \subset L_0(X)$ is said to be *strongly continuous* if $S_0 = I$ and

(i) For every $x \in X$ the mapping $x \to S_t x$ is continuous with respect to t;

(ii) There exists a limit

$$Dx = \lim_{t \to +0} \frac{1}{t}(S_t - I)x \quad \text{for all } x \in X_0 \subset X. \tag{8.2.1}$$

It is easy to verify that D is a linear operator with dom $D = X_0$. The operator D is said to be an *infinitesimal generator for the semigroup* $\{S_t\}_{t \in A(\mathbf{R})}$.

By definition and properties of strongly continuous semigroups we have

$$\overline{\text{dom } D} = X \quad \text{and } DS_t = S_t D \quad \text{on dom } D \quad \text{for all } t \in A(\mathbf{R}). \tag{8.2.2}$$

Moreover, if X is a Banach space then there exist $M > 0$ and $\beta < +\infty$ such that

$$\|S_t x\| \leqslant M e^{\beta|t|} \|x\| \quad \text{for all } t \in A(\mathbf{R}) \text{ and } x \in X. \tag{8.2.3}$$

THEOREM 8.2.3. *Suppose that X is a Banach space, $D \in R(X)$, dim ker $D \neq 0$, F is a bounded initial operator for D corresponding to an $R \in \mathscr{R}_D \cap \cap V(X)$, $S_{A(\mathbf{R})} = \{S_h\}_{h \in A(\mathbf{R})}$ is a strongly continuous semigroup (group) of D-shifts. If D is an infinitesimal generator for $S_{A(\mathbf{R})}$ then $x^\wedge \in C^1(A(\mathbf{R}))$ for all $x \in \text{dom } D$ and*

$$\frac{d}{dh} x^\wedge(h) = (Dx)^\wedge(h) \quad \text{for all } x \in \text{dom } D, h \in A(\mathbf{R}) \tag{8.2.4}$$

(recall that $x^\wedge(h) = FS_h x$ for all $x \in X$ and $h \in A(\mathbf{R})$).

Proof. Suppose that D is an infinitesimal generator for $S_{A(\mathbf{R})}$. It means that $Dx = \lim_{h \to +0} h^{-1}(S_h - I)x$ for all $x \in \operatorname{dom} D$. Write $y = Dx$ for all $x \in \operatorname{dom} D$. Then for all $h \in A(\mathbf{R})$ inequality (8.2.3) implies

$$
\left\| \frac{x^{\wedge}(h+r) - x^{\wedge}(h)}{r} - y^{\wedge}(h) \right\| = \left\| \frac{1}{r}(FS_{h+r}x - FS_h x) - FS_h y \right\|
$$

$$
= \left\| FS_h \frac{1}{r}(S_r - I)x - Dx \right\| \leqslant \|F\| \cdot \|S_h\| \cdot \left\| \frac{1}{r}(S_r - I)x - Dx \right\|
$$

$$
\leqslant Me^{\beta r}\|F\| \cdot \left\| \frac{1}{r}(S_r - I)x - Dx \right\| \to 0 \quad \text{as } r \to +0.
$$

This implies that

$$
y^{\wedge}(h) = \lim_{r \to +0} \frac{x^{\wedge}(h+r) - x^{\wedge}(h)}{r} = x^{\wedge\prime}(h) \quad \text{for all } h \in A(\mathbf{R}),
$$

i.e. $x^{\wedge\prime}(h) = FS_h Dx = (Dx)^{\wedge}(h)$ for all $x \in \operatorname{dom} D$, $h \in A(\mathbf{R})$. ∎

PROPOSITION 8.2.1. *Let X be a complete linear metric locally convex space, $D \in R(X)$, $\ker D \neq \{0\}$ and let $S_{A(\mathbf{R})}$ be a strongly continuous semigroup of D-shifts. If A is an infinitesimal generator for $S_{A(\mathbf{R})}$ then*

$$
D|_{E(\mathbf{R})} = A|_{E(\mathbf{R})}. \tag{8.2.5}
$$

Proof. Indeed, by definition, $E(R) \subset \operatorname{dom} D$. Let $0 \neq \lambda \in \mathbf{C}$ and $0 \neq z \in \ker D$ be arbitrarily fixed. Since the multiplication of elements by scalars is continuous in the space under consideration, we find

$$
\left[\frac{1}{h}(S_h - I) - D\right] e_\lambda(z) = \left[\frac{1}{h}(e^{\lambda h} - 1) - \lambda\right] e_\lambda(z)
$$

$$
= \lambda\left(\frac{e^{\lambda h} - 1}{\lambda h} - 1\right) e_\lambda(z) \to 0 \quad \text{as } h \to +0.
$$

This means that $\lim_{h \to +0} h^{-1}(S_h - I)x$ exists for $x \in E(R)$ and is equal to D, which proves Formula (8.2.5). ∎

THEOREM 8.2.4. *Suppose that X is complete linear metric locally convex space, $D \in R(X)$ is closed, $\dim \ker D \neq 0$, $R \in \mathcal{R}_D \cap V(X)$, $\overline{E(R)} = X$*

and $S_{A(\mathbf{R})} = \{S_h\}_{h \in A(\mathbf{R})}$ *is a strongly continuous semigroup (group) of D-shifts. Then D is an infinitesimal generator for $S_{A(\mathbf{R})}$, so that* $\overline{\text{dom } S} = X$ *and* $S_h D = D S_h$ *on dom D for all* $h \in A(\mathbf{R})$ *(i.e. D-shifts are D-invariant).*

Proof. Since $S_{A(\mathbf{R})}$ is strongly continuous, we conclude that $S_{A(\mathbf{R})}$ has an infinitesimal generator A and $\overline{\text{dom } A} = X$. Proposition 8.2.1 implies that $D|_{E(R)} = A|_{E(R)}$. But $\overline{E(R)} = X$ and both, A and D, are closed. Thus $D = A$ on dom D. Since $\overline{E(R)} = X$ and $\overline{E(R)} \subset \text{dom } D$, we have $\overline{\text{dom } D} = X$. We therefore conclude that D is an infinitesimal generator for $S_{A(\mathbf{R})}$ and $S_h D = D S_h$ on dom D for all $h \in A(\mathbf{R})$. ∎

PROPOSITION 8.2.2. *Suppose that X is a complete linear metric locally convex space, $D \in R(X)$, $\dim \ker D \neq 0$ and $S_{A(\mathbf{R})} = \{S_h\}_{h \in A(\mathbf{R})}$ is a strongly continuous semigroup of R-shifts. If A is an infinitesimal generator for $S_{A(\mathbf{R})}$ then*

$$D|_{P(R)} = A|_{P(R)}. \tag{8.2.6}$$

Proof. Indeed, by definition, $P(R) \subset \text{dom } D$. Let $k \in \mathbb{N}$ and $0 \neq z \in \ker D$ be arbitrarily fixed. Since the multiplication by scalars is continuous in the space X we find

$$\left[\frac{1}{h}(S_h - I) - D \right] R^k z = \frac{1}{h}(S_h R^k z - R^k z) - D R^k z$$

$$= \frac{1}{h}\left(\sum_{j=0}^{k} \frac{h^{k-j}}{(k-j)!} R^j z - R^k z \right) - R^{k-1} z$$

$$= \frac{1}{h}\sum_{j=0}^{k-1} \frac{h^{k-j}}{(k-j)!} R^j z = \sum_{j=0}^{k-1} \frac{h^{k-1-j}}{(k-j)!} R^j z - R^{k-1} z$$

$$= \sum_{j=0}^{k-2} \frac{h^{k-1-j}}{(k-j)!} R^j z = h \sum_{j=0}^{k-2} \frac{h^{k-2-j}}{(k-j)!} R^j z \to 0 \quad \text{as } h \to +0.$$

This means that $\lim\limits_{h \to +0} \frac{1}{h}(S_h - I)$ exists on the set $P(R)$ and is equal to D, which proves Formula (8.2.6). ∎

THEOREM 8.2.5. *Suppose that X is a complete linear metric locally convex space, $D \in R(X)$ is closed, $\ker D \neq \{0\}$, $R \in \mathcal{R}_D$, $\overline{P(R)} = X$ and $S_{A(\mathbf{R})} = \{S_h\}_{h \in A(\mathbf{R})}$ is a strongly continuous semigroup (group) of R-shifts. Then D is an infinitesimal generator for $S_{A(\mathbf{R})}$, so that $\overline{\mathrm{dom}\, D} = X$ and $S_h D = D S_h$ on $\mathrm{dom}\, D$ for all $h \in A(\mathbf{R})$ (i.e. R-shifts are D-invariant).*

Proof. The proof proceeds along the same lines as the proof of Theorem 8.2.4 by means of an application of Proposition 8.2.2. ∎

THEOREM 8.2.6. *Suppose that X is a Banach space, $D \in R(X)$ is closed, $\dim \ker D \neq 0$, $R \in \mathcal{R}_D$ is bounded, $\overline{P(R)} = X$ and $S_{A(\mathbf{R})} = \{S_h\}_{h \in A(\mathbf{R})}$ is a family of R-shifts. If the operator R has the following property*

$$\|R^k\| \leqslant M \frac{\|R\|^k}{k!} \quad \text{for all } k \in \mathbf{N} \quad (M > 0) \tag{8.2.7}$$

then S_h are bounded R-shifts and there exists a constant $c > 0$ such that

$$\|S_h x\| \leqslant c e^{|h|}\|x\| \quad \text{for all } x \in X \text{ and } h \in A(\mathbf{R}). \tag{8.2.8}$$

Proof. Write $r = \|R\|$. Let $z \in \ker D$ be arbitrarily fixed. Then for all $k \in \mathbf{N}$ and $h \in A(\mathbf{R})$ we have

$$\|S_h R^k z\| = \left\| \sum_{j=0}^{k} \frac{|h|^{k-j}}{(k-j)!} \, ^j z \right\| \leqslant M \sum_{j=0}^{k} \frac{|h|^{k-j}}{(k-j)!} \frac{r^j}{j!} \|z\|$$

$$\leqslant M \frac{\|z\|}{k!} \sum_{j=0}^{k} \binom{k}{j} |h|^{k-j} r^j = M \frac{\|z\|}{k!} (|h| + r)^k$$

$$\leqslant M e^{|h|+r} \|z\|.$$

By our assumptions, $\overline{P(R)} = X$. Hence if we put $M e^r = c$ then $\|S_h x\| \leqslant M e^{|h|+r}\|x\| = c e^{|h|}\|x\|$ for all $x \in X$. We therefore conclude that S_h are bounded R-shifts. ∎

Theorems 8.2.4 and 8.2.5 together imply the following

THEOREM 8.2.7. *Suppose that X is a complete linear metric locally convex space, $D \in R(X)$ is closed, $\dim \ker D \neq 0$, F is a continuous initial operator for D corresponding to an $R \in \mathcal{R}_D$ and either*

(i) $\overline{P(R)} = X$ and $S_{A(R)} = \{S_h\}_{h \in A(R)}$ is a strongly continuous semigroup (group) of R-shifts

or

(ii) $R \in V(X), \overline{E(R)} = X$ and $S_{A(R)} = \{S_h\}_{h \in A(R)}$ is a strongly continuous semigroup (group) of D-shifts.

Then, for all $t \in A$ (R),

$$(Dx)^{\wedge}(t) = \frac{d}{dt} x^{\wedge}(t) \quad for \ x \in dom \ D,$$

$$where \ x^{\wedge}(t) = FS_t x, \quad (8.2.9)$$

$$(Rx)^{\wedge}(t) = \int_0^t x^{\wedge}(s)ds, \quad (Fx)^{\wedge}(t) = x^{\wedge}(0) \quad for \ all \ x \in X.$$

So that

$$\varkappa D = \frac{d}{dt}\varkappa, \quad \varkappa R = \int_0^t \varkappa, \quad (\varkappa Fx)(t) = (\varkappa x)(0) \quad for \ x \in X.$$

$$(8.2.10)$$

Proof. By our assumptions D is an infinitesimal generator for $S_{A(R)}$. This, and the continuity of the operator F together imply that for all $x \in dom \ D$, $t \in A(R) = R^+$ we have

$$(Dx)^{\wedge}(t) = FS_t Dx = FS_t \lim_{h \to +0} \frac{1}{h}(S_h - I)x$$

$$= \lim_{h \to +0} \frac{1}{h} S_t(S_h - I)x = \lim_{h \to +0} \frac{1}{h}(FS_{t+h}x - FS_t x)$$

$$= \lim_{h \to +0} \frac{x^{\wedge}(t+h) - x^{\wedge}(t)}{h} = \frac{d}{dt}x^{\wedge}(t).$$

Note that in the case $A(R) = R$ we have to consider $\lim_{h \to 0}$ and we will obtain the same result. The arbitrariness of $x \in X$ implies $\varkappa D = \frac{d}{dt}\varkappa$.

Let $x \in X$ be arbitrarily fixed and let $y = Rx$. Hence $x = Dy$. Observe that $y^{\wedge}(0) = FS_0 Rx = FRx = 0$. This means that $\int_0^t x^{\wedge}(s)ds$

$$= \int_0^t (Dy)^\wedge(s)\,ds = \int_0^t y^{\wedge\prime}(s)\,ds = y^\wedge(t) - y^\wedge(0) = y^\wedge(t) = (Rx)^\wedge(t).$$ The ar-

bitrariness of $x \in X$ implies $\varkappa R = \int_0^t \varkappa.$ Similarly, if $x \in$ dom D is arbitrarily

fixed then $(Fx)^\wedge(t) = [(I - RD)x]^\wedge(t) = x^\wedge(t) - (RDx)^\wedge(t) = x^\wedge(t) -$

$$-\int_0^t (Dx)^\wedge(s)\,ds = x^\wedge(t) - \int_0^t x^{\wedge\prime}(s)\,ds = x^\wedge(t) - x^\wedge(t) + x^\wedge(0) = x^\wedge(0).$$

The equality $(Fx)^\wedge(t) = x^\wedge(0)$ holds also for $x \notin$ dom D, as follows from properties of initial operators. The arbitrariness of $x \in X$ implies $(\varkappa Fx)(t) = (\varkappa x)(0).$ ∎

COROLLARY 8.2.1. *Suppose that all assumptions of Theorem 8.2.7 are satisfied. Then*

$$(D^k x)^\wedge(t) = x^{\wedge(k)}(t) \quad \text{for } x \in \text{dom } D^k,$$

$$\text{hence } \varkappa D^k = \frac{d^k}{dt^k}\varkappa \quad \text{for } k \in \mathbf{N}, \quad (8.2.11)$$

$$(R^k z)^\wedge(t) = \frac{t^k}{k!}z \quad \text{for } k \in \mathbf{N}, z \in \ker D. \quad (8.2.12)$$

If $R \in V(X)$ then

$$e_\lambda(z)^\wedge(t) = e^{\lambda t}z \quad \text{for all } z \in \ker D,\ \lambda \in \mathbf{C}. \quad (8.2.13)$$

Moreover, for all $h \in A(\mathbf{R})$

$$(S_h x)^\wedge(t) = x^\wedge(t+h) \quad \text{for all } x \in X,$$

$$\text{hence } (\varkappa S_h x)(t) = (\varkappa x)(t+h). \quad (8.2.14)$$

Proof. The first of Formulae (8.2.9) is just Formula (8.2.11) for $k = 1$. Suppose Formula (8.2.11) to be true for an arbitrarily fixed $k \in \mathbf{N}$. Then

$$\varkappa D^{k+1} = (\varkappa D^k)D = \frac{d^k}{dt^k}\varkappa D = \frac{d^k}{dt^k}\frac{d}{dt}\varkappa = \frac{d^{k+1}}{dt^{k+1}}\varkappa.$$

We therefore have proved by induction that Formula (8.2.11) holds for all $k \in \mathbf{N}$. We shall prove now also by induction Formula (8.2.12). For $k = 1$ we find $(Rz)^\wedge(t) = \int_0^t z^\wedge(s)\,ds = z \int_0^t ds = tz,$ for $z^\wedge(t) = FS_t z$

$= Fz = z$. Suppose Formula (8.2.12) to be true for arbitrarily fixed $k \in \mathbf{N}$. Then

$$(R^{k+1}z)^{\wedge}(t) = (R(R^k z))^{\wedge}(t) = \left(R \frac{t^k}{k!} z\right)(t) = z \int_0^t \frac{s^k}{k!} \, ds$$

$$= \frac{t^{k+1}}{(k+1)!} z.$$

If $R \in V(X)$ then the operators $e_\lambda = (I - \lambda R)^{-1}$ exist for all $\lambda \in \mathbf{C}$. Since we have $\varkappa(I - \lambda R) = \left(1 - \lambda \int_0^t\right) \varkappa$, Formulae (2.2.39) and (2.2.40) together imply that

$$e_\lambda(z)^{\wedge}(t) = [(I - \lambda R)^{-1}z]^{\wedge}(t) = z + \lambda \int_0^t e^{\lambda(t-s)} z \, ds$$

$$= \left(1 + \lambda \int_0^t e^{\lambda(t-s)} ds\right) z = e^{\lambda t} z.$$

Suppose now that S_h are R-shifts for $h \in A(\mathbf{R})$. Then for all $k \in \mathbf{N} \cup \{0\}$ and $z \in \ker D$ Formula (8.2.12) implies

$$(S_h R^k z)^{\wedge}(t) = \sum_{j=0}^k \frac{h^{k-j}}{(k-j)!} (R^j z)^{\wedge}(t)$$

$$= \left(\sum_{j=0}^k \frac{h^{k-j}}{(k-j)!} \frac{t^j}{j!}\right) z = \frac{z}{k!} \sum_{j=0}^k \binom{k}{j} h^{k-j} t^j$$

$$= \frac{z}{k!} (t+h)^k = (R^k z)^{\wedge}(t+h).$$

Since $\overline{P(R)} = X$ by our assumption we conclude that $(S_h x)^{\wedge}(t) = x^{\wedge}(t+h)$ for all $x \in X$, $h \in A(\mathbf{R})$. Suppose now that $R \in V(X)$ and S_h are D-shifts for $h \in A(\mathbf{R})$.

Formula (8.2.13) implies that for all $\lambda \in \mathbf{C}$, $z \in \ker D$

$$[S_h e_\lambda(z)]^{\wedge}(t) = e^{\lambda h}[e_\lambda(z)]^{\wedge}(t) = e^{\lambda h} e^{\lambda t} z = e^{\lambda(t+h)} z$$

$$= [e_\lambda(z)]^{\wedge}(t+h).$$

Since $\overline{E(R)} = X$ by our assumption, we conclude that $(S_h x)^\wedge(t)$ $= x^\wedge(t+h)$ for all $x \in X$ and $h \in A(\mathbf{R})$. ∎

COROLLARY 8.2.2. *Suppose that all assumptions of Theorem 8.2.7 are satisfied. Then*:

(i) *The operators* $F_h = FS_h$ *for* $h \in A(\mathbf{R})$ *are initial operators for* D *corresponding to right inverses* $R_h = R - F_h R = R - FS_h R$ *and* $F_h \neq F$ *if* $h \neq 0$;

(ii) *If* $R \in V(X)$ *then* $R_h \in V(X)$ *for all* $h \in A(\mathbf{R})$;

(iii) *If* $a, b \in A(\mathbf{R})$ *are arbitrary fixed then for all* $x \in X$, $t \in A(\mathbf{R})$

$$(F_a R_b x)^\wedge(t) = \int_a^b x^\wedge(s)\,ds, \quad \text{hence } \varkappa F_a R_b = \int_a^b \varkappa. \tag{8.2.15}$$

Proof. Let $h \in A(\mathbf{R})$ be arbitrarily fixed. Then $F_h^2 = (FS_h)^2 = FS_h\,FS_h = F^2 S_h$ $= FS_h = F_h$, for $S_h F = F$. Hence F_h are projections. Since for all $z \in \ker D$ we have $F_h z' = FS_h z = Fz = z$, we conclude that F_h are projections onto ker D, hence are initial operators for D corresponding to right inverses $R_h = R - F_h R = R - FS_h R$, respectively (cf. Theorem 2.2.4). If $R \in V(X)$ then exponentials $e_\lambda(z) = (I - \lambda R)^{-1} z$ are well-defined ˙for all $\lambda \in \mathbf{C}$ and $z \in \ker D$. Let $z \neq 0$. Since $F_h e_\lambda(z) = FS_h e_\lambda(z) = Fe^{\lambda h} e_\lambda(z)$ $= e^{\lambda h} Fe_\lambda(z) = e^{\lambda h} z \neq 0$ for all $\lambda \in \mathbf{C}$, Theorem 2.3.4 implies that $R_h \in V(X)$. Let $a, b \in A(\mathbf{R})$ be arbitrarily fixed. Then $F_a R_b = R_a - R_b$ $= R - F_a R - (R - F_b R) = F_b R - F_a R = (F_b - F_a)R$ and

$$(F_b R_a x)^\wedge(t) = [(F_b - F_a) Rx]^\wedge(t) = (F_b Rx)^\wedge(t) - (F_a Rx)^\wedge(t)$$
$$= (FS_b Rx)^\wedge(t) - (FS_a Rx)^\wedge(t)$$
$$= \int_0^{t+b} x^\wedge(s)\,ds|_{t=0} - \int_0^{t+a} x^\wedge(s)\,ds|_{t=0}$$
$$= \int_0^b x^\wedge(s)\,ds - \int_0^a x^\wedge(s)\,ds = \int_0^b x^\wedge(s)\,ds + \int_a^0 x^\wedge(s)\,ds$$
$$= \int_a^b x^\wedge(s)\,ds.$$

The arbitrariness of $x \in X$ implies $\varkappa F_b R_a = \int_a^b \varkappa$. Observe that in the proof of Point (ii) we did not use metric assumptions. ∎

An immediate consequence of Formula (8.2.11) is

COROLLARY 8.2.3. *Suppose that all assumptions of Theorem 8.2.7 are satisfied. Write*

$$X_k^\wedge = \{x: x^\wedge \in C^{(k)}\big(A(\mathbf{R}), \ker D\big) \text{ and } |x^{\wedge(i)}(t)| < a_j \cdot j!, \ 0 < a_j \in \mathbf{R},$$

$$j = 0, 1, ..., k\}, \qquad (8.2.16)$$

$$X_\infty^\wedge = \bigcap_{k \in \mathbf{N} \cup \{0\}} X_k^\wedge, \qquad (8.2.17)$$

where, as before, $x^\wedge(t) = FS_t x$. Assume that the metric on the space $X_\infty^\wedge$ is defined by the almost uniform convergence on compact sets.

Let the function

$$f(t) = \sum_{k=0}^{\infty} a_k t^k \qquad (8.2.18)$$

be entire. Then $[f(D)x]^\wedge(t) = \sum_{k=0}^{\infty} a_k x^{\wedge(k)}(t)$ for $x \in X$, hence

$$\varkappa f(D) = f\left(\frac{d}{dt}\right)\varkappa \quad \text{on } X_\infty^\wedge. \qquad (8.2.19)$$

Examples and Exercises

EXAMPLE 8.2.1. Suppose that X is a Banach space, $D \in R(X)$ is closed, dim $\ker D \neq 0$, F is a bounded initial operator corresponding to a bounded right inverse R and $\{S_h\}_{h \in \mathbf{R}}$ is a strongly continuous family of R-shifts. All assumptions of Theorem 8.2.7 are satisfied. Since the series $\sum_{n=0}^{\infty} \lambda^n \|R\|^n / n!$ is convergent for all $\lambda \in \mathbf{C}$, we conclude that the operator $\sum_{n=0}^{\infty} \lambda^n R^n / n!$ is well-defined and we shall write

$$e^{\lambda R} = \sum_{n=0}^{\infty} \frac{\lambda^n}{n!} R^n \quad (\lambda \in \mathbf{C}). \qquad (8.2.20)$$

The operator $e^{\lambda R}$ is invertible and its inverse is $e^{-\lambda R}$. Corollary 8.2.1 implies that $(S_h e^{\lambda R} x)^\wedge(t) = (e^{\lambda R} x)^\wedge(t+h)$. In particular

$$(e^{\lambda R} z)^\wedge(t) = z \sum_{n=0}^{\infty} \frac{(\lambda t)^n}{(n!)^2} \quad \text{for all } z \in \ker D, \ \lambda \in \mathbf{C}.$$

EXAMPLE 8.2.2. Suppose that $X = C_0(\mathbf{R})$ is the space of all functions bounded and uniformly continuous for $t \in \mathbf{R}$ and $D = d/dt$.

The operator D is not right invertible in X. Indeed, constants are bounded functions and a primitive function of a constant is not a bounded function: $\int_0^t c \, ds = ct$. Hence the operator $R = \int_0^t$ does not preserve the space X. However, in the space X there is defined a strongly continuous semigroup $\{S_h\}_{h \in \mathbf{R}}$ defined by means of *Gauss kernels*

$$(S_h x)(t) = \begin{cases} (2\pi h)^{1/2} \int_{-\infty}^{+\infty} e^{-(t-s)^2/2h} x(s) \, ds & \text{for } h > 0, \\ x(t) & \text{for } h = 0. \end{cases}$$

The infinitesimal generator for this semigroup is $A = \dfrac{1}{2} \dfrac{d^2}{dt^2}$ (cf .Yosida, 1965, Section IX.5, Example 2).

EXAMPLE 8.2.3. Suppose that all assumptions of Theorem 8.2.7 are satisfied, R is continuous and $A(\mathbf{R}) = \mathbf{R}$. Consider a linear equation with scalar coefficients:

$$W(D)x = \sum_{k=0}^{n} \sum_{j=0}^{m} a_{kj} D^k S_{-h_j} x = 0, \qquad 0 = h_0 < h_1 < \dots < h_m.$$
$$(8.2.21)$$

Then the *quasi-polynomial* for the operator $W(D)$ is of the form

$$W^\wedge(t) = \sum_{k=0}^{n} \sum_{j=0}^{m} a_{kj} t^k e^{-h_j t}. \qquad (8.2.22)$$

Indeed, Corollary 8.2.1 implies that

$$[W(D)e_\lambda(z)]^\wedge(t) = W^\wedge(t) e^{\lambda t} z \quad \text{for all } z \in \ker D, \; \lambda \in \mathbf{C}$$

and

$$[W(D)x]^\wedge(t) = \sum_{k=0}^{n} \sum_{j=0}^{m} a_{kj} (D^k S_{-h_j} x)^\wedge(t)$$

$$= \sum_{k=0}^{n} \sum_{j=0}^{m} a_{kj} x^{\wedge(k)}(t - h_j).$$

This means that to Equation (8.2.21) there corresponds a differential-difference equation of the form

$$\sum_{k=0}^{n}\sum_{j=0}^{m}a_{kj}x^{\wedge(k)}(t-h_j) = y^{\wedge}(t), \quad t \in \mathbf{R}, \tag{8.2.23}$$

and both have the same quasi-polynomial $W^{\wedge}(t)$ defined by Formula (8.2.22). Hence both Equations, (8.2.21) and (8.2.22), have similar properties (cf. Przeworska-Rolewicz, 1982). We can study in this way properties of solutions of Equation (8.2.21), in particular, their stability.

EXAMPLE 8.2.4. The following question has been posed by Przeworska-Rolewicz, 1980a: Are all Volterra right inverses of the operator D $= d/dt$ in the space $C[0, 1]$ of the form $R = \int_{a}^{t}, 0 < a < 1$. The answer is positive in the space $C_c[0, 1]$ over complexes and negative in the space $C_r[0, 1]$ over reals (cf. the author and S. Rolewicz, 1985). Namely, we have the following theorem:

(i) *Every continuous Volterra right inverse R for the operator $D = d/dt$ in the space $C_c[0, 1]$ is of the form $R = \int_{a}^{t}$ for an $a \in [0, 1]$.*

This implies the following corollaries:

(ii) *If a continuous right inverse R of the operator $D = d/dt$ is not a Volterra operator and has at least two eigenvalues then it has an infinite number of eigenvalues.*

(iii) *If F is an initial operator for $D = d/dt$ corresponding to a non-Volterra right inverse then the corresponding initial value problem*

$$x' - \lambda x = 0, \quad Fx = x_0 \quad (x_0 \text{ is a given constant})$$

is ill-posed and has infinitely many complex eigenvalues, i.e. infinitely many eigenvectors.

In the real case we can prove the following:

(iv) *Let $a, b \in [0, 1]$ and let*

$$(Rx)(t) = \int_{a}^{t}x(s)\,ds - \frac{1}{2}\int_{a}^{t}x(s)\,ds \quad \text{for } x \in C_r[0, 1].$$

Then R is a Volterra operator.

The result (i) can be extended to the space $C_c^\infty [0,1]$ with the classical metric of uniform convergence of all derivatives (cf. Przeworska-Rolewicz and S. Rolewicz, 1986). This result together with Theorem 8.2.7 permits us also to characterize Volterra right inverses in a complex space. Recall that Corollary 8.2.2, Point (i) gives a sufficient condition for a right inverse to be a Volterra operator and Theorem 2.3.4 gives a necessary and sufficient condition for a right inverse to be a Volterra operator (cf. also Heinig and Rost, 1984).

EXAMPLE 8.2.5. A consequence of results presented in Example 8.2.4 is: Suppose that all assumptions of Theorem 8.2.7 are satisfied. If $\mathscr{F} = \mathbf{C}$ then the only Volterra right inverses are $R_n = R - FS_h R$, where $R \in \mathscr{R}_D \cap V(X)$ (cf. Corollary 8.2.2). This is not true when $\mathscr{F} = \mathbf{R}$, as for $D = \mathrm{d}/\mathrm{d}t$.

EXERCISE 8.2.1. Suppose that X is the space $C(\mathbf{R})$ of all functions continuous on the real line, $D = \mathrm{d}/\mathrm{d}t$, $R = \int_0^t$, $(Fx)(t) = x(0)$, $(S_h x)(t) = x(t+h)$ for all $x \in X$, $h \in \mathbf{R}$. We know that $\{S_h\}_{h \in \mathbf{R}}$ is a family of R-shifts and D-shifts simultaneously. Write $\tilde{D} = D^2$, $\tilde{R} = R^2$. Prove that:

(i) $\tilde{D} \in R(X)$, $\tilde{R} \in \mathscr{R}_{\tilde{D}} \cap V(X)$, $\tilde{F} = F + RFD$ is an initial operator for $\tilde{D}$ corresponding to $\tilde{R}$ and $\ker \tilde{D} = \{c_0 + c_1 t : c_0, c_1 \in \mathbf{R}\}$;

(ii) S_h are neither $\tilde{R}$-shifts nor $\tilde{D}$-shifts;

(iii) Observe that X is a locally convex linear metric space with the metric induced by the countable family of pseudonorms $\|x\|_n = \sup_{\|t\| \leqslant n} |x(t)|$ for $n \in \mathbf{N}$. Prove that $P(\tilde{R}) = X$ and $\tilde{R}$-shifts and $\tilde{D}$-shifts coincide.

EXERCISE 8.2.2. Suppose that all assumptions of Theorem 8.2.7 are satisfied and $A \in L_0(X)$ is stationary. Write $\tilde{D} = D + FA = D + AF$. Prove that:

(i) $\tilde{D} \in R(X)$, $\tilde{R} = R \in \mathscr{R}_{\tilde{D}}$, $\tilde{F} = (I - AR)F$ is an initial operator for $\tilde{D}$ corresponding to R and $\ker \tilde{D} = (I - AR) \ker D$;

(ii) If $\{S_h\}_{h \in A(\mathbf{R})}$ is a family of R-shifts for D then $\{\tilde{S}_h\}_{h \in A(\mathbf{R})}$ is a family of $\tilde{R}$-shifts for $\tilde{D}$ where $\tilde{S}_h|_{P(\tilde{R})} = S_h|_{P(R)}$ and $P(\tilde{R}) = (I - AR)P(R)$;

(iii) If X is a complete linear metric locally convex space, $\overline{P(R)} = X$, D is closed, F and A are continuous and $\{S_h\}_{h \in A(\mathbf{R})}$ is a strongly continuous semigroup of R-shifts, $\{\tilde{S}_h\}_{h \in A(\mathbf{R})}$ is strongly continuous family of $\tilde{R}$-shifts, then

$$\tilde{S}_h = (I - AR) S_h \quad \text{on } X.$$

EXERCISE 8.2.3. Suppose that X is as in Point (iii) of Exercise 8.2.1, D, R, F, S_h ($h \in \mathbf{R}$) are defined, as in that Exercise and $A = dI$ where $d \neq 0$. Applying Exercise 8.2.2 prove that:

(i) $(\tilde{D}x)(t) = x'(t) + dx(0)$ for $x \in C^1(\mathbf{R})$, $(\tilde{F}x)(t) = (1 - dt)x(0)$ for $x \in C(\mathbf{R})$, $\ker \tilde{D} = \{c(1 - dt): c \in \mathbf{R}\}$;

(ii) $\tilde{R}$-shifts for $\tilde{D}$ are of the form

$$(\tilde{S}_h x)(t) = x(t+h) - d \int_0^t x(s+h)\,ds$$

for all $t, h \in \mathbf{R}$, $x \in X$.

EXERCISE 8.2.4. Suppose that $X = C(\mathbf{R} \times \mathbf{R})$ with pseudonorms defined as follows: $\|x\|_n = \sup\limits_{|t| \leq n, |s| \leq n} |x(t, s)|$ for $n \in \mathbf{N}$, $D = \partial/\partial t$,

$$(Rx)(t, s) = \int_0^t x(u, s)\,du, \quad (Fx)(t, s) = x(0, s),$$

$$(S_h x)(t, s) = x(t+h, s)$$

and $(Ax)(t, s) = a(s)x(t, s)$, $a \in C(\mathbf{R})$ is arbitrarily fixed, for all $x \in X$, $t, s, h \in \mathbf{R}$. Prove by applications of results of Exercise 8.2.2 that:

(i) A is stationary, $(\tilde{D}x)(t, s) = x_t'(t, s) + a(s)x(0, s)$, $\text{dom } \tilde{D} = \{x \in C(\mathbf{R}): x \in C^1(\mathbf{R})$ for every fixed $s \in \mathbf{R}\}$, $(\tilde{F}x)(t) = [1 - ta(s)]x(0, s)$ for $x \in X$ and $\ker \tilde{D} = \{[1 - ta(s)]g(s): g \in C(\mathbf{R})\}$;

(ii) $\tilde{R}$-shifts for the operator $\tilde{D}$ are of the form:

$$(\tilde{S}_h x)(t, s) = x(t+h, s) - a(s) \int_0^t x(u+h, s)\,du.$$

EXERCISE 8.2.5. Suppose that X, D, F, R, S_h are defined as in Exercise 8.2.2 and $R \in V(X)$. Determine the family of D-shifts.

EXERCISE 8.2.6. Let $X = C(\mathbf{R}^+)$, $D = t\dfrac{d}{dt}$, $(Rx)(t) = \int_a^t s^{-1}x(s)\,ds$ for an arbitrarily fixed $a > 0$, $A = dI$, where $d \neq 0$.

Applying results of Exercise 8.2.2 prove that D-shifts are of the form:

$$(S_h x)(t) = x(rt) - d\int_a^t s^{-1}x(rs)\,ds \quad \text{for } x \in X, \text{ where } h = \ln r, r > 0.$$

EXERCISE 8.2.7. Suppose that $D \in R(X)$, dim ker $D \neq 0$, F is an initial operator for D corresponding to an $R \in \mathcal{R}_D$, $\{S_h\}_{h \in \mathbf{R}}$ is a family of R-shifts (D-shifts, respectively), $A, B \in L_0(X)$ are stationary, the operator $I - AR = I - RA$ is invertible and $E_A = (I - AR)^{-1}$. Consider again the equation

$$Dx = Ax + BS_{-h}x + y, \quad y \in X \tag{8.2.24}$$

(cf. Section 5.5) together with the conditions

$$FS_t x = FS_t x_0 \quad \text{for } t \in [-h, 0], \quad \text{where } x_0 \in X \text{ is given,} \tag{8.2.25}$$

$$Fx = Fx_0 = x_0^0, \quad x_0^0 \in \ker D \text{ is given.} \tag{8.2.26}$$

Suppose, moreover, that all assumptions of Theorem 8.2.7 are satisfied. Our assumptions and Corollary 8.2.2 together imply that Equation (8.2.24) together with Conditions (8.2.25), (8.2.26) can be rewritten as

$$x^\wedge(t) = Ax^\wedge(t) + Bx^\wedge(t-h) + y^\wedge(t) \quad \text{for } t > 0, \tag{8.2.27}$$

$$x^\wedge(t) = x_0^\wedge(t) \quad \text{for } t \in [-h, 0] \quad \text{and } x^\wedge(0) = x_0^0. \tag{8.2.28}$$

The system (8.2.27)–(8.2.28) can be solved by the so-called *step-by-step method*, i.e. we are solving sequentially equations:

$$x_n^{\wedge\prime}(t) = Ax_n^\wedge(t) + y_n^\wedge(t) \tag{8.2.29}$$

with condition

$$x_n^\wedge\big((n-1)h\big) = x_{n-1}^\wedge\big((n-1)h\big) \quad \text{for } t \in [(n-1)h, nh],$$
$$y_n^\wedge(t) = y^\wedge(t) + Bx_{n-1}^\wedge(t-h), \quad n \in \mathbf{N}. \tag{8.2.29'}$$

The corresponding system is

$$(D-A)x_n = y_n, \quad y_n = y + BS_{-h}x_{n-1}, \quad x_n = FS_{-nh}x_{n-1}$$
$$(n \in \mathbf{N}). \tag{8.2.30}$$

Since $S_{-h}D = DS_{-h}$ on dom D, A is stationary and $E_A R(D-A) + E_A F = I$ for $(I-AR)[R(D-A)+F] = (I-AR)^{-1}(RD-RA+I-RD) = (I-AR)^{-1}(I-AR) = I$, then

$$S_{-nh}x_n = E_A R(D-A)S_{-nh}x_n + E_A + S_{-nh}x_n$$
$$= E_A RS_{-nh}(D-A)x_n + E_A FS_{nh}x_{n-1}$$
$$= E_A RS_{-nh}y_n + E_A FS_{-nh}x_{n-1}$$
$$= E_A[RS_{-nh}(y+BS_{-h}x_{n-1})+FS_{-nh}x_{n-1}]$$

and

$$x_n = S_{nh}E_A[RS_{nh}(y+BS_{-h}x_{n-1})+FS_{-nh}x_{n-1}] \quad \text{for } n \in \mathbb{N},$$
$$(8.2.31)$$

which is a solution of (8.2.30). Recall that x_0 is given by Condition (8.2.25). Write

$$R_{A,h} = S_h E_A RS_{-h}, \quad F_{A,h} = S_h E_A FS_{-h} \quad \text{for } h \in \mathbb{R}. \quad (8.2.32)$$

Then (8.2.31) can be rewritten as

$$x_n = R_{A,nh}(y+BS_{-h}x_{n-1})+F_{A,nh}x_{n-1} \quad \text{for } n \in \mathbb{N}. \quad (8.2.32a)$$

Let U be a linear space, let $u \in U$ and C be a linear operator with dom $C \subset U$ and with the range in X. Put $y = S_{-h}Cu$. Then

$$x_n = R_{A,nh}(S_{-h}C\hat{u}_n + BS_{-h}x_{n-1})+F_{A,nh}x_{n-1} \quad \text{for } n \in \mathbb{N},$$
$$(8.2.33)$$

where $\hat{u}_n(t) = FS_t u$ for $t \in [(n-1)h, nh]$.

Suppose that we consider a boundary condition

$$F_1 x = x^1, \quad \text{where } x^1 \in \ker D, \quad F_1 \neq F \quad (8.2.34)$$

(cf. Section 4.6). We say that F_1 is of the *local type* if

$$(F_1 x)(t) = x(T) \text{ for a } T > 0, \quad x(t) = 0 \text{ for } t > T. \quad (8.2.35)$$

If F_1 satisfies this condition then there exists an $n_0 \in \mathbb{N}$ such that $n_0 h \leqslant T \leqslant (n_0+1)h$. Then for $n = n_0$ we have

$$x^1 = F_1 x_{n_0} = F_1[R_{A,n_0h}(S_h Cu_{n_0} + BS_h x_{n_0-1})+F_{A,n_0h}x_{n_0-1}].$$

If there exists a $u_{n_0} \in U$ such that

$$R_{A,n_0h}(S_h Cu_{n_0} + BS_h x_{n_0-1})+F_{A,n_0h}x_{n_0-1} \in F_1^{(-1)}(x^1) \quad (8.2.36)$$

(i.e. belongs to the inverse image of x_1 by F_1) then the problem (8.2.30)–(8.2.34) has a solution of the form (8.2.33), where $u_0, \ldots, u_{n_0-1} \in U$

are arbitrarily fixed and $u_{n_0} \in U$ satisfies Condition (8.2.36) (cf. Prze-worska-Rolewicz, 1984).

EXERCISE 8.2.8. Suppose that all assumptions of Theorem 8.2.7 are satisfied, X is a Banach space, and D is bounded. Prove that:
  (i) $\cos \lambda D + i \sin \lambda D = e^{\lambda D}$ for $\lambda \in \mathbf{R}$;
  (ii) $[\cos \lambda D]^2 + [\sin \lambda D]^2 = I$ for $\lambda \in \mathbf{R}$.

EXERCISE 8.2.9. It is well-known (cf. Arscott, 1964; Ince, 1926), that every homogeneous ordinary differential equation with periodic coeffi-cients has at least one exponential-periodic solution. Suppose that all assumptions of Theorem 8.2.7 are satisfied, X is a Leibniz D-algebra, $R \in V(X)$, and

$$Q(D) = \sum_{k=0}^{n} Q_k D^k, \quad \text{where } Q_0, \ldots, Q_n \text{ are } S_h \text{ periodic}$$

and

$$(Q_k x)^\wedge(t) = q_k(t) x^\wedge(t) \quad \text{for all } t \in A(\mathbf{R}), \ x \in X$$
$$(k = 0, 1, \ldots, n),$$

where q_k are h-periodic functions. Prove that the equation $Q(D)x = 0$ has at least one exponential-periodic solution (cf. also Exercise 6.2.12).

EXERCISE 8.2.10. Let $X = (s)$ be the space of all sequences $\{x_n\} \subset \mathscr{F}$, where either $\mathscr{F} = \mathbf{R}$ or $\mathbf{C}$, $n \in \mathbf{N}$, with the usual addition and multi-plication by scalars. Consider the difference operator: $D\{x_n\} = \{x_{n+1} - x_n\}$. The operator F_1 defined as $F_1\{x_n\} = x_1\{e_n\}$, where $e_n = 1$ for $n \in \mathbf{N}$, is an initial operator for D corresponding to the Volterra right inverse R_1, where $R_1\{x_n\} = y_n$, $y_1 = 0$, $y_n = \sum_{k=1}^{n-1} x_k$ for $n \geqslant 2$ (cf. Example 2.1.4 and Exercise 2.2.10). Prove that:
  (i) The operators F_m defined by means of the equality:

$$F_m\{x_n\} = \left[\left(1 - \frac{a}{m^2}\right) x_{2m+1} + \frac{a}{m^2} x_1\right]\{e_n\},$$

where $a \in \mathscr{F}$, $m \in \mathbf{N}$, are initial operators for D;

(ii) If $\mathcal{F} = \mathbf{R}$ and $0 < a \leqslant m^2$ then the corresponding inverses $R_m = R_1 - F_m R_1$ are Volterra operators;

(iii) If $\mathcal{F} = \mathbf{R}$ and either $a \leqslant 0$ or $a > m^2$ then R_m are not Volterra operators;

(iv) If $\mathcal{F} = \mathbf{C}$ and $a \neq m^2$ then R_m are Volterra operators (note that for $a = m^2$ we have $F_m = F_1$ and $R_m = R_1$).

Are operators F_m defined by the equality

$$F_m\{x_n\} = \left[\left(1 - \frac{a}{m^2}\right)x_{2m+1} + \frac{b}{m^2}x_1\right]\{e_n\},$$

where $a, b \in \mathcal{F}$, $m \in \mathbf{N}$, initial operators for D if $b \neq a$? (cf. Kalfat, 1986), also 1986a (on p. 602)).

8.3. PERTURBATIONS OF PERIODIC PROBLEMS

We shall now consider perturbations of equations with shifts. Without loss of generality we can consider equations of order 1, because we have seen that equations of a higher order could be studied in a similar way as in preceding sections under appropriate assumptions. We have

THEOREM 8.3.1. *Suppose that $D \in R(X)$, $\dim \ker D \neq 0$, F is an initial operator for D corresponding to an $R \in \mathcal{R}_D$, S_h is an R-shift (on $0 \neq h \in A(\mathbf{R})$). Write*

$$A(S_h) = \sum_{k=0}^{n-1} A_k S_h^k, \quad \text{where } A_k S_h = S_h A_k \quad \text{for } k = 0, 1, \ldots, n-1.$$

$$(8.3.1)$$

Then a unique S_{nh}-periodic solution of the equation

$$Dx = A(S_h)x + y, \quad y \in X_{S_{nh}}, \tag{8.3.2}$$

satisfying the initial condition

$$Fx = x_0, \quad x_0 \in \ker D, \tag{8.3.3}$$

is of the form

$$x = [I - R_n^0 A(S_h)]^{-1} R_n^0\left(y + \frac{1}{nh}x_0\right), \tag{8.3.4}$$

provided that the operator $I - R_n^0 A(S_h)$ is invertible, where

$$R_n^0 = \left(I + FS_{nh} - \frac{1}{nh} RFS_{nh}\right) R. \tag{8.3.5}$$

Proof. Our assumptions and Theorem 5.2.3 together imply that $S_{nh}x = S_h^n x = x$ and $FS_{nh}x = Fx = x_0$, the operator $R_n^0 = \left(I + FS_{nh} - (1/nh) RFS_{nh}\right) R$ is invertible on the space $X_{S_{nh}}$ and

$$x = R_n^0 D_n x = R_n^0 \left(D + \frac{1}{nh} FS_{nh}\right)x = R_n^0 \left(Dx + \frac{1}{nh} Fx\right)$$

$$= R_n^0 A(S_h)x + y + \frac{1}{nh} x_0,$$

i.e.

$$I - R_n^0 A(S_h)x = R_n^0 \left(y + \frac{1}{nh} x_0\right). \tag{8.3.6}$$

Since the operator $I - R_n^0 A(S_h)$ is invertible by our assumptions, we conclude that x is of the required form. ∎

COROLLARY 8.3.1. *Suppose that all assumptions of Theorem 8.3.1 are satisfied and the operator $I - R_n^0 A(S_h)$ is invertible. Then all S_{nh}-periodic solutions of Equation (8.3.2) are of the form*

$$x = [I - R_n^0 A(S_h)]^{-1} R_n^0 \left(y + \frac{1}{nh} z\right), \quad \text{where } z \in \ker D \text{ is arbitrary}. \tag{8.3.7}$$

THEOREM 8.3.2. *Suppose that X is a Banach space, $D \in R(X)$, dim ker $D \neq 0$, F is a bounded initial operator for D corresponding to a compact $R \in \mathcal{R}_D \cap V(X)$, $\{S_h\}_{h \in A(\mathbf{R})}$ are bounded D-shifts. Suppose, moreover, that $\overline{E(R)} = X$. Write*

$$R_n^0 = \left(I + FS_{nh} - \frac{1}{nh} RFS_{nh}\right) R, \tag{8.3.8}$$

$$\tilde{R}_n^0 = \left(I + FS_{nh'} - \frac{1}{nh} RFS_{nh'}\right) R, \quad h \neq h' \in A(\mathbf{R}). \tag{8.3.9}$$

Then

$$\lim_{h' \to h} ||(S_{h'} - S_h)x|| = 0 \quad \textit{for all } x \in X, h \in A(\mathbf{R}), \tag{8.3.10}$$

$$\lim_{h' \to h} ||(\tilde{R}_n^0 - R_n^0)x|| = 0 \quad \textit{for all } x \in X, n \in \mathbf{N}, h \in A(\mathbf{R}).$$
$$\tag{8.3.11}$$

Proof. Since $\{S_h\}_{h \in A(\mathbf{R})}$ are D-shifts and the multiplication of elements by scalars is continuous, we find for all $e_\lambda(z)$, where $\lambda \in \mathbf{C}$, $z \in \ker D$ are arbitrary

$$(S_{h'} - S_h)e_\lambda(z) = (e^{\lambda h'} - e^{\lambda h})e_\lambda(z)$$

$$= e^{\lambda h'}[1 - e^{\lambda(h - h')}]e_\lambda(z) \to 0 \quad \text{as } h' \to h.$$

Since the set $E(R)$ is dense in X, we conclude that

$$||(S_{h'} - S_h)x|| \to 0 \quad \text{as } h' \to h \quad \text{for all } x \in X.$$

An immediate consequence of Formula (8.3.10) is that

$$\lim_{h' \to h} ||(S_{nh'} - S_{nh})x|| = 0 \quad \text{for all } x \in X \ (n \in \mathbf{N}, h \in A(\mathbf{R})).$$
$$\tag{8.3.12}$$

This, and our assumptions together imply for all $x \in X$ that we have

$$||(\tilde{R}_n^0 - R_n^0)x||$$

$$= \left|\left|\left[\left(I + FS_{nh} - \frac{1}{nh} RFS_{nh}\right) - \left(I + FS_{nh'} - \frac{1}{nh} RFS_{nh'}\right)\right]Rx\right|\right|$$

$$= \left|\left|F(S_{nh} - S_{nh'}) - \frac{1}{nh} RF(S_{nh} - S_{nh'})Rx\right|\right|$$

$$\leqslant ||F|| \cdot ||(S_{nh} - S_{nh'})Rx|| + \frac{1}{n|h|} ||R|| \cdot ||F|| \cdot ||(S_{nh} - S_{nh'})Rx||$$

$$\leqslant \left[||F||\left(1 + \frac{1}{n|h|} ||R||\right)\right] ||(S_{nh} - S_{nh'})Rx|| \to 0 \quad \text{as } h' \to h. \blacksquare$$

THEOREM 8.3.3. *Suppose that X is a Banach space, $D \in R(X)$, dim ker D*

$\neq 0$, F is a bounded initial operator for D corresponding to a compact $R \in \mathcal{R}_D$, $\{S_h\}_{h \in A(\mathbf{R})}$ are bounded R-shifts. Suppose moreover that $\overline{P(R)} = X$ and that R_n^0 and $\tilde{R}_n^0$ are defined by Formulae (8.3.8), (8.3.9). Then Formulae (8.3.10) and (8.3.11) hold.

Proof. Since $\{S_h\}_{h \in A(\mathbf{R})}$ are R-shifts and the multiplication of elements by scalars is continuous, we find for all $z \in \ker D$, $k \in \mathbf{N} \cup \{0\}$:

$$(S_{h'} - S_h) R^k z = \sum_{j=0}^{k} \frac{1}{(k-j)!} \left((h')^{k-j} - h^{k-j}\right) R^k z$$

$$= (h' - h) \sum_{j=0}^{k} \frac{1}{(k-j)!} \left[\sum_{m=0}^{k-j-1} (h')^m h^{k-j-m} \right] R^k z \to 0$$

$$\text{as } h' \to h.$$

Since the set $P(R)$ is dense in X we conclude that

$$\|(S_{h'} - S_h) x\| \to 0 \quad \text{as } h' \to h \quad \text{for all } x \in X.$$

The remainder of the proof goes along the same lines as the proof of Theorem 8.3.2. ∎

COROLLARY 8.3.2. *Suppose that either the assumptions of Theorem 8.3.2 or the assumptions of Theorem 8.3.3 are satisfied. Write*:

$$A(t) = \sum_{k=0}^{n-1} A_k t^k, \quad \text{where } A_k \text{ are bounded}, \tag{8.3.13}$$

$$S_h A_k = A_k S_h \quad (k = 0, 1, ..., n-1),$$

$$B_h = R_n^0 A(S_h), \quad B_{h'} = \tilde{R}_n^0 A(S_{h'}). \tag{8.3.14}$$

Then

$$\lim_{h' \to h} \|(B_{h'} - B_h) x\| = 0 \quad \text{for all } x \in X \quad (h \in A(\mathbf{R})). \tag{8.3.15}$$

Proof. We have for all $x \in X$

$$\|(B_{h'} - B_h) x\| = \|[\tilde{R}_h^0 A(S_h) - R_h^0 A(S_h)] x\|$$

$$\leqslant \|R_h^0 [A(S_{h'}) - A(S_h)] x\| + \|(\tilde{R}_h^0 - R_h) A(S_h) x\|.$$

Write

$$b = \left(a + ||F|| \cdot ||S_{nh'}|| + \frac{1}{n|h|} ||R|| \cdot ||F|| \cdot ||S_{nh'}||\right) ||R||. \quad (8.3.16)$$

Since $S_{h'}$ and S_h are either D-shifts or R-shifts we have $S_h^k = S_{kh}$, $S_{h'}^k = S_{kh'}$ for all $k \in \mathbb{N}$.

This, and Formulae (8.3.10), (8.3.11) together imply that for all $x \in X$ we have, if we write $\tilde{x} = A(S_h)x \in X$,

$$||(B_{h'} - B_h)x|| \leqslant b||[A(S_{h'}) - A(S_h)]x|| + ||(\tilde{R}_n^0 - R_n^0)\tilde{x}||$$

$$= b \left\|\sum_{k=0}^{n-1} A_k (S_{h'}^k - S_h^k)x\right\| + ||(\tilde{R}_n^0 - R_n^0)\tilde{x}||$$

$$\leqslant b \sum_{k=0}^{n-1} ||A_k|| \cdot ||(S_{kh'} - S_{kh})x|| + ||(\tilde{R}_n^0 - R_n^0)\tilde{x}|| \to 0$$

$$\text{as } h' \to h. \quad \blacksquare$$

THEOREM 8.3.4. *Suppose that all assumptions of Theorem 8.3.1 are satisfied. Let operators $A(t)$, B_h, $B_{h'}$ be defined by Formulae (8.3.13) and (8.3.14). Consider the equation*

$$Dx = A(S_h)x + y, \quad x \in X_{S_{nh}} \quad (8.3.17)$$

and a perturbed equation

$$D\tilde{x} = A(S_{h'})\tilde{x} + y, \quad y \in X. \quad (8.3.18)$$

If the operator $B_h = I - R_n^0 A(S_h)$ is invertible then Equation (8.3.17) with the condition $Fx = x_0$ has a unique S_{nh}-periodic solution

$$x = [I - R_n^0 A(S_h)]^{-1} R_n^0 \left(y + \frac{1}{nh} x_0\right). \quad (8.3.19)$$

Moreover, there exists a $\delta > 0$ such that for $|h' - h| < \delta$ the operator $B_{h'}$ is invertible, and the perturbed equation (8.3.18) with the condition $Fx = x_0$ also has a unique S_{nh}-periodic solution $\tilde{x}$ of the form

$$\tilde{x} = [I - \tilde{R}_n^0 A(S_{h'})]^{-1} \tilde{R}_n^0 \left(y + \frac{1}{nh} x_0\right), \quad (8.3.20)$$

such that $\lim_{h' \to h} ||\tilde{x} - x|| = 0.$

Proof. If the operator $B_h = I - R_n^0 A(S_h)$ is invertible then Theorem 5.3.3 implies that Equation (8.3.17) together with the condition $Fx = x_0$ has a unique S_{nh}-periodic solution which is of the form

$$[I - R_n^0 A(S_h)]x = R_n^0 \left(y + \frac{1}{nh} x_0 \right) \tag{8.3.21}$$

or more simply

$$(I - B_h)x = R_n^0 y_h, \quad \text{where } y_h = y + \frac{1}{nh} x_0. \tag{8.3.22}$$

The perturbed equation (8.3.18) together with the condition $Fx = x_0$ can be rewritten as follows:

$$[I - \tilde{R}_n^0 A(S_{h'})]\tilde{x} = \tilde{R}_n^0 \left(y + \frac{1}{nh} x_0 \right) \tag{8.3.23}$$

or more simply:

$$(I - B_{h'})\tilde{x} = \tilde{R}_n^0 y_h. \tag{8.3.24}$$

Since R is compact by our assumption, we conclude that the operator $R_n^0 = \left(I + FS_{nh} - \frac{1}{nh} RFS_{nh} \right) R$ is compact. Then the operator $B_h = R_n^0 A(S_h)$ is also compact. The operator $I - B_h = I - R_n^0 A(S_h)$ is invertible by our assumption. We have

$$I - B_{h'} = I - B_h + B_h - B_{h'}$$
$$= (I - B_h)[I + (I - B_h)^{-1}(B_h - B_{h'})],$$

i.e.

$$I - B_{h'} = (I - B_h)(I - T_{h'}), \quad \text{where } T_{h'} = (I - B_h)^{-1}(B_{h'} - B_h). \tag{8.3.25}$$

Corollary 8.3.2 implies that there exists a $\delta > 0$ such that for $|h' - h| < \delta$ we have $\|T_{h'}\| \leqslant q < 1$. Thus the operator $I - B_h$ is an invertible operator. Rewrite Equation (8.3.24) applying Equality (8.3.25)

$$(I - T)\tilde{x} = (I - B_h)^{-1} \tilde{R}_n^0 y_h. \tag{8.3.26}$$

Since the norm of $T\tilde{x}$ is less than 1, the Banach Theorem 8.1.1 implies that Equation (8.3.18) has a unique solution

$$\tilde{x} = (I-T)^{-1}(I-B_h)^{-1}\tilde{R}_n^0 y_h. \tag{8.3.27}$$

But $(I-B_h)^{-1}R_n^0 y_h = x$, and thus we have

$$\begin{aligned}
(I-T)\tilde{x} &= (I-B_h)^{-1}\tilde{R}_n^0 y_h \\
&= (I-B_h)^{-1}(\tilde{R}_n^0 - R_n^0)y_h + (I-B_h)^{-1}R_n^0 y_h \\
&= (I-B_h)^{-1}(\tilde{R}_n^0 - R_n^0)y_h + x.
\end{aligned}$$

Since $T = (I-B_h)^{-1}(B_{h'} - B_h)$, Theorem 8.3.1 and Corollary 8.3. together imply that

$$\begin{aligned}
\|\tilde{x} - x\| &= \|(I-B_h)^{-1}(\tilde{R}_n^0 - R_n^0)y_h\| + \|Tx\| \\
&\leqslant \|(I-B_h)^{-1}\|[\|(\tilde{R}_n^0 - R_n^0)y_h\| + \|(B_{h'} - B_h)x\|] \to 0 \\
&\qquad\qquad\qquad\qquad\qquad\qquad\qquad\qquad\qquad \text{as } h' \to h.
\end{aligned}$$

We therefore conclude that the perturbed equation together with the condition $Fx = x_0$ has a unique S_{nh}-periodic solution, for $|h'-h|$ sufficiently small. ∎

This theorem generalizes theorems on small perturbations of deviations preserving periodic solutions of linear differential-difference equations with periodic coefficients (cf. Przeworska-Rolewicz, 1973 and S. Rolewicz, 1973).

We shall consider now a non-linear periodic problem.

THEOREM 8.3.5. *Suppose that*

(i) *$D \in R(X)$, $0 \neq \dim \ker D < \infty$, F is a bounded initial operator for D corresponding to an $R \in \mathcal{R}_D$;*

(ii) *$\{S_h\}_{h \in A(\mathbf{R})}$ are either D-shifts (provided that $R \in V(X)$) or R-shifts and the semigroup (group) $\{S_h\}_{h \in A(\mathbf{R})}$ is strongly continuous, moreover, D is its infinitesimal generator;*

(iii) *G is a non-linear mapping of X onto $X \times \mathbf{R}$ such that the functions $G^{\wedge}(h, x^{\wedge}(h), \mu)$, $(\partial G^{\wedge}/\partial x^{\wedge})(h, x^{\wedge}(h), \mu)$ are continuous in both variables and ω-periodic with respect to the variable h, where $x^{\wedge}$, $G^{\wedge}$ are defined by the canonical mapping*

$$x^{\wedge}(h) = FS_h x, \quad G^{\wedge}(h, x^{\wedge}, \mu) = FS_h G(x^{\wedge}(h), \mu) \tag{8.3.28}$$

for all $x \in X$, $\mu \in \mathbf{R}$, $h \in A(\mathbf{R})$.

If the linearized equation

$$\frac{dy^{\wedge}(h)}{dh} = \frac{\partial}{\partial y^{\wedge}} G^{\wedge}(h, y^{\wedge}(h), \mu)|_{(h, y^{\wedge}(h), 0)} \tag{8.3.29}$$

has a unique ω-periodic solution $y^{\wedge} = 0$ then the equation

$$Dx = G(x, \mu), \tag{8.3.30}$$

for sufficiently small $|\mu|$, has an S_ω-periodic solution $x(\mu)$ such that

$$x^{\wedge}(h, 0) = y^{\wedge}(h) \quad \text{for all } h \in A(\mathbf{R}) \tag{3.3.31}$$

and $x^{\wedge}(h, \mu) = FS_h x(\mu)$ is continuous with respect to (h, μ).

Proof. Theorem 8.2.7 implies that, by our assumptions (i) and (ii) we have $\dfrac{d}{dh} x^{\wedge}(h) = (Dx)^{\wedge}(h)$ for all $x \in \operatorname{dom} D$. Thus, instead of Equation (8.3.30) we can consider the equation

$$x^{\wedge\prime}(h) = G^{\wedge}\big(h, x^{\wedge}(h), \mu\big), \tag{8.3.32}$$

where $G^{\wedge}\big(h, x^{\wedge}(h), \mu\big)$ is defined by Formulae (8.3.28). It is well-known (cf. Coddington and Levinson, 1955, Chapter XIV, § 1) that under assumption (iii), if the linearized equation has a unique ω-periodic solution $y^{\wedge}(h) \equiv 0$ then Equation (8.3.32) has for sufficiently small $|\mu|$ an ω-periodic solution $x^{\wedge}(h, \mu)$ such that Equality (8.3.31) holds and $x^{\wedge}(h, \mu)$ is continuous with respect to (h, μ). But this means that there exists an $x \in X$ which is S_ω-periodic and such that $x^{\wedge}(h, 0) = y^{\wedge}(h)$, where $x^{\wedge}(h) = FS_h x, h \in A(\mathbf{R})$. Indeed, for all $h \in A(\mathbf{R})$ we have $x^{\wedge}(h+\omega) = FS_{h+\omega} x = FS_h S_\omega x = FS_h x = x^{\wedge}(h)$ for $S_\omega x = x$. ■

Examples and Exercises

EXAMPLE 8.3.1. Consider a linear differential-difference equation with scalar coefficients and with *non-commensurable* delays h_1 and h_2 $(h_1, h_2 > 0, b^2+c^2 > 0)$:

$$x'(t) = ax(t)+bx(t-h_1)+cx(t-h_2)+y(t) \quad (t \in \mathbf{R}) \tag{8.3.33}$$

in the space of continuous functions. Without loss of generality we may assume that $0 < h_1 < h_2$. Then, by our assumption, there is no positive integers m and $n > m$, such that $h_1/h_2 = m/n$. However, by properties of real numbers, there exist $0 < h \in \mathbf{R}$, positive integers m and $n > 0$ and positive reals r_1, r_2 less than 1 such that $h_1 = mh+r_1$, $h_2 = nh+r_2$, hence $h_1-h_2 = (n-m)h+r_1-r_2$.

Let M be the smallest common multiple of numbers m and n. Consider the equation

$$x_h'(t) = ax_h(t)+bx(t-mh)+cx(t-nh)+y(t), \qquad (8.3.34)$$

where $y(t)$ is an Mh-periodic given function.

Applying Theorem 8.3.1 we may find Mh-periodic solutions of Equation (8.3.34). Thus, if this equation has a unique Mh-periodic solution and Equation (8.3.33) has a unique Mh-periodic solution then we may apply Theorem 8.3.4. In other words, Theorem 8.3.5 enables us to reduce problems with non-commensurable delays to problems with commensurable delays which are of a much more simple structure.

EXERCISE 8.3.1. Does the equation

$$x'(t) = ax\left(t-\frac{\pi}{2}\right)+bx(t-1)+\sin t$$

have periodic solutions?

EXERCISE 8.3.2. Does the equation

$$x'(t) = x^2(t+\mu^2)$$

have periodic solution for $0 \neq \mu \in \mathbf{R}$ small enough?

EXERCISE 8.3.3. Does the equation

$$t\frac{dx(t)}{dt} = \mu x^2(rt)+\mu r$$

have solutions invariant under *rotation* on r (i.e. such that $x(rt) = x(t)$ for $0 < \mu \in \mathbf{R}$ small enough? (Cf. Example 5.2.6).

8.4. BIELECKI METHOD AND ITS APPLICATIONS

In order to extend the classical Picard–Lindelöf result, Bielecki (1956) has proved the following theorem:

Suppose that $N(t, s, x)$ is a bounded real functions defined for $0 \leqslant t$, $s \leqslant T$ and $x \in \mathbf{R}$ satisfying the Lipschitz condition with respect to x:

$$|N(t, s, x)-N(t, s, y)| \leqslant L(s)|x-y| \quad \text{for all } x, y \in \mathbf{R}, \quad (8.4.1)$$

where L(t) is a non-negative locally integrable function over the interval $0 \leqslant t \leqslant T$. *Then the equation*

$$x = G(x) + y, \quad \text{where } G(x)(t) = \int_0^t N(t, s, x) \, ds,$$

$$y \in C[0, T] \tag{8.4.2}$$

has a unique solution, which is the limit of a uniformly convergent sequence $\{x_n\}$

$$x(t) = \lim_{n \to 0} x_n(t), \quad \text{where } x_0 = y$$

and

$$x_n(t) = y(t) + \int_0^t N(t, s, x_{n-1}(s)) \, ds \quad (n = 1, 2, \ldots).$$

The proof is based on an introduction of new norms $\| \ \|_p$ $(0 < p < +\infty)$ in the space $C[0, T]$ equivalent to the standard one $\|x\|$ $= \sup_{0 \leqslant t \leqslant T} |x(t)|$. The norms $\| \ \|_p$ are given by the following formula:

$$\|x\|_p = \max_{0 \leqslant t \leqslant T} \left\{ \exp\left[-p \int_0^t L(s) \, ds \right] |x(t)| \right\}. \tag{8.4.3}$$

It can be shown that

$$\|G(x) - G(y)\|_p \leqslant \frac{1}{p} \|x - y\|_p \quad \text{for } x, y \in C[0, T] \tag{8.4.4}$$

(and could be found, for instance in Przeworska-Rolewicz, 1973, cf. also the proof of Theorem 8.4.1).

Therefore this method makes it possible to apply the Banach fixed-point theorem without restrictions on the modulus of the function $N(t, s, x)$ of type "if $N(t, s, x)$ is small enough ...".

Inequality (8.4.4) shows that taking p greater, we obtain a faster approximation.

This theorem could be also formulated for $T = +\infty$; in this case instead of the space $C[0, T]$ we consider

$$X_p = \left\{ x: \exp\left[-p \int_0^t L(s) \, ds \right] |x(t)| < \text{const} \right\}$$

for a $p > 1$, provided that the function L is locally integrable for $t \geqslant 0$.

We shall show that the Bielecki Theorem can be extended for a class of non-linear operators in Banach spaces.

Let E be a Banach space with the norm $\| \ \|_E$. Let $X = C([a, b], E)$ be the Banach space of all functions determined for $a \leqslant t \leqslant b$, with values in E equipped with the norm

$$\|x\| = \sup_{a \leqslant t \leqslant b} \|x(t)\|_E \quad \text{for } x \in X. \tag{8.4.5}$$

THEOREM 8.4.1. *Suppose that*:

(i) *The function $N(t, s, u)$ determined and continuous for $0 \leqslant a \leqslant t, s \leqslant b, x \in E$ and with values in E satisfies the Lipschitz condition*

$$\|N(t, s, u) - N(t, s, v)\|_E \leqslant L(s)\|u - v\|_E$$
$$\text{for all } t, s \in [a, b], u, v \in E, \tag{8.4.6}$$

where L is a locally integrable non-negative function;

(ii) *The function $h \in C[a, b]$ satisfies conditions*:

$$h(a) = a \quad \text{and } a \leqslant h(t) \leqslant t \quad \text{for } t \in [a, b]. \tag{8.4.7}$$

Write

$$\|x\|_p = \sup_{a \leqslant t \leqslant b} \left\{ \exp\left[-p \int_a^{h(t)} L(s)\,\mathrm{d}s \right] \|x(t)\|_E \right\} \tag{8.4.8}$$

for all $x \in X$ and $p \in \mathbf{R}^+$.

Then the equation

$$x(t) = \int_a^{h(t)} N(t, s, x(s))\,\mathrm{d}s + y(t), \quad y \in X, \tag{8.4.9}$$

has a unique solution which is a limit in the norm $\| \ \|_p$ of the sequence of successive approximations:

$$x = \lim_{n \to \infty} x_n, \quad \text{where } x_0 = y,$$

$$x_{n+1}(t) = \int_a^{h(t)} N(t, s, x_n(s))\,\mathrm{d}s + y(t) \quad (n = 0, 1, \ldots). \tag{8.4.10}$$

Proof. It is going on the same lines as the original Bielecki's proof. Observe that $||x||_0 = ||x||$ for $p = 0$ and all norms $|| \ ||_p$ for $p \geqslant 0$ are equivalent. The mapping G defined by means of the equality:

$$G(u)(t) = \int_a^{h(t)} N(t, s, u(s)) \, ds + y(t), \quad u \in X \qquad (8.4.11)$$

maps the space X into itself.

We shall show that

$$||G(u) - G(v)||_p \leqslant \frac{1}{p} ||u - v||_p \quad \text{for } u, v \in X, p > 1. \qquad (8.4.12)$$

Indeed, observe that the function

$$L_1(t) = \int_0^t L(s) \, ds \qquad (8.4.13)$$

is non-negative. Hence for $p > 0$ we have

$$\exp\left[p \int_a^{h(t)} L(s) \, ds\right] = \exp[pL_1(h(t))] \geqslant 1$$

and for all $u, v \in X$

$$||u(t) - v(t)||_E \leqslant \exp[pL_1(h(t))] ||u - v||_p.$$

Since $L_1'(t) = L(t)$, $L_1(h(a)) = L_1(a) = 0$ and $1 - e^{-u} \leqslant 1$ for $u \geqslant 0$ we find

$$\exp\left[-pL_1(h(t))\right] ||G(u) - G(v)||_E$$

$$= \exp\left[-pL_1(h(t))\right] \left\| \int_a^{h(t)} [N(t, s, u(s)) - N(t, s, v(s))] \, ds \right\|_E$$

$$\leqslant \exp\left[-pL_1(h(t))\right] \int_a^{h(t)} L(s) ||u(s) - v(s)||_E \, ds$$

$$\leqslant \exp\left[-pL_1(h(t))\right] \int_a^{h(t)} L(s) \exp[pL_1(s)] ||u - v||_p \, ds$$

$$\leqslant \exp\left[-pL_1(h(t))\right] \int_a^{h(t)} L_1'(s) \exp[pL_1(s)] \, ds ||u - v||_p$$

$$= \frac{1}{p} \exp[-pL_1(h(t))] \exp[pL_1(s)]_a^{h(t)} \|u-v\|_p$$

$$= \frac{1}{p} \exp[-pL_1(h(t))] [\exp pL_1(h(t))-1] \|u-v\|_p$$

$$= \frac{1}{p} \|u-v\|_p \{1-\exp[-pL_1(h(t))]\} \leqslant \frac{1}{p} \|u-v\|_p.$$

Therefore for $p > 1$ the mapping G has a unique fixed point which is a limit in the norm $\| \ \|_p$ of the sequence of successive approximations (cf. Banach Theorem 8.1.1). But all norms $\| \ \|_p$ for $p \in \mathbf{R}^+$ are equivalent. This finishes the proof of our theorem. ∎

In the same manner we can consider Equation (8.4.9) in the spaces $C(\mathbf{R}, E)$, $C(\mathbf{R}^+, E)$ etc. We have only to assume that the function $h(t) \leqslant t$ on $\mathbf{R}$ (or $\mathbf{R}^+$, respectively).

We shall prove now the Bielecki Theorem for right invertible operators.

THEOREM 8.4.2. *Suppose that X is a Banach space, $D \in R(X)$ is closed, $\dim \ker D \neq 0$, F is a bounded initial operator for D corresponding to a bounded right inverse R, $\overline{P(R)} = X$ (resp. R is Volterra and $\overline{E(R)} = X$) and $\{S_h\}_{h \in \mathbf{R}}$ is a strongly continuous group of R-shifts (resp. D-shifts). Suppose, moreover, that $G: X \to X$ is a non-linear mapping satisfying the following conditions*:

$$G(FS_t x) = FS_t G(x) \quad \text{for all } t \in \mathbf{R}, x \in X, \qquad (8.4.14)$$

$$\|G(x)-G(y)\| \leqslant M\|x-y\| \quad \text{for all } x, y \in X. \qquad (8.4.15)$$

Then the problem

$$Dx = G(x), \quad Fx = x_0, \quad x_0 \in \ker D \qquad (8.4.16)$$

has a unique solution, which is the limit (in norm) of sequence of successive approximations:

$$x = \lim_{n \to \infty} x_n, \quad x_{n+1} = RG(x_n)+x_0 \quad (n = 0, 1, 2, \ldots). \ (8.4.17)$$

Proof. By our assumptions, Formulae (8.2.9), (8.2.10) hold and we have $G^{\wedge}(x) = FS_h G(x) = G(FS_h x) = G(x^{\wedge})$. Moreover, since

$$\|S_h x\| \leqslant Ce^{|h|}\|x\| \quad \text{for all } h \in \mathbf{R}, x \in X \qquad (8.4.18)$$

we have for $x, y \in X$

$$||G^{\wedge}(x) - G^{\wedge}(y)|| = ||G(x^{\wedge}) - G(y^{\wedge})|| \leqslant CM||F||e^{|t|}||x - y||.$$
(8.4.19)

Indeed,

$$||G^{\wedge}(x) - G^{\wedge}(y)|| = ||G(x^{\wedge}) - G(y^{\wedge})||$$
$$= ||G(FS_t x) - G(FS_t y)|| \leqslant M||FS_t x - FS_t y||$$
$$\leqslant CM||F||e^{|t|}||x - y||.$$

Observe that the function

$$L(t) = CM||F||e^{|t|} \quad (t \in \mathbf{R})$$
(8.4.20)

is a non-negative locally integrable function of real variable.

On the other hand the problem (8.4.16) is equivalent to the equation

$$x = RG(x) + x_0, \quad x_0 \in \ker D.$$
(8.4.21)

Apply to both sides of Equation (8.4.21) the canonical mapping $\varkappa$. Then by our assumptions

$$x^{\wedge}(t) = FS_t RG(x) + FS_t x_0 = \int_0^t FS_\tau G(x) \, d\tau + Fx_0$$

$$= \int_0^t G(FS_\tau x) \, d\tau + x_0 = \int_0^t G(x^{\wedge}(\tau)) \, d\tau + x_0,$$

where $x_0 = x^{\wedge}(0)$. All assumptions of Theorem 8.4.1 are satisfied with $h(t) \equiv t$, $a = 0$ and $N = \varkappa G$. We therefore conclude that the equation

$$x^{\wedge}(t) = \int_0^t G(x^{\wedge}(\tau)) \, d\tau + x_0$$
(8.4.22)

has a unique solution which is the limit (in norm) of the sequence of successive approximations

$$x^{\wedge} = \lim_{n \to \infty} x_n^{\wedge}, \quad \text{where } x_{n+1}^{\wedge}(t) = \int_0^t G(x_n^{\wedge}(\tau)) \, d\tau + x_0$$

$$(n = 0, 1, 2, \ldots) \quad (8.4.23)$$

for $t \in [0, T]$, where $T > 0$ is arbitrarily fixed.

But the canonical mapping separates points. This means that Equation (8.4.21), hence also the problem (8.4.16), we started with, has a unique solution, which is the limit (in norm) of a sequence of successive approximations

$$x = \lim_{n \to \infty} x_n, \quad \text{where } x_{n+1} = RG(x_n) + x_0$$

$$(n = 0, 1, 2, ...). \quad (8.4.24)$$

∎

Examples and Exercises

EXAMPLE 8.4.1. Let $E = \mathbf{R}$. Consider a non-linear ordinary differential equation

$$x' = f(t, x) \quad (8.4.25)$$

with an initial condition

$$x(t_0) = x_0, \quad t_0 \in [a, b] \text{ is arbitrarily fixed,} \quad (8.4.26)$$

where the function $f(t, x)$, determined and continuous for $t \in [a, b]$, $x \in \mathbf{R}$ satisfies the Lipschitz condition with respect to the variable x:

$$|f(t, x) - f(t, y)| \leqslant M|x - y| \quad \text{for all } x, y \in \mathbf{R}, t \in [a, b].$$

$$(8.4.27)$$

In our case $\|x\|_E = |x|$. Equation (8.4.25) together with Condition (8.4.26) is equivalent to the following *non-linear Volterra integral equation of the second kind*:

$$x(t) = x_0 + \int_{t_0}^{t} f(s, x(s)) \, ds. \quad (8.4.28)$$

Indeed, if x is a solution of Equation (8.4.25) then integrating this equation we obtain $x(t) = \int_0^t f(s, x(s)) \, ds + c$, where c is an arbitrary constant. If we require Condition (8.4.26) to be satisfied then we find $c = x_0$. Conversely, if x is a solution of Equation (8.4.28) then $x(t_0) = x_0$ and $x'(t) = f(t, x(t))$. Condition (8.4.27) implies that f satisfies Condition (8.4.1) with $L(s) \equiv M$ for all $x, y \in X$. All conditions of Theorem 8.4.1 are satisfied. Then Equation (8.4.28) (and, simultaneously Equation (8.4.25) with Condition (8.4.26)) has a unique solution in the

space $C([a, b], \mathbf{R}) = C[a, b]$ which is a limit of uniformly and absolutely convergent sequence of successive approximations:

$$x = \lim_{n \to \infty} x_n, \quad \text{where } x_n(t) = x_0 + \int_{t_0}^{t} f(s, x_{n-1}(s))ds \quad (n \in \mathbf{N}).$$

(8.4.29)

EXAMPLE 8.4.2. Let, as before, $E = \mathbf{R}$. Suppose that $N(t, s, x)$
$= K(t, s)x$ for all $x \in \mathbf{R}$, $t, s \in [a, b]$. Then Equation (8.4.25) becomes
a linear equation

$$x(t) = \int_{a}^{t} K(t, s)x(s)ds + y(t), \quad y \in C[a, b], \tag{8.4.30}$$

which is called a *linear Volterra integral equation of the second kind*.
The function $K(t, s)$ is called a *kernel* of an integral operator K defined
as follows:

$$(Kx)(t) = \int_{a}^{t} K(t, s)x(s)ds. \tag{8.4.31}$$

Since $|N(t, s, x_1) - N(t, s, x_2)| = |K(t, s)| \cdot |x_1 - x_2|$, it is enough
to assume that $K(t, s)$ is continuous for $a \leqslant t, s \leqslant b$. Indeed, writing
$L(s) = \max_{a \leqslant t \leqslant b} |K(t, s)|$ we conclude that $|K(t, s)| \leqslant L(s)$ and Condition (8.4.1) is satisfied. The function $L(s)$ is evidently non-negative and
integrable for $a \leqslant s \leqslant b$. All conditions of Theorem 8.4.1 are satisfied.
Thus Equation (8.4.30) has a unique solution $x(t)$ which is a limit of
a uniformly and absolutely convergent sequence of successive approximations:

$$x(t) = \lim_{n \to \infty} x_n(t), \quad \text{where } x_0(t) = y(t),$$

$$x_n(t) = y(t) + \int_{a}^{t} K(t, s)x_{n-1}(s)ds \quad (n \in \mathbf{N}).$$

The sequence $\{x_n\}$ of successive approximations can be written in
a slightly different way. Namely, write

$$K_0(t, s) \equiv 1, \quad K_1(t, s) = K(t, s),$$

$$K_{n+1}(t, s) = \int_a^t K_n(t, u)K(u, s)\,ds, \tag{8.4.32}$$

$$K_0 = I, \quad (K_n x)(t) = \int_a^t K_n(t, s)x(s)\,ds \quad \text{for } x \in C[a, b]$$

$$(n \in \mathbf{N}).$$

Observe that

$$(K^2 x)(t) = \int_a^t K(t, u)\left[\int_a^u K(u, s)x(s)\,ds\right]du$$

$$= \int_a^t \left[\int_s^t K(t, u)K(u, s)\,du\right]x(s)\,ds$$

$$= \int_a^b K_2(t, s)x(s)\,ds = (K_2 x)(t).$$

In a similar way, by an easy induction, we conclude that

$$K^n = K_n \quad (n = 0, 1, 2, \ldots). \tag{8.4.33}$$

We therefore can rewrite the sequence $\{x_n\}$ of successive approximations as follows:

$$x_1 = y + Kx_0 = y + Ky = (I + K_1)y,$$
$$x_2 = y + Kx_1 = y + K(y + Ky) = y + Ky + K^2 y = (I + K_1 + K_2)y,$$
$$\cdots \cdots \cdots \cdots \cdots \cdots \cdots \cdots \cdots \cdots \cdots \cdots \cdots$$
$$x_n = y + Kx_{n-1} = y + Ky + \ldots + K^n y = (I + K_1 + \ldots + K_n)y$$

$$(n \in \mathbf{N}).$$

Since the sequence $\{x_n\}$ is convergent in the norm of the space $C[a, b]$, writing

$$\mathscr{K}(t, s) = \sum_{n=1}^{\infty} K_n(t, s) \tag{8.4.34}$$

we conclude that a unique solution of Equation (8.4.30) has the form

$$x(t) = y(t) + \int_a^t \mathscr{K}(t, s)y(s)\,ds, \tag{8.4.35}$$

where the series (8.4.34) is uniformly and absolutely convergent. The function $\mathscr{K}(t, s)$ determined by Formulae (8.4.34) and (8.4.32) is called a *resolvent kernel* of the operator K determined by Formula (8.4.31). Define the operator $\mathscr{K}$ as follows:

$$(\mathscr{K}x)(t) = \int_a^t \mathscr{K}(t, s)x(s)\,ds \quad \text{for } x \in C[a, b]. \qquad (8.4.36)$$

The operator $I+\mathscr{K}$ is called a *resolvent* of the operator K. Since Equation (8.4.30) can be rewritten in the form

$$(I-K)x = y \qquad (8.4.37)$$

and this equation has a unique solution for every $y \in C[a, b]$ which can be rewritten in the form

$$x = (I+\mathscr{K})y \qquad (8.4.38)$$

we conclude that

$$(I+\mathscr{K})(I-K)x = (I+\mathscr{K})y = x,$$
$$(I-K)(I+\mathscr{K})y = (I-K)x = y.$$

Thus the operator $I-K$ is invertible in the space $C[a, b]$ and

$$(I-K)^{-1} = I+\mathscr{K}. \qquad (8.4.39)$$

This means that having already known the resolvent kernel of the operator K we can determine the operator $(I-K)^{-1}$ which exists by our assumptions. The resolvent of the operator $R = \int_a^t$ has been determined by Formulae (2.2.39) and (2.2.40).

EXAMPLE 8.4.3. Suppose that $h \in C^1[a, b]$, h maps the interval $[a, b]$ onto itself, $h(a) = a$, $0 \leqslant a \leqslant h(t) \leqslant t$ and $h'(t) > 0$ for $t \in [a, b]$. Suppose that the $\mathbf{R}^n$-valued function $N(t, x)$ is determined and continuous for $t \in [a, b]$, $x \in \mathbf{R}^n$ and satisfies the Lipschitz condition

$$\|N(t, u)-N(t, v)\|_{\mathbf{R}^n} \leqslant L(t)\|u-v\|_{\mathbf{R}^n} \quad \text{for } u, v \in \mathbf{R}^n, \quad (8.4.40)$$

where L is a function such that the function $\tilde{L}(t) = L(t)/h'(h^{-1}(t))$ is a non-negative function integrable over $[a, b]$, where h^{-1} denotes the inverse function. Consider a *differential equation in $\mathbf{R}^n$ with transformed argument*

$$x'(t) = N[t, x(h(t))] \tag{8.4.41}$$

with the initial condition

$$x(a) = x_0. \tag{8.4.42}$$

The system (8.4.41)–(8.4.42) is equivalent to an integral equation

$$x(t) = \int_a^t N[s, x(h(s))]ds + x_0. \tag{8.4.43}$$

If we change the variable $s \to h^{-1}(u)$ and we write

$$\tilde{N}(t, x) = N(h^{-1}(t), x)/h'(h^{-1}(t)),$$

we can rewrite Equation (8.4.43) in the form

$$x(t) = \int_a^{h(t)} \tilde{N}(u, x(u))du + x_0. \tag{8.4.44}$$

The functions $h, \tilde{L}, \tilde{N}$ satisfy all assumptions of Theorem 8.4.1. We therefore conclude that Equation (8.4.44), hence the initial problem (8.4.41)–(8.4.42), has a unique solution which is a limit of the sequence of successive approximations (in the norm $\| \ \|_p, p > 1$):

$$x = \lim_{n \to \infty} x_n, \quad \text{where } x_{n+1}(t) = \int_a^{h(t)} \tilde{N}(u, x_n(u))du + x_0,$$

for $n = 0, 1, 2, \ldots$

EXAMPLE 8.4.4. Consider a *non-linear problem of the Darboux type*:

$$\frac{\partial^2 x(t, s)}{\partial t \partial s} = G(t, s, x(t, s)) \quad \text{in } \Omega = [0, a] \times [0, b], \tag{8.4.45}$$

$$x(t, 0) = \sigma(t), \quad x(0, s) = \omega(s) \quad \text{for } t \in [0, a], s \in [0, b] \tag{8.4.46}$$

(cf. Section 4.2), where the function $G(t, s, x)$ determined for $t, s \in \Omega$, x belonging to a Banach space E satisfies the Lipschitz condition:

$$\|G(t, s, x) - G(t, s, y)\|_E \leqslant L(s)\|x - y\|_E \quad \text{for } x, y \in E, \tag{8.4.47}$$

L is a non-negative, locally integrable function, $\sigma \in C([0, a], E)$, $\omega \in C([0, b], E)$ and $\sigma(0) = \omega(0) = 0$.

The operator $\partial^2/\partial t\,\partial s$ is right invertible and closed in the space $C(\Omega)$. The conditions (8.4.46) induce an initial operator F of the form

$$(Fx)(t, s) = x(t, 0) + x(0, s) - x(0, 0) \qquad (8.4.48)$$

corresponding to a Volterra right inverse $R = \int\limits_{0}^{t}\int\limits_{0}^{s}$. Since $C(\Omega)$ is a Banach space and R is quasinilpotent, we can consider a strongly continuous group of R-shifts (which are simultaneously D-shifts) and

$$\|S_h x\|_E \leqslant C e^{|h|} \|x\|_E \qquad \text{for } x \in X,\ h \in \mathbf{R} \qquad (8.4.49)$$

(cf. Theorems 8.1.7 and 8.2.6).

It is not difficult to verify that

$$S_h F = \left[\exp\left(t\int\limits_{0}^{s} - hI\right)\right] F_1 + \left[\exp\left(s\int\limits_{0}^{s} - hI\right)\right]\int\limits_{0}^{t} F_2 \qquad \text{for } x \in C(\Omega)$$

$$(8.4.50)$$

where $(F_1 x)(t, s) = x(t, 0)$, $F_2(t, s) = x(0, s)$ (cf. Example 4.7 in the author's book, 1980). All assumptions of Theorem 8.4.1 are satisfied. We therefore conclude that the problem (8.4.45)–(8.4.46) has a unique solution which is a limit of an absolutely and uniformly convergent sequence of successive approximations:

$$x = \lim_{n\to\infty} x_n, \quad \text{where } x_0(t, s) = \sigma(t) + \omega(s),$$

$$x_{n+1}(t, s) = \int\limits_{0}^{t}\int\limits_{0}^{s} G\big(u, v, x_n(u, v)\big)\,dv\,du + x_0 \qquad \text{for } n = 0, 1, 2, \ldots$$

EXERCISE 8.4.1. Suppose that the function $N(t, s, x)$ satisfies all conditions of Theorem 8.4.1 and the mapping G is defined by Formula (8.4.11). Prove that for every scalar λ the mapping $I - \lambda G$ is invertible in $C[a, b]$, i.e. the equation $(I - \lambda G)x = y$ has a unique solution $x \in C[a, b]$ for every $y \in C[a, b]$.

EXERCISE 8.4.2. Consider an *ordinary linear differential equation of order n with variable coefficients*

$$x^{(n)} + a_{n-1}(t)x^{(n-1)} + \ldots + a_0(t)x = y, \qquad (8.4.51)$$

where $a_0, \ldots, a_{n-1}, y \in C[a, b]$ are given, with the initial conditions

$$x^{(k)}(t_0) = x_k, \quad x_k \in \mathbf{R} \quad (k = 0, 1, \ldots, n-1), \quad t_0 \in [a, b].$$

$$(8.4.52)$$

Prove that:

(i) the resolvent kernel for this problem is

$$K(t, s) = \sum_{n=1}^{\infty} Q_n^0(t, s),$$
(8.4.53)

where

$$Q_1^0(t, s) = Q^0(t, s) = \sum_{k=0}^{n-1} a_k(t) \frac{(t-s)^{n-k-1}}{(n-k-1)!},$$

$$Q_{n+1}^0(t, s) = \int_s^t Q_n^0(t, s) Q(u, s) du \quad (n \in \mathbb{N});$$

(ii) The operator $I+Q^0$, where $(Q^0 x)(t) = \int_{t_0}^t Q^0(t, s) x(s) ds$ for $x \in X$, is invertible and $(I+Q^0)^{-1} = I + \mathcal{K}$ (cf. Formula (8.4.36)).

(iii) The initial value problem (8.4.51)–(8.4.52) has a unique solution of the form

$$x(t) = \int_{t_0}^t \frac{(t-s)^{n-1}}{(n-1)!} \left\{ y(s) - \sum_{k=0}^{m} a_k(s) \frac{(s-t)^{m-k}}{(m-k)!} + \right.$$

$$+ \int_{t_0}^s K(s, u) \left[y(u) - \sum_{k=0}^{m} a_k(u) \frac{(u-t_0)^{m-k}}{(m-k)!} \right] du \right\} ds +$$

$$+ \sum_{k=0}^{n-1} \frac{(t-t_0)^k}{k!} x_k.$$
(8.4.54)

Observe that in the case when $x_0, \ldots, x_{n-1}$ are arbitrary constants, Formula (8.4.54) gives the general solution of Equation (8.4.51).

EXERCISE 8.4.3. Suppose that all coefficients of Equation (8.4.51) are scalars, i.e. in this case $a_k(t) \equiv a_k \in \mathbb{R}$ for $t \in [a, b]$, $k = 0, 1, \ldots, n-1$. Prove that in this case the general solution of Equation (8.4.51) is of the form:

$$x(t) = \int_{t_0}^t \frac{(t-s)^{n-1}}{(n-1)!} \left[y(s) + \int_{t_0}^s \mathcal{K}(s, u) y(u) du \right] ds -$$

$$-\sum_{m=0}^{n-1} x_m \sum_{k=0}^{m} a_k \int_{t_0}^{t} \frac{(t-s)^{n-1}}{(n-1)!} \left[\frac{(s-t_0)^{m-k}}{(m-k)!} + \right.$$

$$\left. + \int_{t_0}^{s} \frac{(u-t_0)^{m-k}}{(m-k)!} \mathscr{K}(s,u)\,du \right] ds.$$

EXERCISE 8.4.4. Consider a *linear Volterra integral equation of the first kind*

$$\int_{a}^{t} M(t,s)x(s)\,ds = y(t), \qquad\qquad (8.4.55)$$

where $y \in C^1[a,b]$ and $y(a) = 0$, $M(t,s)$ is determined, continuous and has a continuous derivative with respect to t for $a \leqslant t, s \leqslant b$. Prove that:

  (i) Equation (8.4.55) is equivalent to the equation

$$M(t,t)x(t) + \int_{a}^{t} M'_t(t,s)x(s)\,ds = y'(t); \qquad\qquad (8.4.56)$$

  (ii) If $M(t,t) \neq 0$ then Equation (8.4.55) is equivalent to the equation

$$x(t) = \int_{a}^{t} K(t,s)x(s)\,ds, \quad \text{where } K(t,s) = -\frac{M'_t(t,s)}{M(t,t)}.$$

Since the function $K(t,s)$ satisfies all assumptions of Example 8.4.2, a unique solution of this equation is given by Formula (8.4.35).

EXERCISE 8.4.5. Consider a *system of Volterra integral equations*

$$x_k(t) = \int_{a}^{t} N_k(t, s, x_1(s), \ldots, x_m(s))\,ds + y_k(t)$$

$$(k = 1, 2, \ldots, m), \qquad (8.4.57)$$

where the real-valued functions $N_j(t, s, x_1, \ldots, x_m)$ are determined and continuous for $a \leqslant t, s \leqslant b$, $x_1, \ldots, x_m \in \mathbf{R}$ and satisfy the following

Lipschitz condition with respect to variables $x_1, \ldots, x_m$:

$$|N_k(t, s, x_1, \ldots, x_m) - N_k(t, s, x_1', \ldots, x_m')| \leqslant L_k(s) \sum_{j=1}^{m} |x_j - x_j'|$$

for $a \leqslant t, s \leqslant b, x_1, \ldots, x_m, x_1', \ldots, x_m' \in \mathbf{R}$ ($k = 1, \ldots, m$), where the non-negative functions $L_k(t)$ are integrable for $a < t < b$. We assume that $y_1, \ldots, y_m \in C[a, b]$. Prove that the system (8.4.57) has a unique solution in the space $C([a, b], \mathbf{R}^m)$.

EXERCISE 8.4.6. Prove that a system of non-linear differential equations $x_k'(t) = f_k(t, x_1, \ldots, x_m)$ ($k = 1, 2, \ldots, m$) with the initial conditions $x_k(t_0) = x_{k0} \in \mathbf{R}$, $t_0 \in [a, b]$ ($k = 1, 2, \ldots, m$), where the functions $f_k(t, x_1, \ldots, x_m)$ determined and continuous for $a \leqslant t \leqslant b, x_1, \ldots, x_m \in \mathbf{R}$ satisfy the Lipschitz condition with respect to variables $x_1, \ldots, x_n$:

$$|f_k(t, x_1, \ldots, x_m) - f_k(t, x_1', \ldots, x_m')| \leqslant M \sum_{k=1}^{m} |x_k - x_k'|$$

has a unique solution in the space $C([a, b], \mathbf{R}^m)$.

EXERCISE 8.4.7. Suppose that all assumptions of Theorem 4.4.1 are satisfied. Prove that:

The operator $I + Q^0$ is a Volterra operator in the space $X_1 = \{x \in C[t_0, T]: x(t_0) = 0\}$, since we have

$$|[(Q^0)^n x](t)| \leqslant M_0^n \frac{(T-t_0)^n}{n!} \|x\| \quad \text{for } x \in X \ (n \in \mathbf{N}),$$

where

$$M_0 = \begin{cases} M \cdot N \cdot M_a & \text{for } T - t_0 \leqslant 1, \\ M \cdot N \cdot (T - t_0)^N M_a & \text{for } T - t_0 > 1, \end{cases}$$

$$M_a = \max_{0 \leqslant k \leqslant N-1; \, 0 \leqslant j \leqslant M} \|a_{jk}\|$$

(cf. Przeworska-Rolewicz, 1974).

EXERCISE 8.4.8. Suppose that X is a Banach space, $D \in R(X)$, F is a bounded initial operator for D corresponding to a bounded right inverse R. $\overline{P(R)} = X$ (resp. R is Volterra and $\overline{E(R)} = X$) and $\{S_h\}_{h \in \mathbf{R}}$ is a strongly continuous group of R-shifts (D-shifts, respectively). Suppose, moreover, that $G: X \times X \to X$ is a non-linear mapping sytisfying the conditions:

$$G(FS_t x, FS_t y) = FS_t G(x, y) \quad \text{for all } x, y \in X, t \in \mathbf{R},$$
$$\|G(x_1, y_1) - G(x_2, y_2)\| \leqslant M[\|x_1 - x_2\| + \|y_1 - y_2\|],$$
$$x_1, x_2, y_1, y_2 \in X.$$

Prove that

(i) The mapping $G_0(u) = \varkappa G(Ru + x_0, u)$ satisfies assumptions of Theorem 8.4.2 with the function $L(t) = CM\|F\|(\|R\| + 1)e^{|t|}$, C is a constant;

(ii) The equation
$$Dx = G(x, Dx), \quad Fx = x_0, \quad x_0 \in \ker D$$
has a unique solution $x = \lim_{n \to \infty} x_n$, where $x_{n+1} = RG(x_n, Dx_n) + x_0$
for $n \in \mathbf{N}$.

Added in proof:

Let X be a complete linear metric space and let $D \in R(X)$. Write:

$$A_R(D) = \left\{ x \in D_\infty : \sum_{n=0}^{\infty} R^n FD^n x \text{ is convergent and} \right.$$

$$\left. x = \sum_{n=0}^{\infty} R^n FD^n x \right\}$$

where F is an initial operator for D corresponding to an $R \in \mathscr{R}_D$. Elements of $A(D) = \bigcup_{R \in \mathscr{R}_D} A_R(D)$ are said to be *D-analytic*. We find

$$A_R(D) = \{ x \in D_\infty : \lim_{n \to \infty} R^n D^n x = 0 \} \subset D_\infty \quad \text{for } R \in \mathscr{R}_D$$

(cf. Exercise 3.4.9 (iv)).

Suppose that X is a Banach space over $\mathbf{C}$ and there is a quasinilpotent $R \in \mathscr{R}_D$. Then

(i) $H_D[A_R(D)] = A_R(D) \subset A(D)$, in particular $H_D(0) \subset A(D)$;

(ii) $H_D[A_R(D) \oplus R_\infty] = A_R(D) \oplus R_\infty$, where $R_\infty = \bigcup_{i=0}^{\infty} R^i X$.

The proof is based on von Trotha Theorem 3.4.3, Corollary 3.4.2 and Proposition 3.4.6 on p. 218 added in proof (cf. Przeworska-Rolewicz, 1986).

Chapter 9

Miscellanea

9.1. ACCELERATING CONVERGENCE OF ORTHONORMAL SERIES

A linear space X over $\mathbf{C}$ is said to be a *pre-Hilbert* space if for every $x, y \in X$ there exists a complex valued function (x, y), called the *inner product* of x and y, such that

$$(x, x) \geqslant 0 \text{ and } (x, x) = 0 \quad \text{if and only if } x = 0, \qquad (9.1.1)$$

$$(ax, y) = a(x, y) \quad \text{for all } a \in \mathbf{C}, \qquad (9.1.2)$$

$$(y, x) = \overline{(x, y)} \text{ (where } \bar{a} \text{ denotes the conjugate number}$$
$$\text{for } a \in \mathbf{C}), \qquad (9.1.3)$$

$$(x+y, z) = (x, z)+(y, z) \quad (z \in X). \qquad (9.1.4)$$

Properties (9.1.2) and (9.1.3) imply that $(x, ay) = \bar{a}(x, y)$ and $(x, y+z) = (x, y)+(x, z)$ for $x, y, z \in X$ and $a \in \mathbf{C}$. Note that in the real case, i.e. if the field under consideration is $\mathbf{R}$, the inner product is, by definition, a real number and $(y, x) = (x, y)$ for $\bar{a} = a$ if $a \in \mathbf{R}$. This means that the inner product in the real case is a bilinear symmetric mapping of $X \times X$ into $\mathbf{R}$. The non-negative number

$$\|x\| = \sqrt{(x, x)} \quad (x \in X), \qquad (9.1.5)$$

is called the *norm* of an element $x \in X$. The fundamental role in the theory of pre-Hilbert spaces plays the so-called *Schwarz inequality*

$$|(x, y)| \leqslant \|x\| \cdot \|y\| \quad \text{for } x, y \in X. \qquad (9.1.6)$$

Indeed, if $(x, y) = 0$ this inequality is trivially satisfied. Suppose that $(x, y) \neq 0$. Without loss of generality we may assume that $y \neq 0$, hence $(y, y) \neq 0$. Then $0 \leqslant (x+ay, x+ay) = (x, x)+a(x, y)+a(y, x)+ +|a|^2(y, y)$ for all $a \in \mathbf{C}$. If we put $a = -(x, y)/(y, y)$ then we find $(x, x)-|(x, y)|^2/(y, y) \geqslant 0$ which implies (9.1.6).

The Schwarz inequality implies the triangle inequality:

$$||x+y|| \leqslant ||x|| + ||y|| \quad \text{for } x, y \in X.$$

This, and Properties (9.1.1), (9.1.2), (9.1.3) together imply that a pre-Hilbert space is a normed space with the norm induced by the inner product. A complete pre-Hilbert space is said to be a *Hilbert space*. Every Hilbert space is a Banach space, but not conversely.

EXAMPLE 9.1.1. Consider a function defined and measurable on the interval [0, 1]. We identify all functions which are equal almost everywhere and we consider the set $L^2(0, 1)$ of all such functions which are integrable (in the Lebesgue sense) together with their square. It is well-known that $L^2(0, 1)$ is a Hilbert space with the inner product

$$(x, y) = \int_0^1 x(t)\bar{y}(t)\mathrm{d}t \quad \text{for } x, y \in L^2(0, 1); \tag{9.1.7}$$

the norm in $L^2(0, 1)$ is given by

$$||x|| = \left[\int_0^1 |x(t)|^2 \mathrm{d}t\right]^{1/2}. \tag{9.1.8}$$

EXAMPLE 9.1.2. Denote by l^2 the set of all (complex or real) sequences $x = \{x_n\}$ such that $\sum_{n=1}^{\infty} |x_n|^2 < +\infty$. The set l^2 is a Hilbert space with the inner product $(x, y) = \sum_{n=1}^{\infty} x_n \bar{y}_n$. The norm in l^2 is defined by $||x|| = \left(\sum_{n=1}^{\infty} x_n^2\right)^{1/2}$.

Let X be a Hilbert space. Elements $x, y \in X$ are said to be *orthogonal* if $(x, y) = 0$. If Y is a subspace of X then always there exists a subspace $Z \subset X$ such that

$$X = Y \oplus Z \quad \text{and } (y, z) = 0 \quad \text{for all } y \in Y, z \in Z. \tag{9.1.9}$$

The subspace Z is called an *orthogonal complement* of Y and $Y \oplus Z$ is said to be an *orthogonal sum*.

A system $\{e_n\}_{n\in\mathbb{N}}$ of elements of a Hilbert space X is said to be *orthonormal* if $(e_j, e_k) = \delta_{jk}$ for all $j, k \in \mathbb{N}$. One can prove that every system of linear independent elements in a Hilbert space can be transformed into an orthonormal system.

An orthonormal system $\{e_n\}_{n\in\mathbb{N}}$ in a Hilbert space is said to be *complete* if

$$\sum_{i=1}^{\infty} |(x, e_i)|^2 = ||x||^2 \quad \text{for every } x \in X. \tag{9.1.10}$$

The numbers (x, e_i) are called *Fourier coefficients* of $x \in X$. If $\{e_n\}_{n\in\mathbb{N}} \subset X$ is a complete orthonormal system in X then there is no elements $0 \neq y \in X$ such that y is orthogonal to every e_n $(n \in \mathbb{N})$.

EXAMPLE 9.1.3. The system $\{e^{2\pi i n t}\}$ where $n = 0, \pm 1, \pm 2, \ldots$, is a complete orthonormal system in the space $L^2(0, 1)$ (cf. Example 9.1.1).

A Hilbert space X is said to be *separable* if in X there exists a dense countable set Y. In every separable Hilbert space X there exists a complete orthonormal system whose linear span is dense in X.

Let X be a Hilbert space. Operators A and B are said to be *adjoint* if

$$(Ax, y) = (x, By) \quad \text{for all } x \in \text{dom } A, y \in \text{dom } B. \tag{9.1.11}$$

If A and B are adjoint then B is usually denoted by A^*. An operator $A \in L(X)$ is said to be *selfadjoint* if it is closed and

$$(Ax, y) = (x, Ay) \quad \text{for all } x, y \in \text{dom } A. \tag{9.1.12}$$

It is easy to verify that $(A+B)^* = A^*+B^*$, $(AB)^* = B^*A^*$ (if these sum and superposition exist) and $(A^*)^* = A$. However, $(\lambda A)^* = \bar{\lambda}A^*$ for $\lambda \in \mathbb{C}$.

Note that eigenvalues of a selfadjoint operator are reals and that eigenvectors corresponding to different eigenvalues are orthogonal.

THEOREM 9.1.1 (Tasche, 1979). *Suppose that X is a separable Hilbert space over complexes, $D \in L(X)$, $\overline{\text{dom } D} = X$, $\dim \ker D > 0$. Suppose, moreover, that there exists a dense subspace $X_0 \subset X$ such that the restriction $D_0 = D|_{X_0}$ of D is a selfadjoint operator. Let $\{\lambda_k\}_{k\in\mathbb{N}}$ be eigenvalues* *of D_0 of finite multiplicities (each enumerated so many times as its multiplicity indicates) with the properties:*

$$0 < |\lambda_1| \leqslant |\lambda_2| \leqslant \ldots \leqslant |\lambda_k| \leqslant \ldots,$$
$$|\lambda_k| \to +\infty \quad \text{as } k \to +\infty. \tag{9.1.13}$$

Let $\{x_k\}_{k\in\mathbb{N}} \subset X$ be the corresponding complete orthonormal system of eigenvectors:

$$D_0 x_k = \lambda_k x_k \quad (k \in \mathbb{N}). \tag{9.1.14}$$

Then

(i) $X_0 = \left\{ x \in X: \sum_{k=1}^{\infty} |\lambda_k|^2 |(x, x_k)|^2 < +\infty \right\}; \tag{9.1.15}$

(ii) $D_0 \in L(X_0 \to X)$ *can be represented in the form*

$$D_0 x = \sum_{k=1}^{\infty} \lambda_k (x, x_k) x_k \quad \text{for } x \in X_0; \tag{9.1.16}$$

(iii) *The operator R defined by*

$$Rx = \sum_{k=1}^{\infty} \lambda_k^{-1} (x, x_k) x_k \quad \text{for } x \in X, \tag{9.1.17}$$

is compact and selfadjoint with the norm $\|R\| = |\lambda_1|^{-1}$ and

$$R^n x = \sum_{k=1}^{\infty} \lambda_k^{-n} (x, x_k) x_k \quad \text{for } x \in X \text{ and } n \in \mathbb{N}; \tag{9.1.18}$$

(iv) *The operator R is a right inverse of the given operator D, so that $D \in R(X)$.*

Proof. Points (i), (ii), (iii) follow from general properties of self-adjoint operators acting in a Hilbert space with eigenvalues satisfying Condition (9.1.13). To prove Point (iv) observe that D_0 is a restriction of D. This implies that $DR = D_0 R = I$ on X. Hence $D \in R(X)$ and $R \in \mathcal{R}_D$. ∎

THEOREM 9.1.2. *Suppose that all assumptions of Theorem 9.1.1 are satisfied. Then for all $x \in \operatorname{dom} D^n$, $n \in \mathbb{N}$,*

$$x = \sum_{k=0}^{n} R^k F D^k x + \sum_{k=1}^{\infty} \lambda_k^{-n} (D^n x, x_k) x_k \tag{9.1.19}$$

and for $m \geqslant 1$

$$\left\| x - \sum_{k=0}^{n-1} R^k F D^k x - \sum_{k=1}^{m} \lambda_k^{-n} (D^n x, x_k) x_k \right\| \leqslant |\lambda_{m+1}|^{-n} \|D^n x\|.$$

(9.1.20)

Proof. From the Taylor Formula we get

$$x = \sum_{k=0}^{n-1} R^k F D^k x + R^n D^n x \quad \text{for } x \in \text{dom } D^n.$$

This, and Formula (9.1.8) together imply (9.1.19). Therefore, for $m \geqslant 1$,

$$\left\| x - \sum_{k=0}^{n-1} R^k F D^k x - \sum_{k=1}^{m} \lambda_k^{-n} (D^n x, x_k) \right\|^2$$

$$= \sum_{k=m+1}^{\infty} |\lambda_k|^{-2n} |(D^n x, x_k)|^2$$

$$\leqslant |\lambda_{m+1}|^{-2n} \sum_{k=m+1}^{\infty} |(D^n x, x_k)|^2 \leqslant |\lambda_{m+1}|^{-2n} \|D^n x\|^2. \qquad \blacksquare$$

EXAMPLE 9.1.4 (Tasche, 1979). Suppose that $X = L^2(0, 1)$. Define the operator D by the equality $Dx = x''$ with the domain consisting of all $x \in X$ such that x and x' are continuous and differentiable almost everywhere on $[0, 1]$ and $x'' \in X$. Let

$$X_0 = \{x \in \text{dom } D : x(0) = x(1) = 0\}.$$

Clearly, $X_0 \subset \text{dom } D$. Then the restriction $D_0 = D|_{X_0}$ is a self-adjoint operator with single eigenvalues $\lambda_k = k^2 \pi^2$ ($k \in \mathbb{N}$). The corresponding normed (i.e. with the norm equal 1) eigenvectors are $x_k(t) = \sqrt{2} \sin k\pi t$. The right inverse R of D_0 is defined as

$$(Rx)(t) = \int_0^1 K(t, s) x(s) \, ds \quad \text{for } x \in X,$$

(9.1.21)

with

$$K(t, s) = \begin{cases} (1-t)s & \text{if } 0 \leqslant s \leqslant t \leqslant 1, \\ (1-s)t & \text{if } 0 \leqslant t < s \leqslant 1. \end{cases}$$

(9.1.22)

It is easy to verify that

$$\ker D = \text{lin } \{y_1, y_2\},$$

where $y_1(t) = 1 - t$, $y_2(t) = t$ and $y_1(0) = y_2(1) = 1$, $y_1(1) = y_2(0) = 0$. This implies that the initial operator F corresponding to R is of the form

$$(Fx)(t) = x(1)t + x(0)(1 - t) \quad \text{for } x \in X_0. \tag{9.1.23}$$

Putting for $k \in \mathbf{N}$

$$Dy_{2k+1} = y_{2k-1}, \quad Dy_{2k+2} = y_{2k},$$
$$y_{2k+1}(0) = y_{2k+1}(1) = y_{2k+2}(0) = y_{2k+2}(1) = 0, \tag{9.1.24}$$

we find

$$y_{2k+1}(t) = R^k(1 - t) = (-1)^k A_k(1 - t),$$
$$y_{2k+2}(t) = R^k(t) = (-1)^k A_k(t), \qquad (k \in \mathbf{N})$$

where A_k denotes the k-th *Lidstone polynomial*, i.e. a polynomial of degree $2k + 1$ defined by a recurrence formulae

$$A_0(t) = t, \quad A_k''(t) = A_{k-1}(t), \quad A_k(0) = A_k(1) = 0$$
$$(k \in \mathbf{N}). \tag{9.1.25}$$

Consequently, for $x \in C^{2n}[0, 1]$, we find

$$\sum_{k=0}^{n-1} (R^k FD^k x)(t) = \sum_{k=0}^{n-1} [x^{(2k)}(0) A_k(1 - t) +$$
$$+ x^{(2k)}(1) A_k(t)] \quad \text{for } x \in X_0.$$

This, and Formula (9.1.19) together imply that

$$x(t) = \sum_{k=0}^{n-1} [x^{(2k)}(1) A_k(t) + x^{2k}(0) A_k(1 - t)] +$$
$$+ \sum_{k=1}^{\infty} e_k \sin k\pi x, \tag{9.1.26}$$

where

$$e_k = \frac{2(-1)^n}{(k\pi)^{2n}} \int_0^1 x^{(2n)} \sin k\pi s \, ds. \tag{9.1.27}$$

The rate of convergence in the norm of $L_2(0, 1)$ follows from Formula (9.1.22)

$$\left\| x(t) - \sum_{k=0}^{n-1} [x^{(2k)}(1) A_k(t) + x^{(2k)}(0) A_k(1-t)] + \sum_{k=1}^{m} e_k \sin k\pi x \right\|$$

$$\leqslant \frac{1}{[(m+1)\pi]^{2n}} \|x^{(2n)}\|.$$

In a similar way we may obtain an expansion of a function from $L^2(0, 1)$ into orthonormal series consisting both, sine and cosine terms. This example and others can be found in the Tasche's papers, 1979, 1986.

9.2. VON TROTHA PRINCIPLE OF CONTRACTIVE MAPPINGS

Let Z be a linear space over a field $\mathscr{F}$. Denote by $X(Z)$ the Cartesian product of infinitely many copies of Z by itself:

$$X(Z) = \left\{ x = \{z_i\}: z_i \in Z, i \in \mathbf{N}_0 = \{0\} \cup \mathbf{N} \right\}$$

with the addition of elements and multiplication by scalars defined as follows:

$$\{x_i\} + \{y_i\} = \{x_i + y_i\}; \quad \{\lambda x_i\} = \lambda \{x_i\},$$
$$x = \{x_i\}, \quad y = \{y_i\} \in X(Z), \quad \lambda \in \mathscr{F}.$$

The space $X(Z)$ together with the shift operators defined by means of the equalities

$$D_0\{z_i\} = z_{i+1}, \quad R_0\{z_i\} = \{z_{i-1}\} \tag{9.2.1}$$

(where we admit here and in the sequel $x_{-1} = 0$ for all $x \in X(Z)$) is a D_0–R_0 space (cf. Definition 3.4.1). This space will be denoted by $X_0(Z)$ or briefly X_0.

Observe that $\ker D_0 = \left\{ \{z_i\} \subset Z: z_i = 0 \text{ for } i \geqslant 1 \right\}$. If we identify z_i with the sequence $\{z_i, 0, ...\} \in \ker D_0$ then the i-th component $\{0, ..., 0, z_i, 0, ...\}$ for $z = \{z_i\}$ can be written as $R_0^i z$ and

$$(R_0^i D_0^i - R_0^{i+1} D_0^{i+1}) x = R_0^i F_0 D_0^i x = \{0, ..., 0, z_i, 0, ...\}$$
$$(i \in \mathbf{N}_0),$$

where $F_0 = I - R_0 D_0$ projects X_0 onto ker D_0. Evidently, the operators $R_0^i D_0^i$ and $P_{0,i} = R_0^i F_0 D_0^i$ are projections and

$$\ker D_0 \oplus \ldots \oplus R_0^{i-1} \ker D_0 = \big\{ \{z_k\} \subset Z : z_k = 0$$
$$\text{for } 0 \leqslant k \leqslant i-1 \big\}, \quad i \geqslant 1,$$

$$R_0^i X_0 = \big\{ \{z_k\} \subset Z : z_k = 0 \text{ for } 0 \leqslant k \leqslant i-1 \big\}, \quad i \geqslant 1,$$

$$(9.2.2)$$

$$X_0 \supset R_0 X_0 \supset \ldots \supset R_0^i X_0 \supset \ldots \supset \bigcap_{i=0}^{\infty} R_0^i X_0 = \{0\}$$

(cf. Section 3.4.1).

We shall introduce in X_0 a simple but nontrivial metric, compatible with both, the linear and D_0–R_0 structure. Write:

$$|x|_{d_0} = \sum_{i=0}^{\infty} 2^{-i} \frac{|z_i|}{1+|z_i|} \quad \text{for } x = \{z_i\} \in X_0, \qquad (9.2.3)$$

where

$$|z_i| = \begin{cases} 0 & \text{if } z_i = 0, \\ 1 & \text{if } z_i \neq 0, \end{cases} \quad z_i \in Z.$$

It is easy to verify the following:

PROPERTY 9.2.1. *The space X_0 with the metric $d_0(x, y) = |x-y|_{d_0}$ is a complete metric space and the metric $d_0(x, y)$ (briefly: d_0-metric) is translation invariant. However, X_0 is not a linear metric space in the sense defined in Section 8.1.*

PROPERTY 9.2.2. *The subspace*

$$S_0 = \big\{ \{z_i\} \subset Z : z_i = 0 \text{ for almost all } i \in N_0 \big\} \qquad (9.2.4)$$

is dense in X_0.

PROPERTY 9.2.3. *The sets $R_0^i X_0$ defined by (9.2.2) are both open and closed. Moreover*

$$|x|_{d_0} \leqslant 2^{-i} \quad \text{if and only if } x \in R^i X_0 \quad (i \in N_0). \qquad (9.2.5)$$

A sequence $\{x_n\} \subset X_0$ $(n \in \mathbf{N})$ is said to be d_0-*convergent* to an $x \in X_0$ if and only if for every $i \in \mathbf{N}_0$ there is an $N(i) \in \mathbf{N}_0$ such that
$$x_n - x \in R_0^i X_0 \quad \text{for all } n \geqslant N(i). \tag{9.2.6}$$

This convergence can be also described by means of components. Namely, we have

LEMMA 9.2.1. *Let* $x_n = \{z_{n,i}\} \in X_0$ *for* $n \in \mathbf{N}$. *Then the sequence* $\{x_n\}$ *is* d_0-*convergent to an* $x = \{z_i\} \in X_0$ *if and only if for every* $i \geqslant 1$ *there exists an* $N(i)$ *such that* $z_{n,i-1} = z_{N(i),i}$ *for all* $n \geqslant N(i)$.

Proof. Sufficiency. Suppose that $x_n \to x = \{z_i\}$. By (9.2.6) we have $z_{n,j} - z_j \in R_0^i X_0$ for $i \geqslant 1$ and $n \geqslant N(i)$, $j \in \mathbf{N}_0$. This implies $z_{n,i-1} = z_{i-1} = z_{N(i),i-1}$ for all $n \geqslant N(i)$.

Necessity. Let $x = \{z_{N(i),i-1}\}$ and let $i \geqslant 1$ be given. Write $M(i) = \max_{1 \leqslant j \leqslant i} N(j)$. Clearly, $x_n - x \in R_0^i X_0$ for $n \geqslant M(i)$. ∎

The space X_0 is also a nontrivial example of a linear space, where the convergence of series is equivalent to the convergence of their terms to zero. This is shown by the following:

THEOREM 9.2.1. *A series* $\sum_{n=0}^{\infty} x_n$ *is* d_0-*convergent in* X_0 *if and only if* $x_n \to 0$ *(in* d_0-*metric).*

Proof. From Lemma 9.2.1 we infer that $x_n \to 0$ in d_0-metric if and only if for every $i \in \mathbf{N}$ there is an $N(i) \in \mathbf{N}$ such that
$$z_{n,i-1} = 0 \quad \text{for all } n \geqslant N(i). \tag{9.2.7}$$

Let $x_n = \{z_{n,i}\}$, $x = \{z_i\} \in X_0$ and let $x_n' = \left\{ \sum_{j=0}^{n} z_{j,i-1} \right\}$.

Sufficiency. If $x_n \to 0$ (in d_0-metric) then by (9.2.7) for a given $i \geqslant 1$ there is an $N(i)$ such that
$$\sum_{j=0}^{n} z_{j,i-1} - \sum_{j=0}^{N(i)} z_{j,i-1} = 0 \quad \text{for } n \geqslant N(i).$$

Hence, by Lemma 9.2.1 the sequence $\{x_n'\}$ is d_0-convergent, i.e. the series $\sum_{n=0}^{\infty} x_n$ is d_0-convergent.

Necessity. If $x'_n \to x$ (in d_0-metric) then by Lemma 9.2.1 for every $i \geqslant 1$ there exists an $N(i)$ such that

$$\sum_{j=0}^{N(i)} z_{j,\,i-1} + \sum_{j=N(i)+1}^{n} z_{j,\,i-1} = \sum_{j=0}^{N(i)} z_{j,\,i-1} \quad \text{for all } n \geqslant N(i).$$

Thus $z_{n,\,i} = 0$ for all $n \geqslant N(i)+1$, which implies $x_n \to 0$ in d_0-metric. ∎

By Property 9.2.2 the subspace S_0 is dense in X_0. Every point $x = \{z_i\} \in X_0$ is then the limit of the sequence of its n-th sections:

$$x_n = \{z_0, \ldots, z_n, 0, \ldots\} \to x = \{z_n\}. \tag{9.2.8}$$

This means that every series $\sum\limits_{i=0}^{\infty} R_0^i z_i$, where $z_i \in \ker D_0$, is d_0-convergent and $\sum\limits_{i=0}^{\infty} R_0^i z_i = \{z_i\}$. In particular, we have

$$\sum_{i=0}^{\infty} P_{0,\,i}\, x = \sum_{i=0}^{\infty} R_0^i F_0 D_0^i x = x \quad \text{for all } x \in X_0. \tag{9.2.9}$$

In other words: *The Taylor expansion of every element $x \in X_0$ is d_0-convergent to this element.*

Let H be an arbitrary mapping (non necessarily linear) of X_0 into X_0. We say that H is d_0-continuous at the point $x^0 \in X_0$ if and only if for every $i \geqslant 1$ there is an $N(i) \in \mathbf{N}$ such that for all $n \geqslant N(i)$

$$x - x^0 \in R_0^{N(i)} X_0 \text{ implies } H(x) - H(x^0) \in R_0^i X_0. \tag{9.2.10}$$

H is d_0-continuous in X_0 if and only if for every $i \geqslant 1$ there is an $N(i) \in \mathbf{N}$ such that for all $n \geqslant N(i)$,

$$x - y \in R_0^{N(i)} X_0 \text{ implies } H(x) - H(y) \in R_0^i X_0$$
$$\text{for all } x, y \in X_0. \tag{9.2.11}$$

If, in addition, H is additive (the homogeneity of H does not play any role) then the structure of the sets $R_0^i X_0$ implies that H is d_0-continuous in X_0 if and only if for every $i \geqslant 1$ there is an $N(i) \in \mathbf{N}$ such that

$$H(R^{N(i)} X_0) \subset R_0^i X_0 \quad \text{(provided that } H \text{ is additive).} \tag{9.2.12}$$

The shift operators D_0 and R_0 maps the set $\{R_0^i X_0\}_{i \in \mathbf{N}_0}$ into itself.

This, Formulae (9.2.2) and the linearity of the shifts together imply that the d_0-metric is compatible with the D_0–R_0 structure of X_0.

Since X_0 is a complete metric space, the Banach Fixed Point Theorem (Theorem 8.1.1) can be applied to mappings which are contractive in the d_0-metric. A mapping H of X_0 into X_0 is said to be d_0-*contractive* if and only if for every $x, y \in X_0$ and $i \geqslant 1$ there is an $N = N(i, x, y) \in \mathbf{N}$ such that

$$H^n(x) - H^n(y) \in R_0^i X_0 \quad \text{for all } n \geqslant N. \qquad (9.2.13)$$

This definition is equivalent to the following: H is d_0-contractive if

$$H^n(x) - H^n(y) \to 0 \quad \text{for any fixed } x, y \in X_0 \quad \text{(in } d_0\text{-metric)}. \qquad (9.2.14)$$

Observe that the d_0-contractivity does not imply the d_0-continuity and conversely.

To apply this definition we shall use the following obvious

PROPOSITION 9.2.1 (Fixed Point Criterion). *Let X be a complete metric space and let H be a mapping of X into itself. Then the following conditions are equivalent*:
   (i) $\{H^n(x)\}$ *is convergent for at least one $x \in X$*;
   (ii) $H^n(x) - H^n(y) \to 0$ *for all $x, y \in X$*;
   (iii) H *has a unique fixed point x^0, i.e. $x^0 = H(x^0)$, and $H^n(x) \to x^0$ for all $x \in X$.*

Proof. If we write $y = H(x)$ then Conditions (i) and (ii) are mutually related by the equality

$$x - H^{n+1}(x) = \sum_{i=0}^{n} [H^i(x) - H^{i+1}(x)] \quad (x \in X).$$

Hence the sequence $\{H^n(x)\}$ is convergent if and only if the series $\sum_{i=0}^{\infty} [H^i(x) - H^{i+1}(x)]$ is convergent and

$$x - \lim_{n \to \infty} H^n(x) = \sum_{i=0}^{\infty} [H^i(x) - H^{i+1}(x)] \quad (x \in X), \qquad (9.2.15)$$

provided that one of these limits exists. In this case $x^0 = \lim\limits_{n \to \infty} H^n(x)$ is a fixed point of H, according to our criterion.   ∎

THEOREM 9.2.2 (d_0-Fixed Point Theorem, von Trotha, 1981a). *Let H be a d_0-continuous and d_0-contractive mapping of the space $X_0 = X_0(Z)$ into itself. Then H has a unique fixed point.*

Proof. Let $x \in X_0$ be arbitrarily fixed. Write $y = H(x)$. Since, by our assumption, H is a d_0-contraction, we conclude that $H^n(x) - H^{n+1}(x) \to 0$ (in d_0-metric). Theorem 9.2.1 implies that

$$\sum_{i=1}^{\infty} [H^i(x) - H^{i+1}(x)] = K(x) \in X.$$

Formula (9.2.15) and Proposition 9.2.1 together imply that $x - K(x)$ is a unique fixed point of H. Moreover, $x - K(x) = y - K(y)$ for all $x, y \in X$.   ∎

A mapping H of X_0 into itself is said to be *d_0-shrinking* if for every $x \in X_0$ and $i \geqslant 1$ there is an $N = N(i, x) \in \mathbf{N}$ such that

$$H^n(x) \in R_0^i X_0 \quad \text{for all } n \geqslant N. \tag{9.2.16}$$

In other words: *H is d_0-shrinking if and only if $H^n_{\cdot}(x) \to 0$ for every $x \in X_0$ (pointwise).*

PROPOSITION 9.2.2. *If H is d_0-shrinking then it is d_0-contractive and has only zero as a fixed point.*

Proof. The sequences $\{H^n(x)\}$ and $\{H^n(y)\}$ are d_0-convergent to zero for all $x, y \in X_0$. Thus the sequence $\{H^n(x) - H^n(y)\}$ is also d_0-convergent to zero. Hence, by (9.2.14), H is d_0-contractive. Thus H has a fixed point x^0, i.e. $x^0 = H(x^0)$. This implies $x^0 - H^n(x^0) \to 0$. We therefore conclude that $x^0 = 0$.   ∎

COROLLARY 9.2.1. *Let H be a d_0-continuous mapping of X_0 into itself. Then H is d_0-shrinking if and only if H is d_0-contractive with zero as a fixed point.*

Proof. Sufficiency follows from Proposition 9.2.2. Necessity follows from Theorem 9.2.2.   ∎

Note that the notions of a d_0-shrinking mapping and d_0-contraction coincide if the mapping under consideration is additive.

The following theorem slightly generalizes the classical Banach principle of contractive mappings (Theorem 8.1.1):

THEOREM 9.2.3 (von Trotha, 1981a). *Let H be a continuous mapping of a complete metric space with the metric $d(x, y)$ into itself. If the following condition of contractivity is satisfied:*

$$\sum_{i=0}^{\infty} d\big(H^i(x), H^i(y)\big) = K(x, y) < +\infty \quad \textit{for all } x, y \in X,$$

$$(9.2.17)$$

then H has a unique fixed point.

Proof. Since the space X is complete and H is continuous, Condition (9.2.17) implies

$$\sum_{i=0}^{\infty} d\big(H^i(x), H^{i+1}(x)\big) = K\big(x, H(x)\big) < +\infty \quad \text{for } x \in X.$$

Since this series is convergent, we conclude that the sequence $\{H^n(x)\}$ is fundamental for an arbitrarily fixed $x \in X$. Hence there exists the limit $x^0 = \lim_{n \to \infty} H^n(x)$ and $H(x^0) = H\big(\lim_{n \to \infty} H^n(x)\big) = \lim_{n \to \infty} H^{n+1}(x) = x^0$. We therefore conclude that x^0 is a fixed point of H. Condition (9.2.17) also implies that x^0 is a unique fixed point of H. ■

An immediate consequence of this theorem is

COROLLARY 9.2.2. *If X is a complete metric space with the metric $d(x, y)$ and H is a contraction, i.e.*

$$d\big(H(x), H(y)\big) < qd(x, y) \quad \textit{for all } x, y \in X, \text{ where } 0 < q < 1,$$

then H is continuous and satisfies Condition (9.2.17) with $K(x, y)$

$$= \frac{1}{1-q} d(x, y).$$

In particular, we have shown that *the contractivity in d_0-metric implies both, the d_0-continuity and the d_0-contractivity.*

Condition (9.2.17) for the space X_0 can be also formulated as follows:

$$\sum_{n=0}^{\infty} |H^n(x) - H^n(y)|_{d_0} = K(x, y) < +\infty \quad \text{for all } x, y \in X_0.$$

$$(9.2.18)$$

Now it is easy to compare the contractivity conditions in Theorems 9.2.2 and 9.2.3. Namely, Theorem 9.2.1 implies that for a d_0-continuous H and for all $x, y \in X_0$ we have

(i) $H^n(x) - H^n(y) \to 0$ implies the series $\sum_{n=0}^{\infty} [H^n(x) - H^n(y)]$ to be d_0-convergent, but does not imply $\sum_{n=0}^{\infty} |H^n(x) - H^n(y)|_{d_0} \to 0$;

(ii) The convergence of the series $\sum_{n=0}^{\infty} |H^n(x) - H^n(y)|_{d_0}$ implies the d_0-convergence of the series $\sum_{n=0}^{\infty} [H^n(x) - H^n(y)]$, which implies $H^n(x) - H^n(y) \to 0$.

EXAMPLE 9.2.1. Let Z be a linear space (over a field $\mathscr{F}$) and let $A_k \in L_0(Z)$ for $k \in \mathbf{N}$. We admit here $A_0 = 0$. Consider the space $X_0 = X_0(Z)$ and the mapping H defined as follows:

$$H\{z_k\} = \{0, A_1 z_0, A_2 z_1, \ldots\} = \{A_k z_{k-1}\} \quad \text{for } x = \{z_k\} \in X_0.$$

It is easy to verify that H is d_0-continuous and d_0-contractive. We therefore conclude that the system $z_{k+1} = A_{k+1} z_k$ ($k \in \mathbf{N}_0$) with $z_0 \in Z$ arbitrarily fixed has a unique solution

$$x^0 = \{B_l z_0\}, \quad \text{where } B_0 = 0, B_l = A_l \ldots A_1 \text{ for } i \geqslant 1.$$

9.3. LINEAR SYSTEMS. F_1-CONTROLLABILITY

To begin with, we shall define what is meant by a linear system.

DEFINITION 9.3.1. Let X, Y and U be linear spaces (all over $\mathbf{R}$ or $\mathbf{C}$). Suppose that $D \in R(X)$, $\dim \ker D \neq 0$, F is an initial operator for D corresponding to an $R \in \mathscr{R}_D$, $A \in L_0(X)$, $A_1 \in L_0(X \to Y)$, $B \in L_0(U \to X)$,

$B_1 \in L_0(U \rightarrow Y)$. A *linear system* (shortly: (LS)) is, in general, of the form:

$$Dx = Ax + Bu \quad \text{with } BU \subset (D-A)X, \tag{9.3.1}$$

$$\text{(LS)} \quad y = A_1 x + B_1 u, \tag{9.3.2}$$

$$Fx = x_0, \quad \text{where } x_0 \in \ker D. \tag{9.3.3}$$

The spaces X and U are called: the *space of states* and the *space of controls*, respectively. So that, elements $x \in X$ and $u \in U$ are called *states* and *controls*, respectively. The element $x_0 \in \ker D$ is called an *initial state*. A pair $(x_0, u) \in (\ker D) \times U$ is called an *input* and a y determined by (9.3.2) is called an *output* of a linear system under consideration corresponding to that input. Therefore the space $(\ker D) \times U$ is called the *inputs space* and the corresponding set of y's in Y—the *outputs space*. Very often there are considered systems with $A_1 = I$ and $B_1 = 0$, i.e. with $Y = X$ and the output $y = x$. We shall denote these systems by (LS)_0.

The properties of linear systems depend in an essential way on properties of resolving operators $I - RA$ and $I - AR$, respectively. Nguyen Dinh Quyet in a series of papers (1977, 1978, 1978a, 1979, 1981) studied some properties of linear systems in the case of an invertible resolving operator $I - RA$. His results concerning controllability of linear systems are generalized by A. Pogorzelec (1983a, 1984) for resolving operators $I - RA$ and $I - AR$ which are either left or right invertible and for an invertible resolving operator $I - AR$ (cf. also the next section).

Suppose now that we are given a linear system (LS) with an invertible operator $I - RA$. Since the condition $BU \subset (D-A)X$ is satisfied by our assumptions, we conclude that for every fixed $u \in U$ the initial value problem (9.3.1), (9.3.3) is well-posed and its unique solution is of the form:

$$x = E_A(RBu + x_0) = \Phi(x_0, u), \quad \text{where } E_A = (I - RA)^{-1} \tag{9.3.4}$$

(cf. Section 4.1). Therefore, according to Equality (9.3.2), the output y is uniquely determined for every $u \in U$ and $x_0 \in \ker D$ and is of the form:

$$y = A_1 \Phi(x_0, u) + B_1 u, \tag{9.3.5}$$

where $\Phi(x_0, u)$ is defined by (9.3.4).

If we consider a linear system of the type $(LS)_0$ then we have $A_1 = I$, $B_1 = 0$ and $y = x = \Phi(x_0, u)$.

A matrix operator

$$G = (G_0, G_1), \quad \text{where } G_0 = A_1 E_A, G_1 = G_0 RB + B_1,$$
$$E_A = (I - RA)^{-1} \tag{9.3.6}$$

defined on the inputs space $(\ker D) \times U$ is said to be a *transfer operator* for a linear system with an invertible resolving operator $I - RA$. We therefore conclude that to every input (x_0, u) there corresponds a uniquely determined output y (provided that the resolving operator is invertible) which can be written, using the transfer operator, as

$$y = G(x_0, u) = G_0 x_0 + G_1 u, \quad (x_0, u) \in (\ker D) \times U. \tag{9.3.7}$$

Transfer operators determine in a sense corresponding linear systems.

DEFINITION 9.3.2. Suppose, we are given two linear systems: (LS) determined by (9.3.1)–(9.3.2) and (LS)' determined by equalities:

$$Dx = A'x + B'u, \quad B'U \subset (D - A')X, \tag{9.3.1'}$$
$$y = A_1'x + B_1'u \tag{9.3.2'}$$

and the initial condition (9.3.3), where $A', A_1' \in L_0(X)$ and B', B, $\in L_0(U \to X)$. Then the systems (LS) and (LS)' are said to be *equivalent* if there exist operators K, L, M, N such that M and N are invertible and the following identities hold:

$$\begin{aligned}
I - RA' &= M(I - RA)N, \\
RB' &= M[(I - RA)L + RB], \\
A_1' &= [A_1 - K(I - RA)]N, \\
B_1' &= K[(I - RA)L + RB] - A_1 L + B_1.
\end{aligned} \tag{9.3.8}$$

The systems (LS) and (LS)' are said to be *similar*, if

$$N = M^{-1}, \quad K = L = 0. \tag{9.3.9}$$

THEOREM 9.3.1 (Nguyen Dinh Quyet, 1977). *Suppose that all assumptions of Definitions 9.3.1 and 9.3.2 are satisfied. Consider the systems* (LS) *and* (LS)' *with the property that the resolving operators $I - RA$ and $I - RA'$ are invertible. Let G' be the transfer operator for* (LS)', *i.e. $G' = (G_0', G_1')_1'$*

where $G_0' = A_1'(I - RA')^{-1}$, $G_1' = G_0' RB' + B_1$. *If the systems* (LS) *and* (LS)$'$ *are equivalent then their transfer operators satisfy the equalities*:

$$G_0 = G_0' M + K, \quad G_1 = G_1'. \tag{9.3.10}$$

If the systems (LS) *and* (LS)$'$ *are similar then their transfer operators satisfy the equalities*:

$$G_0 = G_0' M, \quad G_1 = G_1'. \tag{9.3.11}$$

Proof. By definitions and our assumptions we have

$$
\begin{aligned}
G_0' &= A_1'(I - RA')^{-1} = [A_1 - K(I - RA)]N[M(I - RA)N]^{-1} \\
&= [A_1 - K(I - RA)]NN^{-1}(I - RA)^{-1}M^{-1} \\
&= [A_1 - K(I - RA)](I - RA)^{-1}M^{-1} \\
&= [A_1(I - RA)^{-1} - K]M^{-1} = (G_0 - K)M^{-1}
\end{aligned}
$$

which implies $G_0 = G_0' M + K$. Similarly,

$$
\begin{aligned}
G_1' &= G_0' RB' + B_1' = A_1'(I - RA')^{-1}RB' + B_1' \\
&= [A_1 - K(I - RA)]N[M(I - RA)N]^{-1}M[(I - RA)L + RB] + \\
&\quad + K[(I - RA)L + RB] - A_1 L + B_1 \\
&= [A_1 - K(I - RA)]NN^{-1}(I - RA)^{-1}M^{-1}M[(I - RA)L + \\
&\quad + RB] + K[(I - RA)L + RB] - A_1 L + B_1 \\
&= [A_1 - K(I - RA)](I - RA)^{-1}[(I - RA)L + RB] + \\
&\quad + K[(I - RA)L + RB] - A_1 L + B_1 \\
&= A_1(I - RA)^{-1}[(I - RA)L + RB] - K[(I - RA)L + RB] + \\
&\quad + K[(I - RA)L + RB] - A_1 L + B_1 \\
&= A_1 L + A_1(I - RA)^{-1}RB - A_1 L + B_1 \\
&= A_1(I - RA)^{-1}RB + B_1 = G_0 RB + B_1 = G_1.
\end{aligned}
$$

The systems (LS) and (LS)$'$ are similar, if $N = M^{-1}$, $K = L = 0$. This, and Formulae (9.3.10), together imply that the transfer operator for similar systems (LS) and (LS)$'$ satisfy Equalities (9.3.11). ∎

Similar results can be obtained for the invertible operator $I - AR$, also for left and right invertible resolving operators $I - RA$, $I - AR$ (cf. Examples and Exercises).

We shall consider now linear systems of the type $(LS)_0$, i.e. with $A_1 = I$, $B_1 = 0$, what means that the corresponding outputs y are equal to x (cf. Definition 9.3.1):

$$Dx = Ax + Bu, \quad \text{where } BU \subset (D-A)\text{dom } D,$$
$$Fx = x_0.$$

This system can be written in equivalent forms as follows: either

$$(I - RA)x = RBu + x_0 \tag{9.3.12}$$

or

$$(I - AR)Dx = Bx + Ax_0, \quad A \ker D \subset (I - AR)X \tag{9.3.13}$$

(cf. Section 3.1, in particular, Formulae (3.1.3), (3.1.5)).

DEFINITION 9.3.3. Consider $(LS)_0$. Suppose that either $I - RA \in R(X)$ or $I - RA \in \Lambda(X)$ or $I - RA$ is invertible or $I - AR \in R(X)$ or $I - AR \in \Lambda(X)$ or $I - AR$ is invertible and $R_A \in \mathcal{R}_{I-RA}$, $L_A \in \mathcal{L}_{I-RA}$, $E_A = (I - RA)^{-1}$, $R^A \in \mathcal{R}_{I-AR}$, $L^A \in \mathcal{L}_{I-AR}$, $E^A = (I - AR)^{-1}$, respectively. Denote by Φ_i $(i = 1, 2, 3, 4, 5, 6)$ the following sets defined for all $x_0 \in \ker D$, $u \in U$:

(i) if $I - RA \in R(X)$ then
$$\Phi_1(x_0, u) = \{R_A(RBu + x_0) + z : z \in \ker(I - RA)\},$$
$$R_A \in \mathcal{R}_{I-RA};$$

(ii) if $I - RA \in \Lambda(X)$ then
$$\Phi_2(x_0, u) = \{L_A(RBu + x_0)\}, \quad L_A \in \mathcal{L}_{I-RA};$$

(iii) if $I - RA$ is invertible then
$$\Phi_3(x_0, u) = \{E_A(RBu + x_0)\}, \quad E_A = (I - RA)^{-1};$$

(iv) if $I - AR \in R(X)$ then
$$\Phi_4(x_0, u) = \{R[R^A(Bu + Ax_0) + z] + x_0 : z \in \ker(I - AR)\},$$
$$R^A \in \mathcal{R}_{I-AR};$$

(v) if $I - AR \in \Lambda(X)$ and $Bu + Ax_0 \in (I - AR)X$ then
$$\Phi_5(x_0, u) = \{RL^A(Bu + Ax_0) + x_0\}, \quad L^A \in \mathcal{L}_{I-AR};$$

(vi) if $I - AR$ is invertible then
$$\Phi_6(x_0, u) = \{RE^A(Bu + Ax_0) + x_0\}, \quad E^A = (I - AR)^{-1}.$$

The sets Φ_i $(i = 1, ..., 6)$ are sets of all solutions of the system $(LS)_0$ in the corresponding cases. Observe that, according to Corollaries 3.1.2, 3.1.3, 3.1.5, 3.1.6, these sets are independent of the choice of right and left inverses R_A, L_A, R^A, L^A, respectively. Therefore, in the cases under considerations to every fixed input (x_0, u) there corresponds an output of the form: $x = \Phi_i(x_0, u)$, where $i = 1, 2, 3, 4, 5, 6$, respectively.

DEFINITION 9.3.4. Consider $(LS)_0$ and the sets $\Phi_i(x_0, u)$ $(i = 1, ..., 6)$ determined in Definition 9.3.3. A state $x \in X$ is said to be *reachable from the initial state* $x_0 \in \ker D$ if for every $R_A \in \mathcal{R}_{I-RA}$ $(L_A \in \mathcal{L}_{I-RA}, E_A = (I-RA)^{-1}, R^A \in \mathcal{R}_{I-AR}, L^A \in \mathcal{L}_{I-AR}, E^A = (I-AR)^{-1}$, respectively) there exists a control $u \in U$ such that $x \in \Phi_i(x_0, u)$ $(i = 1, 2, 3, 4, 5, 6$, respectively). A state $x \in X$ is said to be R_A-reachable $(L_A$-reachable, R^A-reachable, L^A-reachable, respectively) from the initial state $x_0 \in \ker D$ if there exists a control $u \in U$ such that $x = \Phi_i(x_0, u)$, $i = 1, 2, 4, 5$, respectively.

For instance, for $i = 1$ a state $x \in X$ is R_A-reachable if there exists a control $u \in U$ and also a constant $z \in \ker(I-RA)$ such that $x = R_A(RBu + x_0) + z$. For $i = 2$ a state x is L_A-reachable if there exists a $u \in U$ such that $x = L_A(RBu + x_0)$.

An immediate consequence of Definition 9.3.4 is

PROPOSITION 9.3.1. *Suppose that all assumptions of Definition 9.3.4 are satisfied. A state* $x \in X$ *is reachable from a given initial state* $x_0 \in \ker D$ *if and only if there exists an* $R_A \in \mathcal{R}_{I-RA}$ $(L_A \in \mathcal{L}_{I-RA}, E_A, R^A \in \mathcal{R}_{I-AR}, L^A \in \mathcal{L}_{I-AR}, E^A$, respectively*), such that the state* x *is* R_A*-reachable* $(L_A$*-reachable,* E_A*-reachable,* R^A*-reachable,* L^A*-reachable,* E^A*-reachable, respectively) from the initial state* x_0.

Write:

$$\text{Rang}_{U, x_0} \Phi_i = \bigcup_{u \in U} \Phi_i(x_0, u)$$

$$\text{for } i = 1, 2, 3, 4, 5, 6, \ x_0 \in \ker D, \quad (9.3.14)$$

$\text{Rang}_{U, x_0} \Phi_i$, as sets of all solutions of the system (9.3.1), (9.3.3) for a fixed control space U, are sets reachable from the initial state x_0 by means of controls $u \in U$. Theorems 3.1.3 and 3.1.4 imply that all these sets are contained in the domain of D. Namely, we have

PROPOSITION 9.3.2. *Consider the linear system* (9.3.1), (9.3.3) *of the type* $(\text{LS})_0$.

(i) *If* $I - RA \in R(X)$, $BU \subset (D-A)\operatorname{dom} D^k$ *then*

$$\operatorname{Rang}_{U, x_0} \Phi_1 \subset (I - F_A)\operatorname{dom} D^k \oplus F_A X \quad (k \in \mathbb{N}),$$

where F_A *is an initial operator for* $I - RA$;

(ii) *If* $I - RA \in \Lambda(X)$, $BU \subset (D-A)\operatorname{dom} D^k$, $\ker D \subset (I-RA)\operatorname{dom} D^k$ *then*

$$\operatorname{Rang}_{U, x_0} \Phi_2 \subset \operatorname{dom} D^k \quad (k \in \mathbb{N});$$

(iii) *If* $I - RA$ *is invertible and* $BU \subset (D-A)\operatorname{dom} D^k$ *then*

$$\operatorname{Rang}_{U, x_0} \Phi_3 \subset \operatorname{dom} D^k \quad (k \in \mathbb{N});$$

(iv) *If* $I - AR \in R(X)$ *then*

$$\operatorname{Rang}_{U, x_0} \Phi_4 \subset \operatorname{dom} D;$$

(v) *If* $I - AR \in \Lambda(X)$ *and* $A \ker D \subset (I-AR)X$ *then*

$$\operatorname{Rang}_{U, x_0} \Phi_5 \subset \operatorname{dom} D;$$

(vi) *If* $I - AR$ *is invertible then*

$$\operatorname{Rang}_{U, x_0} \Phi_6 \subset \operatorname{dom} D.$$

THEOREM 9.3.2 (Pogorzelec, 1983a, 1984). *Consider a linear system* $(\text{LS})_0$ *defined by Formulae* (9.3.1), (9.3.3). *Let* X', U' *be the conjugate spaces for* X, U, *respectively* (*i.e. according to the definition in Section* 1.5—*spaces of all linear functionals over* X *and* U, *respectively*). *Write*:

$$T = \begin{cases} RR^A & \text{if } I - AR \in R(X) \text{ and } R^A \in \mathcal{R}_{I-AR}, \\ RL^A & \text{if } I - AR \in \Lambda(X), \, A \ker D \subset (I-AR)X \text{ and} \\ & \qquad\qquad L^A \in \mathcal{L}_{I-AR}, \\ RE^A = R(I-AR)^{-1} & \text{if } I - AR \text{ is invertible}, \\ R_A R & \text{if } I - RA \in R(X) \text{ and } R_A \in \mathcal{R}_{I-RA}, \\ L_A R & \text{if } I - RA \in \Lambda(X) \text{ and } L_A \in \mathcal{L}_{I-RA}, \\ E_A R = (I-RA)^{-1} R & \text{if } I - RA \text{ is invertible}. \end{cases}$$

$$(9.3.15)$$

Suppose that $B \in L_0(U \to X, X' \to U')$, $A, D, R, T \in L_0(X, X')$. Then the generalized Kalman condition

$$\ker B^* T^* = \{0\} \tag{9.3.16}$$

holds if and only if for every initial state $x_0 \in \ker D$ every state $x \in RX \oplus$ $\oplus \{x_0\} \oplus V_0$ is reachable from x_0, where

$$V_0 = \ker(I - RA) \quad \text{if } I - RA \in R(X) \text{ and}$$
$$V_0 = \{0\} \quad \text{otherwise.} \tag{9.3.17}$$

Proof. Observe that, by definition,

$$T^* = \begin{cases} (R^A)^* R^* & \text{if } I - AR \in R(X) \text{ and } R^A \in \mathscr{R}_{I-AR}, \\ (L^A)^* R^* & \text{if } I - AR \in \Lambda(X) \text{ and } L^A \in \mathscr{L}_{I-AR}, \\ (E^A)^* R^* & \text{if } I - AR \text{ is invertible}, \\ R^* R_A^* & \text{if } I - RA \in R(X) \text{ and } R_A \in \mathscr{R}_{I-RA}, \\ R^* L_A^* & \text{if } I - RA \in \Lambda(X) \text{ and } L_A \in \mathscr{L}_{I-RA}, \\ R^* E_A^* & \text{if } I - RA \text{ is invertible.} \end{cases} \tag{9.3.18}$$

Let $x_0 \in \ker D$ be an arbitrarily fixed initial state. Let $I - AR \in R(X)$ and let $R^A \in \mathscr{R}_{I-AR}$ be fixed. The condition $BU \subset (D - A) \operatorname{dom} D$ implies

$$RR^A BU \subset \{x = RR^A(I - AR)v - As_0 : v \in X, s_0 \in \ker D\}.$$

Indeed,

$$RR^A(D - A)\operatorname{dom} D$$
$$= RR^A(D - A)\{x = Rv + s_0 : v \in X, s_0 \in \ker D\}$$
$$= RR^A\{x = (I - AR)v - As_0 : v \in X, s_0 \in \ker D\}$$
$$= \{x = RR^A[(I - AR)v - As_0] : v \in X, s_0 \in \ker D\}.$$

Hence the operator $RR^A B$ maps the control space U into the set $\{x = RR^A(I - AR)v - As_0 : v \in X, s_0 \in \ker D\}$. We therefore conclude that Condition (9.3.16) holds if and only if for every $v \in X$ and $s_0 \in \ker D$ there exists a $u \in U$ such that $RR^A B = RR^A[(I - AR)v - As_0]$. This means that for every $v \in X$, $s_0 \in \ker D$, $z \in \ker(I - AR)$ there exists a $u \in U$ such that

$$R[R^A(Bu + As_0) + z] + x_0 = R[R^A(I - AR)v + z] + x_0.$$

In particular, let $s_0 = x_0$ and let $z = F^A v$, where F^A is an initial operator for the right invertible operator $I - AR$, i.e. $F^A = I - R^A(I - AR)$.

Then we find for every $v \in X$ a $z \in \ker(I-AR)$ and a $u \in U$ such that

$$R[R^A Bu + Ax_0 + z] + x_0 = R[(I-F^A)v + F^A v] + x_0.$$

Since F^A is a projection, we have a decomposition onto the direct sum: $X = (I-F^A)X \oplus F^A X$. This implies that for every $v \in X$ there exist $z \in \ker(I-AR)$ and $u \in U$ such that $R[R^A(Bu + Ax_0) + z] + x_0 = Rv + x_0$ and $\mathrm{Rang}_{U, x_0} \Phi_4 = RX \oplus \{x_0\}$. The arbitrariness of $x_0 \in \ker D$ implies the first point of our theorem. The proofs in other cases under consideration are similar. ∎

Note that the generalized Kalman condition (9.3.16) in the case of an invertible $I-RA$ was introduced and applied by Nguyen Dinh Quyet (1977, 1978).

DEFINITION 9.3.5. Consider the linear system $(LS)_0$ defined by Definition 9.3.1. Let $F_1 \neq F$ be an initial operator for D. The state $x_1 \in \ker D$ is said to be F_1-*reachable from the initial state* $x_0 \in \ker D$ if there exists a control $u \in U$ such that $x_1 \in F_1 \Phi_i(x_0, u)$ $(i = 1, 2, 3, 4, 5, 6$, respectively, cf. Definition 9.3.3). The state x_1 will be called a *final state*. The linear system $(LS)_0$ is said to be F_1-*controllable* if for every initial state $x_0 \in \ker D$

$$F_1 \mathrm{Rang}_{U, x_0} \Phi_i = \ker D \quad (i = 1, 2, 3, 4, 5, 6, \quad \text{respectively}).$$
$$(9.3.19)$$

The system $(LS)_0$ is said to be F_1-*controllable to zero* if

$$0 \in \mathrm{Rang}_{U, x_0} \Phi_i \quad (i = 1, 2, 3, 4, 5, 6, \text{ respectively}), \quad (9.3.20)$$

for every initial state $x_0 \in \ker D$.

Clearly, F_1-controllability of a linear system implies its F_1-controllability to zero.

PROPOSITION 9.3.3. *Let be given a linear system* $(LS)_0$ *and an initial operator* $F_1 \neq F$ *for* D. *Suppose that the system* $(LS)_0$ *is* F_1-*controllable to zero and that*

$$F_1(RR^A A + I)\ker D = \ker D \quad \text{if } I-AR \in R(X),\ R^A \in \mathscr{R}_{I-AR},$$
$$(9.3.21)$$

$$F_1(RL^A A + I)\ker D = \ker D \quad \text{if } I - AR \in A(X),\ L^A \in \mathscr{L}_{I-AR},$$

(9.3.22)

$$F_1(RE^A A + I)\ker D = \ker D \quad \text{if } I - AR \text{ is invertible and}$$
$$E^A = (I - AR)^{-1}. \quad (9.3.23)$$

The every final state $x_1 \in \ker D$ is F_1-reachable from zero.

Proof. Suppose that $I - AR \in R(X)$. Let $x_1 \in \ker D$ be arbitrarily fixed. Since, by our assumption, the system $(LS)_0$ is F_1-controllable to zero, we conclude that $0 \in \mathrm{Rang}_{U, x_0} \Phi_4$ for every initial state $x_0 \in \ker D$. This means that there exists a control $u_0 \in U$ such that $0 \in F_1 \Phi_4(x_0, u_0)$. Thus there is a $z_0 \in \ker(I - AR)$ such that $F_1 \{R[R^A(Bu_0 + Ax_0) + z_0] + x_0\} = 0$. This last equality can be written in the equivalent form:

$$F_1[R(R^A Bu_0 + z_0)] = -F_1[(RR^A A + I)x_0].$$

Our assumption (9.3.21) implies that there exists a constant $-\tilde{x}_0 \in \ker D$ such that $F_1[(RR^A A + I)\tilde{x}_0] = x_1$. Then there are a control $u \in U$ and a constant $z_0 \in \ker(I - AR)$ such that $F_1 \{R[R^A(Bu + A\tilde{x}_0) + z_0] + \tilde{x}_0\} = x_1$. This proves that the final state x_1 is reachable from the initial state 0. The arbitrariness of $x_1 \in \ker D$ implies the first point of our proposition. If either $I - AR \in A(X)$ or $I - AR$ is invertible then the proof is similar. ∎

THEOREM 9.3.3 (Pogorzelec, 1983a). *Suppose that all assumptions of Proposition 9.3.3 are satisfied. Then the linear system $(LS)_0$ is F_1-controllable.*

Proof. Suppose that $I - AR \in R(X)$ and $R^A \in \mathscr{R}_{I-AR}$. We have to prove that $F_1 \mathrm{Rang}_{U, x_0} \Phi_4 = \ker D$ for every initial state $x_0 \in \ker D$, i.e. that for every pair $x_0, x_1 \in \ker D$ there exist a control $u_0 \in U$ and a $z_0 \in \ker(I - AR)$ such that

$$F_1 \{R[R^A(Bu_0 + Ax_0) + z_0] + x_0\} = x_1.$$

Since, by our assumption, the system $(LS)_0$ is F_1-controllable to zero, we conclude that there exist a control $u_1 \in U$ and a $z_1 \in \ker(I - AR)$ such that

$$F_1 \{R[R^A(Bu_1 + Ax_0) + z_1] + x_0\} = 0. \quad (9.3.24)$$

On the other hand, Proposition 9.3.3 implies that there exist a control u_2 and a $z_2 \in \ker(I - AR)$ such that

$$F_1\{R[R^A(Bu_2 + 0) + z_2] + 0\} = x_1. \tag{9.3.25}$$

If we add Equalities (9.3.24) and (9.3.25) then we get

$$F_1\{R[R^A(B(u_1 + u_2) + Ax_0) + (z_1 + z_2)] + x_0\} = x_1.$$

We therefore have shown that for $u_0 = u_1 + u_2$ and $z_0 = z_1 + z_2$ the final state is reachable from the initial state x_0. The arbitrariness of $x_0, x_1 \in \ker D$ implies the first point of our theorem. The proof for $I - AR \in \Lambda(X)$ and an invertible $I - AR$ are similar. ∎

Note that the assumptions of Theorem 9.3.3 are sufficient for the F_1-controllability of the system (LS)$_0$ but not necessary, as it will be shown by Examples 9.3.3 and 9.3.4.

PROPOSITION 9.3.4. *Let a linear system* (LS)$_0$ *and an initial operator* $F_1 \neq F$ *for* D *be given. Suppose that the system* (LS)$_0$ *is* F_1-*controllable to zero and that*

$$F_1 R_A \ker D = \ker D \quad \text{if } I - RA \in R(X) \text{ and } R_A \in \mathcal{R}_{I - RA},$$
$$\tag{9.3.26}$$

$$F_1 L_A \ker D = \ker D \quad \text{if } I - RA \in \Lambda(X) \text{ and } L_A \in \mathcal{L}_{I - RA},$$
$$\tag{9.3.27}$$

$$F_1 E_A \ker D = \ker D \quad \text{if } I - RA \text{ is invertible and}$$
$$E_A = (I - RA)^{-1}. \tag{9.3.28}$$

Then every final state $x_1 \in \ker D$ *is* F_1-*controllable.*

Proof. Suppose that $I - RA \in \Lambda(X)$ and that $L_A \in \mathcal{L}_{I - RA}$ is fixed. In a similar way, as in the proofs of Proposition 9.3.3 and Theorem 9.3.3, we can prove the following fact: if the system (LS)$_0$ is F_1-controllable to zero and Condition (9.3.27) is satisfied then this system is F_1-controllable. Theorem 3.1.6 immediately implies that the system (LS)$_0$ is F_1-controllable to zero for $I - RA \in \Lambda(X)$ if and only if is F_1-controllable for $I - AR \in \Lambda(X)$. Since, by Theorem 3.1.6, we have $L^A = DL_A R \in \mathcal{L}_{I - AR}$, we conclude that the following equality holds:

$$F_1(RL^A A + I) = F_1(RDL_A RA + I) = F_1[(I - F)L_A RA + I].$$

The condition $A \ker D \subset (I - AR)X$ (assumed in (9.3.13)) implies $RA \ker D \subset (I - RA) RX \subset (I - RA)X$. It is easy to verify also that $FL_A y = Fy$ for $y \in (I - RA)X$ (cf. Section 3.1). We therefore conclude that $(I - F) L_A RA = L_A RA - FRA = L_A RA$ for $FR = 0$ on X. Hence $F_1(RL^A A + I) = F_1(L_A RA + I) = F_1[-L_A(I - RA) + L_A + I] = F_1(-I + + L_A + I) = F_1 L_A$. Finally, we find that both conditions, the F_1-controllability to zero and (9.3.27) for the case $I - RA \in \Lambda(X)$ together are equivalent to F_1-controllability to zero and (9.3.22) in Proposition 9.3.3 for $I - AR \in \Lambda(X)$. The proofs for $I - RA \in R(X)$ and an invertible $I - RA$ are similar. ∎

Theorem 3.1.6 immediately implies that the system $(LS)_0$ is F_1-controllable to zero for $I - RA \in R(X)$ if and only if is F_1-controllable to zero for $I - AR \in R(X)$. On the other hand, Example 9.3.2 shows that conditions

$$F_1(RR^A A + I) \ker D = \ker D \quad \text{and} \quad F_1 R_A \ker D = \ker D$$

(where $R^A = DR_A R \in \mathcal{R}_{I-AR}$, $R_A \in \mathcal{R}_{I-RA}$) are, in general, not equivalent. Observe that the inclusion $R_A RX \subset RX$ implies that

$$F_1(RR^A A + I) = F_1(RDR_A RA + I) = F_1[(I - F) R_A RA + I]$$
$$= F_1(R_A RA + I) = F_1[-R_A(I - RA) + R_A + I] = F_1(F_A + R_A),$$

where $F_A = I - R_A(I - RA)$ is an initial operator for $I - RA$.

THEOREM 9.3.4 (Pogorzelec, 1983a, 1984). *Let a linear system* $(LS)_0$ *and an initial operator* $F_1 \neq F$ *for D be given. Let T be defined by Formulae* (9.3.15). *Suppose that* $B \in L_0(U \to X, X' \to U')$, $D \in L(X, X')$, $R, A, T \in L_0(X, X')$.

(i) *Suppose that* $I - AR \in R(X)$ *and* $R^A \in \mathcal{R}_{I-AR}$. *If*

$$\ker B^* T^* F_1^* = \{0\} \tag{9.3.29}$$

then the system $(LS)_0$ *is* F_1-*controllable*;

(ii) *Suppose that* $A \ker D \subset (I - AR)X$ *and either* $I - AR \in \Lambda(X)$ *and* $L^A \in \mathcal{L}_{I-AR}$ *or* $I - AR$ *is invertible and* $E^A = (I - AR)^{-1}$. *Then Condition* (9.3.29) *is necessary and sufficient for the system* $(LS)_0$ *to be* F_1-*controllable*.

Proof. Observe that in all cases under consideration the operator $F_1 TB$ maps U onto ker D. Let $x_0, x_1 \in$ ker D be arbitrarily fixed. Suppose that $I - AR \in R(X)$ and that $R^A \in \mathcal{R}_{I-AR}$ is fixed. Condition (9.3.29) is equivalent to the condition

$$F_1 TBU = \text{ker } D. \tag{9.3.30}$$

Since, by our assumption, $BU \subset (D-A)$ dom D, we get

$$TBU = RR^A BU \subset RR^A(D-A)\text{dom } D = T(D-A)\text{dom } D.$$

This implies $F_1 T(D-A)$ dom $D = $ ker D. Hence for every $v \in X$ and $s_0 \in$ ker D there exists a $u \in U$ such that

$$F_1 T(D-A)(Rv+s_0) = F_1 TBu.$$

This equality can be written as follows:

$$F_1 T[(I-AR)v - As_0] = F_1 TBu$$

and

$$F_1(Rv+s_0) = F_1 \{R[R^A(Bu+As_0)+F^A v] + s_0\},$$

where $F^A = I - R^A(I-AR)$ is an initial operator for $I-AR$.

By definition, for every $x_2 \in$ ker D there exists a $v \in X$ such that $F_1 Rv = x_1$. In particular, take $x_2 = x_1 - x_0$ and $s_0 = x_0$. We find that for the state x_1 there exist a control $u \in U$ and $z \in$ ker$(I-AR)$ such that $F_1 \{R[R^A(Bu+Ax_0)+z]+x_0\} = x_1$, where $z = F^A v = x_1 - x_0$. The arbitrariness of $x_0, x_1 \in$ ker D implies F_1 Rang$_{U,x_0} \Phi_4 = $ ker D for every $x_0 \in$ ker D.

Suppose now that $I - AR \in \Lambda(X)$ and that $L^A \in \mathscr{L}_{I-RA}$ is fixed. Again Condition (9.3.29) is equivalent to Condition (9.3.30). Since, by our assumption, $BU \subset (D-A)$ dom D and A ker $D \subset (I-AR)X$, we conclude that $TBU = RL^A BU \subset RL^A(D-A)$ dom $D = T(D-A)$ dom $D = RX$. Hence $F_1 TBU = F_1 RX = $ ker D. This implies that for every $v \in X$ and $s_0 \in$ ker D there exists a $u \in U$ such that $F_1 TBu = F_1 T(D-A)(Rv+s_0)$. Moreover, $F_1[T(Bu+As_0)+s_0] = F_1 Rv+s_0$. Put $s_0 = x_0$ and take such $v \in X$ that $F_1 Rv = x_1 - x_0$. We find

$$F_1[T(Bu+Ax_0)+x_0] = x_1.$$

The arbitrariness of $x_0, x_1 \in$ ker D implies F_1 Rang$_{U,x_0} \Phi_5 = $ ker D for every $x_0 \in$ ker D. For an invertible $I - AR$ the proof is the same.

We shall prove now the necessity of Condition (9.3.29). Suppose then that for every $x_0 \in \ker D$ we have $F_1 \mathrm{Rang}_{U, x_0} \Phi_5 = \ker D$. In particular, if $x_0 = 0$ then we get $F_1 \mathrm{Rang}_{U, 0} \Phi_5 = \ker D$, i.e. $F_1 TBU = \ker D$. This means that the operator $F_1 TB$ maps U onto $\ker D$ which is equivalent to Condition (9.3.29).   ∎

A linear system $(\mathrm{LS})_0$ is said to be *stationary* if A and B are stationary.

THEOREM 9.3.5 (Nguyen Dinh Quyet, 1977, 1978). *Let a stationary linear system with an invertible resolving operator $I - RA$ be given. Then*
(i) *Condition (9.3.29) is of the form*:

$$\ker B^* E_A^* F_1^* = \{0\}, \quad E_A = (I - RA)^{-1}, \tag{9.3.31}$$

provided that $B \in L_0(U \to X, X' \to U')$, $R, F_1, A, E \in L_0(X, X')$, $D \in L(X, X')$;

(ii) *Suppose that the system $(\mathrm{LS})_0$ is F_1-controllable and $BU \subset (I - RA)X$. Then for an arbitrary initial operator $F_2 \in \mathscr{F}_D$ the system $(\mathrm{LS})_0$ is F_2-controllable.*

Proof. Condition (9.3.31) for stationary A and B is an immediate consequence of (9.3.29) since $I - AR = I - RA$, $E_A B = B E_A$ and $\ker R = \{0\}$. In order to prove (ii) observe that for a stationary A we have $(I - RA)R = R(I - RA)$ and $E_A R = R E_A$. By our assumption that the system $(\mathrm{LS})_0$ is F_1-controllable for every $z \in \ker D$ we find a $u \in U$ such that $F_1 \Phi(x_0, u) = F_1 E_A (RBu + x_0) = z$. This implies that

$$(F_1 - F_2) \Phi(x_0, u) = (R_1 - R_2)[DRE_A Bu + DE_A x_0]$$
$$= F_1 R_2 (E_A Bu + DE_A x_0),$$

where F_2 is an initial operator for D corresponding to an $R_2 \in \mathscr{R}_D$. This last equality can be written as

$$F_2 R E_A Bu = F_2 E_A RBu = z - z_2,$$
$$\text{where } z_2 = F_1 R_2 (E_A Bu + DE_A x_0) - F_2 E_A x_0.$$

By our assumption there exist $y \in X$ and $w \in U$ such that $F_2 Ry = z_2$ and $Bw = (I - RA)y \in X$. Since $I - RA$ is invertible, we have $y = E_A Bw$ and we conclude that $F_2 R E_A Bw = F_2 Ry = z_2$. Let $u_1 = u + w$. Then $u_1 \in U$ and $F_2 E_A RBu_1 = F_2 E_A RB(u + w) = F_2 E_A RBu + F_2 E_A RBw$

$= z - z_2 + z_2 = z$. The arbitrariness of $z \in \ker D$ implies Point (ii) of our theorem.* ∎

DEFINITION 9.3.6. Denote by $(LS)_H$ a linear system (LS) with the output $y = Hx$, i.e. with $A_1 = H \in L_0(X \to Y)$, $B_1 = 0$:

$$Dx = Ax + Bu, \quad y = Hx, \quad Fx = x_0, \quad x_0 \in \ker D. \quad (9.3.32)$$

Suppose that the resolving operator $I - RA$ is invertible. The system $(LS)_H$ is said to be *observable* if for every given output $y \in Y$ and input $u \in U$ there exists a unique initial state $x_0 \in \ker D$ such that $y = H\Phi(x_0, u)$, where $\Phi(x_0, u) = E_A(RBu + x_0) = (I - RA)^{-1}(RBu + x_0)$ $= x$ (cf. (9.3.4), (9.3.5)). Let $(\ker D)'$ be the conjugate space to the space $\ker D$, i.e. the space of all linear functionals defined on $\ker D$ (cf. Section 1.5). A functional $f \in (\ker D)'$ is said to be *observable* if there exists a $\varphi \in Y'$ such that $f = E_A^* H^* \varphi$ (provided that $D \in L(X, X')$, $R, A, F, E_A \in L_0(X, X')$, $B \in L_0(U \to X, X' \to U')$, $H \in L_0(X \to Y, Y' \to X')$). Note that $E_A = (I - RA)^{-1}$ maps $\ker D$ into X, i.e. for every $x_0 \in \ker D$ the element $x = E_A x_0$ is a unique solution of the system:

$$(D - A)x = 0, \quad Fx = x_0, \quad x_0 \in \ker D. \quad (9.3.33)$$

THEOREM 9.3.6. (Nguyen Dinh Quyet, 1977, 1978). *Let a linear system* $(LS)_H$ *with the invertible resolving operator* $I - RA$ *be given. The following conditions are equivalent:*

(i) *The system* $(LS)_H$ *is observable;*

(ii) *The following condition is satisfied:*

$$\ker HE_A = \{0\}; \quad (9.3.34)$$

(iii) *All functionals* $f \in (\ker D)'$ *are observable.*

Proof. In order to prove (i) → (ii) observe that the output y corresponding to $x_0 \in \ker D$ and $u \in U$ is of the form $y = HE_A x_0 + HE_A RBu$. Hence the equation $HE_A x_0 = y - HE_A RBu$ has a unique solution x_0 for every $u \in U$ and $y \in Y$ if and only if $\ker HE_A = \{0\}$. A functional

---

* The additional condition for F_2 admitted by Nguyen Dinh Quyet that $F_2 \in \{\tilde{F} \in \mathcal{F}_D: \underset{z \in \ker D}{\forall} \underset{y \in X}{\exists} \tilde{F}Ry = z\}$ is automatically satisfied. His proof was restricted to the case $x_0 = 0$ what is also not essential.

$f \in (\ker D)'$ is observable if and only if Condition (9.3.34) is satisfied. Indeed, the equation $f = E_A^* H^* \varphi$ has a solution φ if and only if $\ker HE_A = \{0\}$ (cf. for instance, the author and Rolewicz, 1968). This implies that all functionals $f \in (\ker D)'$ are observable if and only if Condition (9.3.34) is satisfied. ∎

F_1-controllability and observability of linear systems with shifts has been considered by the author (1980, Chapter 12).

Examples and Exercises

EXAMPLE 9.3.1. Let $X = C[0, T]$ over $\mathbf{C}$. Let $D = d/dt$, $R = \int_{t_0}^{t}$, $(Fx)(t) = x(t_0)$ for a $t_0 \in [0, T]$ and $x \in X$. Consider a linear system $(LS)_0$:

$$Dx = \lambda x + Bu, \quad Fx = x_0, \qquad (9.3.35)$$

where x_0 is a constant, $u \in U$, U is a linear space over $\mathbf{C}$, $\lambda \in \mathbf{C}$, $B \in L_0(U \to X)$ and $BU \subset (D - \lambda I)\,\mathrm{dom}\,D$. The system (9.3.35) has a resolving operator $I - \lambda R \in V(X)$ and its inverse is

$$e_\lambda = (I - \lambda R)^{-1},$$

$$[(I - \lambda R)^{-1} x](t) = x(t) + \lambda \int_{t_0}^{t} e^{\lambda(t-s)} x(s)\,ds \quad \text{for } x \in X.$$

Let F_1 be an initial operator for D defined by the equality: $(F_1 x)(t) = x(t_1)$ for a $t_1 \in [0, T]$, $t_1 \neq t_0$. It is easy to verify that $F_1 e_\lambda \ker D = \ker D$.

The system (9.3.35) is F_1-controllable to zero if and only if to every initial state x_0 there is a control $u \in U$ such that $(F_1 x)(t) = x(t_1) = 0$, where $x = Re_\lambda(Bu + \lambda x_0) + x_0$. By simple calculations we can prove that such controls u exist for a set U satisfying the condition

$$BU \supset \{x(t) = c_1 t + c_2 : c_1, c_2 \in \mathbf{C}\}. \qquad (9.3.35')$$

Proposition 9.3.3 implies that every final state $x_1 \in \ker D$ is F_1-reachable from zero. Theorem 9.3.3 implies that the system (9.3.35) is F_1-controllable for controls satisfying Condition (9.3.35').

EXAMPLE 9.3.2. Let $X = (s)$ be the space of all sequences $x = \{x_n\} \subset \mathbf{R}$ (over $\mathbf{R}$) with the addition and multiplication by scalars: $\{x_n\} + \{y_n\}$

$= \{x_n + y_n\}$, $\lambda\{x_n\} = \{\lambda x_n\}$ for $\{x_n\}$, $\{y_n\} \in X$, $\lambda \in \mathbf{R}$. Consider the following linear system:

$$Dx = Ax + Bu, \quad Fx = x_0, \qquad\qquad (9.3.36)$$

where $D\{x_n\} = \{x_{n+1} - x_n\}$, $F\{x_n\} = x_1\{e_n\}$, where $e_n = 1$ for $n \in \mathbf{N}$, $x_0 \in \ker D = \{c\{e_n\}: c \in \mathbf{R}\}$, $A\{x_n\} = \{x_{n+2}\}$, $U = \ker D$, $B = \lambda I$, $0 \neq \lambda \in \mathbf{R}$, $R\{x_n\} = \{y_n\}$, $y_1 = 0$, $y_n = \sum_{k=1}^{n-1} x_k$ for $n \geqslant 2$. We know that $D \in R(X)$ and that F is an initial operator for D corresponding to $R \in \mathscr{R}_D$ (cf. Examples 2.1.4 and Exercice 2.2.10). The condition $BU \subset (D-A)\operatorname{dom} D$ is satisfied for $BU = \ker D$, $\operatorname{dom} D = X$ and $(D-A)X = X$. The operator $I - RA$ is right invertible. Indeed, we have

$$\ker(I-RA) = \big\{x = \{x_n\} \in X\colon x_{6n+1} = 0, x_{6n+2} = x_{6n+3} = c,$$

$$x_{6n+4} = 0, x_{6n+5} = x_{6n+6} = -c$$

$$\text{for } n \in \mathbf{N}, c \in \mathbf{R}\big\} \neq \{0\}.$$

On the other hand, for every $y \in X$ the equation $(I-RA)x = y$ has a solution $x = \{x_n\}$, where $x_1 = y_1, x_2 = y_2 + a$, $x_3 = a$, $x_4 = -y_3$, $x_5 = -y_4 - a$, $x_6 = y_3 - y_5 + a$, and so on, where $a \in \mathbf{R}$ is arbitrarily fixed. All solutions of this equation are of the form: $x = y + z$, where $z \in \ker(I-RA)$. This implies that $(I-RA)X = X$ and $\dim X/(I-RA)X = 0$, hence $I-RA$ is a mapping onto. A right inverse R_A of $I-RA$ is:

$$R\{x_n\} = \{y_n\} \quad \text{where } y_n = x_n \quad \text{for } n = 1, 2, \quad y_3 = 0,$$

$$y_n = -x_{n-1} \quad \text{for } n = 4, 5 \text{ and}$$

$$y_n = -x_{n-1} + x_{n-2} + y_{n-1} - y_{n-2} \quad \text{for } n \geqslant 6.$$

All solutions of the system (9.3.36) are then of the form: $x = R_A(RBu + x_0) + z$, where $z \in \ker(I-RA)$. Define the operator $F_1 \in \mathscr{F}_D$ in the following way: $F_1\{x_n\} = d\{e_n\}$, where $d = \frac{1}{2}(x_1 + x_2)$. We shall prove that F_1 satisfies the condition $F_1 R_A \ker D = \ker D$ and the system (9.3.36) is F_1-controllable to zero.

Let $\tilde{x}_1 = c\{e_n\} \in \ker D$, $c \in \mathbf{R}$. Then $R_A\tilde{x}_1 = \{y_n\}$, where $y_1 = y_2 = c$, $y_3 = 0$, $y_4 = y_5 = -c$, $y_6 = -2c$, and so on. Moreover, $F_1 R_A \tilde{x}_1 = \frac{1}{2}(y_1 + y_2)\{e_n\} = c\{e_n\} = \tilde{x}_1$. The arbitrariness of $c \in \mathbf{R}$ implies that $F_1 R_A \ker D = \ker D$. On the other hand, for every $x_0 \in \ker D$

the equation $F_1[R_A(RBu+x_0)+z] = 0$ has a solution (u, z) in the set $U \times \ker(I-RA)$, for instance $(u, 0)$, where $u = \lambda^{-1}(I-AR_1)v$, $v \in X$. Indeed, $B = \lambda I$, $\lambda \neq 0$. Since F_1 is an initial operator for D corresponding to a right inverse R_1, we have $\ker F_1 = R_1 X$, for $F_1(R_1 v) = 0$ for all $v \in X$. We are looking then for such $v \in X$ that $R_1 v = R_A(\lambda Ru+x_0) = R_A(\lambda RBu+x_0)$. Since $x_0 \in \ker D$ and $R_A \in \mathcal{R}_{I-RA}$ we find $\lambda Ru+x_0 = (I-RA)R_A(\lambda Ru+x_0) = (I-RA)R_1 v$ and

$$u = \lambda DRu = D[(I-RA)R_1 v - x_0]$$
$$= (D-A)R_1 v - Dx_0 = (I-AR_1)v.$$

We conclude that

$$0 \in F_1\{x = R_A(RBu+x_0)+z: z \in \ker(I-RA), u = U\}$$

$$= F_1 \operatorname{Rang}_{U,x_0} \Phi_1,$$

which means that the system (9.3.36) is F_1-controllable.

Define: $F_2\{x_n\} = (x_1 - x_3)\{e_n\}$. By definition, $F_2 \in \mathscr{F}_D$. We have the following equalities:

$$F_2 R_A \ker D = \ker D \quad \text{and} \quad F_2(F_A + R_A)\ker D = \{0\}, \quad (9.3.37)$$

where $F_A = I - R_A(I-RA)$ is an initial operator for $I - RA$ corresponding to $R_A \in \mathcal{R}_{I-RA}$. Indeed, $F_2 R_A \tilde{x}_1 = \tilde{x}_1$ for $\tilde{x}_1 = c\{e_n\} \in \ker D$, $F_2(F_A+R_A)\tilde{x}_1 = 0$ for an arbitrarily fixed $c \in \mathbf{R}$. The second of Equalities (9.3.37) implies that

$$F_2(RR^A A + I)\ker D = \{0\}, \quad \text{where } R^A = DR_A R \in \mathcal{R}_{I-AR}.$$

Hence, in general, equalities $F_2 R_A \ker D = \ker D$ and $F_2(RR^A A + I)\ker D = \ker D$ do not hold simultaneously.

EXAMPLE 9.3.3. This example shows that the conditions of Theorem 9.3.3 are sufficient for the F_1-controllability, but not necessary. Consider the system described in Example 9.3.2. Define $F_1\{x_n\} = x_3\{e_n\}$, $R^A\{x_n\} = \{y_n\}$, where $y_1 = b \in \mathbf{R}$ is fixed, $y_n = -x_n - \sum_{k=1}^{n-1} y_k$ for $n > 1$. Let $\tilde{x}_1 = c_1\{e_n\} \in \ker D$, where $c_1 \in \mathbf{R}$ is arbitrarily fixed. We find

$$F_1(RR^A A + I)\tilde{x}_1 = b\{e_n\} \in \ker D.$$

The arbitrariness of $\tilde{x}_1 \in \ker D$ implies that the equality $F_1(RR^A A + I)\ker D = \ker D$ is not satisfied. But this equality has been assumed in Theorem 9.3.3. On the other hand, for every $x_0 = c_0\{e_n\}$ and $\tilde{x}_1$ $= c_1\{e_n\} \in \ker D$ there is a control $u \in U$ and $z \in \ker(I - RA)$ satisfying the equation

$$F_1\{R[R^A(Bu + Ax_0) + z] + x_0\} = \tilde{x}_1.$$

For instance, we can admit $z = 0$ and $u = \lambda^{-1}(I - AR)(w - \tilde{x}_1) - \tilde{x}_1$, where $w \in X$ is an arbitrary element satisfying the condition $F_1 Rw = \tilde{x}_1$. The arbitrariness of x_0 and x_1 implies that the system (9.3.36) is F_1-controllable.

EXAMPLE 9.3.4. Suppose that X, D, R, F are defined as in Example 9.3.2, $A\{x_n\} = \{x_{n+1}\}$, $U = X$ and $B = I$. Consider the system (9.3.36): $Dx = Ax + Bu$, $Fx = x_0$. The operator $I - RA$ is left invertible and has a left inverse L_A, where $L_A\{x_n\} = \{y_n\}$, $y_1 = x_1$, $y_2 = -x_3$, $y_n = x_n - -x_{n+1}$ for $n \geqslant 3$. Define an operator $F_1 \in \mathscr{F}_D$ as follows: $F_1\{x_n\} = (x_1 + +x_2)\{e_n\}$. The system (9.3.36) is F_1-controllable, however, the condition $F_1 L_A \ker D = \ker D$ is not satisfied. Indeed, let $x_0 = c_0\{e_n\} \in \ker D$, where $c_0 \in \mathbf{R}$ is arbitrarily fixed. Then $L_A x_0 = \{y_n\}$, where $y_1 = c_0$, $y_2 = -c_0$, $y_n = 0$ for $n \geqslant 3$ and $F_1 L_A x_0 = x_0$.

Let $x_0 = c_0\{e_n\}$, $\tilde{x}_1 = c_1\{e_n\} \in \ker D$, where $c_0, c_1 \in \mathbf{R}$ are arbitrary. The equation $F_1 L_A(RBu + x_0) = x_1$ has always a solution $u \in U$ $= X$, for instance, $u = \{u_n\}$, where $u_1 + u_2 = -\tilde{c}_1$, $u_n = c \in \mathbf{R}$ for $n \geqslant 3$, c is arbitrarily fixed. The arbitrariness of x_0 and x_1 implies the F_1-controllability of the system (9.3.36) for $I - RA \in \Lambda(X)$, hence the F_1-controllability of this system for $I - AR \in \Lambda(X)$. However, our example shows that the assumptions admitted in Theorem 9.3.3 are sufficient but not necessary.

EXAMPLE 9.3.5. Let $X = C([0, T], \mathbf{C})$. Consider the following linear system:

$$x' = Ax, \quad x(t_0) = x_0, \tag{9.3.38}$$

where $t_0 \in [0, T]$, $x_0 = (x_{01}, x_{02})$, $x_{0i} \in \mathbf{R}$ ($i = 1, 2$), $x(t) = (x_1(t),$ $x_2(t))$, $A = (a_{ij})_{i,j=1,2}$, $a_{ij} \in \mathbf{R}$ and $\det A \neq 0$. This system can be written as

$$Dx = Ax, \quad Fx = x_0, \quad x_0 \in \ker D, \tag{9.3.38a}$$

where

$$(Fx)(t) = x(t_0),$$

$$D = \begin{bmatrix} d/dt & 0 \\ 0 & d/dt \end{bmatrix} \in R(X), \quad R = \begin{bmatrix} \int_{t_0}^{t} & 0 \\ 0 & \int_{t_0}^{t} \end{bmatrix}$$

and F is an initial operator for D corresponding to its Volterra right inverse R (cf. Section 3.5). The operator A is, by definition, stationary: $DA = AD$ on dom D and $RA = AR$. The operator $I - RA$ is invertible and its inverse is $(I - RA)^{-1} = [A_{ij}]_{i,j=1,2}$, where

$$A_{12} = a_{12}A_0, \quad A_{21} = a_{21}A_0,$$

$$A_{11} = (I_0 - a_{11}R_0)^{-1}(I_0 + a_{12}A_{12}R_0),$$

$$A_{22} = (I_0 - a_{22}R_0)^{-1}(I + a_{21}A_{21}R_0),$$

$$A_0 = (I_0 - \lambda_1^{-1}R_0)^{-1}(I_0 - \lambda_2^{-1}R_0)^{-1}R_0,$$

$$R_0 = \int_{t_0}^{t},$$

$\lambda_1, \lambda_2 \neq 0$ are roots of the equation det $A \cdot \lambda^2 - (a_{11} + a_{22})\lambda + 1 = 0$ and I_0 is the identity on $C[0, T]$. The solution of the system (9.3.38a), hence of (9.3.38), is then of the form: $x = (I - RA)^{-1}x_0$. For every $x_0 \in \ker D$ and for every $t_1 \in [0, T]$ we have

$$F_1(I - RA)^{-1}x_0 = \Psi(t_1, t_0)x_0,$$

$$\text{where } (F_1 x)(t) = x(t_1) \text{ for } x \in X.$$

The fundamental matrix $\Psi(t_1, t_0)$ is said to be the *transfer matrix* since transforms the state $x(t_0)$ at the moment t_0 into the state $x(t_1)$ at the moment t_1. So that, we can say that $F_1(I - RA)^{-1}$ is the *transfer operator from the initial state Fx to the final state $F_1 x$*.

Suppose that $B = (b_{ij})_{i,j=1,2}$, $b_{ij} \in R$ and U satisfies the condition: $BU \subset X$. Consider the system:

$$Dx = Ax + Bu, \quad Fx = x_0. \tag{9.3.39}$$

This system has for all $u \in U$ a unique solution of the form: $x = (I-RA)^{-1}(RBu+x_0)$, which can be written using the fundamental matrix in the following form:

$$x(t) = \Psi(t, t_0)x_0 + \int_{t_0}^{t} \Psi(t, \tau)(Bu)(\tau)d\tau.$$

Since $RA = AR$, we may write this solution in an equivalent form:

$$x = (I-RA)^{-1}x_0 + (I-RA)^{-1}RBu$$

(cf. also Formulae (9.3.6), (9.3.7)).

EXAMPLE 9.3.6.   Let $X = C([0, T], \mathbf{R}^n)$. Let $A = (a_{ij})_{i,j=1,...,n}$, $B = (b_{ij})_{i,j=1,...,n}$, $a_{ij}, b_{ij} \in \mathbf{R}$ $(i, j = 1, 2, ..., n)$, $x = (x_1, ..., x_n)^T$ be an n-dimensional vector transposed to $(x_1, ..., x_n) \in X$, $x_0 = (x_{10}, ..., x_{n0})^T$. Let $Dx = (dx_1/dt, ..., dx_n/dt)^T$ for $x_j \in C^1[0, T]$ $(j = 1, 2, ..., n)$, $(Fx)(t) = x(t_0)$ and $R = (\int_{t_0}^{t}, ..., \int_{t_0}^{t})$ for $x \in X$ and $t_0 \in [0, T]$. Then $D \in R(X)$ and F is an initial operator for D corresponding to $R \in \mathcal{R}_D$. Moreover, the operator $I-RA = I-AR$ is invertible (cf. Section 3.5). Consider a set of initial operators for D defined as follows:

$$\tilde{\mathscr{F}}_D = \{F_1 \in \mathscr{F}_D : (F_1x)(t) = x(t_1) \text{ for a } t_1 \in [0, T], x \in X\},$$

i.e. the set describing so-called *local conditions*. Observe that the F_1-controllability for the system

$$Dx = Ax+Bu, \quad Fx = x_0 \tag{9.3.40}$$

does coincide with the classical notion of controllability (cf. for instance, Rolewicz, 1987). Indeed, this system is said to be *controllable* (in the classical sense) if for arbitrary $x_0, x_1 \in \mathbf{R}^n$ there exists a control $u \in U$ such that the corresponding state x satisfies the conditions: $x(t_0) = x_0$, $x(t_1) = x_1$ for a finite time $t_1 \in [0, T]$. On the other hand, the system (9.3.40) is F_1-controllable if and only if for every $x_0 \in \ker D$ we have

$$F_1\{x = (I-RA)^{-1}(RBu+x_0): u \in U\} = \ker D. \tag{9.3.41}$$

This condition is equivalent to the equality

$$\ker B^*R^*[(I-RA)^{-1}]^* = 0.$$

Since there is a one-to-one correspondence between ker D and $(F_1 x)(t) = x(t_1)$, we conclude that Condition (9.3.41) holds if and only if the matrix $(B, AB, \dots, A^{n-1}B)$ is of the rank n:

$$\operatorname{rank}(B, AB, \dots, A^{n-1}B) = n.$$

This is the classical *Kalman Condition* of controllability of the system (9.3.40).

If the condition (9.3.41) is satisfied for an $F_1 \in \tilde{\mathscr{F}}_D$ then it is satisfied for every initial operator belonging to $\tilde{\mathscr{F}}_D$. This fact does coincide with the following consequence of Kalman Theorem: *If the system (9.3.40) (which is stationary by our assumptions) is controllable on an interval $[0, t_1]$ then it is controllable on every interval $[0, t_2]$* (cf. also Theorem 9.3.5).

EXAMPLE 9.3.7. Let X, D, A, B, x_0 be defined as in Example 9.3.5. Let

$$\tilde{F} = \begin{bmatrix} F_0 & 0 \\ 0 & F_1 \end{bmatrix}, \quad \tilde{R} = \begin{bmatrix} R_0 & 0 \\ 0 & R_1 \end{bmatrix}, \quad \text{where } R_0 = \int_{t_0}^{t}, \ R_1 = \int_{t_1}^{t},$$

$(F_0 x)(t) = x(t_0)$, $(F_1 x)(t) = x(t_1)$ for $t_0, t_1 \in [0, T]$, $t_1 \neq t_0$, $x \in C[0, T]$. Then $\tilde{F}$ is an initial operator for D corresponding to the Volterra right inverse. Moreover, $\tilde{R}A \neq A\tilde{R}$, the operator $I - \tilde{R}A$ is invertible and

$$(I - \tilde{R}A)^{-1} = \begin{bmatrix} (I - a_{11}R_0)^{-1} & a_{12}(I - a_{11}R_0)^{-1}R_0(I - a_{22}R_1)^{-1} \\ 0 & (I - a_{22}R_1)^{-1} \end{bmatrix}.$$

Since $BU \subset (D - A) \operatorname{dom} D = X$, for every $x_0 \in \ker D$ and $u \in U$ we have a unique solution of a linear system

$$Dx = Ax + Bu, \quad \tilde{F}x = x_0 \tag{9.3.42}$$

which is of the form: $x = (I - \tilde{R}A)^{-1}(\tilde{R}Bu + x_0)$. Put $a_{ij} = a \cdot \delta_{ij}$, $b_{ij} = b \cdot \delta_{ij}$ for $i, j = 1, 2$, where $0 \neq a, b \in \mathbf{R}$. Let

$$\hat{F} = \begin{bmatrix} F_1 & 0 \\ 0 & F_2 \end{bmatrix}, \quad \text{where } (F_2 x)(t) = \frac{1}{T}\int_0^T x(s)\,ds \quad \text{for } x \in C[0, T]$$

and

$$e^{a(T - t_1)} - e^{-at_1} \neq T \quad \left(\text{i.e. } \frac{1}{T}(e^{aT} - 1) \neq e^{at_1}\right) \tag{9.3.43}$$

Clearly, $\hat{F}$ is an initial operator for D, since F_2 is an initial operator for the operator d/dt. In order to prove that the system is $\hat{F}$-controllable, we have to verify (according to Theorem 9.3.3) that

(i) $\hat{F}(I - \tilde{R}A)^{-1} \ker D = \ker D$;

(ii) the system (9.3.42) is $\hat{F}$-controllable to zero.

Indeed, Condition (9.3.42) implies that for every $c = (c_1, c_2) \in \ker D$, $c_1, c_2 \in \mathbf{R}$, the equation $\hat{F}(I - \tilde{R}A)^{-1}z = c$ has a solution $z = (\tilde{c}_1, \tilde{c}_2)$ $\in \ker D$, namely

$$\tilde{c}_1 = c_1 e^{a(t_1 - t_0)}, \quad \tilde{c}_2 = c_2 aT(e^{aT} - 1)^{-1} e^{at_1}.$$

Hence $\hat{F}(I - \tilde{R}A)^{-1} \ker D = \ker D$. The equation

$$\hat{F}(I - \tilde{R}A)^{-1}(\tilde{R}Bu + x_0) = 0, \quad \text{i.e.} \quad \hat{F}(I - \tilde{R}A)^{-1}\tilde{R}Bu = z,$$

where $z = -\hat{F}(I - \tilde{R}A)^{-1}x_0 \in \ker D$

has a solution $u = \dfrac{1}{b}(I - \tilde{R}A)^{-1}w$, where w is an element satisfying the equality: $\hat{F}\tilde{R}w = z$. This means that the system (9.3.42) is $\hat{F}$-controllable to zero, hence $\hat{F}$-controllable.

EXAMPLE 9.3.8. Suppose that X is defined as in Example 9.3.5. Consider a linear system

$$Dx = Ax + Bu, \quad Fx = x_0, \tag{9.3.44}$$

where

$$D = \begin{bmatrix} d/dt & 0 \\ 0 & d/dt \end{bmatrix}, \quad F = \begin{bmatrix} F_0 & 0 \\ 0 & F_1 \end{bmatrix},$$

$$A = (a_{ij})_{i,j=1,2}, \quad B = (b_{ij})_{i,j=1,2},$$

$$x_0 = (x_{10}, x_{20}), \quad a_{12} = a_{21} = 0, \quad a_{22} = \frac{2\pi i}{T},$$

$$a_{11}, b_{ij}, x_{i0} \in \mathbf{R} \quad (i, j = 1, 2),$$

$$BU \subset X, \quad (F_0 x)(t) = x(t_0), \quad (F_1 x)(t) = \frac{1}{T}\int_0^T x(s)\, ds$$

for $x \in C[0, T], \quad t_0 \in [0, T]$.

Clearly, $D \in R(X)$ and F is an initial operator for D corresponding to a right inverse

$$R = \begin{bmatrix} R_0 & 0 \\ 0 & R_1 \end{bmatrix}, \quad \text{where } R_0 = \int_{t_0}^{t},$$

$$(R_1 x)(t) = \int_{t_0}^{t} x(s)\,ds + \frac{1}{T}\int_{0}^{T}(T-s)x(s)\,ds \quad \text{for } x \in C[0, T].$$

Observe that R_0 is a Volterra right inverse for d/dt but R_1 is not a Volterra operator. However, the operator $I - a_{22}R_1$ is right invertible. Then also the operator $I - RA$ is right invertible and has a right inverse, for instance of the form:

$$R_A = \begin{bmatrix} (I-a_{11}R_0)^{-1} & 0 \\ 0 & R_a \end{bmatrix}, \quad \text{where } R_a \in \mathcal{R}_{I-a_{22}R_1}.$$

Hence the initial value problem for the linear system (9.3.44) is ill-posed and all its solutions are of the form:

$$x = R_A(RBu + x_0) + z,$$

where $z = (0, z_1)$, $z_1 \in \ker(I - a_{22}R_1) = \{ce^{a_{22}t} : c \in \mathbf{C}\}$.

In a similar way, as in Examples 9.3.5 and 9.3.7 we can examine the F-controllability for $F \in \mathcal{F}_D$.

Note that Examples 9.3.1–9.3.8 were given by Pogorzelec, 1983a.

EXERCISE 9.3.1. Which results of this section can be generalized for operators $Q(D)$ and $Q\langle D\rangle$ defined by Formulae (3.2.1)? (cf. Nguyen Dinh Quyet, 1977, 1978a for $Q(D)$ with an invertible resolving operator)

EXERCISE 9.3.2. Suppose that all assumptions of Theorem 9.3.4 are satisfied and there is a $z_0 \in \ker(I - AR)$ such that

$$F_1 \{x = R[R^A(Bu + Ax_0) + z_0] + x_0 : u \in U\} = \ker D.$$

Prove that this equality implies

$$F_1 \{x = R[R^A(Bu + Ax_0) + z] + x_0 : u \in U\} = \ker D,$$
$$\text{for every } z \in \ker(I - AR)$$

(cf. Pogorzelec, 1983a).

EXERCISE 9.3.3. Is Theorem 9.3.6 true if the considered linear system has an invertible resolving operator $I-AR$?

EXERCISE 9.3.4. Do Theorems 9.3.2, 9.3.4, 9.3.5, 9.3.6 hold if we consider instead of spaces X' and U' arbitrary conjugate spaces $\Xi \subset X', \Sigma \subset U'$ (in the sense used in Section 1.5)?

EXERCISE 9.3.5. Formulate and prove conditions for linear systems to be equivalent (respectively: similar) analogous to these given in Theorem 9.3.1 if the resolving operator
  (i) $I-RA$ is either left invertible or right invertible;
  (ii) $I-AR$ is either invertible or left invertible or right invertible.

EXERCISE 9.3.6. Suppose that $(LS)_0$ is a linear system with an invertible resolving operator. Let $F_1 \neq F$ be an initial operator for D. Write:

$$A(x_0, F; F_1) = \{x = F_1 \Phi(x_0, u): u \in U\}, \qquad (9.3.45)$$

where $\Phi(x_0, u)$ is defined by Formula (9.3.4). Prove that:
  (i) $A(x_0, F; F_1)$ is a non-empty subset of ker D;
  (ii) $A(x_0, F; F_1) = \{x_0\}$;
  (iii) $A[A(x_0, F; F_\alpha), F_\alpha; F_\beta] = \bigcup_{x \in A(x_0, F; F_\alpha)} A(x, F_\alpha; F_\beta)$;
  (iv) If X, U are Banach spaces, $I-RA$, RB, F are bounded, F_1 is closed then $A(x_0, F; F_1)$ is closed;

  (v) Suppose that conditions of Point (iv) are satisfied; then for an arbitrary $\varepsilon > 0$ there exists a $\delta > 0$ such that for every bounded $F_1, F_2 \in \mathscr{F}_D$ the condition $||F_1 - F_2|| < \delta$ implies that the *Hausdorff distance* $\varrho(A(x_0, F; F_1), A(x_0, F; F_2)) = \inf[\varrho(x, y): x \in A(x_0, F; F_1), y \in A(x_0, F; F_2)]$ is less than ε (cf. Nguyen Dinh Quyet, 1979).

EXERCISE 9.3.7. Suppose that all conditions of Point (iv) in Exercise 9.3.6 are satisfied. A linear system $(LS)_0$ is said to be F_1-*stable* if for a set $S \subset \ker D$ and for arbitrary $\varepsilon > 0$ there exists a $\delta > 0$ such that for any x_0 the condition $||x_0 - v|| < \delta$ for some $v \in S$ implies that for every $x \in A(v, F; F_1)$ there is a $w \in S$ such that $||x - w|| < \varepsilon$.

The system $(LS)_H$ with the output $y = Hx$, where H is bounded, is said to be *output F_1-stable* if for an $S \subset \ker D$ and for arbitrary $\varepsilon > 0$ there exists a $\delta > 0$ such that for every x_0 the condition $\|x_0 - v\| < \delta$ for some $v \in S$ implies that for every $x \in A_H(v, F; F_1)$ there is a $w \in HS$ such that $\|x - w\| < \varepsilon$, where

$$A_H(x_0, F; F_1) = \{Hx: x \in A(x_0, F; F_1)\}.$$

Prove that:

(i) If a linear system $(LS)_0$ is F_1-stable then the linear system $(LS)_H$ is output F_1-stable;

(ii) If a linear system $(LS)_H$ is observable and output F_1-stable then is F_1-stable (cf. Nguyen Dinh Quyet, 1979).

EXERCISE 9.3.8. Generalize for systems $(LS)_H$ all results of this section formulated for $(LS)_0$.

EXERCISE 9.3.9. Is it possible to introduce similar notions as transfer operators, reachability, F_1-controllability and observability for linear systems with left invertible operators?

9.4. DUALITIES AND CONJUGATE SYSTEMS

In the present section we shall consider again linear systems (cf. Definition 9.3.1). Several problems lead us to a notion of a "dual system" which should be defined in an appropriate sense in order to prove required properties. A kind of duality has been defined in Section 6.5 by means of the Green Formula using operators of definite integral. To have a coincidence with classical results, the duality which has to be introduced, should satisfy the following conditions:

(i) A conjugate to a conjugate is the given operator;

(ii) The following are dual with respect to that conjugation:

(ii') left and right invertible operators;

(ii'') initial and final states;

(ii''') retarded and advanced argument;

(iii) A conjugate to the operator $D = \mathrm{d}/\mathrm{d}t$ is the operator $D^* = -\mathrm{d}/\mathrm{d}t$.

Condition (ii'') and (iii) together imply that a conjugate space should be chosen in such a way that the operator $D^* = - \, d/dt$ will be invertible on this space. Indeed, both operators, d/dt and $-d/dt$ are right invertible and the operator $-d/dt$, as a conjugate of a right invertible operator has to be left invertible, hence invertible. This is the crucial point when we have to study conjugate problems.

We shall construct conjugate systems having all required properties. Moreover, it will be shown that, under appropriate assumptions, the conjugation considered is up to the canonical mapping $\varkappa$ identical with that introduced in Section 6.5 by means of the Green Formula.

In order to find a conjugate system for a linear system in a natural way we shall consider an optimization problem.

DEFINITION 9.4.1. Suppose that X and Y are Banach spaces and J is a mapping of X into Y. Let $x \in X$. Suppose that there exists a linear operator $A \in L(X \to Y)$ such that

$$J(x+h) - J(x) = Ah + a(x, h), \qquad (9.4.1)$$

where

$$\frac{\|a(x, h)\|}{\|h\|} \to 0 \quad \text{as } h \to 0. \qquad (9.4.2)$$

Then A is said to be the *Fréchet differential of J at x* (or: *strong differential*). We shall write: $Ah = J_x'(x)(h)$ for $x, h \in X$. If J has a Fréchet differential we say that J is *Fréchet differentiable*. Note that for a linear J the Fréchet differential is equal to J for $a(x, h) = 0$ for all $x, h \in X$.

DEFINITION 9.4.2. Suppose that X and U are Banach spaces and $\varXi \subset X', \varSigma \subset U'$ are total spaces of bounded linear functionals over X and U, respectively, i.e. conjugate spaces in the sense defined in Section 1.5. Suppose that $J(x, u)$ is a nonlinear functional defined on $X \times U$, Fréchet differentiable with respect to each variable and its Fréchet differentials $J_x'(x, u), J_u'(x, u)$ belong to $\varXi$ for every fixed $u \in U$ and belong to $\varSigma$ for every fixed $x \in X$. Consider a linear system $(LS)_0$ in the sense of Definition 9.3.1, i.e.

$$Dx = Ax + Bu, \quad Fx = x_0, \qquad (9.4.3)$$

where $D \in R(X)$, F is an initial operator for D corresponding to an $R \in \mathcal{R}_D$, $BU \subset (D-A)\operatorname{dom} D$ and $x_0 \in \ker D$. We assume that $D \in L(X, \Xi)$, A, R, $F \in L_0(X, \Xi)$, $B \in L_0(X \to U, \Sigma \to \Xi)$ (cf. Formulae (1.5.2), (1.5.3) and a convention admitted in Section 1.5). Assume also that the resolving operator $I-RA$ is invertible and $E_A = (I-RA)^{-1} \in L_0(X, \Xi)$. Write, according to Formula (9.3.4):

$$W = \{(x, u) \in X \times U: x = E_A(RBu+Fx)\}$$
$$= \{(x, u) \in X \times U: x = \Phi(Fx, u)\}. \tag{9.4.4}$$

By definition, W consists of all solutions of the problem (9.3.4) with all possible initial data $x_0 \in \ker D$. Moreover, also by definition, if $(x, u) \in W$ then $x \in \operatorname{dom} D$ for x is a solution of (9.3.4). Hence $\Phi(Fx, u) \subset \operatorname{dom} D$ for all $x \in \operatorname{dom} D$. The set W is said to be *admissible* for the system $(LS)_0$ under consideration. A point $(x^+, u^+) \in W$ is said to be a *local minimum point for the functional* $J(x, u)$ *with the constraints* (9.4.3) if there exists an open set $U_0 \subset U$ such that

$$J(u^+) = \min_{u \in U_0} J(u), \quad \text{where } J(u) = J(\Phi(x_0, u), u)$$
and $Fx^+ = x_0, \quad x^+ = \Phi(x_0, u). \tag{9.4.5}$

If (x^+, u^+) is a local minimum point for J then u^+ is said to be an *optimal control* for the problem under consideration.

THEOREM 9.4.1 (Necessary conditions). *Suppose that all conditions of Definition 9.4.2 are satisfied. If* $(x^+, u^+) \in X \times U_0$ *is a local minimum point for the functional* $J(x, u)$ *with the constraints* (9.4.3) *then there exists a unique solution of the conjugate problem*:

$$D^*\xi = A^*\xi + J_x', \tag{9.4.6}$$
$$F^*A^*\xi = \xi_0, \quad \text{where } \xi_0 = -F^*J_x', \tag{9.4.7}$$

namely

$$\xi = R^*E_A^*J_x', \quad B^*\xi = -J_u'. \tag{9.4.8}$$

Proof. If $(x^+, u^+) \in X \times U$ is a local minimum point for the functional $J(x, u)$ with the constraints (9.4.3) then $J'(u^+) = 0$ and on the set

$$W_0 = \{(x, u) \in X \times U_0: x = \Phi(Fx, u)\} \subset W$$

we have

$$0 = J'(u^+) = J'_x(x^+, u^+)E_A(RBu + Fx) + J'_u(x^+, u^+)u$$
$$= [(B^*R^*E_A^*J'_x + J'_u)(x^+, u^+)]u + F^*E_A^*J'_x(x^+, u^+)x$$

which implies

$$B^*R^*E_A^*J'_x + J'_u = 0, \tag{9.4.9}$$

$$F^*E_A^*J'_x = 0. \tag{9.4.10}$$

Write, as in Formulae (9.4.8), $\xi = R^*E_A^*J'_x$. It is clear that $\xi \in \Xi$. By (9.4.9) we obtain $B^*\xi = -J'_u$. Equation (9.4.6) consists of a left invertible operator D^* and, by our assumption, of $I - RA \in L_0(X, \Xi)$. Hence the operator $E_A^* = (I^* - A^*R^*)^{-1}$ is well-defined and

$$D^*\xi = D^*R^*E_A^*J'_x = (RD)^*E_A^*J'_x = (I - F)^*E_A^*J'_x$$
$$= (I^* - F^*)E_A^*J'_x = E_A^*J'_x - F^*E_A^*J'_x = E_A^*J'_x.$$

On the other hand we find

$$A^*\xi = A^*R^*E_AJ'_x = E_A^*J'_x - (I^* - A^*R^*)(I^* - A^*R^*)^{-1}J'_x$$
$$= D^*\xi - J'_x,$$

which proves that ξ is a solution of Equation (9.4.6). This solution is unique (cf. Section 3.6). Moreover, since $DF = 0$ we have

$$0 = (DF)^*\xi = F^*D^*\xi = F^*(A^*\xi + J'_x) = F^*A^*\xi + F^*J'_x,$$

which implies the condition (9.4.9). This is, which was to be proved. ∎

COROLLARY 9.4.1. *Suppose that all assumptions of Definition 9.4.2 are satisfied, $(x^+, u^+) \in X \times U$ is a local minimum point for $J(x, u)$ with the constraints (9.4.3) and the operator D^* is invertible on Ξ. Then there exists a unique solution ξ of the conjugate problem*

$$D^*\xi = A^*\xi + J'_x, \tag{9.4.6}$$

$$F^*J'_x = 0, \tag{9.4.11}$$

where ξ is defined by (9.4.8): $\xi = R^*E_A^*J'_x$ and $B^*\xi = -J'_u$, $\xi_0 = -F^*J'_x = 0$.

Proof. Indeed, by our assumptions, $R^*D^* = (DR)^* = I^*$ and $I^* = D^*R^*$

$= (RD)^* = (I-F)^* = I^*-F^*$ on Ξ which implies $F^* = 0$. This and (9.4.10) together imply the condition (9.4.11). ∎

REMARK 9.4.1. In the linear case we can reject all topological assumptions. Namely, let X and U be linear spaces (over the same field of scalars) and let $\Xi \subset X'$, $\Sigma \subset U'$ be conjugate spaces. Suppose that $J(x, u) = f(x)+g(u)$ for some $f \in \Xi$, $g \in \Sigma$.

If $(x^+, u^+) \in X \times U$ is a local minimum point (with respect to a subset $U_0 \subset U$) for the functional $U(x, u) = f(u)+g(u)$ with constraints

$$Dx = Ax+Bu, \quad Fx = x_0, \quad x_0 \in \ker D, \qquad (9.4.12)$$

then there exists a unique solution $\xi = R^*E_A^*f$ of the conjugate problem

$$D^*\xi = A^*\xi+f, \quad F^*A^*\xi = \xi_0, \quad \text{where } \xi_0 = -F^*f \quad (9.4.13)$$

with the property: $B^*\xi = -g$. Then we prove in the same way as before that $f(x)+g(u) = 0$ implies the existence of a unique solution of the conjugate problem (9.4.13). ∎

An immediate consequence of Theorem 9.4.1 is the following

COROLLARY 9.4.2. *Suppose that X and U are Hilbert spaces (i.e. we can admit $\Xi = X$, $\Sigma = U$) with scalar products $\langle \ , \ \rangle_X$ and $\langle \ , \ \rangle_U$ respectively. Suppose that $P \in L_0(X)$, $Q \in L_0(U)$ and that $J(x, u)$ is a quadratic functional defined as follows:*

$$J(x, u) = \langle Px, x \rangle_X + \langle Qu, u \rangle_U. \qquad (9.4.14)$$

If $(x^+, u^+) \in X \times U$ is a local minimum point of the functional $J(x, u)$ defined by (9.4.14) with the constraints (9.4.3) then there exists a unique solution ξ of the conjugate problem

$$D^*\xi = A^*\xi+P, \quad F^*A^*\xi = -F^*P, \qquad (9.4.15)$$

where

$$\xi = R^*E_A^*P, \quad B^*\xi = -Q \qquad (9.4.16)$$

(cf. also Nguyen Dinh Quyet, 1978, 1978a).

COROLLARY 9.4.3. *Suppose that all assumptions of Definition 9.4.2 are satisfied and $(x^+, u^+) \in X \times U$ is a local minimum point for $J(x, u)$.*

If $D^* = -D|_{\mathcal{Z}}$ then the conjugate system (9.4.6)–(9.4.7) is of the form:

$$D^*\xi = -A^*\xi - J'_x, \quad F^*A^*\xi = 0, \tag{9.4.17}$$

$$F^*J'_x = 0 \tag{9.4.18}$$

and its unique solution is

$$\xi = -R^*E_A^*J'_x, \quad E_A^* = (I^*+A^*R^*)^{-1}. \tag{9.4.19}$$

Proof. Indeed, in this case the operator D^* is invertible on $\mathcal{Z}$. ∎

We shall give an example only of sufficient conditions for the functional $J(x, u)$ considered in Theorem 9.4.1, subjected to the constraints (9.4.3). One can consider other conditions even weaker which assure the existence of a local minimum point, provided that necessary conditions are satisfied.

THEOREM 9.4.2 (Sufficient conditions). *Suppose that all conditions of Definition 9.4.2 are satisfied and*

(i) $J(x, u)$ *is twice Fréchet differentiable with respect to x and u;*

(ii) *The necessary conditions are satisfied for a $(x^+, u^+) \in W$, i.e. the conjugate system (9.4.6)–(9.4.7) has a unique solution ξ;*

(iii) *The operator*

$$H = T^*J'_{xx}T + T^*J'_{xu} + J'_{ux}T + J'_{uu}, \quad \text{where } T = E_A RB \tag{9.4.20}$$

is positively defined for $x^+ = \Phi(x_0, u^+) = Tu^+ + E_A x_0$, i.e. there exists a bilinear form $\langle \,, \rangle_U$ defined on $U \times U$ such that $\langle Hu, u \rangle_U > 0$ for $u \in U_0$, in particular, for $u = u^+$. Then (x^+, u^+) is a local minimum point for $J(x, u)$ with the constraints (9.4.3).

Proof. By our assumptions,

$$J(u) = J(\Phi(x_0, u), u) = J(Tu + E_A x_0, u) \quad \text{for } u \in U_0.$$

In particular, we have

$$J(u^+) = J(Tu^+ + E_A x_0, u^+).$$

Hence for all $u \in U_0$

$$J''(u) = \langle J'_{xx}(Tu^+ + E_A x_0, u^+)Tu, Tu \rangle_U +$$
$$+ \langle J'_{xu}(Tu^+ + E_A x_0, u^+)u, Tu \rangle_U +$$

$$+\langle J'_{ux}(Tu^+ + E_A x_0, u^+)Tu, u\rangle_U +$$
$$+\langle J'_{uu}(Tu^+ + E_A x_0, u^+)u, u\rangle_U$$

and

$$J''(u) = \langle T^* J'_{xx} Tu, u\rangle_U + \langle T^* J'_{xu} u, u\rangle_U +$$
$$+\langle J'_{ux} Tu, u\rangle_U + \langle J'_{uu} u, u\rangle_U$$
$$= \langle (T^* J'_{xx} T + T^* J'_{xu} + J'_{ux} T + J'_{uu})u, w\rangle_U = \langle Hu, u\rangle_U.$$

Since H is positively defined at u^+ we conclude that $J''(u^+)$ $= \langle Hu^+, u^+\rangle > 0$ which proves that (x^+, u^+) is a local minimum point for $J(x, u)$ subjected to the constraints (9.4.3). ∎

Consider now problems with shifts.

THEOREM 9.4.3. *Suppose that all assumptions of Theorem 8.2.7 are satisfied. Let* $\Xi \subset X'$ *be a conjugate space. Assume that* $D \in L(X, \Xi)$, R, F, S_h $\in L_0(X, \Xi)$ *for all* $h \in A(\mathbf{R})$ *and* $D^* = -D|_\Xi$. *Then the operator* $D^* = -D|_\Xi$ *is an infinitesimal generator for*

$$S^*_{A(\mathbf{R})} = \{S^*_h\}_{h \in A(\mathbf{R})}, \quad where \ S^*_h = S_{-h}|_\Xi^{!}. \tag{9.4.20}$$

Proof. Consider the operators $A_h = D - \dfrac{1}{h}(S_h - I)$ for $h \in A(\mathbf{R})$. Since D is an infinitesimal generator for $\{S_h\}_{h \in A(\mathbf{R})}$, we have

$$Dx = \lim_{h \to 0} \frac{1}{h}(S_h - I)x \quad \text{for all } x \in \text{dom } D, \ \overline{\text{dom } D} = X.$$

This implies that $A_h \to 0$ pointwise. Then for all $h \in A(\mathbf{R})$

$$A^*_h = D^* - \frac{1}{h}(S^*_h - I^*) = -D|_\Xi - \frac{1}{h}(S^*_h - I^*)$$
$$= -\left[D|_\Xi - \frac{1}{-h}(S^*_h - I^*)\right]$$

and for all $x \in \text{dom } D$, $\xi \in \Xi$ we find

$$0 = \lim_{h \to 0} \xi(A_h x) = \lim_{h \to 0}(A^*_h \xi)(x)$$
$$= -\lim_{h \to 0}\left[D\xi - \frac{1}{-h}(S^*_h - I^*)\xi\right](x).$$

The arbitrariness of $x \in \text{dom } D$ and $\xi \in \mathcal{E}$ implies that $D^* = -D|_{\mathcal{E}}$ is an infinitesimal generator for $\{S_h^*\}_{h \in A(\mathbf{R})}$, where $S_h^* = S_{-h}|_{\mathcal{E}}$.   ■

This theorem implies, in particular, the duality between problems with *retarded* and *advanced argument*.

Suppose that X is a D-algebra and F, F_1, $F_1 \neq F$, are initial operators for D corresponding to R, $R_1 \in \mathcal{R}_D$. Recall that a definite integral $F_1 R$ is said to be *total* on the set

$$E_m = \{u \in D_m: F_1 D^j u = FD^j u = 0 \quad \text{for } j = 0, 1, \ldots, m\}$$
$$(m = 0, 1, 2, \ldots, +\infty), \quad (9.4.21)$$

where $D_\infty = \bigcap_{k=0}^{\infty} D_k$, $D_0 = X$, $D_k = \text{dom } D^k$ for $k \in \mathbf{N}$, if

$$F_1 R(xh) = 0 \text{ for all } h \in E_m \quad \text{implies } x = 0 \qquad (9.4.22)$$

(cf. Definition 6.5.1).

Suppose again that all assumptions of Theorem 8.2.7 are satisfied. Write, as before

$$F_h = FS_h \quad \text{for } h \in A(\mathbf{R}). \qquad (9.4.23)$$

Then by Corollary 8.2.2 we get

$$\varkappa F_b R_a = \int_a^b \varkappa \quad \text{for } a, b \in A(\mathbf{R}), \qquad (9.4.24)$$

where $\varkappa$ is the canonical mapping for $\{S_h\}_{h \in A(\mathbf{R})}$.

Example 6.5.3 shows that classical integrals $\int_a^b$ are total (by principal lemma of the variational calculus) on sets

$$E_m^0 = \{x \in C^m[0, T]: x^{(j)}(b) = x^{(j)}(a) = 0 \text{ for } j = 0, 1, \ldots, m\}$$
$$(m = 0, 1, \ldots, +\infty). \qquad (9.4.25)$$

Write:

$$\Xi_m = \left\{\xi \in X': \xi(x) = \int_a^b y(t)x(t)\,dt \text{ for } y \in E_m^0, \ x \in X\right\}$$

$$(m = 0, 1, \ldots, +\infty). \qquad (9.4.26)$$

All spaces $\mathcal{E}_m$ are total, hence conjugate spaces for X. We also have the following inclusions:

$$X' \supset \mathcal{E}_0 \supset \mathcal{E}_1 \supset \ldots \supset \mathcal{E}_m \supset \ldots \supset \mathcal{E}_\infty. \qquad (9.4.27)$$

Suppose now that the canonical mapping is *multiplicative*, i.e. $\varkappa(xy) = (\varkappa x)(\varkappa y)$ for $x, y \in X$. It is so when the inducing initial operator F and all shifts are multiplicative, for $\varkappa x = \{FS_h x\}_{h \in A(\mathbf{R})} = \{\hat{x}(h)\}_{h \in A(\mathbf{R})}$, $x \in X$. Then to the conjugate spaces $\mathcal{E}_m$ defined by Formula (9.4.26) there correspond by Equality (9.4.24) the following sets:

$$Z_m = \{\zeta : \zeta(u) = F_b R_a(uv) \text{ for } v \in E_m, u \in X\}$$
$$(m = 0, 1, \ldots, +\infty), \qquad (9.4.28)$$

where the sets E_m are defined by Formula (9.4.21).

If integrals $F_b R_a$ are total for all $a, b \in A(\mathbf{R})$ then the spaces Z_m are total, i.e.

$$\zeta(u) = 0 \text{ for all } \zeta \in Z_m \quad \text{implies } u = 0 \quad (m = 0, 1, \ldots, +\infty).$$
$$(9.4.29)$$

By Formula (9.4.26) we find

$$\varkappa(uv) = (\varkappa u)(\varkappa v) = \{\hat{u}(h)\hat{v}(h)\}_{h \in A(\mathbf{R})}$$

and for all $\zeta \in Z_m$

$$\varkappa\zeta(u) = \varkappa F_b R_a(uv) = \left\{\int_a^b \hat{u}(t)\hat{v}(t)\,\mathrm{d}t\right\}_{h \in A(\mathbf{R})},$$

i.e.

$$\int_a^b \hat{u}(t)\hat{v}(t)\,\mathrm{d}t = 0 \text{ for all } v \in E_m \quad \text{implies } u = 0.$$

But this means that to $\varkappa\zeta$ there corresponds a functional belonging to a conjugate space of the form (9.4.26).

On the other hand, consider the Green Formula (6.5.3) in the simplest case. Suppose then that $x, y \in \operatorname{dom} D$ and $D^+ = -D$. Then we have

$$F_b R_a(yDx - xD^+y) = c_D^{-1}(F_b - F_a)(xy) - c_D^{-1}F_b R_a f_D(x, y).$$

If X is an almost Leibniz algebra, $y \in \ker D^+ = \ker D$, F_a, F_b are almost averaging then we have $D^+ y = 0$, $f_D(x, y) = 0$, $(F_b - F_a)(xy)$ $= F_b(xy) - F_a(xy) = yF_bx - yF_ax = y(F_b - F_a)x$ and in this case we have the Green Formula in the form:

$$F_b R_a(yDx) = c_D^{-1} y(F_b - F_a)x, \quad y \in \ker D^+. \tag{9.4.30}$$

Consider the equation

$$Dx = u \quad \text{with the condition } F_b x = F_a x.$$

The Green Formula (9.4.30) implies that

$$F_b R_a(y \cdot u) = 0 \quad \text{for all } y \in \ker D^+. \tag{9.4.30'}$$

This means that the right hand side of the equation $Dx = u$ is "orthogonal" in that operator sense to every solution y of the homogeneous dual equation $D^+ y = 0$. But we have seen that, under appropriate assumptions, conditions of the type (9.4.30') can be expressed in terms of conjugate equations using properties of the canonical mapping.

Examples and Exercises

EXAMPLE 9.4.1. Let X be a linear space and let $\Xi \subset X'$ be a conjugate space. Suppose that $D \in R(X)$. Consider the equation

$$Q(D)x = y, \quad y \in X, \tag{9.4.31}$$

where

$$Q(D) = \sum_{k=0}^{N} Q_k D^k, \quad Q_0, \ldots, Q_{N-1} \in L_0(X, \Xi), \quad Q_N = I$$

with the initial conditions

$$F_k D^k x = x_k, \quad x_k \in \ker D \quad (k = 0, 1, \ldots, N-1), \tag{9.4.32}$$

where $F_0, \ldots, F_{N-1}$ are initial operators for D corresponding to right inverses $R_0, \ldots, R_{N-1}$, respectively. We assume here and in the sequel that $D \in L(X, \Xi)$, $F_0, \ldots, F_{N-1}$, $R_0, \ldots, R_{N-1} \in L_0(X, \Xi)$. Write:

$$R_N = R_0 \ldots R_{N-1}, \quad F_N = R_0 + \sum_{k=1}^{N-1} R_0 \ldots R_{k-1} F_k D^k.$$

By the Taylor–Gontcharov Formula we have $F_N = I - R_N D^N$ on dom D^N. On the other hand $D^N R_N = I$. Hence $D^N \in R(X)$ and F_N is an initial operator for D^N corresponding to $R_N \in \mathcal{R}_D$. Moreover,

$$Q(D) R_N = I + Q^0, \quad \text{where } Q^0 = \sum_{k=0}^{N-1} Q_k R_k \ldots R_{N-1} \quad (9.4.33)$$

and the following similarity holds:

$$(I + Q^0) D^N = Q(D)(I - F_N) \quad (9.4.34)$$

(cf. Section 3.2).

Consider the problem (9.4.31)–(9.4.32). If x is a solution of this problem then the element $x_N = Q(D)(I - F_N)x$ is given. Indeed, we have

$$x_N = Q(D)(I - F_N)x$$

$$= Q(D)x - Q(D)\left[F_0 x + \sum_{k=1}^{N-1} R_0 \ldots R_{k-1} F_k D^k x\right]$$

$$= y - Q(D)\left(x_0 + \sum_{k=1}^{N-1} R_0 \ldots R_{k-1} x_k\right).$$

Assume that the operator $I + Q^0$ is invertible and belongs to $L(X, \mathcal{E})$, where Q^0 is defined by (9.4.33). Then by Similarity (9.4.34) the system (9.4.31)–(9.4.32) is equivalent to the following

$$D^N x = y_N, \quad \text{where } y_N = (I + Q^0)^{-1} x_N \quad (9.4.35)$$

with the initial conditions (9.4.32). The system (9.4.35)–(9.4.32) has a unique solution

$$x = R_N y_N + x_0 + \sum_{k=1}^{N-1} R_0 \ldots R_{k-1} x_k.$$

We shall write:

$$[Q(D)]^* = Q^*(D). \quad (9.4.36)$$

Observe that

$$Q^*(D) = \sum_{k=0}^{N} D^{*k} Q_k^*, \quad Q^{0*} = \sum_{k=0}^{N-1} R_{N-1}^* \ldots R_k^* Q_k^*,$$

$$F_N^* = F_0^* + \sum_{k=1}^{N-1} D^{*k} F_k^* R_{k-1}^* \ldots R_0^*.$$

Then Formulae (9.4.33) and (9.4.34) together imply

$$(I^* + Q^{0*})D^{*N} = (I^* - F_N^*)Q^*(D) = (I^* - F_N^*)Q\langle D^*\rangle,$$
$$R_N^* Q^*(D) = I^* + Q^{0*}. \tag{9.4.37}$$

The conjugate problem

$$Q^*(D)\xi = \eta, \tag{9.4.38}$$

has a unique solution $\xi = (I^* + Q^{0*})^{-1} R_N^*$, since $(I + Q^0)^* \xi = R_N^* Q^*(D)\xi = R_N^* \eta$.

Assume, as in Corollary 9.4.1 that D^* is invertible on Ξ. Then $F_i^* = 0$ on Ξ for $i = 1, 2, \ldots, N-1$. This implies $F_N^* = 0$. But F_N is an initial operator for D^N. Hence, arguing as before, we conclude that $R_N^* = R_0^{*N}$ on Ξ and

$$Q^{0*} = \sum_{k=0}^{N-1} R_0^{*k} Q_k^* = Q\langle I, R_0^*\rangle - I \quad \text{on } \Xi.$$

On the other hand, $(R_0 \ldots R_{N-1})^* = R_{N-1}^* \ldots R_0^*$, i.e. to the operator R_k there correspond the operator R_{N-k}. This implies that to the operator F_k there correspond the operator F_{N-k} $(k = 0, 1, \ldots, N-1)$ and

$$R_k^* = -R_{N-k}|_\Xi \quad \text{and} \quad F_k^* = F_{N-k}|_\Xi = 0$$
$$(k = 0, 1, \ldots, N-1). \tag{9.4.39}$$

If, moreover, as in the classical case, $D^* = -D|_\Xi$, then

$$D^{*N} = (-1)^N D^N|_\Xi, \quad R_N^* = (-1)^N R_0^N|_\Xi.$$

Consider a linear system of the form

$$Q(D)x = Bu, \tag{9.4.40}$$
$$FD^k x = x_k, \quad x_k \in \ker D \quad (k = 0, 1, \ldots, N-1), \tag{9.4.41}$$

where $u \in U$, $B \in L_0(U \to X, X' \to U')$. Here we have $F_0 = \ldots = F_{N-1} = F$, hence $R_0 = \ldots = R_{N-1} = R$. Assume, as before, that the operator $I + Q^0 = \sum_{k=0}^{N-1} Q_k R^{N-1}$ is invertible and belongs to $L_0(X, X')$. Let $F_1 \neq F$ be an initial operator for D. The system (9.4.40)–(9.4.41) is F_1-controllable if and only if

$$\ker B^*[(I + Q^0)^*]^{-1}(R^*)^{N-k}F_1^* = \{0\} \quad \text{for } k = 0, 1, \ldots, N-1$$

(Nguyen Dinh Quyet, 1978). The proof is going on the same lines as the proof of Theorem 9.3.4.

In a similar way, as Corollary 9.4.1, we may prove a necessary condition of optimality. Namely, we have the following theorem, provided that all conditions of Definition 9.4.1 are satisfied with respect to the operator $\hat{D} = Q(D)$:

If the operator $Q^*(D)$ is invertible on Ξ and (x^+, u^+) is a local minimum point of the functional $J(x, u)$ with the constraints (9.4.40)–(9.4.41) then there exists a unique solution of the conjugate problem

$$Q^*(D)\xi = J'_x, \quad F^*_N J'_x = 0 \tag{9.4.42}$$

which is of the form:

$$\xi = (R^N)^* [(I + Q^0)^*]^{-1} J'_x, \quad B^*\xi = -J'_u.$$

If $Q^*(D)$ is invertible on Ξ we can apply Corollary 9.4.1. Sufficient conditions follow from Theorem 9.4.2 if we put $T = (I + Q^0)^{-1} R^N B$.

We may consider in the same way a linear system even of a more complicate form, namely put in (9.4.40)

$$B = \tilde{Q}(D), \quad \text{where } \tilde{Q}(D) = \sum_{j=1}^{M} \tilde{Q}_j D^j, \quad \tilde{Q}_j \in L_0(U, U')$$

(provided that $\mathrm{dom}\, D \subset U \subset X$). Then we have a linear system of the form:

$$Q(D)x = Q(D)u, \quad FD^k x = x_k, \quad x_k \in \ker D$$
$$(k = 0, 1, ..., N-1) \tag{9.4.43}$$

which can be examined in the same manner.

Consider now an operator of a more complicated form:

$$P(D) = \sum_{k=0}^{N} (-1)^k D^k P_k D^k, \tag{9.4.44}$$

where

$$P_k \in L_0(X, \Xi) \quad (k = 0, 1, ..., N-1), \quad P_N = I,$$

X, Ξ, D are as before. If $D^* = -D|_\Xi$ is invertible on Ξ and $P^*_k = P_k|_\Xi$

$(k = 0, 1, ..., N-1)$ then the operator $P(D)$ is *self-adjoint*. Indeed, for all $\xi \in \varXi$ we have

$$P^*(D)\xi = \left[\sum_{k=0}^{N}(-1)^k D^k P_k D^k\right]^* \xi = \left[\sum_{k=0}^{N}(-1)^k D^{*k} P_k^* D^{*k}\right]\xi$$

$$= \left[\sum_{k=0}^{N}(-1)^k(-1)^{2k} D^k P_k D^k\right]\xi = P(D)\xi,$$

which implies $P^*(D) = P(D)|_{\varXi}$.

Suppose that the operator

$$I+P^0 = \sum_{k=0}^{N}(-1)^k R^{N-k} P_k R^{N-k}$$

is invertible. It is easy to verify that the operator $P(D)$ is right invertible and has a right inverse of the form:

$$R_{P(D)} = R^N(I+P^0)^{-1}R^N.$$

Hence the operator $P^*(D)$ is, as self-adjoint, invertible and again we may apply Corollary 9.4.1.

EXAMPLE 9.4.2. Let $X = C[a, b]$, $D = \mathrm{d}/\mathrm{d}t$, $R_c = \int_c^t$, $(F_c x)(t) = x(c)$ for $x \in X$, where $c \in [a, b]$ is arbitrary. It is clear that F_c is an initial operator for D corresponding the right inverse R_c. Suppose that we are given n points:

$$a = a_0 < ... < a_{n-1} = b. \tag{9.4.45}$$

In a similar way, as in (9.4.25), write for $m = 0, 1, ..., +\infty$

$$E_m^0 = \{y \in C^m[a, b]: y^{(i)}(a_j) = 0 \text{ for }$$

$$j = 0, 1, ..., n-1, i = 0, 1, ..., m\} \tag{9.4.46}$$

and, as in (9.4.26),

$$\varXi_m = \left\{\xi \in X': \xi(x) = \int_a^b x(t)y(t)\,\mathrm{d}t \text{ for } y \in E_m^0, x \in X\right\}. \tag{9.4.47}$$

By the same arguments, $\varXi_m$ are conjugate spaces and we have inclusions (9.4.27). Therefore the space $\varXi_\infty$ is the smallest one and we restrict all our considerations to that space (cf. also theorems on reduction of conjugate spaces in Section 1.5).

For all $y \in E_\infty$, $x \in \mathrm{dom}\, D = C^1[a, b]$ (resp.: $x \in X$), $c \in [a, b]$, since $y(b) = y(a) = 0$ by our assumption, we find

$$\xi(Dx) = \int_a^b y(t)x'(t)\,dt = [y(t)x(t)]_a^b - \int_a^b x(t)y'(t)\,dt$$

$$= -\int_a^b x(t)y'(t)\,dt,$$

$$\xi(R_a x) = \int_y^b y(t)\left(\int_a^t x(s)\,ds\right) dt = \int_a^b x(s)\left(\int_s^b y(t)\,dt\right) ds$$

$$= \int_a^b x(s)\left(-\int_b^s y(t)\,dt\right) ds,$$

$$\xi(R_b x) = \int_a^b x(s)\left(-\int_a^s y(t)\,dt\right) ds,$$

which implies

$$D^* = -D|_\Xi = -\frac{d}{dt}\bigg|_{\Xi_\infty}, \tag{9.4.48}$$

$$R_a^* = -\int_b^t \bigg|_{\Xi_\infty} = -R_b|_{\Xi_\infty}, \quad R_b^* = -\int_a^t \bigg|_{\Xi_\infty} = -R_a|_{\Xi_\infty} \tag{9.4.49}$$

and

$$y(t) - (F_a^* y)(t) = [(I^* - F_a^*)y](t) = [(R_a D)^* y](t)$$

$$= (D^* R_a^* y)(t) = -\frac{d}{dt}\left(-\int_b^t y(s)\,ds\right) = y(t)$$

and similarly, $y - F_b^* y = y$. Hence

$$F_a^* = 0, \quad F_b^* = 0, \tag{9.4.50}$$

$$R_a^* D^* = I^* - F_a^* = I^*, \quad R_b^* D^* = I^* - F_b^* = I^*.$$

We therefore conclude that D^* is *invertible* on Ξ_∞.

Observe that for all $v \in E_m^0$ $(m = 0, 1, ..., +\infty)$ we have $(R_a v)(a) = 0$ but, in general, $(R_a v)(c) \neq 0$ for $c \in [a, b]$. Thus to have operators preserving a conjugate space, we should consider the set

$$E = \{v \in E_\infty^0: (R_a v)(b) = (R_b v)(a) = 0\}$$

and as the conjugate space

$$\Xi = \left\{\xi \in X': \xi(x) = \int_a^b x(t) y(t) \, dt \text{ for } y \in E, x \in X\right\}.$$

Clearly, D^* is invertible on Ξ.

Formulae (9.4.49) show the duality between *initial* and *final states*. Since D^* is invertible on Ξ, Condition (9.4.7) is in our case of the form: $F_a^* J_x' = 0$.

Consider now initial operators corresponding to some *boundary* and *non-local conditions*:

$$(F_2 x)(t) = \frac{1}{b-a} \int_a^b x(s) \, ds,$$

$$(F_3 x)(t) = dx(a) + (1-d) x(b),$$

$$(F_4 x)(t) = dx(a) + \frac{1-d}{b-a} \int_a^b x(s) \, ds,$$

$$(F_5 x)(t) = x(a) + dx'(b),$$

where $x \in X$ and $0 \neq d \in \mathbf{R}$ is either fixed or a parameter. Observe that

$$F_2 = \frac{1}{b-a} F_b R_a, \quad F_3 = dF_a + (1-d) F_b,$$

$$F_4 = DF_a + (1-d) F_2 = dF_a + \frac{1}{b-a} F_b R_a,$$

$$F_5 = F_a + dF_b D.$$

The corresponding right inverses can be written in the following form:

$$R_2 = R_a - F_2 R_a,$$

$$R_3 = R_a - F_3 R_a = R_a - dF_a R_a - (1-d) F_b R_a$$
$$= R_a + (d-1) F_b R_a,$$
$$R_4 = R_a - F_4 R_a = R_a - dF_a R_a - (1-d) F_2 R_a$$
$$= R_a + (d-1) F_2 R_a,$$
$$R_5 = R_a - F_5 R_a = R_a - F_a R_a - dF_1 DR_a = R_a - dF_b$$

(since $F_a R_a = 0$ and $DR_a = I$).

Since $F_a^* = F_b^* = 0$, Formulae (9.4.49) imply that on $\varXi$

$$F_2^* = \frac{1}{b-a} R_a^* F_b^* = 0, \quad F_3^* = dF_a^* + (1-d) F_b^* = 0,$$

$$F_4^* = dF_a^* + (1-d) F_2^* = 0, \quad F_5^* = F_a^* + dD^* F_b^* = 0$$

and that for $i = 2, 3, 4, 5$

$$R_i^* = R_a^* + (F_i R_a)^* = R_a^* - R_a^* F_i^* = 0 \quad \text{on } \varXi.$$

This, in particular, means that for all problems inducing initial operators F_2, F_3, F_4, F_5 the operator D^* is invertible and has an inverse $R_a^* = -R_b|_\varXi$. Therefore for all these problems Condition (9.4.7) is of the form:

$$F_i^* J_x' = 0 \quad (i = 2, 3, 4, 5).$$

Observe now that

$$F_c^* = (I - R_c D)^* = I^* - \left(\int_c^b \frac{d}{dt} \right)^* = I^* - I^* = 0, \quad c \in [a, b],$$

$$R_c^* = (R_a - F_c R_a)^* = R_a^* - R_a^* F_c^* = R_a^* = -R_b|_\varXi.$$

Hence

$$R_{a_k}^* = -R_b|_\varXi, \quad F_{a_k}^* = 0 \quad (k = 0, 1, \ldots, n-1). \tag{9.4.51}$$

Write now

$$F = \sum_{k=0}^{n-1} d_k F_{a_k}, \quad \text{where } d_k \in \mathbf{R}, \quad \sum_{k=0}^{n-1} d_k = 1. \tag{9.4.52}$$

By definition, F is an initial operator for D corresponding to the right inverse

$$R = R_a - FR_a = \left(I - \sum_{k=0}^{n-1} d_k F_{a_k}\right) R_a.$$

The initial operator F corresponds to the so-called *multipoints boundary value problems* for

$$(Fx)(t) = \sum_{k=0}^{n-1} d_k x(a_k) \quad \text{for } x \in X.$$

Formulae (9.4.51) imply that

$$F^* = \sum_{k=1}^{n-1} d_k F_{a_k}^* = 0,$$

$$R^* = (R_a - FR_a)^* = R_a^* - R_a^* F^* = R_a^* = -R_b|_{\Xi},$$

i.e. again Condition (9.4.7) for the problems with the initial operator F will be of the form: $F^* J_x' = 0$.

EXAMPLE 9.4.3. Suppose that we have a stationary linear system $(LS)_H$ with a stationary output H, i.e.

$$Dx = Ax + Bu, \quad y = Hx, \quad Fx = x_0. \tag{9.4.53}$$

We assume that X, U are Hilbert spaces, i.e. we identify* X and X^*, U and U^*, $D \in R(X)$, F is an initial operator for D corresponding to an $R \in \mathcal{R}_D$, $x_0 \in \ker D$, $A, H \in L_0(X)$, $B \in L_0(U \to X)$ and A, H are stationary. We also assume that the resolving operator $I - RA = I - AR$ is invertible and that $E_A = (I - RA)^{-1} \in L_0(X)$.

Consider the *conjugate system* $(LS)_H^*$:

$$D^* \xi = A^* \xi + H^* v, \quad F^* \xi = 0. \tag{9.4.54}$$

We shall prove the following theorem: *Suppose that D^* is invertible. A necessary and sufficient condition for the system* (9.4.53) *to be observable is that for the conjugate system* (9.4.54) *every state $\xi \in RX \oplus \{x_0\}$ is reachable from every initial state $x_0 \in \ker D$.*

---

* We denote by $X^* \subset X'$ the space of all bounded functionals.

Indeed, by Theorem 9.3.6 the system (9.4.53) is observable if and only if $\ker HE_A = \{0\}$. On the other hand, by Theorem 9.3.2 for the conjugate system (9.4.54) we find that every $\xi \in RX \oplus \{x_0\}$ is reachable from every initial state $x_0 \in \ker D$ if and only if

$$\{0\} = \ker(H^*)^*(E_A^*)^* = \ker HE_A$$

for $E_A^* = (I^* - A^*R^*)^{-1} = (I^* - R^*A^*)^{-1}$ is the resolving operator for the system (9.4.54).

Theorem 9.3.6 implies that the system (9.4.53) is observable if for every functional $f \in (\ker D)'$ there is a $\varphi \in Y' = X$ such that $E_A^*H^*\varphi = f$. Since H is stationary, H^* is also stationary and this last equation can be written in the following form:

$$H^*\varphi = (I - RA)^*f = (I^* - A^*R^*)f = (I^* - R^*A^*)f.$$

Since $D^*R^* = R^*D^* = (DR)^* = I$, we find

$$H^*D^*\varphi = D^*H^*\varphi = D^*(I^* - R^*A^*)\varphi = (D^* - A^*)\varphi,$$

i.e.

$$D^*f = A^*f + H^*v, \quad \text{where } v = D^*\varphi.$$

The condition $F^*f = 0$ is automatically satisfied, because D^* is invertible, hence $F^* = 0$. We therefore conclude that f is a solution of the conjugate system (9.4.54).

EXERCISE 9.4.1. Formulate Definition 9.4.2 and find necessary and sufficient conditions for the problem under consideration in the case when

(i) the resolving operator $I - RA$ is either left or right invertible,

(ii) the resolving operator $I - AR$ is either left or right invertible or invertible.

EXERCISE 9.4.2. Suppose that the system (9.4.53) is F_1-controllable. Is the conjugate system (9.4.54) observable?

EXERCISE 9.4.3. Which results of Sections 9.3 and 9.4 can be generalized for nonlinear systems?

Bibliography

Allan, G. R. (1967): 'On one-side inverses in Banach algebras of holomorhpic vector valued functions', *J. London Math. Soc.* **42**, 460–470.

Anderson, J. (1973): 'On normal derivations', *Proc. Amer. Math. Soc.* **38**, 135–140.

Anderson, J. H., Stampfli, J. G. (1971): 'Commutators and compressions', *Israel J. Math.* **10**, 433–441.

Antosik, P., Mikusiński, J. (1968): 'On Hermite expansions', *Bull. Acad. Pol. Sci.* **16**, 787–791.

Apostol, C., Stampfli, J. (1976): 'On derivation ranges', *Indiana University Math. Journ.* **25**, 857–869.

Arens, R. (1958): 'Dense Inverse Limit Rings', *Michigan Math. J.* **5**, 169–182.

Arscott, F. M. (1964): *Periodic Differential Equations*, Pergamon Press, Oxford.

Atterton, T. W. (1972): 'Definitions of integral elements and quotient rings over noncommutative rings with identity', *J. Austr. Math. Soc.* **13**, 433–466.

Bart, H. (1974): 'Holomorphic relative inverses of operator valued functions', *Math. Annalen*, **208**, 179–194.

Bart, H., Gohberg, I., Kaashoek, M. A. (1982): 'Wiener–Hopf integral equations, Toeplitz matrices and linear systems', in: *Operator Theory: Advances and Applications*, Vol. 3, BirkhäuserVlg., Basel–Boston–Stuttgart.

Bart, H., Gohgerg, I., Kaashoek, M. A. (1982a): 'Convolution equations and linear systems', *Integral Equations and Operator Theory*, **5**, 283–340.

Bart, H., Kaashoek, M. A., Lay, D. (1974): 'Stability properties of finite meromorphic operators', *Proc. Acad. Sci. Amsterdam*, A **77**, 217–259.

Bart, H., Kaashoek, M. A., Lay, D. (1975): 'Relative inverses of meromorphic operators, functions and associated holomorphic projection functions', *Math. Ann.* **218**, 199–210.

Bellert, S. (1957): 'On foundations of operator calculus', *Bull. Acad. Polon. Sci.* **5**, 855–858.

Bellert, S. (1960): 'Operational calculus in linear spaces' (in Polish), *Rozpr. Elektrot.* **6**, 170–212.

Bellman, R., Cooke, K. L. (1963): *Differential-Difference Equations*, Academic Press, New York–London.

Berg, L. (1972): 'Anfangswertprobleme vom Standtpunkt der Algebra', *Sitzungsberichte des Plenums und der Klassen der Akademie der Wissenschaften der DDR*, **11**.

588

Berg, L. (1973): 'Allgemeine Operatorenrechnung', *Überblicke Math. Bibliographisches Institut AG, Mannheim–Wien–Zürich*, 6.

Berg, L. (1974): *Operatorenrechnung*, VEB Deutscher Verlag der Wissenschaften, Berlin. I. *Algebraische Methoden*, 1972; II, *Funktionentheoretische Methoden*, 1974.

Berg, L. (1974a): 'Randwertprobleme in der Operatorenrechnung', *Wiss. Zeitschrift der Universität Rostock*, 2, 621–624.

Berg, L. (1975): *Analysis in geordneten kommutativen Halbgruppen mit Nullelement*, Nova Acta Leopoldina, Halle (Saale).

Berg, L. (1975a): 'Über die Struktur der Lösungen homogener Operatorgleichungen', *Math. Nachr.* 67, 1–11.

Berg, L. (1975b): 'Zur Umkehrung "periodischer" Operatoren', *ZAMM* 55, 65–68.

Berg, L. (1975c): 'Evaluation of the eccessive initial values in the two-dimensional operational calculus', *Ibidem* 55, 701–708.

Berg, L. (1976): 'The method of reducing the order of linear operator equations', *Demonstratio Math.* 9, 1–6.

Berg, L. (1977): 'Solutions of degenerate linear initial value problems', *ZAMM* 57, 65–73.

Berg, L. (1977a): 'The division of distributions by the independent variable', *Math. Nachr.* 78, 327–338.

Berg, L. (1979): 'General iteration methods for the evaluation of linear equations', *Numer. Punct. Anal. a. Optimiz.* 1, 365–381.

Berg, L. (1980): 'Commutable right invertible operators', *Demonstratio Math.* 13, 665–674.

Bergman, S. (1972): *Integral Operators in the Theory of Linear Partial Differential Equations*, III-rd edition, Springer Vlg., Heidelberg–New York.

Bielecki, A. (1956): 'Une remarque sur la méthode de Banach–Cacciopoli–Tikhonov dans la théorie des équations différentielles ordinaires', *Bull. Acad. Polon. Sci.* 4, 261–264.

Bittner, L. (1973): 'Linear equations with an operator polynomial on the left side', *ZAMM* 53, 397–408.

Bittner, R. (1961): 'Algebraic properties of linear derivative equations in linear spaces', *Bull. Acad. Polon. Sci.* 9, 133–139.

Bittner, R. (1961a): 'Operational calculus in linear spaces', *Studia Math.* 20, 1–18.

Bittner, R. (1964): 'Algebraic and analytic properties of solutions of abstract differential equations', *Rozpr. Matem.* 41 (Warszawa).

Bittner, R. (1974): *Operational Calculus in Linear Spaces* (in Polish), PWN, Warszawa.

Bittner, R., Smentek, Z. (1980): 'A theorem on the existence and uniqueness of solutions for abstract quasi-linear differential equations', *Demonstratio Math.* 13, 787–808.

Björk, J. E. (1979): *Rings of Differential Operators*, North-Holland Math. Library 21, Amsterdam–New York.

Bozhinov, N. S. (1976): 'Operational calculus for general linear partial differential operators of the first order', *Comptes Rendus Acad. Bulg. Sci.* **29**, 1261–1264.

Bozhinov, N. S., Dimovski, I. H. (1975): 'Operational calculus for general differential operator of second order', *Ibidem* **28**, 727–730.

Browder, A. (1967): 'Point derivation of functions algebras', *J. Funct. Analysis* **1**, 22–27.

Burton, T. A. (1983): *Volterra Integrals and Differential Equations*, Academic Press, New York–London.

Bykov, Ya. V., Botashev, A. I., Merzl'yakova, G. D. (1974): 'On a symbolic method of solvability of boundary and initial value problems for linear operator equations (in Russian), *Diff. Uravnienija*, **12**, 2193–2207.

Carlitz, L. (1970): 'Some operational formulas', *Math. Nachr.* **45**, 379–389.

Carlitz, L. (1976): 'A theorem on differential operator', *Amer. Math. Monthly* **83'** 351–354.

Carrol, R. (1979): *Transmutation and Operator Differential Equations*, North-Holland, Amsterdam–New York–Oxford.

Cartan, H. (1971): *Differential Calculus*, Herman, Paris.

Chernoff, P. R. (1973): 'Universally commutable operators are scalars', *Michigan Math. J.* **20**, 101.

Chiu, S. C., Ladik, J. F. (1977): 'Generating exactly soluble nonlinear evolution equations by generalized Wronskian technique', *J. Math. Phys.* **18**, 690–700.

Coddington, A., Levinson, N. (1955): *Theory of Ordinary Differential Equations*, Mc Graw–Hill Book Co., New York–Toronto–London.

Cohoon, D. K. (1969): 'Non-existence of a continuous right inverse for parabolic operators', *J. Diff. Equations* **6**, 503–511.

Cohoon, D. K. (1970): 'A continuous right inverse for second order differential operators with variable coefficients', *SIAM J. Appl. Math.* **19**, 648–657.

Cohoon, D. K. (1971): 'Non-existence of a continuous right inverse for surjective linear partial differential operators on the Fréchet space $\gamma^{(\delta)}\Omega$, *J. Diff. Equations* **10**, 291–313.

Conner, P. E., Floyd, E. E. (1964): *Differentiable Periodic Maps*, Springer Vlg., Berlin–Heidelberg–Göttingen.

Cooper, J. L. B. (1951): 'Heaviside and the operational calculus', *The Mathematical Gazette, London Math. Soc.* **18.1**.

Davis, Ch. (1973): 'Explicit functional calculus', *Algebra and its applications* **6**, 193–199.

Delsarte, J. (1937): 'Le calcul linéaire', *Bull. Soc. Math. France.* Comptes rendus de séances.

Delsarte, J. (1938): 'Sur une extension de la formula de Taylor', *J. de Math.* **17**, 213–231.

Delsarte, J. (1971): *Oeuvres de Jean Delsarte*, Edition du Centre National de la Recherche Scientifique, Paris.

Delvos, F.-J., Schempp, W. (1983): 'The method of parametric extension applied to right invertible operators', *Numer. Funct. Analysis and Optimiz.* **6** (2), 135–148.

Dimovski, I. H. (1966): 'Operational calculus for a class of differential equations', *Comptes Rendus Acad. Bulg. Sci.* **19**, 1111–1114.

Dimovski, I. H. (1973): 'Operational calculus for the general linear differential operator of the first order', *Ibidem* **26**, 1579–1582.

Dimovski, I. H. (1975): 'On an operational calculus for vector valued functions', *Mathematica Balcanica* **4**, 11–118.

Dimovski I. H. (1975a): 'Foundations of operational calculi for the Bessel-type differential operator', *Serdica, Bulg. Math. Publicationes* **1**, 51–63.

Dimovski I. H. (1976): 'The new convolutions for linear right-inverse operators of d^2/dt^2', *Comptes Rendus Acad. Bulg. Sci.* **29**, 25–28.

Dimovski I. H. (1978): 'Convolutions of right inverse operators and representation of their multipliers', *Ibidem* **31**, 1377–1380.

Dimovski, I. H. (1982): *Convolutional Calculus*, Publish. House of the Bulg. Acad. Sci., Sofia.

Dimovski, I. H. (1982a): 'Representation formulas for the commutators of Gel'fond-Leont'ev integration operators', in: *Proc. of the 11-th Spring Conference of the Union of Bulg. Mathematicians*, Bulg. Acad. Sci., Sofia, 166–172.

Ditkin, V. A., Prudnikov, A. P. (1962): 'Operational calculus for Bessel operators' (in Russian), *Zh. vych. matem.* **2**, 997–1018.

Ditkin, V. A., Prudnikov, A. P. (1974): *Integral Transforms and Operational Calculus* (in Russian), Moscow, French edition: *Transformations intégrales et calcul opérationnel*, Mir, Moscou, 1982.

Djrbamian, N. N., Caakian, B. A. (1975): 'Class of formulae and expansions of the Taylor–Maclaurin type associated with differential operators of fractional orders (in Russian), *Izv. AN SSR* **38**, 69–122.

Donaldson, J. A. (1972): 'An operational calculus for a class of abstract operator equations', *J. Math. Anal. Appl.* **37**, 167–184.

Dudek, Z. (1978): 'Some properties of Wronskian in D–R spaces of the type Q–L, Pt. I', *Demonstratio Math.* **11**, 1115–1130.

Dudek, Z. (1980): 'Some properties of Wronskian in D–R spaces of the type Q–L, Pt. II', *Ibidem* **13**, 987–993.

Dudek, Z. (1980a): 'Some properties of Wronskian in Q–L algebras', in: *Proc. Intern. Conf. Functional-Differential Systems and Related Topics, I*, Błażejewko, May 1979, Higher College of Engineering, Zielona Góra, 93–98.

Dudek, Z. (1981): 'On decompositions of quasi-Leibniz D–R algebras', *Demonstratio Math.* **14**, 745–757.

Dudek, Z. (1982): 'The relationship between ql-operators and Gelfond-Leontiev derivatives', in: *Proc. Intern. Conf. Functional-Differential Systems and Related Topics, II*, Błażejewko, May 1981, Higher College of Engineering, Zielona Góra, 102–104.

Elsgolc, L. E. (1964): *Introduction into Theory of Differential Equations with Deviating Argument* (in Russian), Izd. Nauka, Moscow.

Fage, M. K. (1958): 'Operator-analytic functions of one independent variable' (in Russian), *Trudy Mosc. matem. obscestva* 7, 227–268.

Fattorini, H. O. (1969): 'Ordinary differential equations in linear topological spaces, I', *J. Diff. Equations* 5, 72–105.

Fattorini, H. O. (1970/71): 'A perturbation theorem for generators of cosine functions' (in Spanish), *L Rev. Un. Mat. Argentina* 27, 191–211.

Fényes, T., Kosik, P. (1972): 'The algebraic derivative and integral in the discrete operational calculus', *Studia Sci. Math. Hungar.* 7, 117–130.

Fortuna, Z. (1974): *Analysis of Properties of Conjugate Gradients in Hilbert Spaces* (in Polish), Ph. D. Diss., Institute of Automatics Technical University of Warsaw, Warszawa.

Förster, K. H. (1972): 'Über Kommutatoren und Störungsmetriken und ihre Beziehung zum Funktionalkalkül', *Manuscripta Math.* 6, 291–308.

Förster, K. H., Garske, G. (1972): *Holomorphic einseitiger Inverser von holomorphen Funktionen mit Werten in eine Banach para-algebra*, Preprint, Math. Institut der Universität Dortmund und Math. Institut der Universität Karlsruhe (TH).

Goldberg, M. A. (1972): 'The derivative of a determinant', *Amer. Math. Monthly* 79, 1124–1126.

Grabiner, S. (1971): 'A formal power series operational calculus for quasinilpotent operators', *Duke Math. J.* 38, 641–658.

Grabowski, P. (1983): *A New Class of Initial Problems for Equations with Right Invertible Operators and its Connections with the Theory of c_0-Semigroups in Banach Spaces* (in Polish), Pt. I, Preprint, Academy of Minning and Metallurgy, Kraków.

Grabowski, P. (1984): *A New Class of Initial....* Pt. II.

Green, C. D. (1969): *Integral Equation Methods*, Nelson, London.

Grothendieck, A. (1957): 'Sur quelques points d'algèbre homologique', *Tohoku Math. J.* 9, 119–221.

Gulick, F. (1970): 'Systems of derivatives', *Trans. Amer. Math. Soc.* 149, 465–488.

Heinig, G., Rost, K. (1984): *Algebraic Methods for Toeplitz-like Matrices and Operators*, Math. Research 19, Akademie-Vlg., Berlin.

Hermite, Ch. (1878): 'Sur la formule d'interpolation de Lagrange', *J. Reine Angew. Math.* 84, 70–79.

Heuser, H. (1975): *Funktionalanalysis*, B. G. Teubner, Stuttgart.

Hutson, V., Pym, J. S. (1970): 'General translations associated with an operator', *Math. Ann.* 187, 241–258.

Ince, E. L. (1926): *Ordinary Differential Equations*, Longmans, Green 1926; London 1927.

Jacobson, N. (1956): *Structure of Rings*, Amer. Math. Soc., Providence, R. I.

Johnson, B. E. (1968): 'Continuity of derivation and a problem of Kaplansky', *Amer. J. Math.* 90, 1067–1073.

Johnson, B. E. (1969): 'Continuity of derivations on commutative algebras', *Amer. J. Math.* **91**, 1–10.

Kaashoek, M. A., Lay, D. C. (1972): 'Ascent, descent and commuting perturbations', *Trans. Amer. Math. Soc.* **169**, 35–47.

Kahane, Ch. (1969): 'On operators commuting with differentiation', *Amer. Math. Monthly* **76**, 171–173.

Kalfat, A. (1986): 'Volterra right inverses for the difference operators', in: *Proc. Intern. Conf. Functional-Differential Systems and Related Topics IV, Jachranka May 26–June 2, 1985*, Higher College of Engineering, Zielona Góra.

Kamen, E. W. (1975): 'On an algebraic theory of systems defined by convolution operators', *Math. Systems Theory* **9**, 57–74.

Kamen, E. W. (1978): 'An operator theory of linear functional-differential equations', *J. Diff. Equations* **27**, 274–297.

Kaplansky, I. (1957): *An Introduction to Differential Algebra*, Paris.

Kaplansky, I. (1970): *Algebraic and analytic aspects of operators algebras. Conference Board of the Mathematical Sciences*, Regional Conference Series in Mathematics, 1, Amer. Math. Soc., Providence, R. I.

Kawada, V. (1951): 'On the derivation in number fields', *Ann. Math.* **56**, 302–314.

Koliha, J. J. (1974): 'Power convergence and pseudoinverse of operators in Banach spaces', *Math. Anal. Appl.* **48**, 446–469.

Koliha, J. J. (1974a): 'Pseudoinverses of operators', *Bull. Amer. Math. Soc.* **80**, 325–328.

Kolodner, I. I. (1975): 'On exp(tA) with A satisfying a polynomial', *J. Math. Anal. Appl.* **52**, 514–524.

Koprinsky, S. C. (1979): 'An operational method for Bessel heat equation', in: *Proc. Conf. Generalized Functions and Operational Calculus, Varna 1975*, Sofia.

Kornacki, H. (1986): 'Non-Leibniz components in non-commutative algebras with unit', *Demonstratio Math.* **18**, 483–497.

Krasnoselskiĭ, M. A. (1966): *Shift Operator along Trajectories of Differential Equations* (in Russian), Izd. Nauka, Moscow.

Kreĭn, S. G. (1967): *Linear Differential Equations in Banach Spaces* (in Russian), Izd. Nauka, Moscow.

Kreĭn, S. G. (1982): *Linear Equations in Banach Spaces*, Birkhäuser, Boston–Basel–Stuttgart.

Kreith, K. (1975): 'Picone's identity and generalizations', *Rendiconti di Mathematica* **8**, 251–262.

Krobeĭnik, Yu. F. (1973): 'The general form of linear operators that commute with the differentiation operator in the space of analytic functions' (in Russian), *Funkcjonal. Anal. i Prilozhen.* **7**, 74–76.

Kuratowski, K. (1969): *Introduction to Calculus*, I-st Polish ed. Monografie Matematyczne 15, PWN, Warszawa 1948, II-nd English ed. Pergamon Press–PWN–Polish Scientific Publishers, Oxford–Warszawa.

Leĭbenzon, Z. L. (1972): 'Algebraic differential transformations of linear differential operators of arbitrary order and their spectral properties that are applicable in the inverse problem. I' (in Russian), *Math. Sbornik* **87** (129), 396–416.

Levitan, B. M. (1973): *Theory of Generalized Shift Operator* (in Russian), Izd. Nauka, Moscow.

Lord, M. E. (1969): 'A discrete algebraic derivative', *J. Math. Anal. Appl.* **25**, 701–709.

Lovass-Nagy, V., Powers, D. L. (1974): 'On under- and over-determined initial value problems', *Int. J. Control* **19**, 653–656.

Machover, M. (1974): 'On the Wronskian Formula', *Math. Mag.* **47**, 89–91.

Maliński, K. (1962): 'Homogeneous differential equations in linear spaces' (in Polish), *Zesz. Nauk. Wyższej Szkoły Pedagogicznej, Gdańsk*, **2**, 13–21.

Maliński, K. (1964): 'Trigonometric quasi-polynomials' (in Polish), *Ibidem* **4**, 11–22.

Maliński, K. (1965): 'Uniqueness of solution of abstract linear differential equations' (in Polish), *Ibidem* **5**, 5–10.

Maliński, K. (1966): 'Solution of abstract non-homogeneous linear differential equations by the method of indetermined coefficients' (in Polish), *Ibidem* **6**, 5–15.

Maliński, K. (1966a): 'Systems of abstract linear differential equations' (in Polish), *Ibidem* **6**, 17–26.

Mażbic-Kulma, B. (1971): 'Differential equations in a linear ring', *Studia Math.* **39**, 157–161.

Mażbic-Kulma, B. (1971a): *Algebraic Derivative and Functional-Differential Equations* (in Polish), Ph. D. Diss. Institute of Mathematics, Polish Academy of Sciences, Warszawa.

März, R. (1979): 'Interpolation in linearen Räumen', *Wissenschaftliche Zeitschrift der Humboldt-Universität zu Berlin, Mat.-Nat. R.* **28**, 515–521.

Meyer, Ch. (1975): *Über einige Klassen von singulären Operatoren und ihre Beziehung zu singulären Integralgleichungen*, Ph. D. Diss. Sektion Mathematik, Technische Hochschule Karl–Marx–Stadt.

Meyer, Ch., Silbermann, B. (1977): 'Die Indexformel für eine Klasse von ausgeartetet singulären Operatoren mit Carlemanscher Verschiebung', *Demonstratio Math.* **10**, 155–167.

Mikusiński, J. G. (1949): 'Sur les fondaments du calcul opératoire', *Studia Math.* **11**, 58–59.

Mikusiński, J. G. (1954): *Operational Calculus*, I-st Polish ed. Monografie Matematyczne 30, PWN, Warszawa, 1953; English ed. Pergamon Press and PWN–Polish Scientific Publishers, Oxford–Warszawa, 1957.

Mikusiński, J. (1958): 'Extension de l'espace linéaire avec dérivation', *Studia Math.* **16**, 156–172.

Miller, J. B. (1966): 'Averaging and Reynolds operators on Banach algebras. I. Representation by derivations and antiderivations', *J. Math. Anal. Appl.* **14**, 527–548.

Miller, J. B. (1967): 'Homomorphisms, higher derivations and derivations on associative algebras', *Acta Scientiarum Math.* **28**, 221–231.

Miller, J. B. (1968): 'Averaging and Reynolds operators on Banach algebras. II. Spectral properties of averaging operators', *J. Math. Anal. Appl.* **23**, 183–197.

Miller, J. B. (1968a): 'A formula for the resolvent of a Reynolds operator', *J. Australian Math. Soc.* **8**, 447–456.

Miller, J. B. (1969): 'Baxter operators and endomorphisms on Banach algebras', *J. Math. Anal. Appl.* **25**, 503–520.

Miller, J. B. (1982): 'The Euler–Maclaurin formula for an inner derivation', *Aequationes Math.* **25**, 42–51.

Miller, J. B. (1983): 'The Euler–Maclaurin formula generated by a summation operator', *Proc. Roy. Soc. of Edinburgh* **95** A, 285–300.

Mingarelli, A. B. (1983): *Volterra–Stieltjes Integral Equations and Generalized Ordinary Differential Expressions*, Lecture Notes in Mathematics 89, Springer Vlg., Berlin–Heidelberg–New York–Tokyo.

Moszner, Z. (1973): 'Structure de l'automate plein, recluit et inversible', *Aequationes Math.* **9**, 46–59.

Moszner, Z. (1973a): 'The translation equation and its application', *Demonstratio Math.* **6**, 309–327.

Moszner, Z., Tabor, J. (1976): 'L'équation de translation sur la structure avec zéro', *Ann. Polon. Math.* **31**, 255–264.

Muhamadev, E. (1967): 'On the theory of periodic completely continuous vector field' (in Russian), *Uspehi Mat. Nauk* **22**, 127–128.

Narkiewicz, W. (1969): 'On a theorem of A. Weil on derivation in number fields', *Colloq. Math.* **20**, 57–58.

Nashed, M. Z. (1976): 'Generalized inverses and applications', in: *Proc. of an Advanced Seminary, University of Wisconsin, Madison* 1973, ed. M. Z. Nashed, Academic Press, New York.

Nguyen Dinh Quyet (1977): *Controllability and Observability of Linear Systems Described by the Right Invertible Operators in Linear Spaces*, Preprint No. 113, Institute of Mathematics, Polish Academy of Sciences, Warszawa.

Nguyen Dinh Quyet (1978): 'On linear systems described by right invertible operators acting in a linear space', *Control and Cybernetics* **7**, 33–45.

Nguyen Dinh Quyet (1978a): 'A minimizing problem of a quadratic functional for a system described by right invertible operators in Hilbert space', *Ibidem* **7**, 27–36.

Nguyen Dinh Quyet (1979): 'On the stability and observability of $(D–R)$-systems in Banach spaces', *Demonstratio Math.* **12**, 203–209.

Nguyen Dinh Quyet (1981): *On the F_1-controllability of the system described by the right invertible operators in linear spaces*, Methods of Mathematical Programming, System Research Institute, Polish Academy of Sciences, PWN–Polish Scientific Publishers, Warszawa, 223–226.

Nguyen Van Mau (1983): 'Characterization of polynomial operators with constant coefficients', *Demonstratio Math.* **16**, 375–405.

Nguyen Van Mau (1983a): 'Characterization of commutators with algebraic operators', *Ibidem* **16**, 843–856.

Nguyen Van Mau (1984): 'Regularization of polynomials in algebraic and almost algebraic operators', *Ibidem* **17**, 15–57.

Nowak, H. (1982): *Finite-Dimensional Perturbations of Equations with Right Invertible Operators* (in Polish), Ph. D. Diss. Institute of Mathematics, Technical University of Warsaw; also *Proc. Intern. Conf. Functional-Differential Systems and Related Topics III*, Błażejewko 1983, Higher College of Engineering, Zielona Góra, 1984, 181–187.

Oldham, K. B. (1974): *The Fractional Calculus*, Academic Press, New York.

Öre, O. (1931): 'Linear equations in non-commutative fields', *Ann. Math.* **32**, 463–477.

Öre, O. (1933): 'Theory of non-commutable polynomials', *Ibidem* **34**, 480–508.

Ostrowski, A. M. (1973): *Solutions of Equations in Banach Spaces*, Academic Press, New York.

Pietsch, A. (1980): *Operators Ideals*, North-Holland, Amsterdam–New York.

Pogorzelec, A. (1981): 'Ill-determined linear systems described by the right invertible operators', in: *Proc. Intern. Conf. Functional-Differential Systems and Related Topics II*, Błażejewko, May 1981, Higher College of Engineering, Zielona Góra, 277–279.

Pogorzelec, A. (1983): 'Initial value problems with ill-determined linear systems with right invertible operators', *Demonstratio Math.* **16**, 407–420.

Pogorzelec, A. (1983a): *Solvability and Controllability of Ill-Determined Systems with Right Invertible Operators*, Ph. D. Diss. Institute of Mathematics, Technical University of Warsaw, Warszawa.

Pogorzelec, A. (1984): 'Controllability of ill-determined systems with right invertible operators', in: *Proc. Intern. Conf. Functional-Differential Systems and Related Topics III*, Błażejewko, May 1983, Higher College of Engineering, Zielona Góra, 189–195.

Pogorzelski, W. (1966): *Integral Equations and their Applications*, I-st Polish ed. Pt. I—1953, Pt. II—1958, Pt. III—1960; English ed. Pergamon Press and PWN–Polish Scientific Publishers, Oxford–Warszawa.

Pogorzelski, W. (1970): *Integral Equations and their Applications*, Pt. IV (in Polish), PWN, Warszawa.

Procesi, C. (1973): *Ring with Polynomial Identities*, M. Dekker Inc., New York.

Przeworska–Rolewicz, D. (1960): 'Sur les involutions d'ordre *n*', *Bull. Acad. Polon. Sci.* **8**, 735–739.

Przeworska–Rolewicz, D. (1960a): 'Sur les équations involutives d'ordre *n*', *Ibidem* **8**, 741–746.

Przeworska–Rolewicz, D. (1961): 'Sur les équations involutives et leur applications', *Studia Math.* **20**, 95–117.

Przeworska–Rolewicz, D. (1962): 'Sur les opérations satisfaisantes à l'identité polynomiale', *Ibidem* **22**, 43–58.

Przeworska–Rolewicz, D. (1963): 'Équations avec opérations algébraïques', *Ibidem* **22**, 337–367.

Przeworska-Rolewicz, D. (1963a): 'Sur l'unique solution polyharmonique d'équation $\sum_{k=0}^{n-1} a_k \Delta^k u = v$', *Atti dell'Academia Nazionale dei Lincei, Ser. VIII, V,* 34–39.

Przeworska-Rolewicz, D. (1965): 'Sur les équations presque algébriques', *Studia Math.* **25**, 163–180.

Przeworska-Rolewicz, D. (1969): 'On categories and pseudocategories' (in Russian), *Funkcjonal. Analiz. i Prilozhen.* **3**, 93.

Przeworska-Rolewicz, D. (1969a): 'A characterization of algebraic derivative', *Bull. Acad. Polon. Sci.* **17**, 11–13.

Przeworska-Rolewicz, D. (1970): 'Algebraic derivative and abstract differential equations', *Anais da Academia Brasileira de Ciencias* **42**, 403–409.

Przeworska-Rolewicz, D. (1972): 'Algebraic derivative and initial value problems', *Bull. Acad. Polon. Sci.* **20**, 629–633.

Przeworska-Rolewicz, D. (1972a): 'Generalized linear equations of Carleman type', *Ibidem* **20**, 635–639.

Przeworska-Rolewicz, D. (1972b): 'Algebraic derivative and definite integrals' *Ibidem* **20**, 641–644.

Przeworska-Rolewicz, D. (1972c): 'A mixed boundary value problem with an algebraic derivative', *Ibidem* **20**, 645–648.

Przeworska-Rolewicz, D. (1972d): 'Concerning left invertible operators', *Ibidem* **20**, 837–839.

Przeworska-Rolewicz, D. (1972e): 'Algebraic theory of right invertible operators', *Studia Math.* **48**, 129–144.

Przeworska-Rolewicz, D. (1973): *Equations with Transformed Argument. An Algebraic Approach,* Elsevier Scientific Publishing Co. and PWN–Polish Scientific Publishers, Amsterdam–Warszawa.

Przeworska-Rolewicz, D. (1973a): 'Algebraic theory of partial differential equations with variable coefficients, in: *Ergebnisse einer Tagung über Funktionentheoretische Methoden bei partiellen Differentialgleichungen,* Gesellschaft für Mathematik und Datenverarbeitung mbH, Bonn 77, 109–127.

Przeworska-Rolewicz, D. (1973b): *Pseudocategories, Para-Algebras and Perturbations of Linear Operators,* Preprint No. 60, Institute of Mathematics, Polish Academy of Sciences, Warszawa.

Przeworska-Rolewicz, D. (1973c): 'Right invertible operators and functional-differential equations with involutions', *Demonstratio Math.* **5**, 165–177.

Przeworska-Rolewicz, D. (1973d): 'Extension of operational calculus', *Control and Cybernetics* **2**, 5–14.

Przeworska-Rolewicz, D. (1974): 'On linear differential equations with transformed argument solvable by means of right invertible operators', *Ann. Polon. Math.* **29**, 141–148.

Przeworska-Rolewicz, D. (1974a): 'Concerning boundary value problems for equations with right invertible operators', *Demonstratio Math.* **7**, 365–380.

Przeworska-Rolewicz, D. (1975): 'Concerning non-linear equations with right invertible operators', *Ibidem* **3**, 313–321.

Przeworska–Rolewicz, D. (1976): 'Admissible initial operators for superpositions of right invertible operators', *Ann. Polon. Math.* **33**, 113–120.

Przeworska–Rolewicz, D. (1976a): 'Remarks on boundary value problems and Fourier method for right invertible operators', *Math. Nachr.* **72**, 109–117.

Przeworska–Rolewicz, D. (1976b): 'Perturbations of linear equations with right invertible operators', *Beiträge zur Analysis* **8**, 115–118.

Przeworska–Rolewicz, D. (1978): 'A generalization of Wroński theorems', *Math. Nachr.* **85**, 47–55.

Przeworska–Rolewicz, D. (1978a): 'On trigonometric identity for right invertible operators', *Commentationes Math.*, Special Volume I, 267–277.

Przeworska–Rolewicz, D. (1979): *Introduction to Algebraic Analysis and its Applications* (in Polish), WNT–Publishers in Sciences and Technology, Warszawa.

Przeworska–Rolewicz, D. (1979a): 'Algebraic analysis and operational calculus', in: *Proc. Conf. Generalized Functions and Operational Calculus, Varna* 1975, Publish. House Bulg. Acad. Sci., Sofia, 178–181.

Przeworska–Rolewicz, D. (1979b): 'New approach to systems with transformed argument', in: *Proc. Conf. Game Theory and Related Topics, Bonn–Hagen* 1978, North-Holland, Amsterdam–New York–Oxford, 243–246.

Przeworska–Rolewicz, D. (1980): *Shifts and Periodicity for Right Invertible Operators*, Research Notes in Mathematics 43, Pitman Advanced Publish. Program, Boston–London–Melbourne.

Przeworska–Rolewicz, D. (1980a): 'Right inverses and Volterra operators', *J. Integral Equations* **2**, 45–56.

Przeworska–Rolewicz, D. (1980b): 'Picone's identity for right invertible operators', in: *Proc. Conf. Special Topics of Applied Mathematics, Bonn* 1979, North-Holland, Amsterdam–New York–Oxford, 27–32.

Przeworska–Rolewicz, D. (1980c): 'Some remarks on shifts for right invertible operators', in: *Proc. Intern. Conf. Functional-Differential Systems and Related Topics I*, Błażejewko 1979, Higher College of Engineering, Zielona Góra, 274–279.

Przeworska–Rolewicz, D. (1981): 'D-shifts and generalized periodic solutions of equations with right invertible operators', in: *Proc. Conf. Methods of Mathematical Programming, Zakopane* 1977, PWN–Polish Scientific Publishers, Warszawa, 261–269.

Przeworska–Rolewicz, D. (1981a): 'Some properties of right inverses', *Serdica, Bulg. Math. Publicationes* **7**, 298–300.

Przeworska–Rolewicz, D. (1981b): 'Integration of unit in linear rings with right invertible operators', *Comptes Rendus. Math. Reports Acad. Sci. Canada* **1**, 227–230.

Przeworska–Rolewicz, D. (1981c): 'Concerning Euler–Lagrange equations in algebras with right invertible operators', in: *Proc. Conf. Game Theory and Economics, Hagen–Bonn* 1980, North-Holland, Amsterdam–New York–Oxford, 435–447.

Przeworska–Rolewicz, D. (1982): 'Integral characterization of Volterra right inverses', *J. Integral Equations* **4**, 89–93.

Przeworska–Rolewicz, D. (1982a): 'Theorems on polynomials in right invertible operators', *Zeitschrift für Analysis und ihre Anwendungen* **1**, 7–9.

Przeworska–Rolewicz, D. (1982b): 'Quasi-polynomials for linear equations with right invertible operators and shifts', in: *Proc. Intern.Conf.Functional-Differential Systems and Related Topics II*, Błażejewko 1981, Higher College of Engineering Zielona Góra, 290–292.

Przeworska–Rolewicz, D. (1982c): 'Picone's identity in non-Leibniz algebras', *Demonstratio Math.* **15**, 1105–1110.

Przeworska–Rolewicz, D. (1983): 'Green's formula for right invertible operators', *Ann. Polon. Math.* **42**, 285–297.

Przeworska–Rolewicz, D. (1983a): 'Non-Leibniz algebras', *Studia Math.* **77**, 69–79.

Przeworska–Rolewicz, D. (1983b): 'Conjugate problems for linear systems with right invertible operators', *Math. Nachr.* **114**, 237–254.

Przeworska–Rolewicz, D. (1984): 'On some boundary value problems for stationary linear systems with shifts', in: *Proc. Intern. Conf. Functional-Differential System. and Related Topics III*, Błażejewko 1983, Higher College of Engineering, Zielona Góra, 95–99.

Przeworska–Rolewicz, D. (1985): 'Generalized Bielecki Theorem', *Ann. Univ. M Curie-Skłodowska* **38**, 125–133.

Przeworska–Rolewicz, D., Rolewicz, S. (1968): *Equations in Linear Spaces*, Monografie Matematyczne 47, PWN–Polish Scientific Publishers, Warszawa.

Przeworska–Rolewicz, D., Rolewicz, S. (1968a): 'On periodic solutions of nonlinear differential-difference equations', *Bull. Acad. Polon. Sci.* **16**, 577–580.

Przeworska–Rolewicz, D., Rolewicz, S. (1969): 'On the control of linear periodic time lag systems', *Studia Math.* **32**, 149–152.

Przeworska–Rolewicz, D., Rolewicz, S. (1985): 'The only continuous Volterra right inverses in $C_c[0, 1]$ for the operator d/dt are $\int_a^t$, *Colloq. Math.* **51**, 279–283.

Przeworska–Rolewicz, D., Rolewicz, S. (1986): 'On a representation of continuous Volterra right inverses to the derivative in the spaces of infinitely differentiable functions', *J. Australian Math. Soc.* **41**, 138–142.

Przeworska–Rolewicz, D., von Trotha, H. (1981): 'Right inverses in D–R algebras with unit', *J. Integral Equations* **3**, 245–259.

Reynerts, H. (1979): 'The linear functional equation $\sum_{i=1}^{n} h_i \times \varphi \circ g_i = h$', *Demonstratio Math.* **12**, 131–139.

Reckziegel, I., Tasche, M. (1980): 'Über die Lösungentartenen Operatorengleichungen mit Matrixkoeffizienten', *Rostock Math. Kolloq.* **13**, 81–96.

Roach, G. F. (1983): 'On a class of non-homogeneous multiparameters problems', in: *Proc. Conf. Dynamical Problems in Math Physics, Oberwolfach* 1982, Lang, Frankfurt, 183–193.

Rolewicz, S. (1973): 'On perturbations of deviations of periodic differential-difference equations in Banach spaces', *Studia Math.* **47**, 31–35.

Rolewicz, S. (1985): *Metric Linear Spaces*, 2-nd revised and extended edition, PWN–Polish Scientific Publishers and D. Reidel, Dordrecht.

Rolewicz, S. (1987): *Functional Analysis and Control Theory*, PWN–Polish Scientific Publishers and D. Reidel, Dordrecht.

Roman, S. M., Rota, G.-C. (1978): 'The umbral calculus', *Advances in Mathematics* **27**, 95–188.

Rota, G.-C. (1964): 'Reynolds operators', in: *Proc. Symposia in Applied Mathematics XVI, Stochastic Process in Mathematics, Physics and Engineering*, Amer. Math. Soc., 70–83.

Rota, G.-C. (1975): *Finite Operator Calculus*, Academic Press, New Yorrk

Saks, S., Zygmund, A. (1965): *Analytic Functions*, II-nd enlarged ed., Monografie. Matematyczne 28, PWN–Polish Scientific Publishers, Warszawa.

Schultz, H. J. (1974): 'Systems of derivations on topological algebras of power series', *Notices Amer. Math. Soc.* **21**, A-177.

Schwabik, S., Tvrdý, M., Vejvoda, O. (1979): *Differential and Integral Equations Boundary Value Problems and Adjoints*, Academia, Praha.

Serbin, H. (1973): 'The operational calculus of Heaviside', *J. Inst. Math. Appl.* **11**, 131–143.

Shapiro, J., Schechter, M. (1973): 'A generalized operational calculus developed from Fedholm operator theory', *Trans. Amer. Math. Soc.* **175**, 439–467.

Shtokalo, I. Z. (1972): *Operational Calculus* (in Russian), Naukova Dumka, Kiev.

Sieńczewski, G. (1982): *Deterministic Models of Stationary Linear Systems with Right Invertible Operators and Identification of their Parameters*, Preprint No. 20, Institute of Mathematics, Polish Academy of Sciences, Warszawa.

Sikorski, R. (1958): 'On Mikusiński's algebraic theory of differential equations', *Studia Math.* **16**, 230–236.

Sikorski, R. (1974): 'Quasi-inverses of morphisms', *Fundamenta Math.* **81**, 343–358.

Sinclair, A. H. (1969): 'Continuous derivations on Banach algebras', *Proc. Amer. Math. Soc.* **20**, 166–170.

Singer, I. M., Wermer, J. (1955): 'Derivation on commutative normed algebras', *Math. Ann.* **129**, 260–264.

Skórnik, K. (1981): 'On tempered integrals and derivatives of non-negative orders', *Ann. Polon. Math.* **40**, 47–57.

Sova, M. (1966): 'Cosine operator functions', *Rozprawy Matematyczne* **49**, Warszawa.

Sprössig, W. (1977): 'Räumliches Analogon zur Volterraschen Integralgleichungen', *Wiss. Z. Techn. Hochschule Karl–Marx–Stadt* **19**, 535–537.

Straškraba, I. (1982): 'Existence and uniqueness of periodic solutions of differential equations in Banach spaces', *Czechoslovak Math. Journ.* **32** (107), 53–73.

Targonski, G. I. (1967): *Seminar on Functional Operators and Equations*, Lecture

Notes in Mathematics 33, Springer Vlg., Berlin–Heidelberg–New York.

Tasche, M. (1974): 'Algebraische Operatorenrechnung für einen rechtsinvertirbaren Operator', *Wiss. Zeitschr. d. Universität Rostock* 23, 735–744.

Tasche, M. (1975): 'Abstrakte Differentialgleichungen mit algebraischen Operatoren', *Ibidem* 24, 1231–1236.

Tasche, M. (1976): 'Operatorenrechnung in einer Algebra', *Beiträge zur Analysis* 9, 125–130.

Tasche, M. (1977): 'Abstrakte lineare Differentialgleichungen mit stationären Operatoren', *Math. Nachr.* 78, 21–36.

Tasche, M. (1977a): 'Randprojektoren und Greenische Operatoren von abstrakten Randwertproblemen und deren Anwendung', *ZAMM* 57, 579–589.

Tasche, M. (1978): *Funktionanalytische Methoden in der Operatorenrechnung*, Nova Acta Leopoldina 49, Halle (Saale).

Tasche, M. (1979): 'Zur Konvergenzbeschleunigung von Fourier-Reihen', *Math. Nachr.* 90, 123–134.

Tasche, M. (1982): 'A unified approach to interpolation methods', *J. Integral Equations* 4, 55–75.

Tasche, M. (1986): 'Accelerating convergence of Fourier expansions', in: *Proc. Intern. Conf. Functional-Differential Systems and Related Topics, IV*, Jachranka, 1985, Higher College of Engineering, Zielona Góra, 121–129.

Tasche, M., Wrase, B. (1979): 'Über lineare Differenzialgleichungen gebrochener Ordnung und verallgemeinerte Abelsche Integralgleichungen', *Demonstratio Math.* 12, 803–820.

Trèves, J. F. (1970): *Linear Partial Differential Equations*, Gordon and Breach, New York–London–Paris.

Trotha von, H. (1981): 'Structure properties of D–R vector spaces', *Dissertationes Math.* 184, Warszawa.

Trotha von, H. (1981a): 'Contractivity in certain D–R spaces', *Math. Nachr.* 101, 207–213.

Turner, T. R. (1972): 'Double commutants of algebraic operators', *Proc. Amer. Math. Soc.* 33, 415–419.

Turner, T. R. (1974): Erratum: "Double commutants ...", *Ibidem* 45, 466.

Viner, I. Ya. (1969): 'Differential equations with involutions' (in Russian), *Differencjalnyje Uravnienija* 5, 1131–1137.

Viner, I. Ya. (1970): 'Partial differential equations with involutions' (in Russian), *Ibidem* 6, 1320–1322.

Voiculescu, D. (1974): 'Norm limits of algebraic operators', *Rev. Roumaine Math. Pur. Appl.* 19, 371–378.

Wermer, J. (1960): 'Dirichlet algebras', *Duke Math. J.* 27, 373–382.

Włodarska-Dymitruk, A. (1975): 'Polynomial-periodic solutions of differential-difference equations', *Demonstratio Math.* 8, 49–65.

Wolfersdorf von, L. (1970): 'Sjöstrandsche Probleme der Richtungsableitung bei einer Gleichung von zusammen Typ', *Math. Nachr.* 45, 263–277.

Wolfersdorf von, L. (1970a): 'Sjöstrandsche Probleme bei einem System von partiellen Differentialgleichungen vom zusammegesetzten Typ', *Math. Nachr.* **46**, 321–339.

Wyman, B. F. (1974): 'Linear systems over rings of operators', in: *Proc. First Intern. Symp. Category Theory Appl. Comp. and Control*, San Francisco.

Yamamuro, S. (1974): *Differential Calculus in Topological Linear Spaces.* Springer Vlg., Berlin–Heidelberg–New York.

Yamamuro, S. (1979): 'A theory of differentiation in locally spaces', *Memoirs of Amer. Math. Soc.* **17**, № 212, Providence, R. I.

Yaohua, Deng (1981): 'The index formula for equations with algebraic operators', *Kexue Tongbao* **26**, 193–199.

Yosida, K. (1983): 'The algebraic derivative and Laplace's differential equation', *Proc. Japan. Acad. Ser. A Math. Sci.* **59**, 1–14.

Yosida, K. (1965): *Functional Analysis*, Springer Vlg., Berlin–Göttingen–Heidelberg.

Zahar–Itkin, M. H. (1973): 'Conditions for the splitting of matrix telegraph equations' (in Russian), *Differencjalnyje Uravnienija* **9**, 565–567, 591.

Zaidman, S. D. (1979): *Abstract differential equations*, Research Notes in Mathematics 36, Pitman Advanc. Publish. Program, San Francisco–London–Melbourne.

Zariski, O., Samuel, P. (1953): *Commutative Algebra*, Van Nostrand, Princeton.

Zima, K. (1968): 'Certain equations in an abstract ring', Wyższa Szkoła Pedagogiczna, Katowice, *Zeszyty Naukowe Sekc. Mat.* **6**, 93–100.

Added in proof:

Kalfat, A. (1986a): 'Volterra right inverses for weighted difference operators', *Demonstratio Math.* (to appear).

Lausch, A., Przeworska-Rolewicz, D. (1986): 'Pseudocategories, Para-algebras and Linear operators', *Math. Nachr.* (to appear).

Przeworska-Rolewicz, D. (1986): 'Remarks on D_∞-spaces', In: *Proc. Intern. Conf. Functions Spaces*, Poznań 1986 (to appear).

Przeworska-Rolewicz, D. (1986a): 'On "neutral" equations with right invertible operators', in: *Proc. Intern. Conf. Functional-Differential Systems and Related Topics, IV, Jachranka* 1985, Higher College of Engineering, Zielona Góra, 95–99.

Authors Index

Subject Index

List of Symbols

ERRATA

Page, line	For:	Read:
37_6	$[A_1]$	$[[A_1]]$
$328_{12,18}$	$\hat{DR} = DAA^{-1}R = DR = I$	$\hat{DR} = A^{-1}DAA^{-1}RA = A^{-1}DRA = I$
352_{11}	$E_A R(BS_{-h}x+y)+x_0 = y_0$	$E_A[R(BS_{-h}x+y)+x_0] = y_0$
420_{18}	$c_{D_1} = c_D$	$c_{D_1} = c_{D_2}$
504_{11}	$x^{\wedge}(t) =$	$x^{\wedge}(t) =$

Danuta Przeworska-Rolewicz, Algebraic Analysis